Innovative Wastewater Treatment & Resource Recovery Technologies

Innovative Wastewater Treatment & Resource Recovery Technologies

Impacts on Energy, Economy and Environment

Edited by

Juan M. Lema and Sonia Suarez

IWA PUBLISHING

Published by **IWA Publishing**
Alliance House
12 Caxton Street
London SW1H 0QS, UK
Telephone: +44 (0)20 7654 5500
Fax: +44 (0)20 7654 5555
Email: publications@iwap.co.uk
Web: www.iwapublishing.com

First published 2017

British Library Cataloguing in Publication Data
A CIP catalogue record for this book is available from the British Library

ISBN: 9781780407869 (Paperback)
ISBN: 9781780407876 (eBook)

Contents

List of contributors

Manuel Alarcón
Biotechnology Centre,
University of Concepción, Chile
alarconvivero@gmail.com

Teresa Alvarino
Department of Chemical Engineering,
Institute of Technology,
Universidade de Santiago de Compostela, Spain
teresa.alvarino@usc.es

Elena Anayiotou
Department of Environmental Science and Technology,
Cyprus University of Technology
elena.anayiotou@gmail.com

Zouhayr Arbib
FCC aqualia Gestión Integral del Agua, S.A, Spain
Zouhayr.Arbib@fcc.es

Eduardo Ayesa
CEIT and Tecnun (University of Navarra), San Sebastián, Spain
eayesa@ceit.es

Janis Baeten
Department of Biosystems Engineering,
Ghent University, Belgium
janis.baeten@ugent.be

Juan A. Baeza
GENOCOV, Departament d'Enginyeria Química, Biològica i Ambiental
Escola d'Enginyeria
Universitat Autònoma de Barcelona, Spain
JuanAntonio.Baeza@uab.cat

Predrag Banković
Institute of Chemistry, Technology and Metallurgy, Center for Catalysis and Chemical Engineering,
University of Belgrade, Serbia
predragb@nanosys.ihtm.bg.ac.rs

Damià Barceló
Catalan Institute for Water Research (ICRA), Girona, Spain.
Department of Environmental Chemistry, IDAEA-CSIC Barcelona, Spain
dbcqam@cid.csic.es

Damien Batstone
Advanced Water Management Centre (AWMC),
The University of Queensland, Australia
d.batstone@awmc.uq.edu.au

Jamie Benidickson
Faculty of Law,
University of Ottawa, Canada
Jamie.Benidickson@uOttawa.ca

Giorgio Bertanza
Department of Civil, Environmental, Architectural Engineering and Mathematics, University of Brescia, Italy
giorgio.bertanza@unibs.it

David Bolzonella
Department of Biotechnology, University of Verona, Italy
david.bolzonella@univr.it

Hande Bozkurt
CAPEC-PROCES Research Center, Department of Chemical and Biochemical Engineering, Technical University of Denmark hboz@kt.dtu.dk

Camilla Maria Braguglia
Water Research Institute, National Research Council, Rome, Italy
braguglia@irsa.cnr.it

Gianluigi Buttiglieri
Catalan Institute for Water Research ICRA, Girona, Spain gbuttiglieri@icra.cat

Raúl Cano
FCC aqualia Gestión Integral del Agua, S.A, Spain
raulcano7@gmail.com

Marta Carballa
Department of Chemical Engineering, Institute of Technology, Universidade de Santiago de Compostela, Spain
marta.carballa@usc.es

Hélène Carrère
INRA, Laboratoire de Biotechnologie de l'Environnement (LBE), Narbonne, France
helene.carrere@supagro.inra.fr

Gilda Carvalho
UCIBIO, REQUIMTE, Departamento de Química, Faculdade de Ciências e Tecnologia, Universidade Nova de Lisboa, Portugal
gs.carvalho@fct.unl.pt

Alba Castillo
Laboratory of Chemical and Environmental Engineering (LEQUIA), University of Girona, Spain
alba.castillo@lequia.udg.cat

Paula ML Castro
Escola Superior de Biotecnologia, Universidade Católica Portuguesa Porto, Portugal
plcastro@porto.ucp.pt

Grzegorz Cema
Environmental Biotechnology Department, Faculty of Power and Environmental Engineering, Silesian University of Technology, Poland
grzegorz.cema@polsl.pl

Zeynep Cetecioglu Gurol
Department of Chemical Engineering, School of Chemical Science and Engineering, KTH - Royal Institute of Technology, Sweden
zeynepcg@kth.se

Jean-Marc Choubert
Wastewater Treatment and Recovery Ecotechnologies Research Team, Irstea, Research Institute of Science and Technology for Environment and Agriculture, France
jean-marc.choubert@irstea.fr

Joan Colón
BETA Tech. Center. (TECNIO Network), U Sciente Tech., University of Vic-Central University of Catalonia, Spain. AXIA Environmental Technologies
joan.colon@uvic.cat

Jesús Colprim
Institute of the Environment LEQUiA, University of Girona, Spain
J.Colprim@lequia.udg.cat

Joaquim Comas
Laboratory of Chemical and Environmental Engineering (LEQUIA), University of Girona, Spain. Catalan Institute for Water Research (ICRA), Girona, Spain
joaquim.comas@udg.edu

Lluis Corominas
Catalan Institute for Water Research (ICRA), Girona, Spain
lcorominas@icra.cat

Sarah Cotterill
School of Civil Engineering and Geosciences, Newcastle University, UK
sarah.cotterill@newcastle.ac.uk

Dafne Crutchik
Department of Biotechnology, University of Verona, Italy
dafne.crutchik@gmail.com

Justyna Czajkowska
Water Supply and Sewage System Division, Warsaw University Life Sciences, Poland
justyna_czajkowska@sggw.pl

Ignacio de Godos
FCC aqualia Gestión Integral del Agua, S.A, Spain
idgodos@gmail.com

Wim de Schepper
VITO, Belgium
wim.deschepper@vito.be

Angela J. Dean
School of Communication and Arts, The University of Queensland. Cooperative Research Centre for Water Sensitive Cities, Monash University, Clayton. Behaviour Works, Monash Sustainability Institute, Monash University, Clayton, Australia
a.dean@uq.edu.au

Guido del Moro
Water Research Institute, National Research Council, Bari, Italy guido.delmoro@ba.irsa.cnr.it

Marc A. Deshusses
Department of Civil and Environmental Engineering, Duke University Durham, North Carolina, USA
marc.deshusses@duke.edu

Claudio Di Iaconi
Water Research Institute, National Research Council, Bari, Italy
claudio.diiaconi@ba.irsa.cnr.it

Rubén Díez-Montero
Department of Water and Environmental Sciences and Technologies, University of Cantabria, Spain
ruben.diezmontero@unican.es

Xochitl Dominguez-Benetton
VITO, Belgium
xoch@vito.be

Carlos G. Dosoretz
Division of Environmental, Water and Agricultural Engineering, Faculty of Civil and Environmental Engineering, Technion - Israel Institute of Technology
carlosd@tx.technion.ac.il

Jörg E. Drewes
Chair of Urban Water Systems Engineering Technical University of Munich, Germany
jdrewes@tum.de

Jan Dries
Research group BioGEM, Faculty of Applied Engineering, University of Antwerp, Belgium
jan.dries2@uantwerpen.be

Ana Lorena Esteban-García
Department of Sciences and Techniques of Water and the Environment, University of Cantabria, Spain
analorena.esteban@unican.es

Francesco Fatone
Department of Material, Environmental and Urban Planning Science and Engineering, Polytechnic University of Marche, Italy
f.fatone@univpm.it

Fernando Fdz-Polanco
Department of Chemical Engineering and Environmental Technology, University of Valladolid, Spain
ffp@iq.uva.es

Gumersindo Feijoo
Department of Chemical Engineering, Institute of Technology, Universidade de Santiago de Compostela, Spain
gumersindo.feijoo@usc.es

Tamara Fernández-Arévalo
CEIT and Tecnun (University of Navarra), San Sebastián, Spain
tfernandez@ceit.es

Kelly S. Fielding
School of Communication and Arts, The University of Queensland. Cooperative Research Centre for Water Sensitive Cities, Monash University, Australia
k.fielding@uq.edu.au

Xavier Flores-Alsina
CAPEC-PROCES Research Center, Department of Chemical and Biochemical Engineering, Technical University of Denmark
xfa@kt.dtu.dk

Isabel Fonts
Centro Universitario de la Defensa-AGM. Aragon Institute of Engineering Research (I3A), University of Zaragoza, Spain
isabelfo@unizar.es

Andrea Fra-Vázquez
Department of Chemical Engineering, Institute of Technology, Universidade de Santiago de Compostela, Spain
andrea.fra@usc.es

David Gabriel
Department of Chemical, Biological and Environmental Engineering, Universitat Autònoma de Barcelona, Spain
david.gabriel@uab.cat

Xavier Gamisans
Department of Mining, Industrial and ITC Engineering, Universitat Politècnica de Catalunya, Spain
xavier.gamisans@upc.edu

Alex Gali
Cetaqua, Spain
agali@cetaqua.com

Manel Garrido-Baserba
Department of Civil and Environmental Engineering, University of California, USA
manelg@uci.edu

Krist V. Gernaey
CAPEC-PROCES Research Center, Department of Chemical and Biochemical Engineering, Technical University of Denmark
kvg@kt.dtu.dk

Andrea Gianico
Water Research Institute, National Research Council, Rome, Italy
gianico@irsa.cnr.it

Noemí Gil-Lalaguna
Aragon Institute of Engineering Research (I3A), University of Zaragoza, Spain
noemigil@unizar.es

Emine Gozde Ozbayram
Department of Environmental Engineering, Faculty of Civil Engineering
Istanbul Technical University, Turkey
gozbayram@itu.edu.tr

Paloma Grau
CEIT and Tecnun (University of Navarra), San Sebastián, Spain
pgrau@ceit.es

Andrea Guerrini
Department of Management, University of Verona, Italy
andrea.guerrini@univr.it

Albert Guisasola
GENOCOV, Departament d'Enginyeria Química, Biològica i Ambiental, Escola d'Enginyeria, Universitat Autònoma de Barcelona, Spain
albert.guisasola@uab.cat

Zhen He
Department of Civil and Environmental Engineering, Virginia Polytechnic Institute and State University, USA
zhenhe@vt.edu

Mark Gerard Healy
Civil Engineering, National University of Ireland, Galway
mark.healy@nuigalway.ie

Elizabeth Heidrich
School of Civil Engineering and Geosciences, Newcastle University, UK
elizabeth.heidrich@newcastle.ac.uk

Brigitte Helmreich
Urban Water Systems Engineering, Technical University of Munich, Germany
b.helmreich@tum.de

Sarah Hendry
Centre for Water Law, Policy and Science (under the auspices of UNESCO), University of Dundee, UK
s.m.hendry@dundee.ac.uk

Lucía Hernández Leal
European Centre of Excellence for Sustainable Water Technology, Wetsus, Netherlands
Lucia.Hernandez@wetsus.nl

Francesc Hernández-Sancho
Department of Applied Economics, University of Valencia, Spain
francesc.hernandez@uv.es

Dang Ho
Trojan Technologies, London, Ontario, Canada. Department of Chemical and Biochemical Engineering, Western University, Canada
dho@trojanuv.com

Nils Horstmeyer
Chair of Urban Water Systems Engineering Technical University of Munich, Germany
nils.horstmeyer@tum.de

Almudena Hospido
Department of Chemical Engineering, Institute of Technology, Universidade de Santiago de Compostela, Spain
almudena.hospido@usc.es

Tim Huelsen
Advanced Water Management Centre (AWMC), The University of Queensland, Australia
t.huelsen@awmc.uq.edu.au

Bahar Ince
Institute of Environmental Sciences, Bog˘aziçi University, Turkey
bahar.ince@boun.edu.tr

Orhan Ince
Department of Environmental Engineering, Faculty of Civil Engineering
Istanbul Technical University, Turkey
inceor@itu.edu.tr

Ulf Jeppsson
Division of Industrial Electrical Engineering and Automation (IEA), Lund University, Sweden
ulf.jeppsson@iea.lth.se

Sara Johansson
Institute of the Environment LEQUiA, University of Girona, Spain
johansson@lequia.udg.cat

Evina Katsou
Department of Mechanical, Aerospace and Civil Engineering, Institute of Environment, Health and Societies, Brunel University, UK
Evina.katsou@brunel.ac.uk

Stuart Khan
Technical University of Munich, Germany. University of New South Wales, Australia
s.khan@unsw.edu.au

Dimitrios Komilis
Department of Environmental Engineering, Democritus University of Thrace, Greece
dkomilis@env.duth.gr

Pawel Krzeminski
Section of Systems Engineering and Technology, Norwegian Institute for Water Research (NIVA), Norway
pawel.krzeminski@niva.no

Javier Lafuente
Department of Chemical, Biological and Environmental Engineering, Universitat Autònoma de Barcelona, Spain
javier.lafuente@uab.cat

Alette A.M. Langenhoff
Department of Environmental Technology, Wageningen University & Research, the Netherlands
alette.langenhoff@wur.nl

Enrique Lara
FCC aqualia Gestión Integral del Agua, S.A, Spain
elarac@fcc.es

Henrik Fred Larsen
Danish Road Directorate, Denmark
hfl@vd.dk

Raquel Lebrero
Department of Chemical Engineering and Environmental Technology, University of Valladolid, Spain
raquel.lebrero@iq.uva.es

Lance Leverette
Industrial Strategic Market Consultant, Rhode Island, United States
Lance.leverette@gmail.com

Yago Lorenzo- Toja
Department of Chemical Engineering, Institute of Technology, Universidade de Santiago de Compostela, Spain
yago.lorenzo@usc.es

Gerasimos Lyberatos
School of Chemical Engineering, National Technical University of Athens, Greece
lyberatos@chemeng.ntua.gr

Miguel M. Iglesias
Department of Chemical Engineering, Institute of Technology, Universidade de Santiago de Compostela, Spain
miguel.mauricio@usc.es

Simos Malamis
Department of Water Resources and Environmental Engineering, School of Civil Engineering, National Technical University of Athens, Greece
malamis.simos@gmail.com

Adrien Marchi
Aquafin, Belgium

Desirée Marín-Navarro
CETaqua, Water Technology Centre, Barcelona, Spain
dmarin@cetaqua.com

Alma Masic
EAWAG, Switzerland
alma.masic@eawag.ch

Nedal Massalha
The Galilee Society Institute of Applied Research, Israel
nidal@gal-soc.org

Andrew McLeod
Cranfield Water Science Institute, School of Applied Sciences, Cranfield University, UK
a.j.mcleod@cranfield.ac.uk

Süreyya Meriç
Namık Kemal University, Turkey
smeric@nku.edu.tr

Steffen Metzger
Institute of Sanitary Engineering, Water Quality and Solid Waste Management, University of Stuttgart, Germany
steffen.metzger@koms-bw.de

Philipp Michel
Chair of Urban Water Systems Engineering Technical University of Munich, Germany
philipp.michel@tum.de

Maria Teresa Moreira
Department of Chemical Engineering, Institute of Technology, Universidade de Santiago de Compostela, Spain
maite.moreira@usc.es

María Molinos-Senante
Hydraulic and Environmental Engineering Department, Institute of Urban and Territorial Studies, Pontificia Universidad Católica de Chile.
Center of Sustainable Urban Development
mmolinos@uc.cl

Anuska Mosquera-Corral
Department of Chemical Engineering, Institute of Technology, Universidade de Santiago de Compostela, Spain
anuska.mosquera@usc.es

Raúl Muñoz
Department of Chemical Engineering and Environmental Technology, University of Valladolid, Spain
mutora@iq.uva.es

Dragana Mutavdžić Pavlović
Department of Analytical Chemistry, Faculty of Chemical Engineering and Technology, University of Zagreb, Croatia
dmutavdz@fkit.hr

Ayten Namli
Department of Soil Science, Faculty of Agriculture, Ankara University, Turkey
aytenkrc@gmail.com

Valentin Nenov
Department of Water Treatment Technologies, Burgas University, Bulgaria
vnenov@btu.bg

Ioanna Ntaikou
Institute of Chemical Engineering Sciences, Patras, Greece
ntaikou@iceht.forth.gr

Adrian Oehmen
UCIBIO, REQUIMTE, Departamento de Química, Faculdade de Ciências e Tecnologia, Universidade Nova de Lisboa, Portugal
a.oehmen@fct.unl.pt

Piotr Oleskowicz-Popiel
Institute of Environmental Engineering, Faculty of Civil and Environmental Engineering, Poznan University of Technology, Poland
piotr.oleskowicz-popiel@put.poznan.pl

Tania Palmeiro-Sánchez
Department of Chemical Engineering, Institute of Technology, Universidade de Santiago de Compostela, Spain
tania.palmeiro@usc.es

Matteo Papa
Department of Civil, Environmental, Architectural Engineering and Mathematics, University of Brescia, Italy
matteo.papa@unibs.it

Sara Isabel Pérez-Elvira
Department of Chemical Engineering and Environmental Technology, University of Valladolid, Spain
sarape@iq.uva.es

Mira Petrovic
Catalan Institute for Water Research (ICRA). Catalan Institution for Research and Advanced Studies (ICREA), Barcelona, Spain
mpetrovic@icra.cat

Maite Pijuan
Catalan Institute for Water Research (ICRA), Girona, Spain
mpijuan@icra.cat

Elzbieta Plaza
Department of Sustainable Development, Environmental Science and Engineering, Royal Institute of Technology (KTH), Sweden
elap@kth.se

Manel Poch
Laboratory of Chemical and Environmental Engineering (LEQUIA), University of Girona, Spain
manuel.poch@udg.edu

Marie-Noëlle Pons
Laboratory of Chemical Engineering Sciences, University of Lorraine, France
marie-noelle.pons@univ-lorraine.fr

Sergio Ponsá
BETA Tech. Center. U Sciente Tech., University of Vic-Central University of Catalonia, Spain
sergio.ponsa@uvic.cat

Jose Porro
Cobaltwater, Girona, Spain
jose.porro@cobaltwater-us.com

Narcis Pous
Institute of the Environment LEQUiA, University of Girona, Spain
narcis.pous@lequia.udg.cat

Sebastià Puig
Institute of the Environment LEQUiA, University of Girona, Spain
sebastia@lequia.udg.cat

Christian Remy
Berlin Centre of Competence for Water (KWB), Berlin, Germany
christian.remy@kompetenz-wasser.de

Daniele Renzi
Alto Trevigiano Servizi SRL, Treviso, Italy
drenzi@altotrevigianoservizi.it

Maria A.M. Reis
UCIBIO, REQUIMTE, Departamento de Química, Faculdade de Ciências e Tecnologia, Universidade Nova de Lisboa, Portugal
amr@fct.unl.pt

Joan Rieradevall
Institute of Environmental Science and Technology, Universitat Autònoma de Barcelona, Spain
joan.rieradevall@uab.cat

Ángel Robles
Institut Universitari d'Investigació d'Enginyeria de l'Aigua i Medi Ambient (IIAMA), Universitat Politècnica de València, Spain
angel.robles@uv.es

Ignasi Rodríguez-Roda
Laboratory of Chemical and Environmental Engineering (LEQUIA), University of Girona. Catalan Institute for Water Research (ICRA), Girona, Spain
irodriguezroda@icra.cat

Frank Rogalla
FCC aqualia Gestión Integral del Agua, S.A, Spain
frogalla@fcc.es

Mael Ruscalleda
Institute of the Environment LEQUiA, University of Girona, Spain
mael@lequia.udg.cat

Isam Sabbah
The Galilee Society Institute of Applied Research.
Department of Biotechnology Engineering, Braude College, Israel
isabbah@braude.ac.il

José Luis Sánchez
Aragon Institute of Engineering Research (I3A), University of Zaragoza, Spain
jlsance@unizar.es

F. Dilek Sanin
Department of Environmental Engineering, Middle East Technical University, Turkey
dsanin@metu.edu.tr

Domenico Santoro
Trojan Technologies, London, Ontario, Canada. Department of Chemical and Biochemical Engineering, Western University, Canada
dsantoro@trojanuv.com

Siva Sarathy
Trojan Technologies, London, Ontario, Canada Department of Chemical and Biochemical Engineering, Western University, Canada
ssarathy@trojanuv.com

Zachary Scott
Trojan Technologies, London, Ontario, Canada zscott@trojanuv.com

Aurora Seco
Departament d'Enginyeria Química, Escola Tècnica Superior d'Enginyeria, Universitat de València, Spain
aurora.seco@uv.es

Gürkan Sin
CAPEC-PROCES Research Center, Department of Chemical and Biochemical Engineering, Technical University of Denmark
gsi@kt.dtu.dk

Tadeusz Siwiec
Water Supply and Sewage System Division Warsaw University Life Sciences, Poland
tadeusz_siwiec@sggw.pl

Ana Soares
Cranfield Water Science Institute, School of Applied Sciences, Cranfield University, UK
a.soares@cranfield.ac.uk

Mª Eugenia Suárez-Ojeda
GENOCOV Research Group, Department of Chemical, Biological and Environmental Engineering, Universitat Autònoma de Barcelona, Spain
MariaEugenia.Suarez@uab.cat

Eoin Syron
School of Chemical and Bioprocess Engineering, University College Dublin, Ireland
eoin.syron@ucd.ie

Carlota Tayà
BETA Tech. Center. U Sciente Tech., University of Vic-Central University of Catalonia, Spain
carlota.taya@uvic.cat

Iñaki Tejero
Department of Water and Environmental Sciences and Technologies, University of Cantabria, Spain
juan.tejero@unican.es

Carmen Teodosiu
Department of Environmental Engineering and Management, "Gheorge Asachi" Technical University of Iasi, Romania
cteo@ch.tuiasi.ro

Taina Tervahauta
European Centre of Excellence for Sustainable Water Technology, Wetsus, Netherlands
taina.tervahauta@wetsus.nl

Maria Concetta Tomei
Water Research Institute, National Research Council, Rome, Italy
tomei@irsa.cnr.it

Mark van Loosdrecht
Department of Biotechnology, Delft University of Technology, The Netherlands
M.C.M.vanLoosdrecht@tudelft.nl

Ángeles Val del Río
Department of Chemical Engineering, Institute of Technology, Universidade de Santiago de Compostela, Spain
mangeles.val@usc.es

Ian Vázquez-Rowe
Peruvian LCA Network, Department of Engineering, Pontificia Universidad Católica del Perú
ian.vazquez@pucp.pe

Paola Verlicchi
Department of Engineering, University of Ferrara, Italy
paola.verlicchi@unife.it

Eveline Volcke
Department of Biosystems Engineering, Ghent University, Belgium
eveline.volcke@ugent.be

Nick Voulvoulis
Centre for Environmental Policy, Imperial College London, UK
n.voulvoulis@imperial.ac.uk

Darko Vrecko
Department of Systems and Control, Jožef Stefan Institute, Ljubljana, Slovenia
darko.vrecko@ijs.si

Ioannis Vyrides
Department of Environmental Science and Technology, Cyprus University of Technology
Ioannis.vyrides@cut.ac.cy

Elena Zambello
Department of Engineering, University of Ferrara, Italy
elena.zambello@unife.it

Grietje Zeeman
Sub-department Environmental Technology, Wageningen University, Netherlands
Grietje.Zeeman@wur.nl

About the editors

Prof. Juan M. Lema. Professor of Chemical Engineering at the University of Santiago de Compostela (USC) since 1984. He holds a degree in Industrial Chemistry (1971) and a PhD in Science (1975) by USC. He is an IWA Fellow. His scientific production can be summarized as follows: Supervisor of 51 PhD Theses; Publication of 352 papers in peer-reviewed SCI journals (Scopus) (H factor: 51) and more than 8,000 citations and 25 book chapters; Author of 3 European and 8 Spanish patents; More than 100 invited Conferences in National and International Congresses and Universities in Europe, China and Latinoamerica; Participation in 37 projects funded by European Commission (13), Spanish Ministry of Science and Technology (16) and Galician Government (8) and in 32 relevant research contracts with companies. He was the Coordinator of the Novedar_Consolider Project (consortium of 11 Spanish and Dutch Universities) and the Chair of the COST Action ES1202 "Conceiving Wastewater treatment in 2020" (Water, 2020). His main research areas include: a) Removal of organic micropollutants from wastewater; b) Anaerobic (co) digestion; c) Development of novel bioreactors for wastewater treatment; d) Biorefinery.

Dr. Sonia Suárez. Is Project Manager in the Group of Environmental Engineering and Bioprocesses at the Department of Chemical Engineering of the University of Santiago de Compostela (USC) since 2008. She is a Chemical Engineer (2002) and obtained her PhD (2008) at USC. Her scientific production can be summarized as follows: Supervisor or 1 PhD Thesis; Publication of 16 papers in peer-reviewed SCI journals (Web of Science) (H factor: 9 and more than 500 citations) and 6 book chapters; Editor of the book "Innovative Technologies for Urban Wastewater Treatment Plants (ISBN:13-978-84-693-3992-3)"; Author of 1 European patent; 36 presentations at international Conferences; Participation in 15 Projects (4 international and 11 national) and in 4 relevant contracts with Companies. She was the Grant Holder Manager of the COST Action ES1202 "Conceiving Wastewater treatment in 2020" (Water, 2020). Her main research interest is related to the removal of organic micropollutants (pharmaceutical and personal care products) from municipal and hospital wastewaters by innovative physico-chemical and biological processes.

Preface

From Sanitary to Environmental Engineering. The 3R concept

Even before, but especially during the Roman Empire, many innovative solutions including aqueducts for water supply, sewers like the "*cloaca maxima*" or water reservoirs, were conceived to provide a certain level of comfort for the population. Impressive works such as the Segovia aqueduct in Spain, the Basilica Cistern in Istanbul or the "forica" (public toilets) in Pompeia are some selected examples from this period.

Later, in the Middle Ages, some Benedictine and Cistercian convents applied innovative and very simple solutions for wastewater management (Wiesmann, 2006). Their concept implied not only water supply but also a simple wastewater treatment and a further treatment/use as fertiliser by irrigation (Figure 1).

The dramatic growth of size and population of cities during the industrial revolution caused very serious problems, such as the cholera epidemic in London, which greatly alarmed the population and made it necessary the installation of sewers in the largest cities in Europe. For example, in the middle of the XIX Century, more than 340 km of sewers were constructed in Paris to derive the wastewater discharges downstream the Seine River, with the aim of warding off the problem rather than solving it (Figure 2).

FROM SANITARY TO ENVIRONMENTAL ENGINEERING

In the second part of the 19th Century, engineers realised that Nature was incapable to of coping with water pollution and that some measures should be taken to restore the equilibrium in rivers and lakes. A very efficient solution based on the ideas of Benedictine monasteries was implemented to control the pollution of the Speer river in Berlin: After a first settling system, wastewater was distributed by a radial system with 12 channels for land irrigation, attaining a very high efficiency even though it had a low treatment capacity (Wiesmann, 2007).

Along the following years, the concept of trickling filter was developed in systems such us that installed in The Hague at the end of the XIX century. The use of this technology, in combination with settling units and post-irrigation, allowed the capacity to be impressively improved from 0.3 to more than 500 m^3/Ha h (Seeger, 1999) (Figure 3).

However, the great leap took place a hundred years ago when Arden and Lockett (1914) discovered that a microorganism mass (sludge) sufficiently aerated was able to further increase the efficiency of carbon removal and even for ammonia oxidation. This concept, "activated sludge", shaped the development of most solutions for wastewater treatment for the next century (Jenkins & Wanner, 2014). From then on, thousands of Sewage Treatment Plants (STP) were built up with the same principles used in the design of the STP of Essen, Germany, in 1925 (Figure 4).

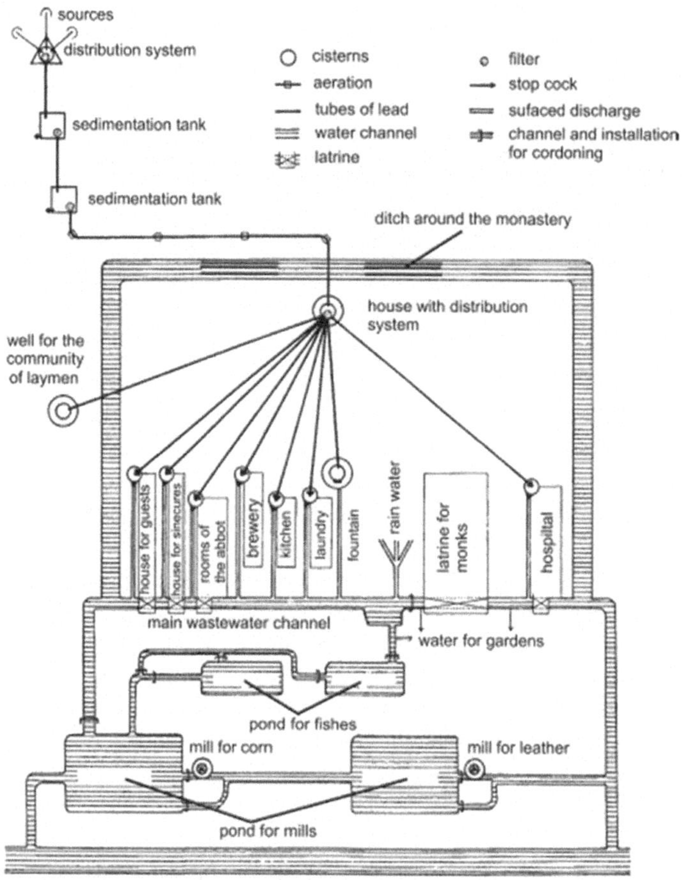

Figure 1 Water and wastewater management in Cistercian monasteries in the middle ages (Wiesmann *et al.* 2006).

All those developments took place into the "Sanitary Engineering" framework (from Latin *Sanitas* i.e. health), a specific branch of Civil Engineering whose major goal was to find technical solutions to preserve human health, to minimise harmful effects to humans and preserve all things in Nature that were useful to humans, under an anthropocentric view of Nature as existing for the service of the human being.

The environment has not been a major concern of Society until a few years ago. In the second half of the XX Century two major impacts greatly influenced the change of mentality of Society concerning its relationship with Nature. The first was triggered by the publication of the book "The Silent Spring" by Rachel Carson (1962), in which the effect of pesticides (particularly DDT) massively applied on living organisms in cultivation areas in the USA was analysed. The book shocked very much the American society and had an enormous social and even political repercussion that contributed to making the

Society more sensitive towards the environment. This fact could be considered as the starting point of *Environmentalism*, a scientific, political, cultural and economic movement that developed quickly, mostly in industrialised countries, since 1970.

Figure 2 Paris sewer workers (XIX Century). *http://assets.atlasobscura.com/article_images/10670.*

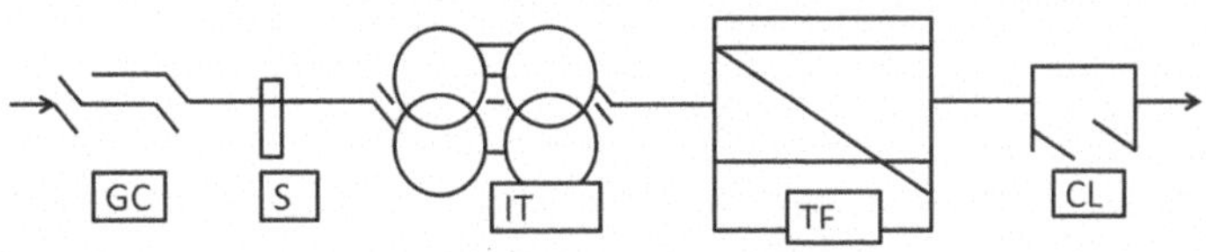

Figure 3 A wastewater treatment system based on Immhof tank (IT) and Trickling filter (TF) with pretreatment (GC: grid chamber, S: settling) and postreatment (CL: Clarifier) units. *Adapted from (Seeger, 1999).*

A second important impact emerged from Europe. In April 1968, by the initiative of Aurelio Pecci, an Italian economist, and the Scottish chemist Alexander King, a group of 36 scientists, economists and politicians from different countries, under the sponsorship of the Agnelli Foundation, met in Rome to debate about the changes in our Planet as a consequence of human activity. Two years later they founded the "Club of Rome" which in 1972 published its first report, "The limits to growth" (Meadows *et al.* 1979), written by specialists in Theory of Systems from MIT. In it, which some scenarios and options available to harmonise sustainable progress and environmental limitations were analysed. The conclusions evidenced the contradiction of the uncontrolled use of Nature in a World with limited resources. The impact was tremendous at the international level and somehow it can be considered as the origin of the *Sustainability* concept, which was clearly stated further in the "Report of the World Commission on Environment and Development: Our Common Future" (http://www.un-documents.net/wced-ocf.htm) known as "The Brundland Report" issued by the UN in 1987.

Figure 4 Rellinghausen Sewage Treatment Plant in Essen (Germany) in 1925 for 45,000 inhabitants. *http://www.ruhrverband.de/fileadmin/con_images/abwasser/historie_bild9_g.jpg.*

These two concepts, *Environmentalism* and *Sustainability* are progressively modifying the perception about the role of humans on Earth from an *Egocentric* view, characterised by the submission of Nature to the interests of humans, to *Ecocentrism*, in which Nature is the focus of attention (Figure 5).

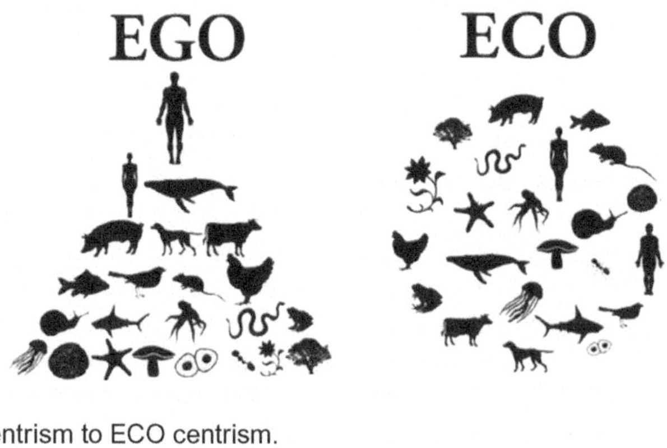

Figure 5 From EGO centrism to ECO centrism.

Accordingly, *Sanitary Engineering*, whose major goal was to preserve the health and the goods useful to humans, evolved quickly towards the concept of *Environmental Engineering*, which pretends to protect Nature, including all living organisms, and to preserve the natural resources. Wastewater treatment conceived under this new paradigm implies that new technologies should not only target the "technical efficiency" but also other environmental, social and economic impacts should be taken into account, under a holistic approach. This constituted the main objective of the *COST Action Water_2020* that was launched in November 2012, in the frame of which this book was shaped.

THE 3 R CONCEPT IN WASTEWATER TREATMENT

The 3R is a well-established concept in the context of waste management. In this book we propose a re-definition and adaptation of this concept to wastewater treatment (Figure 6): *Reducing Requirements and Impacts (Section I)*, focussing on advanced technologies that demand less energy (Chapter 1–5) and space (Chapter 6–8), as well as on options to reduce the impacts related to the emission of organic micropollutants (Chapter 9–10), gases and odours (Chapter 11) and sludge (Chapter 12); *Reusing Water and Sludge (Section II)*, by discussing options for producing high-quality municipal wastewater effluents for reclamation and reuse purposes (Chapter 13) and by analysing the agricultural use and value of sludge after different treatment strategies (Chapter 14); and *Recovering Resources (Section III)*, being the added value by-products considered in the book energy (Chapter 15), metals (Chapter 16), nutrients (Chapter 17), and organic substances (Chapter 18).

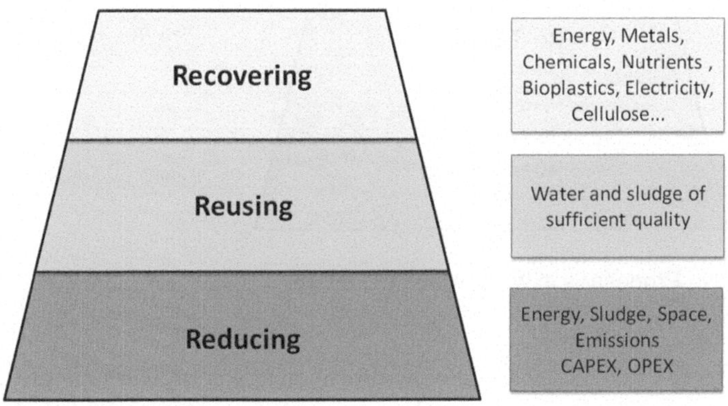

Figure 6 3R Objectives in wastewater treatment.

In the book, the 3R Objectives are approached by advanced technologies currently available at different Technology Readiness Levels (TRL). In all cases a specific conclusion section with a mini-assessment in terms of environmental and socio-economic impact has been included.

The evaluation of technologies cannot be based exclusively on technical aspects (i.e. removal efficiencies, operational conditions, reliability, etc.), the *Economic, Environmental, Legal and Social Impacts (Section IV)* derived from their implementation, in accordance with the sustainability concept, should also be considered. In this sense, several tools are presented for the assessment of the impact of innovation on wastewater treatment Economics (Chapter 19), the analysis of environmental impacts mainly based on Life Cycle Assessment (Chapter 20–21), and the impacts caused by the emission of emerging pollutants such as Greenhouse Gases and Odours (Chapter 23) and Micropollutants (Chapter 24). Transversal factors related to water innovation actions, such as public perceptions of recycled water (Chapter 22) and Legal and Policy Frameworks (Chapter 25) are also considered as relevant boundaries.

The 3R concept in wastewater treatment can be defined not only for *Objectives* potentially attained by using innovative *Technologies* but also for innovative *Processes* in which several technologies, traditional or innovative, can be combined to reach an optimised flowsheet (Figure 7). Thus, several solutions implying several levels of innovation can be conceived: *Retrofitting*, meaning the inclusion of some new units into conventional treatment scheme to improve some process parameters (e.g. Nutrient removal in the centrate (Chapter 1) or pre-treatment of sludge (Chapter 12)); *Re-Thinking* the STP, with a substantial modification of flowsheets, in which innovative technologies are included (e.g. Anaerobic treatment of municipal wastewater (Chapter 2), Aerobic granular sludge reactors (Chapter 6), Enhanced primary treatment (Chapter 8)); and *Re-Imagining* which implies a new conception, completely changing the current flowsheets (e.g. Algal systems (Chapter 4), Bioelectrochemical systems (Chapter 5)), and even switching the main role of such installations from treating wastewater at the lowest impact, to recovering valuable and finite resources (water, energy, phosphorus, etc.).

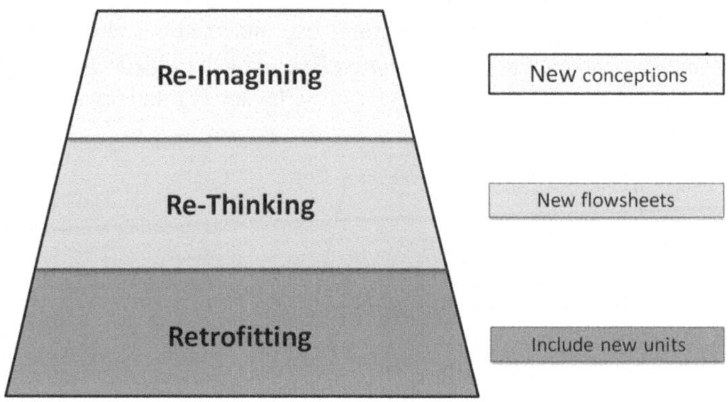

Figure 7 3R Innovative Processes in wastewater treatment.

It is important to realise that the inclusion or modification of a particular unit into a process affects the overall efficiency. For example, when a thermal sludge pre-treatment is included, the COD and N mass flow are substantially modified and the expected benefits might not be sufficient to offset new requirements in other units. Therefore a holistic approach is required to assess the suitability of a particular process for a particular problem.

In *Section V* different tools for *Conceiving, Comparing and Selecting Efficient Processes* are presented, whose aim is to model and select the best STP layout for each particular situation from a combined economic and environmental point of view. This selection process is of increased complexity due to the co-existence of many technological alternatives and restrictions. Automatic optimization and/or decision-making systems, such as Environmental Decision Support Systems (Chapter 26) and Superstructure-based optimization tools (Chapter 27), reduce the time needed to make decisions and improve their quality and consistency. The STP is approached in this section from a global perspective, considering the impact of implementing innovative unit technologies on the global plant efficiency and sustainability, considering the whole STP in terms of energy, mass (C, nutrients, EPs, GHGs) and cost balances thanks, to the development of plant wide modelling approaches (Chapter 28).

We would like to thank the authors for their enthusiastic participation in the writing of this book and for their commitment to advancing the holistic assessment of innovative processes. We also want to acknowledge the COST Action Water_2020, in which most of the authors have participated. It allowed us to set-up a network of researchers advancing in this new perspective about wastewater treatment innovation.

REFERENCES

Ardern E. and Lockett W. T. (1914). "Experiments on the oxidation of sewage without the aid of filters". *J. Soc. Chem. Ind.*, **33**, 523 and (Part II) *J. Soc. Chem. Ind.*, **33**, 1122.

Carson R. (1962). "The silent spring". Houghton Mifflin.

Jenkins D. and Wanner J. (2014). "Activated sludge 100 years and counting". IWA Publishing.

Meadows D., Meadows D., Randers J. and Behrens W. (1979). "The limits to growth". Potomak Ass. Book

Seeger H. (1999). "The history of the german wastewater treatment". *European Water Management*, **2**(5), 51–56

Wiesmann U. *et al.* (2006). "Historical development of wastewater collection and treatment". In: Fundamentals of Biological Wastewater Treatment. Wiley-VCH Verlag, Weinheim.

Juan M. Lema and Sonia Suárez

Part 1

Reducing Requirements and Impacts

Part 1a: *Reducing Energy Requirements*

Chapter 1

Nutrient removal

Francesco Fatone, Juan A. Baeza, Damien Batstone, Grzegorz Cema, Dafne Crutchik, Rubén Díez-Montero, Tim Huelsen, Gerasimos Lyberatos, Andrew McLeod, Anuska Mosquera-Corral, Adrian Oehmen, Elzbieta Plaza, Daniele Renzi, Ana Soares and Iñaki Tejero

1.1 INTRODUCTION

1.1.1 Nutrient management regulation and implications on energy consumptions

After decades from the Urban Wastewater Treatment Directive (271/91/EEC), nutrient pollution resulting from excess nitrogen (N) and phosphorus (P) is still a leading cause of degradation of water quality in Europe (European Commission – JRC, 2014). More stringent nutrient management practices and regulations are therefore needed and have been undertaken. Considering for example the recently identified "ecoregions" in the USA (WERF, 2010), it is clear that current trends are establishing very low standard for in-stream concentrations of N and P which will result in standard for nutrient discharge in sensitive watersheds much lower than 10 mgN/L and 1 mgP/L set by the Directive 271/91/EEC. Technology-based nutrient limits at or near the limit of technology (LOT) are being considered in several regions in the United States and abroad. The LOT for total nitrogen (TN) is typically defined as 3.0 mg/L and total phosphorus (TP) of 0.1 mg/L or the mass-load-based equivalent at the design capacity of the wastewater treatment plant. In some regions, especially sensitive watersheds or ecosystems, TP limits much less than 0.1 mg/L are being considered.

In Europe a recent survey carried out within the Water_2020 network (ES1202 COST Action) concerned the most sensitive areas, where special local nutrient management legislation is applied (Table 1.1). The Water_2020 partners pointed out that the lowest limits on both total nitrogen and phosphorus are set in Finland for the Helsinki Region wastewater treatment plant. Here, the standards of 4.5 mgN/L and 0.3 mgP/L must be achieved to discharge into the eutrophicated Baltic Sea. On the other hand, standard for P discharge in very sensitive watershed are already as low as 0.1 mgP/L and further lowering around Europe is planned.

Table 1.1 Standard for nutrient discharge in sensitive watersheds lower than European legal requirements.

Country	TN (mg/L)	TP (mg/L)	NH_4-N (mg/L)	NO_2-N (mg/L)	NO_3-N (mg/L)	NH_3 (mg/L)
Italy	10	0.5	2	0.3		
Austria	>70% removal (for Temp > 12°C)	1	5 (for Temp > 8°C)	1	–	n/a
Finland	>70% removal and max 20 mg/L (for T > 12°C) Current average 4.5 mg/L	<0.3				
Germany	10	0.1	5	–	–	–
Greece		1	2	0.15	7	0.16
Ireland	15	2				
Netherlands	20	1	10			
Norway	– (70% removal)	0.25 (95% removal)				
Poland	10	0.25	3	0.03	7	
Sweden	8	0.4				
United Kingdom	>80% removal, 10 mg/L	0.5	3			
Switzerland	>70% removal	0.8	$NH_4 + NH_3$: 2 mg/L N (for T > 10°C)	0.3		

When considering the questions "how low can we go" and "what is stopping us from going lower" (WERF, 2010), we must consider that the nutrient challenge consists in striking the balance between nutrient removal, greenhouse gas emissions, receiving water quality, and costs, so a triple bottom line (TPL) analysis is needed to include environmental, economic, and social pillars (Falk *et al.* 2013).

To achieve the new, lower effluent limits that are close to the technology-best-achievable performance, facilities have begun to look beyond traditional treatment technologies (U.S. EPA, 2007). Nutrient removal processes could be classified in three "levels" of effluent concentration: i) achievable with conventional nutrient removal technologies (8 mgN/L and 1 mgP/L); ii) enhanced removal requires tertiary treatment and chemical addition to achieve low concentrations (3 mgN/L and 0.1 mgP/L); iii) requires state-of-the-art technology and enhanced/optimized treatment operation, especially to simultaneously achieve both the very low N and P levels (1 mgN/L and 0.01 mgP/L).

The more is the nutrient removal technology complexity, the more is the energy consumption and the Greenhouse gas (GHG) emissions, which largest contributors were found to be energy related (Falk *et al.* 2013) (Table 1.2).

Therefore, energy efficiency in nutrient removal in wastewater treatment plants (WWTPs) is clearly one of the key pillar to consider for the water-energy-carbon nexus.

Table 1.2 Energy consumptions and GHG emissions estimated by Falk *et al.* (2013) for a treated flowrate of 40000 m^3/d municipal wastewater.

TN Limit (mgN/L)	TP Limit (mgP/L)	Specific Consumption kWh/m^3 (Increase %)	GHG Emissions (tonCO_2/year)
>10	>1	0.14 (baseline)	4590
8	1	0.17 (+20%)	5570
8	0.1–0.3	0.18 (+27%)	6600
2	0.1	0.20 (+41%)	7570
<2	<0.02	0.38 (+169%)	12950

1.1.2 Biological Nutrients Removal processes: microbial and energy overview

In recent times, there has been an increased emphasis on increasing the efficiency of BNR processes and reducing the operational costs. One means of improving the cost-effectiveness is by employing short-cut nitrogen removal, or nitrogen removal via the nitrite pathway (Table 1.3). This involves aerobic nitritation by AOBs coupled with anoxic denitritation by denitrifiers, thus necessitating the limitation of NOB growth and activity. Some WWTP operational conditions are known to favour AOB at the expense of NOB, such as the higher growth rate of AOB at temperatures higher than 25°C (Hellinga *et al.* 1998), as well as the lower affinity of NOB for oxygen, where a low dissolved oxygen (DO) concentration will favour nitrite accumulation instead of nitrate. Short-cut nitrogen removal reduces the oxygen demand of the WWTP by 25% through eliminating the need to oxidise nitrite to nitrate, while simultaneously reducing the COD needed for denitrification by 40% through eliminating the need to reduced nitrate to nitrite. Aeration is widely considered to be one of the main energetic costs associated with WWTP operation, while the external dosing of COD sources also increases costs due to the expense associated with the COD supply as well as the increased sludge production, where sludge processing and disposal also represents one of the main operational costs associated with WWTPs.

Table 1.3 Comparison of the conventional BNR with the advanced BNR processes.

Process	Oxygen Requirements ($kgO_2/kgN_{rem.}$)	COD Consumption ($kgCOD/kgN_{rem.}$)	Sludge Produced ($kgVSS/kgN_{rem.}$)	Total Treatment Cost[a] (€/kgN_{rem})
Nitrification/ denitrification	4.33	2.86	1–1.2	3–5
Nitritation/denitritation	3.26	1.72	0.8–0.9	1.5–2.5
Deammonification	1.71	0	<0.1	1–2

[a]Total cost includes both capital and O&M cost.

In Table 1.3, a comparison is made between the biomass production, COD and oxygen requirements associated with wastewater treatment plant processes performing COD, N and P removal, as well as their respective nitrogen and phosphorus removal levels (standardized per mg of nitrogen removed). It is clear from Table 1.3 the savings in COD and oxygen requirements as well as the reduced sludge production achievable through short-cut nitrogen removal as compared to conventional nitrification/denitrification.

The anaerobic ammonia oxidation (Anammox) process has also attracted much attention in recent years, since it achieves N removal from wastewater with even further reductions in aeration and COD requirements, as well as sludge production (Strous *et al.* 1997, 1999), as can be observed in Table 1.3. In Anammox, ammonium is oxidized directly to dinitrogen gas using nitrite as the electron acceptor. Anammox bacteria are autotrophic and require low oxygen concentrations to survive, or are otherwise rapidly outcompeted by comparatively faster-growing nitrifiers. Partial nitritation is often employed prior to the Anammox process in order to generate sufficient nitrite. The molar nitrite/ammonia ratio is about 1.3 in the Anammox reaction, due to the simultaneous production of a small amount of nitrate (Strous *et al.* 1997). Anammox is becoming increasingly employed during the treatment of high strength ammonium-containing wastewaters with low COD content, including the supernatant from sludge digesters, landfill leachates and industrial wastewaters (Wett 2006; van der Star *et al.* 2007; Ganigue *et al.* 2009; Lackner *et al.* 2014) due to its high potential to increase the cost-effectiveness in WWTPs, although it remains a sensitive process to operate in practice, since it is prone to inhibition by toxic compounds and the slow biomass growth rate results in lengthy start-up/recovery periods.

Phosphorus is another key nutrient that stimulates the growth of toxic cyanobacteria (blue-green algae), and has been found to often be the limiting nutrient leading to eutrophication (Mainstone & Parr 2002). While P can be removed via chemical precipitation, enhanced biological phosphorus removal (EBPR) processes promote the removal of phosphorus from wastewater without the need for the addition of these chemicals, leading to a more cost-effective option for phosphorus removal when operated successfully. The addition of chemicals not only increases the operational costs due to the demand of reagents, but also increases the sludge production. In the EBPR process, the group of organisms primarily responsible for phosphorus removal are known as the PAOs. In order to promote the development of PAOs and, consequently, P removal, anaerobic followed by anoxic and/or aerobic conditions are generally employed, thus combining very well with BNR processes designed for biological nitrogen removal. PAOs are able to take up carbon sources such as volatile fatty acids (VFA) anaerobically and store them as polyhydroxyalkanoates (PHA), providing them a selective advantage over most ordinary heterotrophs. However, a competitor group of organisms known as glycogen accumulating organisms (GAOs) are also capable of anaerobic VFA uptake and therefore can also be enriched under similar conditions as PAOs, consuming the generally limited VFA supply without contributing to P removal.

While the majority of the P is generally taken up under aerobic conditions in most conventional EBPR processes, simultaneous denitrification and P removal can save on aeration, minimise sludge production and reduce the demand of readily biodegradable COD, which is often-limiting (Table 1.4). The combination of nitritation with EBPR via denitritation can lead to further savings in both COD and oxygen demands (Table 1.4). Nevertheless nitrite accumulation (in the form of free nitrous acid) is known to inhibit P uptake by Polyphosphate Accumulating Organisms (PAOs) when present at high levels (Saito *et al.* 2004; Zhou *et al.* 2007), and can lead to the undesirable production of N_2O (a powerful greenhouse gas) (Zhou *et al.* 2008). This is of particular relevance for sludge unacclimatized to high nitrite levels (Zhou *et al.* 2011). BNR processes applying denitritation and P removal should avoid excessive levels of nitrite accumulation to prevent N_2O accumulation and maximize P removal. While the potential for increasing the cost-effectiveness of EBPR systems through increasing the P fraction removed anoxically is high, this still remains a challenge to achieve in practice since the aerobic zone can only be eliminated in segregated sludge systems and PAOs grow more quickly aerobically than anoxically, lowering their denitrification capacity.

Furthermore it should be noted that the estimates presented in Table 1.4 neglect the growth of Glycogen Accumulating Organisms (GAOs), which would increase the COD demand and therefore sludge production in systems where they proliferate. Some factors favouring the growth of GAOs include high

Table 1.4 Comparison between the nitrogen and phosphorus removal achieved by different BNR processes, with the corresponding COD and oxygen requirements and biomass production (X) indicated. Estimates obtained from the default yield coefficients of the ASM1 model (Henze *et al.* 2000), the metabolic models for Enhanced Biological Phosphorus Removal (EBPR) (Smolders *et al.* 1994; Kuba *et al.* 1996; Oehmen *et al.* 2010c) and a model of the anammox process (Ni *et al.* 2014). Estimates are normalized per mg of nitrogen removed and neglect maintenance/decay processes.

C, N, P			C, N			C	
Nitritation D-EBPR Nitrite	**Nitrification D-EBPR Nitrate**	**O_2-EBPR+ Nitrification Denitrification**	**Partial Nitritation Anammox**	**Nitritation Denitritation**	**Nitrification Denitrification**	**Aerobic C Oxidation**	**Units**
2.23	3.71	5.90	0.54	3.66	5.34	17.36	mgCOD[a]
0.94	0.94	0.75	0.95	0.77	0.68	0.00	mgN (via Nit/Denit and Anammox)
1.00	1.00	1.00	1.00	1.00	1.00	1.00	mgN (tot including assimilation)
0.16	0.16	0.17	0.01	0.05	0.07	0.23	mgP
3.06	4.08	4.86	1.63	3.72	4.70	5.73	mgO_2
0.94	0.94	3.04	0.54	2.64	3.74	11.63	mgX
14.1	23.6	34.1	50.0	69.4	71.4	74.6	COD/P

[a]COD assumed to be present as either readily biodegradable COD or acetate in the case of EBPR processes.

temperature, low pH and a very high acetate/propionate ratio (Lopez-Vazquez *et al.* 2009), as well as high DO concentrations (Carvalheira *et al.* 2014). In this sense, the increasing practice of operating BNR plants at low DO levels can be beneficial not only for reducing aeration costs, but also minimizing the growth of GAOs. Furthermore, there are differences between the different groups of PAOs and GAOs related to their capacity to denitrify. Experimental studies have shown that some clades of Accumulibacter (a common PAOs in BNR plants) are able to denitrify from nitrate onwards, while essentially all Accumulibacter clades denitrify from nitrite onwards (Carvalho *et al.* 2007; Flowers *et al.* 2009; Guisasola *et al.* 2009; Oehmen *et al.* 2010a). With respect to different microbial groups of GAOs present in WWTPs, some Competibacter sub-groups have been found to be able to reduce nitrate and nitrite, where others are able to reduce nitrate only or not denitrify at all, while Defluviicoccus Cluster I was able to reduce nitrate, but not nitrite and Defluviicoccus Cluster II were unable to denitrify (Kong *et al.* 2006; Burow *et al.* 2007). Overall, it is clear that comparatively few GAOs groups are able to metabolise nitrite as compared to PAOs, suggesting that combining nitrogen removal via the nitrite pathway with denitrifying EBPR is a potential means of eliminating GAOs (Taya *et al.* 2013), leading to further savings in COD, aeration and sludge production as compared to P removal via nitrate or oxygen.

1.2 REDUCING ENERGY FOOTPRINT NOW, BY RETROFITTING

Water_2020 members have been working to retrofit existing WWTP by applying innovative energy efficient nutrient removal technologies.

1.2.1 Sidestream technologies/systems

1.2.1.1 ELAN system: Pilot study and full scale retrofitting

The ELAN® (Autotrophic Nitrogen Removal in Spanish) process has been developed by the company FCC Aqualia (Spain) with the know-how of the University of Santiago de Compostela by means of collaborative work that was started in the year 2009 (Vázquez-Padín *et al.* 2014a). It is based on combining the partial nitrification and anammox processes (1.1), for nitrogen removal from wastewater, in a single aerobic unit where the biomass is grown in the form of granules (Vázquez-Padín *et al.* 2014b).

$$NH_4^+ + 0.85O_2 + 1.11HCO_3^- \rightarrow 0.44N_2 + 0.11NO_3^- + 2.56H_2O + 1.11CO_2 \quad (1.1)$$

This process has been successfully applied, at pilot scale, for the treatment of the reject water from WWTPs characterized by an ammonia concentration of 0.5–1.5 g NH_4^+-N/L and a temperature in a range of 18–31°C. Two locations have been tested in two WWTPs located in Galicia. In one case the used reject water was collected from an anaerobic digester treating the sludge from the plant (WWTP of Vigo) and in the other case sludge was co-digested together with agricultural wastes (WWTP of Guillarei).

At the moment a full scale plant of 115 m^3 is being started up for the treatment of the reject water in the STP of Guillarei. This reactor has been designed to treat 67 kg N/d corresponding to 23% of the total nitrogen treated in the WWTP (Vázquez-Padín *et al.* 2014b). Once this reactor is implemented, an improvement of the quality of the produced effluent is expected by means of the decrease from 15 to 13 g TN/m^3 in the effluent of the plant.

Special features of the process

The ELAN® process is specially appropriated for the removal of nitrogen from wastewater streams characterized by the low BOD_5/TN content which would require external addition of carbon source for the performance of the conventional nitrification-denitrification processes. Taking into account that the anammox process performs in optimal conditions at temperatures in the range of the mesophilic values, the effluents produced from sludge anaerobic digesters in STPs are appropriated to be treated by the ELAN® process. Its application to treat this sidestream will allow for the obtaining of a better quality effluent, in terms of nitrogen content, of the STP as less nitrogen has to be treated in the mainstream treatment system. This process can be easily integrated in the plant as an extra unit after the anaerobic digester provided with a previous equalization tank.

The ELAN® process is carried out in sequencing batch reactors (SBR) operated in cycles comprising feeding, reaction (aerobic), settling and withdrawal periods. The feeding can be performed throughout the reaction phase.

AOB are placed in the external layers (aerobic) of the granular biomass while the anammox is in the inner (anoxic) part. The operation of the system is based on the control of the dissolved oxygen concentration in the bulk liquid which affects the oxygen depth penetration inside the granules (Morales *et al.* 2015a).

The tested hydraulic retention time (HRT) applied to this system ranged from 0.8 to 1.5 days. The DO concentration was maintained around 1–2 mg O_2/L according to the proposed control strategy in Vázquez-Padín *et al.* (2010) based on the modification of HRT and DO concentration. Later, this strategy allowed for the speed up of the start-up process of a single unit where the partial nitrification and anammox processes took place (Vázquez-Padín *et al.* 2012). This combined strategy was used to avoid limiting or inhibitory conditions for the process caused by concentrations of ammonia and nitrite, either low or high, respectively. Furthermore, from the experimental results, a correlation between the conductivity values measured in the reactor and the amount of nitrogen removed in an operational cycle was estimated which allows for the estimation of the nitrogen removal capacity of the system. From the stoichiometry corresponding to the ELAN® process [1], the fact that the maximum achievable nitrogen removal percentage cannot be higher than 90% is clear, as around 10% of the fed ammonia is transformed to nitrate.

Summary of results obtained from pilot plant studies and roadmap to full scale application

Experiments performed in pilot plants indicated that reject water with loads of 1.1 kg N/$m^3 \cdot$ d can be treated with efficiencies larger than 80% in an ELAN system (Vázquez-Padín *et al.* 2014b). The amount of produced biomass is reduced by 90% compared to conventional nitrification-denitrification systems. In this unit no nitrite oxidation to nitrate was observed in the media. From previous calculations it has been estimated that the full scale ELAN in the STP of Guillarei will remove about 25% of the total nitrogen reducing the footprint and reaction volume of more than 90%. In addition, in case of co-digestion the application of an anammox based process seems to be the only option if the aim is to valorise as much organic matter as possible for biogas production (Vázquez-Padín *et al.* 2014b).

The application of the ELAN® process for the treatment of the mainstream o STP is being under extensive study at the moment. However certain identified drawbacks need to be solved before its application at full scale. These are the maintaining of an appropriated balance between AOB and anammox bacteria activity and the complete suppression of the NOB (Al-Omari *et al.* 2015). The maximization of the biomass retention in the system is crucial to guarantee the stability of the process by means of the control of the minimum settling velocity of the biomass fixed by means of the settling time in the operational cycle. Regarding the produced effluent besides the presence of nitrate, previously mentioned, and considering the mass transfer diffusion inside the granules also minimum ammonia concentrations will be present. A minimum amount of this compound (0.48 mg NH_4^+-N/mg O_2) has to be present in the bulk liquid to assure the consumption of all the DO by AOB which guarantee the removal of nitrite in the anoxic zone (Morales *et al.* 2015b).

1.2.1.2 Attached growth biofilm (RBC-MBBR) pilot study and full scale retrofitting

The process of partial nitritation/anammox was first observed in a rotating biological contactor (RBC) for the treatment of landfill leachate in Mechernich, Germany (Hippen & Rosenwinkel 1997), Kölliken, Switzerland (Siegrist *et al.* 1998) and Pitsea, Great Britain (Schmid *et al.* 2003). Then the research on the use of RBC were mainly investigated in lab-scale. Generally, the RBC technology is popular because of its simple construction and operation. Additionally, the system provides stability and high resistance to load changes and toxic substances inflow. However, the occurrence of the anammox process in these installations was rather not an intended action, but the process developed spontaneously in the oxygen limited conditions. For the successful operation of the partial nitritation/anammox process, except of very efficient biomass retention, a good balance between aerobic Ammonium Oxidizing Bacteria (AerAOB) and Anaerobic Ammonium Oxidizing Bacteria (AnAOB) is needed. In the RBC such a balance is obtained spontaneously due to the limited penetration of the oxygen in the biofilm (De Clippeleir *et al.* 2009). In the RBC units, the aeration can only be controlled by the immersion level and rotation speed, and in this way the oxygen supply plays a crucial role in the performance of partial nitritation/anamox in RBC. Courtens and coworkers (2014) demonstrated that nitratation could be controlled by the discs immersion level and not the rotational speed. Pynaert *et al.* (2003) showed that this system can be highly loaded, while Cema *et al.* (2007) demonstrated that nitrogen can be effectively removed in RBC unit even at temperatures below 20°C. Recently, it was indicated about full scale implementation of partial nitritation/ anammox RBC for treatment of decentralized wastewater. This investment was led by DeSah BV, Sneek (Netherland), for digested black water, with currently a 0.5 m^3 RBC serving 64 population equivalents (PE), and a 6 m^3 reactor for 464 PE. In this case, the process control was based on the rotation speed (1–4 rpm) and by setting the pH at 7.0–7.5 (Lackner *et al.* 2014).

One of the first dedicated implementation of the partial nitritation process was realized in Moving Bed Biofilm Reactor (MBBR) at the Hattingen WWTP near Bochum (Germany). The MBBR process is based on the biofilm principle that take advantage of both activated sludge process and conventional fixed film systems without their disadvantages. The reactor can be operated at very high load and the process is insensitive to load variations and other disturbances (Odegaard *et al.* 1994). Additionally, unlike most biofilm reactors, the reactor volume in the MBBR is totally mixed and consequently there is no dead or unused space in the reactor. The process performed in MBBR reactor was called DeAmmon®. This process is realized in biofilm system for two purposes: to ensure biomass storage in the reactor and to achieve necessary sludge age, and to provided a biofilm with different layers for nitritation bacteria at outer aerobic layer and anammox bacteria in the inner layer (Ling, 2009).

The first full scale partial nitritation/anammox pilot plant was erected in cooperation of the institute of Water Quality and Waste Management, University of Hanover (ISAH), the Ruhrverband and the PURAC GmbH (Rosenwinkel & Cornelius 2005; Gustavsson, 2010). The reactor had got an overall volume of 319 m^3, and was one reactor biofilm system, with three stages, filled in 40% with Kaldnes carriers K1 (effective area is equal to 500 m^2/m^3). To retain the media within each stage, the sieves were provided between stages. The reactor was at first operated as nitritation denitritation process with external carbon source, as a necessary step to build the biofilm structure on the carrier material. As nitritation control by pH appear too expensive (huge amounts of chemicals were required), the aeration control was used to ensure stable nitritation. At first the low oxygen level was adjusted (below 1 mg/l), then intermittent aeration was tested. 18 months later the nitrogen removal reached 70–80% at a load between 100–160 kg N/d. It must be also stated, that this first full scale application was characterized by quite high energy consumption equal to 5.6 kWh/kgN removed, and such high energy consumption was explained by low nitrogen load (Gustavsson, 2010).

Generally, the experience form the first full scale application as well other studies at semi technical scale (Rosenwinkel & Cornelius 2005; Szatkowska *et al.* 2007; Cema *et al.* 2010) indicated that one stage MBBR with simultaneous partial nitritation/anammox process was better option with higher nitrogen removal rates with lower operation temperature. However, as for the overall process performance, the nitrite production is the rate-limiting step (Szatkowska *et al.* 2007), the dissolved oxygen control is one of the crucial factor impacting the nitrogen removal efficiency as well as nitrogen removal rates. At Hattingen it was discovered that during the intermittent aeration, the mechanism of lag phases of nitrosomonas and nitrobacter under anoxic condition was useful for nitrobacter suppression and high process stability. According to Gaul *et al.* (2005) for fast start-up of partial nitritation anammox process the short HRT is required in order to wash up the suspended organisms competing with mass transfer limited biofilm cells for alkalinity as limiting substrate. Other mandatory precondition is an oxygen flux adapted to the surface loading rate prevent complete nitrification.

Based on the experience at Hattingen and the Royal Institute of Technology (KTH), the second DeAmmon® plant was started at Himmerfjärden WWTP (260000 P.E.) in Stockholm, Sweden (Ling, 2009). Two existing pre-sedimentation tanks were converted into MBBR tanks. The DeAmmon® plant is designed for a nitrogen loading of 600 kg/d. The start-up times for two lines were 9 and 12 month, however the effective start up time has been equal to 6–7 months. In this plant, the designed energy consumption was to 2.3 kWh/kgN removed. Since the penetration of oxygen into the biofilm in MBBRs limits the reduction rates, higher DO concentration is needed in such systems at Himmerfjärden WWTP has a DO set-point between 3–4 mg O_2/L.

One-stage partial nitritation/anammox process in MMBR system called ANITA™Mox was developed by AnoxKaldnes/Veolia. It was shown that the use of new carriers materials such as BiofilmChip™ M (1200 m^2/m^3), Anox™ K5 (800 m^2/m^3) or new Chip K3 (500 m^2/m^3) allows for compact design and simple operation with very high biomass retention ability (Lemaire *et al.* 2011; WERF, 2014). The very slow start-up of the partial nitritation/anammox process due to the very slow growth of the anammox bacteria was considered as a big drawback of the biofilm process. During the start-up period bacteria are very sensitive to high oxygen concentration, nitrite and free ammonia. Additionally, studies performed by Schneider *et al.* (2009) showed, that for the biofilm system, seeding with active biomass is not a sufficient start-up strategy. However, the research performed at Sjölunda WWTP showed that the seeding with small fraction of colonized carriers can significantly reduce the start-up period from up to year down to 2–3 months (Depending on the amount of seeding) (Lemaire *et al.* 2013; Christensson *et al.* 2013). However, further study performed by Kanders *et al.* (2014) showed that the addition of seeding carriers enabled immediate nitrogen removal performance, having no influence on the total start-up time. The results indicate that anammox bacteria contained in the inflow reject water were the relevant source for enrichment on the virgin carriers, while anammox originating from the seeding biofilm did not contribute significantly.

The first AnitaTMox full scale application was started up in 2010 at Sjölunda WWTP in Malmö, Sweden. This unit was made of four parallel 50 m^3 reactors in order to test different carrier filling degree (from 40–50%) and to test different types of carriers. The reactor was operated at temperature of 27–30°C with 40% media fill volume and designed load of 200 kgN/d, pH of 6.7–8.0 without any control and with DO in range of 0.5 to 1.5 with continuous aeration mode. In this condition, after 4 months of operation (with 3% of seeding medium) the ammonium nitrogen removal rate reached 1.3 kgN/m^3d with 90% of ammonium removal efficiency. These was achieved without any pretreatment, without any chemical addition (external carbon, acid or base solution), without no need of mechanical mixing (continuous aeration strategy) and without any heating system. In this system, the energy consumption was equal to 1.45–1.75 kWh/kgN removad. Additionally, this system is characterized with small carbon footprint as the N_2O emission is in the range of 0.2–0.9% of reduced nitrogen (Christensson *et al.* 2013). Table 1.5 reports the overview of the main biofilm full scale partial nitritation/anammox systems.

Table 1.5 An overview of the volumetric removal rates in full scale partial nitritation/anammox Rotating Biological Contactors (RBC) and Moving Bed Biofilm Reactors (MBBR).

Reactor Type	Plant	Volume (m^3)	Nitrogen Removal Rate (kg N $m^{-3}d^{-1}$)	Reference
RBC	Kölliken	33	0.4	Siegrist *et al.* (1998)
RBC	Pitsea	240	1.7	Schmid *et al.* (2003)
MBBR	Himmerfjärden	1400 (2 × 700)	0.6	Ling (2009)
MBBR	Sjölunda	200 (4 × 50)	1.2	Christensson *et al.* (2013)
MBBR	Sundet	300	0.63[a]	Christensson *et al.* (2013)

[a]The N loading limited by reject water supply (200 kgN/d instead of 320 kgN/n initially expected) (Lemaire *et al.* 2013).

1.2.1.3 Short-Cut Enhanced Nutrient Abatement (S.C.E.N.A): pilot study and full scale retrofitting

Completely autotrophic nitrogen removal is becoming the most attractive biological process for the treatment of sludge reject waters in municipal WWTPs with several full-scale applications (Lackner *et al.* 2014). This process cannot enhance the phosphorus bioaccumulation and should be followed by struvite crystallisation for sustainable phosphorus recovery. However, if the economic and life cycle costs are taken into account, phosphate recovery as struvite is not considered the only sustainable alternative, and the use of composted sludge was considered another efficient phosphate based compounds for fertilisation (Hao *et al.* 2013). In consolidated schemes for EBPR, PAOs take up excess phosphorus and store it as polyphosphate in the cell mass using the energy from the aerobic heterotrophic oxidation of organic materials. If PAOs are exposed to anaerobic conditions, they obtain energy from the hydrolysis of the accumulated polyphosphate to take up VFAs as PHAs. On the other hand, the novel denitrifying biological phosphorus removal via nitrite (DBPRN) offers the potential to integrate phosphorus and nitrogen removal in a robust process in which ammonium is oxidised to nitrite under aerobic conditions, while under anoxic conditions denitrification via nitrite and enhanced biological phosphorus uptake occur simultaneously by the DPAOs. The DBPRN process is enhanced when high percentages of propionic and butyric acids are used as carbon source (Frison *et al.* 2013), which can be obtained in-situ by alkaline fermentation of sewage sludge (Longo *et al.* 2015). The sludge derived alkaline fermentation liquid was rich in propionic and butyric acid and has been found to be a high added value carbon source which drives the competitive advantage towards PAOs against the GAOs. Several studies demonstrated that using sludge-derived Short Chain Volatile Fatty Acids (SCFAs) resulted in superior BNR performance than using synthetic acetate (Tong *et al.* 2007; Zheng *et al.* 2010). Higher phosphorus removal efficiency was achieved with the use of SCFAs derived from waste activated sludge (WAS) compared to acetate (Tong *et al.* 2007). The authors explained that the presence of propionate was probably the reason for better phosphorus removal, while the higher nitrogen removal efficiency might be due to the better use of exogenous denitrification pathway for nitrogen removal. In addition, sludge fermentation liquid was reported to significantly reduce the NO and N_2O emissions from biological nitrogen and phosphorus removal during anaerobic-low DO process.

Further to these background, the University of Verona developed the S.C.E.N.A. (Short-Cut Enhanced Nutrients Abatement) system for reject water treatment in collaboration with the public-owned water utility Alto Trevigiano Servizi. S.C.E.N.A. can be described according to the following key processes: (1) alkaline fermentation of sewage sludge for the production of the best available carbon source (BACS) which is a VFA-mixture with the high content of propionic acid; (2) nitritation in aerobic conditions (DO > 1.5 mg/L, so as to also minimize N_2O emissions); (3) denitritation and via-nitrite biological phosphorus uptake achieved through the dosage of the BACS; (4) process control on the basis of low-cost sensors of pH, conductivity and redox potential. Initially the SCENA scheme was developed at pilot-scale at the municipal treatment plant of Treviso for nitrifying and denitrifying biological phosphorus removal via nitrite the anaerobic co-digestate of waste activated sludge and the organic fraction of municipal solid waste (OFMSW). Here nutrient removal was successfully tested at low volumetric nitrogen loading rates (vNLR = 0.2 kgN m^3/d), at the vNLR of the system's nitrifying capacity (0.8 kgN m^3/d) and above its nitrifying capacity (1.1 kgN m^3/d). The phosphorus uptake rate under anoxic conditions was, on average, 10.7 and 9.2 and mg P/VSS · h and occurred at nitrite concentrations exceeding 10 mg/L and up to140 mg/L.

After the successful pilot-scale application in the municipal wastewater and OFMSW treatment plant Treviso, the SCENA pilot plant was tested in a conventional municipal wastewater treatment plant and applied in the Carbonera WWTP (Veneto region, Italy), where the BACS was produced from the alkaline fermentation of sewage sludge. In the sludge fermentation process within SCENA, the production of SCFAs by alkaline fermentation has proved to be highly dependent on pH and temperature. The use of wollastonite was successfully tested and the fermentation liquid consisted mainly of acetic, propionic and butyric acid (37, 34 and 15% respectively). At the same time the addition of wollastonite decreased the capillary suction time and time to filtration (by 51% and 59% respectively), resulting in more favorable dewatering potentials. The fermentation liquid produced was tested as a carbon source for nutrient removal into scSBR leading to the nutrient removal rates reported in Table 1.6.

Table 1.6 Specific nutrient removal rates.

sAUR (mgN/gVSS · h)	10–15
$sNUR_{BACS}$ (mgN/gVSS · h)	45–70
sPUR (mgP/gVSS · h)	4.5–8

According to the best parameters derived from the pilot-scale experimentation, the first full scale S.C.E.N.A. system was designed and realized to treat 7,66 kgN/h and 0,35 kgP/h which are respectively equivalent to 26,2% and 12,4% of the total loading influent to the mainstream of the Carbonera WWTP. In order to optimize the revamping of existing facilities, the SCENA system was realized underground in a storage tank for septic tanks dismissed more than 15 years ago. The existing tank was divided in: (1) reject water equalization (40 m^3); (2) Short-cut Sequencing Batch Reactor (70 m^3); (3) Alkaline sludge fermentation (50 m^3); (4) Storage sludge fermentation liquid (BACS) (50 m^3). In addition, the separation of the fermentation liquid was carried out in a energy-effective screw press. Finally, the SCENA system of Carbonera is described in Figure 1.2.

On the basis of conservative assumptions of average nitrogen removal of 30 kgN/d, the O&M costs for nitrogen removal were calculated and are reported in Table 1.7.

The S.C.E.N.A. O&M cost of 1.6 €/kgN removed are much lower than the O&M cost of the mainline (3.5 €/kgN removed, Renzi *et al.* 2015) even not taking into account the economic advantages in terms of enhanced phosphorus uptake. Current research efforts are dedicated to the optimization of sludge fermentation, including the fermentation of cellulosic sludge in order to optimize the VFA content of the fermentation liquid. In addition, further investigations on the full scale SCENA system will deal with the NO and N_2O emissions which should be lower than micro-aerobic processes thanks the carbon source and the complete aerobic conditions. Finally, the installation of the dynamic thickener of the sewage sludge before the fermentation will optimize the VFA content of the BACS and reduce the O&M costs up to 1.2–1.4 €/m^3.

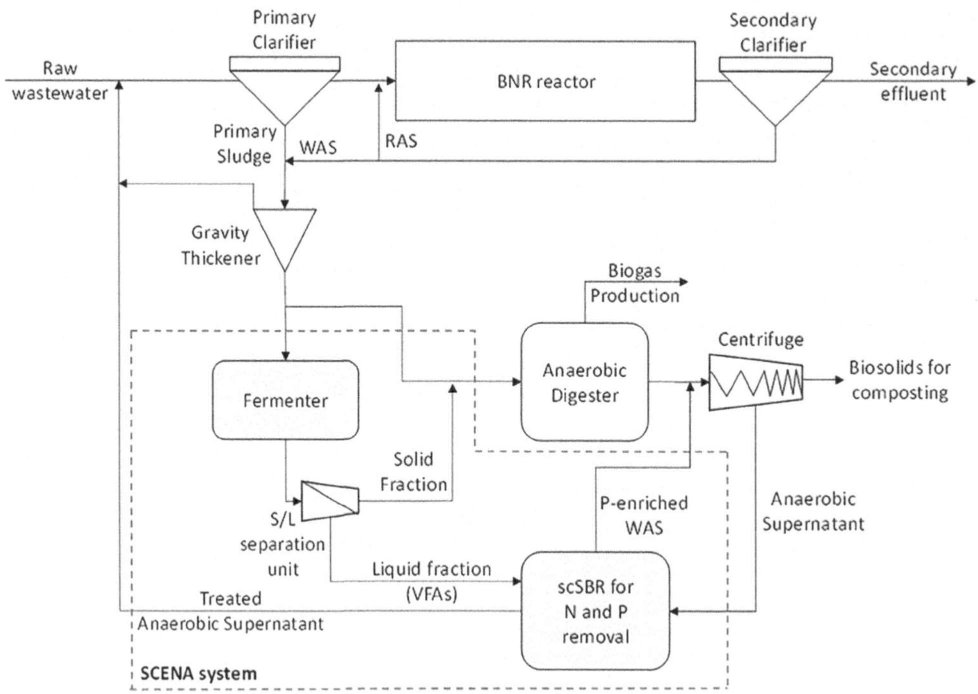

Figure 1.1 The S.C.E.N.A. system integrated in Carbonera municipal WWTP.

Table 1.7 Cost for the nitrogen removal in the anaerobic supernatant with S.C.E.N.A system.

Operation Unit	Unit	Value	Percentage	O&M (€/kgN Removed)
Reject water equalization	kwh/d	9	7%	0.05
Sequencing Batch Reactor	kwh/d	56	43%	0.31
Sludge fermentation	kwh/d	48	37%	0.26
Solid/Liquid separation	kwh/d	17	13%	0.09
Energy consumption	kwh/d	130	100%	0.72
Polyelectrolite	kg/d	8.5		0.41
Excess sludge	ton/d	0.07		0.16
Personnel				0.21
Maintenance				0.10
Total				1.60

1.2.2 Mainstream technologies/systems

1.2.2.1 Innovative (anoxic) BNR (BioP) including control automation

As for the mainstream, major focus for the technologies ready-to-retrofit is given to the process control automation potential, which is considered a crucial action towards energy saving by retrofitting existing WWTPs.

Using novel benchmarking scenarios, some recent works have developed, evaluated and optimised novel control strategies for simultaneous organic matter, N and P removal considering different performance criteria in order to increase stability of EBPR process, reducing operational costs and improving effluent quality (Guerrero *et al.* 2014, 2013, 2012, 2011b; Machado *et al.* 2009; Ostace *et al.* 2013). The main control loops that have been studied for the typical anaerobic/anoxic/oxic (A2/O) configuration are detailed below.

Total suspended solids (TSS) is usually considered an inventory variable and hence it is controlled to be maintained in a fixed value in all the control scenarios for comparison purposes. A proportional-integral (PI) feedback control loop using the purge flow-rate as manipulated variable, and TSS in a reactor as measured variable is the typical configuration. This control loop guaranties a proper sludge retention time (SRT) where PAO and nitrifiers can coexist (Machado *et al.* 2009).

Dissolved oxygen in the aerobic reactors is usually measured and controlled by manipulating the aeration flow-rate or the oxygen global mass transfer coefficient. Typical PI controllers with a fixed DO setpoint are used for this purpose. A more efficient alternative is a cascade control structure with a PI algorithm controlling the ammonium concentration in the last aerobic reactor (primary control loop) by modifying the DO setpoint of the secondary DO control loop (Guerrero *et al.* 2012, 2011b; Ostace *et al.* 2013).

Nitrate control loops are implemented to control the amount of nitrate recycled to the anoxic reactor. This is a key parameter to maintain a good EBPR activity. If an excess of electron acceptor is introduced in the anoxic reactor, less organic matter will be fermented to VFA and hence lower PAOs activity would be observed. The recommended control loop is based on a PI feedback controller using the internal recycle flow-rate from the aerobic reactor to the anoxic reactor as manipulated variable and the nitrate concentration in the last anoxic reactor as the measured variable. Alternative measurement points as the last aerobic reactor have been proposed for nitrate, but the anoxic reactor provides the better decoupling of control loops, avoiding the generation of internal disturbances because of the effect of different control loops (Machado *et al.* 2009; Ostace *et al.* 2013).

The implementation of some of these control strategies can improve N/P removal in full scale WWTP. Automatic control of the WWTP can greatly improve the operational costs of the plant while maintaining low pollutant effluent concentrations and achieving a more stable performance even under intense operation. Improvements for the operational costs depend on the specific scenario of WWTP conditions and influent flow-rate, composition and variability. Improvements in operational costs with respect to non-optimized reference operation are reported as 13%–14% (Machado *et al.* 2009; Ostace *et al.* 2013). However, different operational points can be selected to balance different criteria such as effluent quality, operational costs, microbiological risks (Guerrero *et al.* 2012) or GHG emissions (Flores-Alsina *et al.* 2011a). Regarding GHG emissions, N_2O production can have a high impact on the overall carbon footprint of the WWTP (Daelman *et al.* 2013), where some operational factors are known to increase N_2O production, including nitrite accumulation, DO level, COD/N ratio and operational transitions. Validating the impact of online control strategies on the global GHG emissions experienced by WWTPs is still needed in order to improve our ability to simultaneously limit GHG emissions while optimising process performance and minimising energy expenditures.

In addition to the control strategies that can improve nutrient removal, other new operational strategies and configurations have been also proposed for full scale application. Some of them are based on taking profit of PAO denitrification capacities for removing nitrite and nitrate using external carbon sources or using the internal substrate stored during the anaerobic phase. DPAO activity is usually considered to have lower sludge production and hence its treatment cost can be reduced. Guerrero *et al.* (2013) compare the performance of different EBPR configurations and operational costs under different model assumptions. Inclusion of nitrite as state variable allows a more realistic description of N removal, better accounting of the organic matter needed to denitrify (i.e. denitritation requires less COD than denitratation). It enables a better description and understanding of the competition between PAO and OHO for the carbon source, especially in systems operated to achieve nitritation for short-cut nitrogen removal. Regarding the effect of the plant configurations on C/N/P removal, the highest P removal is obtained for Johannesburg (JHB) and Modified University of Cape Town (MUCT). UCT and BarDenPho-5-stage (BDP-5-stage) configurations resulted in the lowest P removal because of the high amounts of NOX entering the anaerobic zone, favouring heterotrophic denitrification instead of EBPR. On the other hand, if effluent quality and operational costs are considered simultaneously Anaerobic-Anoxic-Oxic (A2/O) results in the most balanced configuration. It does not require excessive carbon source addition to meet P discharge limits and presents lower operational costs than other configurations.

Experimental studies evaluating new alternatives for EBPR

Pilot studies have shown that anoxic P removal can be maintained at long term using configurations as MUCT (Zeng *et al.* 2016, 2013). Nitritation is established through short hydraulic retention time (6 h) and low dissolved oxygen concentration (0.3–0.5 mg/L). Anoxic P removal around 88% was achieved in the pilot MUCT process treating municipal wastewater during a year. Change of operation from complete nitrification to nitritation demonstrated changes in the microbial community, being PAOII the dominant PAO when operating with nitritation. Moreover, a positive correlation with the nitrite accumulation ratio was observed for P removal efficiency, which was 30% higher than that under complete nitrification, suggesting that nitrite was appropriate as electron acceptor for denitrifying P removal when treating carbon limited wastewater. In addition, simultaneous nitritation/denitritation and anaerobic ammonia oxidation (anammox) possibly contribute to nitrogen removal in the aerobic zone (Zeng *et al.* 2014). Another pilot-scale process combined an anaerobic baffled reactor and an aerobic membrane bioreactor (Wu *et al.* 2015). The combination of both reactors allowed nitritation, simultaneous nitrification/denitrification and denitrifying P removal, making it a process with low energy demand and efficient nutrient removal.

The previous conventional WWTP configurations have the drawback of being single sludge systems, where heterotrophs, nitrifiers and PAO must coexist alternating different operational conditions as anaerobic, anoxic and aerobic reactors. Two-sludge systems are proposed to obtain better conditions for the different groups. Marcelino *et al.* (2011) propose a combination of two SBRs (a heterotrophic SBR with denitrifying PAOs for P removal and an aerobic SBR for N removal) to achieve simultaneous biological C, N and P removal with DPAO activity, but also implementing the nitrite pathway. Partial nitrification is obtained in the autotrophic SBR by controlling aerobic phase length, while anoxic dephosphatation activity of DPAO using nitrite is obtained in the heterotrophic SBR. This configuration overcomes the potential drawbacks of one-sludge systems for the treatment of low COD/N wastewaters, where the aeration phase disfavours denitrifying phosphorus removal and a more significant fraction of COD is consumed aerobically. Integration of nitritation with PAO activity should be implemented taking care to avoid high nitrite concentrations in the anoxic reactor, because free nitrous acid (FNA) is an important inhibitor of biological activity. Although PAO are capable of denitritation (Oehmen *et al.* 2010a), aerobic and anoxic PAO inhibition due to FNA has been widely reported, especially when nitrite concentration is high and pH decreases. Then, continuous stirred tank reactors are recommended or long feeding phases in SBR configurations. Slow feeding of nitrified wastewater with high nitrite concentration is recommended because it allows maintaining low FNA concentration in the anoxic reactor. Other two-sludge configurations have been also reported for continuous systems (Zou *et al.* 2014), although the separation of nitrifiers and heterotrophs imply the need of additional settlers that are not required in conventional configurations. On the other hand a real anoxic zone with very low oxygen input is obtained, favouring DPAO denitrifying activity using internal storage polymers.

Other innovative P-removal WWTP configurations are being proposed for P-recovery in addition to removal (Acevedo *et al.* 2015; Kodera *et al.* 2013; Lv *et al.* 2014; Valverde-Pérez *et al.* 2015; Zou *et al.* 2014). Most of these configurations are based on the PAO capacity to produce an effluent with much higher P concentration than the influent to the WWTP. Effluent from the anaerobic reactor (or anaerobic phase in SBRs) after VFA uptake multiplies P concentration and provides an enriched stream with high ammonium and phosphate concentration which is more suitable for recovering as precipitates of struvite or other P salts. The only additional requirement is the need of a supplementary settling phase to separate the enriched P stream. On the other hand, P recovery as a secondary enriched stream will have an additional benefit for the process, as it would decrease the amount of COD required for P removal, which can be of interest in carbon-limited wastewaters.

Finally, a new P removal process in association with sulphur cycle has been recently reported (Wu *et al.* 2013). This process is of particular interest in regions as Hong Kong, where it has practiced seawater toilet flushing since 1958 for freshwater savings, resulting in high sulphate concentration in the wastewater. Sulfate reduction Autotrophic denitrification and Nitrification Integrated (SANI) process was expanded with EBPR, describing a novel sulfur cycle associated to EBPR processes. The new EBPR process is supported by observations as anaerobic phosphate release associated with acetate uptake, poly-P hydrolysis, PHA synthesis and aerobic P-uptake associated to PHA (and poly-S2-/S0) degradation, and poly-P formation; biomass with high P content; absence of chemical precipitated P crystals; P sludge with more than 90% polyP and microscopic observation of PHA, poly-P and S globules in the biomass. Recently, a further development of the anaerobic/anoxic denitrifying sulphur cycle-associated EBPR, named as DS-EBPR, has been reported for an alternating anaerobic/anoxic SBR with simultaneous COD/N/P removal (Wu *et al.* 2014). It was confirmed the sulphur cycle-associated biological P uptake utilizing nitrate as electron acceptor. This new bioprocess reduces operation time and enhances volumetric loading of SBR compared with the alternating anaerobic/oxygen-limited aerobic. Interestingly, no conventional PAO are detected in the sludge. A synergistic relationship may exist between sulphur cycle and biological phosphorus removal, but further research is required in this new configuration before considering its full-scale application.

1.2.2.2 Ion exchange for P recovery: pilot study and full scale retrofitting

The recent development of a new generation of NH_4-N and PO_4-P selective media, possessing enhanced exchange capacity and selectivity, has boosted the prospect of ion exchange for wastewater treatment.

Ion exchange is a process unit in which ions of a given species are displaced from an insoluble exchange material by ions of a different species in solution, hence it can be viewed a separation process based on the mass transfer between liquid-solid phases. The exchange of ions is reversible between phases, meaning that there is no permanent change to the resin and this can be regenerated using a brine solution. This involves the transfer of one or more ions from the aqueous phase to the solid by exchange or ion displacement with the same charge that are united by electrostatic forces to superficial functional groups. The process efficiency depends in the liquid-solid equilibrium and the mass transfer velocity. Ion exchange processes offer a number of advantages including concentration of the nutrients into small volumes, ability to handle shock loads, insensitivity to temperature, an ability to reach near zero effluent concentrations and the ability to recover the nutrients for reuse/sell as a product.

Selective ion exchange is a promising alternative to biological nitrification for the removal of ammoniacal nitrogen (NH_4-N) at WWTP, where it has been applied to the treatment of full-flow secondary effluent and return-flow anaerobic sludge liquors (Mercer *et al.* 1970; Koon & Kaufman 1975; Svetich, 1993; Mackinnon *et al.* 2003; Aiyuk *et al.* 2004; Hankins *et al.* 2005; Thornton *et al.* 2007). This includes full-scale (ca.15,000 m^3 d^{-1}) operation, where a zeolite media (clinoptilolite) was used as the primary NH_4-N treatment at a WWTPs over a period of 30 years (Svetich, 1993, 2015). Ion exchange media operates by a physico-chemical mechanism that selectively removes dissolved ammonium ions (NH_4^+) upon contact with wastewater and substitutes them for ions initially residing within the media (e.g. sodium, Na^+); thereby maintaining charge balance. Ion exchange can readily achieve <1 mg NH_4-N L-1 in the effluent and operates consistently at low temperatures and variable NH_4-N loads that may be detrimental to conventional biological nitrification for NH_4-N removal (Thornton *et al.* 2007). Therefore IEX technology is resilient to the strict limits of consent for discharge proposed by legislation, i.e. the European Water Framework Directive (WFD), which are unlikely to managed technically or cost effectively by intensification of current methods (Georges *et al.* 2009). Ion exchange may significantly reduce the energy associated with NH_4-N removal by eliminating the aeration demanded for nitrification via the activated sludge process (ASP), which typically represents >50% of the total electrical cost for WWTP (Tchobanoglous *et al.* 2003; McAdam *et al.* 2011). It is suggested that the removal of nutrients by ion exchange, in tandem with emerging full-flow anaerobic carbonaceous treatments (i.e. anaerobic membrane bioreactors, AnMBR), may enable energy neutral STW by totally supplanting ASP (Martin-Garcia *et al.* 2013).

Recently, ion exchange was highlighted as an alternative full-flow PO4-P treatment method due to the emergence of commercially available, PO4-P selective, hybrid anion exchange (HAIX) media (Blaney *et al.* 2007; Kumar *et al.* 2007; Martin *et al.* 2009; Boyer *et al.* 2011). HAIX can readily achieve 0.1 mg PO4-P L-1 in the effluent, with a PO4-P capacity of 10–15 mg PO4-P g-1, demonstrating the promise of ion exchange as a future-proof technology (Martin, 2009). Prior to the introduction of HAIX the application of ion exchange for PO4-P removal was impractical due to poor media selectivity in the presence of competing anions, such as sulfate, nitrate and bicarbonate (Liberti *et al.* 1979). However, HAIX contains embedded iron nanoparticles with a specific affinity for PO4-P that provide generally superior removal versus alternative media (Boyer *et al.* 2011). Once the removal capacity of exchange media is approached and the PO4-P concentration in the exchanger effluent exceeds the desired limit, known as breakthrough, the media must be regenerated by passing several bed volumes (BV) of chemical solution through the exchanger to restore the initial removal capability. For example; HAIX uses a 2% sodium hydroxide (NaOH) solution to efficiently liberate media-bound PO4-P in to the caustic solution so that it may treat further anaerobic effluent.

1.2.2.3 Via-nitrite mainstream biological nitrogen removal

In the recent years, special attention has been given on partial nitrification denitrification (PND) as a mainstream nutrient removal perspective. In a PND process, ammonium oxidation takes place up to the level of nitrite via AOB and then nitrite is denitrified by denitrifying bacteria (DB) to molecular nitrogen, bypassing nitrate formation.

Compared to conventional N-removal, PND presents significant advantages, as it theoretically can save up to 25% of the required oxygen for complete nitrification, 40% of organic carbon for denitrification, and can achieve a lower sludge production (Abeling & Seyfried 1992; Chen *et al.* 1991; Katsogiannis *et al.* 2003; Turk & Mavinic 1987). In addition, it has been reported (Turk & Mavinic 1987; Peng & Zhu 2006) that nitrite reduction enhances the denitrification rate by 63% with a 33–35% lower sludge production during nitrification and 55% during denitrification. In addition, PND leads to reduced CO_2 emissions by 20% (Peng & Zhu 2006).

Several approaches have been proposed towards the accomplishment of partial nitrification-denitrification, such as variations in DO concentration, pH, temperature, free ammonia (concentration, and free nitrous acid concentration.

Using a frequent enough switching between aerobic and anoxic conditions in a SBR and an aerobic to anoxic phase ratio specific to the treated wastewater, it is possible to achieve PND, i.e., partially oxidize ammonia to nitrite (aerobically) and then subsequently reduce nitrite to nitrogen gas (under anoxic conditions), without producing nitrate (Katsogiannis *et al.* 2002, 2003; Kornaros *et al.* 2008). It was later proved (Kornaros *et al.* 2010) that what is exploited here, is the time lag exhibited by the NOB when anoxic conditions are changed abruptly to aerobic.

Among the various methods that have been proposed for partial nitrification, the proposed exploitation of the time-lag exhibited by nitrite oxidizing bacteria under alternating aerobic/anoxic conditions, seems most promising, as it does not require any addition of chemicals or extreme growth conditions (e.g., temperature), but a simple manipulation of the operating conditions. The key point, according to this strategy, is to enhance the nitritification process, while inhibiting or suppressing, at the same time, the nitratification process to obtain a mixed liquor biomass enriched in AOB and poor in NOB. If nitratification is inhibited to an extent that the minimum SRT required or NOB growth is higher than the current SRT of the system, then the NOB will be washed out and the effluent will not contain nitrate.

The proved methodology, with the aid of ENBIO Ltd (a spin-off company) was successfully transferred from the lab-scale SBR treating synthetic wastewater to a 2 m^3 pilot-scale SBR system (Katsogiannis *et al.* 2002), which was constructed and operated in the Wastewater Treatment Plant of Patras (Western Greece) treating raw municipal wastewater.

It was demonstrated that three pairs of aerobic/anoxic phases with a relative duration of 1:2 (8-hour cycle: fill 30 min, react 360 min, settle 60 min, draw/idle 30 min) secures the desired by-pass of nitrate production in the pilot SBR. The wastewater feed characteristics were 24.3 mg NH_4-N/L, 101 mg/L soluble COD. The mean temperature was 19.4°C and the Sludge Retention Time was 16.2 d.

The SBR operation may be optimized. To this end the models developed in (Katsogiannis *et al.* 2002; Kornaros *et al.* 2010) may prove quite useful. In addition, it would be interesting to develop a continuous WWTP process that will exploit the NOB time-lag concept, in other words to achieve in space, what is achieved by the SBR in time.

1.3 REDUCING ENERGY FOOTPRINT TOMORROW BY RE-THINKING

1.3.1 Mainstream systems

1.3.1.1 Systems with AnAmmOx for mainstream application: pilot scale results and full scale perspective

The number of full-scale applications of the Anammox process for nitrogen removal from anaerobic digestion of reject water (with high nitrogen concentrations and high temperatures) showed that energy and chemicals requirements can be substantially lowered, comparing to the traditional nitrification/denitrification process, when treating wastewaters with a low C:N ratio. A new challenge is to apply the partial nitritation/Anammox process for nitrogen removal in a main stream of the WWTP, at low temperatures and low ammonium concentrations. Application of the Anammox process is especially interesting in A-B systems where organic matter is removed in A stage and nitrogen in B stage.

Proposed system configurations

By application of the new process configuration with separate stages for organic carbon removal and nitrogen removal improved nitrogen removal and increased biogas production can be obtained. Application of this new solution will lead towards resource effective wastewater treatment plant in future.

Application of the deammonification process for nitrogen removal from mainstream municipal wastewater will allow maximize biogas production from organic matter content in wastewater and convert wastewater treatment plants into energy producing facilities without compromising the treatment efficiency. In these new systems a first step (stage A) can be based on organic matter removal by use of high rate activated sludge process (HRAS) or anaerobic treatment in upflow anaerobic sludge blanket (UASB) reactor. Different system configurations are presented in Figure 1.2. Pretreated in stage A wastewater is characterized by a low COD/N ratio, what is advantageous for application of a partial nitritation/anammox for nitrogen removal. The main challenges are anammox bacteria retention in biomass and suppression of NOB growth. There are a number of studies which prove that the effective nitrogen removal can be reached in conditions close to mainstream with low temperature or/and low nitrogen concentration, where the biomass of nitrifiers and anammox bacteria is in a form of biofilm (De Clippeleir *et al.* 2013; Malovanyy *et al.* 2015a) or granules (Lotti *et al.* 2014; Wett *et al.* 2013).

Nitrite production is the limiting step in deammonification process. Earlier studies for reject water treatment have shown that AOB of suspended biomass contribute significantly to nitrite production and up to 75% of nitrite production can be attributed to suspended biomass activity (Malovanyy *et al.* 2015b). It was shown that a combination of moving bed biofilm (mainly responsible for anammox activity) and suspended sludge (mainly responsible for aerobic ammonium oxidation) in an integrated fixed film activated sludge (IFAS) configuration could increase 3-fold the nitrogen removal capacity of the deammonification process treating anaerobic digestion reject water compared with a system without suspended sludge retention (Veuillet *et al.* 2014).

Results of the IFAS reactor operation with pre-treated in UASB mainstream wastewater presented by Malovanyy *et al.* (2015b) also clearly showed the advantages of IFAS compared with biofilm system in MBBR for nitrogen removal with deammonification process. In IFAS configuration advantages of both activated sludge and biofilm could be used. The ratio of AOB/NOB activity was much higher in suspended sludge than in biofilm. Suitable aeration strategy should allow reaching the highest total AOB activity while maintaining NOB activity on low level. When the MLSS concentration in the reactor was increased to over 800 mg TSS/L and after recirculation of suspended sludge started, the nitrite production was only 28% lower than what was required for utilizing all the available anammox bacteria capacity. The nitrogen removal efficiency of $70 \pm 4\%$ was reached with nitrogen removal rate of 55 ± 6 g N/($m^3 \cdot d$) (Malovanyy *et al.* 2015b).

Temperature influence on deammonification process performance

The temperature of wastewater is one of the most important technological parameter for deammonification process because of the influence on the activity of autotrophic bacteria, which take part in ammonium nitrogen oxidation (AOB, NOB) and nitrite reduction (anammox). The optimum temperature for operation of anammox process is from about 35°C to 40°C. Usually the temperature of supernatant from sludge dewatering after anaerobic digestion is above 25°C and because of this, most of the full-scale side stream deammonification systems work under such conditions. Decreasing the temperature of the treated wastewater has generally resulted in significantly decreased nitrogen conversion rate both in short term of alteration and long-term reactor operation (Szatkowska *et al.* 2006). Better understanding of how to control the deammonification process to promote the required biomass activity at low temperatures is necessary in order to apply deammonification process successfully for mainstream treatment.

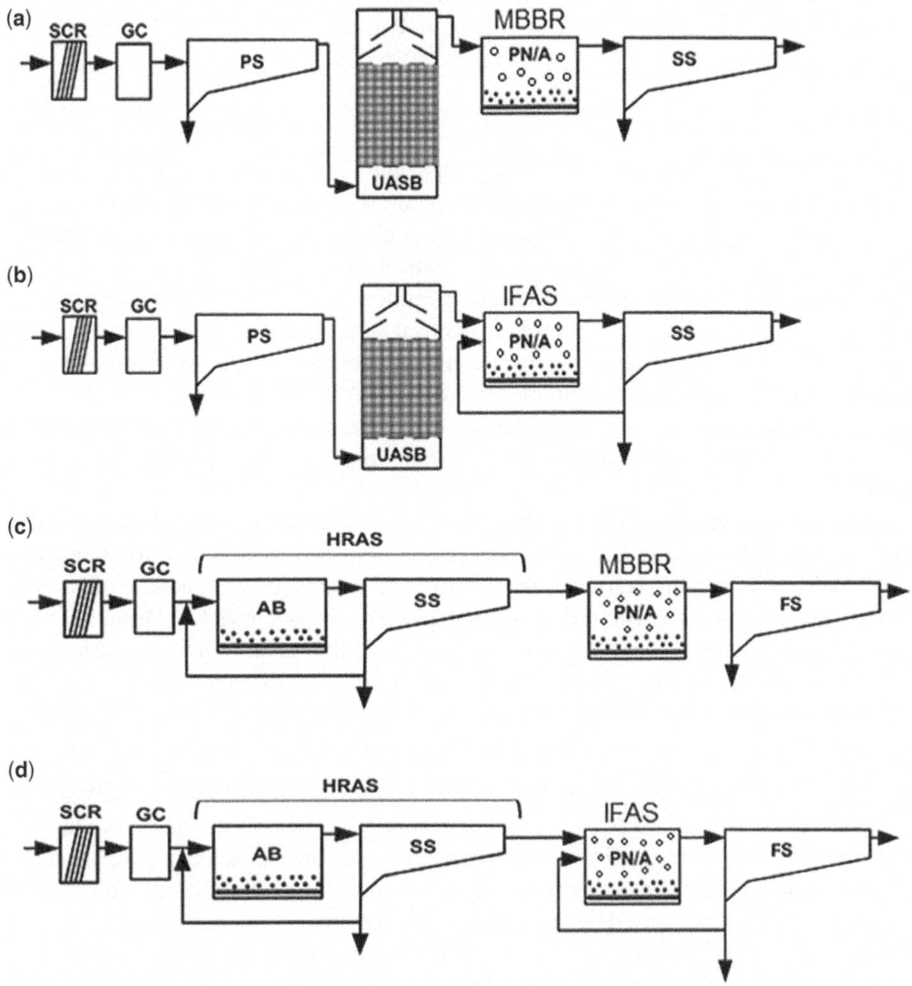

SCR	Screens	PS	Primary sedimentation	AB	Activated sludge reactor
GC	Grit chamber	SS	Secondary sedimentation	PN/A	Partial nitritation/Anammox
FS	Final sedimentation	UASB	Upflow Anaerobic Sludge Blanket reactor		

Figure 1.2 Process configuration with mainstream deammonification: (a) Deammonification in MBBR after UASB; (b) Deammonification in IFAS after UASB; (c) Deammonification in MBBR after High Rate Activated Sludge process (HRAS); (d) Deammonification in IFAS after High Rate Activated Sludge process (HRAS).

Pilot scale experiments with decreasing temperature from 25°C to 10°C showed that the threshold temperature below which capacity of deammonification process sharply decreased was 16°C. The microbial community composition, as well as the biofilm structure in the MBBR was stable throughout the experiments with decreasing temperatures indicating that the MBBR carriers provided a stable and protected environment for the microorganisms. Anammox bacteria (Brocadia sp.) were dominating the biofilm community in the thick biofilms, with a much lower abundance of AOB (Nitrosomonas sp.) located near the oxygenated biofilm-water interface, where also NOB (Nitrobacter sp. and Nitrospira sp.) were situated (Persson *et al.* 2014).

Benefits of technology and roadmap to increase the technology readiness level

If as proposed in Figure 1.2 alternatives for mainstream applications are used with separate organic matter removal more substrate is available for biogas production and wastewater treatment plant not only gives treatment results with high quality effluent, but also becomes energy positive plant. Using MBBR based technology gives lower sludge production during deammonification process.

The most sustainable and energy positive wastewater treatment plants are on the way to be developed with the application of Anammox process for nitrogen removal. The main challenge for the success of the new technology is maintaining stable and effective deammonification (partial nitritation and anammox processes) at low temperature and low ammonium concentration. An effective retention of the Anammox biomass in a reactor is required. Inflow nitrogen concentration in mainstream wastewater is low (25–50 mg NH_4^+-N/L) and together with low yield and growth rate of Anammox bacteria this leads to low Anammox biomass production

NOB may compete with AOB and with anammox bacteria. The competition can lead to reducing the efficiency of the wastewater treatment. Therefore the scope of ongoing research on nitritation/Anammox in a main stream at a low temperature is to find the effective and feasible control method of NOB growth and then optimization of the process. The following strategies and factors have been tested: intermittent aeration strategy; pH increase; influence of inorganic carbon concentration on Anammox bacteria activity; interactions between ammonia nitrogen concentrations, pH and dissolved oxygen concentrations and their influence on NOB suppression.

1.3.1.2 AnoxAn: A novel anaerobic-anoxic reactor for biological nutrient removal

The AnoxAn reactor was conceived and patented by Tejero *et al.* (2010) with the objective of unifying the anoxic and anaerobic zones in a single upflow sludge blanket reactor, aimed at achieving high compactness and efficiency. Its application is envisaged in those cases where retrofitting of existing WWTPs for BNR, or the construction of new ones, is limited by the available surface area.

The AnoxAn reactor is a continuous upflow anaerobic-anoxic sludge blanket reactor inside which the environmental conditions are vertically divided up with an anaerobic zone at the bottom prior to an anoxic zone above (Figure 1.3). This setup avoids the use of chemicals and the need for additional source of organic matter for BNR by means of EBPR and anoxic pre-denitrification, as it is in conventional configurations such as A2/O, Modified Bardenpho, UCT and VIP. A clarification zone at the top of the reactor avoids the escape of large amounts of biomass, thus promoting high sludge concentration in a sludge blanket reactor type.

The biological anaerobic-anoxic functioning of AnoxAn is coupled with an aerobic reactor (for the removal of residual organic matter, phosphate uptake, and nitrification) and a secondary sedimentation unit (or a final filtration step), in order to complete the treatment train. A nitrate rich stream is recycled to the anoxic zone of AnoxAn, providing the conditions for denitrification.

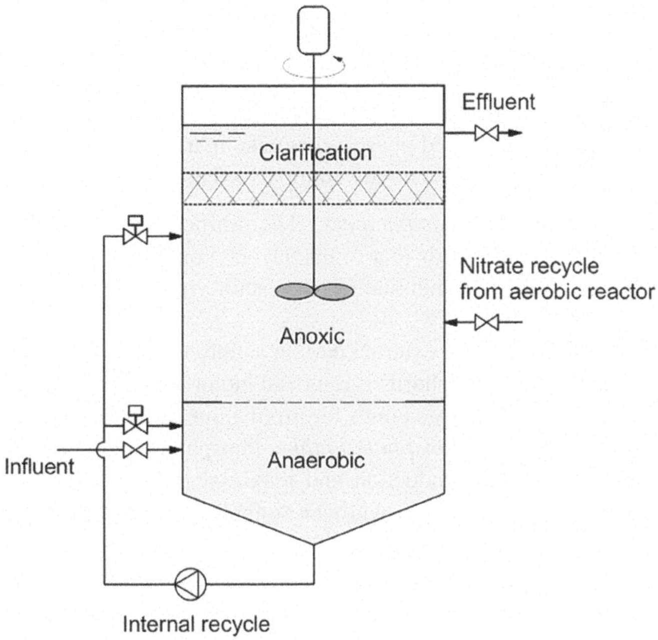

Figure 1.3 AnoxAn reactor scheme.

The main specific features of the AnoxAn reactor are: (i) upflow operation; (ii) hydraulic separation between the anoxic and anaerobic zones; and (iii) suspended solids retention. Such characteristics allow for a reduced footprint requirement, providing high compactness and efficiency. First of all, the upflow operation contributes to energy saving for mixing, plug-flow and sustainable high sludge concentration (Lettinga *et al.* 1980). Regarding the hydraulic separation, it is required in order to establish separate anoxic and anaerobic conditions, that is to keep negligible nitrate concentration in the anaerobic zone. The desired hydraulic separation between the anoxic and anaerobic zones is achieved through specific mechanical mixing devices and baffles, while keeping the influent flow up-way through the reactor. Independent mixing devices are implemented for the anaerobic and anoxic zones, by means of top entry or side entry dry-installed agitators, submersible mixers, and/or recirculation pumps. The targets of those devices are to keep the biomass in suspension reducing the extent of sludge settling and to provide good contact between the wastewater and biomass in each zone. Excessive mixing energy should be avoided in order to allow for the hydraulic separation between both zones, which can be performed through intermittent operation of the mixing devices. In addition, in order to limit the flow exchange and to improve the hydraulic separation, a baffle is introduced between the anoxic and anaerobic zones. This baffle could be implemented as a perimeter frame along the wall or by means of a rigid horizontal net whose voids allow for wastewater and biomass flow. The upflow setup results in biomass retention to some extent, due to suspended solids settling, and it is assisted by means of an additional baffle at the top of the reactor. This baffle consists of a set of rigid horizontal nets, or a set of lamellas, providing favourable conditions for suspended solids settling. In this way, an upper clarification zone is established so that large biomass escape is prevented, achieving high biomass concentration inside the reactor. Additionally, a periodic recirculation of suspended solids from the anaerobic to the anoxic zone is carried out in order to avoid excessive biomass accumulation in the anaerobic zone and to enhance biomass circulation inside the reactor being exposed to alternating anaerobic-anoxic conditions. In this way, phosphate uptake using nitrate as electron acceptor by means of denitrifying phosphate accumulating organisms (DPAOs) is promoted. Nevertheless, some escape of suspended solids is expected in order to provide alternating anaerobic-aerobic conditions to complete biological phosphorus removal by means of PAOs. Overall, the novel configuration claims anaerobic phosphate release, anoxic denitrification and phosphate uptake in a single reactor.

The main advantages of the AnoxAn reactor are summarized as follows:

- *Simplicity, high efficiency and compactness.* The unification of the anaerobic and anoxic compartments in a single reactor leads to a simple layout, compared to conventional configurations for BNR. Additionally, a better exploitation of the reactor volume is achieved due to high biomass concentration.
- *No need for chemicals addition.* An external carbon supply for denitrification is not needed due to pre-anoxic denitrification, and phosphorus is removed biologically without the need for chemicals.
- *Reduced energy requirement.* Energy savings for mixing due to upflow operation.
- *Simultaneous denitrification and phosphate uptake.* Phosphate uptake by DPAOs leads to energy savings for aeration, less sludge production and maximal influent organic substrate exploitation (Vlekke *et al.* 1988; Kuba *et al.* 1993), providing a suitable alternative for influent wastewaters with low C/N ratio.

Pilot scale studies

The capability of the AnoxAn configuration to establish two hydraulically separated zones inside the single reactor, while achieving adequate mixing conditions in the two zones and keeping the continuous influent flow up-way through it, was assessed by means of hydraulic characterization experiments and model simulations (Díez-Montero *et al.* 2015a). The feasibility assessment of the desired hydraulic behaviour, prior to the evaluation of its biological performance treating wastewater, was considered essential and was addressed in that study. Residence time distribution (RTD) experiments in clean water were performed in a bench-scale (48.4 L) AnoxAn prototype. The observed behaviour was described by a hydraulic model consisting of continuous stirred tank reactors and plug-flow reactors. The impact of the denitrification process in the anoxic zone on the hydraulic separation was subsequently evaluated through model simulations. The desired hydraulic behaviour proved feasible, involving little mixing between the anaerobic and anoxic zones (mixing flowrate 40.2% of influent flowrate) and negligible nitrate concentration in the anaerobic zone (less than 0.1 mg/L) when denitrification was considered (Figure 1.8).

The same AnoxAn prototype was coupled with an aerobic hybrid membrane bioreactor for the performance evaluation of AnoxAn in the removal of organic matter and nutrients from municipal wastewater without primary settling (Díez-Montero *et al.* 2016). The overall average removal efficiencies of TN and TP reached 75% and 89%, respectively, with a HRT of 10 hours. The development of a sludge blanket allowed several purposes in the single multi-environment AnoxAn reactor: biomass retention; hydrolysis of influent particulate organic matter; phosphate release in the anaerobic zone with an HRT of 1.1 hours; and nearly complete denitrification with an anoxic HRT of 2.7 hours. Phosphate uptake in the anoxic zone resulted virtually negligible under the conditions of the study, in spite of the potential denitrifying phosphate accumulating activity evaluated through batch tests. This was attributed to the influent wastewater characteristics, with no limiting organic matter availability (C/N > 10> g COD/gN) for both PAO and conventional denitrifying heterotrophs. Regarding nitrate removal, it was observed that only 5% of the nitrate recycled from the aerobic reactor was removed in the anaerobic zone, thus confirming the success of the anoxic zone performing denitrification and the feasibility of the hydraulic separation between the anoxic and the anaerobic zones of the AnoxAn reactor.

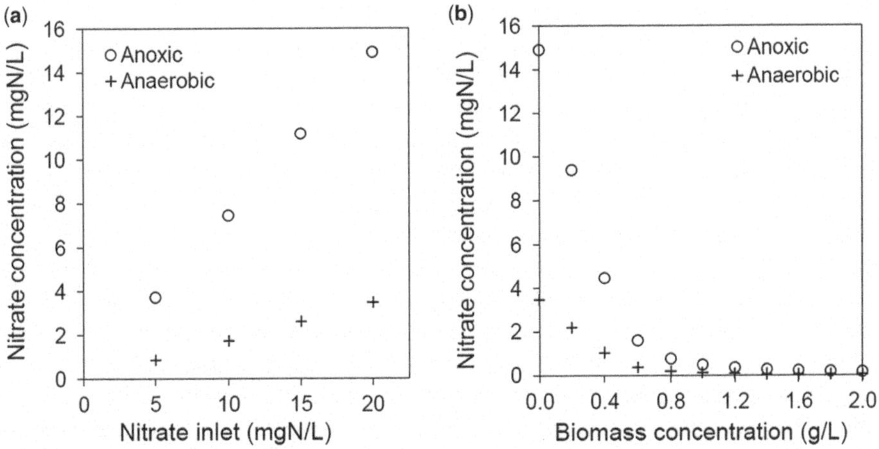

Figure 1.4 Tracer (nitrate) concentration in the anoxic and anaerobic zones: (a) for different tracer (nitrate) injections in the nitrate recycle inlet not taking into account denitrification and (b) for different biomass concentrations including denitrification model in the anoxic zone with a tracer (nitrate) injection in the nitrate recycle inlet of 20 mgN/L.

Economic assessment and roadmap for full scale application

Cost estimates are dependent on local requirements and specific application and economy of scale applies. Nevertheless, in order to assess the potential economic savings of the implementation of the AnoxAn reactor, an economic analysis of a hypothetical realization has been carried out (Díez-Montero *et al.* 2015b). An AnoxAn reactor has been designed based on a 16,500 m^3/d average daily flow, and compared with the equivalent anaerobic and anoxic stages of a conventional BNR treatment system. The economic study has considered the investment and operational costs of the resulting AnoxAn reactor, and the investment and operational costs of the anaerobic and anoxic stages of a UCT treatment system. The investment cost included construction works, electrical and mechanical equipment, electrical facilities, instrumentation and control. The operational cost included the energy consumption corresponding to the operation of the electrical devices. The economic assessment did not include: (i) pretreatment, primary treatment, aerobic stage, and sludge handling and treatment; (ii) land cost, buildings and urbanization; and (iii) staff, maintenance and chemicals consumption. The result has been expressed as the total annualized equivalent cost (TAEC) of both alternatives (AnoxAn vs. UCT anaerobic-anoxic), as shown in Table 1.8, assuming an expected life of the proposed treatment systems of 20 years and an interest rate of 3%.

Table 1.8 Investment, operational and total annualized equivalent costs of the hypothetical AnoxAn realization compared to the equivalent anaerobic and anoxic stages of a UCT type BNR process.

	Unit	AnoxAn		UCT	
Investment cost	€	652,885		528,918	
Electricity cost	€/kWh	0.10	0.14	0.10	0.14
Operational cost	€/year	17,713	24,798	41,045	57,464
TAEC	€/year	61,597	68,682	76,597	93,015

The results of the economic assessment show remarkable differences between both alternatives. The investment cost of the AnoxAn reactor was estimated 23% higher than that of the equivalent UCT system, mainly due to the additional cost of lamellas or baffles. However, the energy savings of the AnoxAn reactor lead to an operational cost lower than half of that of the UCT system. Eventually, the TAEC of the AnoxAn reactor resulted from 20 to 26% lower than the one of the equivalent UCT system, considering an electricity cost from 0.10 to 0.14 €/kWh. This indicates the significance of the potential energy savings and the corresponding economic benefit of the AnoxAn reactor.

It should be pointed out that the pilot scale studies have been performed in an AnoxAn prototype with specific dimensions. According to the setup, it is expected that such type of reactor could be applied for small-sized wastewater treatment. The implementation in medium and large-scale WWTPs would entail the construction of multiple modular units of the AnoxAn reactor, which could be far from the optimum from the technical and economic points of view. Further research is required in order to develop new AnoxAn configurations, maintaining the same features but with different dimensions. The study of the hydrodynamics of these specific new configurations by means of experimental tests and model simulations is considered a crucial step in order to assess its feasibility and scalability. Such new configurations and its shapes could mimic typical primary clarifiers, activated sludge reactors, etc., aimed at making the AnoxAn concept readily applicable at full scale, for instance for existing WWTPs upgrade to BNR. The upgrades based on AnoxAn attempt to use the existing facilities, thus reducing the capital expenditure for new reactors, and will provide an energy efficient process for BNR. Nevertheless, AnoxAn could also be applied for the construction of new WWTPs for BNR, in cases of limited available surface area.

Despite the fact that there are no full scale installations of the AnoxAn reactor, some of its fundamentals have been applied in several proposals for existing WWTPs upgrade for BNR. In one specific case study, two similar trickling filter WWTPs were asked to be upgraded to achieve nitrogen and phosphorus effluent standards. The proposed upgrade aimed to use the existing primary clarifier to host an anaerobic-anoxic reactor for BNR, with suspended solids retention, based on the AnoxAn setup. However, due to the shape and dimensions of the primary clarifier in such case study, a concentric configuration was proposed instead of a vertically compartmentalized upflow reactor. Several scenarios were simulated to preliminarily design and to optimize the anaerobic-anoxic reactor, and eventually several of them were found to successfully achieve both nitrogen and phosphorus removal, using the existing facilities without the need for new reactors (Díez-Montero *et al.* 2015b).

1.3.1.3 Domestic wastewater treatment with purple phototrophic bacteria

Purple phototrophic bacteria and domestic wastewater treatment

Purple phototrophic bacteria (PPB) are widely found in soil, freshwater, marine, and wastewaters and can be readily isolated from these sources (Zhang *et al.* 2003). PPB are characterized by versatile metabolic modes including (1) photoautotrophic growth with light as energy source, H_2 as the electron donor and CO_2 as the electron acceptor and carbon source (Koku *et al.* 2002), (2) chemoautotrophic growth in the dark with H_2 as energy source, O_2 as electron acceptor and CO_2 as carbon source (Qian & Tabita 1996), (3) photoheterotrophic growth with various organic compounds (e.g. butyrate, propionate, acetate, malate, succinate) as electron donor (Kim *et al.* 2004), (4) chemoheterotrophic in the dark with organics as electron donor and carbon source and O_2 as electron acceptor (Dubbs *et al.* 2000) and finally (5) fermentative growth with sugars as electron donor in the dark under anaerobic conditions (Tabita, 1988).

PPB grow through anoxygenic phototrophism (Bryant *et al.* 2007) whereby light energy, rather than chemical energy is used for ATP generation (Basak & Das 2007). Bacteriochlorin (BChls) and carotenoids function as light harvesting complexes. Compared to algae and most cyanobacteria, PPB contain BChl a and/or BChl b, which enables them to absorb light in the near infra-red (NIR) (Bertling *et al.* 2006). Carotenoids are less effective in light harvesting but extend the range of wavelength to drive anoxygenic photosynthesis (Hurse & Abeydeera 2002) and fulfil photo-oxidative protection and accessory light harvesting functions (Cogdell *et al.* 1981).

Mixed culture PPB have recently been proposed as an alternative platform for wastewater treatment utilising photoheterotrophic growth (Batstone *et al.* 2014). The concept is based on non-destructive assimilation and accumulation of organics (COD), nitrogen (NH_4-N) and phosphorous (PO_4-P) whereby resources are preserved in the biomass, rather than dissipated e.g. into the atmosphere.

The ATP generation via light avoids catabolic oxidative phosphorylation and results in biomass yields close to 1 (Suhaimi *et al.* 1987). Basically, all degradable COD is assimilated without dissipation of carbon e.g. as CO_2. This maximises the utilisation of macronutrients N and P and results in a potential recovery of close to 100%. This creates new options for resource recovery from diluted wastewater streams.

Batch tests using domestic wastewater as inoculum showed that infrared could selectively enrich PPB from domestic wastewater alone, and that domestic enriched mixed PPB could remove soluble COD, NH_4-N and PO_4-P from domestic wastewater to current effluent limits by assimilation (Hülsen *et al.* 2014). Recent research has identified that for every 100 mg of soluble COD around 6.0 ± 1.0 mg of nitrogen (N) and 0.8 ± 0.2 mg phosphorous (P) can be assimilated into the solid phase (Hülsen *et al.* 2015 in submission).

Laboratory scale (2 L) photo anaerobic membrane bioreactors (PAnMBR) at ambient (20°C) as well as cold (10°C) temperatures were operated for 2 years and supported these findings from batch in continuous operation. A schematic set-up of the PAnMBR is shown in Figure 1.9.

At ambient temperatures and anaerobic illuminated conditions the continuous PAnMBR treated real domestic wastewater in a single stage, without biological oxidation of organics or nitrogen to discharge limits (TCOD < 100 mgL^{-1}; TN < 10 mg L^{-1} and TP < 1 mg L^{-1}) (Hülsen *et al.* 2015 in submission). This was at a SRT of <3 d, and HRT of <1 d. Operation at 10°C was almost identical to that at 20°C, indicating that operation at cold temperatures could extend the application of PPB for wastewater treatment from tropical to temperate regions.

Wastewater treatment at low temperatures is challenging and the plant design needs to accommodate for reduced growth rates of e.g. nitrifies which is effectively done in BNR plants by adjusting the minimum SRT, with around 45% longer SRTs required at 10°C compared to 20°C (Kos, 1998). For anaerobic biomass the effects are more prolonged, with HRTs approximately doubling for every 10°C decrease and sludge rentention times of up to 100 d (Zeeman & Lettinga 1999). For the anammox process low temperatures are know to cause activity drops of anammox in combination with selective advantages for nitrite oxidising bacteria (Hu *et al.* 2013), which creates practical limitations for the nitrite supply for the anammox (Hendrickx *et al.* 2014).

Although PPB are adaptable to cold temperatures a major drawback remains. PPB cannot utilise the whole spectrum of organics in the wastewater but are limited to lower molecular substances such as organic acids and alcohols (Inui *et al.* 1995) and some sugars (Inui *et al.* 1999) which means that COD (around 200 mg L^{-1}) has to be added in order to remove N and P to below discharge limits. Therefore, additional COD (such as VFA) needs to be added to achieve N and P discharge limits. Adding this from commodity resources would increase the operational costs to infeasible levels. The other option is the COD supply directly from the PPB biomass generated during the release phase of treatment.

The PPB biomass generated in the partitioning step contains close to 100% of the wastewater COD, N and P in a small concentrated volume. A critical element is the release and recovery. Anaerobic digestion is a key candidate due to its relatively low cost, good understanding of the underlying process (Mata-Alvarez *et al.* 2000), ability to recover organics as methane or via fermentation to organic acids while retaining ammonia as NH_4-N and phosphates as soluble PO_4-P for subsequent chemical recovery (Demirel & Scherer 2008). Several anaerobic digestion configurations, such as mesophilic, thermophilic, temperature phased can be combined with various sludge pre-treatment methods, to maximise the biogas production and the N and P release.

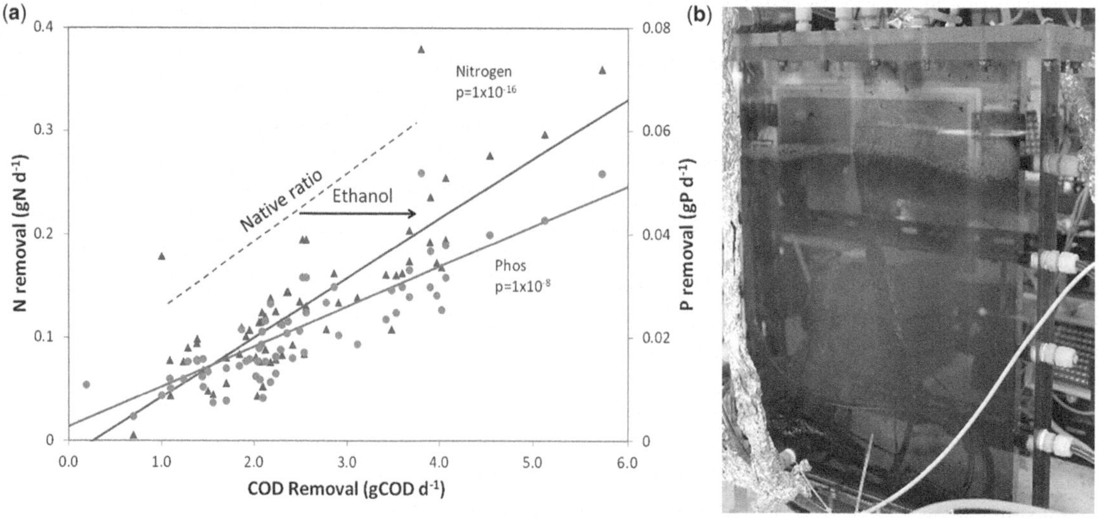

Figure 1.5 COD:N:P ratio of the process (Left) and photo of the reactor (Right) (Hülsen *et al.* 2014, 2015 submitted).

Another option is the direct use of the PPB biomass. PPB contain up to 65% of proteins (Shipman *et al.* 1975; Ponsano *et al.* 2004) and were reported to be an excellent food additive for fish, chicken (Kobayashi & Tchan 1973) and poultry farming (Ponsano *et al.* 2004). The use of PPB as organic fertilizer increased production of citrus fruits (Kobayashi & Tchan 1973) and improved the soil quality, growth and yield of crops (Xu, 2001). These studies clearly show the potential value of PBB biomass.

However, the applicability of this biomass depends very much on the wastewater treated and the local legislations. The direct use as food additive or the application as organic fertilizer would make this process a non-waste product producing one.

The application of PPB for wastewater treatment is an entirely new approach for mainline domestic as well as for industrial wastewater treatment. Instead of reducing the biomass/sludge of the wastewater treatment process, this approach aims to produce biomass as carrier for COD, N and P. The feasibility of this platform depends on the utilisation of this biomass. There is different value related to different utilisation options. The assimilated resources have to be recovered either as energy, heat and fertiliser or applied on fields and for livestock. In any case a product has to balance the operational costs of the resource partitioning.

1.4 CONCLUDING REMARKS AND SUSTAINABILITY INDICATORS

This chapter outlined the potential of new sustainable technologies to revamp, renew and rethink municipal wastewater treatment plants towards the reduction of energy footprint. The technical, economic and environmental sustainability was finally preliminary quantified in Table 1.9 by indicators used in the Water_2020 Environmental Decision Support System (EDSS).

Table 1.9 Mini-assessment of the technologies.

Parameters	ELAN System	RBC-MBBR	SCENA System	Mainstream AnAmmOx UASB + MBBR	Mainstream AnAmmOx UASB + IFAS	Ion Exchange for P Recovery	Ion Exchange for N Recovery	Purple Phototropic Bacteria	AnoxAn	Via-nitrite SBR Mainstream
1. N removal efficiencies	80%	90% NH_4-N; 80% TN	90%NH_4-N; 80% TN	36%	70%		0.99	90–96%[a]	75%[c]	99%
2. P removal efficiencies	10%[a]	–	60–65%	n/a	n/a	0.99		90–99%[a]	89%[c]	
3. Sludge per kg of N removed	0.11 g VSS/g N	No data	0.40–0.45 kgVSS/kgN removed	n/a	n/a		0	~0.8 gVSS/gCOD; ~0.08 gVSS/gN	1.1 gVSS/gN[c]	
4. Sludge per kg of P removed	ND	–	ND	n/a	n/a	0		ND	7.4 gVSS/gP[c]	
5. kWh per kg of N removed	1[b]	1.45–1.75	1.6–1.7	n/a	n/a		OPEX including: electricity, chemicals, maintenance and staff = 6.5 euro/kg N removed	~1–2 kWh/gCOD, simultaneously removing TN and TP to <10 mgTN/L, <1 mgP/L	1.1 kWh/kgN[d]	
6. kWh per kg of P removed	ND	–	10–15[a]	n/a	n/a	OPEX including: electricity, chemicals, maintenance and staff = 7.5 euro/kg P removed			4.9 kWh/kgP[d]	
7. N_2O (and GHG) per kg of N removed	6 kg N_2O-N/kg N removed[b]	0.2–0.9%	0.50%	n/a	n/a		0	none	ND	
8. Maximum N specific loading (with reference to the reaction volume and MLVSS) to achieve the max removal aforementioned	1 kg N/(m³ · d) and 0.06 g N/(g VSS · d)	0.7–1.2 kg N/m³/day	0.7–0.8 kg N/m³/day	51 g N/m³ d	79 g N/m³ d	not applicable, as this is chemical physical process		0.3 kgN/(m³ d) main stream	75.9 gN/m³ · d[c]	72 gN/m³ · d

(Continued)

Table 1.9 Mini-assessment of the technologies (*Continued*).

	Parameters	ELAN System	RBC-MBBR	SCENA System	Mainstream AnAmmOx UASB + MBBR	Mainstream AnAmmOx UASB + IFAS	Ion Exchange for P Recovery	Ion Exchange for N Recovery	Purple Phototropic Bacteria	AnoxAn	Via-nitrite SBR Mainstream
9.	Maximum P specific loading (with reference to the reaction volume and MLVSS) to achieve the max removal aforementioned	ND		0.05–0.10 kg $P/m^3/day$	51 g N/m^3 d	79 g N/m^3 d		not applicable, as this is chemical physical process	0.04 kgP/(m3 d)*main stream	9.6 $gP/m^3 \cdot d$[c]	
10.	Flowrate range able to be treated	The flow rate range is not relevant; the design of the unit is performed based on the nitrogen loading rate	125–960 m^3/d	65 m^3/d (around 40 kgN/d)	scalable	scalable	no limits	no limits		ND	
11.	Capital cost (€/m^3)	ND		Estimated 1.45–1.55 €/kgN removed			1,300,000 euros assuming a 10,000 PE site, for 40 years	2,350,000 euros assuming a 10,000 PE site, for 40 years		0.022 €/m^3[e]	
	Operational parameters:										
12.	HRT	The one to achieve a nitrogen removal rate of 1 kg N/($m^3 \cdot d$)	10–20 h	1–1.1 h	0,6 days	0,4 days	3 min empty bed contact time. Regeneration once every 2 weeks for 5–6 hours that enables CaP recovery	10–20 min empty bed contact time. Regeneration once every 10 days for 5–6 hours. NH_4SO_4 recovery	12–24 h	4.2 h, only AnoxAn reactor	12 h
	SRT	>50 d		12–15 d	n/a	25 for suspended biomass	not applicable	not applicable	2–3 days	39 days[c]	16 d

	RAS MLSS	None	–	None	no RAS	n/a	not applicable	not applicable	–	Up to 5 gTSS/L in the anoxic zone Up to 10 gTSS/L in the anaerobic zone	
	T	18–35°C	25–30°C (20 lowest)	25–28°C	25	25	Should not be affected by temperature	Should not be affected by temperature	10–40	18.0 ± 3.2°C[f]	19.4°C
	Conductivity	2–12 mS/cm	–	1.8–2.6 mS/cm (Typical range in the reactor)	n/a	n/a	ok at ww pH	ok at ww pH	up to 20 mS/cm	Up to 10.4 mS/cm[g]	
	pH	7.5–8.5	6.7–7.5	7.5–8.2	7	6,9	ok at ww pH	ok at ww pH	6.5–9	6.5–8.5	7–8
13.	Reactants consumption	None	0	Polyelectrolyte (38–40 gN/kgN removed)	no chemicals	no chemicals	NaOH	NaCl	COD	No need for reactants	No need for reactants
14.	Is this technology applied for the mainstream or sidestream?	sidestream	yes	sidestream	mainstream	mainstream	usually mainstream, but can be applied for side stream too	usually mainstream, but can be applied for side stream too	mainstream	mainstream	mainstream
	Qualitative operational data:										
15.	Need of specialized staff	During the design and start-up processes, specialized staff is necessary. Once the system started up, the same maintenance as the one required for the rest of the WWTP is enough, i.e., cleaning of probes, refilling engine oil, etc.		Specialized operators are required to assist the reactor during the 20–30 days of start-up. Then, conventional operations are required (maintainance of the pumps and blowers, cleaning sensors). Estimated time for mantainance around 3 days per years.			Yes	Yes		ND	

(Continued)

Table 1.9 Mini-assessment of the technologies (*Continued*).

Parameters	ELAN System	RBC-MBBR	SCENA System	Mainstream AnAmmOx UASB + MBBR	Mainstream AnAmmOx UASB + IFAS	Ion Exchange for P Recovery	Ion Exchange for N Recovery	Purple Phototropic Bacteria	AnoxAn	Via-nitrite SBR Mainstream
Problem frequency	Almost never	Rare	Rare			not yet fully understood	not yet fully understood	to be determined	ND	Rare
Control over the process	Yes	DO control	Yes (process controll based on DO, conductivity and pH. ORP and MLSS installed for monitoring			Needs development	Needs development	Yes	ND	Needs development
Safety and health risk	None		None			H&S risks assciated with chemical storage	H&S risks assciated with chemical storage	None	ND	None
Robustness	Yes	High	High			very	very	Yes	ND	High
Flexibility	Yes	High	High			very	very	Yes	ND	High
Stability	Yes	High	High			very	very	Yes	ND	High
16. State of development (emergent, adaptive use, innovative, established)	Innovative/ Established	Established	Innovative/ Established	innovative	innovative	Pilot-plant and demonstration plant currently	Pilot-plant and demonstration plant currently	Innovative	Emergent	Emergent
17. Compatibility with other technologies	Combined to a previous anaerobic digestion reactor	?	Possible installation using existing tanks.	yes	yes	very compatible	very compatible	stand alone or after anaerobic with bypass	Compatibility and need for an aerobic nitrifying reactor	stand-alone

[a]P consumed for biomass growth is not considered.
[b]Compared to the STP without ELAN® in the sidestream.
[c]Average value for the overall system, obtained at bench-scale coupled with an aerobic hybrid MBR.
[d]Energy requirement estimated for the AnoxAn reactor, not including the aerobic reactor.
[e]Equivalent annual capital cost (€/year) / Flowrate (m^3/year); calculated by project.
[f]Average ± standard deviation, measured during the bench-scale operation aforementioned.
[g]Maximum of 24 h composite samples, measured during the bench-scale operation aforementioned.

1.5 REFERENCES

Abeling U. and Seyfried C. F. (1992). Anaerobic-aerobic treatment of high strength ammonium wastewater nitrogen removal via nitrite. *Water Science and Technology*, **26**, 1007–1015.

Acevedo B., Camiña C., Corona J. E., Borrás L. and Barat R. (2015). The metabolic versatility of PAOs as an opportunity to obtain a highly p-enriched stream for further p-recovery. *Chemical Engineering Journal*, **270**, 459–467.

Aiyuk S., Xu H., van Haandel A. and Verstraete W. (2004). Removal of ammonium nitrogen from pretreated domestic sewage using a natural ion exchange. *Environmental Technology*, **25**(11), 1321–1330.

Al-Omari A. A., Han M., Lemaire R., Morales N., Vázquez-Padín J. R., Stewart H., Val del Río A. and Wett B. (2015). Mainstream Deammonification. In: Shortcut Nitrogen Removal-Nitrite Shunt and Deammonification. Water Environment Federation, 107–127. ISBN: 978-1-57278-313-3.

Basak N. and Das D. (2007). The prospect of purple non-sulfur (PNS) photosynthetic bacteria for hydrogen production: the present state of the art. *World Journal of Microbiology and Biotechnology*, **23**(1), 31–42.

Batstone D. J., Hülsen T., Mehta C. M. and Keller J. (2014). Platforms for energy and nutrient recovery from domestic wastewater: a review. *Chemosphere*, **140**, 2–11.

Bertling K., Hurse T. J., Kappler U. and Rakic A. D. (2006). Lasers – an effective artificial source of radiation for the cultivation of anoxygenic photosynthetic bacteria. *Biotechology Bioengineering*, **94**(2), 337–345.

Blaney L. M., Cinar S. and Sengupta A. K. (2007). Hybrid anion exchanger for trace phosphate removal from water and wastewater. *Water Research*, **41**, 1603–1613.

Bodirsky B. L., Popp A., Lotze-Campen H., Dietrich J. P., Rolinski S., Weindl I., Schmitz C., Müller C., Bonsch M., Humpenöder F., Biewald A. and Stevanovic M. (2014). Reactive nitrogen requirements to feed the world in 2050 and potential to mitigate nitrogen pollution. *Nature Communications*, **5**, article number 38–58.

Booker N. A., Priestley A. J. and Fraser I. H. (1999). Struvite formation in wastewater treatment plants: opportunities for nutrient recovery. *Environmental Technology*, **20**(7), 777–782.

Boyer T. H., Persaud A., Banerjee P. and Palomino P. (2011). Comparison of low cost and engineered materials for phosphorus removal from organic rich surface water. *Water Research*, **45**(16), 4803–4814.

Bryant D. A., Costas A. M. G., Maresca J. A., Chew A. G. M., Klatt C. G., Bateson M. M., Tallon L. J., Hostetler J., Nelson W. C., Heidelberg J. F. and Ward D. M. (2007). Candidatus Chloracidobacterium thermophilum: an aerobic phototrophic acidobacterium. *Science*, **317**(5837), 523–526.

Burow L. C., Kong Y. H., Nielsen J. L., Blackall L. L. and Nielsen P. H. (2007). Abundance and ecophysiology of Defluviicoccus spp., glycogen accumulating organisms in full-scale wastewater treatment processes. *Microbiology-Sgm*, **153**, 178–185.

Carvalheira M., Oehmen A., Carvalho G., Eusebio M. and Reis M. A. M. (2014). The impact of aeration on the competition between polyphosphate accumulating organisms and glycogen accumulating organisms. *Water Research*, **66**, 296–307.

Carvalho G., Lemos P. C., Oehmen A. and Reis M. A. M. (2007). Denitrifying phosphorus removal: linking the process performance with the microbial community structure. *Water Research*, **41**, 4383–4396.

Cema G. (2010). Comparative Study on Different Anammox Systems. PhD thesis, TRITA-LWR, 1053.

Cema G., Wiszniowski J., Żabczyński S., Zabłocka-Godlewska E., Raszka A. and Surmacz-Górska J. (2007). Biological nitro gen removal from landfill leachate by deammonification assisted by heterotrophic denitrification in a rotating biological contactor. *Water Science and Technology*, **55**(8–9), 35–42.

Chen S. K., Juaw C. K. and Cheng S. S. (1991). Nitrification and denitrification of high-strength ammonium and nitrite wastewater with biofilm reactor. *Water Science and Technology*, **23**, 1417–1425.

Christensson M., Ekström S., Andersson Chan A., Le Vaillant E. and Lemaire R. (2013). Experience from start-ups of the first ANITA Mox plants. *Water Science and Technology*, **67**, 2677–2684.

Cogdell R. J., Hipkins M. F., MacDonald W. and Truscott T. G. (1981). Energy transfer between the carotenoid and the bacteriochlorophyll within the B-800-850 light-harvesting pigment-protein complex of Rhodopseudomonas sphaeroides. *BBA – Bioenergetics*, **634**, 191–202.

Courtens E. N. P., Boon N., De Clippeleir H., Berckmoes K., Mosquera M., Seuntjens D. and Vlaeminck S. E. (2014). Control of nitritation in an oxygen-limited autotrophic nitrification/denitrification rotating biological contactor through disc immersion level variation. *Bioresource Technology*, **155**, 182–188.

Daelman M. R. J., van Voorthuizen E. M., van Dongen L. G. J. M., Volcke E. I. P. and van Loosdrecht M. C. M. (2013). Methane and nitrous oxide emissions from municipal wastewater treatment - results from a long-term study. *Water Science and Technology*, **67**, 2350–2355.

De Clippeleir H., Siegfried E. Vlaeminck S. E., Carballa M. and Verstraete W. (2009). A low volumetric exchange ratio allows high autotrophic nitrogen removal in a sequencing batch reactor. *Bioresource Technology*, **100**(21), 5010–5015.

De Clippeleir H., Vlaeminck S. E., De Wilde F., Daeninck K., Mosquera M., Boeckx P., Verstraete W. and Boon N. (2013). One-stage partial nitritation/anammox at 15°C on pretreated sewage: feasibility demonstration at lab-scale. *Applied Microbiology and Biotechnology*, **97**, 10199–10210.

Demirel B. and Scherer P. (2008). The roles of acetotrophic and hydrogenotrophic methanogens during anaerobic conversion of biomass to methane: a review. *Reviews in Environmental Science and Bio/Technology*, **7**(2), 173–190.

Díez-Montero R. (2015a). AnoxAn: a Novel Anaerobic-Anoxic Reactor for Biological Nutrient Removal from Wastewater. PhD thesis, University of Cantabria.

Díez-Montero R., De Florio L., González-Viar M., Volcke E. I. P. and Tejero I. (2015b). Feasibility of hydraulic separation in a novel anaerobic-anoxic upflow reactor for biological nutrient removal. *Bioprocess and Biosystems Engineering*, **38**(1), 93–103.

Díez-Montero R., De Florio L., González-Viar M., Herrero M. and Tejero I. (2016). Performance evaluation of a novel anaerobic-anoxic sludge blanket reactor for biological nutrient removal treating municipal wastewater. *Bioresource Technology*, **209**, 195–204.

Dubbs J. M., Bird T. H., Bauer C. E. and Tabita F. R. (2000). Interaction of CbbR and RegA* Transcription Regulators with the Rhodobacter sphaeroides cbb I Promoter-Operator Region. *Journal of Biological Chemistry*, **275**(25), 19224–19230.

EEC (1991a). Council Directive 91/271/EEC of 21 May 1991 Concerning Urban Wastewater Treatment. European Economic Community, Brussels, Belgium.

European Commission (2014). Towards Macro-Regional Integrated Nutrient Management Managing Nitrogen and Phosphorus Loads to Water Bodies: Characterisation and Solutions. Joint Research Centre Technical Reports, Institute for Environment and Sustainability.

Falk M. W., Reardon D. J., Neethling J. B., Clark D. L. and Pramanik A. (2013). Striking the Balance between Nutrient Removal, Greenhouse Gas Emissions, Receiving Water Quality, and Costs. *Water Environment Research*, **85**(12), 2307–2316.

Flores-Alsina X., Rodriguez-Roda I., Sin G. and Gernaey K. V. (2009). Uncertainty and sensitivity analysis of control strategies using the benchmark simulation model No1 (BSM1). *Water Science and Technology*, **59**, 491–499.

Flores-Alsina X., Corominas L., Snip L. and Vanrolleghem P. A. (2011). Including greenhouse gas emissions during benchmarking of wastewater treatment plant control strategies. *Water Research*, **45**, 4700–4710.

Flores-Alsina X., Gernaey K. V. and Jeppsson U. (2012). Global sensitivity analysis of the BSM2 dynamic influent disturbance scenario generator. *Water Science and Technology*, **65**, 1912–1922.

Flowers J. J., He S., Yilmaz S., Noguera D. R. and McMahon K. D. (2009). Denitrification capabilities of two biological phosphorus removal sludges dominated by different 'Candidatus Accumulibacter' clades. *Environmental Microbiology Reports*, **1**, 583–588.

Frison N., Di Fabio S., Cavinato C., Pavan P. and Fatone F. (2013). Best available carbon sources to enhance the via-nitrite biological nutrients removal from supernatants of anaerobic co-digestion. *Chemical Engineering Journal*, **215–216**, 15–22.

Ganigue R., Gabarro J., Sanchez-Melsio A., Ruscalleda M., Lopez H., Vila X., Colprim J. and Dolors Balaguer M. (2009). Long-term operation of a partial nitritation pilot plant treating leachate with extremely high ammonium concentration prior to an anammox process. *Bioresource Technology*, **100**, 5624–5632.

Gaul T., Märker S. and Kunst S. (2005). Start-up of moving bed biofilm reactors for deammonification: the role of hydraulic retention time, alkalinity and oxygen supply. *Water Science and Technology*, **52**(7), 127–133.

Georges K., Thornton A. and Sadler R. (2009). Transforming wastewater treatment to reduce carbon emissions, Report: SC070010/R2, Environment Agency Evidence Directorate.

Gernaey K. V. and Jørgensen S. B. (2004). Benchmarking combined biological phosphorus and nitrogen removal wastewater treatment processes. *Control Engineering Practice*, **12**, 357–373.

Guerrero J., Guisasola A. and Baeza J. A. (2011a). The nature of the carbon source rules the competition between PAO and denitrifiers in systems for simultaneous biological nitrogen and phosphorus removal. *Water Research*, **45**, 4793–4802.

Guerrero J., Guisasola A., Vilanova R. and Baeza J. A. (2011b). Improving the performance of a WWTP control system by model-based setpoint optimisation. *Environmental Modelling & Software*, **26**, 492–497.

Guerrero J., Guisasola A., Comas J., Rodríguez-Roda I. and Baeza J. A. (2012). Multi-criteria selection of optimum WWTP control setpoints based on microbiology-related failures, effluent quality and operating costs. *Chemical Engineering Journal*, **188**, 23–29.

Guerrero J., Flores-Alsina X., Guisasola A., Baeza J. A. and Gernaey K. V. (2013). Effect of nitrite, limited reactive settler and plant design configuration on the predicted performance of simultaneous C/N/P removal WWTPs. *Bioresource Technology*, **136**, 680–688.

Guerrero J., Guisasola A. and Baeza J. A. (2014). A novel control strategy for efficient biological phosphorus removal with carbon-limited wastewaters. *Water Science and Technology*, **70**, 691–697.

Guerrero J., Guisasola A. and Baeza J. A. (2015). Controlled Crude Glycerol Dosage to Prevent EBPR Failures IN C/N/P Removal WWTPS. *Chemical Engineering Journal*, **271**, 114–127.

Guisasola A., Qurie M., Vargas M. D., Casas C. and Baeza J. A. (2009). Failure of an enriched nitrite-DPAO population to use nitrate as an electron acceptor. *Process Biochemistry*, **44**, 689–695.

Hankins N. P., Pliankarom S. and Hilal N. (2004). An equilibrium ion-exchange study on the removal of NH_4^+ ion from aqueous effluent using clinoptilolite. *Separation Science and Technology*, **39**(15), 3639–3663.

Hao X., Wang C., van Loosdrecht M. C. M. and Hu Y. (2013). Looking beyond struvite for P-Recovery. *Environmental Science & Technology*, **47**, 4965–4966.

Hellinga C., Schellen A., Mulder J. W., van Loosdrecht M. C. M. and Heijnen J. J. (1998). The SHARON process: an innovative method for nitrogen removal from ammonium-rich waste water. *Water Science and Technology*, **37**, 135–142.

Hendrickx T. L. G., Kampman C., Zeeman G., Temmink H., Hu Z., Kartal B. and Buisman C. J. N. (2014). High specific activity for anammox bacteria enriched from activated sludge at 10°C. *Bioresource Technology*, **163**, 214–222.

Henze M., Gujer W., Mino T. and van Loosdrecht M. C. M. (2000). Activated Sludge Models ASM1, ASM2, ASM2d and ASM3: Scientific and Technical Report No. 9. IWA Publishing, London.

Hippen A. and Rosenwinkel K.-H. (1997). Aerobic deammonification: a new experience in the treatment of waste waters. *Water Science and Technology*, **35**, 111–120.

Hülsen T., Batstone D. J. and Keller J. (2014). Phototrophic bacteria for nutrient recovery from domestic wastewater. *Water Research*, **50**, 18–26.

Hülsen T., Barry E. M., Lu Y., Puyol D., Keller J., Batstone D. J. (2015 in submission). Domestic wastewater treatment with purple phototrophic bacteria using a novel continuous photo anaerobic membrane bioreactor. *Water Research*, manuscript number: WR32005.

Hurse T. J. and Abeydeera W. P. P. (2002). Quantification of sulphur and sulphur-containing compounds in wastewaters by means a combination of liquid chromatographic methods. *Journal of Chromatography A*, **942**(1–2), 201–210.

Inui M., Momma K., Matoba R., Ikuta M., Yamagata H. and Yukawa H. (1995). Characterization of alcohol-assimilating photosynthetic purple non-sulfur bacteria and cloning of molecular chaperones from a purple non-sulfur bacterium. *Energy Conversion and Management*, **36**(6–9), 767–770.

Inui M., Nakata K., Roh J. H., Zahn K. and Yukawa H. (1999). Molecular and functional characterization of the Rhodopseudomonas palustris no. 7 phosphoenolpyruvate carboxykinase gene. *Journal of Bacteriology*, **181**, 2689–2696.

Kanders L., Areskoug T., Schneider Y., Ling D., Punzi M. and Beier M. (2014). Impact of seeding on the start-up of the one-stage deammonification MBBRs. *Environmental Technology*, **35**(21–24), 2767–2773.

Katsogiannis A. N., Kornaros M. and Lyberatos G. (2002). Long-term effect of total cycle time and aerobic/anoxic phase ratio on nitrogen removal in a sequencing batch reactor. *Water Environment Research*, **74**, 324–337.

Katsogiannis A. N., Kornaros M. and Lyberatos G. (2003). Enhanced nitrogen removal in SBRs bypassing nitrate generation accomplished by multiple aerobic/anoxic phase pairs. *Water Science and Technology*, **47**, 53–59.

Kim M. K., Choi K.-M., Yin C.-R., Lee K.-Y., Im W. -T., Lim J. H. and Lee S.-T. (2004). Odorous swine wastewater treatment by purple non-sulfur bacteria, Rhodopseudomonas palustris, isolated from eutrophicated ponds. *Biotechnology Letters*, **26**(10), 819–822.

Kobayashi M. and Tchan Y. T. (1973). Treatment of industrial waste solutions and production of useful by-products using a photosynthetic bacterial method. *Water Research*, **7**(8), 1219–1224.

Kodera H., Hatamoto M., Abe K., Kindaichi T., Ozaki N. and Ohashi A. (2013). Phosphate recovery as concentrated solution from treated wastewater by a PAO-enriched biofilm reactor. *Water Research*, **47**, 2025–2032.

Koku H., Eroğlu İ., Gündüz U., Yücel M. and Türker L. (2002). Aspects of the metabolism of hydrogen production by Rhodobacter sphaeroides. *International Journal of Hydrogen Energy*, **27**(11–12), 1315–1329.

Kong Y. H., Xia Y., Nielsen J. L. and Nielsen P. H. (2006). Ecophysiology of a group of uncultured Gammaproteobacterial glycogen-accumulating organisms in full-scale enhanced biological phosphorus removal wastewater treatment plants. *Environmental Microbiology*, **8**, 479–489.

Koon J. H. and Kaufman W. J. (1975). Ammonia removal from municipal wastewaters by ion exchange. *Journal of the Water Pollution control Federation*, **47**(3), 448–465.

Kornaros M., Marazioti C. and Lyberatos G. (2008). A pilot scale study of a sequencing batch reactor treating municipal wastewater operated via the UP-PND process. *Water Science and Technology*, **58**(2), 435–438.

Kornaros M., Dokianakis S. N. and Lyberatos G. (2010). Partial nitrification/denitrification can be attributed to the slow response of nitrite oxidizing bacteria to periodic anoxic disturbances. *Environmental Science & Technology*, **44**(19), 7245–7253.

Kuba T., Smolders G., van Loosdrecht M. C. M. and Heijnen J. J. (1993). Biological phosphorus removal from wastewater by anaerobic-anoxic sequencing batch reactor. *Water Science and Technology*, **27**(5/6), 241–252.

Kuba T., Murnleitner E., van Loosdrecht M. C. M. and Heijnen J. J. (1996). A metabolic model for biological phosphorus removal by denitrifying organisms. *Biotechnology and Bioengineering*, **52**, 685–695.

Kumar M., Badruzzaman M., Adham S. and Oppenheimer J. (2007). Beneficial phosphate recovery from reverse osmosis (RO) concentrate of an integrated membrane system using polymeric ligand exchanger (PLE). *Water Research*, **41**(10), 2211–2219.

Lackner S., Gilbert E. M., Vlaeminck S. E., Joss A., Horn H. and van Loosdrecht M. C. M. (2014). Full-scale partial nitritation/anammox experience - An application survey. *Water Research*, **55**, 292–303.

Lemaire R., Liviano I., Ekström S., Roselius C., Chauzy J., Thornberg D. Thirsing S. and Deleris S. (2011). 1-stage Deammonification MBBR process for reject water sidestream treatment: investigation of start-up strategy and carrier design. In: Proceedings of the WEF Nutrient Recovery and Management Conference, 9–12 January 2011, Miami.

Lemaire R., Christensson M., Zhao H., Le Noir M. and Voon C. (2013). Experience from Start-up and Operation of Deammonification MBBR Plants, and Testing of a New Deammonification IFAS Configuration. In: Proceedings of the Water Environment Federation, WEFTEC 2013, Chicago, IL, USA, pp. 1926–1947.

Lettinga G., van Velsen A. F. M., Hobma S. W., de Zeeuw W. J. and Klapwijk A. (1980). Use of the Upflow Sludge Blanket (USB) reactor concept for biological wastewater treatment. *Biotechnology and Bioengineering*, **22**, 699–734.

Liberti L. Boari G. and Passino R. (1979). Phosphates and ammonia recovery from secondary effluents by selective ion exchange with production of a slow-release fertilizer. *Water Research*, **13**(1), 65–73.

Ling D. (2009). Experience from commissioning of full-scale DeAmmon® plant at Himmerfjärden (Sweden). In: Proceedings of 2nd IWA Specialized Conference on Nutrient Management in Wastewater Treatment Processes, 6–9 September 2009, Kraków, 403–410.

Longo S., Katsou E., Malamis S., Frison N., Renzi D. and Fatone F. (2015). Recovery of volatile fatty acids from fermentation of sewage sludge in municipal wastewater treatment plants. *Bioresource Technology*, **175**, 436–444.

Lopez-Vazquez C. M., Oehmen A., Hooijmans C. M., Brdjanovic D., Gijzen H. J., Yuan Z. and Van Loosdrecht M. C. M. (2009). Modeling the PAO-GAO competition: effects of carbon source, pH and temperature. *Water Research*, **43**, 450–462.

Lotti T., Kleerebezem R., van Erp Taalman Kip C., Hendrickx T. L. G., Kruit J., Hoekstra M. and van Loosdrecht M. C. M. (2014). Anammox growth on pretreated municipal wastewater. *Environmental Science & Technology*, **48**, 7874–7880.

Lv J., Yuan L., Chen X., Liu L. and Luo D. (2014). Phosphorus metabolism and population dynamics in a biological phosphate-removal system with simultaneous anaerobic phosphate stripping. *Chemosphere*, **117**, 715–721.

Machado V. C., Gabriel D., Lafuente J. and Baeza J. A. (2009). Cost and effluent quality controllers design based on the relative gain array for a nutrient removal WWTP. *Water Research*, **43**, 5129–5141.

Machado V. C., Lafuente J. and Baeza J. A. (2014). Activated sludge model 2d calibration with full-scale WWTP data: comparing model parameter identifiability with influent and operational uncertainty. *Bioprocess and Biosystems Engineering*, **37**, 1271–1287.

Mackinnon I. D. R., Barr K., Miller E., Hunter S. and Pinel T. (2003). Nutrient removal from wastewaters using high performance materials. *Water Science and Technology*, **47**(11), 101–107.

Mainstone C. P. and Parr W. (2002). Phosphorus in rivers - ecology and management. *Science of the Total Environment*, **282**, 25–47.

Malovanyy A., Trela J. and Plaza E. (2015a). Mainstream wastewater treatment in integrated fixed film activated sludge (IFAS) reactor by partial nitritation/anammox process. *Bioresource Technology*, **198**, 478–487.

Malovanyy A., Yang J., Trela J. and Plaza E. (2015b). Combination of UASB reactor and partial nitritation/Anammox MBBR for municipal wastewater treatment. *Bioresource Technology*, **180**, 144–153.

Marcelino M., Wallaert D., Guisasola A. and Baeza J. A. (2011). A two-sludge system for simultaneous biological C, N and P removal via the nitrite pathway. *Water Science and Technology*, **64**, 1142–1147.

Martin B. D., Parsons S. A. and Jefferson B. (2009). Removal and recovery of phosphate from municipal wastewaters using polymeric anion exchanger bound with hydrated ferric oxide nanoparticles. *Water Science and Technology*, 2637–2645.

Martin Garcia I., Mokosch M., Soares A., Pidou M. and Jefferson B. (2013). Impact on reactor configuration on the performance of anaerobic MBRs: treatment of settled sewage in temperate climates. *Water Rsearch*, **47**(14), 4853–4860.

Mata-Alvarez J., Macé S. and Llabrés P. (2000). Anaerobic digestion of organic solid wastes. An overview of research achievements and perspectives. *Bioresource Technology*, **74**(1), 3–16.

McAdam E. J., Luffler D., Martin-Garcia N., Eusebi A. L., Lester J. N., Jefferson B. and Cartmell E. (2011). Integrating anaerobic processes into wastewater treatment. *Water Science and Technology*, **63**(7), 1459–1466.

Mercer B. W., Ames L. L., Touhill C.J., Van Slyke W. J. and Dean R. B. (1970). Ammonia removal from secondary effluents by selective ion exchange. *Journal of the Water Pollution Control Federation*, **42**(2), 95–107.

Morales N., Val del Río A., Vázquez-Padín J. R., Gutiérrez R., Fernández-Gonzalez R., Icaran P., Rogalla F., Campos J. L., Méndez R. and Mosquera-Corral A. (2015a). Influence of dissolved oxygen concentration on the startup of the anammox based process: ELAN®. *Water Science and Technology*, **72**(4), 520–527.

Morales N., Val del Río A., Vázquez-Padín J.R., Méndez R., Mosquera-Corral A. and Campos J. L. (2015b). Integration of the Anammox process to the rejection water and main stream lines of WWTPs. *Chemosphere*, **140**, 99–105.

Ni B.J., Peng L., Law Y., Guo J. and Yuan Z. (2014). Modeling of nitrous oxide production by autotrophic ammonia-oxidizing bacteria with multiple production pathways. *Environmental Science & Technology*, **48**, 3916–3924.

Ødegaard H., Rusten B. and Westrum T. (1994). A new moving bed biofilm reactor – application and results. *Water Science and Technology*, **29**, 157–165.

Oehmen A., Carvalho G., Freitas F. and Reis M. A. M. (2010a). Assessing the abundance and activity of denitrifying polyphosphate accumulating organisms through molecular and chemical techniques. *Water Science and Technology*, **61**, 2061–2068.

Oehmen A., Carvalho G., Lopez-Vazquez C. M., van Loosdrecht M. C. M. and Reis M. A. M. (2010b). Incorporating microbial ecology into the metabolic modelling of polyphosphate accumulating organisms and glycogen accumulating organisms. *Water Research*, **44**, 4992–5004.

Oehmen A., Lopez-Vazquez C. M., Carvalho G., Reis M. A. M. and van Loosdrecht M. C. M. (2010c). Modelling the population dynamics and metabolic diversity of organisms relevant in anaerobic/anoxic/aerobic enhanced biological phosphorus removal processes. *Water Research*, **44**, 4473–4486.

Ostace G. S., Baeza J. A., Guerrero J., Guisasola A., Cristea V. M., Agachi P. Ş. and Lafuente J. (2013). Development and economic assessment of different WWTP control strategies for optimal simultaneous removal of carbon, nitrogen and phosphorus. *Computers & Chemical Engineering*, **53**, 164–177.

Persson F., Sultana R., Suarez M., Hermansson M., Plaza E. and Wilen B. M. (2014). Structure and composition of biofilm communities in a moving bed biofilm reactor for nitritation-anammox at low temperatures. *Bioresource Technology*, **154**, 267–273.

Ponsano E., Pinto M., Garcia-Neto M. and Lacava P. (2004). Performance and color of broilers fed diets containing Rhodocyclus gelatinosus biomass. *Revista Brasileira de Ciência Avícola*, **6**, 237–242.

Pynaert K., Smets B. F., Wyffels S., Beheydt D., Sciliano S.D. and Verstraete W. (2003). Characterization of an Autotrophic Nitrogen-Removing Biofilm from a Highly Loaded Lab-Scale Rotating Biological Contactor. *Applied and Environmental Microbiology*, **69**(6), 3626–3635.

Qian Y. and Tabita F. R. (1996). A global signal transduction system regulates aerobic and anaerobic CO_2 fixation in Rhodobacter sphaeroides. *Journal of Bacteriology*, **178**(1), 12–18.

Renzi D., Longo S., Frison N., Malamis S., Katsou E. and Fatone F. (2015). Chapter 16: Short-cut enhanced nutrient removal from anaerobic supernatants: pilot scale results and full scale development of the S.C.E.N.A. process. In: Sewage Treatment Plants: Economic Evaluation of Innovative Technology for Energy Efficiency. Integrated Environmental Technology Series. Stamatelatou and Tsagarakis (eds), IWA Publishing, London, UK, pp. 295–310.

Rosenwinkel K.-H. and Cornelius A. (2005). Deammonification in the Moving-Bed Process for the Treatment of Wastewater with High Ammonia Content. *Chemical Engineering & Technology*, **28**(1), 49–52.

Saito T., Brdjanovic D. and van Loosdrecht M. C. M. (2004). Effect of nitrite on phosphate uptake by phosphate accumulating organisms. *Water Research*, **38**, 3760–3768.

Schmid M., Walsh K., Webb R., Rijpstra W., van de Pas-Schoonen K., Verbruggen M., Hill T., Moffett B., Fuerst J., Schouten S., Damste J., Harris J., Shaw P., Jetten M.S.M. and Strous M. (2003). Candidatus "Scalindua brodae", sp. nov., Candidatus "Scalindua wagneri", sp. nov., Two New Species Anaerobic Ammonium Oxidizing Bacteria. *Systematic and Applied Microbiology*, **26**, 529–538.

Shipman R. H., Kao I. C. and Fan L. T. (1975). Single-cell protein production by photosynthetic bacteria cultivation in agricultural by-products. *Biotechnology and Bioengineering*, **17**(11), 1561–1570.

Siegrist H., Reithaar S., Koch G. and Lais P. (1998). Nitrogen loss in a nitrifying rotating contactor treating ammonium-rich wastewater without organic carbon. *Water Science and Technology*, **38**, 241–248.

Smolders G. J. F., Vandermeij J., Van Loosdrecht M. C. M. and Heijnen J. J. (1994). Stoichiometric model of the aerobic metabolism of the biological phosphorus removal process. *Biotechnology and Bioengineering*, **44**, 837–848.

Strous M., VanGerven E., Zheng P., Kuenen J.G. and Jetten M. S. M. (1997). Ammonium removal from concentrated waste streams with the anaerobic ammonium oxidation (anammox) process in different reactor configurations. *Water Research*, **31**, 1955–1962.

Strous M., Fuerst J. A., Kramer E. H. M., Logemann S., Muyzer G., van de Pas-Schoonen K. T., Webb R., Kuenen J. G. and Jetten M. S. M. (1999). Missing lithotroph identified as new planctomycete. *Nature*, **400**, 446–449.

Suhaimi M., Liessens J. and Verstraete W. (1987). NH4+-N assimilation by Rhodobacter capsulatus ATCC 23782 grown axenically and non-axenically in N and C rich media. *Journal of Applied Microbiology*, **62**(1), 53–64.

Svetich R. (1993). Long-term use of clinoptilolite in the treatment of sewage at Tahoe-Truckee Sanitation Agency, Truckee, California. In: International conference on the Occurrence, Properties and Utilisation of Natural Zeolites, 20–28 June 1993, Red Lion Hotel-Riverside, Boise, Idaho, pp. 197–201.

Svetich T. (2015). Personal communication, 10th January 2015.

Szatkowska B. and Plaza E. (2006). Temperature as a factor influencing the Anammox process performance. In: Water and Environmental Management Series, R. Stuetz & L. Teik-Thye (eds), Young Researchers, WEMS No. 12, London, UK, pp. 51–58.

Szatkowska B., Cema G., Plaza E., Trela J. and Hultman B. (2007). One-stage system with partial nitritation and Anammox processes in moving Bed biofilm reactor. *Waster Science and Technology*, **55**(8–9), 19–26.

Tabita F. (1988). Molecular and cellular regulation of autotrophic carbon dioxide fixation in microorganisms. *Microbiological Reviews*, **52**(2), 155–189.

Taya C., Kumar Garlapati V., Guisasola A. and Baeza J. A. (2013). The selective role of nitrite in the PAO/GAO competition. *Chemosphere*, **93**, 612–618.

Tchobanoglous G., Burton F. L. and Stensel H. D. (2003). Wastewater Engineering: Treatment and Reuse. Metcalf and Eddy, McGraw Hill, New York, USA.

Tejero I., Díez R., Esteban A. L., Lobo A., Temprano J. and Rodríguez L. (2010). Reactor biológico anóxico-anaerobio para la eliminación de nutrientes de aguas residuales (Anoxic-anaerobic biological reactor for nutrient removal from wastewater). Spanish Patent ES2338979 (in Spanish).

Thornton A. (2007). Application of Ion Exchange for Ammonium Removal from Municipal Wastewaters. PhD thesis, Cranfield University.

Tong J. and Chen Y. (2007). Enhanced biological phosphorus removal driven by short-chain fatty acids produced from waste activated sludge alkaline fermentation. *Environmental Science & Technology*, **41**, 7126–7130.

USEPA (2007). Biological Nutrient Removal Processes and Costs.

USEPA (2008). Municipal Nutrient Removal Technologies Reference Document.

Valverde-Pérez B., Ramin E., Smets B. F. and Plósz B. G. (2015). EBP2R – An innovative enhanced biological nutrient recovery activated sludge system to produce growth medium for green microalgae cultivation. *Water Research*, **68**, 821–830.

van der Star W. R. L., Abma W. R., Blommers D., Mulder J. -W., Tokutomi T., Strous M., Picioreanu C. and van Loosdrecht M. C. M. (2007). Startup of reactors for anoxic ammonium oxidation: experiences from the first full-scale anammox reactor in Rotterdam. *Water Reseacrh*, **41**, 4149–4163.

van Voorthuizen E., Zwijnenburg A., van der Meer W. and Temmink H. (2008). Biological black water treatment combined with membrane separation. *Water Research*, **42**(16), 4334–4340.

Vázquez-Padín J. R., Campos J. L., Mosquera-Corral A., Méndez R. and Revsbech N. P. (2010). Microbial community distribution and activity dynamics of granular biomass in a CANON reactor. *Water Research*, **44**, 4359–4370.

Vázquez-Padín J. R., Fernández R., Rogalla F., Campos J. L., Mosquera-Corral A. and Méndez R. (2012). Fast start-up of a CANON pilot scale on reject wáter – influence of DO and HRT control. IWA Nutrient Removal and Recovery 2012: Trends in NNR, 23–25 September 2012, Harbin, China.

Vázquez-Padín J. R., Fernández González R., Campos J. L., Mosquera-Corral A. and Méndez P. (2014a). Method for starting up and controlling a biological process for ammonium removal by the action of autotrophic bacteria in wastewater. European Patent: EP2740713 (A1).

Vázquez-Padín J. R., Morales N., Gutiérrez R., Fernández R., Rogalla F., Barrio J. P., Campos J. L., Mosquera-Corral A. and Méndez R. (2014b). Implications of full-scale implementation of an anammox based process as post-treatment of a municipal anaerobic sludge digester operated with co-digestion. *Water Science and Technology*, **69**(6), 1151–1158.

Verstraete W., Van de Caveye P. and Diamantis V. (2009). Maximum use of resources present in domestic "used water". *Bioresource Technology*, **100**(23), 5537–5545.

Veuillet F., Lacroix S., Bausseron A., Gonidec E., Ochoa J., Christensson M. and Lemaire R. (2014). Integrated fixed-film activated sludge ANITATMMox process – a new perspective for advanced nitrogen removal. *Water Science and Technology*, **69**, 915–922.

Vlekke G. J. F. M., Comeau Y. and Oldham W. K. (1988). Biological phosphate removal from wastewater with oxygen or nitrate in sequencing batch reactors. *Environmental Technology Letters*, **9**, 791–796.

Water Environment Research Foundation (WERF) (2010a). Proceedings of the Workshop on Nutrient Removal: How Low Can We Go & What Is Stopping Us from Going Lower? http://www.werf.org (accessed 04 December 2015)

Water Environment Research Foundation (WERF). (2010b). Nutrient Management: Regulatory Approaches to Protect Water Quality Volume 1 – Review of Existing Practices, D.L. Clark *et al.* (eds)

Water Environment Research Foundation (WERF) Nutrient Challenge. (2014). Deammonification, J.B. Neethling (ed.), Water Environment Research Foundation (WERF), http://www.werf.org (accessed 30 August 2014)

Wett B. (2006). Solved upscaling problems for implementing deammonification of rejection water. *Water Science and Technology*, **53**, 121–128.

Wett B., Omari A., Podmirseg S. M., Han M., Akintayo O., Gómez Brandón M., Murthy S., Bott C., Hell M., Takács I., Nyhuis G. and O'Shaughnessy M. (2013). Going for mainstream deammonification from bench to full scale for maximized resource efficiency. *Water Science and Technology*, **68**, 283–289.

Wu D., Ekama G. A., Lu H., Chui H. -K., Liu W. -T., Brdjanovic D., van Loosdrecht M. C. M. and Chen G. -H. (2013). A new biological phosphorus removal process in association with sulfur cycle. *Water Research*, **47**, 3057–69.

Wu D., Ekama G. A., Wang H. G., Wei L., Lu H., Chui H. K., Liu W. T., Brdjanovic D., van Loosdrecht M. C. M. and Chen G. H. (2014). Simultaneous nitrogen and phosphorus removal in the sulfur cycle-associated Enhanced Biological Phosphorus Removal (EBPR) process. *Water Research*, **49**, 251–264.

Wu P., Xu L., Wang J., Huang Z., Zhang J. and Shen Y. (2015). Partial Nitrification and Denitrifying Phosphorus Removal in a Pilot-Scale ABR/MBR Combined Process. *Applied Biochemistry and Biotechnology*, **177**(5), 1003–1012.

Xu H. -L. (2001). Effects of a Microbial Inoculant and Organic Fertilizers on the Growth, Photosynthesis and Yield of Sweet Corn. *Journal of Crop Production*, **3**(1), 183–214.

Zeeman G. and Lettinga G. (1999). The role of anaerobic digestion of domestic sewage in closing the water and nutrient cycle at community level. *Water Science and Technology*, **39**(5), 187–194.

Zeng W., Wang X., Li B., Bai X. and Peng Y. (2013). Nitritation and denitrifying phosphorus removal via nitrite pathway from domestic wastewater in a continuous MUCT process. *Bioresource Technology*, **143**, 187–195.

Zeng W., Li B., Wang X., Bai X. and Pengm Y. (2014). Integration of denitrifying phosphorus removal via nitrite pathway, simultaneous nitritation-denitritation and anammox treating carbon-limited municipal sewage. *Bioresource Technology*, **172C**, 356–364.

Zeng W., Li B., Wang X., Bai X. and Peng Y. (2016). Influence of nitrite accumulation on "Candidatus Accumulibacter" population structure and enhanced biological phosphorus removal from municipal wastewater. *Chemosphere*, **144**, 1018–1025.

Zhang D., Yang H., Zhang W., Huang Z. and Liu S. J. (2003). Rhodocista pekingesis sp. Nov., a cyst-forming phototrophic bacterium from a municipal wastewater treatment plant. *International Journal of Systematic and Evolutionary Microbiology*, **53**(4), 1111–1114.

Zheng X., Chen Y. and Liu C. (2010). Waste activated sludge alkaline fermentation liquid as carbon source for biological nutrients removal in anaerobic followed by alternating aerobic–anoxic sequneching batch reactors. *Chinese Journal of Chemical Engineering*, **18**, 478–485.

Zhou Y., Pijuan M. and Yuan Z. G. (2007). Free nitrous acid inhibition on anoxic phosphorus uptake and denitrification by poly-phosphate accumulating organisms. *Biotechnology and Bioengineering*, **98**, 903–912.

Zhou Y., Pijuan M., Zeng R. J. and Yuan Z. (2008). Free Nitrous Acid Inhibition on Nitrous Oxide Reduction by a Denitrifying-Enhanced Biological Phosphorus Removal Sludge. *Environmental Science & Technology*, **42**, 8260–8265.

Zhou Y., Oehmen A., Lim M., Vadivelu V. and Ng W. J. (2011). The role of nitrite and free nitrous acid (FNA) in wastewater treatment plants. *Water Research*, **45**, 4672–4682.

Zou H. M., Lu X. W. and Li T. (2014). Effect of side-stream phosphorus recovery on biological phosphorus removal performance investigated by chemical and microbial analyses in a novel BNR-IC process. *Water Science and Technology*, **70**(9), 1441–1447.

Chapter 2

Anaerobic treatment of municipal wastewater

Orhan Ince, Zeynep Cetecioglu Gurol, E. Gozde Ozbayram, Miguel M. Iglesias, Bahar Ince, Nedal Massalha, Ángel Robles, Isam Sabbah and Aurora Seco

2.1 INTRODUCTION

Meeting the growing energy demand while taking into consideration the necessary environmental aspects is one of the main challenges modern societies are faced with. Thus, due to the depletion of fossil fuels and the increase in the greenhouse gas emissions, renewable energy production has a crucial role for meeting the requirements.

Anaerobic treatment is a degradation process of complex organic materials through serial reactions of different metabolic groups of microorganisms, namely hydrolytic, acidogenic, acetogenic bacteria and methanogenic archaea to the final products methane (CH_4) and carbon dioxide (CO_2). An acceptable organic matter removal in an anaerobic reactor depends upon the presence of an adequate level of methanogenic activity (Ince *et al.* 1994). In recent years, anaerobic treatment technologies have received a great attention, since the production of methane-rich biogas is considered as a green energy source, replacing fossil fuels. If the system operation is managed properly, it can be a net energy producer, while reducing organic material at the same time. The main advantages are: lower nutrient requirements, small land requirements, resistance to shock organic loadings, lower operational costs due to less digested sludge production, and no aeration requirement. Since municipal wastewater already contains macronutrients (nitrogen and phosphorus) and micronutrients, additional chemicals are not required for the process. However, the requirement for higher investment costs, a higher sensitivity to toxic compounds, and the fact that hydrogen sulphide is contained in the biogas, which leads to odour and corrosion, are the major drawbacks of anaerobic treatment systems. In addition, as they are usually not sufficient to meet the effluent discharge standards, they need to be coupled with another post-treatment method. Moreover, the low growth rate of methanogens extends the start-up period when compared with aerobic systems (Skouteris *et al.* 2012; McCarty *et al.* 2011). Temperature is one of the most important operational parameters in anaerobic processes and heating of reactors requires energy and capital cost; hence, relatively high biogas production can be anticipatedat ambient temperatures only in tropical regions (Martinez-Sosa *et al.* 2011).

During the anaerobic treatment of municipal wastewater, organic nitrogen and phosphorus are decomposed to ammonia and phosphate, which are not removed from the system, while their concentration in the liquid phase increases. The effluent of an anaerobic system treating municipal wastewater includes ammonia nitrogen concentration in the range 30–50 mg/L and a phosphorus concentration of about 10–17 mg/L (Foresti *et al.* 2006). Thus, anaerobic treatment systems offer a great economical potential for nutrient recovery in the future, turning thus the conventional municipal treatment plants into resource factories. Innovative technologies are being developed to achieve maximum resource reuse.

2.1.1 Energy nexus: Is anaerobic treatment a feasible way for municipal wastewater?

Municipal wastewater is typically characterized by low organic strength and high particulate organic matter content (Ozgun *et al.* 2013). These characteristics lead to low biomass growth rate and make maintaining biomass concentration under defined levels difficult, often causing biomass wash-out. Additionally, the particulate fraction of organic matter has to be broken down and solubilized through hydrolysis before the conversion to final products. Hydrolysis is accepted as the rate-limiting step for the treatment of wastewaters that contain a high amount of particulate material. Thus, kinetically, anaerobic treatment consists of two relatively slow steps, where the first step is decomposition of particulate matter into soluble compounds (Martinez-Sosa *et al.* 2011) and the second is the conversion of dissolved organic matter to biogas. Another important issue in the anaerobic treatment of municipal wastewater is temperature. At lower temperatures, the microbial activity declines causing a decrease in biogas production and poor effluent quality. The technological developments on anaerobic processes enable to treat municipal wastewater with a temperature at above 15°C. However, more research studies should be conducted for revealing the optimum treatment conditions at lower temperatures.

Generally, anaerobic processes are applied to high-strength industrial wastewaters and they are installed for digesting the primary and secondary sludge produced by conventional aerobic biological systems in municipal wastewater treatment plants (Foresti *et al.* 2006). However, there are some full-scale applications treating municipal wastewaters in tropical regions such as India, Colombia, and Brazil. Because of the climate change, increasing energy demand and water scarcity, municipal wastewaters are considered as a resource rather than a waste to handle and anaerobic treatment processes are then naturally considered as an alternative treatment technology for municipal wastewater (Gao *et al.* 2014).

The installation of an anaerobic treatment process as a core system for municipal wastewater treatment has been suggested in recent years, given the possibility for energy and fertilizer production (Gao *et al.* 2014). Improvements in anaerobic biotechnology became more important due to the possibility of sustainable energy production from municipal wastewater (Smith *et al.* 2012).

In case biomass synthesis is ignored, one gram of O_2 is required for the aerobic removal of one gram of chemical oxygen demand (COD) equivalent organic matter from wastewater, stoichiometrically. However, one gram COD equivalent organics produce 0.35 L of methane under anaerobic conditions. Also, if anaerobic treatment is installed instead of activated sludge, up to 1 kWh (fossil energy) kg COD^{-1} removed is conserved, depending on the system used for aeration of the activated sludge. Furthermore, 13.5 MJ CH_4 energy can be produced by removal of 1 kg COD under anaerobic conditions giving 1.5 kWh electric through cogeneration systems (assuming 40% electric conversion efficiency) (van Lier *et al.* 2015).

The application of anaerobic systems to municipal wastewater treatment is still limited, because the amount of methane produced in the system cannot cover the heating requirements. The organic matter removal efficiencies, construction and operational costs are the main parameters for anaerobic treatment unit implementation in municipal treatment plants. Therefore, these technologies are easily installed in countries with warmer climates (Skouteris *et al.* 2012). Thus, the anaerobic process could be an alternative municipal wastewater treatment for minimizing sludge production and optimizing energy use. It should be noted that, in case of cold weather the profits of anaerobic treatment are not clear. For this reason, many studies are conducted to increase efficiencies, to reveal solutions for enhancing effluent quality, to shorten start-up period and to minimise costs.

2.2 ANAEROBIC REACTOR TYPES FOR MUNICIPAL WASTEWATER TREATMENT

During the past ten years, anaerobic treatment technologies have been applied to municipal wastewaters in various configurations for enhancing treatment quality at lower hydraulic retention times. The first continuous flow anaerobic reactor was developed in 1905 by Karl Imhoff, who designed a single tank for enhanced settling and simultaneous digestion of the settled solids. This system was especially implemented for municipal wastewaters and is still operated in tropical regions (van Lier *et al.* 2015).

The type of the reactor configuration applied for the treatment plants depends on the wastewater characteristics, required effluent quality, ambient temperature, etc. Recently, many studies have been conducted on the treatment of municipal wastewater at ambient temperatures using higher technology anaerobic reactor types, such as the upflow anaerobic sludge blanket (UASB) and expanded granular sludge bed (EGSB), anaerobic membrane bioreactor (AnMBR) etc. Moreover, a considerable amount of full-scale anaerobic treatment plants have been constructed and have resulted in sufficient removal efficiencies due to their high rate characteristics (Feng *et al.* 2008).

2.2.1 Anaerobic membrane bioreactor (AnMBR)

Membrane bioreactor (MBR) technology combines the biological degradation process of organic matter with a direct solid–liquid separation process by membrane filtration. By using micro or ultrafiltration membrane technology (with pore sizes usually ranging from 0.03 to 0.4 μm), MBR technology allows achieving a complete retention of the microorganisms inside the system. This complete retention of microorganisms allows high SRT to be obtained with reduced working volumes (Ince *et al.* 1993, 1997; Judd & Judd, 2011). In this respect, MBR applied to municipal wastewater treatment is a promising alternative to obtain high biomass concentrations in the system by decoupling both HRT and SRT. Depending on the configuration of the filtration process, MBRs can be classified in submerged/immersed MBRs and side-stream MBRs. Several MBR variants exist in the market nowadays, but they are all originally variants of the two configurations mentioned above. Currently, the submerged/immersed MBR is one of the most promising technologies in municipal wastewater field.

The first application of membranes in anaerobic wastewater treatment was reported in 1978 (Grethlein, 1978). It consisted of an external cross-flow membrane applied to the treatment of a septic tank effluent. On the other hand, the first commercially-available AnMBR was developed by Dorr-Oliver in the early 1980s, which was known as membrane anaerobic reactor system (MARS). It was designed for high-strength whey processing wastewater treatment and consisted of a completely mixed suspended growth anaerobic reactor and a crossflow membrane filtration loop. Nevertheless, this system was only tested at pilot scale due to the high membrane investment and operating costs. Since then, many pilot- and laboratory-scale investigations about the application of this technology to industrial wastewater treatment have been carried out. Nowadays, several full-scale plants are treating industrial wastewaters (Dereli *et al.* 2012). However, the application of AnMBR technology for municipal wastewater treatment has been less reported (Lin *et al.* 2013). As commented before, two main MBR configurations can be defined, based on how the membranes are integrated with the bioreactor: sidestream and submerged/immersed.

In sidestream MBRs, the membrane modules are placed outside the reactor, and the mixed liquor circulates at high velocity over a recirculation loop that contains the membrane. The membranes are operated in cross-flow configuration (they are not immersed in the mixed liquor). The main constraint of the sidestream configuration is the high energy requirements, due to the high operational transmembrane pressures (TMP) required to achieve the desired cross-flow velocities. The main advantage of the sidestream configuration is that the membranes are physically cleaned due to the cross-flow velocity applied over the membrane surface.

In submerged/immersed MBRs, the membranes are directly immersed in the mixed liquor (either in the reactor or in a separate membrane tank), thus they are operated in a dead-end configuration. The submerged MBR configuration involves lower energy requirements in comparison with sidestream configurations, since permeate is obtained by vacuum filtration. However, submerged MBRs operate at lower transmembrane fluxes (J), thus requiring larger membrane surfaces for a given treatment flow. In order to minimise fouling in submerged configurations, an amount of gas is introduced/recycled to the system from the bottom of the membranes.

Side-stream AnMBRs are mostly applied to the treatment of high-loaded wastewaters (e.g., industrial wastewater), whilst the use of submerged/immersed AnMBRs is usually limited to the treatment of low-loaded wastewaters. This is mainly due to the low cost and suitable fouling control that is achieved by combining low-pressure filtration and membrane scouring by air/gas sparging (Judd & Judd, 2011). Studies on the treatment of municipal wastewater with AnMBRs mostly utilised submerged configurations (Ozgun *et al.* 2013), which can be divided in two types: internal submerged MBR and external submerged MBR system (see Figure 2.1). A membrane externally connected to an anaerobic bioreactor is the configuration most commonly used. The bioreactor can be a continuously stirred tank reactor (CSTR) (Giménez *et al.* 2011), an upflow anaerobic sludge blanket (UASB) (Gouveia *et al.* 2015) an expanded granular sludge bed (EGSB) (Chu *et al.* 2005), or a fluidised bed (Shin *et al.* 2014) reactor coupled to membrane filtration.

Figure 2.1 shows an example of the possible configurations of a submerged AnMBR.

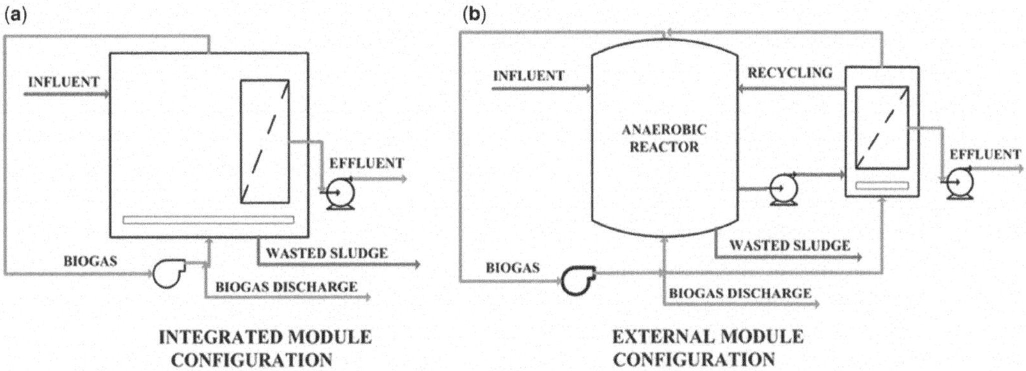

Figure 2.1 Configurations of a submerged anaerobic membrane bioreactor: (a) integrated and (b) external (Robles, 2013).

Hence, AnMBR technology gathers the before-mentioned advantages of anaerobic treatment processes (e.g., minimum sludge production due to low biomass yield of anaerobic organisms; low energy demand since no aeration is required; and methane production that can offset or even outweigh process energy requirements jointly with main benefits of using MBR technology for urban wastewater treatment:

- Obtaining solid- and microorganism-free effluent (high effluent quality).
- Generating a high-quality effluent that can be suitable for reuse. Typical output quality of membrane systems includes suspended solid (SS) <1 mg/L.
- Reducing the footprint of the WWTP.

- Increasing hydraulic capacity due to decoupling hydraulic retention time (HRT) and SRT (i.e. high SRT can be achieved without increasing the reactor volume).
- Intensifying the organic matter removal process due to a higher microbial biodiversity in the reactor (retention of the low-growth rate microorganisms).
- Good disinfection capability.
- Reducing the energy demand, cost and environmental impacts of the WWTP (Pretel *et al.* 2016).

As mentioned above, AnMBR technology has shown a high capability for obtaining high effluent quality. This is particularly important when considering reuse options. Moreover, controlling the SRT of the process allows the growth of species that would be simply washed out in other systems without increasing the operating temperature. This fact increases the possibility of application of MBR technologies for a high diversity of wastewaters, including wastewaters with persistent pollutants.

Although there are still some uncertainties around AnMBR performance, this technology is becoming increasingly popular for municipal wastewater treatment. Indeed, although this technology still has not been applied to full-scale municipal wastewater treatment, recent publications (e.g., Lin *et al.* 2013; Ozgun *et al.* 2013) have reported increasing interest in its use by the scientific community.

This interest is mainly focused on the opportunity to develop an MBR concept with the advantage of a clarified effluent with improved energy balance and very low sludge production. AnMBR is a potential technology to accomplish the challenges that municipal wastewater treatment will face in the near future. Concerning sludge production, reported reductions in the sludge production when using AnMBR technology of up to 90%. Pretel *et al.* (2014) reported sludge productions when treating municipal wastewater at different operating conditions from 0.16 to 0.5 kg TSS kg^{-1} $COD_{removed}$. Moreover, depending on the operating conditions, the sludge produced in AnMBR could achieve suitable conditions, which allows its direct disposal on farmland without requiring further digestion.

Regarding energy consumption, Pretel *et al.* (2014) reported power requirements (when methane capture is not considered) in AnMBR technology treating municipal wastewater of around 0.22 $kWh \cdot m^{-3}$. This value is lower than the power requirements reported in literature for other wastewater treatment processes. For instance, Judd and Judd (2011) reported membrane and total aeration energy demands of around 0.34 and 0.55 $kWh \cdot m^{-3}$ for the full-scale aerobic MBR from Peoria (USA). Higher values were reported by the same authors for other full-scale aerobic MBRs (e.g., Running Springs MBR WWTP, USA, consuming around 1.3–3 $kWh \cdot m^{-3}$). Therefore, from an energy perspective, AnMBR operating at ambient temperature could be a promising sustainable system compared to other existing wastewater treatment technologies, since methane can be used to cover energy requirements.

However, being an emerging technology, there are still some barriers or challenging issues that limit the widespread practical application of AnMBRs. Thus, further breakthroughs with respect to the following challenges should be pursued:

- *Membrane fouling and cleaning*: Membrane fouling is the result of the interaction between membrane surface and sludge suspension. This phenomenon usually decreases system productivity, causes frequent membrane chemical cleaning, which might reduce membrane lifespan whilst increasing replacement costs, and increases energy requirements for sludge recirculation or gas scouring (Lin *et al.* 2013). In this respect, membrane fouling and cleaning issues remain a critical obstacle limiting the widespread application of membrane systems in WWTPs (Lin *et al.* 2013; Jeison, 2007).
- *Methane production*: methane production in AnMBR treating municipal wastewater is lower than that in aerobic membrane bioreactor (MBR) when treating high-strength wastewaters. For instance, a recent review on AnMBR treating different industrial wastewaters (Dereli *et al.* 2012) reported specific methane productions similar to the theoretical maximum value in standard conditions of

temperature and pressure (0.35 m^3 $CH_4 \cdot kg^{-1}$ $COD_{removed}$). However, methane production decreases from 0.06 to 0.27 m^3 $CH_4 \cdot kg^{-1}$ $COD_{removed}$ when treating low-strength wastewaters such as municipal wastewater (Ozgun *et al.* 2013) mainly due to the loss of dissolved methane with the effluent and the possible competition between Methanogenic Archaea (MA) and Sulphate Reducing Bacteria (SRB) for the available substrate.

- *Sulphide production*: When municipal wastewater containing sulphate is anaerobically treated, sulphate is reduced to sulphide. The production of this end product can cause some disadvantages. For instance, the amount of produced biogas is reduced because some of the influent COD (approx. 2 g COD per g SO_4-S) is consumed by SRB (see, for instance, Giménez *et al.* 2011). Moreover, the presence of hydrogen sulphide in the biogas and the mixed liquor causes some technical problems such as: 1) toxicity to anaerobic microorganisms; 2) reduction of the quality of the produced biogas; 3) corrosion in pipes, engines and boilers, entailing higher maintenance and replacement costs; and 4) downstream oxygen demand for oxidising hydrogen sulphide.
- *Temperature*: Low ambient temperatures have been normally considered a barrier for anaerobic treatment because the energy requirements associated with heating large quantities of wastewater outweigh the energy recovery potential (Lettinga *et al.* 2001). Although anaerobic processes are most often operated at high/moderate temperatures to increase microorganism growth rate, AnMBRs have recently been shown to perform adequately at lower temperatures (e.g., 15–20°C) (Martínez-Sosa *et al.* 2012; Pretel *et al.* 2014). However, the lower the temperature, the higher the proportion of the produced methane that is dissolved in the mixed liquor (Giménez *et al.* 2012). Indeed, Pretel *et al.* (2014) reported that up to 85% of the methane produced in AnMBR is extracted with the effluent when operating at low SRT and temperature (30 days and around 15°C, respectively). This methane dissolved in the effluent could strip out in downstream open-air steps thus being emitted to the atmosphere.
- *Lack of direct nutrient removal capability*: A post-treatment is required to produce an effluent suitable for being discharged directly into the aquatic environment. However, some approaches can be applied according to McCarty *et al.* (2011): 1) chemical precipitation or its conversion into struvite for recovery as fertiliser; 2) anammox process, which oxidises ammonia with nitrite to produce harmless N_2 gas; 3) source-separation of urine so that it does not become part of the municipal wastewater; and 4) crop or landscape irrigation of the AnMBR effluent.

Hence, submerged AnMBRs are considered as a candidate technology to improve sustainability in the wastewater treatment field, broadening the range of application of anaerobic biotechnology to low-strength wastewaters (e.g., urban ones) or extreme environmental conditions (e.g., low operating temperatures). This alternative technology is more sustainable (rather than aerobic-based technologies) because wastewater turns into a renewable source of energy and nutrients, whilst providing a recyclable water resource. AnMBRs not only have the main advantages of MBRs (i.e. high quality effluent, and reduced space requirements) but also the main advantages of anaerobic processes. In this regard, low sludge production can be obtained due to the low biomass yield of anaerobic organisms, low energy demand is required since aeration is not needed, and methane is produced – a renewable energy source that improves the energy balance in this system. Mention must also be made of the potential nutrient recovery from wastewater either when the effluent can be used for direct irrigation or when it must be further treated by using nutrient recovery technologies.

2.2.2 Upflow anaerobic sludge blanket Reactor (UASB)

The upflow anaerobic sludge blanket (UASB) reactor is the most commonly applied high-rate anaerobic system for municipal wastewaters. This reactor configuration was designed in the Netherlands by Lettinga and his co-workers in the early 70's. The number of UASB reactors operated worldwide is more than 700 which corresponds to nearly 65% of all the existing anaerobic treatment systems (Nnaji *et al.* 2014).

The main advantages of this system are the simple design, operational simplicity, low capital and operational costs and ability to tolerate fluctuation in pH, temperature and influent substrate concentration (Alvarez *et al.* 2006). Also in UASB systems, microorganisms can naturally form granules, which enhance the settling quality and prevent biomass washout from the reactor (Ince *et al.* 2001; Seghezzo *et al.* 1998). Furthermore, this system enables a good contact between the wastewater and the microorganisms by natural turbulence resulting from gas bubbles and the system achieves high organic compound removal efficiency at shorter retention times. Thus, the system does not require mechanical mixing, which decreases the energy consumption and the operational costs. Due to the granulation in UASB systems, the hydraulic retention time can be different from the sludge retention time, so the hydraulic retention time can be designated independently reducing the treatment times from days (typical for conventional digesters) to hours. The effectiveness and stability of a UASB reactor depends strongly on the initial start-up, which in turn is mainly affected by numerous physical, chemical and biological parameters (Chong *et al.* 2012). The effects of operating parameters on microbial community structures of UASB were established in detail by advanced molecular tools in the literature (Akarsubasi *et al.* 2006; Kolukirik *et al.* 2006; Ayman Oz *et al.* 2012).

There are some drawbacks of UASB systems. The most important issue is the poor effluent quality of a UASB reactor which cannot meet with the discharge limits (Ince *et al.* 2005). Also, the effluent may include high nutrient concentrations and pathogens and requires post-treatment to avoid the contamination of receiving environments. Thus, complete pathogen elimination, highest organic material removal and maximum nutrient recovery must be taken into account, while selecting the proper post-treatment option (Khan *et al.* 2011). Also if the system is exposed to high organic loading, volatile fatty acids accumulation can occur and cause inhibition (Dogan *et al.* 2005).

2.2.3 Expanded granular sludge bed reactor (EGSB)

The expanded granular sludge blanket (EGSB) is a high-rate system designed by de Man *et al.* (1988) as an improved UASB reactor. Due to the higher upflow velocities inside the EGSB reactor, induced by a high height/diameter ratio and recycle rate, sludge is expanded through the reactor and the contact of microorganisms with the influent increases, which leads to a better effluent quality and reduction of dead volumes (Fuentes *et al.* 2011). The removal performance and stability of the EGSB system are strictly dependent on the bed expansion and biomass washout. Thus, accurate prediction of the bed expansion plays a crucial role during the design and operation phases of EGSB reactors.

The EGSB system offers many benefits over UASB reactors. Due to the high upflow velocity, higher mixing is provided in the system, which enhances the mass transfer and biomass activity as well as transportation of substrates into flocs. Furthermore, the reactor can be operated under higher organic and hydraulic loadings as well as under psychrophilic conditions. The wastewaters containing lipid and toxic compounds can also be treated by EGSB systems. Besides, EGSB reactors are efficient in treatment of low strength wastewaters such as municipal wastewaters. Nevertheless, complete removal of suspended solids does not take place in these systems (Mao *et al.* 2015).

2.2.4 Anaerobic sequencing batch reactor (ASBR)

Anaerobic sequencing batch reactors are suspended growth systems and all the biochemical reactions as well as separation occur in a single tank. These systems are suitable treatment options for low-flow streams and can be implemented for a wide variety of wastewaters. Due to the operational simplicity and high removal performances, these systems are applied in many plants. The main difference between ASBR and continuous flow system is the system design in which equalization, reaction and sedimentation takes place in a single tank in a defined time sequence. Furthermore, ASBR systems are flexible to influent wastewater volumes and can tolerate shock loadings which is differ from continuous systems (Mao *et al.* 2015). Also, capital and operational expenditure of these systems are cost-effective.

The main challenges of ASBRs are: requirement of high reactor volumes which can lead to dead zones, channelling and clogging. Also, biogas can affect biomass settling properties and dilution of toxic compounds cannot be realised in this type of reactor. There are many studies in the literature investigating the effects of toxic compounds on ASBR systems (Aydin *et al.* 2015a, 2015b, 2015c).

2.2.5 Anaerobic baffled reactor (ABR)

One of the high-rate anaerobic reactors, the anaerobic baffled reactor was designed by McCarty and colleagues in the early 1980's. There is a series of cells in one tank and the influent wastewater goes through a series of downflow and upflow compartments, passing through a sludge blanket in the latter. The system can be modified according to the wastewater characteristics and required effluent quality by changing the baffles. In recent years, the ABRs became attractive due to their advantages. ABRs show very stable performance under shock loadings regarding to compartmentalized design. Due to the unique design of this system, it can be operated as a two-stage anaerobic reactor in which acidogenesis and methanogenesis run separately and provide the favourable conditions for responsible microbial groups. As in the UASB system, HRT can be different from SRT, which allows achieving efficient organic matter removal, lower sludge generation and small footprint. Furthermore, using the granulating sludge enhances the system stability and tolerating the extreme conditions. Due to the mechanical simplicity, it requires low capital and operation costs (Hahn *et al.* 2015; Mao *et al.* 2015). The challenges of ABRs, such as biomass washout from the system, insufficient mixing and low loading capacity limit their full-scale application.

2.2.6 Full scale applications

Due to the fact that anaerobic treatment systems are more suitable for high-strength wastewaters, they have been mostly applied to industrial wastewaters in full scale. However, there are many municipal treatment plants that include anaerobic treatment units not only for the treatment of sludge, but also wastewater.

In Brazil, a lot of anaerobic treatment units, mainly of the UASB type, have been implemented and many more are under construction or in the design stage. In Mangueira – Brazil, the UASB reactor with an 810 m^3 is operated with a COD removal efficiency of 60%–90%. During the influent COD fluctuations, the COD removal efficiency does not change due to the adjustment of organic loading rates. The reactor shows stable performance (Florencio *et al.* 2001). In a study carried out by Nada *et al.* (2011), the removal efficiency of a UASB reactor with 4187 m^3 total volume implemented in Sanhour, Fayoum in Egypt was monitored. The reactor showed stable performance after reaching steady-state conditions (20 weeks). The total COD, total biochemical oxygen demand ($tBOD_5$) and total suspended solids (TSS) removal performances of the effluent appeared to be 70% ± 11, 70% ± 13 and 85% ± 7, respectively, which is found satisfactory by the authors. According to the results, because of the lower energy demand of this system, operation and maintenance costs are 30–60% lower than conventional systems. In another study, Heffernan *et al.* (2011), reviewed the performance of 10 full-scale upflow anaerobic sludge blanket reactors in municipal treatment plants located in warm regions; 7 plants were located in Brazil, 2 in India and 1 in the United Arab Emirates. The removal efficiencies were different between plants and COD, BOD_5 and TSS removal performances ranged from 44–77%, 37–80% and 45–84%, respectively. Also, 15 full-scale UASB systems are implemented in the Yamuna River basin under its Yamuna Action Plan (Sato *et al.* 2006). According to the effluent characteristics, all plants are deficient in meeting the regulation limits. The reason of this situation is related with the high pollutant content of influent wastewater, which is contaminated with industrial wastewaters. In these plants, the operation and maintenance practices should be revised and the sources of municipal wastewater should be determined in order to reach better quality of effluent.

2.2.7 Pilot scale applications

There are many pilot-scale applications of various types of anaerobic treatment systems for municipal wastewater all around the world. Researchers tested the performance of different reactor configurations and combinations under certain conditions.

In the study of Alvarez *et al.* (2006), a UASB reactor with a 35 m^3 total volume treating municipal wastewater was operated under psychrophilic conditions. In the effluent, total suspended solids (TSS) and total chemical oxygen demand (tCOD) removal were 75–85% and 63–73%, respectively. In another study also carried out by Alvarez *et al.* (2008), the authors operated a two-stage pilot plant consisting of a hydrolytic upflow sludge bed (HUASB) reactor (2.5 m^3) and an UASB reactor (20 m^3). This configuration showed a good performance for influent wastewater COD value higher than 250 mg/L; however, the removal efficiency fell down during the extreme dilutions caused by rainfalls. Tandukar *et al.* (2007) operated a combined UASB (1 m^3) and down-flow hanging sponge (DHS) reactor (0.5 m^3). The pilot plant was installed in the municipal wastewater treatment plant of Nagaoka City, Japan. Under certain circumstances, the BOD_5 removal was 92%. The authors proposed that, this combined system can be a potential and cost-effective solution for municipal wastewaters over the activated sludge process.

According to the resource recovery concept from municipal wastewater, El-Shafai and his colleagues (2014) operated an UASB reactor (40 L) followed by a series of duckweed ponds. During the summer season, the COD, BOD and TSS removal efficiencies were very high (93%, 96% and 91%, respectively) and the system achieved nearly complete faecal coliform removal. On the other hand, removal performance was reduced during the winter and the system was not efficient in faecal coliform removal.

In a study carried out by Dai *et al.* (2011), municipal wastewater was treated by an expanded granular sludge bed (EGSB) reactor with 23 m^3 total volume and 12.5 m^3 active volumes, removing soluble organic matter and followed by an electrochemical (EC) reactor to oxidize the NH_3–N (16 L). The pilot plant was set up in a municipal wastewater treatment plant in Shanghai, China. After iron shavings addition, the removal efficiency of COD, SS and total phosphorus (TP) were increased to 81.5%, 89% and 62.4%, respectively. In the effluent of the EC, NH_3–N was measured as 12 mg/L. The results were very promising for possible full-scale application.

In another study, a four-compartment anaerobic baffled reactor (ABR) with 1 m^3 volume was tested for two years treating raw municipal wastewater (Hahn *et al.* 2015). The reactor was operated at ambient water and air temperatures, and implemented in Plum Creek Water Reclamation Authority (PCWRA)'s wastewater treatment plant. The average TSS and BOD_5 removal was 83 ± 10% and 47 ± 15%, respectively. The authors concluded that, the ABR can be an alternative primary treatment option in cold climate regions.

Shin and his colleagues (2014) tested the performance of the pilot-scale staged anaerobic fluidized membrane bioreactor (SAF-MBR) system integrated to the Bucheon municipal wastewater treatment plant in the Republic of Korea. The total reactor volume was 0.99 m^3 and was operated with a working volume of 0.77 m^3. Prior to the acclimation period, the reactor reached chemical oxygen demand (COD) and biochemical oxygen demand (BOD5) removals of 81% and 85%, respectively, during the winter. After the reactor was fully acclimated, the COD and BOD_5 removal efficiencies increased to 90%. The performance of AnMBR treating municipal wastewater at pilot-scale has been evaluated in different studies over the last years (see Table 2.1).

Table 2.1 Operating conditions and treatment performance of membrane coupled anaerobic technologies for municipal wastewater treatment at pilot scale.

Reactor Configuration	Pore Size (μm)	Membrane Area (m^2)	Reactor Type	Volume (L)	HRT (h)	SRT (d)	OLR (kgCOD/m^3/d)	MLSS (g/L)	T (°C)	J (LMH)	CODi (mg/L)	COD_{REM} (%)	CH_4 Yield L_{CH4}/gCOD	Biosolids Wasted kgSS/kgCOD	Reference
Submerged	0.05	31	CSRT	1500	6–21	70		8–22	33	10	445	87	0.294 (0.069[b])		Giménez *et al.* (2011)
Submerged	0.04	3.5	CSRT	350	19		0.5–1.1		20–35	7	630	82–90	0.23–0.27		Martinez-Sosa *et al.* (2011)
Side-stream	100[a]	5.1	UASB	849	6	180			22	45–50	445	93			Herrera-Robledo *et al.* (2011)
Submerged	0.05	31	CSRT	2100	6–21	74–76			20–33		410–720			0.117–0.228	Giménez *et al.* (2012)
Submerged	0.04	3.5	CSRT	350	17.5–26.5		0.4–0.9	9.5–17.3	20	7–12	612	84–94	0.24–0.29		Martinez-Sosa *et al.* (2012)
Submerged	0.05	31	CSRT	2100	5.5–24.5	70		7.5–31.5	20–33	10–13.3	459	85		0.16–0.23	Robles *et al.* (2013)
Submerged	0.05	31	CSRT	2100	15.7	30–70		12.5–32.5	17–33	9–20	650			0.16–0.55	Pretel *et al.* (2014)
Submerged	0.03	39.5	FB	1000	4.6–6.8			0.6–1.2	8–30	4.1–7.5	223–362	81–94		0.026–0.11[c]	Shin *et al.* (2014)
Submerged	0.045	0.93	UASB	310	7		2–2.5		18	10–15	892	87	0.18–0.23		Gouveia *et al.* (2015a)
Submerged	0.045	0.93	UASB	459	12.8–14.2		1.6–2.0		18	10–14	978	90		0.05–0.083[c]	Gouveia *et al.* (2015b)
Submerged	0.04	5.4	CSRT	550	8.5	70		11.5–23	23	17		80–94	0.091–0.102	0.13–0.25	Dong *et al.* (2015)

[a]kDa; [b]observed; [c]SV.

A recent study dealing with the application of membrane coupled fluidised reactors is the one described in Shin *et al.* (2014). The anaerobic fluidized membrane bioreactor used in this study was operated at ambient temperatures ranging between 8–30°C for 485 days with an HRT ranging between 4.6 and 6.8 hours. Although operating at considerably low temperatures (8°C in wintertime), the reactor reached an average COD removal efficiency of 81%, increasing to up to 98% in summertime. As a result of GAC fluidisation, the membrane fouling propensity was low. This fact allowed reaching low energy requirements (around 0.23 kW h/m^3), which could be met by energy input from the produced methane. Moreover, biosolids production varied from 0.023 to 0.11 gVS/gCOD (yearly average biosolids production of 0.051 gVS/gCOD), which is significantly low compared to aerobic systems.

As regards UASB coupled to membrane filtration for municipal wastewater treatment, Herrera-Robledo *et al.* (2011) and Gouveia *et al.* (2015a) operated pilot-scale systems with working volumes of 0.85 and 0.35 m^3, respectively. As a result of operating with attached biomass, low HRT (6 and 7 hours, respectively) were possible (see Table 2.1). Herrera-Robledo *et al.* (2011) evaluated the performance of an AnMBR system treating raw sewage at ambient temperature. The system, which was operated at an SRT of 180 days and an average temperature of 22 ± 3°C, achieved COD removal efficiencies of 93% on the average. The low MLSS levels in the membrane tank allowed operating at fluxes of around 45–50 L/m^2/h On the other hand, Gouveia *et al.* (2015a) evaluated the long-term performance of an AnMBR system consisting of an UASB reactor coupled to an external submerged membrane module. The results of this study proved the long-term reliability of AnMBR technology for municipal wastewater treatment at psychrophilic temperatures. COD removal resulted was 87% even though the OLR was high (2–2.5 kgCOD/m^3/d), whilst the methane production ranged around 0.18–0.23 LCH_4/gCOD. The same authors (Gouveia *et al.* 2015b) evaluated the long-term performance of an AnMBR system consisting of an UASB reactor coupled to an integrated ultrafiltration unit. The system, which treated municipal wastewater at 18 ± 2°C, resulted in stable operation for more than 3 years. Higher COD removal efficiencies were achieved (90%) compared to the external membrane configuration, although the OLR was lower in this case (1.6–2.0 kgCOD/m^3/d). In addition, low biosolids productions were obtained (around 0.05–0.08 g VS/g $COD_{removed}$).

Concerning CSTR based systems; Martinez-Sosa *et al.* (2011) evaluated the performance of an anaerobic CSTR coupled to a submerged hollow-fibre membrane for municipal wastewater treatment at mesophilic and psychrophilic temperature conditions. The system was operated for 100 days at 35, 28 and 20°C, obtaining COD removals close to 90% in all cases. Nevertheless, an increase in fouling rate was observed at 20°C probably due to the accumulation of solids in the reactor. Reasonable methane yields (0.23–0.27 LCH_4/gCOD) were achieved during the whole experimental period. The same pilot plant was later operated at psychrophilic conditions (20°C) for 90 days (Martinez-Sosa *et al.* 2011), achieving similar methane yields and COD removals.

Another example of CSTR-based AnMBR pilot-plant is the one reported by Giménez *et al.* (2011, 2012). The main characteristic of this system was that it was equipped with two industrial-scale hollow-fibre membrane modules. It is worth to point out that the effect of the main operating conditions on membrane performance depends heavily on the membrane size (HF length is a critical parameter). This plant was fed with the effluent from the pre-treatment of a full-scale municipal WWTP located in Valencia (Spain). Giménez *et al.* (2011) aimed at assessing the effect of several operational variables on both biological and separation process performance. The plant was operated under mesophilic conditions (at 33°C), at a 70 days solid retention time, and variable hydraulic retention time ranging from 20 to 6 hours. The pilot plant performance demonstrated that AnMBR can be a promising technology for municipal wastewater treatment. The competition between Methanogenic Archaea (MA) and Sulphate Reducing Bacteria (SRB) must be considered, especially when the influent wastewater presents a low COD/SO_4-S ratio. Almost 90% of COD removal was achieved, but with a low methane yield, which was mainly due to the COD removed by SRB. On the other hand, no irreversible fouling problems were detected, even for Mixed Liquor Total Solids (MLTS) above 22 $g \cdot L^{-1}$. Robles *et al.* (2013) evaluated the effect of temperature (20, 25 and 33°C) on the performance of this AnMBR system. The results from this study revealed that MLTS is a key factor affecting membrane permeability (K). K was higher under psychrophilic than mesophilic conditions when operating at similar transmembrane fluxes and MLTS. It was related to the decrease in biomass activity observed during psychrophilic operation. Thus, lower extracellular polymeric substances (EPS) and soluble microbial products (SMP) levels were observed at psychrophilic conditions, which affected not only the three-dimensional floc matrix, but also the fouling propensity. Nevertheless, after almost one year of operation, no irreversible fouling problems were detected. In order to assess the feasibility of the system, Pretel *et al.* (2014) evaluated the operating cost of the system when treating sulphate-rich municipal wastewater at ambient temperature (ranging from 17 to 33°C). Energy consumption, methane production, and sludge handling and recycling to land were evaluated. The results revealed that optimising specific gas demand with respect to permeate volume (SGDP) and sludge retention time (for given ambient temperature conditions) is essential to maximise energy savings (minimum energy demand: 0.07 $kWh \cdot m^{-3}$). Moreover, low/moderate sludge productions were obtained (minimum value: 0.16 kg $TSS \cdot kg^{-1}$ $COD_{removed}$), which further enhanced the overall operating cost of the plant (minimum value: €0.011 per m^3 of treated water). On the other hand, the sulphate content in the municipal wastewater significantly affected the final production of methane and thereby affected the overall energy consumption. Indeed, this study revealed that AnMBR technology is likely to be a net energy producer when treating low/non sulphate-loaded wastewaters in warm/hot climates: theoretical maximum energy productions of up to 0.11 $kWh \cdot m^{-3}$ could be achieved.

2.2.8 Different lab-scale options – immobilization

In practice, operational stability obstacles still limit wide application of commercialized anaerobic technologies for wastewater treatment (Dupla *et al.* 2004). Moreover, anaerobic processes are highly vulnerable to organic and hydraulic load fluctuation, suffer active biomass washout from reactors, are sensitive to inhibitors, and require lengthy periods of acclimation to achieve efficient biodegradation (Dereli *et al.* 2012). To overcome the principal limitations of conventional and unstable processes, granular-based anaerobic systems incorporating an immobilized biomass have been proposed as an alternative technology for complex wastewater treatment (Wu *et al.* 2008). Systems with immobilized biomass facilitate the use of compact units operating without recirculation or separation systems. Compared to suspended growth processes, immobilized microorganisms possess several advantages, including high metabolic activity rates and strong resistance to toxic chemicals (Dwyer *et al.* 1986; Massalha *et al.* 2007; Wu *et al.* 2008). The effective control of sludge retention time, potentially high biomass concentration and, consequently, the option of applying low hydraulic retention times have encouraged the adoption of immobilized biomass technology (Fazolo *et al.* 2007). Moreover, immobilization of biomass increases its tolerance to hydraulic or quality shocks, providing a secure environment for efficient activity (Gao *et al.* 2011; Massalha *et al.* 2015).

Based on previous developments of biomass drying procedures and unique immobilization techniques using hydrophilic polyurethane foam (HPUF) (Massalha *et al.* 2015a), the influence of biomass immobilization within a matrix enriched with powdered activated carbon (PAC) was examined (Massalha *et al.* 2015b). This system was successfully operated at a lab-scale system (0.5 L) to test the effects of OLR increase, phenol shock load (5 g/L for three days), and continuous exposure to phenol. HPUF matrix enriched with PAC has added advantage of securing the biomass from sudden inhibitory shock loads, and enables fast recovery and adaptation to phenol biodegradation (Masslaha *et al.* 2015b).

This technology was also tested at a pilot-scale level with 4 m^3 of high-rate upflow reactor with upper zone of immobilized matrix. This technology is based on a unique, patented process for preparing "pre-treated" biomass, immobilized in a polymeric matrix that has high handling capability and fast acclimatization to wastewater streams (Sabbah *et al.* 2011). The technology was developed specifically for treatment of high-load agro-industrial waste water. The pilot is operated in Karmiel's Wastewater treatment plant (North of Israel) in collaboration with Mekorot the Israeli National Water Company. The implementation of this process guarantees a sustainable wastewater treatment technology to meet existing and future effluent quality requirements at energy-saving approach of about 25% compared to the activated sludge process and with a significant reduction of the secondary sludge (more than 35%).

The information obtained here is expected to be useful for development of tolerant, stable and environmentally-effective anaerobic systems for the treatment of organic wastes containing problematic constituents. However, further research is needed on optimizing process performance at low temperatures and demonstrating performance at pilot and full scale.

2.3 MODELING OF ANAEROBIC TREATMENT SYSTEMS

Anaerobic digestion consists of a complex network of biological and chemical reactions. The main biological reactions include disintegration of particulates, hydrolysis of complex organic matter, acidogenesis, acetogenesis and methanogenesis. Other significant reactions would be sulphate reduction, nitrate reduction or acetate oxidation. These biological reactions are impacted by physicochemical phenomena occurring in the reactor medium such as gas stripping (CH_4, CO_2 and H_2), precipitation, acid/base equilibria which, in turn, lead to changes on pH, temperature variations, etc. Although it is possible to describe all the aforementioned phenomena in a single mathematical model, the resulting set of equations and parameters would be too complex for universal application. On the contrary, a mathematical model is often defined as a "partial description of a system", indicating that not all the observed phenomena are accounted for in a mathematical model. As a conclusion, each application of anaerobic digestion will require a model that fits the objectives of the application rather than systematically using a comprehensive model.

2.3.1 Review of models

Among other characteristics, models are classified into i) data-driven models (also called black box models), which are built inductively as relations between input and output data and ii) mechanistic models (also called first-principle or white box models), which are built deductively from conservation laws, kinetic equations, etc.

Data driven models include, among others, fuzzy logic models, artificial neural networks and partial least square. The development of data driven models is, in general, simpler than mechanistic models and can be more accurate than mechanistic models if the knowledge about the process is scarce (e.g., the reaction kinetics or the model parameters are uncertain). This is, in particular, the case when accounting for phenomena other than the digestion of COD, such as the fate of NH_3, H_2S or the cycle of phosphorous. Among data driven models, artificial neural networks (ANN) stand out due to their high predicting power. ANN consist of connecting inputs and outputs by several layers of nodes (neurons) with elementary mathematical functions (e.g., hyperbolic tangent), that combined and calibrated, can accurately predict nonlinear relationships. On the other hand, it must be taken into account that training such an ANN model can require a large amount of data Holubar *et al.* (2003) used 500 days of operation for training an (ANN). Furthermore, data driven models do not provide insight about the process and extrapolation to other processes must be done with care as they are prone to overfitting.

Mechanistic models are built deductively by using physical, chemical and biological knowledge about the system. The first AD models were developed by describing only the step that limits the overall process.

In systems with high solids content, such as primary and activated sludge digesters, hydrolysis is rate limiting of the overall process (Eastman & Ferguson, 1981), in particular if the fraction of particulates is large. On the other hand, acetoclastic methanogenesis is limiting for diluted systems with a large amount of dissolved organic material, which was the basis of the first attempts to model AD (Andrews, 1969).

As the AD models tried to describe with more detail the biological reactions involved, the IWA Task Group for Mathematical Modelling of Anaerobic Digestion Processes presented a consensus model called the Anaerobic Digestion Model No. 1 (ADM1) with the aim of providing a generic model that can be widely applicable for anaerobic processes (Batstone *et al.* 2002). The ADM1 consists of five reaction steps (disintegration, hydrolysis, acidogenesis, acetogenesis and methanogenesis) and uses 26 state variables in an attempt to generically cover anaerobic processes. For more specific applications, it is possible to extend the standard ADM1 by including sulphate reduction, nitrate reduction or more detailed disintegration and hydrolysis in function of the type of substrate.

When the anaerobic stage is modelled together with other units within a plant, it is necessary to consider how to connect the ADM1 model with other models, in particular with the widespread activated sludge models (ASM). Rosen *et al.* (2006) connected the ADM1 with the ASM by using an interface that makes an assignation between the variables of both models. A rather different approach was followed by Grau *et al.* (2007) who integrated the ADM1 into a plant wide model, i.e. all the processes and variables are merged into a single comprehensive model of all the units in the plant, thereby avoiding the need of interfaces.

The reader looking for more information is advised to check Lauwers *et al.* (2013) for an excellent and comprehensive review on the existing AD models.

2.3.2 Model selection for a given application

The selection of a model depends on the application, the purpose, the accuracy needed, the available data, etc. As mentioned, ADM1 provides a good trade-off of genericity and complexity, being extendable for accounting for more complex phenomena. However, the complexity of ADM1 and the need of many input parameters, together with the expansion of AD of animal waste in farms, have led to the development of simple calculators. These calculators are better suited for existing and standard processes, such as biogas production in farms, and do not require particular technical knowledge. As the calculators are not aimed at universal applications, but rather focus on, e.g., animal waste AD, the use of simpler correlations is possible. Finally, these calculators provide estimates of biogas production, energy production and a financial assessment. For example, the Irish Environmental Protection Agency provides a Microsoft Excel based tool in their website. However, the estimations can differ notably; Kythreotou *et al.* (2014) compared six of these calculators and found differences of more than 130% in the estimations.

Technology development and analysis of a given process operation are applications that need very realistic and complex models, based on factual knowledge. Important issues here deal with minimising inhibition, ensuring non-limiting gas transfer, etc. Therefore, ADM1 or extensions are the most suitable choice.

Upcoming critical issues in design of AD units, in particular for UASB reactors, are the behaviour of substrate particles and how the hydraulics impacts the sludge retention (Batstone, 2006). Any attempt to model the hydraulics of the reactor will greatly increase the complexity and the computational requirements; therefore, a simple reaction model should be preferred for design applications, considering the hydrolysis and the limiting bioconversion stage.

Online model-based control requires monitoring the state variables of the model, either by direct measurement, and either by estimation by an observer which can be challenging for the most complex AD models. As a result, a number of reduced complexity models have been developed for different applications such as manipulating feed to avoid wash-out, or maximizing the biogas production (Simeonov *et al.* 2007)

2.4 PROBLEMS AND FUTURE PERSPECTIVES

2.4.1 Problems

As discussed in the first part of the chapter, anaerobic technologies are mainly preferred for energy recovery from waste (water). Today's technology causes increasing energy demand and fossil fuels, which are associated with climate change concern, are still the main energy source. So we need to design a next generation wastewater treatment plant which produces energy more than its consumption. Basically, efforts need to be made toward a better efficiency and a more sustainable use of wastewater's energy potential (McCarty *et al.* 2011).

Anaerobic treatment is the most sustainable way to recover energy from the wastewater. However 60–70% of the energy potential is lost during the conversion into electricity (U.S. Environmental Protection Agency, 2007). During conversion of organic materials the municipal wastewater to the methane gas, 8% of the potential energy is lost. Also 7% of energy is lost as maintenance of microbial cells. The total lost increased up to 20% with some engineering fails and approx. 80% of methane gas can be used as a potential energy source. Another important issue is that only 65% of the methane can be converted into the electricity. Thus, totally 28% of the energy obtained from anaerobic degradation of municipal wastewater can be used (McCarty *et al.* 2011).

Another advantage of the anaerobic treatment application in municipal wastewater treatment, which has been already discussed above, is nutrient recovery. Ammonia and orthophosphate are directly used in the agriculture as ferti-irrigation (Ozgun *et al.* 2013). Using the municipal wastewater as a source of ferti-irrigation has an added value because production of N and P required energy. According to this knowledge, using wastewater as a nutrient source is a good opportunity instead of energy spending as seen in Table 2.2.

Table 2.2 Energy characteristics of a typical municipal wastewater.

Constituent	Typical Concentrations (mg/L)	Energy (kWh/m^3)		
		Maximum Potential from Organic Oxidation	Required to Produce Fertilizing Elements	Thermal Heat Available for Heat-Pump Extraction
Organics (COD)				
Total	500			
Refractory	180			
Suspended	80	0.31		
Dissolved	100	0.39		
Biodegradable	320			
Suspended	175	0.67		
Dissolved	145	0.56		
Nitrogen				
Organic	15		0.29	
Ammonia	25		0.48	
Phosphorus	8		0.02	
Water				7.0
Totals		1.93	0.79	7.0

Source: Ozgun *et al.* (2013).

In spite of the energy and nutrient recovery opportunities, anaerobic treatment for municipal and/or municipal wastewater is still discussed because of doubts regarding its feasibility. Low COD contents and consequently low energy production are the main concerns nowadays.

2.4.2 Suggestions

2.4.2.1 Source separation and anaerobic treatment of black water stream

A new hot topic for sanitation and wastewater management has been discussed and new technologies have been developed, which is referred to as Resources Management Sanitation, Ecological Sanitation (EcoSan), or Decentralised Sanitation and Reuse (DESAR) in the recent years (Lens *et al.* 2001; Otterpohl & Oldenburg, 2007). These new technologies are based on source separation of wastewater flows and organic waste on a household level, followed by an appropriate treatment of each stream in decentralized or semi-centralized systems and consequent reuse of water and nutrients in spite of centralized conventional wastewater treatment The basic approach is separated wastewater generated on household level as black water and grey water. The main components of black water are urine, faeces, toilet paper and flushing water coming from toilets. Black water contains most of the nutrients, around half of the municipal COD load, the major part of the pathogens (Vinnerås *et al.* 2006). As seen in Table 2.3, blackwater is suitable for anaerobic digestion because of high strength nature. Also anaerobic treatment gives an opportunity to recycle the nutrients in the black water for agricultural purposes. New approaches have been reported in the literature for anaerobic treatment of black water collected with vacuum toilets however mainly four types of the reactors are used as a CSTR (continuously stirred tank reactor), an accumulation system, a UASB-septic tank, anaerobic MBR (Upflow anaerobic sludge blanket) (van Voorthuizen *et al.* 2008).

Also combining black water with kitchen waste increases the methane production Zeeman *et al.* (2008) reported that methane production can be doubling by addition of kitchen waste. Luostarinen *et al.* (2007) showed that anaerobic digestion of black water and kitchen waste combination is efficient at low temperatures such as 10°C and 20°C. Kujawa-Roeleveld *et al.* (2006) a similar approach for treatment of black water and kitchen refuse under low temperature in an accumulation system and they reported that accumulation system is appropriate for more concentrated system; it is not efficient or sole black water. van Voorthuizen *et al.* (2008) studied membrane systems through the treatment of black water and they reported the methane production was lower than expected however nutrient recovery is beneficial.

Table 2.3 Black water characteristics.

Parameter	Unit	Black Water from Vacuum Toilets (Kujawa-Roeleveld *et al.* 2006)	Black Water from Vacuum Toilets (Zeeman *et al.* 2007)
Total COD	mg/L	9500–12300	19000
Dissolved COD	mg/L	1400–2800	5000
VFA-COD	mg/L	500–1900	1300
Particulate COD	mg/L	7000–9600	14000
NH_4-N	mg/L	600–1000	1400
Total P	mg/L	90–140	280
Particulate COD /Total COD	–	76%	74%
COD/N/P	–	95/10/1	68/5/1

2.4.2.2 A hybrid system: algae combined with anaerobic digester

In the recent years, algal studies to treat nutrients in municipal wastewater increase. Also microalgae hold a great promise as feedstock for biofuel production. Anaerobic digestion is an established technology to obtain energy. However, this new combination is still under investigation and a hybrid system can be developed: While microalgae remove the nutrient, they serve as a substrate for anaerobic digestion to produce methane and to recover nutrients. In the literature, there are limited studies on this (Alcántara *et al.* 2013; Mahdy *et al.* 2014; Mooij *et al.* 2013; Posadas *et al.* 2013). These studies showed that algal biomass is a potential co-substrate for biogas production. However there are still limitations in terms of both technical and economical in the cultivation and methane production of microalgae for full-scale application and to overcome these limitations, the potential for nutrient and energy recovery in the overall biofuel production process should be quantified by mass-energy balances (Alcántara *et al.* 2013; Ras *et al.* 2011).

2.5 FUTURE ASPECTS

Anaerobic treatment is a core technology for energy and nutrient recovery from the wastewater and all approaches discussed above imply that anaerobic technology will have a place for the next generation wastewater treatment plants. Improvement of present technology and combination of different technologies with anaerobic treatment are required.

Anaerobic membrane technology is the most suitable for the treatment of low COD content wastewaters. So, newer membrane processes coupled with complete anaerobic treatment of wastewater offer the potential for wastewater treatment to become a net generator of energy, rather than the large energy consumer that it is today (McCarty *et al.* 2011). Also effluents of anaerobic membrane reactors are pathogen free and conserving nutrients in the liquid phase, which make them suitable for agricultural use. Additionally membrane fouling problems are still waiting for a solution. More research on the development of both economical and effective technologies is necessary for the rapid implementation of anaerobic membrane reactor technology, especially at full-scale (Ozgun *et al.* 2013).

Another important issue is the separated collection of municipal wastewater in terms of black and grey water. Organic components are more concentrated in this stream and it is suitable for anaerobic digestion. Also addition of kitchen waste and/or sewage sludge increases the methane production.

Hybrid systems, which serve algae as a feedstock for methane production, provide an opportunity to increase energy yield and to recover nutrient and also a low effect to decrease CO_2 release from treatment plants. More studies on this system should be conducted to increase the efficiency.

Table 2.4 Environmental/economic/technical indicators.

Category		AnMBR	UASB
Environmental	Savings in GHG emission	−0.50–5.5 kg CO_2e/m^3	0.14–0.35 kg CO_2e/m^3
Economic	Energy consumption	−0.15 to 0.21 $kWh \cdot m^{-3}$	0.08 kWh/kg $COD_{removed}$
	Other operational costs	€0.07–€0.101/m^3	0.17–0.2$/$m^3$
Technical	COD removal	>90%	60%–90%
	Sludge production	0.16 kg TSS · kg–1 COD removed	0.03 to 0.2 kg SS/kg COD removed
	Methane production	0.06 to 0.27 m^3 $CH_4 \cdot kg^{-1}$	0,02–0,3 m^3 $CH_4 \cdot kg^{-1}$
Social	Awareness/Acceptability	Market on the rise	It is a core method

Also lab-scale studies on improvement of anaerobic technologies are still going on. Kim *et al.* (2015) studied a system of membrane distillation combined with an anaerobic moving bed biofilm reactor for the treatment of municipal wastewater. In another study, bamboo carrier anaerobic baffled reactor was used for municipal wastewater (Feng *et al.* 2008). Integrated anaerobic fluidized-bed membrane bioreactor was developed for the same purpose (Gao *et al.* 2014). Lab-scale studies should be continued however, the tested and successful systems should be improved and implemented into full-scale.

2.6 CONCLUSIONS

In this chapter, we summarized profits and limitations of anaerobic treatment systems for municipal wastewater in the framework of energy production and nutrient recovery. The treatment performance of anaerobic systems vary significantly according to the whole process schemes employed to the plant and the technical, economic and environmental sustainability of two common processes was presented in the Table 2.4.

2.7 REFERENCES

Akarsubasi A. T., Ince O., Ayman Oz N., Kirdar B. and Ince B. (2006). Evaluation of performance, acetoclastic methanogenic activity and archaeal composition of full-scale UASB reactors treating alcohol distillery wastewaters. *Process Biochemistry*, **1**(41), 28–35.

Alcántara C., García-Encina P. and Muñoz R. (2013). Evaluation of mass and energy balances in the integrated microalgae growth-anaerobic digestion process. *Chemical Engineering Journal*, **221**, 238–246.

Álvarez J. A., Ruiz I., Gómez M., Presas J. and Soto M. (2006). Start-up alternatives and performance of an UASB pilot plant treating diluted municipal wastewater at low temperature. *Bioresource Technology*, **97**, 1640–1649.

Álvarez J. A., Armstrong E., Gómez M. and Soto M. (2008). Anaerobic treatment of low-strenght municipal wastewater by a two-stage pilot plant under psychrophilic conditions. *Bioresource Technology*, **99**, 7051–7062.

Andrews J. (1969). Dynamic model of the anaerobic digestion model. *Journal of the* Sanitary Engineering Division, **1**, 95–116.

Aydin S., Ince B. and Ince O. (2015a). Development of antibiotic resistance genes in microbial communities during long-term operation of anaerobic reactors in the treatment of pharmaceutical wastewater. *Water Research*, **83**, 337–344.

Aydin S., Ince B. and Ince O. (2015b). Application of real-time PCR to determination of combined effect of antibiotics on Bacteria, Methanogenic Archaea, Archaea in anaerobic sequencing batch reactors. *Water Research*, **76**, 88–98.

Aydin S., Ince B., Cetecioglu Z., Arikan O., Ozbayram E. G., Shahi A. and Ince O. (2015c). Combined effect of erythromycin, tetracycline and sulfamethoxazole on performance of anaerobic sequencing batch reactors. *Bioresource Technology*, **186**, 207–214.

Ayman Oz N., Ince O., Turker G. and Ince B. (2012). Effect of seed sludge microbial community and activity on the performance of anaerobic reactors during the start-up period. *World Journal of Microbiology and Biotechnology*, **2**(28), 637–647.

Batstone D. J. (2006). Mathematical modelling of anaerobic reactors treating domestic wastewater: rational criteria for model use. *Reviews in Environmental Science and Bio/Technology*, **5**, 57–71.

Batstone D. J., Keller J., Angelidaki I., Kalyuzhnyi S. V., Pavlostathis S. G., Rozzi A., Sanders W. T. M., Siegrist H. and Vavilin V. A. (2002). The anaerobic digestion model no 1 (ADM1). *Water Science and Technology*, **45**(10), 65–73.

Chong S., Sen T. K., Kayaalp A. and Ang H. M. (2012). The performance enhancements of upflow anaerobic sludge blanket (UASB) reactors for domestic sludge treatment a state-of-the-art review. *Water Research*, **46**, 3434–3470.

Chu L. B., Yang F. L. and Zhang X. W. (2005). Anaerobic treatment of domestic wastewater in a membrane-coupled expended granular sludge bed (EGSB) reactor under moderate to low temperature. *Process Biochemistry*, **40**, 1063–1070.

Dai R., Liu Y., Liu X., Zhang X., Zeng C. and Li L. (2011). Investigation of a sewage-integrated technology combining an expanded granular sludge bed (EGSB) and an electrochemical reactor in a pilot-scale plant. *Journal of Hazardous Materials*, **192**, 1161–1170.

de Man A. W. A., van der Last A. R. M. and Lettinga G. (1988). The use of EGSB and UASB anaerobic systems or low strength soluble and complex wastewaters at temperatures ranging from 8 to 30°C. In: Proceedings of the 5th International Symposium on Anaerobic Digestion, Bologna, Italy, E. R. Hall and P. N. Hobson (eds), pp. 197–208.

Dereli R. K., Ersahin M. E., Ozgun H., Ozturk I., Jeison D., van der Zee F. and van Lier J. B. (2012). Potentials of anaerobic membrane bioreactors to overcome treatment limitations induced by industrial wastewaters. *Bioresource Technology*, **122**, 160–170.

Dogan T., Ince O., Ayman Oz N. and Ince B. (2005). Inhibition of volatile fatty acid production in granular sludge from an UASB reactor. Journal of Environmental Science and Health, *Part A*, **3**(40), 633–644.

Dong Q., Parker W. and Dagnew M. (2015). Impact of FeCl3 dosing on AnMBR treatment of municipal wastewater. *Water Research*, **80**, 281–293.

Dupla M., Conte T., Bouvier J. C., Bernet N. and Steyer J. P. (2004). Dynamic evaluation of a fixed bed anaerobic digestion process in response to organic overloads and toxicant shock loads. *Water Science and Technology*, **49**, 61–68.

Dwyer D. F., Krumme M. L., Boyd S. A. and Tiedje J. M. (1986). Kinetics of phenol biodegradation by an immobilized methanogenic consortium. *Applied and Environmental Microbiology*, **52**, 345–351.

Eastman J. and Ferguson J. (1981). Solubilization of particulate organic carbon during the acid phase of anaerobic digestion. *Journal Water Pollution Control Federation*, **53**(3), 352–366.

El-Shafahi S., El-Gohary F., Nasr F. A., van der Steen P. and Gijzen H. J. (2007). Nutrient recovery from domestic wastewater using a UASB-Duckweed Ponds system. *Bioresource Technolology*, **98**(4), 798–807.

Fazolo A., Foresti E. and Zaiat M. (2007). Removal of nitrogen and organic matter in a radial-flow aerobic-anoxic immobilized biomass reactor used in the posttreatment of anaerobically treated effluent. *Applied Biochemistry and Biotechnology*, **142**, 44–51.

Feng H., Hu L., Mahmood Q., Qiu C., Fang C. and Shen D. (2008). Anaerobic domestic wastewater treatment with bamboo carrier anaerobic baffled reactor. *International Biodeterioration & Biodegradation*, **62**, 232–238.

Florencio L., Kato M. T. and Morals J. C. (2001). Domestic sewage treatment in full-scale UASBB plant at Mangueira, Recife, Pernambuco. *Water Science and Technology*, **44**(4), 71–77.

Foresti E., Zaiat M. and Vallero M. (2006). Anaerobic processes as the core technology for sustainable domestic wastewater treatment: consolidated applications, new trends, perspectives, and challenges. *Reviews in Environmental Science and Biotechnology*, **5**, 3–15.

Fuentes M., Scenna N. J. and Aguirre P. A. (2011). A coupling model for EGSB bioreactors: hydrodynamics and anaerobic digestion processes. Chemical Engineering & Process, **50**, 316–324.

Gao D., An R., Tao Y., Li J., Li X. and Ren N. (2011). Simultaneous methane production and wastewater reuse by a membrane-based process: evaluation with raw domestic wastewater. *Journal of Hazardous Materials*, **186**, 383–389.

Gao D. W., Hu Q., Yao C., Ren N. Q. and Wu W. M. (2014). Integrated anaerobic fluidized-bed membrane bioreactor for domestic wastewater treatment. *Chemical Engineering Journal*, **240**, 362–368.

Giménez J. B., Robles A., Carretero L., Durán F., Ruano M. V., Gatti M. N., Ribes J., Ferrer J. and Seco A. (2011). Experimental study of the anaerobic urban wastewater treatment in a submerged hollow-fibre membrane bioreactor at pilot scale. *Bioresource Technology*, **102**, 8799–8806.

Giménez J. B., Martí N., Ferrer J. and Seco A. (2012). Methane recovery efficiency in a submerged anaerobic membrane bioreactor (SAnMBR) treating sulphate-rich urban wastewater: evaluation of methane losses with the effluent. *Bioresource Technology*, **118**, 67–72.

Gouveia J., Plaza F., Garralon G., Fdz-Polanco F. and Peña M. (2015a). Long-term operation of a pilot scale anaerobic membrane bioreactor (AnMBR) for the treatment of municipal wastewater under psychrophilic conditions. *Bioresource Technology*, **185**, 225–233.

Gouveia J., Plaza F., Garralon G., Fdz-Polanco F. and Peña M. (2015b). A novel configuration for an anaerobic submerged membrane bioreactor (AnSMBR). Long-term treatment of municipal wastewater under psychrophilic conditions. *Bioresource Technology*, **198**, 510–519.

Grethlein H. E. (1978). Anaerobic digestion and membrane separation of domestic wastewater. *Journal Water Pollution Control Federation*, **50**, 754–763.

Hahn M. J. and Figueroa L. A. (2015). Pilot scale application of anaerobic baffled reactor for biologically enhanced primary treatment of raw municipal wastewater. *Water Research*. http://dx.doi.org/10.1016/j.watres.2015.09.027

Heffernan B., van Lier J. B. and van der Lubbe J. (2011). Performance review of large scale up-flow anaerobic sludge blanket sewage treatment plants. *Water Science and Technology*, **63**(1), 100–107.

Herrera-Robledo M., Cid-León D. M., Morgan-Sagastume J. M. and Noyola A. (2011). Biofouling in an anaerobic membrane bioreactor treating municipal sewage. *Separation and Purification Technology*, **81**, 49–55.

Holubar P., Zani L., Hager M., Fröschl W., Radak Z. and Braun R. (2003). Start-up and recovery of a biogas-reactor using a hierarchical neural network-based control tool. *Journal of Chemical Technology and Biotechnology*, **78**(8), 847–854.

Ince O., Anderson G. K. and Ince B. (1992). Improvement in performance of an anerobic contact digester using a crossflow ultrafiltration membrane technique for brewery wastewater treatment. In: 28th Purdue Industrial Waste Conference Proceedings, Lews Publishers, Chelsea, Michigan, printed in USA.

Ince O., Anderson G. K. and Ince B. (1994). Use of specific methanogenic activity test for controlling stability and performance in anaerobic digestion of brewery wastewater. In: 29th Purdue Industrial Waste Conference Proceedings, Lews Publishers, Chelsea, Michigan, printed in USA.

Ince O., Anderson G. K. and Ince B. (1997). Composition of the microbial population in a membrane anaerobic reactor system during start-up. *Water Research*, **31**(1), 1–10.

Ince O., Ince B. and Yenigün O. (2001). Determination of potential loading capacity of an upflow anaerobic sludge blanket reactor using specific methanogenic activity test. *Journal of Chemical Technology and Biotechnology*, **6**(76), 573–578.

Ince O., Kolukirik M., Ayman N. and Ince B. (2005). Comparative evaluation of full-scale UASB reactors treating alcohol distillery wastewaters in terms of performance and methanogenic activity. *Journal of Chemical Technology and Biotechnology*, **80**, 138–144.

Judd S. and Judd C. (2011). The MBR Book: Principles and Applications of Membrane Bioreactors for Water and Wastewater Treatment, 2nd edn. Elsevier, ISBN: 978-0-08-096682-3.

Khan A. A., Gaur R. Z., Tyagi V. K., Khursheed A., Lew B., Mehrotra I. and Kazmi A. A. (2011). Sustainable options of post treatment of UASB effluent treating sewage: a review. *Resources Conservation and Recycling*, **55**, 1232–1251.

Kim H., Shin J., Won S., Lee J., Maeng S. K. and Song K. G. (2015). Membrane distillation combined with an anaerobic moving bed biofilm reactor for treating municipal wastewater. *Water Research*, **71**, 97–106.

Kolukirik M., Ince O. and Ince B. (2006). Methanogenic Community Change in a Full-Scale UASB Reactor Operated at a Low F/M Ratio. *Journal of Environmental Science and Health, Part A*, **42**, 903–910.

Kujawa-Roeleveld K., Elmitwalli T. and Zeeman G. (2006). Enhanced primary treatment of concentrated black water and kitchen residues within DESAR concept using two types of anaerobic digesters. *Water Science and Technology*, **53**, 159–168.

Kythreotou N., Florides G. and Tassou S. A. (2014). A review of simple to scientific models for anaerobic digestion. *Renewable Energy*, **71**, 701–714.

Lauwers J., Appels L., Thompson I. P., Degrève J., Van Impe J. F., Dewil R. (2013). Mathematical modelling of anaerobic digestion of biomass and waste: power and limitations. *Progress in Energy and Combustion Science*, **39**, 383–402.

Lens P., Zeeman G. and Lettinga G. (2001). *Decentralised Sanitation and Reuse – Concepts, Systems and Implementation.* IWA Publishing, London, UK.

Lettinga G., Rebac S. and Zeeman G. (2001). Challenge of psychrophilic anaerobic wastewater treatment. *Trends in Biotechnology.*, **19**, 363–370.

Lin H., Peng W., Zhang M., Chen J., Huachang H. and Zhang Y. (2013). A review on anaerobic membrane bioreactors: applications, membrane fouling and future perspectives. *Desalination*, **314**, 169–188.

Luostarinen S., Sanders W., Kujawa-Roeleveld K. and Zeeman G. (2007). Effect of temperature on anaerobic treatment of black water in UASB-septic tank systems. *Bioresource Technology*, **98**, 980–986.

Mahdy A., Mendez L., Ballesteros M. and González-Fernández C. (2014). Algaculture integration in conventional wastewater treatment plants: anaerobic digestion comparison of primary and secondary sludge with microalgae biomass. *Bioresource Technology*, **184**, 236–244.

Mao C., Feng Y., Wang X. and Ren G. (2015). Review on research achievements of biogas from anaerobic digestion. *Renewable and Sustainable Energy Reviews*, **45**, 540–555.

Martinez-Sosa D., Helmreich B., Netter T., Paris S., Bischof F. and Horn H. (2011). Anaerobic submerged membrane bioreactor (AnSMBR) for municipal wastewater treatment under mesophilic and psychrophilic temperature conditions. *Bioresource Technology*, **102**, 10377–10385.

Martinez-Sosa D., Helmreich B. and Horn H. (2012). Anaerobic submerged membrane bioreactor (AnSMBR) treating low-strength wastewater under psychrophilic temperature conditions. *Process Biochemistry*, **47**, 792–798.

Massalha N., Basheer S. and Sabbah I. (2007). Effect of adsorption and bead size of immobilized biomass on the rate of biodegradation of phenol at high concentration levels. *Industrial & Engineering Chemistry Research*, **46**, 6820–6824.

Massalha N., Brenner A., Sheindorf C., Haimov Y. and Sabbah I. (2015). Enriching composite hydrophilic polyurethane foams with PAC to enhance adsorption of phenol from aqueous solutions. *Chemical Engineering Journal*, **280**, 283–292.

McCarty P. L., Bae J. and Kim J. (2011). Domestic wastewater treatment as a net energy producer-can this be achieved? Environ*mental Science and Technology*, **45**, 7100–7106.

Mooij P. R., Stouten G. R., Tamis J., van Loosdrecht M. C. M. and Kleerebezem R. (2013). Survival of the fattest. *Energy & Environmental Science*, **6**, 3404.

Nnaji C. C. (2014). A review of the upflow anaerobic sludge blanket reactor. Desalination and Water Treatment, **52**(22–24), 4122–4143.

Otterpohl R. and Oldenburg M. (2007). Ecological sanitation: innovative sanitation systems for urban and peri-urban areas; high- and low-tech solutions with resources recovery. In: Global Change: Enough Water for All? J. L. Lozán, H. Graßl, P. Hupfer, L. Menzel and C. -D. Schönwiese (eds), Wissenschaftliche Auswertungen und GEO-Verlag, Hamburg, Germany.

Ozgun H., Dereli R. K., Ersahin M. E., Kinaci C., Spanjers H. and Van Lier J. B. (2013). A review of anaerobic membrane bioreactors for municipal wastewater treatment: integration options, limitations and expectations. *Separation and Purification Technology*, **118**, 89–104.

Posadas E., García-Encina P. -A., Soltau A., Domínguez A., Díaz I. and Muñoz R. (2013). Carbon and nutrient removal from centrates and domestic wastewater using algal-bacterial biofilm bioreactors. *Bioresource Technology*, **139**, 50–58.

Pretel R., Robles A., Ruano M. V., Seco A. and Ferrer J. (2014). The operating cost of an anaerobic membrane bioreactor (AnMBR) treating sulphate-rich urban wastewater. *Separation and Purification Technology*, **126**, 30–38.

Pretel R., Robles A., Ruano M. V., Seco A. and Ferrer J. (2016). Economic and environmental sustainability of submerged anaerobic MBR-based (AnMBR-based) technology as compared to aerobic-based technologies for moderate-/high-loaded urban wastewater treatment. *Journal of Environmental Management*, **166**, 45–54.

Ras M., Lardon L., Bruno S., Bernet N. and Steyer J. P. (2011). Experimental study on a coupled process of production and anaerobic digestion of Chlorella vulgaris. *Bioresource Technology*, **102**, 200–206.

Robles A. (2013). Modelling, simulation and control of the filtration process in a submerged anaerobic membrane bioreactor treating urban wastewater. PhD Thesis, Dept. of Hydraulic Engineering and Environment, Universitat Politècnica de València.

Robles A., Ruano M. V., Ribes J. and Ferrer J. (2014). Performance of industrial scale hollow-fibre membranes in a submerged anaerobic MBR (HF-SAnMBR) system at mesophilic and psychrophilic conditions. *Separation and Purification Technology*, **104**, 290–296.

Rosen C., Vrecko D., Gernaey K. V., Pons M. N. and Jeppsson U. (2006). Implementing ADM1 for plantwide benchmark simulations in Matlab/Simulink. *Water Science and Technology*, **54**(4), 11–19.

Sato N., Okubo T., Onodera T., Ohashi A. and Harada H. (2006). Prospects for a self-sustainable sewage treatment system: a case study on full-scale UASB system in India's Yamuna River Basin. *Journal of Environmental Management*, **80**, 198–207.

Seghezzo L., Zeeman G., van Lier J. B., Hamerles H. V. M. and Lettinga G. (1998). A review: the anaerobic treatment of sewage in UASB and EGSB reactors. *Bioresource Technology*, **65**, 175–190.

Shin C., McCarty P. L., Kim J. and Bae J. (2014). Pilot-scale temperate-climate treatment of domestic wastewater with a staged anaerobic fluidized membrane bioreactor (SAF-MBR). *Bioresource Technology*, **159**, 95–103.

Simeonov I., Noykova N. and Gyllenberg M. (2007). Identification and extremum seeking control of the anaerobic digestion of organic wastes. *Cybernet Information Technology*, **7**(2), 73–84.

Skouteris G., Hermosilla D., López P., Negro C. and Blanco Á. (2012). Anaerobic membrane bioreactors for wastewater treatment: a review. *Chemical Engineering Journal*, **198–199**, 138–148.

Smith A. L., Stadler L. B., Love N. G., Skerlos S. J. and Raskin L. (2012). Perspectives on anaerobic membrane bioreactor treatment of domestic wastewater: a critical review. *Bioresource Technology*, **122**, 149–159.

Tandukar M., Ohashi A. and Harada H. (2007). Performance comparison of a pilot-scale UASB and DHS system and activated sludge process for the treatment of municipal wastewater. *Water Research*, **41**, 2697–2705.

U.S. Environmental Protection Protection Agency (2007). Opportunites for and Benefits of Combined Heat and Power at Wastewater Treatment Facilities. EPA-430-R-07-003. Washington, DC.

van Lier J. B., van der Zee F. P., Frijters C. T. M. J. and Ersahin M. E. (2015). Celebrating 40 years anaerobic sludge bed reactors for industrial wastewater treatment. *Reviews in Environmental Science and Biotechnology*, **14**(4), 681–702, DOI: 10.1007/s11157-015-9375-5.

van Voorthuizen E., Zwijnenburg A., van der Meer W. and Temmink H. (2008). Biological black water treatment combined with membrane separation. *Water Research*, **42**, 4334–4340.

Vinnerås B., Palmquist H., Balmer P. and Jönsson H. (2006). The characteristics of household wastewater and biodegradable solid waste – A proposal for new Swedish design values. *Urban Water Journal*, **3**, 3–11.

Wu S., Hung C., Lin C., Lin P., Lee K., Chang F. and Chang J. (2008). HRT-dependent hydrogen production and bacterial community structure of mixed anaerobic microflora in suspended, granular and immobilized sludge systems using glucose as the carbon substrate. *International Journal of Hydrogen Energy*, **33**, 1542–1549.

Zeeman G., Kujawa K., de Mes T., Hernandez L., de Graaff M., Abu-Ghunmi L., Mels A., Meulman B., Temmink H., Buisman C., van Lier J. and Lettinga G. (2008). Anaerobic treatment as a core technology for energy, nutrients and water recovery from source-separated domestic waste(water). *Water Science and Technology*, **57**, 1207–1212.

Chapter 3

Resource recovery from source separated domestic wastewater; energy, water, nutrients and organics

Lucía Hernández Leal, Taina Tervahauta and Grietje Zeeman

3.1 INTRODUCTION

The world is increasingly facing resource shortages, including water, energy and nutrients, especially phosphorus (Cordell *et al.* 2009). Domestic waste and wastewater contains each of these resources. Most nutrients and organics, a potential energy source, in domestic waste (water) originate from faeces, urine (so called black water, BW) and kitchen waste (KW). Their separated collection with a minimum amount of water allows for a more effective and efficient recovery. This can be attained by applying low flush toilets like vacuum toilets. Alternatively also urine can be separated from the remaining domestic wastewater streams (Jonsson *et al.* 1997; Lienert *et al.* 2006), containing major fraction of the nutrients (Zeeman *et al.* 2011). In parallel, the separated discharges from washing machines, bath, shower, hand basins and kitchen sink (also known as grey water, GW) represent a less polluted potential source for water reuse. Human pharmaceuticals, hormones and personal care products reach the environment primarily via household wastewater. Separated collection of BW, GW and urine has shown options for the more effective removal of these compounds. Several new technologies for resource recovery and MP removal from source separated domestic streams have been developed in The Netherlands over the last 10 years. The present chapter gives an overview of the achievements on 'New Sanitation', on laboratory, pilot and demonstration scale.

3.2 RESOURCES AND POLLUTANTS IN DOMESTIC WASTEWATER

The volumes of waste(water) from households in the Netherlands are 0.1, 1.4, 0.2 and 79 L/cap/d of faeces, urine, kitchen refuse and grey water, respectively (Table 3.1). By diverting black water (BW) from grey water (GW) and adding kitchen waste (KW) to BW, 80–92% of the nutrients N, P and K are separated in a concentrated stream with a small volume (1.5–2 litres $p^{-1}d^{-1}$) (Zeeman, 2012). The BW plus KW mixture also contains ca. 70% of organics, contained in domestic wastewater plus kitchen waste (Zeeman *et al.* 2011). The remaining 30% is in the GW.

Table 3.1 Mean resource load in faeces, urine, kitchen refuse and grey water.

Parameter	Unit	Faeces	Urine	Kitchen Refuse	Grey Water
Volume	L/cap/d	0.1	1.4	0.2	79
COD	g/cap/d	50	11	59	52
TN	g/cap/d	1.8	9	1.7	1.2
TP	g/cap/d	0.5	0.8	0.2	0.4
PO_4^{3}–P	g/cap/d	0.2	0.3	–	0.1
K	g/cap/d	0.9	2.8	0.2	0.8

Source: Adapted from Tervahauta *et al.* (2013).

Aside from the resources present in domestic wastewater, pollutants such as organic micropollutants and heavy metals must be removed prior reuse. Pharmaceuticals in black water and personal care products in grey water were present in higher concentrations than in conventional sewage. For example diclofenac (1.9–66 μg/L) and ibuprofen (55–258 μg/L) were measured in concentrations 10 times higher than in sewage (de Graaff *et al.* 2011). Concentrations of triclosan in grey water were 6.3–35.7 μg/L, also higher than reported concentrations in sewage (0.05–5.2 μg/L). Furthermore, discharges of pharmaceuticals and hormones via GW are not irrelevant. About 70% of diclofenac and 20% of ibuprofen load into domestic wastewater has been measured in GW from a new housing estate Waterschoon, in the Netherlands (Butkovskyi *et al.* 2015). A nursery home is part of the housing estate. The concept of source separation of waste streams results in higher concentrations of micropollutants due to the lower water use, which may increase treatment efficiencies. However, collection of micropollutants in only one stream cannot be achieved by source separation. For practical reasons it is recommended to combine GW and BW after nutrients and COD have been recovered for removal of micropollutants in one treatment unit. The type of water use will influence latter decision as combining the BW and GW effluent will increase the pathogen content of the water, to be reused. Selection of technologies for micropollutant removal should also be based on pathogen removal efficiency.

3.3 ANAEROBIC TREATMENT CORE TECHNOLOGY IN 'NEW SANITATION'

Anaerobic treatment of vacuum collected BW in respectively a UASB-septic-tank and a UASB at 25°C, was studied by Kujawa-Roeleveld *et al.* (2005) and de Graaff (2011). High removal efficiencies of 78% are reported in both studies. The UASB can be operated at a relatively low HRT of 8.7 days, in comparison to 30 days in the UASB septic tank. A UASB septic tank is, unlike a UASB, also designed for the storage of sludge. This system can therefore be attractive for use in small communities or house-on-site, where minimal maintenance is required (Zeeman, 2012). The UASB reactor for BW treatment generates 10 L CH_4 $p^{-1}d^{-1}$, a liquid comprising 91% of the nitrogen and 61% of the phosphorus and an organic sludge.

3.3.1 Organic sludge and heavy metals

Tervahauta *et al.* (2014) investigated the quality of the organic sludge with respect to heavy metal content for use in agriculture and concluded that BW sludge has superior quality in comparison to conventional sewage sludge and moreover promotion of soil application of BW sludge over livestock manure and artificial fertilizers could further reduce the heavy metal content in the soil/food cycle.

Co-treatment of GW sludge, produced in a bio-flocculation unit, and BW in a UASB can increase the CH_4 production with 23–73% (Hernandez Leal *et al.* 2010; Tervahauta *et al.* 2014). However GW sludge addition deteriorates the organic sludge quality (Tervahauta *et al.* 2014). Tervahauta *et al.* (2014) determined an extra 36% (Cu), 32% (Zn) and 19% (Ni) contribution to the heavy metal input in the UASB reactor on a solid matter basis. When agricultural use of organic sludge prevails over increased energy production, a bio-flocculation system for grey water, characterised by the production of a non-stabilised sludge is not feasible within a new sanitation concept. Alternatively an aerobic sequencing batch reactor, with or without anaerobic pre-treatment, could be applied for GW treatment (Hernandez Leal *et al.* 2010).

3.3.2 Recovery of phosphorus during or after UASB treatment?

Recovery and reuse of phosphorus from waste and wastewater is crucial as phosphorus is a finite resource and the phosphate stock is predicted to be exhausted at the end of the 21st century (i.e. Driver *et al.* 1999; Cordell *et al.* 2009). De Graaff (2011) report a recovery of 0.22 kg P/p/y, representing 10% of the artificial phosphorus fertiliser production in the world, via struvite precipitation from the liquid produced in a UASB treating BW. The difference with the recovery potential of 27% of the global phosphorus consumption, as calculated by Zeeman (2012), can be attributed to the 40% of the P accumulated in the organic sludge of the UASB, the absence of KW in the experiments of de Graaff *et al.* (2011) and the P produced via faeces and urine outside houses (at work etc.). Recent developments show the combined anaerobic treatment of BW and production of 1–2 mm calcium phosphate granules within a UASB reactor (Tervahauta *et al.* 2014). At normal operational conditions of the UASB, 7 g P/person/year can be recovered as calcium phosphate granules, representing 2% of the incoming phosphorus in the UASB reactor. Local pH gradient formed due to a biofilm around the granules is hypothesized to play an important role in the precipitation of calcium phosphate. Follow-up research focuses on the mechanisms behind the calcium phosphate precipitation and development of a methodology to enhance the recovery.

3.3.3 Removal or recovery of nitrogen?

Unlike phosphorus, nitrogen is not a finite resource and is abundantly available in the atmosphere. However, the large scale production of nitrogen fertilizers from the air via the Haber-Bosch process, consumes a considerable amount of energy (37–45 kJ/g N; (Maurer *et al.* 2003)). Presently available technology enables the application of energy efficient removal of nitrogen from the liquid produced in a UASB treating BW, via nitritation and anammox processes (Vlaeminck *et al.* 2009; de Graaff *et al.* 2010, 2011). A breakthrough nitrogen recovery technology is reported by (Kuntke *et al.* 2011, 2012) for urine. A MFC (Microbial Fuel Cell) is applied for a migrational ion flux (i.e. NH_4^+-N) to the cathode chamber, driven by the electron production from anaerobic degradation of organic matter in urine. The prevailing high pH at the cathode enables efficient recovery of the ammonia. A second very promising technique is the use of algae growth for combined recovery of N and P. Tuantet *et al.* (2014) reported for the first time the growth of *Chlorella sorokiniana* in a continuously operated (>8 months) short light-path (5 mm) photo-bioreactor (HRT = 1 day) with concentrated human urine as a substrate. At optimised conditions with respect to light path (5 mm), N:P molar ratio and magnesium content, productivity increased to 14.8 g biomass L^{-1} d^{-1}, matching a removal of 1.3 g N L^{-1} d^{-1} and 0.15 g P L^{-1} d^{-1}. Fernandes *et al.* (2015) disclosed the possibility of applying a photo-bioreactor for recovering N and P not only from urine but also from the liquid produced in a UASB treating BW.

3.4 REMOVAL OF MICROPOLLUTANTS FROM BLACK AND GREY WATER

Only a few micropollutants are biodegraded during biological treatment of both grey and black water (Hernandez Leal *et al.* 2010; de Graaff *et al.* 2011). Concentrations of carbamazepine and diclofenac remain the same after a lab-scale treatment train of UASB reactor, partial nitritation and anammox (see Table 3.2) (de Graaff *et al.* 2010). Physical-chemical post treatment is required to remove these compounds and protect the aquatic environment (Eggen *et al.* 2014).

Table 3.2 Concentration of selected organic micropollutants in black and grey water (in μg/L).

Compound	Blackwater[a]	Grey water[b]
Carbamazepine	0.1–2.5	
Cetirizine	0.2–3	
Ibuprofen	55–258	
Diclofenac	1.9–66	
Doxycycline	1.8–9.7	
Galaxolide		5.7–19.1
4-Methylbenzylidene camphor		8.9–20
Triclosan		6.3–35.7

[a]de Graaff *et al.* (2011).
[b]Hernandez Leal *et al.* (2010).

Addition of granular activated carbon (GAC) in a lab-scale UASB reactor (5.7 g GAC/L reactor) resulted in lower concentrations of diclofenac, ibuprofen, metoprolol, galaxolide and triclosan in the effluent and UASB sludge (Butkovskyi, 2015). Other performance parameters of the UASB reactors such as biogas production and COD removal were unaffected by the GAC addition. A disadvantage of this process is, however, the separation of the GAC from waste sludge prior to its application on land. Furthermore, the required amount of GAC can be about 3 times higher within the UASB reactor than as an UASB effluent treatment unit (Butkovskyi, 2015).

Algae have been successfully applied for the removal and recovery of nutrients from UASB effluent (Fernandes *et al.* 2015). The fate of micropollutants during this process was studied by de Wilt *et al.* (2016). Degradation of diclofenac, ibuprofen, metoprolol, trimethoprim, carbamazepine, paracetamol, estrone and 17β-estradiol ranged from 20–100%. Except for metoprolol, removal was mostly due to photolysis or biodegradation by organisms other than the algae (de Wilt *et al.* 2016). No sorption to algal biomass was observed for ibuprofen, paracetamol, metoprolol, estrone and β-estradiol. For diclofenac, trimethoprim, carbamazepine, and ethylinestradiol adsorption ranged from 0.2 μg/g (ethylinestradiol) to 1.4 μg/g (diclofenac). De Wilt *et al.* (2016) compared direct urine applied on land with algae biomass. Recovery of nutrients via algae treatment can result in a strongly decreased load of micropollutants on land, when algae are used as an organic fertiliser; for example, diclofenac loads are decreased from 2.9 kg/ha to 0.0055 kg/ha.

Co-composting of UASB excess sludge and wood waste indicated high conversion of spiked micropollutants (800–1100 ng/g), ranging from 88% for carbamazepine, 93% for triclosan, to 99.9% for estrone and diclofenac (Butkovskyi *et al.* 2016). Removals were similar both at 35°C and 50°C. A fraction of triclosan was transformed into methyltriclosan (12–24%), which accumulated in the compost. The other compounds were not monitored for transformation products.

Three different configurations of biological reactors (anaerobic, anaerobic + aerobic and aerobic alone) were tested for grey water treatment. The aerobic reactor alone lead to the highest removal of micropollutants, varying from >90% for triclosan, around 80% for galaxolide and 0 for UV filter PBSA (Hernandez Leal *et al.* 2010). Yet, most compounds were still measured in the effluent, such as galaxolide (2 μg/L). For that reason different technologies were tested to remove the remaining micropollutants. Ozonation and activated carbon were equally effective in removing micropollutants from aerobically treated grey water and eliminate toxicity to water fleas (Hernandez Leal *et al.* 2011, 2012). Additionally, electrochemical treatment was also tested but resulted in the formation of chlorinated compounds which increased the toxicity of the effluent (Butkovskyi *et al.* 2014).

The fate of 14 pharmaceuticals (paracetamol, ibuprofen, diclofenac, naproxen, trimethoprim, ciprofloxacin, metformin, hydrochlorothiazide, gemfibrozil, atenolol, propranolol, metoprolol, carbamazepine and oxazepam) and 3 of their transformation products (desmethylnaproxen, guanyl urea and carbamazepine-diol) was monitored at a full scale sanitation system in which BW and GW are treated separately for organics and nutrients removal/recovery (see Section 2.8, Figure 3.4) (Butkovskyi *et al.* 2015). Two configurations of treatment were tested, were effluents of BW and GW are separated or combined; the concentration of pharmaceuticals in the effluents is shown in Table 3.3. The reduced concentrations on the combined effluent were only due to the dilution factor. Concentrations in the effluents are above calculated no effect concentrations and therefore post treatment is necessary to decrease the concentration of pharmaceuticals (Butkovskyi *et al.* 2015). Sludge from the UASB reactor contained ciprofloxacin, metoprolol and propranolol in the range of μg/g. Concentration of all studied micropollutants in struvite was below quantification limits (Butkovskyi *et al.* 2015).

Table 3.3 Concentration of pharmaceuticals in the full scale new sanitation system in Sneek, The Netherlands. n is the number of samples out of the 15 sampling events, with the concentration above the quantification limit.

Compound	Concentration in the Effluent of Black Water Treatment System, μg/L		Concentration in the Effluent of Grey Water Treatment System, μg/L
	1st Sampling Period (Separated Treatment System)	2nd Sampling Period (Combined Treatment System)	2nd Sampling Period (Combined Treatment System)
Paracetamol	<LOD	<LOD	<LOD
Ibuprofen	2.7 ± 2.4 (n = 7)	<LOD	<LOD
Diclofenac	3.9 ± 2.0 (n = 14)	1.7 ± 1.0 (n = 10)	1.5 ± 0.9 (n = 9)
Naproxen	0.4 ± 0.2 (n = 8)	0.8 ± 0.6 (n = 10)	<LOD
Desmethylnaproxen	N.D.	<LOQ	<LOQ
Trimethoprim	<LOD	<LOD	<LOD
Ciprofloxacin	3.6 ± 0.8 (n = 15)	1.8 ± 0.3 (n = 15)	0.4 ± 0.3 (n = 6)
Metformine	7.8 ± 0.9 (n = 15)	22.0 ± 10.0 (n = 15)	8.7 ± 8.6 (n = 15)
Guanyl urea	N.D.	6.4 ± 5.2 (n = 10)	7.4 ± 8.5 (n = 15)
Hydrochlorothiazide	23.6 ± 1.5 (n = 15)	22.4 ± 2.5 (n = 15)	3.8 ± 0.5 (n = 15)
Gemfibrozil	<LOD	<LOD	<LOD
Atenolol	<LOD	<LOD	<LOD
Propranolol	<LOD	<LOD	<LOD
Metoprolol	116 ± 15.7 (n = 15)	62.6 ± 4.8 (n = 15)	7.4 ± 1.2 (n = 15)
Carbamazepine	<LOD	<LOD	<LOD
Carbamazepine-diol	<LOD	<LOD	<LOD
Oxazepam	3.2 ± 0.7 (n = 15)	3.0 ± 0.5 (n = 14)	<LOQ

Source: Butkovskyi *et al.* 2015.

N.D. = Not determined.

3.5 MULTI-CRITERIA ASSESSMENT ON ENVIRONMENTAL AND SOCIAL ASPECTS IN NEW SANITATION

Recovery of energy, water, nutrients and organics in new sanitation is influenced by the degree of wastewater separation and by the chosen treatment scheme. Tervahauta *et al.* (2013) determined the relationship between recovery of different resources and different treatment schemes in three source separated sanitation concepts: (1) urine separation from mixed waste stream (2) separation of BW, GW and KW (3) separation of urine, faeces, GW and KW (Figure 3.1). The main focus was on the comparison of urine and BW separation using toilets with different flushing volumes, and the different GW treatment systems. The comparison was done in terms of energy balance, including the indirect energy gain from water reuse and nutrient recovery, and recovery of water, nutrients and organics. The energy balance was determined by the primary energy consumption from the collection, transport and treatment of wastewater, and by the energy production from methane recovery in the UASB reactor. Water recovery was defined as the reusable GW effluent according to suggested water reuse standards. Nutrients (N, P and K) from urine were considered to be reused via spreading on agricultural land, from BW effluent via struvite precipitation, and from BW and GW sludge via land application. The organics were considered to be recovered from BW and GW sludge.

Separation of urine, faeces, GW and KW (concept 3) had the lowest net primary energy consumption due to the lowest flushing volume of the urine diverting toilet (lower influent volume requires less heating energy in the UASB reactor) and the reduced energy input for nitrogen removal (Figure 3.2). Urine separation from the mixed stream (concept 1) had higher net primary energy consumption due to the high energy input and lack of energy production in the aerobic treatment system. Applying a vacuum urine diverting toilet will require less energy than the gravity urine diverting toilet due to the lower flushing volume (2 L/cap/d compared to 5 L/cap/d). However, currently urine diverting toilets with such low flushing volume are not available in the market.

Direct nutrient recovery from urine via spreading on agricultural land (concepts 1 and 3) results in significant recovery of nitrogen (Figure 3.3), influencing the energy balance through indirect energy gain from nutrient recovery. An improved energy balance can be seen particularly with concept 3 due to the additional indirect energy gain from water reuse (GW effluent). The energy balance of the different GW treatment systems is also influenced by the indirect energy gain from water reuse. Initially, bio-flocculation of GW in the A-trap, followed by a Trickling filter (TF), and sub-sequent GW sludge co-digestion with BW improves energy balance. However, since the A-trap plus TF effluent does not meet the reuse standards, GW treatment in the sequencing batch reactor (SBR) plus TF, with higher effluent quality becomes energetically more favourable.

The nutrient recovery method and the form of the recovery product also influence the energy balance. In direct land application of urine and BW sludge, the transport distance need to be taken into account due to the high water content. Recovery of nutrients from urine and BW as concentrated mineral fertilizers would therefore be more favourable in terms of transportation. Nitrogen could be recovered from urine as NH_4^+-N in a Microbial Fuel Cell (Kuntke *et al.* 2011, 2012), and phosphorus from urine and BW as struvite or as calcium phosphate granules from BW, in a UASB reactor (Tervahauta *et al.* 2014). The choice of recovering phosphorus as struvite or as calcium phosphate also influences the reuse possibilities and its potential to replace artificial fertilizers.

Sensitivity analysis in the multi-criteria assessment provides the boundary conditions for optimal resource recovery. Operational conditions in the treatment scheme, such as retention times and operational temperature, influence the energy balance and recovery rates of water, nutrients and organics. For example, the geographical location of the treatment unit determines the atmospheric temperature, and therefore the heating energy required. The choice between different toilets and flushing volumes significantly influences the wastewater volume, and therefore the energy input required for the treatment units. According to

Thibodeau *et al.* (2011), the water consumption for the vacuum toilet is one of the most critical factors influencing the economic viability of BW and GW source-separation systems. The transport distance plays an important role in the energy balance of the land application of urine and BW sludge. With longer distances it becomes more feasible to recover resources in a form with lower water content.

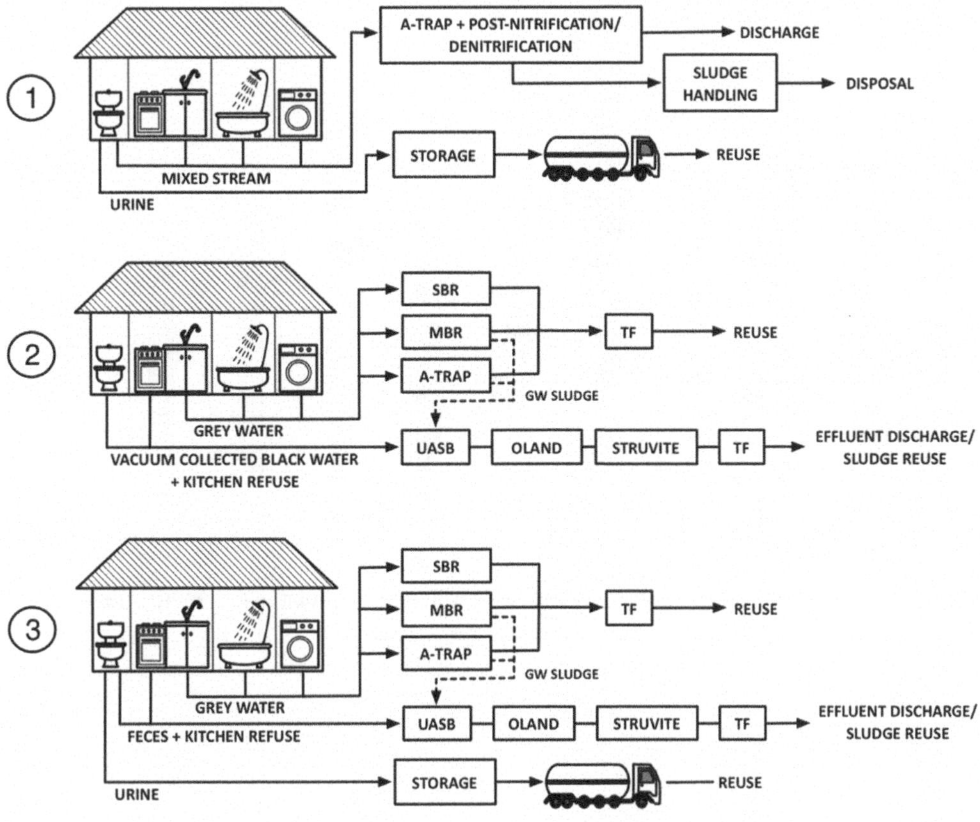

Figure 3.1 New sanitation concepts included in the comparison with wastewater streams and corresponding treatment systems (SBR = sequencing batch reactor, MBR = membrane bioreactor; A-trap = A-stage of AB-process; TF = trickling filter; UASB = up-flow anaerobic sludge blanket reactor; OLAND = oxygen limited anaerobic nitrification denitrification).

To be able to successfully implement new sanitation systems, the social aspect of the end-users need to be involved in the process. More so, the end-users with their practical aspect to sanitation play a key role in the transition from the old wastewater treatment systems to new sanitation systems (Hegger, 2007). According to the survey done in the *Waterschoon* project in Sneek collecting BW and GW, most of the inhabitants consider the new sanitation system practical and hygienic, and are proud of contributing to more environmentally friendly practices (de Graaf & van Hell, 2014). However, some practical aspects, such as the noise from the vacuum toilet, require further considerations to make it more user-friendly. In the study of Lienert *et al.* (2006), most users found the urine separating toilet equal to the conventional toilet in terms of design, hygiene and smell. However, problems with maintenance, such as blockages and toilet paper disposal, created challenges for accepting this early stage technology. Education and communication about the purpose and correct use of new sanitation technology along with improvement in toilet design are crucial for increased implementation of new sanitation systems.

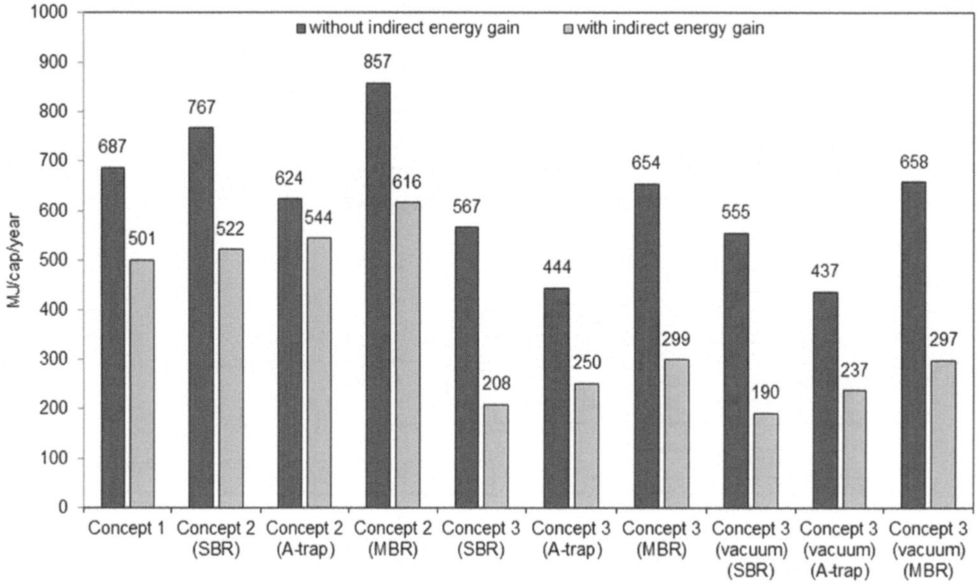

Figure 3.2 Energy balance of new sanitation concepts with and without indirect energy gain.

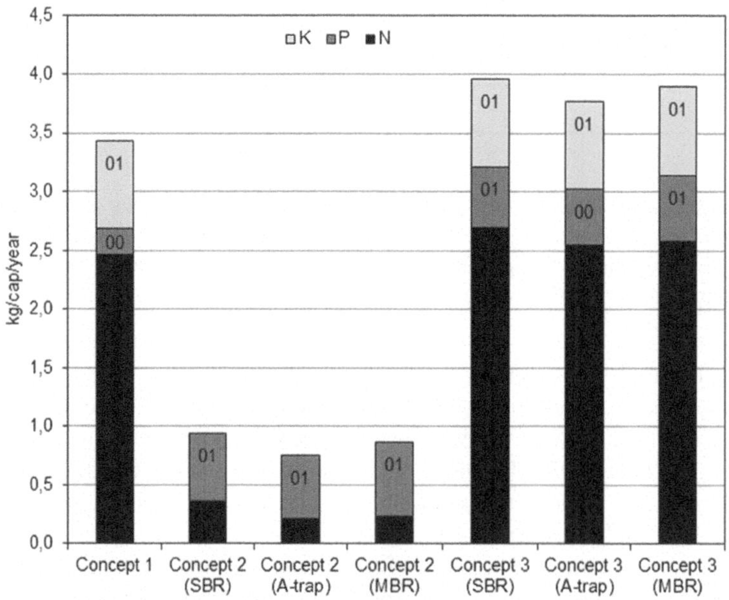

Figure 3.3 Nutrient recovery in new sanitation concepts.

3.6 NEW SANITATION IN PRACTICE IN THE NETHERLANDS

The concept of UASB treatment of BW collected with vacuum, followed by nitrogen removal and phosphorus recovery and GW treatment via bio-flocculation and aerobic post-treatment or a constructed wetland is now applied at full scale in Villa Flora, an office and exhibition building connected to the Floriade in Venlo and at a 250 houses estate (Waterschoon) in Sneek, The Netherlands (de Graaf & van Hell, 2014). In Zutphen, The Netherlands, the Saniphos concept is in operation where urine, mainly collected at festivals, is treated for recovery of phosphate via struvite precipitation and ammonium-sulphate via air stripping. In the sustainable building of the NIOO (Dutch Institute for Ecological Research) in Wageningen a full scale UASB reactor is connected to a pilot photobioreactor for the treatment of black water. After the pilot stage a full scale photobioreactor is foreseen. Vasconcelos Fernandes *et al.* (2015) show the high potential of the growth of algae for closing domestic nutrient cycles. When growing *Chlorella sorokiniana* in batch-wise operated flat-panel photobioreactors, 100% of the phosphorus from anaerobically treated black water was removed within 4 days, while nitrogen uptake continued after P depletion.

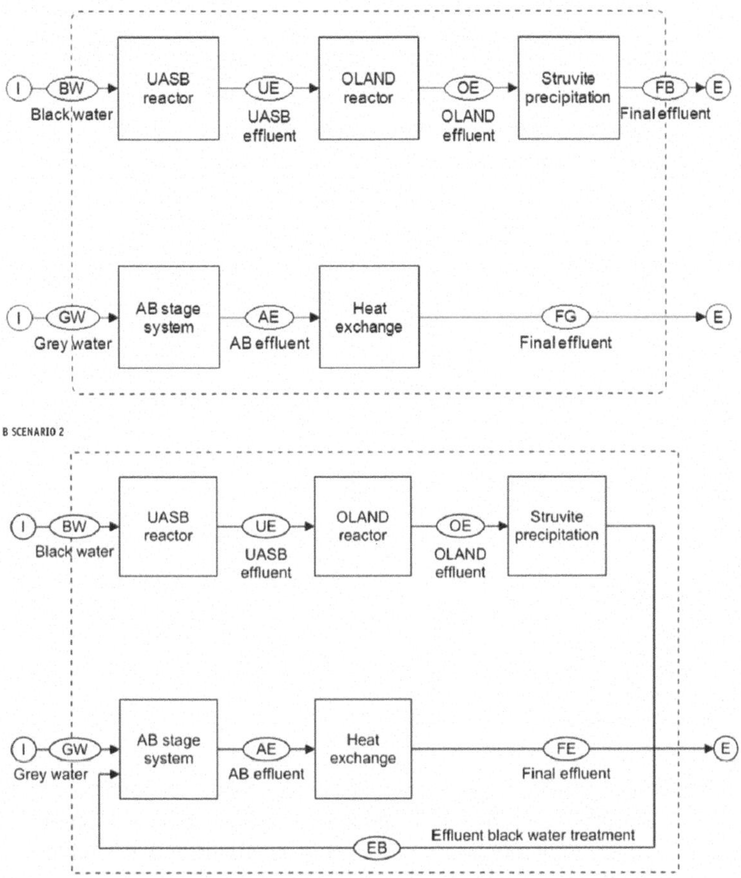

Figure 3.4 Flow scheme of the treatment/recovery system of Sneek, Waterschoon. During the first period BW and GW are separately treated (scenario 1), while during the 2nd period BW effluent is treated together with GW effluent (scenario 2) (de Graaf & van Hell, 2014).

The New Sanitation application in Waterschoon, Sneek, is as compared to the other mentioned applications, most monitored and researched. At the time of the monitoring, 32 apartments and one nursery home were connected to the treatment and recovery concept. All units were equipped with vacuum toilets and kitchen grinders, connected to the same vacuum system as the toilets. The treatment/recovery system is designed for 550 persons, while during the monitoring campaign the built houses were occupied by 79 persons. A flow scheme of the treatment /recovery is shown in Figure 3.4. Two scenarios were subsequently applied. During the first period BW and GW are separately treated (scenario 1), while during the 2nd period BW effluent is treated together with GW effluent (scenario 2) (de Graaf & van Hell, 2014). Results of the long term monitoring program are summarised in Table 3.4.

Table 3.4 Results of a long term monitoring program for the New Sanitation system in Sneek, designed for 550 persons and operated for 79 persons.

Parameter	Unit	BW + KW[a]	UASB effluent[a]	Effluent OLAND[a]	(Effluent) struvite reactor[a]	GW	Effluent GW system[d]	combined Influent (GW + effluent BW)[e]	Effluent combined system[e]
COD_t	mg/l	9500	720	991[b]	254	636	62	528	57 (19)[c]
COD_{ss}	mg/l	5500	120	756[b]	34	100	0	194	0
N_t	mg/l	1000	960	285[b]	276	16.2	4.2	19.7	6.7 (2.6)[c]
NH_4^+-N	mg/l	640	820	29[b]	22	6.5	1.0	2.3	0.8 (0.8)[c]
NO_3-N	mg/l			181[b]	206		0		2.3 (1.7)[c]
NO_2-N	mg/l			12[b]	8		0		0.1 (0.1)[c]
Pt	mg/l	110	86	83[b]	20	13.2	8.7	17.1	13.5 (2.5)[c]
PO_4-P	mg/l	80	76	65[b]	9	11.2	7.9	10.4	12.6 (2.1)[c]
K	mg/l			420[b]					
Mg	mg/l			59[b]	89				
Flow	l/d	1084	–	–				6779	
BW prod.	l/cap/d	13.7	–	–					
CH_4 prod.	l/kgCOD_{inf}		271						
Sludge prod.	kgCOD/d		0.717						
Struvite prod.	kgP/y				24				

[a]15 months monitoring period; weekly sampled.
[b]Sampled before the settler.
[c]Standard deviation in brackets.
[d]3 months monitoring period; weekly sampled.
[e]6.5 months monitoring period; weekly sampled.
Source: Based on de Graaf & van Hell (2014).

Due to the low loading rate applied as compared to the design load, some treatment units performed less than expected, i.e. the OLAND system, the bio-flocculation (A-stage) and the P recovery in the B-stage of the GW treatment. The relatively high total water flow during the monitoring period is due to the large water consumption of the kitchen grinders. Latter also resulted in increased energy consumption for heating the influent of the UASB reactor (Table 3.6). For the new to build houses technical measures are taken to reduce this water consumption. Extrapolation of achieved results shows that the designed system can be applied for 1200 persons (de Graaf & van Hell, 2014). Results of the extrapolation are presented in Table 3.5.

Table 3.5 Measured (78 persons) and extrapolated (1200 persons) influent and effluent load and removal efficiencies of the New Sanitation system Waterschoon, Sneek, The Netherlands, in comparison with a conventional municipal wastewater treatment system in Deventer, The Netherlands (measured).

Parameter	Unit	Waterschoon 79 persons	Waterschoon 1200 persons	Reference, Deventer (100,000 persons)
Flow (GW + BW)	L/cap/d	86	82	194[a]
Influent COD	g/cap/d	175	175	107
Influent TN	g/cap/d	15.3	15.3	9.5
Influent TP	g/cap/d	2.5	2.5	1.4
Effluent COD	g/cap/d	4.9	4.9	6.7
Effluent TN	g/cap/d	0.6	0.6	2.1
Effluent TP	g/cap/d	1.2	0.1	0.3
Removal COD	%	97	97	94
Removal TN	%	96	96	79
Removal TP	%	53	95	77
CH_4 production	m^3/cap/y	13.8	12.2	6.1
Sludge production	kg TS/cap/y	4.2[b]	9.2	16.7
Use Me salt[c]	mol/cap/y	4.0	18.0	5.6

[a]Including rain water.
[b]Due to the low applied loading, hardly any GW sludge is produced.
[c]In Waterschoon Mg is used for struvite recovery, in Deventer Fe is used for P removal.
Source: Based on de Graaf & van Hell, 2014.

Next to a monitoring program of 2.5 years for COD, N, P, energy production and consumption, also an energy balance and a financial evaluation are made. The energy consumption and production for Waterschoon and the reference situation are presented in Table 3.6.

An extensive economic evaluation was executed in which both the actual costs of the New Sanitation system, Waterschoon were calculated, and the extrapolated costs for a similar, optimised, new to build New Sanitation system, for 1200 persons. Table 3.7 presents the investments costs for a New Sanitation system for 1200 persons. The total investment of ca. 2.24 million euros is more or less equally distributed over collection and transport, surplus costs for in-house sewerage and treatment.

The related depreciation, maintenance and exploitation costs and savings are presented in Table 3.8. The results in Table 3.8 illustrate total costs of ca. 72 euros per person per year when a 1200 persons New Sanitation system is applied. In a comparison, the total costs for conventional sanitation (sewerage and treatment) is calculated to be 64.90 euros per person per year when a system for 100,000 persons is considered. Applying price volatility calculations and uncertainty factors in the calculations, a scale of application between 1000 and 1500 persons is estimated to result in equal costs of New Sanitation and conventional Sanitation (de Graaf en van Hell, 2014).

Several new projects are under development in the Netherlands. Within the office building of the Ministry of Infrastructure and Environment in The Hague water free urinals and vacuum toilets are installed. BW collected with the vacuum toilets are treated in a UASB reactor for energy recovery. The nitrogen and phosphorus in the urine and UASB effluent will be recovered applying struvite precipitation followed by a Microbial Fuel Cell for ammonium recovery (Wetsus, 2013).

Table 3.6 Energy production and consumption, measured (78 persons) and extrapolated (1200 persons) of the New Sanitation system Waterschoon, Sneek, The Netherlands, in comparison with a conventional municipal wastewater treatment system in Deventer, The Netherlands (measured).

Parameter	Unit	Waterschoon 79 persons[a]	Waterschoon 1200 persons[b]	Deventer (100,000 persons)[c]
Drinking water production & delivery	Kwh_p/cap/y	−35	−35	−58
Energy consumption treatment	Kwh_p/cap/y	−277	−50	−6
Diesel consumption WKK	Kwh_p/cap/y	0	0	−3
Heat production heat pump	Kwh_p/cap/y	477	477	0
Electricity consumption heat pump	Kwh_p/cap/y	−264	−264	0
Heat production biogas	Kwh_p/cap/y	133	148	6
Electricity consumption treatment	Kwh_p/cap/y	−781	−52	−75
Electricity production WKK	Kwh_p/cap/y	0	0	61
Electricity consumption transport ww	Kwh_p/cap/y	−92	−42	−13
Total energy consumption/ production	Kwh_p/cap/y	−838	184	−88

[a]Measured.
[b]Extrapolated based on measurements and expert judgement.
[c]Measured, conventional municipal wastewater treatment system in Deventer, The Netherlands (Based on de Graaf & van Hell, 2014).

Table 3.7 Investment costs New Sanitation concept, based on Waterschoon, optimised and extrapolated for 1200 persons.

Element	Investment costs (€)	Investment costs per person (€)	Share total investment costs (%)	Owner[a]
Collection/transport	737,000	682	33	Municipality
Surplus costs in-house sewerage	707,000	655	32	Housing cooperation
Treatment	800,000	741	36	Housing cooperation
Total investments	2,244,000	2078	100%	

Source: Based on de Graaf & van Hell (2014).

Table 3.8 Depreciation, maintenance and exploitation costs and savings, New Sanitation concept, based on Waterschoon, optimised and extrapolated for 1200 persons.

Element	Unit	Total (€)	Total per person	Share (%)
Depreciation				
Collection	€/year	16,193	14.99	23
Surplus costs in-house sewerage	€/year	23,578	21.83	33
Treatment	€/year	31,238	28.92	44
Total Depreciation	**€/year**	**71,010**	**65.75**	**100**
Maintenance/exploitation/savings				
Collection	€/year	3217	2.98	46
Surplus costs in-house sewerage	€/year	–	–	–
Treatment	€/year	73,499	68.05	1045
Savings[a]	€/year	−69,683	−64.52	-991
Total Maintenance/exploitation/ savings	**€/year**	**7033**	**6.51**	**100**
Total Depreciation & Maintenance/ exploitation/savings	**€/year**	**78,043**	**72.26**	**100**

[a]Municipality and inhabitants.
Source: Based on de Graaf & van Hell (2014).

3.7 CONCLUSIONS

A New Sanitation concept based on source separation of BW and GW is developed and implemented at several locations in The Netherlands;

Anaerobic treatment of BW and KW in a UASB reactor is the core technology of the developed New Sanitation concept;

Bio-flocculation of GW produces a sludge that can be co-treated in the UASB for increased energy production, but at the expense of the sludge quality with respect to heavy metals;

Co-composting of UASB BW excess sludge and wood waste show high conversion, 88–99.9%, of several MPs like carbamazepine, triclosan, estrone and diclofenac;

Post-treatment of GW & BW is recommended to reduce MPs below no effect concentrations;

New techniques for N and P recovery, viz. MFC, precipitation of calcium phosphate granules within a UASB reactor and growth of algae on BW effluents and urine, are under development for next generation New Sanitation concepts.

3.8 REFERENCES

Butkovskyi A. (2015). Removal of Micropollutants in Source Separated Sanitation. Sub-department Environmental Technology. PhD, Wageningen University, Wageningen, 192.

Butkovskyi A., Jeremiasse A. W., Hernandez Leal L., van der Zande T., Rijnaarts H. and Zeeman G. (2014). Electrochemical Conversion of Micropollutants in Gray Water. *Environmental Science & Technology*, **48**(3), 1893–1901.

Butkovskyi A., Leal L. H., Rijnaarts H. H. M. and Zeeman G. (2015). Fate of pharmaceuticals in full-scale source separated sanitation system. *Water Research*, **85**, 384–392.

Butkovskyi A., Ni G., Leal L. H., Rijnaarts H. H. M. and Zeeman G. (2016). Mitigation of micropollutants for black water application in agriculture via composting of anaerobic sludge. *Journal of Hazardous Materials*, **303**, 41–47.

Cordell D., Drangert J. O. and White S. (2009). The story of phosphorus: global food security and food for thought. *Global Environmental Change-Human and Policy Dimensions*, **19**(2), 292–305.

de Graaf R. and van Hell A. J. (2014). New Sanitation Noorderhoek, Sneek. P. Hermans. STOWA (Dutch Foundation for Applied Water Research), Amersfoort, p. 304.

de Graaff M. S., Zeeman G., Temmink H., van Loosdrecht M. C. M. and Buisman C. J. N. (2010). Long term partial nitritation of anaerobically treated black water and the emission of nitrous oxide. *Water Research*, **44**(7), 2171–2178.

de Graaff M. S., Temmink H., Zeeman G. and Buisman C. J. N. (2011a). Energy and phosphorus recovery from black water. *Water Science and Technology*, **63**(11), 2759–2765.

de Graaff M. S., Temmink H., Zeeman G., van Loosdrecht M. C. M. and Buisman C. J. N. (2011b). Autotrophic nitrogen removal from black water: calcium addition as a requirement for settleability. *Water Research*, **45**(1), 63–74.

de Graaff M. S., Vieno N. M., Kujawa-Roeleveld K., Zeeman G., Temmink H. and Buisman C. J. N. (2011c). Fate of hormones and pharmaceuticals during combined anaerobic treatment and nitrogen removal by partial nitritation-anammox in vacuum collected black water. *Water Research*, **45**(1), 375–383.

de Wilt A., Butkovskyi A., Tuantet K., Leal L. H., Fernandes T. V., Langenhoff A. and Zeeman G. (2016). Micropollutant removal in an algal treatment system fed with source separated wastewater streams. *Journal of Hazardous Materials*, **304**, 84–92.

Driver J., Lijmbach D. and Steen I. (1999). Why recover phosphorus for recycling, and how? *Environmental Technology*, **20**(7), 651–662.

Eggen R. I. L., Hollender J., Joss A., Scharer M. and Stamm C. (2014). Reducing the discharge of micropollutants in the aquatic environment: the benefits of upgrading wastewater treatment plants. *Environmental Science & Technology*, **48**(14), 7683–7689.

Fernandes T. V., Shrestha R., Sui Y. X., Papini G., Zeeman G., Vet L. E. M., Wijffels R. H. and Lamers P. (2015). Closing Domestic Nutrient Cycles Using Microalgae. *Environmental Science & Technology*, **49**(20), 12450–12456.

Hegger D. (2007). Greening Sanitary Systems: An End-user Perspective. PhD thesis, Wageningen University, Environmental Policy, Wageningen, 280.

Hernandez Leal L., Temmink H., Zeeman G. and Buisman C. J. N. (2010a). Bioflocculation of grey water for improved energy recovery within decentralized sanitation concepts. *Bioresource Technology*, **101**(23), 9065–9070.

Hernandez Leal L., Temmink H., Zeeman G. and Buisman C. J. N. (2010b). Comparison of Three Systems for Biological Greywater Treatment. *Water*, **2**(2), 155–169.

Hernandez Leal L., Vieno N., Temmink H., Zeeman G. and Buisman C. J. N. (2010c). Occurrence of Xenobiotics in Gray Water and Removal in Three Biological Treatment Systems. *Environmental Science & Technology*, **44**(17), 6835–6842.

Hernandez Leal L., Temmink H., Zeeman G. and Buisman C. J. N. (2011). Removal of micropollutants from aerobically treated grey water via ozone and activated carbon. *Water Research*, **45**(9), 2887–2896.

Hernandez Leal L., Soeter A. M., Kools S. A. E., Kraak M. H. S., Parsons J. R., Temmink H., Zeeman G. and Buisman C. J. N. (2012). Ecotoxicological assessment of grey water treatment systems with Daphnia magna and Chironomus riparius. *Water Research*, **46**(4), 1038–1044.

Jonsson H., Stenstrom T. A., Svensson J. and Sundin A. (1997). Source separated urine-nutrient and heavy metal content, water saving and faecal contamination. *Water Science and Technology*, **35**(9), 145–152.

Kujawa-Roeleveld K., Fernandes T., Wiryawan Y., Tawfik A., Visser W. and Zeeman G. (2005). Performance of UASB septic tank for treatment of concentrated black water within DESAR concept. *Water Science and Technology*, **52**(1–2), 307–313.

Kuntke P., Geleji M., Bruning H., Zeeman G., Hamelers H. V. M. and Buisman C. J. N. (2011). Effects of ammonium concentration and charge exchange on ammonium recovery from high strength wastewater using a microbial fuel cell. *Bioresource Technology*, **102**(6), 4376–4382.

Kuntke P., Smiech K. M., Bruning H., Zeeman G., Saakes M., Sleutels T., Hamelers H. V. M. and Buisrnan C. J. N. (2012). Ammonium recovery and energy production from urine by a microbial fuel cell. *Water Research*, **46**(8), 2627–2636.

Lienert J. and Larsen T. A. (2006). Considering user attitude in early development of environmentally friendly technology: a case study of NoMix toilets. *Environmental Science & Technology*, **40**(16), 4838–4844.

Maurer M., Schwegler P. and Larsen T. A. (2003). Nutrients in urine: energetic aspects of removal and recovery. *Water Science and Technology*, **48**(1), 37–46.

Tervahauta T., Hoang T., Hernandez L., Zeeman G. and Buisman C. (2013). Prospects of source-separation-based sanitation concepts: a model-based study. *Water*, **5**(3), 1006–1035.

Tervahauta T., Bryant I. M., Leal L. H., Buisman C. J. N. and Zeeman G. (2014a). Improved Energy Recovery by Anaerobic Grey Water Sludge Treatment with Black Water. *Water*, **6**(8), 2436–2448.

Tervahauta T., Rani S., Hernandez Leal L., Buisman C. J. N. and Zeeman G. (2014b). Black water sludge reuse in agriculture: are heavy metals a problem? *Journal of Hazardous Materials*, **274**, 229–236.

Tervahauta T., van der Weijden R. D., Flemming R. L., Hernandez Leal L., Zeeman G. and Buisman C. J. N. (2014c). Calcium phosphate granulation in anaerobic treatment of black water: a new approach to phosphorus recovery. *Water Research*, **48**, 632–642.

Thibodeau C., Monette F., Glaus M. and Laflamme C. B. (2011). Economic viability and critical influencing factors assessment of black water and grey water source-separation sanitation system. *Water Science and Technology*, **64**(12), 2417–2424.

Tuantet K., Temmink H., Zeeman G., Janssen M., Wijffels R. H. and Buisman C. J. N. (2014). Nutrient removal and microalgal biomass production on urine in a short light-path photobioreactor. *Water Research*, **55**, 162–174.

Vlaeminck S. E., Terada A., Smets B. F., Van der Linden D., Boon N., Verstraete W. and Carballa M. (2009). Nitrogen Removal from Digested Black Water by One-Stage Partial Nitritation and Anammox. *Environmental Science & Technology*, **43**(13), 5035–5041.

Wetsus. (2013). Wetsus urine project implemented at the Ministry of infrastructure and environment. https://www.wetsus.nl/home/wetsus-news/press-release-%28in-dutch%29/1 (Retrieved 28 January 2016).

Zeeman G. (2012). New Sanitation. Sub-department Environmental Technology Wageningen University. Inaugural Speech.

Zeeman G. and Kujawa-Roeleveld K. (2011). Resource recovery from source separated domestic waste(water) streams; full scale results. *Water Science and Technology*, **64**(10), 1987–1992.

Chapter 4

Wastewater treatment in algal systems

Ignacio de Godos, Zouhayr Arbib, Enrique Lara, Raúl Cano, Raúl Muñoz and Frank Rogalla

4.1 INTRODUCTION

The term microalgae refers to a polyphyletic group of phototrophic microorganisms that inhabit the illuminated zone of almost all water bodies in the earth. This includes polluted environments where availability of substrates such as nitrogen, phosphorous and carbon is higher than clean water. As a result of microalgae metabolism, oxygen is liberated to the water and used by heterotrophic microorganisms in the breakdown of organic matter. Nitrogen, phosphorous and other nutrients are assimilated into algal-bacterial biomass resulting in a rapid depletion from the water of these pollutants. This natural process has been successfully implemented for the purification of different kinds of polluted effluents, the most common being urban wastewater. Large ponds allow for sunlight exposition of water creating the conditions for microalgae growth and removal of the target contaminants. High quality effluents, with low concentration of nutrients and pathogens, have been reported in full scale and pilot units at different locations worldwide. Regarding the management of urban wastewater, facilities based on these fundamentals are currently in operation. However, most of them are large ponds unmixed without microalgae biomass recovery, namely stabilization ponds. These traditional wastewater treatment systems are out of the topic of the work herein presented based on the fact that microalgae systems are those in which the algal-bacterial biomass generated is harvested and recovered. Therefore, mechanically mixed ponds, namely High Rate Algae Ponds (HRAPs), are the most widespread technology with some real scale installations under operation (Craggs *et al.* 2012). The design of these reactors and the main operational issues has been reviewed by Oswald (1988) and Dodd (1986). Several studies have pointed out the advantages of HRAPs over the conventional activated sludge process in terms of low energy cost and high quality final effluents (García *et al.* 2006). However, compared to the conventional activated sludge tanks, which are the most widespread bioreactor configuration for municipal wastewater treatment in western countries, microalgae based systems are rarely implemented and when installed, the units are small or demonstration-scale. However, a change on this trend has occurred in the last decade (Christenson & Sims, 2011).

Interest on microalgae for management of polluted effluents has been closely related to the attention addressed to the utilization of these organisms for biofuel production. Thus, research and demonstration projects and publication of scientific papers devoted to the potential use of microalgae for wastewater treatment have experienced an increase during the periods of global energy crisis and rise in the price of crude oil. During the 1970's and the early 1980's, large cultivation methods have been tested, several combinations of effluents treatment and energy production have been developed and new strategies, such as the injection of carbon dioxide to promote algae growth, have been considered (Weissman & Goebel, 1987; Benemann *et al.* 1980). Overall, these studies have pointed out the very low operational and installation costs of the microalgae systems and possible synergies between biofuels production and wastewater treatment.

Since 2003 to the end of 2014 the oil price has gradually increased and alternative energy fuels have been reconsidered, including microalgae (Borowitzka & Moheimani, 2013). During these years, research and development projects have been devoted to the optimization of bioprocesses aimed at generating energy. The potential for energy generation of microalgae has been revaluated considering the new energy scenario and the last advances in the cultivation and harvesting technology (Pittman *et al.* 2011; Christenson & Sims, 2011). In spite of the really optimistic studies based on biodiesel production from microalgae with high lipid content published (Chisti, 2007), there is no commercially viable installations based on this concept. Although the potential biomass generation per unit of surface is high compared to conventional crops, the production of an energy positive balance is hard to achieve in a real scale installation. In several life cycle analysis (LCA) and studies carried out, the use of wastewater as a nutrient and water source has been considered as the only possibility for microalgae biomass to compete with other biofuel feedstock (Christenson & Sims, 2011).

Out of the topic of bioenergy production and from the point of view of the sustainability in wastewater treatment, microalgae have to be also revaluated given the considerable lower operational costs involved in the process. Considerable energy savings have been reported at different locations and scales (Garcia *et al.* 2000; Olguín, 2003). In the work herein presented, a summarize energy balance is included in section 4.4 based on the results found in a demonstration plant with a treatment capacity of 150 m^3 of wastewater per day. From this study, it can be concluded that microalgae based systems are economic and energetically competitive compared to conventional WWT facilities based on activated sludge. From the point of view of the sustainability, the recovery of nitrogen and phosphorous by assimilation into biomass results in a more appealing scenario compared to nitrification-denitrification plants where most of the nitrogen is lost as atmospheric N_2 and phosphorous is precipitated with sludge. The assimilation of elements from wastewater is also playing an important role in the carbon footprint. Emissions of CO_2 in a microalgae-based WWT facility are considerable lower (even negative if flue gas is injected) compared to mechanically aerated tanks used in activated sludge processes.

In spite of all these advantages, the number of microalgae-based installations devoted to the management of wastewaters is scarce nowadays. Most of the research has been carried out at laboratory and bench scale, focusing the efforts in tasks such as strain selection or design of new cultivation methods. However, bearing in mind the progress done in this technology, the major challenges are nowadays in the engineering and bioprocess design in order to achieve economically and environmentally attractive solutions to wastewater management using microalgae. This chapter will present and critically discuss the fundamentals, potential and limitations of this promising green-biotechnology based on the 15 years of experience of the Environmental Technology of Valladolid University (Spain) in the field of applied phycology and the know-how of FCC aqualia S.A. acquired within the EU project ALL-GAS (devoted to construct a 10 ha microalgae-based wastewater treatment facility).

4.2 FUNDAMENTALS OF MICROALGAE BASED SYSTEMS

4.2.1 Photosynthetic aeration, symbiosis and algal-bacterial interactions

Pollution abatement in microalgae-based systems depends on the symbiosis between photosynthetic and heterotrophic microorganisms. The oxygen required by organic matter-oxidizing heterotrophs and NH_4^+-oxidizing litotrophs during wastewater treatment is supplied by water photolysis during photosynthesis, which also provides the energy and reducing power necessary to support the reduction of CO_2 into microalgae (Muñoz & Guieysse, 2006). The net O_2 production associated to microalgae growth depends to a large extent on the nitrogen source, varying from 1.5 to 2 gO_2 g biomass^{-1} when NH_4^+ or NO_3^- are present in the cultivation broth (Eqs. 4.1 and 4.2):

$$10CO_2 + 2NH_4^+ + 5H_2O \rightarrow 2C_5H_8O_2N + 2H^+ + 10.5O_2 \quad (4.1)$$

$$10CO_2 + 2NO_3^- + 2H^+ + 7H_2O \rightarrow 2C_5H_8O_2N + 14.5O_2 \quad (4.2)$$

The production of O_2 is therefore intrinsically linked to microalgae productivity, which itself depends on the photobioreactor design and operational/environmental conditions. Thus, volumetric O_2 production rates of up to 4.3 $kgO_2\ m^{-3}\ day^{-1}$ have been reported by Torzillo *et al.* (2003) in a 5 cm diameter tubular photobioreactor during *Spirulina platensis* cultivation. Likewise, Tredici and co-workers (1991) also estimated O_2 production rates ranging from 8.4 to 12 $kgO_2\ m^{-3}\ day^{-1}$ at irradiances of 1500–2600 $\mu E\ m^{-2}s^{-1}$ during the cultivation of *S. platensis* in a vertical alveolar panel reactor. In comparison, oxygenation rates by mechanical surface aerators of ≈3 $kgO_2\ m^{-3}\ day^{-1}$ are typically encountered in aerated tanks. Based on theoretical O_2 mass transfer efficiencies of 8 $kgO_2\ kWh^{-1}$ normally achieved in aerated ponds via membrane diffusers, the energy savings could reach a potential value of 1–1.5 $kWh\ m^{-3}\ day^{-1}$ using photosynthetic oxygenation compared to conventional activated sludge processes. However, photosynthetic oxygenation, despite depending on a free-solar irradiation, entails a significant energy consumption associated to the mixing of the algal-bacterial cultivation broth (0.1–1 $W\ m^{-3}$ in open photobioreactors, 200–1000 $W\ m^{-3}$ in enclosed photobioreactors) (Alcántara *et al.* 2015a). A part from photosynthetic oxygenation, open photobioreactors can receive a significant O_2 supply from atmospheric diffusion during the night as a result of the positive concentration gradient in the cultivation broth.

Pollutant (i.e C, N, P and pathogens) removal from domestic wastewaters in algal-bacterial photobioreactors is mainly based on the hydrolysis-oxidation of organic matter and nutrients to CO_2, H_2O, NO_3^- (or NH_4^+ in the absence of nitrification) and PO_4^{3-} and on the assimilation of part of the organic matter, nutrients and CO_2 into biomass. These processes are symbiotically carried out by microalgae and bacteria, which can play complementary or competitive roles within the consortium during wastewater treatment. In the presence of light, microalgae (or cyanobacteria) produce the O_2 required by heterotrophs to oxidize the organic pollutants using the CO_2 released during the mineralization process (Figure. 4.1). Under certain conditions also nitrifiers can growth concomitantly with heterotrophs and microalgae, where NH_4^+ is oxidized to NO_2^- and NO_3^-. Recently, this microbial synergy has been optimized in advanced photobioreactors resulting in increased nitrogen removal (Alcántara *et al.* 2015a; de Godos *et al.* 2014b). Since the O_2 produced during the photosynthetic assimilation of CO_2 is originated from the biochemical H_2O oxidation, the potential O_2 supply of microalgae-based processes is almost unlimited. Traditionally, heterotrophic activity has been exclusively associated to bacteria, although the predominance of microalgae recently observed in the cultivation broth of open system treating domestic wastewater under certain scenarios suggested the occurrence of a mixotrophic algal metabolism and a key role of microalgae during organic matter removal (Gabriel Acién personal communication). However, this hypothesis still needs to be validated.

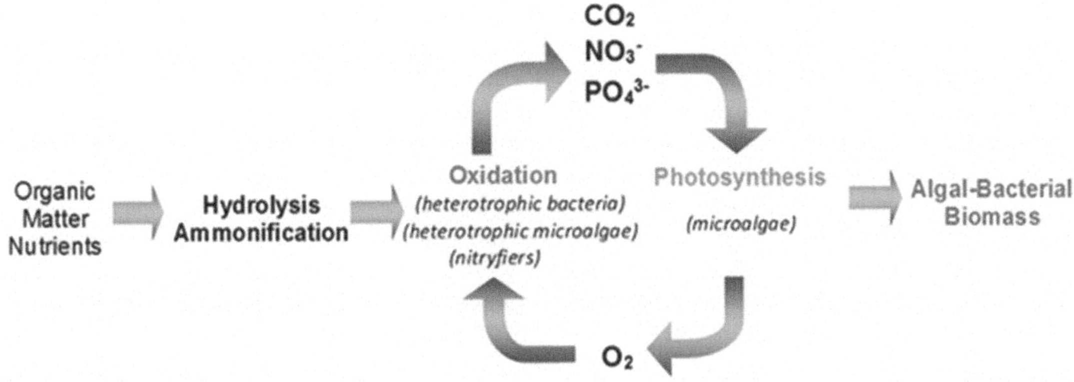

Figure 4.1 Schematic diagram of the symbiosis in algal-bacterial consortia during domestic wastewater treatment.

The interactions between microalgae and bacteria are not only limited to a simple CO_2/O_2 exchange, but can also exert both synergistic and antagonistic mutual effects on their respective activity (Figure 4.2). Thus, microalgae can cause a negative effect on bacterial activity due to (i) the photosynthetically-induced increase in the pH (up to 10–11), dissolved oxygen concentration (up to 30–40 mg L^{-1}) and temperature of the cultivation broth, (ii) the excretion of inhibitory metabolites, and (iii) a more efficient competition for inorganic carbon (Oswald, 2003). In fact, these antibacterial mechanisms are responsible for the removal of 99% of the concentration of pathogens such as *Escherichia coli* in high rate algal ponds treating domestic sewage (Posadas *et al.* 2015a). Likewise, Muñoz and co-workers recently reported a microalgae-mediated inhibition of nitrification activity as a result of the competition for CO_2 between microalgae and nitrifying bacteria (Alcántara *et al.* 2015b; de Godos *et al.* 2014b). On the other hand, microalgae can exert a positive influence

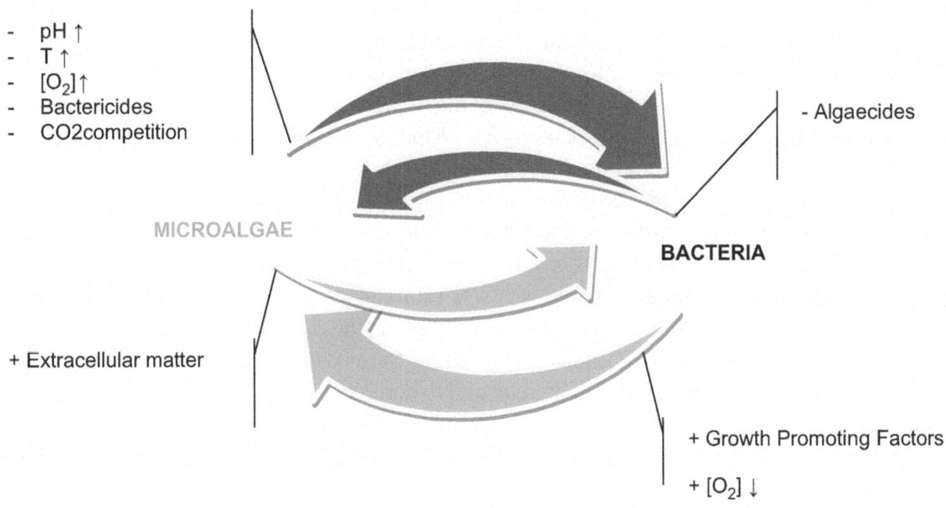

Figure 4.2 Interactions between microalgae and bacteria during wastewater treatment in algal-bacterial photobioreactors.

on bacterial activity by releasing extracellular matter that co-metabolically helps in the bacterial degradation of recalcitrant pollutants (Wolfaardt *et al.* 1994). In addition, bacterial growth can enhance microalgae activity by releasing growth-promoting factors such as indole-3-acetic acid (Gonzalez & Bashan, 2000) and vitamin B12 (Kazamia *et al.* 2012) or by reducing the dissolved oxygen concentration in the cultivation broth and thus alleviating the inhibition of microalgae activity at high O_2 concentrations (Mouget *et al.* 1995). Bacteria can also inhibit microalgae by producing algicidal extracellular metabolites (Fukami *et al.* 1997). The above described interactions between microalgae and bacteria might be even more complex and result in a mutual selection of community clusters and even determine the variations in the structure of the communities. In this context, the implementation of advanced molecular tools such as high throughput shotgun sequencing will allow identifying species that tend to appear together in symbiotic relationships and correlate population structure with the macroscopic functionalities observed in the photobioreactor (Ferrero *et al.* 2012).

4.2.2 Carbon, nitrogen and phosphorous removal mechanisms

Wastewater treatment in algal-bacterial photobioreactors relies on both biotic and abiotic mechanisms, with complementary contributions of these mechanisms to the removal of carbon, nitrogen and phosphorous that ultimately depend on the environmental (temperature, pH and irradiation) and operational (hydraulic retention time and turbulence in the cultivation broth) conditions and photobioreactor design (open *vs.* enclosed systems) (Alcántara *et al.* 2015a). While the removal of biodegradable organic carbon is carried out by aerobic heterotrophs (which convert organic matter into CO_2 and H_2O), inorganic carbon is mainly removed by assimilation into microalgae biomass and stripping (since a negligible removal by assimilation into nitrifying biomass has been recorded). Organic carbon removal currently constitutes the simplest and most cost-effective process during integral wastewater treatment, entailing hydraulic retention times of 1.5–2 days for the complete removal of a BOD concentration of 200 mg O_2 L^{-1} (assuming a microalgae productivity of 26 g m^{-2} d^{-1}). The removal of inorganic carbon is not compulsory in any legislation in the world, but it can entail an increase in the COD concentration in receiving water bodies if converted (via photosynthesis) into freely suspended microalgae. Finally, the anoxic removal of organic carbon via denitrification (with CO_2 and N_2 as the main final products) in novel two-stage denitrification-nitrification photobioreactor configurations has been recently demonstrated (Alcántara *et al.* 2015b).

Nitrogen removal via assimilation into algal-bacterial biomass (nitrogen as a source of proteins) is the preferred removal mechanism over abiotic mechanisms such as NH_3 stripping at the high pH values typically encountered in active microalgae cultures. Among the available nitrogen forms present in wastewater (organic nitrogen, ammonium, nitrite and nitrate), NH_4^+ is the preferred nitrogen source based on its energetically favourable assimilation into aminoacids (since nitrate and nitrite need to be reduced to a N^{3-} state). A complete nitrogen assimilation in microalgae-based wastewater treatment can be achieved at C/N/P ratios of 100/18/2 (Oswald, 1988), although most domestic, livestock and agro-industrial wastewaters present significantly lower C/N ratios (<4). An operational strategy successfully tested to boost nitrogen recovery in the form of algal biomass is the external supplementation of inorganic carbon via flue gas or biogas supply (Posadas *et al.* 2015b; Serejo *et al.* 2015) when locally available. A hydraulic retention time of 3–4 days is typically required to completely deplete the nitrogen from a medium strength domestic wastewater (45 mg N L^{-1}) in photobioreactors operated at a biomass productivity of 26 g m^{-2} d^{-1}). Another innovative strategy to enhance nitrogen removal in wastewaters with a low C/N ratio in scenarios where no flue gas can be supplied to deplete nitrogen from wastewater, is the use of denitrification-nitrification photobioreactor configurations. In this system, ammonium nitrification occurs in the photobioreactor by the action of autotrophic nitrifiers in the presence of sufficient concentrations of NH_4^+, O_2 and CO_2, while the NO_3^- and NO_2^- produced are returned to an anoxic tank receiving the

Table 4.1 Microalgae population structure in experimental open HRAPs treating domestic, livestock and industrial wastewaters under outdoor and indoor conditions.

Experimental System	Microalgae Population Structure	References
1.5 m^2 outdoor HRAP treating domestic wastewater	*Dictyosphaerium, Chlorella, Micractinium, Scenedesmus armatus, Scenedesmus acutus*	Garcia *et al.* (2000)
1 L anoxic reactor interconnected to a 2.7 L stirred tank photobioreactor treating domestic wastewater at 400 μE m^{-2}-s^{-1}	*Scenedesmus ecornis, Pseudanabaena* sp., *Acutodesmus obliquus, Chlorella* sp., *Scenedesmus obtusus, Stigeoclonium setigerum, Planktothrix isothrix, Leptolyngbya benthonica, Chlorella vulgaris, Limnothrix planctonica, Geitlerinema* sp. and others.	Alcántara *et al.* (2015b)
1.5 m^2 outdoor HRAP treating 1:10 and 1:20 diluted swine manure	*Chlamydomonas, Microspora, Chlorella, Nitzschia, Achnanthes, Protoderma, Selenastrum, Oocystis, Ankistrodesmus.*	de Godos *et al.* (2009)
1.2 m^2 Indoor HRAP treating diluted vinasse and raw vinasse	*Geitlerinema sp. Limnothrix planktonica, Pseudoanabaena minima, Stigeoclonium tenue, Leptolyngbya benthonica, Planktolyngvya brevicellularis, Staurosira sp.*	Posadas *et al.* (2015b)

wastewater (located before the photobioreactor). Both nitrate and nitrite are used as electron acceptor in the anoxic tank to support the oxidation of organic matter, resulting in CO_2 and N_2 production and therefore in a dissimilatory nitrogen removal. Thus, Muñoz and co-workers successfully implemented a denitrification-nitrification photobioreactor with biomass recycling, which supported an efficient removal of organic carbon (88%), inorganic carbon (82%) and total nitrogen (75%) at a hydraulic retention time as low as 2 days (Alcántara *et al.* 2015b; de Godos *et al.* 2014b).

Phosphorous removal in algal-bacterial photobioreactors occurs simultaneously via assimilation into biomass and chemical precipitation into hydroxyapatite. Microalgal photosynthesis mediates an increase in the pH of the cultivation broth as a result of CO_2 removal, which induces P-PO_4^{3-} precipitation in the presence of Ca^{2+} in the form of $Ca_5(OH)(PO_4)_3$ according to equation 4.3 (Ruiz-Martinez *et al.* 2015).

$$3HPO_4^{2-} + 5Ca^{2+} + 4OH^- \rightarrow Ca_5(OH)(PO_4)_3 + 3H_2O \tag{4.3}$$

The assimilation of phosphorous, an essential component of cell membranes and genetic material, would require a hydraulic retention time of 4 days to completely remove 5 mg P L^{-1} from a domestic wastewater (assuming a productivity of 26 g m^{-2} d^{-1} and a biomass P content of 1%). However, microalgae biomass can increase its P content up to 4% as a result of a luxury P uptake in the form of intracellular energy-storing polyphosphates (Powell *et al.* 2008). This phenomenon confirms the extraordinary metabolic versatility of microalgae, which apparently exhibit functionalities so far attributed to bacteria.

4.2.3 Strain selection

The microbial population structure (both at a microalgae and bacteria level) is intrinsically determined by variations in the environmental conditions prevailing in the cultivation broth (irradiance, temperature, pH), wastewater characteristics (inorganic carbon, organic matter, phosphorous and NH_4^+ concentrations), operational conditions (HRT, external CO_2 supply, agitation) and photobioreactor configuration (open and enclosed), the latter two influencing the cultivation broth conditions. Tolerance to pollution is often a key selection pressure determining microalgae dominance during microalgae-based wastewater treatment, with *Euglena, Oscillatoria, Chlamydomonas, Scenedesmus* and *Chlorella* ranking in the top 5 of pollution-tolerant genera according to a pioneer study by Palmer (1969). Thus, *Euglena* and *Chlamydomonas* are dominant at high organic loads in wastewater treatment, while *Chlorella* and *Scenedesmus* are the most abundant species at medium loads. On the other hand, *Euglena* and *Scenedesmus* species predominate over *Chlorella* below 15°C due to their higher tolerance to low temperatures (Muñoz & Guieysse, 2006). Table 4.1 shows the most important microalgae genera identified during wastewater treatment in photobioreactors.

Most experimental works conducted have reported very fast changes of microalgae population during wastewater treatment in photobioreactors operated under both indoor and outdoor conditions. Although microalgae dominance of only one genera has been recorded during the treatment of domestic wastewater in HRAPs in Almeria and Cadiz (Spain), *Scendemus* and *Coelastrum*, respectively (Garcia *et al.* 2000; Posadas *et al.* 2015b; Alcántara *et al.* 2015b). In this context, the dominance of slow growing photosynthetic microorganisms in open photobioreactors is often hindered by contamination by the small, rapidly growing microalgae, so enclosed photobioreactors have been proposed by some authors since they support more effective species control (Tredici, 1999). Efforts to sustain a certain microalgal population in open photobioreactors by controlling the operational parameters have not always been successful. For instance, Benemann *et al.* (1980) failed to establish the predominance of *Oscillatoria* sp. and *Micractinium* sp. by microscreening of the biomass and subsequent recirculation into the photobioreactor. Likewise, microalgae sedimentation and recycling did not support the dominance of monoalgal cultures in an anoxic-oxic photobioreactor treating domestic wastewater (Alcántara *et al.* 2015a). However, Wood (1987) successfully maintained the dominance of a *Stigeoclonium* strain by combining a short hydraulic retention time (HRT), to wash the freely-suspended microalgae, with crossflow microscreening of the target strain. At this point, it must be stressed that most common installations are large scale ponds open to the atmosphere and with an important daily input of microorganisms with the wastewater, and therefore, the control of populations could be hard if not impossible. On the other hand, special attention has been devoted in the last years to the selection of strains tolerant to pollution (NH_4^+, organic matter or CO_2) in well controlled laboratory or bench scale experiments (Aravantinou *et al.* 2013). However, due to the high removal efficiency of these systems (e.g. 90% COD removals and 90% N-NH_4^+ removals) the concentration of the contaminants is often very low in the cultivation broth and therefore inhibition is not likely to occur.

4.2.4 Influence of environmental parameters

Temperature, pH, irradiance and the concentration of CO_2, O_2 and inhibitory compounds in the cultivation broth constitute the environmental parameters with the highest influence on microalgae activity and therefore on process oxygenation and on the extent of the different nutrient removal mechanisms (Muñoz & Guieysse, 2006). This section is devoted to briefly describe the influence of these environmental parameters on microalgae-based wastewater treatment.

- *Temperature:* Microalgal and bacterial activity rises at increasing process temperature up to a species-dependent protein denaturalization threshold (≈35–40°C). Optimal temperatures for photoautotrophic growth vary from 20 to 30°C, although thermo-tolerant microalgal and bacterial species are commonly present in microalgae-based wastewater treatment systems. Higher temperatures are very often recorded in algal-bacterial photobioreactors during peak sun hours (especially in enclosed systems), which results in both a reduced photosynthesis and nutrient removal due to a reduction of CO_2 solubility (if wastewater treatment is supplemented with additional CO_2 to boost nutrient assimilation) and an increase in nitrification (with a concomitant reduction in NH_4^+ stripping) (Alcántara *et al.* 2015a).

- *pH:* Microalgae and cyanobacteria grow optimally at a neutral pH, although species such as *Spirulina platensis* and *Chlorococcum littorale* exhibit an optimum pH at 9 and 4, respectively. As a rule of thumb, microalgal CO_2 uptake in photobioreactors leads to pH values reaching 10–11, which itself can mediate a partial inhibition of the algal-bacterial activity (Oswald, 1988). In this context, pH plays also a key role on microalgae inhibition during wastewater treatment as a result of the increase in the fraction of the free NH_3 (Muñoz *et al.* 2006). In this context, nitrification might counter-balanced the increase in pH mediated by photosynthesis as a result of the release of H^+ during NH_4^+ oxidation.
- *Irradiance:* Photosynthetic activity linearly increases up to an irradiance of ≈400 μE m^{-2} s^{-1}, although some species of *Scenedesmus* and *Chlorella* undergo a saturation of their photosynthetic machinery at 100–200 μE m^{-2} s^{-1} (~5–10% solar irradiance) (Muñoz & Guieysse, 2006). Photosynthetic activity remains constant over a species-dependent irradiance interval but gradually decrease at irradiances over 1000 μmol photon m^{-2} s^{-1}. All figures here provided are extremely species-dependent and refer to the irradiance received by individual cells. Hence, microalgae photoinhibition can be significantly alleviated by operating the photobioreactor under high culture densities and a proper mixing regime. Finally, it must be stressed that the sole absorption of photosynthetic active radiation (PAR) and losses due to reflection, respiration, photosaturation and photoinhibition entail a decrease of the maximum efficiency of photosynthesis to values of 1.5–5% of the total impinging solar radiation (Alcántara *et al.* 2015).
- *CO_2 and O_2 concentrations:* Microalgae and cyanobacteria can grow even at CO_2 atmospheric levels (~0.0387% v/v) due to the presence of highly efficient carbon concentrating mechanisms in the vicinity of the RUBISCO enzyme. Despite organic matter mineralization provides inorganic carbon for photosynthesis and nitrification, the supplementation of inorganic carbon (via flue gas or biogas) is often required during domestic wastewater treatment to deplete their nutrient content and prevent from CO_2 competition between the autotrophic communities present in the cultivation broth (Alcántara *et al.* 2015b). On the other hand, high dissolved O_2 concentrations (>20 mg L^{-1}) favour photorespiration and O_2 radicals formation, resulting in a partial inhibition of photosynthesis. Cultivation broth degassing strategies with CO_2-enriched air have been successfully tested in HRAPs to support an enhanced nutrient assimilation while concomitantly decreasing the dissolved oxygen concentrations. The O_2 consumption mediated by organic matter oxidation and nitrification during wastewater treatment entails a decrease in the oxygen concentration, which prevents microalgae inhibition.
- *Toxic compounds:* Heavy metals or organic pollutants from industrial discharges to the sewer network can induce inhibition of the microbial communities supporting wastewater treatment in photobioreactors. Bacteria are often much more tolerant to toxic inhibitory than microalgae, which are severely inhibited in the presence of a few milligrams per litre of toxicants. The most common inhibitory compound during wastewater treatment is NH_3, whose concentration exponentially rises when increasing the pH values in the cultivation broth as a result of photosynthesis. Microalgae from the genus *Chlorella* have shown a high tolerance to NH_3 concentration, which might explain its predominance in microalgae-based wastewater treatment systems (Muñoz & Guieysse, 2006).

4.3 MICROALGAE BASED SYSTEMS USED FOR WASTEWATER TREATMENT

4.3.1 Bioreactors

Microalgae cultivation systems are normally classified into open and closed photobioreactors. Open ponds (namely High Rate Algae Ponds or Raceway ponds) are the most widespread systems, with some commercial installations under operation in different locations. Closed systems are diverse in configuration and they are normally devoted to the production of microalgae biomass for the commercialization of high value products (e.g polyunsaturated fatty acids or pigments). The higher light availability achieved in closed reactors allow higher volumetric productivities. In this context, closed tubular reactors can achieve productivities of 0.4 g $L^{-1}d^{-1}$ (Acién *et al.* 2001), while open ponds achieve maximum values of only 0.1 g $L^{-1}d^{-1}$. However, when compared in terms of areal productivity, the superior performance of closed systems is not that significant, both kind of systems supporting comparable average productivities in the same range (15–25 g $m^{-2}d^{-1}$). Therefore, the potential treatment of wastewater per surface unit must be equivalent in both systems, given that the hydraulic retention times in both kind of systems are similar (between 2 and 6 days). In spite of the considerable efforts given to the design of new photobioreactor configuration for wastewater treatment, mixed open ponds are nowadays the only large scale implemented technology. This is mainly due to the high energy costs involved during the operation of closed photobioreactors. For example, tubular photobioreactors, need an average power consumption for mixing ranging from 1000 to 2000 W per m^3 of culture broth (Acién *et al.* 2001). On the other hand, the energy required for mixing in a large scale conventional HRAPs "equipped with a paddle wheel" ranges from 0.1 and 2 W per m^3, depending on the size and design (Weissmann & Goebel, 1987; Mendoza *et al.* 2013). Based on the fact that the hydraulic retention time is similar in both photobioreactor configuration (open pond and tubular photobioreactor), the energy required in HRAPs can compete with the energy demand reported in activated sludge processes (approximately 0.5 kWh per m^3 of water treated). Despite other kind of closed photobioreactors have lower energy consumptions than tubular reactors, the requirements are still high compared with HRAPs. For instance, power requirements of bubble columns is estimated in 40 W m^{-3} and in flat panels this value is even higher, 53 W m^{-3} (Sierra *et al.* 2008). Likewise, bubble columns and flat panels have some important drawbacks derived from the high installation costs involved. Therefore, several studies have pointed out that raceway ponds or HRAPs are the only technology capable of supporting the large cultivation of microalgae in wastewater (Grobbelar, 2011).

Raceway ponds were initially designed and engineered in the mid 1960s by Professor Oswald and few modifications over the original were reported during the past century. Shallow ponds (0.20–0.40 m deep) mixed by means of a paddle wheel and with a variable surface (from some meters to 1 hectare) have been evaluated for treatment of wastewater (Craggs *et al.* 2012; Benemann *et al.* 1980; Olguín *et al.* 2003). Although this design has been shown to support a reasonable performance, the recent interest in microalgae cultivation has promoted the development of new configurations. The main purpose of these innovations is the reduction of the energy consumption involved in mixing. In this way, the optimization of the HRAPs entailed a reduction in the operational costs that results in a more attractive energy scenario. Other issues such as the reduction in the dead zones and the prevention of biomass settling have been also addressed. Using novel simulation techniques such as Computational Fluid Dynamics (CFD), different configurations of raceway ponds have been simulated in order to elucidate an optimum reactor design with minimum energy consumption and enough mixing to guarantee microalgae growth. These configurations are based on the design of the bends and the mixing device (paddle wheels or propellers). Sompech *et al.* (2012) simulated pond configurations based on different number of flow deflector baffles in the bends and configurations with asymmetrical island bends, and concluded that a bend provided with three baffles and a slightly modified island exhibited the lowest energy consumption. Liffman *et al.* (2012) also used CFD simulations in order to compare novel designs including a modified bends consisting of variations in depth from shallow in the centre to deep in the axis of the curve. According to these simulations, this new configuration resulted in a decrease in 87% of the energy consumption compared to the conventional raceway design. CFD simulations were also used in the evaluation and validation of an innovative mixing system where the paddle wheel was changed by a propeller (Chiaramonti *et al.* 2012). A new approach in raceway design has been recently proposed and patented by the company FCC aqualia

(Lara & Rogalla, 2015). The invention includes a flow acceleration device and depth transition regions in the bends, which results in an overall improvement of the hydraulics and in a considerable decrease in energy consumption (up to 4 times less energy than conventional raceway ponds for 1 ha-HRAPs). This system is currently in operation in a wastewater treatment plant (Chiclana de la Frontera, Spain) and its performance was compared to a conventional raceway mixed by a paddle wheel at industrial scale of 500 m^2 (Figure. 4.3).

Figure 4.3 Raceway ponds installed in Chiclana de la Frontera (South Spain) with two different configurations: conventional mixing by paddle wheels (left) and innovative patented raceway mixed by propeller with modified bends (right).

4.3.2 CO_2 addition, implications in the process

Microalgae are mainly composed of carbon, nitrogen, phosphorous, sulphur, oxygen, hydrogen and other trace elements such as metals. Although biomass composition depends on the cultivation medium, the average C:N ratio ranges from 10:1 to 5:1 depending on the availability of nitrogen (Benemann *et al.* 2003). This is considerable higher than domestic wastewaters where the C:N ratio is approximately 3:1. Therefore, carbon limiting conditions are typically encountered in microalgae-based systems during domestic WWT. This limitation is evidenced by the elevated pH values (up to 10) normally achieved at midday in HRAPs because of the consumption of HCO_3^- ions. The inorganic carbon concentration decreases significantly in photosynthetically active cultures and the availability of CO_2 is further reduced due to the elevated pH since most of the inorganic carbon is in form of bicarbonate and carbonate ions. Some authors have stated that this limitation can result in a reduction of the potential production of biomass and nutrient assimilation and therefore the injection of CO_2 has been considered as an attractive strategy (Wang *et al.* 2008). In wastewater, carbon is present as organic carbonaceous components, which can be consumed by aerobic chemoorganotrophs producing CO_2, and in inorganic form as CO_2, HCO_3^- and CO_3^{2-}. An extra addition of CO_2 can decrease the pH and shift the equilibrium of inorganic carbon towards CO_2 and HCO_3^-, compounds that can be readily consumed by algae (generally CO_3^{2-} is not consumed). The higher availability of inorganic carbon and the pH closer to the optimum for microalgae growth (7–8 for most green microalgae), ultimately increase photosynthetic productivity. Besides this, the control of pH in the neutral range has other positive effects such as the prevention of inhibition due to the presence of NH_3 and the reduction of nitrogen losses by volatilization. According to the published experiences, HRAPs supplemented with carbon dioxide treating wastewater produce a significantly higher amount of biomass per surface unit. Azov and Shelef (1982) reported an enhancement in the productivity between 65–95% when carbon dioxide was supplied into a 1000 m^2 ponds. Productivities more than double have been reported in CO_2 supplemented HRAPs treating domestic wastewater compare to control ponds without CO_2 (Benemann *et al.* 2003; Park & Craggs, 2010). However, considerable lower productivities (30–35%) have been reported in livestock wastewater or secondary treated wastewater (de Godos *et al.* 2010; Arbib *et al.* 2013).

In the context of wastewater treatment, the source of CO_2 must be flue gas or biogas since the cost of commercial CO_2 is not affordable. In addition, HRAPs can be used as a CO_2 abatement technology since the biofixation of this constituent is driven by solar energy. Therefore, flue gases emitted to the atmosphere during industrial processes are the most suitable residual CO_2 source. Depending on the fuel source, flue gas exhibits a carbon dioxide content between 4 and 12%. Bubbling of this gas at the base of the open ponds may not be effective enough since the residence time of the flue gas bubbles might be too short, resulting in a small quantity of carbon dioxide transferred to the cultivation broth. Some systems have been proposed to increase the mass transport of CO_2 to the algal broth. In the engineering studies of Oswald (1988), a sump of around 1 m-deep where the gas was injected is described. A counter-current operation provided by a baffle placed in the middle of this sump was devised to increase the mass transfer coefficients. However, a recent experimentation has reported the disadvantages of counter-current operation in terms of higher head loss in the reactor and a subsequent increase in the power consumption during the mixing (Mendoza *et al.* 2013). In addition, no clear benefits in mass transfer and carbonation potential due to a counter-current operation have been demonstrated. Indeed, an unobstructed sump without counter-current operation can provide a CO_2 mass transfer efficiency of 95% from flue gas to the microalgae culture (de Godos *et al.* 2014a). Other carbonation systems proposed for open ponds are bubble columns where the microalgae culture broth and the flue gas are brought in contact. Putt *et al.* (2011) reported an efficiency of 83% in the carbonation of a raceway reactor coupled to a column for gas exchange. A similar system has been tested for the simultaneous carbonation of microalgae cultures and biogas up-grading (elimination of H_2S and CO_2) producing a high purity of biomethane and maintaining the pH in the neutral range (Serejo *et al.* 2015). The use of biogas as a CO_2 source allow recycling of part of the residual carbon produced during anaerobic digestion process.

In spite of the benefits of CO_2 addition, the supply of this component in real scale facilities must be carefully studied in energy and economic terms. First, the availability of flue gas can be limited in the proximities of the wastewater treatment plants and a customized-transportation of CO_2 through pipelines might be prohibitive. Therefore, flue gas should be generated in the microalgae-based WWT facility using boilers o Combine Heat and Power engines. However, this kind of technology is often limited to large installations with a considerable investment cost and therefore very distant from real microalgae wastewater systems that are limited to small or medium facilities due to the space requirements. In addition, a considerable amount of energy is required for the injection of CO_2 through blower pumps and subsequently the potential energy benefits are further reduced.

4.3.3 Harvesting of biomass

A major challenge in microalgae-based wastewater treatment consist in separating the microalgae from the treated wastewater, namely microalgae harvesting. Biomass concentrations in microalgal cultures are usually low (~0.5 g L^{-1} in open pond reactors up to ~5 g L^{-1} in closed photobioreactors). This means that a large volume of water has to be removed to harvest the biomass. As a result of the small size of microalgal cells (2–20 μm) and their colloidal stability in suspension, harvesting by means of sedimentation or simple screening is not easy to achieve. During microalgae cultivation for high-value product production, harvesting is done in one step process by centrifugation or tangential microfiltration. However, these technologies are too expensive and energy-intensive in case of wastewater treatment where large volumes of culture medium need to be processed. The harvesting has been identified as the bottleneck in the development of the large-scale production systems for biofuel production or wastewater treatment processes (Uduman *et al.* 2010). In mass production units, microalgal biomass recovery accounts with 20–30% to the total operational cost (Molina *et al.* 2003).

Sedimentation by gravidity is the most common and cost effective method of algal biomass removal in wastewater treatment (Nurdogan & Oswald, 1996). The use of settlers has been proposed since the early works of Oswald (1988) and the more recent studies of Craggs *et al.* (2012). However, the algal settling units typically used have relatively long retention times (1–2 days) and only remove 50–80% of the biomass (Nurdogan & Oswald, 1996; Brennan & Owende, 2010; Park *et al.* 2011). These removal rates are not enough to accomplish with discharge limits for total suspended solids (maximum of 35 mg TSS L^{-1} or 90% of removal according to the directive 91/271/CEE). Biomass recycling from settlers to the culture system has been proposed as a method of selection pressure for the enrichment in rapidly settling biomass. This strategy enhances the removal of solids from the final effluent (Park *et al.* 2013). However, removal rates are still far from the targets limits and changes in biomass composition cannot be controlled. Therefore, alternative options have been studied including chemical and mechanical methods (Shen *et al.* 2009; Alabi *et al.* 2009; Brennan & Owende, 2010; Mata *et al.* 2010). However, most of the technologies greatly increase the operational costs (Benemann, 2008). Mechanical centrifugation or filtration could be used for the removal of algal biomass, but the high energy requirement (between 2.1 and 8 kWh per m^3 of effluent processed, according to Danquah *et al.* 2009) makes them only economically viable for a secondary thickening of previously harvested algae (with 1–2% solids), therefore a prior step is required (Benemann, 2008; Alabi *et al.* 2009).

The use of chemical coagulants and flocculants has been proposed as a previous step before sedimentation or flotation. These chemicals aggregate the microalgae cells and increase the effective "particle" size, thus enhancing biomass recovery. Microalgae cells are tiny, smaller than 30 μm, and their density is similar to water, with sedimentation velocity lower than 10^{-6} m s^{-1}. To efficiently settle microalgae cells, the sedimentation velocity must be higher than 10^{-4} m s^{-1}, similar to the velocities of activated sludge flocs in conventional WWT plants. Coagulation-flocculation followed by sedimentation is anticipated to be a relatively inexpensive harvesting technique in terms of energy consumption, however the coagulant-flocculant costs are often a significant portion of the overall process costs. Weissman and Goebel (1987) estimated that chemical costs accounted for 4% to 7% of the total operating costs, respectively. Overdosing of chemicals, specially coagulants such as $FeCl_3$, $Fe(SO_4)_3$ and $Al_2(SO_4)_3$ could affect greatly the pH of the system. Some authors also reported some auto-flotation phenomenon at high oxygen concentration above 16 mg O_2L^{-1} (Bare *et al.* 1975), this behaviour can pose a major problem in the settling tank operation.

In this context, where an elevated separation of cells from liquid must be achieved at low operational cost, the use of solid/liquid separation technologies from water industry is mandatory. Recently, flotation has been proposed as a technological alternative for microalgae harvesting with promising results. This process can be described as a physiochemical type of gravity separation in which air or gas is bubbled through a solid-liquid suspension and the gas molecules are attached to the solid particles. These particles are carried to the surface of the liquid where accumulate, being subsequently removed. Microalgae separation by dissolved air flotation has been often combined with coagulation-flocculation (Bare *et al.* 1975) and it is possible to obtain microalgal slurries of up to 6%, which allows for an easy energy recovery. Beside this, chemical addition is considerable lower than dosing for gravity sedimentation due mainly to the dragging effect of the bubbles and to the fact that hydraulic retention times for flotation are sensitively lower. Energy consumption for flotation units is appropriate in the context of WWT management with values around 0.04 kWh per m^{-3} processed. The slurry generated in flotation units has a concentration between 30 to 60 g L^{-1}, therefore it is suitable for the subsequent stabilization of the biomass in anaerobic digesters (biomethanization). After the harvesting processing, where concentrations between 10 and 60 g TSS L^{-1} are achieved, a conventional dewatering system can be applied in order to re-concentrate this biomass. By filtration or centrifugation, a total biomass concentration up to 250 g TSS L^{-1} can be achieved. Solar dryer has been recently proposed as a final step in order to produce an almost water free microalgae biomass. This solar driven technology is an environmentally friendly alternative compatible with the microalgae technology.

The choice of the harvesting method affects directly the energy recovery of microalgae based treatment. Figure 4.4 shows the main downstream process available and an approximation of the concentration of biomass achieved in one step. Route 1 is normally used in microalgae production units for high value product extraction, and its elevated operational costs avoid its implementation in WWT operations. Route 2 has been proposed for treatment of WW without energy recovery of the microalgae biomass and the process ends in a production of a dry biomass easily to manage. Finally, Route 3 has been proposed and assessed in the simultaneous wastewater treatment and bioenergy production through the revalorization of the microalgae biomass as biomethane through anaerobic digestion. The digestion of the slurry produced within a flotation unit, followed by decanter centrifugation and solar dryer allow for a positive energy balance mainly due to the very low energy cost during the harvesting, approximately 0.05 kWh per m^{-3} treated. This option has been considered in the energy balance of the overall process described in the section 4.4.

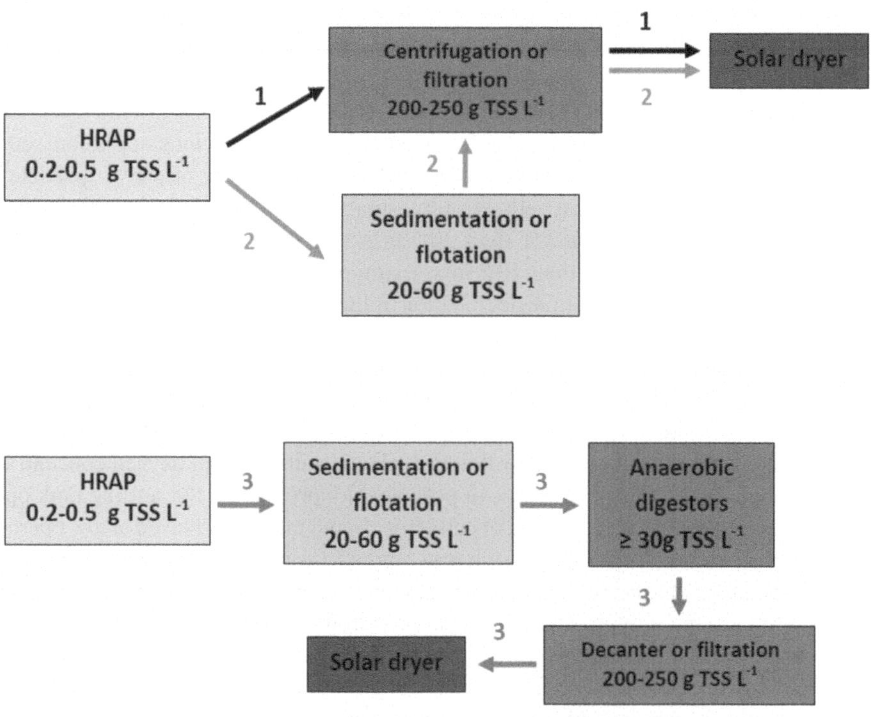

Figure 4.4 Diagram of the alternatives for harvesting of microalgae biomass. Route 1 used in production of microalgae biomass in industry of high-value products; routes 2 and 3 have been proposed for WW processes. The average concentration of the biomass is expressed as Total Suspended Solids (TSS).

4.4 CONSIDERATIONS FOR A REAL SCALE INSTALLATION

The robustness of microalgae based systems for the treatment of wastewater has been reported in several locations and with different kinds of wastewater. As stated in previous section 2, kinetics of microalgae growth is affected by several variables that depend on environmental conditions and wastewater composition. Therefore, the operation of the bioreactors must be done accordingly to the possible variations (mainly light and nutrient availability and organic matter content). In the case of domestic wastewater, experimental units operated in different locations have shown the viability of the process regardless the very different conditions in climate and water composition. In Table 4.2, some of the experiences in large scale HRAP treating domestic wastewater are summarized. Most of the reported works describe experimental or demonstration ponds installed inside or in the proximities of conventional treatment facilities, therefore they do not work as independent units. A pre-treatment of the wastewater before feeding the raceways is often applied in order to reduce turbidity and the concentration of suspended solids. Several of these units use primary treated wastewater, after settling, for feeding the HRAP (Azov & Shelef, 1982; Benemann *et al.* 1980; Canovas *et al.* 1996; García *et al.* 2006). In other cases, anaerobic pre-treatment as a previous step has been assessed (El Hamouri *et al.* 1995). In this case, the use of Upflow Anaerobic Sludge Blanket reactors is one of the most suitable choices. These kind of digesters are currently in operation as primary treatment of domestic wastewater in tropical countries with wastewater temperatures above 20°C, such as Brazil, Colombia and India (Foresti *et al.* 2002). The treatment of UASB effluents by HRAP increases the amount of energy recovered from the process. Banks *et al.* (2011) summarized the main configurations possible of integration of anaerobic processes and microalgae culture, concluding that the pre-treatment of wastewater in UASB reactors and the anaerobic digestion of microalgae, is one of the most promising options. Based on this concept the company, FCC Aqualia is currently developing a demonstration scale unit based in Chiclana de la Frontera (Spain) with a projected treatment capacity of 4000 m^3 per day and estimated biomethane production of vehicle fuel quality (All-Gas project, see website in reference section). In the last years, some research studies have also proposed the use of secondary treated wastewater (after conventional treatment of activated sludge) for microalgae cultivation (Zang *et al.* 2008; Ruiz *et al.* 2013). However, this last option envisaged the production of biomass as a feedstock for biofuel production but did not represent an alternative for wastewater treatment itself.

Table 4.2 Large scale HRAP reported experiences.

Location: Town (Country)	Total Surface (m^2)	Pre-treatment of the Wastewater	References
Christchurch (New Zealand)	14,000	Primary treatment	Craggs *et al.* (2012)
Rawat (Morocco)	790	Anarobic digestion	El Hamouri *et al.* (1995)
St. Helena (USA)	20,000	Primary treatment	Benemann and Oswald (1993)
Hollister (USA)	50,000	Primary treatment	Benemann and Oswald (1993)
Haifa Bay (Israel)	1000	Primary treatment	Azov and Shelef (1982)
Chiclana de la Frontera (Spain)	1000	Sieved Wastewater	This work

The source of wastewater determines the performance, sustainability and the potential energy recovery of the process. The main options, primary or anaerobically treated wastewater, have their own advantages and disadvantages. While the use of primary wastewater has a lower capital cost (since installation of settlers are more economical than anaerobic digesters as UASB), the energy recovery using primary wastewater is

BOX 4.1 TECHNOLOGICAL OPPORTUNITIES

The implementation of this environmentally friendly technology depends on the selection of the appropriate methods for culture, harvesting and energy valorisation. A considerable effort has been done in the improvement of the technology during the last decade and during the previous periods of interest in the utilization of microalgae motivated by the energy crises. Industrially relevant units are in operation at some locations with excellent results in terms of pollution removal. These systems offer a very low energy consumption or even positive production if anaerobic treatments are used for biomethane generation. In exchange, a significant surface is required to install the raceway ponds that allow for the photosynthetic microalgae growth. Therefore, microalgae treatment plants are appropriate for small or medium size towns where conventional systems are unsustainable. The future for the implementation of treatment plants based on this concept depends on the application of the existing technology of water treatment and the knowledge created in demonstration scale projects.

BOX 4.2 IMPACTS (ECONOMIC AND ENVIRONMENTAL ASSESSMENT)

Table 4.3 Environmental/Economic/Technical indicators.

Category	Criteria	Indicator
Environmental	GHG emissions Water quality Production of excess sludge	• Emissions of GHG are negligible or even negative if CO_2 is supply to the microalgae culture. • Microalgae culture is a secondary and tertiary treatment since it provides simultaneously nutrient and organic matter removal and disinfection. • The removal rates of nitrogen and phosphorous reported are between 70 and 90% and between 50 and 95%, respectively. • COD and BOD concentration are normally below of the discharge limits if the microalgae biomass is efficiently separated: 40–100 mgCOD/L and 0–20 mgBOD/L. • Removal of pathogens is very high resulting in *E.Coli* concentrations between 100–1000 UFC/mL. • Production of biomass is very high compare to conventional treatment process. Between 300 and 600 g VSS are produce per m^3 of WW processed.
Economic	Energy consumption Other operational costs	• The microalgae cultivation involves very low electricity consumption between 0.05–0.08 kWh per m^3 of WW. If anaerobic processes are coupled positive energy balances can be achieved. • The cost of the treatment will depend on the selection of the technology of the different steps: culture, harvesting and biomass processing. The studies indicate competitive prices compare to conventional treatment (activated sludge)
Technical	Efficiencies (COD, Nutrient removal) Loading rates	• The percentages of nitrogen and phosphorous removal ranged between 60 to 90% and between 50 to 95%, respectively. In case of COD the removal rate will depend on loading rate with averaging in 0.3–0.8 kg COD per m^{-3} d^{-1}
Social	Awareness Acceptability	• Since the microalgae based systems involve environmental benefits such the neutral carbon footprint and the possible energy recovery, these systems can be easily accepted. However, the implementation also involves unconventional requirements such a considerable surface for the installation of the ponds.

considered lower because less biomethane is produced. In any case, the superior performance of the algae-based systems in terms of energy consumption has been reported. For example, Alcántara *et al.* (2015b) reported a power consumption of 0.048 kWh per m^3 of wastewater treated. Even lower values have been claimed in similar studies, until 0.01–0.03 kWh per m^3 was reported by Garcia *et al.* (1999). However, these calculations are probably only based on the electricity consumption due to algae growth and additional energy inputs and outputs are not considered. In this way, the energy and mass flows of the demonstration plant designed and operated by FCC aqualia in Spain establishes a complete different scenario since every step of the process has been considered including the energy valorisation of the algae biomass and the wastewater through anaerobic digestion. It must be stressed that in this study the biomethane valorisation has been addressed to the use as fuel for vehicles and therefore it includes the electricity consumption due to the up-grading and compression of the biogas. Other energy requirements as the heat for algae digestion, harvesting (based in flotation) and water pumping are also considered. In order to homogenize the energy values of biomethane, electricity and heat requirements, a conversion efficiency factor of 33% (from thermal to electrical power) has been considered. This assumption is necessary given the fact that an algae-based WWT facility based on this concept consumes electricity and thermal power, and produces energy as biomethane. The energy balance shows an average consumption of 0.74 kWh (thermal) per m^3 of water treated and a production of 1.24 kWh (thermal) as biomethane that can be easily placed as biofuel in vehicles. The average balance shows a positive value of 0.5 kWh (thermal) produced per cubic meter of water processed. This is more appealing than the reported negative balance presented in conventional WWT facilities equipped with activated sludge and anaerobic digestion for management of the solids. For instance, Hernández-Sancho *et al.* 2011 determined the average consumption of conventional WWT plants in 0.51 kWh electric per m^3 (equivalent to 1.53 kWh thermal m^{-3}).

The energy recovery through lipid extraction for biodiesel production has been claimed as one of the most promising alternatives. However, in this study we consider that coupling both processes, cleaning water and production high oil content biomass, is not still energetically competitive in view of the low content of lipids of the biomass generated during WWT. The strategies of lipid enrichment are mainly based on the accumulation of these components under stress conditions and specially the nitrogen starvation during relative long terms (Khozin-Goldberg *et al.* 2005). Given that an elevated rate of nitrogen removal per unit of surface or volume is one of the priorities of wastewater management, the deficiency of this nutrient in the culture broth is hard to achieve and oil content is not high enough to consider the biomass generated as a feedstock for biodiesel production. Nevertheless, this topic is still under research and innovative cultures methods are nowadays proposed (Mooij *et al.* 2015).

Besides the energy recovery that can be achieved with the use of microalgae, other important advantages must be highlighted as the alternative uses of algae biomass as biofertilizers. The high nitrogen content, average 6 to 9% of dry matter, and phosphorous, ~1%, makes microalgae biomass a valuable source of nutrients for agriculture. The benefits are not only related to the nutrient content, microalgae can contain compounds that promote germination, leaf or steam growth, flowering or can be used as a biological protection agent against plant diseases (Pulz & Gross, 2004; FAO, 2010). It must be also considered that the nutrient recycling from wastes to the agriculture contributes to a reduction in the fossil fuel consumption and CO_2 emissions, even more when the production of fertilizers is done during the cleaning of water with low impact on the environment.

4.5 CONCLUSIONS

Microalgae have the potential to support a cost-effective wastewater treatment with final effluents containing very low levels of pollution and involving a minimum environmental impact. The current advances in the technology of microalgae culture, harvesting and energy valorisation of microalgae biomass are the result of the increasing interest on the use of these organisms during the periods of high energy and oil prices. The solutions for an efficient production of microalgae, removal of pollution and energy valorisation of the biomass lie on the use of the pre-existing technology of wastewater treatment. In that manner, the recent advances in computational simulations have allowed for improved reactors design and the current solids/liquid separation techniques can provide an efficient biomass harvesting. Beside this, if anaerobic digestion processes are coupled to the microalgae culture a considerable amount of energy can be recovered during the process. Therefore, the main challenges for the implementation of this technology are related to the selection of the technology and bioprocess engineering design. An attractive solution for the treatment of organic polluted effluents is achieved with these organisms since they provide assimilation of carbon, nitrogen, phosphorous and energy of sunlight resulting in a more sustainable technology than the conventional treatments.

4.6 REFERENCES

Acién F. G., Fernández J. M., Sánchez J. A., Molina E. and Chisti Y. (2001). Airlift-driven external-loop tubular photobioreactors for outdoor production of microalgae: assessment of design and performance. *Chemical Engineering Science*, **56**(8), 2721–2732.

Alabi A. O., Tampier M. and Bibeau E. (2009). Microalgae Technologies and Processes for Biofuels/Bioenergy Production in British Columbia: Current Technology, Suitability and Barriers. The British Columbia Innovation Council.

Alcántara C., Dominguez J., García D., Blanco S., Perez R., Garcia-Encina P. and Muñoz R. (2015a). Evaluation of wastewater treatment in a novel anoxic-aerobic algal-bacterial photobioreactor with biomass recycling through carbon and nitrogen mass balances. *Bioresource Technology*, **191**, 173–186.

Alcántara C., Posadas E., Guieysse B. and Muñoz R. (2015b). Microalgae-based waste water treatment. In: Handbook of Microalgae: Biotechnol. Adv, K. Se-Kwon (ed.), Elsevier, pp. 439–455. ISBN: 978-0-12-800776-1.

All-Gas project. http://www.all-gas.eu/Pages/default.aspx

Aravantinou A. F., Theodorakopoulos M. A. and Manariotis I. D. (2013). Selection of microalgae for wastewater treatment and potential lipids production. *Bioresource Technology*, **147**, 130–134.

Arbib Z., Ruiz J., Álvarez P., Garrido C., Barragán J. and Perales J. A. (2013). Effect of pH control by means of flue gas addition on three different photobioreactors treating urban wastewater operating in continuous mode long-term. *Ecological Engineering*, **57**, 226–235.

Azov Y. and Shelef G. (1982). Operation of high-rate oxidation ponds: theory and experiments. *Water Research*, **16**(7), 1153–1160.

Banks C. J., Heaven S., Lara Corona E. S. and Rogalla F. (2011). Energy production from integration of high rate anaerobic and algal wastewater treatment: the EU FP7 All-Gas project. In: 9th IWA Specialist Group Conference on Waste Stabilisation Ponds, Adelaide, 1-3-45.

Bare W. F. R. Jones N. B. and Middlebrooks E. J. (1975). Algae removal USIng dissolved air flotation. *Journal of the Water Pollution Control Federation*, **47**, 153–169.

Benemann J. R. (2003). Biofixation of CO2 and greenhouse gas abatement with microalgae – technology roadmap. Report No. 7010000926 prepared for the U.S. Department of Energy National Energy Technology Laboratory.

Benemann J. R. and Oswald W. J. (1996). Systems and economic analysis of microalgae ponds for conversion of CO2 to biomass. Quarterly Technical Progress Report. Department of Energy, Pittsburgh Energy Technology Center.

Benemann J. R., Koopman B., Weissman J. and Goebel R. (1980). Development of microalgae harvesting and high rate pond technologies in California. In: Algae Biomass Production and Use, G. Shelef and C. J. Soeder (eds.), Elsevier/North-Holland Biomedical Press, Amsterdam, pp. 457–495.

Borowitzka M. A. and Moheimani N. R. (2013). Algae for biofuels and Energy. In: Developments in Applied Phycology, Vol. 5, Springer Science + Business Media, Dordrecht, pp. 17–36.

Brennan L. and Owende P. (2010). Biofuels from microalgae e a review of technologies for production, processing, and extractions of biofuels and co-products. *Renewable and Sustainable Energy Reviews*, **14**(2), 557–577.

Canovas S., Picot B., Casellas C., Zulkifi H., Duboi A. and Bontoux J. (1996). Seasonal development of phytoplankton and zooplankton in a high rate algal pond. *Water Science and Technology*, **33**(7), 199–206.

Chiaramonti D., Prussi M., Casini D., Tredici M. R., Rodolfi L., Bassi N., Zittelli G. C. and Bondioli P. (2012). Review of energy balance in raceway ponds for microalgae cultivation: re-thinking a traditional system is possible. *Applied Energy*, **102**(2), 101–111.

Chisti Y. (2007). Biodiesel from microalgae. *Biotechnology Advances*, **25**, 294–306.

Christenson L. and Sims R. (2011). Production and harvesting of microalgae for wastewater treatment, biofuels and bioproducts. *Biotechnology Advances*, **29**, 686–702.

Craggs R., Sutherland D. and Campbell H. (2012). Hectare-scale demonstration of high rate algal ponds for enhanced wastewater treatment and biofuel production. *Journal of Applied Phycology*, **24**(3), 329–337.

Danquah M. K., Gladman B., Moheimani N. and Gorde G. M. (2009). Microalgal growth characteristics and subsequent influence on dewatering efficiency. *Chemical Engineering Journal*, **151**, 73–78.

Dodd J. C. (1986). Elements of pond design and construction. In: CRC Handbook of Microalgal Mass Culture, A. Richmond (ed.), CRC Press, Boca Raton, FL, pp. 265–283.

El Hamouri B., Jellal J., Outabiht H., Nebri B., Khallayoune K., Benkerroum A., Hajli A. and Firadi R. (1995). The performances of a High-Rate algal Pond in the Moroccan climate. *Water Science and Technology*, **31**(12), 67–74.

European Commision Directive (1998). Amending Council Directive 91/271/EEC with respect to certain requirements established in Annex I, Official Journal of the European Communities, 98/15/EC, 27 February, 29–30.

FAO Aquatic Biofuels Working Group. (2010). Review paper Algae-based biofuels: applications and co products, July 2010 ISBN 978-92-5-106623-2.

Ferrero E. M., de Godos I. Rodriguez E. M., García-Encina P. A., Muñoz R. and Bécares E. (2012). Molecular characterization of bacterial communities in algal–bacterial photobioreactors treating piggery wastewaters. *Ecological Engineering*, **40**, 121–130.

Foresti E. (2002). Anaerobic treatment of domestic sewage: established technologies and perspectives. *Water Science and Technology*, **45**(10), 181–186.

Fukami K., Nishijima T. and Ishida Y. (1997). Stimulative and inhibitory effects of bacteria on the growth of microalgae. *Hydrobiologia*, **358**, 185–191.

García J. and Mujeriego R. (1999). Oxigenación fotosintética en lagunas para depuración de aguas residuales. *Tecnología del Agua*, **195**, 57–64.

Garcia J., Hernández-Marine M. and Mujeriego R. (2000a). Influence of phytoplankton composition on Biomass Removal from High-Rate Oxidation Lagoons by Means of Sedimentation and Spontaneous Flocculation. *Water Environment Research*, **72**, 230–237.

García J., Mujeriego R. and Hernández-Mariné M. (2000b). High rate algal pond operating strategies for urban wastewater nitrogen removal. *Journal of Applied Phycology*, **12**, 331–339.

García J., Green B. F., Lundquist T., Mujeriego R., Hernández-Mariné M. and Oswald W. J. (2006). Long term diurnal variations in contaminant removal in high rate ponds treating urban wastewater. *Bioresource Technology*, **97**, 1709–1715.

de Godos I., Blanco S., García-Encina P., Bécares E. and Muñoz R. (2009). Long Term operation of High Rate Algae Ponds for the Bioremediation of Piggery Wastewaters at High Loading Rates. *Bioresource Technology*, **100**(19), 4332–4339.

de Godos I., Blanco S., García-Encina P., Bécares E. and Muñoz R. (2010). Influence of flue gas sparging on the performance of high rate algae ponds treating agro-industrial wastewaters. *Journal of Hazardous Materials*, **179**, 1049–1054.

de Godos I., Mendoza J. L., Acién F. G., Molina E., Banks C. J., Heaven S. and Rogalla F. (2014a). Evaluation of carbon dioxide mass transfer in raceway reactors for microalgae culture using flue gases. *Bioresource Technology*, **153**, 307–314.

de Godos I., Vargas V. A., Guzmán H. O., Soto R., García B., García P. A. and Muñoz R. (2014b). Assesing carbon and nitrogen removal in a novel anoxic-aerobic cyanobacterial-bacterial photobioreactor configuration with enhanced biomass sedimentation. *Water Research*, **61**, 77–85.

Gonzalez L. E. and Bashan Y. (2000). Increased growth of the microalga Chlorella vulgaris when coimmobilized and cocultured in alginate beads with the plant- growth-promoting bacterium Azospirillum brasilense. *Applied and Environmental Microbiology*, **66**, 1527–1531.

Grobbelar J. U. (2011). Microalgae mass culture: the constrains of scaling-up. *Journal of Applied Phycology*, **24**(3), 315–318.

Hall D. O., Acién Fernández F. G., Guerrero E. C., Rao K. K. and Grima E. M. (2003). Outdoor helical tubular photobioreactors for microalgal production: modeling of fluid-dynamics and mass transfer and assessment of biomass productivity. *Biotechnology and Bioengineering*, **82**(1), 62–73.

Hena S., Fatimah S. and Tabassum S. (2015). Cultivationof algae consortiumin a dairy farm wastewater for biodiesel production. *Water Resources and Industry*, **10**, 1–14.

Kazamia E., Czesnick H., Nguyen T. T., Croft M. T., Sherwood E., Sasso S., Hodson S. J., Warren M. J. and Smith A. G. (2012). Mutualistic interactions between vitamin B12-dependent algae and heterotrophic bacteria exhibit regulation. *Environmental Microbiology*, **14**(6), 1466–76.

Khozin-Goldberg I., Shrestha P. and Cohen Z. (2005). Mobilization of arachidonyl moieties from triacylglycerols into chloroplastic lipids following recovery from nitrogen starvation of the microalga Parietochloris incisa. *Biochimica et Biophysica Acta*, **1738**, 63–71.

Lara E. and Rogalla F. (2015). Open reactor for cultivating microalgae. European Patent number 2875724 Date 4/11/2015.

Liffman K., Paterson D. A., Liovic P. and Bandopadhayay P. (2012). Comparing the energy efficiency of different high rate algal ponds designs using computational fluid dynamics. *Chemical Engineering Research & Design*, **91**(2), 221–226.

Mendoza J. L., Granados M. R., Godos I., Acién F. G., Molina E., Banks C. and Heaven S. (2013). Fluid-dynamic characterization of real-scale raceway reactors for microalgae production. *Biomass & Bioenergy*, **54**, 267–275.

Molina E., Belarbi E.-H., Ancién Fernández F. G., Robles-Medina A. and Chisti Y. (2003). Recovery of Microalgal biomass and metabolites: process options and economics. *Biotechnology Advances*, **20**, 491–515.

Mooij P. R., Stouten G. R., Tamis J., van Loodrecht M. C. M. and Kleerebezem R. (2013). Survival of the fattest. *Energy & Environmental Science*, **6**, 3404.

Mouget J. L., Dakhama A., Lavoie M. C. and De la Noue J. (1995). Algal growth enhancement by bacteria: is consumption of photosynthetic oxygen involved? *FEMS Microbiology Ecology*, **18**, 35–43.

Muñoz R. and Guieysse B. (2006). Algal-bacterial processes for the treatment of hazardous contaminants: a review. *Water Research*, **40**, 2799–2815.

Nurdogan Y. and Oswald W. J. (1995). Enhanced nutrient removal in high rate ponds. *Water Science and Technology*, **31**, 33–43.

Olguín E. J. (2003). Phycoremediation: key issues for cost-effective nutrient removal process. *Biotechnology Advances*, **22**, 81–91.

Oswald W. J. (1988). Micro-algae and waste-water treatment. In: Micro-Algal Biotechnology, M. A. Borowitzka and L. J. Borowitzka (eds.), Cambridge University Press, 305–328.

Oswald W. J. (2003). My sixty years in applied algology. *Journal of Applied Phycology*, **15**, 99–106.

Palmer C. M. (1969). A composite rating of algae tolerating organic pollution. *Journal of Phycology*, **5**, 78–82.

Park J. B. K. and Craggs R. J. (2010). Wastewater treatment and algal production in high rate algal ponds with carbon dioxide addition. *Water Science and Technology*, **61**(3), 633–639.

Park J. B. K., Craggs R. and Shilton A. (2011). Recycling algae to improve species control and harvest efficiency from high rate algae pond. *Water Research*, **45**, 6637–6649.

Park J. B. K., Craggs R. J. and Shilton A. N. (2013). Investigating why recycling gravity harvested algae increases harvestability and productivity in high rate algae ponds. *Water Research*, **47**, 4904–4917.

Pittman J. K., Dean A. P. and Osundeko O. (2011). The potential of sustainable algal biofuel production using wastewater resources. *Bioresource Technology*, **102**, 17–25.

Posadas E., Morales M. M., Gomez C., Acien-Fernandez G. and Muñoz R. (2015a). Influence of pH and CO_2 source on the performance of microalgae-based secondary domestic wastewater treatment in outdoors pilot raceways. *Chemical Engineering Journal*, **265**, 239–248.

Posadas E., Serejo M., Blanco S., Perez R., Garcia-Encina P. and Muñoz R. (2015b). Minimization of Biomethane Oxygen Concentration during Biogas Upgrading in Algal-Bacterial Photobioreactors. *Algal Research*, **12**, 221–229.

Powell N., Shilton A. N., Pratt S. and Chisti Y. (2008). Factors Influencing Luxury Uptake of Phosphorus by Microalgae in Waste Stabilization Ponds. *Environmental Science & Technology*, **42**(16), 5958–5962.

Putt R., Singh M., Chinnasamy S. and Das K. C. (2011). An efficient system for carbonation of high-rate algae pond water to enhance CO_2 mass transfer. *Bioresource Technology*, **102**, 3240–3245.

Ruiz J., Álvarez-Díaz P. D., Arbib Z., Garrido-Pérez, Barragán J. and Perales J. A. (2013). Performance of a flat panel reactor in the continuous culture of microalgae in urban wastewater: prediction from a batch experiment. *Bioresource Technology*, **127**, 456–463.

Ruiz-Martinez A., Serralta J., Romero I., Seco A. and Ferrer J. (2015). Effect of intracellular P content on phosphate removal in Scenedesmus sp. Experimental study and kinetic expression. *Bioresource Technology*, **175**, 325–332.

Serejo M., Posadas E., Boncz M., Blanco S., Garcia-Encina P. and Muñoz R. (2015). Influence of biogas flow rate on biomass composition during the optimization of biogas upgrading in microalgal-bacterial processes. *Environmental Science & Technology*, **49**(5), 3228–3236.

Shen Y., Yuan W., Pei Z., Wu Q. and Mao E. (2009). Microalgae mass production methods. *Transactions of the ASABE*, **52**, 1275–1287.

Sierra E., Acién F. G., Fernández J. M., García J. L., González C. and Molina E. (2008). Characterization of a flat plate photobioreactor for the production of microalgae. *Chemical Engineering Journal*, **138**(1-3), 136–147.

Sompech K., Chisti Y. and Srinophakun T. (2012). Design of raceway ponds for producing microalgae. *Biofuels*, **3**(4), 387–397.

Torzillo G., Pushparaj B., Masojidek J. and Vonshak A. (2003). Biological constraints in algal biotechnology. Biotechnology and Bioprocess Engineering, **8**, 339–348.

Tredici M. R. (1999). Bioreactors, photo. In: Encyclopedia of Bioprocess Technology: Fermentation, Biocatal. Biosep., M. C. Flickinger and S. W. Drew (eds.), John Wiley & Sons.

Tredici M. R., Carlozzi P., Zittelli G. C. and Materassi R. (1991). A Vertical Alveolar Panel (Vap) for Outdoor Mass Cultivation of Microalgae and Cyanobacteria. *Bioresource Technology*, **38**, 153–159.

Uduman N., Qi Y., Danquah M. K., Forde G. M. and Hoadley A. (2010). Dewatering of microalgal cultures: a major bottleneck to algae-based fuels. *Journal of Renewable Sustainable Energy*, **2**, 012701–012715.

Wang B., Li Y., Wu N. and Lan C. Q. (2008). CO_2 bio-mitigation using microalgae. *Applied Microbiology and Biotechnology*, **79**, 707–718.

Weissman J. C. and Goebel R. P. (1987). Design and Analysis of Microalgal Open Pond Systems for the Purpose of Producing Fuels. Report SERI/STR-231-2840. Solar Energy Research Institute, Golden, CO.

Wolfaardt G. M., Lawrence J. R., Robarts R. D. and Caldwell D. E. (1994). The Role of Interactions, Sessile Growth, and Nutrient Amendments on the Degradative Efficiency of a Microbial Consortium. *Canadian Journal of Microbiology*, **40**, 331–340.

Wood A. (1987). Simple Wastewater treatment system incorporating the selective cultivation of a filamentous algae. *Water Science and Technology*, **19**, 1251–1254.

Zhang E., Wang B., Wang Q., Zhang S. and Zhao B. (2008). Ammonia–nitrogen and orthophosphate removal by immobilized *Scenedesmus* sp. isolated from municipal wastewater for potential use in tertiary treatment. *Bioresource Technology*, **99**(9), 3787–3793.

Chapter 5

Niches for bioelectrochemical systems in sewage treatment plants

Sebastià Puig, Juan A. Baeza, Jesús Colprim, Sarah Cotterill, Albert Guisasola, Zhen He, Elizabeth Heidrich and Narcis Pous

5.1 INTRODUCTION

The existence of bacteria that are able to transfer electrons to an electrode was first described in 1911 (Potter, 1911). Although the use of microorganisms for human purposes has a long-life, the use of electroactive microorganisms (able to use an electrode as electron donor/acceptor) for human devices is recent (Schröder, 2011). The different applications that explode the interactions microbe-electrode can be defined as Bioelectrochemical systems (BES) (Rabaey *et al.* 2009).

Normally, BES are composed of an anode and a cathode separated by an ion exchange membrane (Figure 5.1). In the anode compartment, oxidation reactions deliver protons to the media and electrons to the electrode. Protons diffuse to the cathode through an ion exchange membrane, while electrons are transferred by an electric connection. In the cathode, protons and electrons are consumed to carry out reduction reactions. Different applications have been found for bioelectrochemical systems depending on the reactions that occur in each compartment (Rabaey *et al.* 2009).

As an electrochemical process, the operation of each BES is determined by the potential at which each reaction occurs. If reactions with overall positive cell potentials are coupled (Equation 5.1), a thermodynamically favorable process is produced (Equation 5.2) that can generate electrical current.

$$E_{cell} = E_{cathode} - E_{anode} \tag{5.1}$$

where E_{cell} is the cell voltage (V); $E_{cathode}$ is the cathode potential (V) and E_{anode} is the anode potential (V).

$$\Delta G = -n \cdot F \cdot E_{cell} \tag{5.2}$$

where ΔG is the Gibbs free energy (J); n represents the number of electrons involved in the overall electrochemical process; F is Faraday's constant (96485 C · mole^{-1}) (V) and E_{cell} is the cell voltage (V).

If the resulting Gibbs free energy is a negative value, the process is favorable, and energy can be harvested in the form of electricity, which is the case of the so-called **microbial fuel cells (MFC)**. Otherwise it means that the process will not take place spontaneously; therefore energy will need to be applied to drive the process. These kind of systems are known as **Microbial electrolysis cells (MEC)** (Logan & Rabaey, 2012).

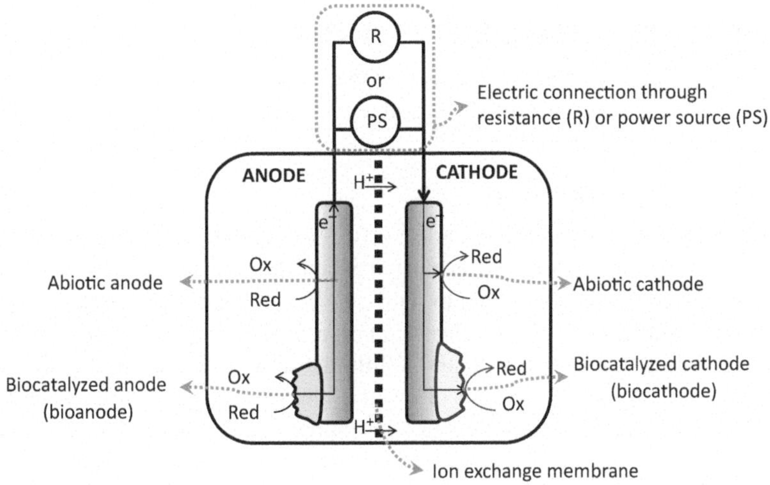

Figure 5.1. Basic schematic of a bioelectrochemical system.

5.1.1 Microbial fuel cells

Microbial fuel cell technology can deal with wastewater treatment while electricity is simultaneously generated (Liu *et al.* 2004; Logan *et al.* 2006). In an MFC, organic substrates are oxidised by exoelectrogenic bacteria, which produces electrons that are transferred to an anode and then flow to a cathode. The anode and cathode are linked by a conductive material containing a resistor. Protons produced at the anode migrate through the solution across a cation exchange membrane (CEM) to the cathode, where they combine with a reducible compound and electrons. MFCs offer the possibility of harvesting electricity from organic waste and renewable biomass (Rabaey & Verstraete, 2005; Lee *et al.* 2008). These are attractive sources of energy because they are carbon-neutral: the oxidation of the organic matter only releases recently fixed carbon back into the atmosphere (Lovley, 2006).

5.1.2 Microbial electrolysis cell

In a Microbial Electrolysis Cell (MEC), external energy is supplied to the system to allow a thermodynamically non-spontaneous process to occur or to stimulate a spontaneous reaction. Biocatalysed electrolysis is an electrolytic process that electrically connects the oxidation of organic material at a biological anode to the reduction of protons at the cathode so that hydrogen gas is formed. The anaerobic oxidation of organic carbon to carbon dioxide and hydrogen is thus split up into two half reactions: (i) the conversion of organic material into bicarbonate, protons, and electrons, and (ii) the conversion of protons and electrons to hydrogen gas.

The theoretical cell voltage (emf) of biocatalysed electrolysis of organic material is about −0.12 V (Rozendal *et al.* 2008). The negative value of this theoretical cell voltage indicates that electrical energy needs to be supplied to the system to drive the biocatalysed electrolysis reactions. This supply of electrical energy is accomplished by including a power supply into the electrical circuit. Essentially, dissolved organic material is thus electrolyzed into bicarbonate and hydrogen gas during the process with the electrochemically active microorganisms acting as the catalyst; hence the name biocatalyzed electrolysis.

5.2 BES IN SEWAGE TREATMENT PLANTS

Bioelectrochemical systems have been explored according to three main concepts: to produce energy from organic substrates, to generate/recovery products and to provide specific environmental services.

5.2.1 Bioelectricity production

Bioelectricity production is a feature that distinguishes BES from other wastewater treatment technologies, and thus must be properly understood and evaluated when developing BES for wastewater treatment. "Electricity" is a general term, which may refer to "current" (I), "voltage" (V), "power" (P), or "energy" (E).

The term "current" indicates the flow of electric charge. It indicates the amount of electrons that are flowing in a electric circuit at a certain moment. While the term "voltage", is the difference in electric potential energy between two points per unit of electric charge. The relation between both is ruled by Ohm's law (Equation 5.3), which includes another relevant parameter, the resistance of the system (R).

$$I = V/R \tag{5.3}$$

Hence, both current and voltage are important parameters to define an electrochemical system. In order to have an electric current flowing in a system, a voltage difference is required. However, in order to evaluate the rate at which the electric energy is being produced or transferred, another parameter should be included, the electric "power". Which is the outcome of both voltage and current (Equation 5.4)

$$P = V \cdot I \tag{5.4}$$

Finally, in order to know the energy produced from an electrochemical system, it is necessary to know the amount of time (t) through which a certain electric power has been produced (Equation 5.5)

$$E = P \cdot t \tag{5.5}$$

One must clearly understand the difference among those terms, because different applications can result in significantly different electricity generation in a BES. In general, there are two types of bioelectricity production, depending on the application purposes: high energy (power) output and high current generation. In BES, the strategy followed to obtain high energy production has been based on the usage of high external resistances, to generate high differences voltage between anode and cathode. On the contrary, another strategy has been allowing high current flows by using low external resistances.

High energy (power) output has been the main goal for BES development and for the majority of the relevant studies. However, "power output" has been mistakenly used to represent "energy production" for a long time (He *et al.* 2013). To properly demonstrate energy performance, the factor of time must be included, and thus a new parameter "normalized energy recovery" (NER) was proposed (Ge *et al.* 2014). NER can be expressed in either kWh m^{-3} (the volume of the treated wastewater) or kWh $kgCOD^{-1}$ (the amount of the removed organics, which it also points the efficiency of energy harvesting). BES can produce NER > 2.0 kWh m^{-3} without significant influence of reactor volume; however, most reported BES produced less than 1.5 kWh m^{-3} (Ge *et al.* 2014). Given the fact that an aerobic treatment process for domestic wastewater requires energy input of about 0.6 kWh m^{-3} (McCarty *et al.* 2011), application of BES may achieve energy neutral or even positive treatment. The highest NER based organic removal is nearly 2.0 kWh $kgCOD^{-1}$ while most results are lower than 1.0 kWh $kgCOD^{-1}$

(Ge *et al.* 2014). Hence, considering that the energy given by 1 mol of COD (1 mol O_2) is equivalent to 430 J, the highest NER achieved (2.0 kWh $kgCOD^{-1}$) meant an energy efficiency of 55%. While the most common result (1.0 kWh $kgCOD^{-1}$) implies an energy efficiency of 28%. The high NER is usually obtained with optimized conditions such synthetic substrates, pH buffer and/or controlled temperature. Energy production in a BES fed with real wastewater could be significantly lower. For example, a long-term study of MFCs operated with actual domestic wastewater and under a non-laboratory condition reports average NER lower than 0.03 kWh m^{-3} or 0.18 kWh $kgCOD^{-1}$ (Zhang *et al.* 2013). Therefore, evaluation of BES for energy recovery from wastewater will require more studies treating actual wastewater outside laboratory. A great challenge for application of the produced electricity is efficient extraction and transfer of electric energy (Wang *et al.* 2015). Despite the low energy production when treating actual wastewater, we should also recognize the advantage of BES in low energy consumption. Because of "anaerobic" treatment and omission of active aeration for organic matter removal, BES could consume less than 0.05 kWh m^{-3} (Zhang *et al.* 2013; Li *et al.* 2014), which makes it possible to achieve an energy-neutral treatment process. However, these calculations should be taken carefully, since in high energy producing BES, oxygen is required as cathode reagent. And, thus, aeration is needed in that compartment. Nevertheless, electricity (or energy) production in a BES needs to be further improved through optimized configuration/operation, materials/catalysts, and better understanding of microbial processes in order to achieve higher power densities.

Table 5.1 Summary of the reported MEC pilot plant performances.

Plant	I. Heidrich	II. Cotterill	III. Baeza		IV. Cusick	V. Escapa	
Applied voltage (V)	1.1	1.2	1.5		0.9	0.7	
Temp (°C)	1–22	3.7–19.4	25		31	19.2	
Operating mode	Continuous	Continuous	Continuous		Continuous	Batch	Continuous
Substrate	Domestic wastewater	Domestic wastewater	Glucose	Diluted crude glycerol	Winery wastewater	Domestic wastewater	
sCOD removal (%)	33	43.6	26.8	35.0	62	80 (average of 2 tests)	20 (average of 5 tests)
CE (%)	41.2	43	48.3	32.2	–	159 (average of 2 tests)	256 (average of 5 tests)
H_2 ($L \cdot d^{-1}$)	0.6	1	4.1	2.3	190 (86 ± 6% CH_4)	0.03	–
H_2 purity (%)	98.5	98.4	76.1	83	trace amount (86 ± 6% CH_4)	87	–
Energy efficiency (%)	49–70				1651	131	97
OLR ($g \cdot L^{-1} \cdot d^{-1}$)	0.47	0.66		0.5	0.7–2.0	0.06	0.27 (3)

High current generation is preferred when BES is used for removing target contaminants or producing valuable chemicals in its cathode. At a higher current, more electrons are flowing through the system, and thus, the oxidative or reductive turnover rates can be higher. As a result of electricity generation, electrons flow from an anode electrode to a cathode electrode, and those electrons can be used to reduce a certain compounds in the cathode. BES has been used to remove nitrogen (especially nitrate) and metals via reduction reactions (Kelly & He, 2014; Wang & Ren, 2014). Value-added compounds such as hydrogen gas, organic compounds, hydrogen peroxide can also be produced through electrochemical or microbial-electrochemical reactions in the cathode (Logan *et al.* 2008; Nevin *et al.* 2010; Rozendal *et al.* 2009). To achieve maximum removal or production of the target compounds, high current generation (and thus more electrons) will be preferred, and energy production under those conditions is not important.

5.2.2 Bioelectrochemical hydrogen production in WWTP

Wastewater treatment is currently an energy intensive process. Thus, research in this field should focus foremost on decreasing its energy requirements and then on recovering part of the chemical energy contained within it. Hydrogen production from industrial and municipal wastewater using bioelectrochemical systems (BES) is arousing scientific interest and could reduce simultaneously the high energetic and economic costs of wastewater treatment.

Bioelectrochemical hydrogen production from domestic or industrial wastewater has been successfully achieved at lab-scale with high energy recovery yields and high organic matter removal. However, these results have been obtained at very small scale (order of ml) and under carefully controlled lab conditions. Scaling-up BES is now the bottleneck, limiting the application of these systems. Few attempts have been made with varying results (Table 5.1; Figure 5.2). This section aims to review the preliminary efforts in scaling-up BES for hydrogen production and detail some of the experimental results obtained.

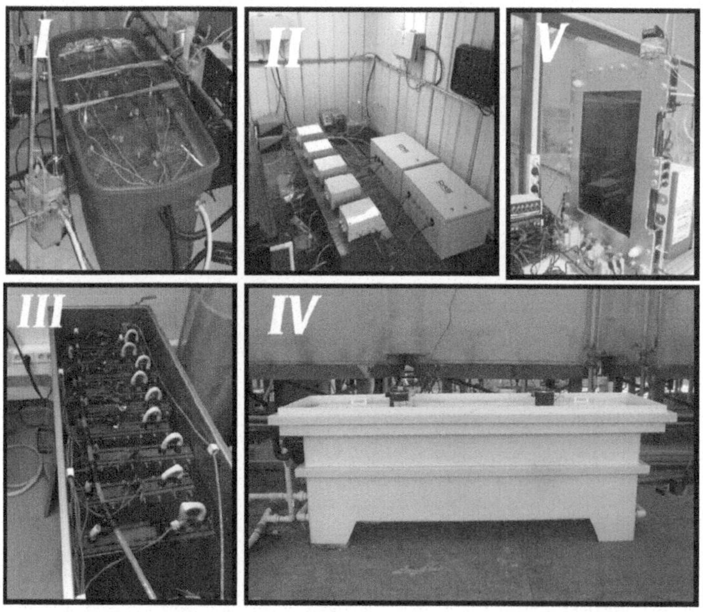

Figure 5.2 Pictures of the reported MEC pilot plants.

Reported pilot plant configurations:

I) **Heidrich *et al.* 2013, 2014 – Howdon Pilot, Northumbrian Water Ltd. (UK)**
A 100L cassette-design reactor with six identical cassettes was installed and run for 12 months at Howdon STW in the city of Newcastle-upon-Tyne, UK. An average of 246,500 m^3 of domestic wastewater is treated daily at this site, covering a population equivalent of 1,000,000. The wastewater was taken from the grit channels after screening, but before settlement, before being pumped through the reactor at 0.07 $L \cdot min^{-1}$ to give a HRT of 1 day. An applied voltage of 1.1 V was supplied to the cells using two bench top variable DC power supplies, and was recorded across a 0.1 Ω resistor using ADC-16 Pico data loggers. The temperature was not controlled, and wastewater tempertures fluctutated between 1 and 22°C throughout the year. In this work, an sCOD removal of 33% could be obtained, with and H2 production of 0.6 $L \cdot d^{-1}$.

II) **Cotterill *et al.*, unpublished – Fishburn STW Pilot, Northumbrian Water Ltd. (UK)**
A 130L cassette-design reactor with 10 identical cassettes was installed on a domestic wastewater treatment site in County Durham (Fishburn STW), and operated for 6 months. The site treats on average 1,000 m^3 of domestic wastewater per day. The wastewater was taken, at a rate of 0.09 $L \cdot min^{-1}$, from the open channel after screening and grit removal, prior to primary settlement. A voltage of 1 V was initially applied, but this was increased to 1.2 V after a week of limited gas production. Voltage was recorded across a 1 Ω resistor using ADC-20 and ADC-24 Pico data loggers. Temperature was not controlled and wastewater temperatures varied between 3.7 and 19.4°C between July and December. By applying these conditions, Cotterill and co-workers reached an sCOD removal of 43.6% and an hydrogen production of 1 $L \cdot d^{-1}$.

III) **Baeza *et al.*, 2017 – Universitat Autònoma de Barcelona (UAB) (Spain)**
A 130 L pilot plant was constructed in the university facilities which included ten cassette-like MECs. The cathodic chambers of the cassettes (5L each) were connected and the catholyte containing the hydrogen produced was cycled in a separate circuit. The plant was fed with a series of synthetic wastewaters with different degrees of complexity (acetate, glucose and diluted crude glycerol) for a period of 4 months. The average COD inlet concentration was 400 mg/L to mimic a real municipal wastewater scenario. The voltage added was 1.5 V and was recorded through a 12 Ω resistor. The pilot was operated at room temperature (around 25°C). The authors reached a sCOD removal of 26.8 and 35.0% when glucose and diluted crude glycerol was used as a substrate, respectively. The highest hydrogen production (4.1 $L \cdot d^{-1}$) was obtained when using glucose.

IV) **Cusick *et al.* 2011 – Napa Valley (USA)**
A 1000L reactor, with 24 electrode modules operating in parallel, was operated outdoors and fed winery wastewater. Each electrode module contained six anodes (thermally pre-treated graphite fibre brushes) and six cathodes (SS 304 mesh). Wastewater was fed at a rate of 1 $m^3 \cdot d^{-1}$ to the reactor. The reactor contained internal substrate recirculation and baffle plates to promote an optimum flow through the reactor. The voltage supplied was 0.9V and recorded through a 0.01 Ω external resistor. The reactor was operated during the harvest season (August to November) with a controlled temperature of 31 ± 1°C. The results obtained presented a high sCOD removal (62%), with a cathodic biogas production of 190 $L \cdot d^{-1}$.

V) **Escapa *et al.* 2015 – University of León (Spain)**
Two twin units working as single-chamber MECs were installed in the university facilities and fed with domestic wastewater. The wastewater was collected from Navalmorales WWTP in Toledo (Spain) where an average of 1,000 m^3 of domestic wastewater is treated daily for a population

equivalent of around 5,000. The wastewater was taken from the primary settler effluent and purged with nitrogen before entering the cell. During the start-up period, the wastewater was amended with 500 mg/L of acetate to enhance biofilm formation. Initially the MECs were run in batch mode conducting four tests in total; two with an applied voltage and two in open circuit mode. After this, both units were operated continuously, with HRTs ranging between 6 and 42 hours. The voltage added was 0.7 V and was recorded through a 16 Ω resistor. Temperature in the room was controlled to 19.2 ± 1.1°C. Under batch mode, the authors reported sCOD removal of 80%, with a hydrogen production of 0.03 $L \cdot d^{-1}$.

5.2.3 Bioelectrochemical denitrification in WWTPs

Bioelectrochemical systems can be an innovative tool to improve conventional technologies for dealing with nitrogen in wastewater treatment plants. Nowadays, two different strategies are being proposed, and both of them have been investigated using BES: nitrogen removal or nitrogen recovery.

5.2.3.1 Nitrogen removal in WWTPs using BES

In terms of nitrogen removal, two main applications could be evaluated. On the one hand, BES could be an alternative to the common secondary treatment of a WWTP by supporting nitrification-denitrification processes. On the other hand, BES could improve the already operating WWTPs by a mean of a polishing step to decrease nitrate and organic matter levels to standards for water reuse.

I) **Nitrification-denitrification in WWTPs**

One of the most ambitious objectives of BES for WWTPs would be the application of BES as an alternative to conventional nitritification-denitrification processes. In this application, ammonium is biologically oxidized to nitrate (nitrification) in an aerobic process, while nitrate is bioelectrochemically reduced to dinitrogen gas in the cathode by autotrophic bacteria. This application could be especially fruitful in those wastewaters with low COD/N ratio (but it has been evaluated for both high and low COD/N ratios), since the use of BES would avoid the requirements of external organic matter dosing. Moreover, the autotrophic nature of bioelectrochemical denitrification would reduce the sludge generation (and thus the cost associated to its disposal) as well.

Most commonly, BES for nitrification-denitrification are operated as MFC. The basis of this operation is organic matter oxidation at the anode and nitrate reduction at the cathode. But one of the key aspects is: where ammonium is oxidized to nitrate (nitrification)? In order to solve this issue different configurations have been reported in literature. The first approach evaluated the use of an external nitrifier reactor (Virdis *et al.* 2008; Vilajeliu-Pons *et al.* 2015). In such system, wastewater is firstly fed to the anode (organic matter is oxidized), then directed to the external nitrifying reactor (ammonium is oxidized to nitrate) and, finally, introduced to the cathode (nitrate is reduced to dinitrogen gas). Hence two reactors are needed for the treatment. A more compact alternative can be found in simultaneous nitrification-denitrification configuration (SND) (Virdis *et al.* 2010; Vilajeliu-Pons *et al.* 2015). In this configuration, the ammonium oxidation is promoted directly in the cathode by bubbling oxygen. According to a comparative study (Vilajeliu-Pons *et al.* 2015), both external nitrifier reactor SND configurations can present similar removal rates. However, in SND strategy, the nitrifier reactor can be suppressed, reducing the space required for the treatment.

In addition, other configurations have been tested. For example, Zhang and He (2012) proposed the usage of a double cathode: One aerobic cathode where ammonium is oxidized but, simultaneously, the excess of oxygen can be electrochemically reduced to get electricity; and a second, anoxic, cathode for bioelectrochemical nitrate reduction (Zhang & He, 2012). A less intensive treatment was also tested by Yan *et al.* (2012). In this case, the authors proposed an air-cathode MFC, where the oxygen diffused through the membrane to the anode was used by nitrifying bacteria to convert ammonium to nitrate. Finally, nitrate was bioelectrochemically reduced to dinitrogen gas using the cathode electrode (Yan *et al.* 2012).

II) **Denitrification in WWTPs as a polishing step**

The purpose of a tertiary treatment in urban wastewater treatment plants is to provide a final treatment stage to polish the effluent quality for reusing purposes. When polishing the nitrogen content, denitrification finds limitations on the low content of biodegradable organic matter in the secondary effluent. For this reason, dosage of organic matter is required. However, if autotrophic denitrification strategies are explored, this extra cost can be suppressed. In this niche of application, denitrifying BES could fit in. In denitrifying BES, different electrochemical configurations can be implemented: i) MFC or ii) MEC.

In a polishing step, where biodegradable organic matter is limited, MFC configuration would be restricted. While a MEC application, where external power is supplied, would be feasible. By applying external energy input, cathode nitrate reduction could be performed without requiring anode organic matter oxidation (Pous *et al.* 2015).

MEC application can be performed following two different strategies: i) two-electrode and ii) three-electrode arrangement.

In a two-electrode arrangement, a fixed cell voltage (Kondaveeti & Min, 2013; Kondaveeti *et al.* 2014; Lee *et al.* 2013) or a fixed current (Park *et al.* 2005; Sakakibara & Kuroda, 1993) is used, which implies the use of a device of low complexity. A DC power is enough to supply the energy required. But it can imply low treatment efficiency because the external energy input is continuously supplied, regardless of the denitrifying process performance.

In a three-electrode arrangement, the cathode potential can be imposed (poised) to enhancee bioelectrochemical denitrification (Gregory *et al.* 2004). It means that the cathode potential value is kept constant by a high complex apparatus, a potentiostat. It has been demonstrated that the cathode potential is a relevant parameter for the denitrification performance (Virdis *et al.* 2009; Cheng *et al.* 2012; Pous *et al.* 2015). By switching the cathode potential, the difference between the standard redox potential for nitrate reduction and the redox potential of the electron donor (the cathode electrode; working electrode) can be modified; thus, the energy gained by bacteria can be switched (Schröder, 2007). In consequence, the cathode potential value controls the whole denitrification process (Virdis *et al.* 2009; Cheng *et al.* 2012; Pous *et al.* 2015). As stated above, this operational strategy requires the use of a complex apparatus, a potentiostat. Thus, the capital cost is higher compared to a two-electrode arrangement strategy. Nevertheless, in a three-electrode configuration, the amount of energy supplied depends on the activity of the biocathode at the given cathode potential. Hence, only the energy required for denitrification is delivered. In consequence, the operational cost of a three-electrode arrangement can be lower than a two-electrode arrangement (Pous *et al.* 2015).

5.2.3.2 Nitrogen recovery in WWTPs using BES

In a context of a worldwide constriction of resources availability, the recovery of nutrients from WWTPs should be taken into account. In the case of nitrogen, BES has been demonstrated as a promising technology for recovering nitrogen, either in form of ammonia or struvite.

I) **Ammonia recovery**

Ammonium is a widely used fertilizer, and it has been conventionally obtained by the energy-intensive Haber-Bosch process. The energy costs associated to both Haber-Bosch process and nitrification-denitrification treatments could be minimized if ammonia was recovered from WWTPs and used as fertilizer. Following this objective, several studies have evaluated the recovery of ammonia from wastewater and, more specifically, from urine using BES (Ledezma *et al.* 2015). In order to recover ammonia, BES can be operated either as MFC (Kuntke *et al.* 2012; Ieropoulos *et al.* 2012) or MEC (Ledezma *et al.* 2015). The recovery of ammonia using BES is based on the charge neutrality principle, and requires the use of a cation exchange membrane. The electron flow between anode and cathode implies that cations (like ammonium NH_4^+) are forced to diffuse from the anode to the cathode compartment through the membrane. Once in the cathode, its high pH allows ammonium to be converted into ammonia (NH_3), which is highly volatile, and thus can be easily recovered through stripping (Kuntke *et al.* 2012).

Table 5.2 Comparison of the microbial electrochemical systems performances.

Plant	I. Heidrich	II. Cotterill	III. Baeza		IV. Cusick	V. Escapa	
Applied voltage (V)	1.1	1.2	1.5		0.9	0.7	
Temp (°C)	1–22	3.7–19.4	25		31	19.2	
Operating mode	Continuous	Continuous	Continuous		Continuous	Batch	Continuous
Substrate	Domestic wastewater	Domestic wastewater	Glucose	Diluted crude glycerol	Winery wastewater	Domestic wastewater	
sCOD removal (%)	33	43.6	26.8	35.0	62	80 (average of 2 tests)	20 (average of 5 tests)
CE (%)	41.2	43	48.3	32.2	–	159 (average of 2 tests)	256 (average of 5 tests)
H_2 ($L \cdot d^{-1}$)	0.6	1	4.1	2.3	190 (86 ± 6% CH_4)	0.03	–
H_2 purity (%)	98.5	98.4	76.1	83	trace amount (86 ± 6% CH_4)	87	–
Energy efficiency (%)	49–70				1651	131	97
OLR ($g.L^{-1} \cdot d^{-1}$)	0.47	0.66	0.5		0.7–2.0	0.06	0.27 (3)

Table 5.3 Comparison of the nitrogen removal performances.

Electrochemical Configuration	**MFC**	**MEC With Poised Cathode Potential**		
BES geometry	Rectangular	Rectangular	Rectangular	Tubular
Water	Groundwater		Synthetic wastewater	
Conductivity ($\mu S \cdot cm^{-1}$)	955 ± 121	918 ± 31	5293 ± 155	4199 ± 282
HRT (h)	11.9	5.0	16.8	0.6
NO_3^- consumption rate ($gN \cdot m^{-3}_{NCC} \cdot d^{-1}$)	47.6 ± 2.4	112.9 ± 8.0	97.4 ± 9.4	699.8 ± 7.0
NO_2^- accumulation (% reduced NO_3^-)	0.6 ± 0.6	0.0 ± 0.0	1.2 ± 1.6	0.0 ± 0.0
N_2O emissions (% reduced NO_3^-)	50	6.4 ± 8.2	13.4 ± 13.8	33.3 ± 9.5
Energy consumption (kWh gN-NO_3-removed^{-1})	–	–	$0.68 \cdot 10^{-2}$	–
Reference	(Pous *et al.* 2013)	(Pous *et al.* 2015b)	Pous *et al.*	Pous *et al.*

II) **Struvite precipitation**

Struvite is a crystalline solid composed of magnesium, ammonia and phosphate at equimolar concentrations ($MgNH_4PO_4 \cdot 6H_2O$). The precipitation of struvite in WWTPs would not only allow nitrogen, but also, phosphorus recovery. Phosphorus is an essential fertilizer, but it has been estimated that accessible phosphorus reserves could be depleted in the next 50 years (Gilbert, 2009). Hence, its recovery has become a research priority. In order to precipitate struvite, alkaline conditions are required, which are conventionally imposed by chemical additions (Doyle & Parsons, 2002). In BES, the reducing processes occurring in cathodes generate alkaline conditions in this compartment. Besides cathode basification is seen as a drawback in most of BES applications, it is worthy for struvite recovery (Ichihashi & Hirooka, 2012). The cathodic alkaline conditons can be provoked using different BES configurations. Accordingly, struvite has been recovered in a plethora of BES as air-cathode MFCs (Ichihashi & Hirooka, 2012; You *et al.* 2016), or single-chamber (Cusick & Logan, 2012) and double-chamber MEC (Cusick *et al.* 2014).

5.3 CONCLUSIONS

Bioelectrochemical systems advances in sewage treatment can improve its energy use and resource recovery. BES possesses a realistic economic potential. However, at the same time further fundamental research is needed and technological hurdles have to be taken to minimize methane production in anodes, hydrogen losses, electrochemical losses. As research and development on BESs is cross-disciplinary it highlights the need for truly integrated process development at all scales.

In terms of energy production, BES can be used for direct electricity production or for an indirect energy recovery through hydrogen or methane production. Both applications have a promising future. The removal or recovery of nutrients in BES has presented promising results. In terms of treatment, it could improve the economic viability of the current WWTP. While the recovery of nutrients would give a second chance to the nutrients presents in the "waste"-water, pushing WWTPs to the circular economy. Ideally, a BES for sewage treatment should couple both energy production and nutrient recovery or removal.

5.4 REFERENCES

Baeza J. A., Martínez-Miró À., Guerrero J., Ruiz Y. and Guisasola A. (2017). Bioelectrochemical hydrogen production at pilot scale from urban wastewater. Submitted to *Journal of Power Sources*.

Cheng K. Y., Ginige M. P. and Kaksonen A. H. (2012). Ano-cathodophilic biofilm catalyzes both anodic carbon oxidation and cathodic denitrification. *Environmental Science & Technology*, **46**(18), 10372–10378.

Cusick R. D. and Logan B. E. (2012). Phosphate recovery as struvite within a single chamber microbial electrolysis cell. *Bioresource Technology*, **107**, 110–115.

Cusick R. D., Bryan B., Parker D. S., Merrill M. D., Mehanna M., Kiely P. D., Liu G. and Logan B. E. (2011). Performance of a pilot-scale continuous flow microbial electrolysis cell fed winery wastewater. *Applied Microbiology and Biotechnology*, **89**, 2053–2063.

Cusick R. D., Ullery M. L., Dempsey B. A. and Logan B. E. (2014). Electrochemical struvite precipitation from digestate with a fluidized bed cathode microbial electrolysis cell. *Water Research*, **54**, 297–306.

Doyle J. D. and Parsons S. A. (2002). Struvite formation, control and recovery. *Water Research*, **36**(16), 3925–3940.

Escapa A., San Martin M.I. and Moran A. (2015). Scaling-up of membraneless microbial electrolysis cells (MECs) for domestic wastewater treatment: Bottlenecks and limitations. *Bioresource Technology*, **180**, 72–78.

Ge Z., Li J., Xiao L., Tong Y. and He Z. (2014). Recovery of Electrical Energy in Microbial Fuel Cells. Environmental Science & Technology Letters, **1**, 137–141.

Gilbert N. (2009). Environment: the disappearing nutrient. *Nature*, **461**(7265), 716–718.

He Z. (2013). Microbial fuel cells: now let us talk about energy. *Environmental Science & Technology*, **47**, 332–333.

Heidrich E.S., Dolfing J., Scott K., Edwards S.R., Jones C. and Curtis T.P. (2013). Production of hydrogen from domestic wastewater in a pilot-scale microbial electrolysis cell. *Applied Microbiology and Biotechnology.*, **97**(15), 6979–6989.

Heidrich E.S., Edwards S.R., Dolfing J., Cotterill S.E. and Curtis T.P. (2014). Performance of a pilot scale microbial electrolysis cell fed on domestic wastewater at ambient temperatures for a 12 month period. *Bioresource Technology*, **173**, 87–95.

Ichihashi O. and Hirooka K. (2012). Removal and recovery of phosphorus as struvite from swine wastewater using microbial fuel cell. *Bioresource Technology*, **114**, 303–307.

Ieropoulos I., Greenman J. and Melhuish C. (2012). Urine utilisation by microbial fuel cells; energy fuel for the future. *Physical Chemistry Chemical Physics: PCCP*, **14**(1), 94–98.

Kelly P. and He Z. (2014). Nutrients removal and recovery in bioelectrochemical systems: a review. *Bioresource Technology*, **153**, 351–360.

Kondaveeti S. and Min B. (2013). Nitrate reduction with biotic and abiotic cathodes at various cell voltages in bioelectrochemical denitrification system. *Bioprocess and Biosystems Engineering*, **36**(2), 231–238.

Kondaveeti S., Lee S.-H., Park H.-D. and Min B. (2014). Bacterial communities in a bioelectrochemical denitrification system: the effects of supplemental electron acceptors. *Water Research*, **51**, 25–36.

Kuntke P., Smiech K. M., Bruning H., Zeeman G., Saakes M., Sleutels T. H. J. A., Hamelers H. V. M. and Buisman C. J. N. (2012). Ammonium recovery and energy production from urine by a microbial fuel cell. *Water Research*, **46**(8), 2627–2636.

Ledezma P., Kuntke P., Buisman C. J. N., Keller J. and Freguia S. (2015). Source-separated urine opens golden opportunities for microbial electrochemical technologies. *Trends in Biotechnology*, **33**(4), 214–220.

Lee H. S., Parameswaran P., Kato-Marcus A., Torres C. I. and Rittmann B. E. (2008). Evaluation of energy-conversion efficiencies in microbial fuel cells (MFCs) utilizing fermentable and non-fermentable substrates. *Water Research.*, **42**(6–7), 1501–1510.

Lee S.-H., Kondaveeti S., Min B. and Park H.-D. (2013). Enrichment of Clostridia during the operation of an external-powered bio-electrochemical denitrification system. *Process Biochemistry*, **48**(2), 306–311.

Li J., Ge Z. and He Z. (2014). Advancing membrane bioelectrochemical reactor (MBER) with hollow-fiber membranes installed in the cathode compartment. *Journal of Chemical Technology and Biotechnology*, **89**, 1330–1336.

Liu H. and Logan B. E. (2004). Electricity generation using an aircathode single chamber microbial fuel cell in the presence and absence of a proton exchange membrane. *Environmental Science & Technology*, **38**(14), 4040–4046.

Logan B. E. and Rabaey K. (2012). Conversion of wastes into bioelectricity and chemicals by using microbial electrochemical technologies. *Science*, **337**(6095), 686–690.

Logan B. E., Aelterman P., Hamelers B., Rozendal R., Schröder U., Keller J., Freguia S., Verstraete W. and Rabaey K. (2006). Microbial fuel cells: methodology and technology. *Environmental Science & Technology*, **40**(17), 5181–5192.

Logan B.E., Call D., Cheng S., Hamelers H.V., Sleutels T.H., Jeremiasse A.W. and Rozendal R.A. (2008). Microbial electrolysis cells for high yield hydrogen gas production from organic matter. *Environmental Science & Technology*, **42**, 8630–8640.

Lovley D. E. (2006). Bug juice: harvesting electricity with microorganisms. *Nature Reviews Microbiology*, **4**(7), 497–508.

McCarty P.L., Bae J. and Kim J. (2011). Domestic Wastewater Treatment as a Net Energy Producer-Can This be Achieved? *Environmental Science & Technology*, **45**, 7100–7106.

Nevin K.P., Woodard T.L., Franks A.E., Summers Z.M. and Lovley D.R. (2010). Microbial electrosynthesis: feeding microbes electricity to convert carbon dioxide and water to multicarbon extracellular organic compounds. *MBio*, **1**.

Park H. I., Kim D. K., Choi Y.-J. and Pak D. (2005). Nitrate reduction using an electrode as direct electron donor in a biofilm-electrode reactor. *Process Biochemistry*, **40**(10), 3383–3388.

Potter M.C. (1911). Effects accompanying the decomposition of organic compounds. *Royal Society of London. Series B*, **84**, 260–276.

Pous N., Puig S., Balaguer M. D. and Colprim J. (2015). Cathode potential and anode electron donor evaluation for a suitable treatment of nitrate-contaminated groundwater in bioelectrochemical systems. *Chemical Engineering Journal*, **263**, 151–159.

Rabaey K. and Verstraete W. (2005). Microbial fuel cells: novel biotechnology for energy generation. *Trends in Biotechnology*, **23**(6), 291–298.

Rabaey K., Angenent L., Schröder U. and Keller J. (2009). Bioelectrochemical Systems: From Extracellular Electron Transfer to Biotechnological Application. International Water Association Publishing, London.

Rozendal R., Hamelers H.V.M., Molenkamp R. J. and Buisman C.J.N. (2007). Performance of single chamber biocatalyzed electrolysis with different types of ion exchange membranes. *Water Research*, **41**, 1984–1994.

Rozendal R.A., Leone E., Keller J. and Rabaey K. (2009). Efficient hydrogen peroxide generation from organic matter in a bioelectrochemical system. *Electrochemistry Communications*, **11**, 1752–1755.

Sakakibara Y. and Kuroda M. (1993). Electric prompting and control of denitrification. *Biotechnology and Bioengineering*, **42**, 535–537.

Schröder U. (2007). Anodic electron transfer mechanisms in microbial fuel cells and their energy efficiency. *Physical Chemistry Chemical Physics: PCCP*, **9**(21), 2619–2629.

Schröder U. (2011). Discover the possibilities: microbial bioelectrochemical systems and the revival of a 100-year–old discovery. *Journal of Solid State Electrochemistry*, **15**(7–8), 1481–1486.

Vilajeliu-Pons A., Puig S., Pous N., Salcedo-Dávila I., Bañeras L., Balaguer M. D. and Colprim J. (2015). Microbiome characterization of MFCs used for the treatment of swine manure. *Journal of Hazardous Materials*, **288**, 60–68.

Virdis B., Rabaey K., Yuan Z. and Keller J. (2008). Microbial fuel cells for simultaneous carbon and nitrogen removal. *Water Research*, **42**(12), 3013–3024.

Virdis B., Rabaey K., Yuan Z., Rozendal R. A. and Keller J. (2009). Electron fluxes in a microbial fuel cell performing carbon and nitrogen removal. *Environmental Science & Technology*, **43**(13), 5144–5149.

Virdis B., Rabaey K., Rozendal R. A., Yuan Z. and Keller J. (2010). Simultaneous nitrification, denitrification and carbon removal in microbial fuel cells. *Water Research*, **44**(9), 2970–2980.

Wang H. and Ren Z.J. (2014). Bioelectrochemical metal recovery from wastewater: a review. *Water Research*, **66**, 219–232.

Wang H., Park J.D. and Ren Z.J. (2015). Practical energy harvesting for microbial fuel cells: a review. *Environmental Science & Technology*, **49**, 3267–3277.

Wei Y., Van Houten R. T., Borger A. R., Eikelboom D. H. and Fan Y. (2003). Minimization of excess sludge production for biological wastewater treatment. *Water Research*, **37**(18), 4453–4467.

Yan H., Saito T. and Regan J. M. (2012). Nitrogen removal in a single-chamber microbial fuel cell with nitrifying biofilm enriched at the air cathode. *Water Research*, **46**(7), 2215–2224.

You J., Greenman J., Melhuish C. and Ieropoulos I. (2016). Electricity generation and struvite recovery from human urine using microbial fuel cells. *Journal of Chemical Technology & Biotechnology*, **91**(3), 647–654.

Zhang F. and He Z. (2012). Simultaneous nitrification and denitrification with electricity generation in dual-cathode microbial fuel cells. *Journal of Chemical Technology & Biotechnology*, **87**(1), 153–159.

Zhang F., Ge Z., Grimaud J., Hurst J. and He Z. (2013). Long-term performance of liter-scale microbial fuel cells installed in a municipal wastewater treatment facility. *Environmental Science & Technology*, **47**, 4941–4948.

Part 1

Reducing Requirements and Impacts

Part 1b: *Reducing Space*

Chapter 6

Aerobic granular sludge reactors

Claudio Di Iaconi, Janis Baeten, Paula ML Castro, Guido Del Moro, Jan Dries, Angeles Val del Rio, Mark van Loosdrecht and Eveline Volcke

6.1 INTRODUCTION

Aerobic granular sludge (AGS) technology is one of the major developments in the field of wastewater treatment of the past fifteen years. Granular sludge is a particulate type of biofilm that displays significantly different metrics and physical-chemical characteristics compared to activated sludge flocs occurring in conventional wastewater treatment plants. AGS is defined as aggregates of microbial cells which do not coagulate under reduced hydrodynamic shear and which subsequently settle significantly faster than activated sludge flocs (de Kreuk *et al.* 2005).

Aerobic granulation is a process of microbial cell-to-cell self-immobilization involving biological, physical, and chemical phenomena, which leads to densification of the aggregates. The process is activated by selective environmental pressures on microorganisms commonly found in wastewater treatment plants using an optimum 'window' of operating conditions (Lee *et al.* 2010).

The existence of a dissolved oxygen (DO) gradient and redox profile within granules leads to a layered structure where nitrifying, denitrifying and facultative anaerobic organisms can coexist. Therefore, due to this structure, AGS can simultaneously remove chemical oxygen demand (COD), nitrogen and phosphorous from wastewater (Pronk *et al.* 2015a).

AGS reactors have been extensively studied for treating a wide variety of industrial wastewater, toxic compounds and municipal wastewater thanks to their many advantages compared to the conventional activated sludge systems consisting in low foot-print, excellent biomass settleability, high and stable rates of metabolic activity, resilience to shocks and toxins due to the protection by layered structure, long biomass residence time, and the possibility for bioaugmentation. Although the high interest and the significant advantages of the aerobic granular bioreactors over the conventional technologies only a limited number of full scale applications have been reported so far after almost two decades from its first development. The overall aim of this chapter is therefore to analyze the granular sludge bioreactors in order to evaluate the maturity level of this technique and at the same time to identify the issues that have prevented or slowed its widespread application at full scale wastewater treatment plants.

6.2 APPLICATIONS OF AEROBIC GRANULATION

6.2.1 Industrial wastewater treatment

The low footprint and the possibility of simultaneous removal of organic matter, nitrogen and phosphorus in a single unit make the aerobic granular technology a compact and good option for treating industrial wastewater. Different types of industrial wastewater were treated with aerobic granular systems: dairy products, malting, abattoir, soybean process, brewery, fish canning, winery, pig manure, palm oil mill, seafood, textile, leachates, soy sauce, pharmaceutical, etc. Table 6.1 summarizes the performances of AGS systems in treating selected industrial effluents. Looking at data reported in this table it is possible to notice that aerobic granules can be obtained independently of pollutants and organic loading rates. In fact, the types of industrial wastewater succesfully tested are numerous and somewhat different. Furthermore, aerobic granules were succesfully formed by applying organic loading rates from 1 to 7 kg COD/m^3 · d. Large variations in generating time of granular biomass can be also noted among the applications listed in Table 6.1. In fact, granules were formed after only a few days (Cassidy & Belia, 2005; Su & Yu, 2005) to several months (Schwarzenbeck *et al.* 2004; Lotito *et al.* 2014). Pre-cultured granules or synthetic wastewater, however, were used for starting-up the reactors in some cases (Yilmaz *et al.* 2008; Lopez-Palau *et al.* 2009; Di Iaconi *et al.* 2011). In all the experiences listed in Table 6.1 large soluble COD removals were achieved. However, the overall COD removal efficiency was often limited by the high concentration of suspended solids (coming from floccular biomass) in the effluent.

Nutrients removal (N and P) was not always significant as squencing batch reactor (SBR) cycle was not optimized. In fact, an anaerobic feeding phase followed by an aerated phase are needed for simultaneous COD, nitrogen and phosphorous removal in a single reactor. The role of the anaerobic feeding phase is the stimulation of slow-growing polyphosphate- and glycogen-accumulating organisms (PAO and GAO, respectively) that accumulate storage polymers which are subsequently consumed in the aerated phase to perform P-uptake coupled to denitrification (de Kreuk & van Loosdrecht, 2004). Such a similar anaerobic-aerobic sequence was adopted only in a limited number of studies listed in Table 6.1. The SBR cycle applied by Cassidy and Belia (2005) for the treatment of abattoir wastewater, consisted of 120 min mixed anaerobic fill followed by 220 min of aerated reaction. Yilmaz *et al.* (2008) applied an anaerobic phase of almost 80 min (including feeding time) followed by 400 min of aeration, for the treatment of nutrient-rich abattoir wastewater. Kishida *et al.* (2009) used a 90 min anaerobic phase followed by 120 min aeration to treat livestock wastewater high in N and P. In these studies, nitrogen removal relies on simultaneous nitrification-denitrification, where the denitrification is executed by denitrifying PAO (DPAO) (Yilmaz *et al.* 2008). In order to have a near-complete nitrogen removal from the wastewater, and to avoid the presence of nitrate at the start of the next anaerobic phase, a post-anoxic phase of 80 to 140 min was included in the SBR processes operated by Yilmaz *et al.* (2008) and Kishida *et al.* (2009).

In contrast to these studies, most of the industrial entries in Table 6.1 do not include an anaerobic phase in the SBR cycle, they instead applied a pre-denitrification phase in the process during and after the rapid feeding phase. Wang *et al.* (2007a) and Liu *et al.* (2015b) applied an anoxic/aerobic sequence of approximately 40 min/300 min for brewery and slaughterhouse wastewater, respectively. Arrojo *et al.* (2004) included an initial 30 min anoxic phase for treatment of dairy wastewater. Val del Río *et al.* (2013) operated various SBR systems without anoxic phase, but relied on the high initial COD concentration in the reactor to allow denitrification at the start of the cycle. As a result, nitrates (and/or nitrites) accumulate at the end of the SBR cycles in many of the industrial laboratory scale reactors and contribute to total nitrogen species in the effluent.

These observations highlight a major conceptual difference between many of the industrial applications at laboratory (and also at pilot scale) and the Nereda process developed by Royal Haskoning DHV that is currently operated at full scale for the treatment of domestic wastewater (Pronk *et al.* 2015a). Most importantly, it seems that the potential of the simultaneous removal of carbon (C), nitrogen (N) and phosphorous (P) from industrial wastewater is not fully exploited yet.

Table 6.1 Performances of lab-scale AGS reactors with selected types of industrial wastewater.

Type of Wastewater	OLR_{max} (kg COD/$m^3 \cdot d$)	Removal Efficiency (%)			Effluent TSS (mg/L)	Time for Granulation (d)	Granule Diameter (mm)	References
		COD[a]	N	P				
Dairy products	7.0	90	70	–	50–800	21	3.5	Arrojo *et al.* 2004
Malting	3.2	80	–	–	–	7	–	Schwarzenbeck *et al.* 2004
Abattoir	2.6	99	97	98	43 (as VSS)	4	1.7	Cassidy and Belia, 2005
Dairy plant	5.9	90	80	67	500 (as VSS)	105	–	Schwarzenbeck *et al.* 2005
Soybean-processing	6.0	98.5	–	–	–	15	1.2	Su and Yu, 2005
Brewery	3.5	88.7	88.9	–	–	63	2–7	Wang *et al.* 2007b
Fish canning	1.7	95 (as TOC)	40	–	30–100	75	3.4	Figueroa *et al.* 2008
Abattoir	2.7	68	86	74	306	Pre-cultured granules	0.7–1.6	Yilmaz *et al.* 2008
Winery	6.0	95	–	–	<100 (as VSS)	40 (with synthetic wastewater)	2.0	Lopez-Palau *et al.* 2009
Pig manure	4.4	87	70	–	1000 (as VSS)	25	5	Figueroa *et al.* 2011
Palm oil mill	6.0	90	–	–		110	0.9	Gobi *et al.* 2011
Seafood	2.0–4.4	90	30	–	300–700 (as VSS)	130	2–11	Val del Rio *et al.* 2013
Soy sauce	5.3	87	76 (as NH_3)	–	–	30	2.0–2.5	Harun *et al.* 2014
Slaughterhouse	2.8	95.1	99.3 (as NH_3)	83.5	–	90	0.6–1.8	Liu *et al.* 2015b
Pharmaceutical industry	5.5	80	–	–	260–350 (as VSS)	15	–	Inizan *et al.* 2005
40% municipal + 60% industrial		80	50	–	20–100	400	0.2–0.9	Liu *et al.* 2010
Mixed municipal-textile	2	82	87.5	–	4	–	–	Lotito *et al.* 2014
Dyeing textile	2.6	78	66		28			Lotito *et al.* 2012
Stabilized municipal leachate	1	54	99		50	Pre-cultured granules		Di Iaconi *et al.* 2011
Livestock	0.2 (as TOC)	97	99.7	99		Pre-cultured granules		Kishida *et al.* 2009

[a]Removal of soluble COD is reported in some cases.

6.2.2 Municipal wastewater treatment

Municipal sewage is generally characterized by lower COD and nutrient concentrations compared to industrial wastewater. The application of AGS systems for treating low-strength wastewater (such as domestic ones) requires long start-up periods since organic load values lower than 1 kg COD/m^3 · d are not particularly indicated for aerobic granules formation (de Kreuk & van Loosdrecht, 2006). The volume exchange ratio and the settling time of the SBR were found to be two key factors with this type of wastewater (Ni *et al.* 2009).

The application of AGS systems for municipal wastewater treatment was first carried out on synthetic urban wastewater and successively on real municipal wastewater.

Lashkarizadeh *et al.* (2015) found that changing the carbon source from readily biodegradable acetate to a municipal wastewater does not have major impacts on aerobic granular sludge characteristics for a 5.7 L SBR reactor. Nonetheless, a readily biodegradable carbon source is needed to obtain biological phosphorus and nitrogen removal. Also, Long *et al.* (2014) have inoculated a SBR (9 L working volume), treating synthetic wastewater, with a mature aerobic granular sludge. They added 1–3 g of sodium acetate at each cycle in order to obtain nitrogen removals above 90%. Also, removal rates of COD, phosphorous and ammonia were 95%, 88% and 99%, respectively.

Wagner *et al.* (2015) found, however, that the performances of a granular biomass reactor treating domestic wastewater were lower than those obtained with a synthetic wastewater. In fact, after granule formation (160 days start-up), nitrogen was principally removed by partial nitrification up to nitrite, followed by denitrification via nitrite assimilation with efficiencies of 86.6%, 59.5% and 60.5% for nitrification, denitrification and total nitrogen, respectively.

Also, Liu and Dong (2011) studied the COD and nitrogen removal from domestic wastewater (collected from the drainage system of a residential area) but differently they employed a micro-aerobic granular sludge reactor having 8 L working volume. The reactor showed stable and effective COD removals during the whole experimental period. Nevertheless, nitrogen removal depended on the operation mode. The optimal settings were 0.20 kg/m^3 · d, 8 h, 11–16 g/L and below 0.10 kg COD/kg$_{MLSS}$ · d for oxygen influx, HRT, MLSS and sludge loading rate, respectively. Also, Wang *et al.* (2009) dealt with granular biomass coming from a lab scale SBR reactor (8 L working volume), treating real domestic wastewater (from a University campus). They found that, even if COD, ammonia an phosphorus in the influent fluctuated, the SBR maintained stable removal performances for COD, TOC, phosphate, NH_3 and TN, with average removal efficiencies of 80%, 70%, 71%, 92%, and 47%, respectively. Moreover, granule analysis showed good simultaneous nitrogen and phosphorus removal performances.

Verawaty *et al.* (2010) obtained a fully granular system from floccular sludge after 50 days of operation with domestic wastewater by progressively decreasing the settling time and HRT of the SBR. The SBR was running for 160 days during which the granules remained stable. Removal efficiencies of 99% and 98% were obtained for nitrogen and phosphorous, respectively, at HRT of 9.6 h thus demonstrating the potential of granular biomass system for achieving good biological nitrogen removal during the treatment of domestic wastewater.

6.2.3 Toxic compounds degradation and biosorption of dyestuffs and heavy metals

The treatment of wastewater containing toxic compounds is also possible with AGS reactors as the layered structure of the granules creating concentration gradients protects microorganisms from the impact of direct acute toxicity associated with these compounds (Maszenan *et al.* 2011). Degradation of typical toxic pollutants occurring in wastewaters such as phenol (Ho *et al.* 2010) and different types of phenol derivates like nitrophenols and chlorinated phenols (Maszenan *et al.* 2011) has been demonstrated. Table 6.2 reports some lab-scale applications for removing selected organic toxic compounds by biodegradation.

Table 6.2 Degradation of selected organic toxic compounds by AGS reactors.

Compound	Inlet Substrate Concentration/Organic Load	Removal Efficiency (%)/ Degradation rate	References
Phenol	1.8 $kg_{phenol}/m^3 \cdot d$	49 $g_{phenol}/kg_{VSS} \cdot h$	Adav *et al.* 2007
Aniline	10–500 mg/L	over 90% 37.5 $g/m^3 \cdot d$	Dai *et al.* 2015
2-Chloroaniline, 3-chloroaniline, 4-chloroaniline and 3,4-dichloroaniline mixture	20–450 mg/L (total inlet concentration; equal amounts of each chloroaniline)	50–100%, 40–100%, 10–100% and 10–100%, respectively	Zhu *et al.* 2008
2-Fluorophenol	25–50 mg/L	up to 100%	Duque *et al.* 2011
4-Chloroaniline	20–200 mg/L	40–98%	Zhu *et al.* 2013
4-Chlorophenol	<50 mg/L	18–100% 3.40 $g/kg_{VSS} \cdot h$	Carucci *et al.* 2009
4-Nitrophenol	<40.1 mg/L	19.3 $g/kg_{VSS} \cdot h$	Yi *et al.* 2006
Methyl tert-butyl ether	25–500 mg/L	0.544 $kg_{MTBE}/kg_{VSS} \cdot d$	Zhang *et al.* 2008
Nitrobenzene	50–1000 mg/L	30–100%	Zhao *et al.* 2011
tert-Butyl Alcohol	100–600 mg/L	7.4 $g_{TBA}/kg_{VSS} \cdot h$	Tay *et al.* 2005
2,4-dichlorophenol	50–100 mg/L	94% 39.6 $g_{2,4\text{-}DCP}/kg_{VSS} \cdot h$	Wang *et al.* 2007b
2,4,6-trichlorophenol	0–15 mg/L 0.05–0.07 kg $COD/m^3 \cdot d$	100% 0.45–2.36 $g_{TCP}/kg_{VSS} \cdot h$	Carucci *et al.* 2008
o-cresol	100 mg/L 0.11 $kg_{o\text{-}cresol}/m^3 \cdot d$	up to 100%	Jemmat *et al.* 2014

Source: Adapted from Amorim, 2013.

Furthermore, due to the high surface area and porosity, aerobic granules can be a good alternative to remove heavy metals and dyes, by bioaccumulation or adsorption, being the maximum adsorption density of aerobic granules three times greater than that of sludge flocs. In fact, the removal by biosorption of heavy metals such as Cd^{+2}, Co^{+2}, Cu^{+2}, Ni^{+2} and Zn^{+2} has been successfully obtained with aerobic granular biomass (Liu *et al.* 2003; Sun *et al.* 2008a; Xu & Liu, 2008). Accumulation of heavy metals in the granules, however, can affect the stability of the granules and process. For example, Marques *et al.* (2013) report that levels of Zn^{2+} as high as 100 mg/L were shown to affect the nutrient removal in the granular sludge reactor, although the effluent still generally complied with admissible legal values concerning organic matter, nitrogen and Zn^{2+}. Furthermore, they also observed that resuming the process to lower levels in the feed led to recovery of the reactor performance. The adsorption of dyes such as Malachite Green, Eriochrome Black T, Reactive Brilliant Blue, Congo Red and Reactive Brilliant Red (Sun *et al.* 2008b; Hailei *et al.* 2010) has also been achieved. More recently, the removal through microbial degradation of an azo dye in a granular sludge reactor was demonstrated with a feed of synthetic textile wastewater supplemented with Acid Red 14 at concentrations of up to 60 mg/L (Franca *et al.* 2015).

In respect to endocrine disrupter compounds (EDCs), Balest *et al.* (2008) compared the performance of an aerobic granular system with that of activated sludge process for the removal of five different EDCs obtaining for all the compounds a better removal efficiency with aerobic granular biomass. Removal of pharmaceuticals and personal care products (PPCPs) (ibuprofen, naproxen, prednisolone, norfloxacin, and sulfamethoxazole, supplied at 50 μg/L) in a AGS reactor was also demonstrated to occur through mechanisms of adsorption to granular biomass although degradation was also hypothesized (Zhao *et al.* 2015). Adsorption was also shown to be a mechanism for removal of fluoroquinolone (FQ) antibiotics ofloxacin, norfloxacin and ciprofloxacin fed at the μM range to a granular SBR reactor (Amorim *et al.* 2014). In this latter study, the SBR was intermittently fed with the FQs for a one-year period. It is noteworthy that the removal of COD was not affected during the shock loadings of the FQs. Furthermore, activity of ammonia oxidizing bacteria and nitrite oxidizing bacteria did not seem to be inhibited by the presence of FQs although during the FQs feeding nitrate accumulation was observed at the effluent suggesting that denitrification was inhibited. A decrease on P removal capacity was also observed. In the long run, exposure to FQs promoted granules disintegration leading to an increase of the effluent solid content. A study focused on the impact of tetracycline (TC) on the performance of reactors with either nitrifying granules or conventional granules has shown that the presence of TC in the feed, although at high concentrations of 10 mg/L, did not affect greatly the latter ones, whereas the nitrifying ones broke down into smaller fractions with consequent nitrite accumulation (Shi *et al.* 2009). In this case the granules produced more exopolymeric substances (EPS) in order to protect themselves from the toxicity of the compound.

When unacclimated granular sludge is not able to readily degrade compounds exhibiting high resistance to biodegradation, as microorganisms present in the native sludge do not always maintain the capacity to degrade pollutants, bioaugmentation of the granules with specialized strains can become an alternative. Bioaugmentation can have some drawbacks which may interfere with its successful application: (i) toxic and/or recalcitrant (micro)pollutants may appear intermittently and/or may be present at very low concentrations, leading to the loss of degrading capacity by the specialized strain, due to the absence of the selective pressure; (ii) the presence of protozoa can affect the success of the bioaugmentation, as the added culture can be grazed by these microorganisms. Nevertheless, the efficacy of bioaugmenting granular sludge with cultures degrading toxic compounds has been observed. Bioaugmentation of granular sludge in a SBR reactor with a specialized bacterial strain able to degrade 2-Fluorophenol (2-FP), a Rhodococcus sp. strain FP1 was reported to lead to full degradation of the compound in a synthetic feed with a stoichiometric release of fluoride from the compound. The 2-FP degrading strain was successfully retained by the aerobic granules, as shown through the recovering of the strain from the granular sludge after more than one year operation, corroborating that granular sludge is capable to incorporate and retain specialized degraders. Overall, the granular SBR was shown to be robust, exhibiting a high performance after bioaugmentation with the 2-FP degrading strain in terms of nutrient removal and 2FP degradation. Furthermore, in the bioaugmented reactor DGGE profiles have shown a wide bacterial diversity with an even distribution of species, and the granular sludge microbial community was quite stable along time, with the dominant bacterial group present being Proteobacteria (Duque *et al.* 2015), demonstrating that a stable reactor performance does not necessarily implies a stable microbial community and vice versa. In the latter case, the stability might have been influenced by the selective pressure exerted by the toxic compound 2-FP. Recently different works have been addressed to the use of selected functional strains as inoculum to cultivate aerobic granules able to treat specific toxic compounds like: 2,4-dichlorophenoxyacetic acid (Quan *et al.* 2010), phenol (Adav *et al.* 2007), pyridine (Liu *et al.* 2015a) and azo-dyes (Kee *et al.* 2015).

6.3 SCALE-UP: FROM THE LAB TO FULL SCALE

AGS reactors have been extensively applied at the laboratory scale for treating different types of wastewater. During the last decade the process has started to be scaled-up to pilot and full scale. Particular attention has been paid to the operation of the plant as the scale-up of granular systems leads to modification of the hydrodynamic conditions, which are very important for the formation and maintenance of aerobic granules stability. Furthermore, some aspects like the way to obtain a quickly start-up and then to speed up the generation of granular biomass have been also investigated.

The first pilot plant based on aerobic granular technology was built and operated in the Netherlands for the treatment of municipal wastewater (de Bruin *et al.* 2005). The plant was designed for simultaneous organic matter, nitrogen and phosphorous removal from municipal wastewater using two parallel bubble columns with a diameter of 0.6 m and height of 6 m. Several operational philosophies were tested to figure out the conditions which lead to the granulation with municipal wastewater as a substrate. Fast formation of granules was observed under conditions of extensive COD removal, extensive biological phosphate removal and low nitrate effluent concentrations.

Ni *et al.* (2009) operated with a reactor of 1 m^3 to treat low strength municipal wastewater (lower than 200 mg COD/L). The first small granules appeared after 80 days from plant start-up. After 300 days of operation, the biomass consisted by 80% of granules having a diameter between 0.2–0.8 mm and a settling velocity of 18–40 m/h. The plant showed COD and ammonia removal efficiencies of 90% and 95%, respectively.

With a wastewater composed of 40% domestic and 60% industrial wastewater Liu *et al.* (2010) needed about 400 days to obtain granule-dominant sludge in the pilot-scale SBR. The plant was able to remove more than 80% and 98% of COD and ammonia content, respectively.

Morales *et al.* (2013) have observed a very quick formation of granular biomass which remained stable during swine slurry treatment at pilot scale. They obtained interesting COD and ammonia removal efficiencies whereas nitrogen and suspended solids removals were poor.

A successful pilot scale application for treating municipal wastewater has been also reported by Rocktäschel *et al.* (2015). The authors have added, however, an easily degradable organic carbon (i.e., acetate) until stable granulation was obtained. They also used an anaerobic phase followed by an aerobic one at the beginning of each SBR according to what reported in section 6.2.1. In order to reduce the start-up time requested by AGS systems it is also possible to apply a strategy similar to the one usually applied in anaerobic reactors that is based on the use of pre-cultured granules seeded into the reactor (Liu *et al.* 2005; Verawaty *et al.* 2012).

A new granular biomass system, known with the acronym SBBGR (Sequencing Batch Biofilter Granular Reactor) has been developed during the last decade by the Water Research Institute of the Italian National Research Council. In this system the granules are not suspended but retained by a filling material. The whole biomass consisting in a mixture of biofilm and granules is completely confined in a dedicated zone of the reactor known as the bed, and a secondary settler is therefore no longer necessary. This feature gives a great advantage in terms of granular biomass stability since a sedimentation phase is no longer required. This system has been scaled-up to pilot scale for treating tannery and municipal primary effluents (Di Iaconi *et al.* 2010a; Di Iaconi *et al.* 2010b). In the case of raw municipal wastewater treatment at pilot scale (see Figure 6.1), 7 months were needed for obtaining a biomass with 50% of granules. Furthermore, the results obtained after a long term operation of 463 days have shown that the plant was able to remove more than 90% of COD, TSS and TKN content from a raw municipal sewage. Nitrogen removal efficiencies of 80% were also achieved thanks to the establishment of simultaneous nitrification-denitrification process favored by the plant's high biomass concentration (Di Iaconi *et al.* 2014). Furthermore, SBBGR treatment followed by sand filtration was able to produce an effluent suitable for agriculture reuse. The first full scale plant based on SBBGR technology is under construction in Apulia (Italy) by retrofitting an existing activated sludge basin. The techno-economic evaluation of the upgrading of an existing wastewater treatment plant based on activated sludge process (15,000 PE equipped with a sludge aerobic stabilization stage) with an SBBGR technology has shown that converting the traditional biological section into an SBBGR allows to achieve a drastic reduction in sludge production (up to 70%), area requirement (up to 50%) and treatment costs (up to 40%) (Di Iaconi *et al.* 2016).

Figure 6.1 Picture of SBBGR pilot plant.

AGS technology has been also scaled up to full scale. Several full-scale AGS plants have been built in the Netherlands, Portugal, South Africa and China in the recent years (almost all based on Nereda® technology) although only general information has been reported in the scientific literature. However, more recently the performances and operating conditions of two full scale AGS plants have been reported (Pronk *et al.* 2015a; Li *et al.* 2014). The first refers to the upgrading of Garmerwolde WWTP, the Netherlands. In particular, the AGS plant, consisting in two reactors having a height of 7.5 m and volume 9,600 m^3 each operating in parallel with the existing plant based on activated sludge process, was designed for treating 41% (28,600 m^3/d) of the total influent received at Garmerwolde WWTP. The authors (Pronk *et al.* 2015a) report the complete achievement of granular biomass after about 5 months from plant start-up. The biomass consisted for more than 60% of granules larger than 1 mm. Nutrient removal was already obtained during plant start-up although not optimized yet. In addition to the good performances obtained operating at HRT of 17 h, the authors report a significant lower energy consumption of AGS plant compared to the conventional activated sludge treatment plant in the Netherlands. The second full scale experience whose detailed operating informations are reported in the scientific literature refers to the upgrading of Yancang WWTP in Haining, a coastal city in Eastern China (Li *et al.* 2014). In this case, a full scale granular sludge SBR (consisting of four parallel reactors of 12,540 m^3 each; length/width/depth: 55/38/6 m) was built for treating 50,000 m^3/d of wastewater (30% domestic and 70% industrial wastewater). Differently from Nereda plant described by Pronk *et al.* (2015a), in this case the AGS plant was fed with the primary effluent and after a hydrolysis step. Furthermore, the effluent quality of AGS plant was improved by a clariflocculation step. Li *et al.* (2014) report the formation of granular biomass (granules with a diameter of 0.5 mm) from activated sludge after only 30 days from plant start-up. Furthermore, the biomass having a SVI of about 43 mL/g remained stable for the next 310 days. The percentage of granular biomass fraction is however not reported as well as the performances of the plant during the start-up period. The data reported by the authors show COD and ammonia removal efficiencies similar to those of an anaerobic/oxic plug flow (A/O) reactor and an oxidation ditch (OD) operating in the same WWTP and under similar organic loading rates (i.e. around 0.6 kg COD/m^3 · d). Granular sludge plant showed, however, a better nitrogen removal efficiency thanks to the presence of a simultaneous nitrification and denitrification process, although the absence of a planned anoxic phase in the treatment cycle of the full SBRs. Finally, phosphorous removal was not obtained due to the absence of an anaerobic phase.

6.4 CRITICAL ASPECTS

The stability of the granular structures at long term operation is one of the key factors for a successful full scale implementation of AGS systems and that is actually the issue that prevented or slowed their widespread application. Liu and Liu (2006) attributed it to the overgrowth of filamentous microorganisms whereas Zheng *et al.* (2006) to an intracellular protein hydrolysis and degradation at the anaerobic granule core. Adav *et al.* (2010) demonstrated that under a high organic load rate (OLR) the microorganisms lose their capability for autoaggregation due to a reduction in the quantity of secreted protein. Lee *et al.* (2010) reviewed the problem of granules stability and indicated the following phenomena responsible: (1) outgrowth of filamentous organisms; (2) hydrolysis of the anaerobic core; (3) function loss of strains; (4) role of EPS; and (5) other mechanisms. These authors proposed the following strategies to enhance granules stability at long-term operation: (1) applying appropriate operational conditions; (2) selection of a slow-growing organism; (3) suppressing activity of anaerobes; and (4) strengthening granule core.

Figure 6.2 shows some examples of loss granules stability. The most common cause of granules destabilization is the proliferation of filamentous organism in their surface, normally provoked by an insufficient supply of oxygen. The limitation of the dissolved oxygen concentration can be due to its low level in the liquid phase (Mosquera-Corral *et al.* 2005; Liu *et al.* 2010) or by an increase in the organic load applied (Isanta *et al.* 2012). The increase of the organic load applied to the reactor is very common in the case of industrial effluents, since they are characterized by high organic contents and variable composition which could affect the stability of aerobic granules. Inizan *et al.* (2005) increased the OLR from 3 to 5.5 kg/m^3 · d with a pharmaceutical effluent and observed the development of filamentous bacteria on the granule surface. As consequence the granules were rapidly transformed in big pellets that were not settling anymore but were floating, inducing a rapid wash out of all biomass and the instability of the process. In a study aimed at evaluating the influence of the increase in organic load applied (3 to 13 kg COD/m^3 · d) on granules stability by using the effluent from a seafood industry, Val del Rio *et al.* (2013) have found that, despite high COD removal efficiency (about 90%), the granules disintegrated at OLRs higher than 4.4 kg COD/m^3 · d due to the high size and the oxygen transfer limitation. These experiences seem to indicate that each type of substrate has its maximum OLR without affecting the granule stability. Therefore, at full scale application when a peak of OLR can occur this need to be avoided or mitigate in order to maintain the granular sludge structure.

De Kreuk and van Loosdrecht (2004) reported that the selection of slow-growing organisms such as phosphate or glycogen accumulating bacteria can enhance granule stability at low DO concentrations.

Verawaty *et al.* (2012) have found that for each operating conditions set (e.g., mixing, reactor geometry, aeration, shear forces, wastewater characteristics, etc.) the granules in the reactor approaches towards a certain critical size beyond which their breakage occurs with consequent impact on reactor performance and final effluent quality. These authors highlight the importance of biomass washout management for obtaining a stable granular sludge.

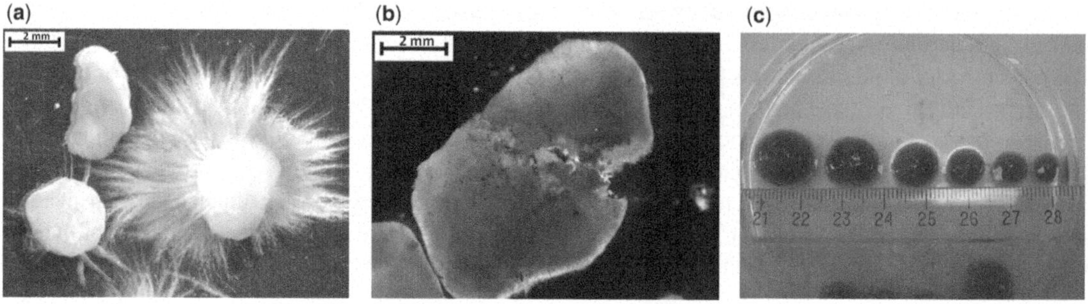

Figure 6.2 Pictures of aerobic granular sludge: (a) with filamentous outgrowth treating synthetic wastewater; (b) with breakage of granules treating pig manure; and (c) with a high size diameter treating fish canning effluents.

Lee and Chen (2015) proposed to promote the internal precipitation with magnesium carbonate to enhance the structural stability of aerobic granules. They tested these strengthened granules in a continuous-flow reactor for 220 days at organic loadings of 6–39 kg/m^3 · d. However, Isanta *et al.* (2012) have studied the long term operation of granular biomass with enhanced biological phosphorus removal and observed that the accumulation of inorganic particles in the granules core, as a result of phosphorous salts precipitation, could be the cause of the mature granules instability. They proposed a regular purge based on the inorganic content rather than the sludge retention time to avoid this destabilization.

Very long times for the development of granular sludge represent an aspect to be taken into account for the practical application of AGS technology also in consideration of the need to maintain good nutrient removal performances during such a period (the operating conditions applied during plant start-up are not suitable for nutrient removal). The challenge is then to rapidly obtain granules treating also low-strength wastewater while simultaneously maintaining nutrient removal. Some strategies have been proposed for shortening the period of granules formation. Ivanov *et al.* (2008) have proposed the use of selected safe starter pure cultures with high cell aggregation ability whereas Pijuan *et al.* (2011) have obtained interesting results by adding a small amount of crushed granules in the seeding sludge (consisting of activated sludge) of the reactor.

The quality of the effluent in terms of solids concentration is another common bottleneck of the aerobic granular systems. Schwarzenbeck *et al.* (2005) stated that the major fraction of solids in the effluent consists of washed out biomass. Inizan *et al.* (2005) with an effluent from the pharmaceutical industry observed that the elevated total COD and SS concentrations in the effluent were mainly due to the suspended solids in the inlet wastewater that are not removed in the reactor and to biomass detachment from the granules surface. Lopez-Palau *et al.* (2009) observed that the solids in the effluent increased with the increase of the applied load, as a result of the high growing rate of the biomass and it washed out. Therefore a pre- or post-treatment is recommended to fulfil the disposal requirements when relevant suspended solids concentrations are present in aerobic granular systems effluents.

In this sense Gobi *et al.* (2011), when treating palm oil mill wastewater with an aerobic granular system, evaluated the effectiveness of a post-treatment based on the use of the own aerobic granules (coming from the biological system) as adsorbent for the turbidity and remaining COD, obtaining 99% and 21% removal, respectively. Arrojo *et al.* (2004) observed that the presence of solids in the effluent was strongly affected by the particulate COD to biomass ratio (CODp/VSS) applied to the system. They obtained lower concentrations of TSS in the effluent when the system was operated with a CODp/VSS ratio lower than 0.12 g COD/g VSS. Figueroa *et al.* (2011) observed an increase of solids concentration in the effluent when the pig manure fed to the aerobic granular system contained a higher amount of solids, thus indicating a poor solids removal capacity of the aerobic granular reactor. These authors also reported episodes of solids wash out when the volume of biomass inside the reactor reached the maximum level of the effluent port. Therefore to avoid this problem of solids concentration peaks in the effluent selectively purges need to be applied in the system. Rocktäschel *et al.* (2015) with municipal wastewater at pilot scale have concluded that the amount of aerobic granules on the total biomass concentration is a key operation tool to control the suspended solids effluent concentration. These authors stated that low solids effluent concentration could be achieved only when the granule content (defined as the percentage of $TSS_{granule}$ on the $TSS_{reactor}$ concentration) was below 80%.

At full scale, Pronk *et al.* (2015a) obtained an effluent with suspended solids content as low as 20–30 mg/L by removing with the surplus sludge the smaller and less settling sludge.

At laboratory scale, treating a mixture of municipal and textile wastewater, Lotito *et al.* (2014) operated with a SBBGR system based on the mixture of biofilm and granules packed in a filling material. This innovative configuration allowed the packed biomass to be acted as a filtering medium to remove suspended particulate matter (and the associated COD) from the wastewater. Furthermore the high age of the biomass allowed the hydrolysis of the captured solids to produce soluble organic compounds which were then removed by the same biomass. Di Iaconi *et al.* (2014) tested SBBGR at pilot scale for treating raw municipal wastewater and observed a high solids removal efficiency (>80%), with residual solids concentration values in the effluent lower than 20 mg SS/L. These works indicate the possibility to improve the design of current aerobic granular systems in order to meet the solids effluent quality.

6.5 MODELLING GRANULAR SLUDGE REACTORS

Several mathematical models have been developed to simulate AGS bioreactors. The biological processes inside granules are determined by concentration gradients of substrates and oxygen. These gradients are a result of the distribution of different types of bacteria, diffusion coefficients, conversion rates, granule size and other factors all influencing each other. Because of these closely interacting processes models are a most useful tool to gain process insight and optimize AGS reactors. Empirical optimisation would require an immensely large amount of experiments, even further complicated by the long time needed to reach steady state conditions (de Kreuk *et al.* 2007).

A mathematical model is always a simplified representation of reality (Henze, 2008). The more it is true to reality, the less calibration is needed to make predictions for specific wastewater treatment plants. On the other hand more realistic models tend to be more complex and therefore are more computationally demanding. Depending on the goal of the model, different phenomena should be taken into account. Since AGS is a relatively new technology no widely accepted modelling approach, like activated sludge models (ASM) for conventional activated sludge (CAS) exists yet (Su & Yu, 2006). In this stage the trade off between generalizability and the complexity is of even greater importance. This section reviews several model features which are important when modelling granular sludge reactors.

6.5.1 Bioconversion processes

The bioconversion processes taking place inside granules (COD-degradation, nitrification, denitrification, phosphorus storage, etc.) are described in terms of reaction stoichiometry and kinetics at the bacterial level. The reaction rates are dependent on the conditions at the location where they take place, irrespective of the reactor configuration (Henze *et al.* 2008). One of the ASMs is often the starting point for the bioconversion model, but modifications are made because of theoretical considerations and to obtain a good agreement with experimental data from AGS reactors.

Different groups of surrogate organisms have been defined in different studies (see Table 6.3). One of the main distinctions is whether or not PAO are included. These are dominant when the reactor operates with a long anaerobic feed period (de Kreuk & van Loosdrecht, 2004). This leads to simultaneous nitrification, denitrification and phosphorus removal (SNDPR). Pulse feeding on the other hand allows less significant biological phosphorus removal. Beun *et al.* (2001) for example, did not monitor nor did they try to predict phosphorus removal by including PAO metabolism. de Kreuk *et al.* (2007) were the first authors to include biological phosphorus removal in their model. These authors however disregarded heterotrophs (OHO) to limit computation time, based on the observation from another simulation study that PAO become dominant at steady-state under most circumstances (Xavier *et al.* 2007). Kagawa *et al.* (2015) also took into account GAO next to PAO and OHO because they were not interested in steady-state values but wanted to simulate the microbial dynamics during the start-up of a reactor. Regarding autotrophic organisms, some authors like Su and Yu (2006) assumed only the surrogate group of the autotrophic nitrifying organisms (ANO) to be responsible for nitrification while others made a distinction between ammonia oxidizing organisms (AOO) and nitrite oxidizing organisms (NOO) to model two-step nitrification, Vázquez-Padín *et al.* (2010) being one example. Denitrification is assumed to be performed by OHO and/or PAO.

Next to COD-oxidation, nitrification and denitrification there are some other reactions that might be important depending on the goal of the model. Beun *et al.* (2001) and many other authors deemed the results to be accurate enough without implementation of hydrolysis of particulates. Ni and Yu (2008) meanwhile used a model that allowed for hydrolysis, a process probably more important in this case since data of a reactor fed by actual soy-bean wastewater was used instead of synthetic wastewater containing only soluble COD. Some authors state more research is needed before hydrolysis can be accurately modelled in AGS-systems because the mechanisms are not entirely understood yet (de Kreuk *et al.* 2007; Pronk *et al.* 2015b). All models consider intracellular storage compounds like polyhydroxyalkanoates (PHA) and glycogen (Gly) inside PAO and/or OHO because they play a crucial role in a discontinuously operating reactor like a SBR. So far only the article of Ni and Yu (2010b) explicitly accounted for production and consumption of EPS and soluble microbial products (SMP).

6.5.2 Intragranule heterogeneity

A granule has a macroscopic volume, allowing physical, chemical and biological heterogeneity within their boundaries. This means diffusion gradients can exist, bacteria are distributed in a non-uniform but often structured way and even physical characteristics like the porosity can differ along the depth of a granule (Ni & Yu, 2010a). To what extent this heterogeneity is taken into account differs between publications. The complexity of modelling this aspect depends largely on the number of spatial dimensions taken into account. Biofilm models usually take into account their heterogeneity in one, two or three dimensions, but also zero-dimensional models that completely ignore the biofilm structure have been proposed (Plattes *et al.* 2008). For AGS we see the same diversity of approaches. A zero-dimensional model was applied by Zhou *et al.* (2013), by using a modified version of ASM3. The one-dimensional approach only considering radial gradients is the most popular (see Table 6.3), but taking into account more dimensions allows for irregular granule boundaries and complex spatial distributions of components (Wanner *et al.* 2006).

Depending on the number of model dimensions different softwares are used. For their 0-D model, Zhou *et al.* (2013) used Matlab. For 1-D models the specific wastewater treatment simulator Aquasim is popular to model simultaneous diffusion and reaction, but Su and Yu (2006) managed to use Matlab by discretizing granules into homogenous slices. General purpose simulators like Matlab allow greater flexibility (Henze, 2008). Regarding two dimensions, software for Individual based Modelling (IbM) (Xavier *et al.* 2005b) has been used by Xavier *et al.* (2007) and Kagawa *et al.* (2015). In IbM biomass is represented by spherical particles that can individually grow, divide and move when overlap with other particles arises. Detachment of the particles can also be included (Xavier *et al.* 2005a).

Table 6.3 Overview of models for AGS with different features.

Surrogate Groups	Supplementary Reactions	Dimensions	Software	Size Distribution	References
OHO ANO		1	Aquasim	Uniform Steady state	Beun *et al.* 2001
OHO ANO	Hydrolysis of particulates	1	Matlab	Heterogeneous Steady state	Su and Yu, 2006
AOO NOO PAO		1	Aquasim	Uniform Steady state	de Kreuk *et al.* 2007
OHO AOO NOO PAO		2	IbM-simulator	Uniform Dynamic	Xavier *et al.* 2007
OHO	Hydrolysis of particulates	1	Aquasim	Uniform Steady state	Ni and Yu, 2008
OHO	Production and utilisation of EPS & SMP	1	Aquasim	Uniform Steady state	Ni and Yu, 2010b
OHO AOO NOO		1	Aquasim	Uniform Steady state	Vázquez-Padín *et al.* 2010
OHO ANO	Hydrolysis of particulates	1	Matlab	Heterogeneous Dynamic	Su *et al.* 2013
OHO AOO NOO	Hydrolysis of particulates	0	Matlab	n/a	Zhou *et al.* 2013
OHO AOO NOO PAO GAO	Hydrolysis of particulates	2	IbM-simulator	Uniform Dynamic	Kagawa *et al.* 2015

6.5.3 Intergranule heterogeneity

The granule size determines the surface to volume ratio, an important parameter for mass transport of solutes. The observed effect of granule size on the performance of a granular sludge reactor for autotrophic nitrogen removal is related to the oxygen penetration depth: an increasing particle size results in less aerobic and more anoxic volume (Volcke *et al.* 2010). The granule size also determines the settling velocity, which can be of importance for SBRs to obtain selective wash-out small granules and flocs (Ni & Yu, 2010a). Even though an AGS reactor contains granules with a wide variety of sizes typically a uniform granule size is assumed for the whole reactor. For instance, Vázquez-Padin *et al.* (2010) characterize granules by their overall volumetric mean diameter, while de Kreuk *et al.* (2007) use a surface-mean diameter and Ni and Yu (2008) rely on an 'average' diameter without specifying its nature.

To which extent the modelling approach to deal with the granule size distribution influences the simulation results was investigated by Volcke *et al.* (2012), by comparing (1) the use of a single characteristic diameter, (2) weighing the simulation results for various characteristic diameters and (3) using a multiple compartment model. Whereas the use of a single characteristic diameter was found sufficient in cases where only the overall reactor behaviour needs to be assessed, when taking into account granule size distributions is required to study the solute exchange between particles of different sizes. With approach (1) and (2) the choice of the characteristic diameter (arithmetic mean, surface-area weighed mean or volumetric mean) had a clear impact on the simulation results, underlining the importance of stating the adopted methodology. The surface-area weighed or volumetric mean diameters are good options because of their physical meaning: the surface area governs the exchange of substrates, while the volume determines the amount of biomass.

The application of Aquasim to simulate multiple particle size classes at once was found limited by several numerical problems (Volcke *et al.* 2012). An example of successful application of a granular sludge model with multiple size classes in Matlab has been reported by Su and Yu (2006). They used 100 size classes with a normal distribution fitted to an experimentally determined distribution. Su *et al.* (2013) went even further by applying a distribution that could change over time, by taking into account breakage, growth and detachment of the granules and modelling the settling process that leads to selective wasting. This led to the proposal of optimization strategies for a fast granulation while maintaining a proper reactor performance by adjusting the settling time.

6.5.4 Flow patterns inside the bulk fluid

Opposed to CAS systems, AGS grows in a reactor with a discontinuous operation. Therefore the flow patterns inside the tank are more complicated and variable in time. Specifically, for the Nereda® process there is an upflow approaching plug flow behaviour through the granule bed during feeding, while the aeration phase is characterised by intensive mixing (Pronk *et al.* 2015a). During the third phase, there is still some turbulence for a while because of the remaining air bubbles and settling granules. Plug flow feeding leads to higher local concentrations in the granular sludge bed compared to a completely mixed bulk. The higher concentrations encountered by the bacteria leads to higher acetate uptake rates. de Kreuk *et al.* (2007) avoided explicit modelling of the plug flow behaviour by assuming complete mixing while temporarily increasing the diffusion constant of acetate artificially. This compensates for the lower concentrations of acetate in the bulk. As mentioned in the introduction of this section: this simplification comes at the cost of a lower generalization. The approach may lead to invalid results when scenario analyses are made for strongly differing operational conditions.

6.6 CONCLUSIONS

This chapter has attempted to give an overview of the practical applications of aerobic granular sludge technology. AGS technology has been extensively researched at the laboratory and pilot scale for treating different types of wastewater. Several experiences have been also carried out at full scale. All these experiences highlight the high potential of AGS systems for simultaneously removing carbon, nitrogen and phosphorous from wastewater. A specific sequence of the reactor cycle phases is however requested for achieving high nutrient removals. The long times requested for obtaining granular biomass and especially its stability for long term operation represent two critical aspects for a successful full scale implementation. This would explain the delay in the diffusion at full scale of this technology also in consideration of its significant advantages over conventional systems. Intensive efforts have been made in recent years, however, for understanding the long term stability of AGS allowing to develop specific remedial actions. Microbial methods have been shown to be effective tools for engineering stable AGS systems whereas modelling and simulation have provided an important support for process optimization. Although the number of full-scale installations is steadily increasing, however, only limited information on their operating data is available in the scientific literature.

6.7 REFERENCES

Adav S. S., Chen M.-Y., Lee D.-J. and Ren N.-Q. (2007). Degradation of phenol by Acinetobacter strain isolated from aerobic granules. *Chemosphere*, **67**(8), 1566–1572.

Adav S. S, Lee D.-J. and Lai J.-Y. (2010). Potential cause of aerobic granular sludge breakdown at high organic loading rates. *Applied Microbiology and Biotechnology*, 85(5), 1601–1610.

Amorim C. L. (2013). Microbial degradation of fluorinated compunds. Studies on biodegradation mechanisms and biotreatment systems. Ph.D thesis, Catholic University of Portugal, Porto, Portugal.

Amorim C. L., Maia A. S., Mesquita R. B., Rangel A. O., van Loosdrecht M. C., Tiritan M. E. and Castro P. M. (2014). Performance of aerobic granular sludge in a sequencing batch bioreactor exposed to ofloxacin, norfloxacin and ciprofloxacin. *Water Research*, **50**, 101–113.

Arrojo B., Mosquera-Corral A., Garrido J. M. and Méndez R. (2004). Aerobic granulation with industrial wastewater in sequencing batch reactors. *Water Research*, **38**(14–15), 3389–3399.

Balest L., Lopez A., Mascolo G. and Di Iaconi C. (2008). Removal of endocrine disrupter compounds from municipal wastewater using an aerobic granular biomass reactor. *Biochemical Engineering Journal*, **41**(3), 288–294.

Beun J. J., Heijnen J. J. and van Loosdrecht M. C. M. (2001). N-Removal in a granular sludge sequencing batch airlift reactor. *Biotechnology and Bioengineering*, **75**(1), 82–92.

Carucci A., Milia S., De Gioannis G. and Piredda M. (2008). Acetate-fed aerobic granular sludge for the degradation of chlorinated phenols. *Water Science and Technology*, **58**(2), 309.

Carucci A., Milia S., De Gioannis G. and Piredda M. (2009). Acetate-fed aerobic granular sludge for the degradation of 4-chlorophenol. *Journal of Hazardous Materials*, **166**, 483.

Cassidy D. P. and Belia E. (2005). Nitrogen and phosphorus removal from an abattoir wastewater in a SBR with aerobic granular sludge. *Water Research*, **39**(19), 4817–4823.

Dai Y., Jiang Y., Su H. (2015). Influence of an aniline supplement on the stability of aerobic. *Journal of Environmental Management*, **162**, 115–122.

de Bruin L. M. M., van der Roest H. F., de Kreuk M. K. and van Loosdrecht M. C. M. (2005). Promising results pilot research aerobic granular sludge technology at WWTP Ede. In: Promising results pilot research aerobic granular sludge technology at WWTP Ede. In: Aerobic Granular Sludge. Water and Environmental Management Series. IWA Publishing, Munich, pp. 135–142.

de Kreuk M. K. and van Loosdrecht M. C. M. (2004). Selection of slow growing organisms as a means for improving aerobic granular sludge stability. *Water Science and Technology*, **49**(11–12), 9–17.

de Kreuk M. K., McSwain B. S., Bathe S., Tay S. T. L., Scwarzenbeck N. and Wilderer P. A. (2005). Aerobic Granular Sludge. In: Discussion Outcomes, Bathe S., de Kreuk M. K., Mc Swain B. S. and Scwarzenbeck N. (eds.), IWA, London, UK, pp. 153–169.

de Kreuk M. K. and van Loosdrecht M. C. M. (2006). Formation of aerobic granules with domestic sewage. *Journal of Environmental Engineering-Asce*, **132**(6), 694–697.

de Kreuk M. K., Picioreanu C., Hosseini M., Xavier J. B. and van Loosdrecht M. C. M. (2007). Kinetic model of a granular sludge SBR: Influences on nutrient removal. *Biotechnology and Bioengineering*, **97**(4), 801–815.

Di Iaconi C., De Sanctis M., Rossetti S. and Ramadori R. (2010a). SBBGR technology for minimising excess sludge production in biological processes. *Water Research*, **44**, 1825–1832.

Di Iaconi C., Del Moro G., De Sanctis M. and Rossetti S. (2010b). A chemically enhanced biological process for lowering operative costs and solid residues of industrial recalcitrant wastewater treatment. *Water Research*, **44**, 3635–3644.

Di Iaconi C., De Sanctis M., Rossetti S. and Mancini A. (2011). Bio-chemical treatment of medium-age sanitary landfill leachates in a high synergy system. *Process Biochemistry*, **46**(12), 2322–2329.

Di Iaconi C., De Sanctis M. and Lopez A. (2014). A single-stage biological process for municipal sewage treatment in tourist areas. *Journal of Environmental Management*, **144**(0), 34–41.

Di Iaconi C., Del Moro G., Bertanza G., Canato M., Laera G., Heimersson S. and Svanström M. (2016). Upgrading small wastewater treatment plants with the sequencing batch biofilter granular reactor technology: Techno-economic and environmental assesment. *Journal of Cleaner Production*, **148**, 606–615.

Duque A. F., Bessa V. S., Carvalho M. F. and Castro P. M. (2011). Bioaugmentation of a rotating biological contactor for degradation of 2-fluorophenol. *Bioresource Technology*, **102**(19), 9300–9303.

Duque A. F., Bessa V. S. and Castro P. M. L. (2015). Characterization of the bacterial communities of aerobic granules in a 2-fluorophenol degrading process. *Plant Biotechnology Reports*, **5**, 98–104.

Wanner O., Eberl H. J., Morgenroth E., Noguera D., Picioreanu C., Rittmann B. E. and van Loosdrecht M. C. M. (2006). IWA Task Group on Biofilm Modeling. Mathematical Modeling of Biofilms. IWA Scientific and Technical Report No.18, IWA Publishing, London, 208 pages. ISBN 1843390876.

Figueroa M., Mosquera-Corral A., Campos J. L. and Mendez R. (2008). Treatment of saline wastewater in SBR aerobic granular reactors. *Water Science and Technology*, **58**(2), 479–485.

Figueroa M., Val del Rio A., Campos J. L., Mosquera-Corral A. and Mendez R. (2011). Treatment of high loaded swine slurry in an aerobic granular reactor. *Water Science and Technology*, **63**(9), 1808–1814.

Franca R. D. G., Vieira A., Mata A. M. T., Carvalho G. S., Pinheiro H. M. Lourenço N. D. (2015). Effect of an azo dye on the performance of an aerobic granular sludge sequencing batch reactor treating a simulated textile wastewater. *Water Research*, **85**, 327–336.

Gobi K., Mashitah M. D. and Vadivelu V. M. (2011). Development and utilization of aerobic granules for the palm oil mill (POM) wastewater treatment. *Chemical Engineering Journal*, **174**(1), 213–220.

Hailei W., Ping L., Guosheng L., Xin L. and Jianming Y. (2010). Rapid biodecolourization of eriochrome black T wastewater by bioaugmented aerobic granules cultivated through a specific method. *Enzyme and Microbial Technology*, **47**(1–2), 37–43.

Harun H., Nor-Anuar A., Ujang Z., Othman I. and Rosman N. H. (2014). Efficiency of Aerobic Granulation Technology in Treating High Strength Soy Sauce Wastewater. *Sains Malaysiana*, **43**(10), 1485–1490.

Henze M., van Loosdrecht M. C. M., Ekama G. and Brdjanovic D. (Eds.) (2008). Biological wastewater treatment. Principles, modelling and design. IWA publishing, London, U.K, 512 pages. ISBN 9781843391883; eISBN9781780401867.

Ho K. L., Chen Y. Y., Lin B. and Lee D. J. (2010). Degrading high-strength phenol using aerobic granular sludge. *Applied Microbiology and Biotechnology*, **85**(6), 2009–2015.

Inizan M., Freval A., Cigana J. and Meinhold J. (2005). Aerobic granulation in a sequencing batch reactor (SBR) for industrial wastewater treatment. *Water Science and Technology*, **52**(10–11), 335–343.

Isanta E., Suárez-Ojeda M. E., Val del Río Á., Morales N., Pérez J. and Carrera J. (2012). Long term operation of a granular sequencing batch reactor at pilot scale treating a low-strength wastewater. *Chemical Engineering Journal*, **198–199**, 163–170.

Ivanov V., Wang X.-H. and Stabnikova O. (2008). Starter culture of Pseudomonas veronii strain B for aerobic granulation. *World J Microb Biot*, **24**(4), 533–539.

Jemaat Z., Suarez-Ojeda M. E., Perez J. and Carrera J. (2014). Sequentially alternating pollutant scenarios of phenolic compounds in a continuous aerobic granular sludge reactor performing simultaneous partial nitritation and o-cresol biodegradation. *Bioresource Technology*, **161**, 354–361.

Kagawa Y., Tahata J., Kishida N., Matsumoto S., Picioreanu C., van Loosdrecht M. C. M. and Tsuneda S. (2015). Modeling the nutrient removal process in aerobic granular sludge system by coupling the reactor- and granule-scale models. *Biotechnology and Bioengineering*, **112**(1), 53–64.

Kee T. C., Bay H. H., Lim C. K., Muda K. and Ibrahim Z. (2015). Development of bio-granules using selected mixed culture of decolorizing bacteria for the treatment of textile wastewater. *Desalination and Water Treatment*, **54**(1), 132–139.

Kishida N., Tsuneda S., Kim J. H. and Sudo R. (2009). *Simultaneous Nitrogen* and *Phosphorus Removal from High-Strength Industrial Wastewater* Using Aerobic Granular Sludge. *Journal of Environmental Engineering*, **135**(3), 153–158.

Lashkarizadeh M., Yuan Q. and Oleszkiewicz J.A. (2015). Influence of carbon source on nutrient removal performance and physical–chemical characteristics of aerobic granular sludge. *Environmental Technology*, **36**(17), 2161–2167.

Lee D. J. and Chen Y. Y. (2015). Magnesium carbonate precipitate strengthened aerobic granules. *Bioresource Technology*, **183**, 136–140.

Lee D. J., Chen Y. Y., Show K. Y., Whiteley C. G. and Tay J. H. (2010). Advances in aerobic granule formation and granule stability in the course of storage and reactor operation. *Biotechnology Advances*, **28**(6), 919–934.

Li J., Ding L.-B., Cai A., Huang G.-X. and Horn H. (2014). Aerobic Sludge Granulation in a Full-Scale Sequencing Batch Reactor. *Bio Med Research International*, 2014, 1–12 (article ID 268789).

Liu Q. S., Liu Y., Tay S. T. L., Show K. Y., Ivanov V., Benjamin M. and Tay J. H. (2005). Startup of pilot-scale aerobic granular sludge reactor by stored granules. *Environmental Technology*, **26**(12), 1363–1369.

Liu X. and Dong C. (2011). Simultaneous COD and nitrogen removal in a micro-aerobic granular sludge reactor for domestic wastewater treatment. *Systems engineering P*, **1**, 99–105.

Liu X., Chen Y., Zhang X., Jiang X., Wu S., Shen J., Sun X., Li J., Lu L. and Wang L. (2015a). Aerobic granulation strategy for bioaugmentation of a sequencing batch reactor (SBR) treating high strength pyridine wastewater. *Journal of Hazardous Materials*, **295**(0), 153–160.

Liu Y. and Liu Q.-S. (2006). Causes and control of filamentous growth in aerobic granular sludge sequencing batch reactors. *Biotechnology Advances*, **24**(1), 115–127.

Liu Y., Xu H., Yang S.-F. and Tay J.-H. (2003). A general model for biosorption of Cd2 + , Cu2+ and Zn2+ by aerobic granules. *Journal of Biotechnology*, **102**(3), 233–239.

Liu Y. Q., Moy B., Kong Y. H. and Tay J. H. (2010). Formation, physical characteristics and microbial community structure of aerobic granules in a pilot-scale sequencing batch reactor for real wastewater treatment. *Enzyme and Microbial Technology*, **46**(6), 520–525.

Liu Y., Kang X., Li X. and Yuan Y. (2015b). Performance of aerobic granular sludge in a sequencing batch bioreactor for slaughterhouse wastewater treatment. *Bioresource Technology*, **190**(0), 487–491.

Long B., Yang C. Z., Pu W., Yang J. K., Shi Y. F., Wang J., Bai J., Zhou X. Y., Jiang G. S., Li C. Y. and Liu F. B. (2014). The stability of aerobic granular sludge treating municipal sludge deep dewatering filtrate in a bench scale sequencing batch reactor. *Bioresource Technology*, **169**, 244–250.

Lopez-Palau S., Dosta J. and Mata-Alvarez J. (2009). Start-up of an aerobic granular sequencing batch reactor for the treatment of winery wastewater. *Water Science and Technology*, **60**(4), 1049–1054.

Lotito A. M., Fratino U., Mancini A., Bergna G. and Di Iaconi C. (2012). Effective aerobic granular sludge treatment of a real dyeing textile wastewater. *International Biodeterioration & Biodegradation*, **69**, 62–68.

Lotito A. M., De Sanctis M., Di Iaconi C. and Bergna G. (2014). Textile wastewater treatment: Aerobic granular sludge vs activated sludge systems. *Water Research*, **54**(0), 337–346.

Marques A. P. G. C., Duque A.F., Bessa V., Mesquita R. Rangel A. O. S. S. and Castro P. M. L. (2013). Performance of an aerobic granular sequencing batch reactor fed with wastewaters contaminated with Zn^{2+}. *Journal of Environmental Management*, **128**, 877–882.

Maszenan A. M., Liu Y. and Ng W. J. (2011). Bioremediation of wastewaters with recalcitrant organic compounds and metals by aerobic granules. *Biotechnology Advances*, **29**(1), 111–123.

Morales N., Figueroa M., Fra-Vázquez A., Val del Río A., Campos J. L., Mosquera-Corral A. and Méndez R. (2013). Operation of an aerobic granular pilot scale SBR plant to treat swine slurry. *Process Biochemistry*, **48**(8), 1216–1221.

Mosquera-Corral A., de Kreuk M. K., Heijnen J. J. and van Loosdrecht M. C. M. (2005). Effects of oxygen concentration on N-removal in an aerobic granular sludge reactor. *Water Research*, **39**(12), 2676–2686.

Ni B.-J. and Yu H.-Q. (2008). Growth and storage processes in aerobic granules grown on soybean wastewater. *Biotechnology and Bioengineering*, **100**(4), 664–672.

Ni B.-J. and Yu H.-Q. (2010a). Mathematical modeling of aerobic granular sludge. *Biotechnology Advances*, **28**, 895–909.

Ni B.-J. and Yu H.-Q. (2010b). Modeling and simulation of the formation and utilization of microbial products in aerobic granular sludge. *AIChE Journal*, **56**(2), 546–559.

Ni B. J., Xie W. M., Liu S. G., Yu H. Q., Wang Y. Z., Wang G. and Dai X. L. (2009). Granulation of activated sludge in a pilot-scale sequencing batch reactor for the treatment of low-strength municipal wastewater. *Water Research*, **43**(3), 751–761.

Pijuan M., Werner U., Yuan Z. (2011). Reducing the start up time of aerobic granular sludge reactors through seeding floccular sludge with crushed aerobic granules. *Water Research*, **45**(16), 5075–5083.

Plattes M., Henry E. and Schosseler P. M. (2008). A zero-dimensional biofilm model for dynamic simulation of moving bed bioreactor systems: Model concepts, Peterson matrix, and application to a pilot-scale plant. *Biochemical* Engineering Journal, **40**(2), 392–398.

Pronk M., de Kreuk M. K., de Bruin B., Kamminga P., Kleerebezem R. and van Loosdrecht M. C. M. (2015a). Full scale performance of the aerobic granular sludge process for sewage treatment. *Water Research*, **84**, 207–217.

Pronk M., Abbas B., Al-zuhairy S. H. K., Kraan R., Kleerebezem R. and van Loosdrecht M. C. M. (2015b). Effect and behaviour of different substrates in relation to the formation of aerobic granular sludge. *Applied Microbiology and Biotechnology*, **99**(12), 5257–5268.

Quan X. C., Tang H., Xiong W.-C. and Yang Z.-F. (2010). Bioaugmentation of aerobic sludge granules with a plasmid donor strain for enhanced degradation of 2,4-dichlorophenoxyacetic acid. *Journal of Hazardous Materials*, **179**(1–3), 1136–1142.

Rocktäschel T., Klarmann C., Ochoa J., Boisson P., Sørensen K. and Horn H. (2015). Influence of the granulation grade on the concentration of suspended solids in the effluent of a pilot scale sequencing batch reactor operated with aerobic granular sludge. *Separation and Purification Technology*, **142**(0), 234–241.

Schwarzenbeck N., Erley R., Mc Swain B. S., Wilderer P. A. and Irvine R. L. (2004). Treatment of malting wastewater in a granular sludge sequencing batch reactor (SBR). *Acta Hydroch Hydrob*, **32**(1), 16–24.

Schwarzenbeck N., Borges J. M. and Wilderer P. A. (2005). Treatment of dairy effluents in an aerobic granular sludge sequencing batch reactor. *App Microbiol Biot*, **66**(6), 711–718.

Shi X.-Y., Yu H.-Q., Sun Y.-J. and Huang X. (2009). Characteristics of aerobic granules rich in autotrophic ammonium-oxidizing bacteria in a sequencing batch reactor. *Chemical Engineering Journal*, 147(2–3), 102–109.

Su K. Z. and Yu H. Q. (2005). Formation and characterization of aerobic granules in a sequencing batch reactor treating soybean-processing wastewater. *Environmental science & technology*, **39**(8), 2818–2827.

Su K.-Z. and Yu H.-Q. (2006). A generalized model for aerobic granule-based sequencing batch reactor. 1. Model development. *Environmental Science & Technology*, **40**(15), 4703–4708.

Su K.-Z., Ni B.-J. and Yu H.-Q. (2013). Modeling and optimization of granulation process of activated sludge in sequencing batch reactors. *Biotechnology and Bioengineering*, **110**(5), 1312–1322.

Sun X.-F., Wang S.-G., Liu X.-W., Gong W.-X., Bao N. and Gao B.-Y. (2008a). Competitive biosorption of zinc(II) and cobalt(II) in single- and binary-metal systems by aerobic granules. *Journal of Colloid and Interface Science*, **324**(1–2), 1–8.

Sun X.-F., Wang S.-G., Liu X.-W., Gong W.-X., Bao N., Gao B.-Y. and Zhang H.-Y. (2008b). Biosorption of Malachite Green from aqueous solutions onto aerobic granules: Kinetic and equilibrium studies. *Bioresource Technology*, **99**(9), 3475–3483.

Tay S. T., Zhuang W.-Q. and Tay J.-H. (2005). Start-up, microbial community analysis and formation of aerobic granules in a tert-butyl alcohol degrading sequencing batch reactor. *Environmental Science & Technology*, **39**, 5774.

Val del Rio A., Figueroa M., Mosquera-Corral A., Campos J. L. and Mendez R. (2013). Stability of Aerobic Granular Biomass Treating the Effluent from A Seafood Industry. *International Journal of Environmental Research*, **7**(2), 265–276.

Vázquez-Padín J. R., Mosquera-Corral A., Campos J. L., Méndez R., Carrera J. and Pérez J. (2010). Modelling aerobic granular SBR at variable COD/N ratios including accurate description of total solids concentration. *Biochemical* Engineering Journal, **49**, 173–184.

Verawaty M., Coma M., Yuan Z., Pijuan M. and Bond P. L. (2010). Application of Aerobic Granular Technology to Treat Domestic Wastewater for Biological Nutrient Removal. AWA Proceeding. Australia's National Water Conference and Exhibition (Ozwater 10), 8–10 March 2010, Brisbane.

Verawaty M., Pijuan M., Yuan Z. and Bond P. L. (2012). Determining the mechanisms for aerobic granulation from mixed seed of floccular and crushed granules in activated sludge wastewater treatment. *Water Research*, **46**(3), 761–771.

Volcke E. I. P., Picioreanu C., De Baets B. and van Loosdrecht M. C. M. (2010). Effect of granule size on autotrophic nitrogen removal in a granular sludge reactor. *Environmental Technology*, **31**(11), 1271–1280.

Volcke E. I. P., Picioreanu C., De Baets B. and van Loosdrecht M. C. M. (2012). The granule size distribution in an anammox-based granular sludge reactor affects the conversion-Implications for modeling. *Biotechnology and Bioengineering*, **109**(7), 1629–1636.

Wagner J., Weissbrodt D. G., Manguin V., Ribeiro da Costa R. H., Morgenroth E. and Derlon N. (2015). Effect of particulate organic substrate on aerobic granulation and operating conditions of sequencing batch reactors. *Water Research*, **85**, 158–166.

Wang S.-G., Liu X.-W., Zhang H.-Y., Gong W.-X., Sun X.-F. and Gao B.-Y. (2007a). Aerobic granulation for 2,4-dichlorophenol biodegradation in a sequencing batch reactor. *Chemosphere*, 69, 769.

Wang S.-G., Liu X.-W., Gong W.-X., Gao B.-Y., Zhang D.-H. and Yu H.-Q. (2007b). Aerobic granulation with brewery wastewater in a sequencing batch reactor. *Bioresource Technology*, **98**(11), 2142–2147.

Wang F., Lu S., Wei Y. and Ji M. (2009). Characteristics of aerobic granule and nitrogen and phosphorus removal in a SBR. *Journal of Hazardous Materials*, **164**(2–3), 1223–1227.

Xavier J. B., Picioreanu C., Rani S. A., van Loosdrecht M. C. M. and Stewart P. S. (2005a). Biofilm-control strategies based on enzymic disruption of the extracellular polymeric substance matrix – a modelling study. *Microbiology*, **151**(12), 3817–3832.

Xavier J. B., Picioreanu C. and van Loosdrecht M. C. (2005b). A framework for multidimensional modelling of activity and structure of multispecies biofilms. *Environmental Microbiology*, **7**(8), 1085–1103.

Xavier J. B., De Kreuk M. K., Picioreanu C. and van Loosdrecht M. C. M. (2007). Multi-scale individual-based model of microbial and bioconversion dynamics in aerobic granular sludge. *Environmental Science & Technology*, **41**(18), 6410–6417.

Xu H. and Liu Y. (2008). Mechanisms of Cd2 +, Cu2+ and Ni2+ biosorption by aerobic granules. *Sep Purif Technol*, **58**(3), 400–411.

Yi S., Zhuang W.-Q., Wu B., Tay S. T.-L. and Tay J.-H. (2006). Biodegradation of p-nitrophenol by aerobic granules in a sequencing batch reactor. *Environmental Science & Technology*, **40**, 2396.

Yilmaz G., Lemaire R., Keller J. and Yuan Z. (2008). Simultaneous nitrification, denitrification, and phosphorus removal from nutrient-rich industrial wastewater using granular sludge. *Biotechnology and Bioengineering*, **100**(3), 529–541.

Zhang L. L., Chen J. M. and Fang F. (2008). Biodegradation of methyl t-butyl ether by aerobic granules under a cosubstrate condition. *Applied Microbiology and Biotechnology*, **78**, 543.

Zhao D., Liu C., Zhang Y. and Liu Q. (2011). Biodegradation of nitrobenzene by aerobic granular sludge in a sequencing batch reactor (SBR). *Desalination*, 281, 17.

Zhao X., Chen Z., Wanga X., Li J., Shen J. and Xu H. (2015). Remediation of pharmaceuticals and personal care products using an aerobic granular sludge sequencing bioreactor and microbial community profiling using Solexa sequencing technology analysis. *Bioresource Technology*, **179**, 104–112.

Zheng Y.-M., Yu H.-Q., Liu S.-J. and Liu X.-Z. (2006). Formation and instability of aerobic granules under high organic loading conditions. *Chemosphere*, **63**(10), 1791–800.

Zhou M., Gong J., Yang C. and Pu W. (2013). Simulation of the performance of aerobic granular sludge SBR using modified ASM3 model. *Bioresource Technology*, **127**, 473–481.

Zhu L., Xu X., Luo W., Cao D. and Yang Y. (2008). Formation and microbial community analysis of chloroanilines-degrading aerobic granules in the sequencing airlift bioreactor. *Journal of Applied Microbiology*, **104**, 152.

Zhu L., Lv M., Dai X., Zhou J.-H. and Xu X.-Y. (2013). The stability of aerobic granular sludge under 4-chloroaniline shock in a sequential air-lift bioreactor (SABR). *Bioresource Technology*, **140**, 126.

Chapter 7

Membranes in wastewater treatment

Pawel Krzeminski, Evina Katsou, Carlos G. Dosoretz, Ana Lorena Esteban-García, Lance Leverette, Simos Malamis, Valentin Nenov, Ángel Robles, Aurora Seco and Eoin Syron

7.1 INTRODUCTION

7.1.1 MBR's when does it make sense?

Membrane bioreactor (MBR) technology combining activated sludge process with a membrane filtration is currently considered a well-established, mature technology with many full-scale MBR plants treating municipal and industrial wastewater. In Europe by the end of 2008, 37 MBR plants having a capacity higher than 5000 m^3/d were operating, while more than 800 commercial MBR applications were in use. It is expected that by 2019 more than 5 million m^3 of wastewater per day will be treated by MBR plants. The Henriksdal wastewater treatment plant (WWTP) in Stockholm will be upgraded to an MBR that will treat 864,000 m^3/d of wastewater, making it the largest MBR plant in the world, when it will be commissioned in 2018. For the comparison, in 2004 the largest MBR plant in the world was Nordkanal MBR plant with a design capacity of 45,000 m^3/d at that time. The difference in the design capacity between the Nordkanal and the Henriksdal WWTPs demonstrate the significant growth of MBR technology.

Recent R&D developments in MBRs in respect to novel configurations have been focused on membrane fouling control, energy demand reduction, enhanced nutrient removal or removal of refractory compounds (Meng *et al.* 2011; Huang & Lee, 2015).

7.1.2 Energy demand reduction

As an energy-saving alternative to energy intensive air scouring used for fouling mitigation application of intermittent electric field was proposed (Akamatsu *et al.* 2010). Different configurations were developed among which: electrochemical MBR (EMBR) coupled with low voltage and low intensity electric field (Liu *et al.* 2012b), submerged membrane electro-bioreactor (MEBR) with electrocoagulation incorporated inside of MBR and intermittent direct current field applied (Bani-Melhem & Elektorowicz, 2011), and electro-MBRs (e-MBRs) with stainless steel or titanium anodes and intermittent application of low-voltage electric-field (Zhang *et al.* 2015). In these systems fouling is suppressed through electro-coagulation, electrophoresis and electrostatic repulsion/rejection mechanisms.

Others combined conventional MBR and microbial fuel cells (MFC) into a bioelectrochemical membrane reactor (BEMR). In BEMR, bacteria oxidize various substrates and produce electricity reducing the energy demand of MBR process, while MBR improves biomass retention and effluent quality which are known drawbacks of MFC systems. Coupling of membrane filtration with MFC is discussed in details in Section 3.1.7.3.3.

Integration of membrane filtration with microalgae in membrane photobioreactor (MPBR) provides organic carbon removal in MBR and nutrients in photobioreactor. The system has a potential to reduce nutrient removal costs in MBR in parallel to reduction in nutrient and primary harvesting costs in the microalgae processing (Bilad *et al.* 2014; Drexler & Yeh, 2014).

7.1.3 Enhanced nutrients and/or refractory compounds removal

To provide enhanced nutrient removal MBR process may be combined with the biological processes, such as batch reactor (SBR), anoxic–oxic (AO or A/O), and anaerobic–anoxic–oxic (A^2O), moving bed bioreactor (MB), modified Ludzacke Ettinger (MLE). For example, A/O vertical MBR (VMBR, VSMBR or IVMBR) compose of lower anoxic and upper oxic zones in one reactor developed to enhance removal of organics and nutrients, reduce membrane fouling and lower sludge production. In A^2O-MBR, anaerobic process increases removal of organics, whereas anoxic-aerobic process enhances removal of nitrogen through the pre-denitrification and aerobic nitrification. In addition, soluble EPS and other potential membrane foulants are removed with a benefit for a downstream MBR process (Sun *et al.* 2015).

Other configurations were also developed to increase, besides nutrients, removal of organics and/or refractory compounds. For example, integration of MBR and ozonation processes increased removal of COD and ozonation biodegradable products at lower ozone doses compared to separate MBR and ozonation systems without affecting MBR filtration or biological processes (Pollice *et al.* 2012). An MBR-TiO_2, combining MBR and UV/TiO_2 photocatalysis process, removed up to 95% of carbamazepine from synthetic wastewater (Laera *et al.* 2011). A continuous flow fungal MBR achieved 80–90% removal of bisphenol A and about 55% removal of diclofenac by utilizing inhibitory compounds as fungi substrate (Yang *et al.* 2013). In osmotic MBRs (OMBR, FO-MBR), forward osmosis (FO) membrane extracts water from activated sludge into concentrated with salts draw solution generating driving force - osmotic pressure. The diluted draw solution is treated by reverse osmosis, producing high-quality water and re-concentrated draw solution that may be reused in the FO process (Achilli *et al.* 2009). OMBRs offer excellent removal of contaminants, reduce membrane fouling but fluxes tends to be low and removal efficiency gradually decrease due to decreasing driving force (Cornelissen *et al.* 2008; Zhang *et al.* 2014).

7.1.4 Synergistic effects utilization

Biofilm processes have been implemented in MBR by addition of media, such as carriers, in moving or fixed bed configurations, or aerated membranes in the bioreactor as a support for biofilm growth. In hybrid growth membrane bioreactors biomass is in suspension and attached to packing media providing reduce membrane fouling rates. Bio-entrapped membrane reactor (BEMR) using bio-ball carriers had reduced membrane fouling, improved removal of COD, NH_4 and phenol, and required less frequent chemical cleanings due to slower TMP increase compared with conventional MBRs (Rafiei *et al.* 2014). According to Cuevas-Rodriguez *et al.* (2015) moving bed MBR provides better membrane filtration performance but lower nitrogen removal compared to conventional, moving bed and fixed bed MBRs. A batch granulation

MBR (BG-MBR) consisting of a sequencing batch airlift reactor (SBAR), settler and submerged MBR allowed for simultaneous organic/nitrogen removal and membrane fouling control (Thanh *et al.* 2013).

Synergetic effects of activated carbon adsorption, biodegradation and membrane filtration have been utilized by addition of powdered activated carbon (PAC), granular activated carbon (GAC) or zeolite, typically used to control membrane fouling, to increase resistant pollutants removal. With the addition of PAC, the biological activated carbon provides uptake and/or entrapment of soluble organics and colloids resulting in improved trace organics removal. However, removal efficiency of some compounds deteriorates gradually and addition of PAC lost with sludge discharge is needed to maintain filtration performance. Application of activated carbon in MBRs have been discussed in detailed in a recent review (Skouteris *et al.* 2015).

Among commercial developments, Microdyn-Nadir introduced in 2008 a BIO-CEL membrane which combines the advantages of hollow fibre and flat-sheet membrane. The BIO-CEL self-supporting membrane is made of two flat sheet membrane sheets laminated on the spacer to provide support and enable backwashing. According to the company, membrane has also a self-healing potential (based on turbidity). In addition, a continuous mechanical cleaning process with granulates has been developed under the Bio-Cell®-MCP name. In 2014, GE introduced MACarrier, i.e. MBR with carriers, to tackle refractory COD, toxicity, phenols during treatment of difficult wastewater streams. Fibracast, part of Anaergia group, developed FibrePlate™ membrane which combines the strengths of hollow fibre and flat-sheet membrane in to one UF hybrid membrane. According to the producer, the membrane couple high packing density and a back-wash capability, with low TMP and ease of operation, while reduced capital and operating costs. In 2015, a new flux enhancement technology for tubular membranes, Helix, was presented by Pentair.

7.2 INNOVATIVE USE OF MEMBRANES IN WASTEWATER TREATMENT

The technologies discussed in the following sub-sections allow to not only reduce required land space but are also well suited to contribute in reducing energy requirements associated with wastewater treatment.

7.2.1 Anaerobic Membrane Bioreactors

Anaerobic digestion (AD) has been successfully applied for wastewater treatment during the last 30 years. Nonetheless, its application to high-rate and/or low-strength wastewater treatment (e.g. urban wastewater – UWW) was limited by the difficulty in retaining slow-growth-rate anaerobic microorganisms when operating at short hydraulic retention times (HRTs). This technology results in a high-quality effluent, while it generates biogas that can be used for energy production. The complexity of the anaerobic process can be balanced by the energy production and the reduction of carbon footprint.

The adoption of AnMBR technology can convert traditional wastewater treatment plants (WWTPs) into water resource recovery facilities (WRRFs). The latter is in line with the new green concept, which considers wastewater as a renewable source of energy, reclaimed water and valuable nutrients. Moreover, it contributes to the reduction of greenhouse gases (GHG) emissions and biosolids production.

AnMBRs can be classified as submerged/immersed MBRs or side-stream MBRs depending on the configuration of the filtration process. Side-stream AnMBRs are mainly applied for the treatment of high-strength wastewaters (e.g. industrial wastewater), while the use of submerged/immersed AnMBRs is usually limited to the treatment of low-strength wastewaters (e.g. UWW). This is mainly due to the low cost and suitable fouling control that is achieved by combining low-pressure filtration and membrane scouring by gas sparging (Judd & Judd, 2011). The bioreactor can be a continuously stirred tank reactor

(CSTR) (Giménez *et al.* 2011), an upflow anaerobic sludge blanket (UASB) (Gouveia *et al.* 2015b), an expanded granular sludge bed (EGSB) (Chu *et al.* 2005), or a fluidised bed (Shin *et al.* 2014) reactor coupled with membrane filtration.

7.2.1.1 Feasibility for the treatment of different wastewater streams

The efficiency of AnMBR technology has been widely demonstrated at bench scale for the treatment of high-strength (Padmasiri *et al.* 2007) and low-strength wastewaters (Ho & Sung, 2009a). Moreover, the performance of AnMBR treating UWW at pilot scale has been evaluated in different studies during the last years, whilst the process has been successfully applied at full-scale for the treatment of high-strength industrial wastewater (Christian *et al.* 2011). However, most of these studies were conducted at elevated temperatures (>25°C), while the application of this process at psychrophilic environments (<20°C) is limited. The latter is essential in order to validate the economic viability of the technology within the context of UWW treatment. Two review papers (O'Flaherty *et al.* 2006; van Haandel *et al.* 2006) have concluded that anaerobic treatment of UWW is possible at low temperature, provided the system is operated at long sludge retention times (SRTs).

The use of AnMBR process for wastewater treatment reduces sludge production, eliminates aeration and generates methane. AnMBR offers the possibility to operate the system in energy neutral or even positive net energy balance due to biogas generation. Moreover, the solid-free effluent (complete biomass retention is normally achieved) can be used for nutrient recovery in downstream units or water reuse (e.g. irrigation). Furthermore, the AnMBR permeate can be reused for agricultural purposes (i.e. fertigation). Macronutrients such as ammonium and orthophosphates are not removed by anaerobic bioprocesses and pathogens can be retained by the membrane unit . However, the nutrient discharge limits for nutrients vary depending in the regulation. The selection of the most suitable wastewater treatment scheme should be performed based on the guidelines or regulations that are imposed by the prevailing legislation (Italy, Spain, etc.).

7.2.1.2 Barriers for widespread application

Despite the above mentioned advantages, there are still some barriers or challenging issues that limit the extensive use of AnMBRs. Thus, further breakthroughs should be pursued. High operating costs related to membrane fouling control and mitigation (e.g. energy requirements for gas/liquid recirculation and chemical reagent consumption) are still limiting factors for the widen application of AnMBRs. However, membrane acquisition and/or replacement costs have significantly decreased due to the reduction of the membrane module costs. The main barriers that limit the wider application of this technology are described in the following sections of the current chapter.

7.2.1.3 Membrane fouling

Fouling is the major drawback of the membrane systems, affecting the operation and performance of the filtration process. Although part of membrane fouling improves organic removal, excessive fouling must be controlled by back flushing and/or biogas sparging.

AnMBRs typically operate at transmembrane fluxes lower than the respective ones in aerobic MBRs. This is mainly due to the reduced flocculation and increased concentrations of fine particulates and colloids under anaerobic conditions. However, some studies that compared aerobic MBR and AnMBR for UWW treatment have revealed similar fouling issues (Achilli *et al.* 2011) or even less fouling propensity in the latter case (Baek & Pagilla, 2006). The impact of several fouling factors on membrane fouling in AnMBRs is summarised in the review of Lin *et al.* (2013), including operating parameters, sludge properties and membrane characteristics. Bérubé *et al.* (2006) focused their review paper on membrane fouling and achievements in fouling control in order to promote the full-scale implementation of AnMBRs within UWW treatment.

7.2.1.4 Fouling mitigation

Fouling mitigation (during operation) and membrane maintenance and replacement are key factors for the AnMBR market exploitation. Further research is required to determine the most effective fouling mitigation strategy per energy input. One commonly applied fouling mitigating strategy entails operating membranes in sub-critical filtration conditions, which are bounded by the so-called critical flux. The operation of the membranes at sub-critical conditions increases membrane lifespan, which reduces maintenance costs. However, this practice increases the investment and/or operating expenses (need for higher membrane areas and/or intensities of the gas sparging for membrane scouring). Robles *et al.* (2012) determined critical fluxes at different operating conditions in an AnMBR system equipped with industrial-scale hollow fibre membranes. The results indicated that it is theoretically feasible to operate membranes sub-critically at high MLSS concentrations without applying prohibitive specific gas demand per unit of membrane area (SGDm) levels (from 0.17 to 0.50 $Nm^3/h.m^2$) while maintaining flux between 10 and 15 LMH. A considerable increase in flux can be achieved in sub-critical filtrations conditions with a slight increase in SGDm. Concerning chemical cleaning, several procedures have been summarised in the work of Dong (2015). A common practice recommended by manufacturers includes extended cleaning durations (2–24 h) with typical reagent concentrations of 2–5 g/L NaOCl coupled with 10–15 mM/L citric acid or 50–100 mM/L oxalic acid.

On-line fouling control

According to Smith *et al.* (2012), future research should focus on the development of control systems for the optimisation of the membrane operation. Jeison and van Lier (2006) developed an on-line cake-layer management protocol that monitored critical flux constantly and prevented the build-up of excessive cake layer on the membrane surface. Robles *et al.* (2014) applied a model-based supervisory controller to optimise filtration in an AnMBR pilot plant with industrial-scale membranes. Energy savings of up to 25% were achieved with the use of gas sparging to scour membranes, while the downtime for physical cleaning was about 2.4% of operating time. The operating cost of the AnMBR system with the use of the supervisory controller was about €0.045 m^3; out of which 53.3% was associated with energy costs. In another study, Robles *et al.* (2015) obtained similar results using a 2-layer control system to measure the flow rate (controlling the HRT), the excess sludge volume (controlling the SRT), the temperature, and the gas sparging intensity in the anaerobic reactor. The authors controlled the permeate flow rate, the TMP, the sludge flow-rate recycled through the membrane tanks, and the gas sparging intensity in the membrane tanks. Hence, the application of on-line fouling control strategies can effectively reduce membrane fouling and thus decrease energy demand and increase membrane lifetime in the AnMBR processes.

7.2.1.5 Mathematical modelling

A critical issue for advancing on AnMBR development is the use of mathematical models capable to accurately predict system performance under different design and operating conditions. The main focus of AnMBR modelling is on fouling prediction and control. Robles *et al.* (2014) developed a mechanistic cake accumulation model and applied both mechanistic and empirical models for the control of cake accumulation. Ho and Sung (2009b) evaluated the effect of the characteristics and rheological properties of anaerobic sludge characterised by different solids content by applying different theoretical filtration laws. In another study, Herrera-Robledo and Noyola (2015) identified the contribution of fouling mechanisms during the ultrafiltration of anaerobic effluent–like mixtures. On the other hand, the effect of hydrodynamics and mixing has been recently modelled in AnMBRs. Trad *et al.* (2015) combined 1D and 3D methodologies including single-phase and two-fluid CFD models for the optimisation of the reactor's design and mixing conditions. Yang *et al.* (2012b) applied fluid dynamic modelling to analyse the mass transfer characteristics in the tubular membrane of a side-stream AMBR that was operating with a gas-lift two-phase flow. In addition, cake accumulation models have been combined with CFD in a multi-dimensional domain (Boyle-Gotla *et al.* 2014).

However, the developed filtration models do not generally consider biological reactions, while reactive models have been used only for the evaluation of the general biological performance of AnMBRs. Thus, there is a need for the development of integrated models that link reactive models that describe the presence and accumulation of specific constituents (e.g. colloidal organics, SMP, particulate matter, etc.) with fouling mechanisms, hydraulics, membrane physical interactions and multiphase reactions. This can facilitate the optimisation of the AnMBR process (Batstone *et al.* 2015).

7.2.1.6 Life Cycle Cost (LCC)

Considering cost aspects, Lin *et al.* (2013) reported that membrane investment and membrane scouring energy accounted for the largest fraction of the total life cycle capital costs and operating costs, respectively. However, the operating costs can be totally offset by the benefits obtained from biogas recovery. On the other hand, Smith *et al.* (2014) found that the total cost of an AnMBR treating medium- and high-strength UWW was €0.124 per m^3. In the same study, energy recovery increased by 60 and 130% of the total energy consumption when the system treated medium- and high-strength wastewater, respectively. Ferrer *et al.* (2015) evaluated the total cost of a full-scale AnMBR treating low-sulphate and sulphate-rich UWW at different operating temperatures; the total cost was between €0.07 and €0.101 per m^3. Pretel *et al.* (2015) evaluated the economic impact of a system that includes primary settling and AD of the waste activated sludge for the design of a submerged AnMBR-based WWTP for the treatment of sulphate-rich and low-sulphate UWW at 15 and 30°C. AnMBR without PS and without subsequent AD of the waste sludge was the most economic option (minimum cost: €0.05 per m^3) for the treatment of low-sulphate UWW at average temperatures (above 15°C). In fact, the authors demonstrated that surplus energy of 0.1 kWh per m^3 can be achieved when the methane is captured. The combination of AnMBR with PS and AD was the most economically sustainable option for the treatment of sulphate-rich UWW (minimum cost of €0.05 per m^3 and a maximum surplus energy of 0.09 kWh per m^3). The total cost of the AnMBR WWTP treating low-sulphate was significantly lower that the respective one for the treatment of sulphate-rich UWW (cost savings up to 45% can be achieved). On the other hand, the cost of an AnMBR-based WWTP including CAS- or MBR-based post-treatment for nutrient removal would result in about €0.126–0.169 per m^3 (Pretel *et al.* 2016).

7.2.1.7 Life Cycle Assessment (LCA)

Several studies have been published on the environmental impact assessment of WWTPs following the LCA. Only few works have been conducted on LCA (and LCC) for the assessment of the environmental performance of AnMBR for UWW treatment. This is mainly attributed to the lack of full-scale data.

Smith *et al.* (2014) used environmental and economic criteria in order to evaluate and compare submerged AnMBRs with alternative aerobic processes. More specifically, the main objective of this study was to compare the AnMBR technology with conventional wastewater processes that have been applied for energy recovery; high rate activated sludge with AD (HRAS + AD), conventional activated sludge with AD

(CAS + AD) and aerobic membrane bioreactor with AD (AeMBR + AD). AnMBR recovered 49% more energy as biogas than HRAS + AD for the treatment of medium-strength UWW at 15°C (baseline scenario). However, the global warming impact of AnMBR was high due to the dissolved methane emissions in the effluent. Pretel *et al.* (2013) studied the energy balances and performed the LCA of an AnMBR system that applied industrial-scale membranes for the treatment of UWW at different temperatures. The results revealed the improved environmental performance of the AnMBR at ambient temperature, recovery of nutrients (environmental impact in eutrophication can be reduced up to 45%) and dissolved methane (positive environmental impact can be obtained) from the effluent. On the other hand, Pretel *et al.* (2016) assessed the economic and environmental sustainability of submerged AnMBR (coupled with aerobic post-treatments) compared to aerobic MBR and CAS for the removal of organic matter, nitrogen and phosphorus from moderate-/high-strength UWW. AnMBR with a CAS as post-treatment for nutrient removal was identified as a sustainable option for moderate-/high-strength UWW treatment: a minimum energy consumption of 0.04 kWh per m^3 and low sludge production was obtained under the given operating conditions. In addition, significant reductions in several environmental impact categories (GWP, marine aquatic ecotoxicity, abiotic depletion and acidification) and LCC (minimum LCC value of around €0.135 per m^3) can be achieved in comparison with other existing UWW treatment processes.

7.2.1.8 Challenges and future perspectives for the use of AnMBRs

The main limitations for widespread application (which is the case for aerobic MBRs), especially in the large-scale UWW treatment, are optimum configuration, operation at low temperature, biogas production, and membrane performance.

- Membrane fouling issues that could be more prevalent in AnMBRs than in aerobic MBRs. Thus, AnMBR would operate at lower transmembrane fluxes. Moreover, another possible constraint could be related with the operating costs of the filtration system. This is mainly due to possible high biogas sparging intensities in submerged AnMBRs and high cross-flow velocities in side-stream AnMBRs for fouling control and/or mitigation. Other issues that should be considered include backwashing and cleaning frequency and use of chemicals. The improvement of the membrane performance and the minimisation of relevant costs remains a challenge that should be addressed for the full-scale expansion of AnMBR technology. AnMBRs are also more prone to inorganic fouling caused by calcium, phosphorus and sulphur precipitates; struvite is the dominant inorganic foulant in the treatment of industrial wastewater streams.
- *Effluent quality and nutrients removal limitations*: more advanced treatment is usually required for nutrients removal; however, this is location specific depending on the effluent quality concerns.
- Continuation of research towards the treatment of 'more difficult' wastewater streams; low-strength wastewaters. Establishment of a low-energy dissolved methane recovery process in order to limit global warming impact (connection with LCA).
- *Nutrients and energy recovery*: Approach wastewater as a source of recovered products moving towards the circular concept and the economic and environmental sustainability of wastewater management of alternate renewable energy sources, and other factors.

7.2.2 Membranes for gas transfer

7.2.2.1 Into what is different about membranes for gas transferring

The most important property of membranes is their ability to control the rate of permeation of different species. The capability of membranes to separate components has been exploited by the water industry mainly for solids separation. In processes such as microfiltration, ultrafiltration, nanofiltration, reverse osmosis or electrodialysis, the membrane acts to reject the pollutants (suspended or dissolved), allowing for the purified water to pass through. Membranes also act as filters in MBRs, in which their function is to rêtain biomass solids in a biological reactor, thereby replacing the sedimentation tank.

The other way to exploit membranes is for solute transfer, and specifically for gas transfer. The most common type of gas needed in biological treatment processes is oxygen, which, is the electron acceptor for aerobic degradation. Hydrogen may act as electron donor for autotrophic denitrification or the reduction of other oxidized contaminants. Finally, methane may support cometabolic oxidation of some organic compounds. Although it has been the technology of choice for many years, using conventional diffusers to dissolve a gas in an aqueous medium has some disadvantages: the low solubility of the gases requires continuous bubbling. Introducing the gas at the bottom of the tank means that the gas must be pressurised to overcome the hydrostatic head resulting in a high-energy consumption, and gas bubbles have a fixed residence time before they reach the surface of the liquid and are lost to the headspace which results in a waste of the undissolved gas. This is particularly costly in the case of expensive gases or when an explosive atmosphere is created, such as hydrogen and methane.

Along with the dissolving the gas in liquid membranes can provide an alternative way to supply the gas to the microorganisms: the gas is diffused through a gas permeable membrane directly to a biofilm that is growing on the other side, in contact with the wastewater. This bubbleless delivery of the gas straight into the biofilm allows for the increase in the gas transfer efficiency to a maximum of 100%. Moreover, large surface areas for the biofilm can be achieved when thin hollow fibre membranes are used. Together with the extended effective sludge age and high biomass concentration typical of biofilm reactors, this configuration allows for small reactor footprints. The general term for this type of reactors is membrane biofilm reactor (MBfR) (Martin & Nerenberg, 2012), and when air or oxygen is delivered they are usually referred to as membrane-aerated biofilm reactors (MABR).

When hydrogen is the supply gas, MBfRs are particularly relevant as a safer way for delivery. Regarding MABRs, the increasing interest is due mostly to two important features: unique biofilm stratification and the exceptional low energy required for oxygen transfer.

In the MABR, oxygen and soluble wastewater constituents diffuse from opposite directions (counter-diffusional), while in a conventional biofilm both oxygen and pollutants diffuse from the same direction (co-diffusional). In a co-diffusional biofilm, the rate of microbial activity is limited at the biofilm-liquid interface, where both substrate and oxygen exist. Bulk liquid dissolved oxygen concentrations are generally maintained at 2 mg/l or less to maximise the oxygen transfer rate. Micro-organism population stratification in a biofilm depends on the relative concentrations of oxygen and substrate in the internal layers; the aerobic, anoxic and anaerobic conditions can all exist within the biofilm. Nitrifying microorganisms will tend to grow near the membrane of a MABR, where oxygen concentration is high and ammoniacal nitrogen concentration is higher than that of carbonaceous substrate, due to the higher diffusion rate of ammonia. Heterotrophic carbon removal predominates further from the membrane surface and this may be via aerobic or anoxic pathways depending on the oxygen concentration. When the oxygen is depleted within the biofilm ideal conditions for denitrification are created inside the biofilm and even in the bulk liquid.

The counter-diffusional biofilm implies two main benefits. On the one hand, slow growing degrading microorganisms, such as ammonia oxidizing bacteria, are located adjacent to the membrane and thus protected from erosion, shocks or inhibitory compounds. Secondly, simultaneous organic carbon removal, nitrification and denitrification can be achieved, which involves compactness avoids the requirement for pH correcting and removes the requirement for separate aerobic and anoxic tanks. However, the fact that oxygen and substrates diffuse from opposite sides implies that biofilms thickness control turns out to be essential for an adequate performance. In every biofilm process, the control of biofilm thickness is an important issue to avoid clogging and liquid flow distribution problems, but for MABRs excessive biofilm thickness results in higher diffusion resistances, which can profoundly affect substrates and oxygen fluxes.

The outstanding oxygen transfer characteristics of MABRs are associated with better oxygen delivery control and a high potential for energy savings compared to conventional systems. In a membrane, the mass transfer coefficient only depends on the membrane characteristics and the surface area is constant, whilst for bubble oxygenation both are sensitive to atmospheric and hydrostatic factors. By just regulating oxygen partial pressure in the membranes (adjusting feed pressure, air flow in open-end membranes or oxygen concentration), a high flexibility in design and response to variations in oxygen demand is possible.

The reduction in energy consumption in a MABR relies on the fact that it is possible to achieve very high oxygen transfer efficiencies (OTEs) with low head losses. While in conventional bubble aeration OTEs are around 12–37% (for a depth of 4.5 m), in some configurations of MABR (dead-end mode) an OTE of up to 100% can be attained. Furthermore, in conventional bubble aeration frictional losses through the piping, hydrostatic pressure of the water and backpressure required to blow bubbles through the diffuser device must be taken into account. In a MABR, only the sufficient pressure to overcome the frictional resistance across the fibre length is required.

Finally, bubbleless aeration reduces the stripping of volatile organic carbons (VOCs), greenhouse gases, aerosols and malodorous compounds from water, also preventing foaming.

7.2.2.2 Types of membranes and configurations

Multiple configurations of MABRs have been proposed. Most research has been carried out with hollow or tubular fibres; with both microporous and dense membranes being studied. Microporous membranes have the advantage of a negligible mass transfer but only low gas pressures can be applied: failures due to flooding (caused by wetting of breaks) have been reported. On the other side, dense membranes, such as silicone or PolyMethylPentane tubing, are more resistant and higher intramembrane pressures can be applied, however their thicker walls offer significantly higher diffusional resistance. Composite membranes (essentially porous membranes with a thin dense layer) might provide an intermediate solution. Gas supply modes are dead-end, flow-through and self-aerated. In a dead-end mode, maximum oxygen transfer efficiency can be achieved but with the accumulation of gases mainly due to nitrogen and carbon dioxide lower overall oxygen transfer rates result with the opposite (higher transfer rates, but lower transfer efficiency) occurring in flow-through MABRs. An intermediate balance is required to maximize the efficiency (such as periodic air venting or controlled air flow rate). Finally, self or passively aerated MABR have also been proposed. In this configuration, both ends of the hollow fibre membranes or one of the sides of a flat sheet (BioGill (a)) are connected with the atmospheric air and so oxygen passive transport across the membrane occurs spontaneously.

7.2.2.3 Potential advantageous uses of gas transferring membranes in an WWTP

The potential advantages of MBfRs previously mentioned have been explored in a number of applications. For instance, the high oxygen transfer efficiencies and the possibility of using pure oxygen offer MABRs a potential for high oxygen demanding wastewaters, such as landfill leachate or food processing wastewater. Other example would be the biodegradation of xenobiotic compounds. The benefits of bubbleless aeration, preventing VOCs from being released, and the protection of slow growing microorganisms from wash-out and toxics, can be exploited.

With a focus exclusively on the use of MABR in municipal wastewater treatment, one of the main attractions of the technology is the ability to achieve low energy consumption and in some cases energy positive wastewater treatment plants. In a typical domestic wastewater treatment, the cost of aeration can account for 50% or even more of the energy costs. With higher OTEs and lower power requirements for blowers, MABRs can substantially reduce energy consumption for aeration. For a system conducting carbon an ammonium removal, Semmens (2005) calculated between four and forty times less power in an MABR than in a CAS reactor. Aybar *et al.* (2014) estimated power savings of 45 to 86% with a hybrid MABR compared to a CAS for COD and TN removal. Energy requirements at pilot or full-scale of 0.17 kWh/m^3 (BioGill), 0.10 kWh/m^3 (Shechter, 2015) or even less than 0.10 kWh/m^3 (Syron *et al.* 2014) have been reported.

The other highly remarkable feature of MABRs for urban wastewater treatment is their suitability for concurrent COD and total nitrogen removal, allowing for reduced tank volumes, elimination of internal water recycle and no need for external carbon and pH correctors. Nitrifying bacteria grow close to the membrane, where the oxygen concentration is the highest and heterotrophic denitrifying bacteria grow in the outer biofilm, close to the biofilm-liquid interface. Ammonium diffuses from the bulk liquid to the surface of the membrane, where nitrification takes place. Nitrites and nitrates are denitrified in the outer layer of the biofilm. Numerous studies have confirmed this microbial stratification and its feasibility for simultaneous nitrification and denitrification (e.g. Terada *et al.* 2003). In this application, the performance seems to depend largely on achieving an optimum biofilm thickness, due to the limitations to substrate diffusion of thick biofilms. An alternative to avoid excessive biofilm thickness is the hybrid MABR: biofilm on the membrane is devoted to nitrification and suspended biomass (Downing & Nerenberg, 2008) or plastic media (Landes *et al.* 2011) in the anoxic bulk liquid for organic removal and denitrification. In MABRs, membranes are arranged in modules or cassettes. They are appropriate both for retrofitting existing WWTP (to cope with higher loads and/or achieve nitrogen removal) and in new treatment plants.

7.2.2.4 Challenges in the use of gas transferring membranes

While the first reports on the MBfR technology concept date back to the 1980s, when the patent of Onishi *et al.* (1980) and the scientific paper of Timberlake *et al.* (1988) were published and pilot studies were carried out in the 1990s (Brindle *et al.* 1999), commercialisation of the MBfR is very recent. The first variant to come to the market was the hydrogen based MBfR by APTwater, Inc. in 2012 named ARoNiteTM. Although its main application is the treatment of groundwater polluted with nitrate and other oxidized compounds to obtain drinking water, an application in WWTP has also been reported. The technology was used to reduce the nitrate concentration of a tertiary effluent intended for reuse below the detectable limit of 0.5 mg-N/L. The aerated version (MABR) is currently being commercialized by a number of different companies throughout the world, BioGill (Australia), Emefcy (Israel), GE (Canada) and Oxymem (Ireland). BioGill and Emefcy both use flat sheet membranes come in modules, the modules are then built up to accommodate the desired flow and load. While GE and Oxymem have taken an alternative approach and have used hollow fibre membranes which are assembled into modules, these modules can then be installed into existing or specially designed tanks and to treat the entire flow or add additional capacity by providing both additional biomass and additional aeration.

The MBfR is a very promising technology combing significant energy savings with the process resilience and low sludge production of a fixed film system. Therefore the MABR has the potential to form a key component of the municipal wastewater treatment plant of the future, but to achieve this there needs to be a significant increase in the number of researchers investigating both the fundamental processes underlying this technology and also the parameters governing the operation and installation of the systems.

Although extensive work has already been done with respect to different aspects of MBfRs, more research and development is needed to fully exploit the potential of the technology. The priorities (Martin & Nerenberg, 2012; Aybar *et al.* 2014), can be categorized into two areas: performance and cost effectiveness.

As previously stated, maintaining an optimum biofilm thickness is a critical point to maximize reaction rates in MBfRs. The technology would benefit from more knowledge in the field of biofilm development and detachment mechanisms, both from basic research and with reference to technological issues (membrane materials, module design, scouring strategies). As for membranes performance, the prevention or removal of water from the lumen, and the effect of gas back-diffusion need to be addressed more in depth. In addition, more pilot-scale studies devoted to compare MBfRs and conventional processes would help to establish the most competitive niches for the technology. Some other aspects of the process, such as the characterization of the produced sludge, the effect of shock loads, start-up strategies or the fluid hydrodynamics in the reactor are also interesting research points.

MBfRs, and specifically MABRs, have the capacity for great operational savings mainly related with low energy needs for oxygenation, but also with less sludge processing and disposal costs. Moreover, the reduction in space requirements due to simultaneous nitrification and denitrification and the high biomass concentration associated with biofilm systems, can lead to a significant reduction in initial investment costs where available space is scarce. On the other side, the membranes cost and (and consequently membranes life expectancy) is the most important factor to determine the cost effectiveness of the technology. Although some preliminary studies have tried to assess the economic savings of MABRs compared to conventional systems (Aybar *et al.* 2014), there is a deep need for further research on the subject. The development of cheaper and more durable membrane materials and configurations, the study of operation modes to maximize energy savings or the impact of operation parameters on the treatment cost are some of the topics to deal with.

7.2.3 Microbial Desalination Cells (MDC) – anionic and cationic exchange membranes

Most bio-electrochemical reactors (BES), often called microbial fuel cells (MFCs), consist of an anode chamber in which biodegradable materials are oxidized by microorganisms generating electrons and cathode chamber with electron acceptors (such as oxygen) provided . The electrons migrate to the cathode through the external circuit. The anode and the cathode are separated by a proton exchange membrane (PEM) that acts as a solid electrolyte bridge. Protons migrate to the cathode from the anodic chamber through the membrane or directly across the wastewater in the case of membrane-less MFCs. Besides the electricity production, the electron flow can be used for reduction of various chemicals (MFC) and electron/proton interaction in the cathodic chamber leading to H_2 formation (MEC).

The membrane separator is one of the main components of the MFC that significantly affects its overall cost and power density. Though membrane free MFCs are accepted in some particular cases, the application of membranes offers higher coulombic efficiency (CE) compared with membrane free MFC in case of usage of mixed culture. It is attributed to the loss of substrate due to the presence of oxygen that is diffused through the cathode and the limited distance between the anode and cathode in a membrane-free MFC because the potential negative effect of oxygen on the activity of the anaerobic culture bacteria on the anode (Fan *et al.* 2007).

Ion exchange membranes separate the biological anode from the cathode reactions, while at the same time facilitating the transport of ions through the membrane in order to maintain electroneutrality in the system (Rozendal *et al.* 2006).

In BES applied for wastewater treatment typically cation exchange membranes (CEMs) are in use. Various studies have shown that in that case mainly cation species other than protons are responsible for the transport of positive charge through the membrane. This fact is due to much higher concentrations of cations in wastewater than protons (typically 105 times higher than the concentration of protons).

The transport of other cation species causes a pH increase at the cathode, which negatively affects the cell performance (Rozendal *et al.* 2006). Nafion membranes as one of the CEM representatives have long been the preferred separators for MFCs because their high proton conductivity attributed to the negatively charged hydrophilic sulfonate group attached to its hydrophobic fluorocarbon backbone. However, the high cost of Nafions forced the usage of other cheaper membranes such as the conventional cation exchange and anion exchange membranes (AEM); and polymer/composite membranes (Kim *et al.* 2007).

In attempting to reduce pH decrease in the anode chamber and the ion-transport resistance in MFCs several conventional CEM/AEM membranes were tested. Better performance of AEM compared to CEM has been shown (Kim *et al.* 2007). This better performance is caused mainly by the much lower internal resistance of the AEM configuration compared to the CEM configuration.

7.2.3.1 Principles and operation of MDCs

MFCs can be modified to contain of pair of anion and cation exchange membranes, resulting in cell that can both desalinate water, a microbial desalination cell (MDC). Basically, an MDC is a combination of an MFC and an electrodialysis (ED) cell. A typical MDC consists of three chambers, (i) an anode, middle (salt), and (ii) a cathode, separated by an anion exchange membrane, between the anode and the middle chambers, and (iii) a cation exchange membrane, between the cathode and the middle chambers, respectively (Figure 7.1). Different configurations have been proposed for MDCs, such as stacked reactor and upflow tubular reactor (Kim & Logan, 2011; Jacobson *et al.* 2011a; Ge *et al.* 2014). Optimized operation through recirculation between the anode and the cathode compartments could achieve pH balance without the addition of costly buffer (Chen *et al.* 2012). Fundamental studies have also been conducted to reveal the diverse microbial community in the anode of MDC (Luo *et al.* 2012a,b).

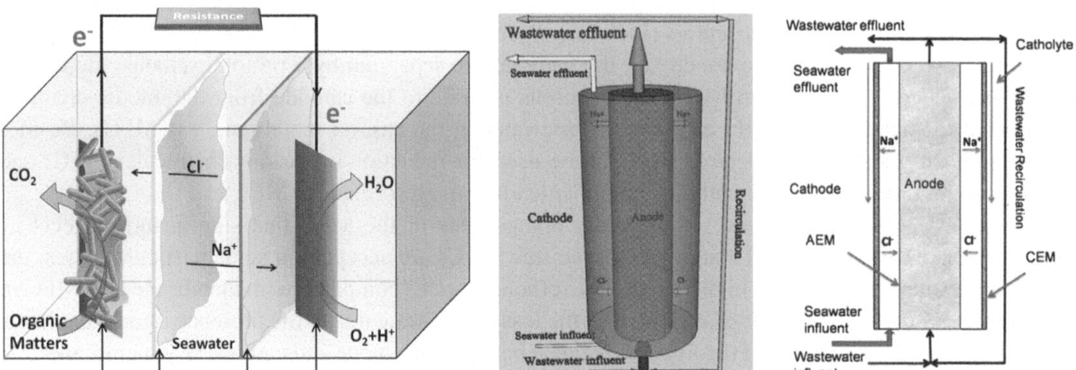

Figure 7.1 Schematic of a microbial desalination cell. Left, conventional configuration (Ping *et al.* 2013). Right, air cathode (concentric) configuration (Jacobson *et al.* 2011b).

It is worth noting that unlike in the MFC, protons released in the anolyte stay in the anolyte (in MFC's they diffuse across a PEM to combine with oxygen and electrons). Thus the pH in the anode cell will drop with time, inhibiting the activity of exoelectrogenic bacteria in the anode. At the same time the pH rises in the cathode cell due to the consumption of protons, in the catholyte. The resulting pH imbalance in MDCs is much greater than that in MFCs, and can limit the reactors functionality unless a buffer or some other method of stabilizing pH is used such as a circulation between the anode and cathode cells (Chen *et al.* 2012). Among BES operating at a fixed resistance, MFCs are set to operate at maximum power density whereas MDCs are set to maximum current intensity.

When electricity is produced through anode bacterial oxidation and cathode reduction, anions migrate into the anode chamber via an anion-exchange membrane and cations transport into the cathode chamber through a cation-exchange membrane. As a result, salts are "relocated" into the wastewater stream and salt concentration in the feed water can be greatly reduced. For every electron moving from the anode electrode to the cathode electrode, e.g., one Na^+ and one Cl^- is required to transport through ion-exchange membranes. The effluent from the middle chamber contains less salt, thereby accomplishing desalination; the anode and cathode effluents are enriched with salts. The addition of the desalination cell to the MFC gives the MDC its dual functionality: simultaneous wastewater treatment and desalination.

7.2.3.2 Performance of MDCs

While the dominant driving force to salt reduction in MDCs is the electrical potential gradient created by the bacterial oxidation of organic matter there are a few other factors that contribute as well. The first of these factors is the ionic concentration difference between salt solutions and wastewaters, also known as junction potential (φ_{jct}). Most wastewaters have an ionic content between 0.5 g/L to 0.9 g/L, whereas seawater has an ionic concentration of 35 g/L. This substantial concentration difference drives ions to move out of the salt solution, into the adjacent anode and cathode cells. Another factor that drives desalination in the MDC reactor is osmotic water transport. The osmotic water transport occurs due to the solvents tendency to move across the IEM from a region of lower solute concentration to a region of higher solute concentration. This phenomenon dilutes the salt solution and improves the water recovery in the three-cell MDCs.

The majority of MDC research has been conducted with synthetic wastewaters which are made of an easily biodegradable substrate, such as acetate or xylose. The resulting COD removals have been relatively high (mostly over 70%). When using actual wastewater the COD removal drops to approx. 50% with 66% salt removal and a CE of 31% (Luo *et al.* 2012a). Salinity reduction in three-cell MDCs has been reported to be above 90% with a starting feed concentration of 30–35 g/L NaCl. Although impressive, this high desalination rate required a salt solution to wastewater ratio of 1:133 (Kim & Logan, 2013). A continuous MDC operating at a hydraulic retention time of 0.8 d achieved a salt removal rate of 1.2 g/L · d from brackish water with an initial salinity of 5.9 g/L (CE varied between 5–10%), with nearly 70% removal of organic compounds in wastewater (Ping *et al.* 2015a). Thus for practical purposes MDC's are more likely to be used for partial salt removal from seawater, or for salt removal from brackish water. For example MDCs used for brackish water desalination achieved up to 99% removal from a starting salt concentration of 10 g/L NaCl, with a salt solution to wastewater concentration of 1:14 (Luo *et al.* 2010). Integrating experiments with different brackish water sources and mathematical models it was concluded that given sufficient HRT, the desalinated brackish water could meet the none-restriction irrigation standard in terms of both salinity and composition of individual ions (Ping *et al.* 2015a).

In either application, back diffusion of inorganic ions and organic substances can reduce the quality of the desalinated water. Having the adjacent anode cell containing wastewater separated from the salt solution by an AEM, MDCs generate concerns regarding the possible leakage of organic or inorganic contaminants from wastewater into the desalinated stream. A previous study of an MDC operated with a high acetate concentration discovered extensive leakage of acetic ions from the anolyte into the salt solution opposite the electrical field, thereby lowering the effluent quality and causing biofouling (Luo *et al.* 2012b). Additionally an MDC fed with wastewater (~260 mg/L total organic carbon) exhibited mild organic leakage (~7 mg/L) in the desalinated effluent (Ping *et al.* 2015a). Many inorganic anions can freely cross AEM, and thus leakage or back diffusion of anions from the anolyte into the salt solution against the electrical field is expected to occur to some degree. A small amount of back diffused organics might lead to microbial contamination of the desalinated water, and post-disinfection may be required to inhibit bacteria growth.

Such back diffusion of contaminants from the anolyte into the desalinated stream was reported to be controlled by two mechanisms, Donnan effect for negatively charged ion exchange and molecule transport driven by osmotic water flux. Current generation effectively suppressed Donnan effect from 68.2% to 7.2%, and then molecule transport became more responsible for back diffusion. Negatively charged or nonpolar organic molecules with molecular weight larger than 350 Da can be retained by the AEM resulting in less back diffusion (Kim & Logan, 2013). The ratio between osmotic water flux and salt solution flow rate was found to be a critical design criterion for MDC development. Desalinating seawater (a high-salinity stream) will likely cause more back diffusion than that brackish water with a lower salinity, because of a higher osmotic water flux resulted from a greater concentration gradient between the anolyte and salt solution. To reduce back diffusion, an MDC with low organic loading rates, small salt concentration gradient between anolyte and salt solutions, and high bioelectricity production should be favoured.

Comparing the a long term AEM/CEM behaviour in MDC, it was observed that the AEM are more sensitive to biofouling, resulting from wastewater and microbial growth on organic compounds migrating across the membrane, while the CEM had substantial inorganic scaling, mainly consisting of calcium and magnesium (Figure 7.2). The membrane resistance of the CEM increased more significantly than that of the AEM, indicating that CEM needs more maintenance during MDC operation (Ping *et al.* 2013).

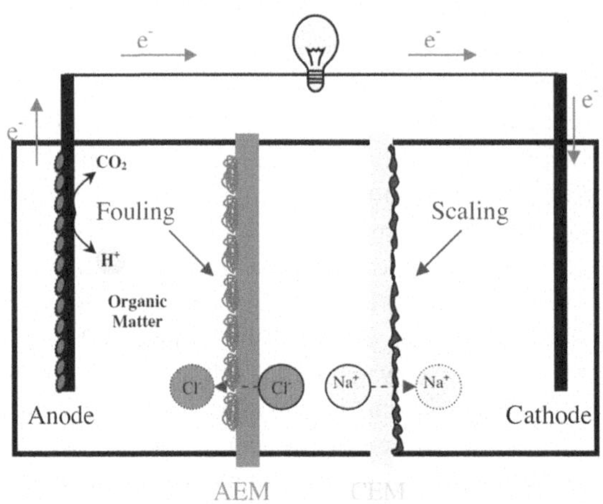

Figure 7.2 Schematic diagram depicting the main membrane fouling mechanism in a MDC (Luo *et al.* 2012a).

7.2.3.3 Anionic and Cationic exchange membranes

Another modification of MDC is the bipolar membrane MDC. The bipolar membrane consists of an anion and cation selective layers laminated (heat-pressed or glued) together as a single membrane. The bipolar membrane is normally placed next to the anode chamber of the MDC, thereby creating a four-chamber MDC (Buck, 2014). The bipolar membrane MDC are applied mainly for treating high salinity water.

Water passing through the membrane splits up into protons and hydroxyl ions at the interface of the bipolar membrane and a high potential gradient is created. As the organic matter is oxidized in the anode chamber, the hydroxide ions are released from the bipolar membrane into the anode chamber, while the hydrogen ions flow into the additional fourth chamber. Membrane properties influence the MDC performance significantly. These properties include high electivity permeability, low water splitting voltage drop and electrical resistance and long-life duration (Buck, 2014). Increasing the ion-exchange capacity of a membrane would increase the desalination efficiency from 50 to 63% (Saito *et al.* 2010). In general, bipolar membranes are more susceptible to organic and biological fouling than the conventional CEM/AEM membranes due to their exposure to wastewater in the anode chamber.

The comparative electrochemical performance of CEM, AEM, mosaic membrane (CMM) and bipolar membrane (BPM), Figure 7.3, used as ion exchange membranes in a bio-catalyzed electrolysis cell the membranes are rated in the order AEM-CEM-CMM-BPM (Rozendal *et al.* 2008). However, with respect to the transport numbers for protons and/or hydroxyl ions and the ability to prevent pH increase in the cathode chamber, the ion exchange membranes are rated in the order BPM-AEM-CMM-CEM. Charge-mosaic membranes represent a subset of bipolar membranes. They contain anion- and cation-permeable domains. Anions and cations can pass through the membrane without violation of microscopic electrical neutrality.

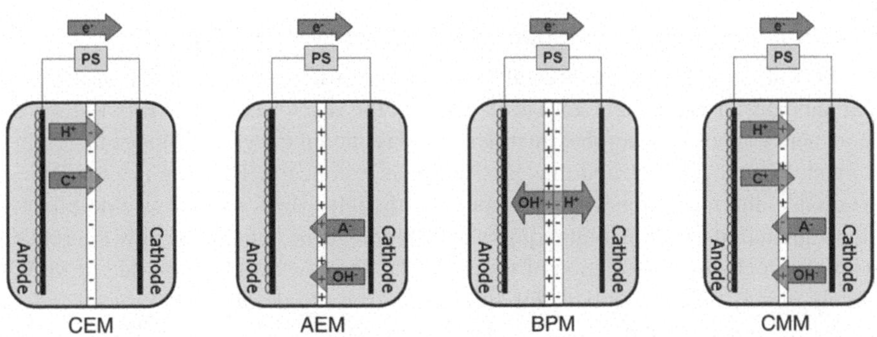

Figure 7.3 Theoretical working principle of membrane charge transport in four different types of ion exchange membranes used in biocatalyzed electrolysis: (A) CEM – through the transport of cations (ideally protons) from anode to cathode; (B) AEM – through the transport of anions (ideally hydroxyl ions) from cathode to anode; (C) BPM – through water splitting into protons and hydroxyl ions inside the membrane; (D) CMM – through the transport of cations (ideally protons) from anode to cathode AND/OR anions (ideally hydroxyl ions) from cathode to anode. PS = power supply, C^- = Cations, A^+ = Anions (Rozendal *et al.* 2008).

The available information related to the application of membranes in MFC shows that the most important properties which membrane separators must possess in order to improve the MFCs performance and enlarging MFC implementation (Leong *et al.* 2013) are:

- Low membrane resistance to facilitate proton transfer from anodic chamber to cathodic chamber,
- Non-porous or dense membranes to prevent the oxygen diffusion from the cathode chamber to the anaerobic anodic chamber and substrate crossover in the opposite direction,

- High biofouling resistance,
- Lower membrane cost.

In aiming to lower the membrane resistance several proton conductive materials, such as polybenzimidazoles, polyamides, polyether imides, polysulphones, polyphenylenesulfides, poly-etheretherketones and polyphenyquinoxalines have been used. Such materials can be protected from being easily fouled by adding hydrophilic nano-particles to increase their hydrophilicity which follows to higher resistant to biofouling. One example of cheaper MFC membranes are the functionalized zeolites synthesized by using 1,3-propane sulfone as a functionalization agent of sulfonic acids to improve the proton transport properties. Sulfonated polymer membranes containing sulfonate groups such as SPEEK membranes and BPSH are now increasingly being used to replace the more widely used Nafions membrane in MFCs to overcome the shortage of Nafions membranes and to lower MFC cost (Chae *et al.* 2007). SPEEK membranes are produced by sulfonating the cheaper, chemically and thermally stable and mechanically strong native PEEK membranes in order to increase proton conductivity. Other examples of composite materials are Zirfon (Pant *et al.* 2010) and Hyflon (Ieropoulos *et al.* 2009) which are alternatives of Nafions. Zirfon, which consists of 85 wt.% of a hydrophilic ZrO_2 powder and 15 wt.% polysulfone, is a macro porous organic-mineral material. In comparison with Nafion, Zirfon is of higher oxygen permeability (penetrability) which is detrimental to anodic reactions, but its specific resistance is much lower.

7.2.3.4 Challenges and future perspectives for the use of MFC's

The fact that MFC's utilize self-replicating catalysts (bacteria) to drive organic oxidation offers a sustainable and affordable method for energy generation. In addition the reactor has the ability to treat wastewaters while saving on aeration costs (anaerobic treatment), and reducing the cost of sludge handling (anaerobic bacteria have a slower growth rate and thus less sludge is produced). Thus the reactor offers an energy efficient, sustainable alternative for water treatment. Although MDCs can achieve desalination without external voltage applied, its desalination requires a fairly long time because of slow microbial oxidation, compared with electrochemical oxidation of water in ED. The low desalination rate of MDCs indicates that this technology may be more suitable for application as a pre-desalination process for conventional desalination such as reverse osmosis or ED. Another potential application of MDC technology is to desalinate low-salinity water, such as brackish water or treated wastewater effluents. Development of AEM and CEM membranes with high ionic transport capabilities and antifouling/biofouling properties will certainly increase the applicability of MDCs in integrated wastewater treatment/desalting or marginal water.

Independently of the MFC applications, serious challenge is the biofouling. Biofouling deteriorates membrane performance expressed by the reduced power output of MFC and leads to change in membrane properties. In aiming to solve the problem attempts have been done to reduce biofouling formation through anti-adhesion and anti- microbial techniques. The membrane internal resistance should be as low as possible, in order to increase current and power densities. In attempting to lower the internal resistance conductive membranes favouring the transportation of ions through the membrane can be utilized. One of the problems in applying CEM/AEM as separators used to reduce oxygen diffusion and substrate crossover in MFC is the pH splitting. The level of pH splitting depends on the types of membrane separator. MFCs using CEM as separators encounter severe pH splitting problems compared to the AEM choice. pH splitting in MFCs can be solved by using buffer solutions (phosphate and bicarbonate buffers), which can stabilize pH and also can increase solution conductivity.

Despite the high cost, the Nafion membrane are currently most popular membrane separator used in MFCs because of its chemical and thermal stability and high proton conductivity. By the other side they are suffering of severe proton (cation) accumulation, oxygen diffusion and loss of substrate. In aiming to lower MFC cost, sulfonated polymer membranes containing sulfonate groups are now increasingly being used to replace the widely used Nafion membrane in MFCs. As substitute of Nafion membrane, proton conductive materials based on polyamides, polyether imides, poly-sulphones, etc. has been synthesized. Recently, nanocomposite membranes have also attracted much interest in MFC applications, because there are proofs that nanocomposite membranes are cheaper and compatible in power production.

7.3 CONCLUSIONS AND PERSPECTIVES

Technological opportunities

The most mature technology is that of aerobic MBR treating urban wastewater with many full-scale applications around the world (TRL = 9). It has been accepted alternative for applications requiring effluent of high quality or reliability, small footprint, stringent nutrients discharge requirements or for water reuse reasons. Although a wide range of energy consumption is reported (0.5–4.2 kWh/m^3), in full-scale plants the typical energy requirements are below 1.5 kWh/m^3.

The AnMBR has much lower energy requirements however anaerobic conditions are less favorable for filtration and more prone to fouling. Operation at low temperature, nutrients removal limitations and optimum configuration are other identified challenges. This is why AnMBR has only very few full-scale applications (TRL = 8). Given the early stage of development and uncertainties around AnMBR performance, it is unclear how detailed design and operational decisions influence the economic and environmental performance of AnMBRs. Moreover, the selection of appropriate layouts for wastewater treatment should take into account not only economic terms (i.e. investment, operation and maintenance) but also environmental terms (e.g. eutrophication, global warming potential (GWP), marine ecotoxicity, etc.).

MBfRs, and specifically MABRs can achieve operational savings mainly related with low energy needs for oxygenation, but also with less sludge processing and disposal costs. Smaller space requirements reduce initial investment costs and process resilience adds to the technology potential. Maintaining an optimum biofilm thickness is of crucial importance for MBfR performance. The first full-scale system was launched in 2012 (TRL = 8).

MFC's offers a sustainable and affordable method for energy generation combined with the ability to treat wastewaters with reduced cost of sludge handling. Until now, MFC have only been applied at bench and pilot scale without full-scale applications (TRL = 5). MDCs can achieve desalination without external voltage applied but low desalination rate indicates application as a pre-desalination process or to desalinate low-salinity water (brackish water or treated wastewater effluents). Attempts have been done to reduce biofouling formation, which reduce power output of MFC and lead to change in membrane properties, through anti-adhesion and anti- microbial techniques.

Electric power requirements

Table 7.1 shows the typical energy consumption of different technologies involved in the urban wastewater treatment, considering preliminary treatment, CAS, high rate activated sludge (HRAS), MBR, AnMBR, MBfR, MABR and MDC processes.

Table 7.1 Typical energy consumption of different technologies involved in the UWW treatment.

System	Energy Consumption (kWh/m^3)	Sources
Preliminary treatment	0.16–0.30	Meda and Cornel 2010
CAS	0.19–1.4	Fenu *et al.* 2010
	0.3–1.4	Meda and Cornel 2010
HRAS	−0.079–0.13	Smith *et al.* 2014
MBR	0.5–2.5	Meda and Cornel 2010
	0.7–1.8	Palmowski *et al.* 2010
	0.4	Tao *et al.* 2010
	0.9–3.9	Judd and Judd 2011
	0.4–4.2	Krzeminski *et al.* 2012
	0.8–2.4	Barillon *et al.* 2013
	0.4–2.1	Gabarron *et al.* 2014
	0.8–3.0	Itokawa *et al.* 2014
AnMBR	−0.15–0.21	Smith *et al.* 2014
	−0.07–0.22	Ferrer *et al.* 2015
	−0.12–0.19	Pretel *et al.* 2015
	0.03–0.23	Pretel *et al.* 2016
MABR	0.10[a]	Casey *et al.* 2008
	0.09–0.12[b]	Casey *et al.* 2008
	0.02	EMEFCY (2011)
	<0.10	Syron *et al.* 2014
	0.10	Shechter, 2015
	0.07	EMEFCY (2016)
	0.3	BioGill (a)
MDC	−0.12– −0.04	Jacobson *et al.* 2011; Jiang *et al.* 2012; He *et al.* 2013

[a]Theoretical comparison between CAS and MABR for 10,000 m^3/d, including oxygenation and other energy consumption. For total N removal with membrane as the sole source of oxygen. The respective value for CAS was 0.35 kWh/m^3.
[b]Theoretical comparison between CAS and MABR for 3,792 and 115,200 m^3/d, including oxygenation and mixing. For BOD and NH_4 removal with membrane as the sole source of oxygen. The respective values for High Purity Oxygen Activated Sludge and CAS were 0.15 and 0.5 kWh/m^3.

According to the recently reported data from full-scale MBRs treating UWW, the typical specific energy consumption is in the range of 0.7–1.6 kWh/m^3 (Palmowski *et al.* 2010; Krzeminski *et al.* 2012; Gabarron *et al.* 2014; Itokawa *et al.* 2014). On the other hand, the energy consumption may be as high as 2.2–3.0 kWh/m^3 (Judd & Judd, 2011; Barillon *et al.* 2013) or as low as 0.4 kWh/m^3 for fully optimised MBRs (Tao *et al.* 2010) and new generation of MBRs (Xiao *et al.* 2014).

The energy consumption of AnMBR and HRAS ranged from −0.15 to 0.21 kWh/m^3 and from −0.08 to 0.13 kWh/m^3, respectively (Smith *et al.* 2014). However, AnMBR energy demand does not usually include the energy needed to remove nutrients as in the case of the rest wastewater treatment processes. Nevertheless, Pretel *et al.* (2016) evaluated the energy demand of an AnMBR system coupled to CAS or aerobic MBR for nutrient removal, resulting in energy demands from 0.03 to 0.23 kWh/m^3. Nonetheless, the results obtained in each case study depend on influent wastewater characteristics and operating conditions.

The main feature of MABRs is their potential to greatly save energy consumption for aeration, due to the possibility to achieve high oxygen transfer efficiencies with low pressure blowers. Some theoretical comparisons between MABRs and CAS gave 45 to 86% power savings with a hybrid MABR for COD and TN removal (Aybar *et al.* 2014), between 15 and 50 times less energy for a nitrification system (Lackner, 2009) and 4 to 40 times inferior for COD and NH_4 removal (Semmens, 2005). In the latter case, the big differences are explained by different OTE assumptions (10 and 30% respectively). Some authors are working on increasing OTE while maintaining high oxygen transfer rates by applying specific membrane venting strategies (Perez-Calleja *et al.* 2015). Pilot results (Côté *et al.* 2015) showed aeration efficiencies at process conditions in a MABR of 4 kgO_2/kWh, with a potential to achieve 6 kgO_2/kWh, compared to typical aeration efficiencies for fine-pore diffusers of 0.7–2.6 kgO_2/kWh (Stenstrom & Rosso, 2008). Also at pilot scale, Syron *et al.* (2014) compared a MABR with an existing bubble-based aeration system in a STP, obtaining 70% reduction in aeration energy. The reported MABR energy requirements are of <0.10–0.3 kWh/m^3. To the authors knowledge, no references on energy consumption or costs for H_2 based MBfR have been reported.

Capital and operational costs

The cost of any system will vary widely both in CAPEX and OPEX depending on a range of factors including capacity, type of system, pretreatment, local personnel and electric power costs, etc. For the MBRs, membrane prices have been shown to be steadily falling and have reached the mature stage of development. Additionally, the price of an MBR system will depend on the negotiation as higher volume systems will result in lower per part prices. Turnkey systems such as those offered by GE are by the nature of their build different than those constructed from various component suppliers. Construction prices will also differ as the cost of land and labor in Rome, Italy will be exponentially higher than Rome, Georgia, US. As far as OPEX, labor and energy costs of operating the plant will mirror that of building the plant as it will depend on the locale of the system.

Many publications discuss experiments in which differing variations on an MBR system are monitored in order to measure performance to cost. However, for market purposes these degrees of granularity do not simplify the decision process or answer the basic question of, "how much does an MBR cost?" Such studies do not have a high enough sample volume of systems or take into account the differing brands of components available on the market, as they are scientific studies focused on specific technologies not market studies focused on average prices across a wide range. Multiple units across multiple locations are, for market purposes, the only method in which to show real world applications in which an average cost can be projected.

One study commissioned by Ovivo in the US, published in 2012, did focus on the average. Although it should be taken into account that they employ flat sheet membranes systems and the study does show that such systems have a significant cost advantage, the size of the sample helps to clarify average CAPEX while adjusting for system type and system capacity variations.

A study that began in 2008, it was said that due to the fact that most systems in the US are below 6 mgd in capacity, the sample size was limited to 84 plants in which 43 were flat sheet and 41 were hollow fibre type systems (Figure 7.4). Due to the roughly comparable costs being from one country, the fact that they were municipal only, and a sample size greater than one to a handful based on a scientific study, this can serve as a comparable tool in which to understand market costs.

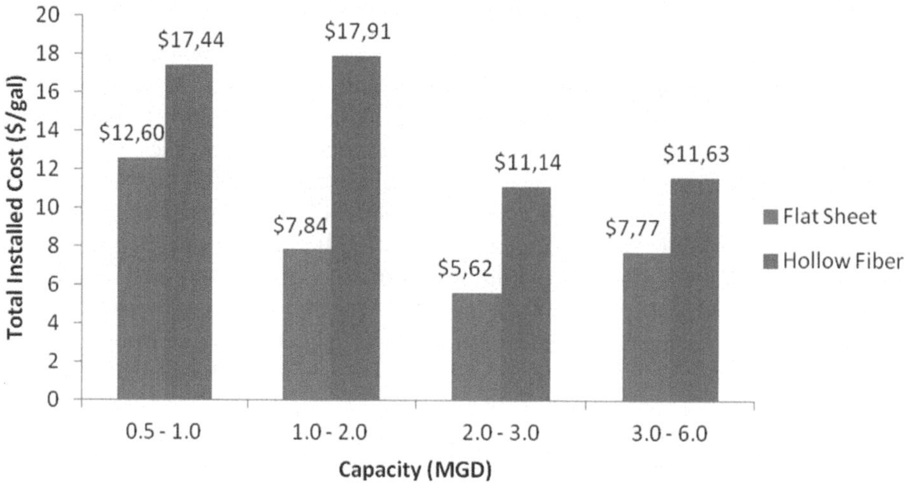

Figure 7.4 Survey of comparative CAPEX of flat sheet and hollow fibre MBR systems. (*Source:* MBRs Don't Have to Break the Bank, Kimberly Mathis, Ovivo, 19th January 2012).

When comparing MABR with conventional technologies, three potential sources of costs savings must be considered: energy cost savings in aeration, less sludge processing and disposal costs and reduction in space requirements. The available theoretical studies that have compared CAS with MABR have shown that the potential economic savings of MABRs strongly depends on membranes price and electricity costs. Casey *et al.* (2008) obtained that MABR offers lower total costs with respect to CAS for COD and total nitrogen removal as long as the price of membrane is lower than 40 €/m^2 and electricity costs higher than 0.1 €/kWh. Aybar *et al.* (2014) claim that savings up to US$/1,000 m^3 of treated water could be achieved depending on the costs of membranes and electricity (comparing hybrid MABRs and CAS for COD and N removal). Membrane lifetime and the oxidation rates of COD and N are also critical factors. More research is needed to better quantify the capital and operational costs of MABRs. With respect to H_2-based MBfRs, no studies for sewage treatment are available to date. A preliminary analysis to treat drinking water sources contaminated with perchlorate found the technology cost competitive (Adham *et al.* 2004).

Table 7.2 shows typical capital and operational costs of different technologies applied for urban wastewater treatment.

A mini-assessment of different technologies applied for urban wastewater treatment discussed in this chapter is presented in Table 7.3.

Challenges and future perspectives

Despite continuous improvements and developments the membrane based technologies applied in wastewater treatment still need to overcome key challenges in order to further expand their applications and broaden their market penetration. The list of key challenges identified is presented below.

Table 7.2 Typical CAPEX and OPEX of different technologies involved in the UWW treatment.

System	CAPEX	OPEX (€/m³)	References
MBR	269–487 €/PE,	0.08–0.12[a]	Coté *et al.* 2004
	1499–2240 €/m³ · d	0.26	Engelhard and Lindner 2006
	345–475€/PE[b]	0.36	De Wever *et al.* (2008)
	250–650 €/PE[c]	0.29	Krzeminski, 2013
	250–450 €/PE[4]		Brepols, 2010
	720–1260 €/m³ · d		Lesjean *et al.* 2005
			Adapted from BioGill (b)
AnMBR	0.048–0.058 €/m³	0.012–0.053	Ferrer *et al.* 2015
	0.070–0.077	0.056–0.057	Pretel *et al.* 2016
MBfR	N.A.	N.A.	
MABR	540–810 €/m³	0.041	Adapted from BioGill (a, b) [e]
	732 €/m³ · d	0.010	Adapted from Casey *et al.* (2008)[f]
MDC	N.A.	N.A.	

[a]OPEX for 38,000 m³/d MBR equipped with hollow fibre membranes excluding personnel cost.
[b]CAPEX including civil, mechanical, electromechanical work and membranes.
[c]CAPEX for 45,000 m³/d, 80,000 PE, for 6430 m³/d, 9000 PE.
[d]CAPEX for 80,000–3000 PE.
[e]OPEX only includes electricity, which is calculated based on $0.15/KWh. 1$ = 0.9€. Costs estimated only for STP and food and beverage processes within the range of 0–500 m³/d.
[f]OPEX includes electricity, sludge disposal and labor for a hypothetical case with 10,000 m³/d wastewater.

Table 7.3 Mini-assessment of different technologies involved in the UWW treatment.

Parameter	MBR	AnMBR	MABR	MDC
COD removal efficiencies	91–96%	>90%	75–96%	50–90%
N removal efficiencies	95–98%	–	80–94% NH_4 66–78% TN	>90%
P removal efficiencies	67–85%[a] 94–96%[b]	–	–	>95%
Flow rates able to be treated	10–50 LMH	15–20 LMH	N.A.	N.A.
HRT	4–28 h	15–20 h	8–28 h	0.3–2 d
SRT	15–42 d	>15 d[c]	n.a.	N.A.

[a]Without chemical P removal.
[b]With chemical P removal.
[c]Depending on ambient operating temperature.

Economics of membrane use aspects

The aspects related to the economics of membrane use covers cost effective membrane capital costs, membrane life expectancy and also energy consumption. The specific challenges related to the economics of membrane use are:

- Energy efficiency and competitive cost per m³ wastewater treated

The operation strategy of the MBR significantly affects the energy consumption; significant energy savings can be achieved by optimizing the main operating parameters. At optimal operating conditions large MBRs can reach as low as 0.4 kWh/m^3 in terms of specific energy requirements. Configurations and operation modes to maximize energy savings in novel MBR configurations such as, Membrane Biofilm Reactors (MBfR) or Anaerobic Membrane Bioreactors (AnMBRs).

- Cost effective membrane materials production

Membrane capital costs, durability against washing chemicals and life expectancy.

- Up-scaling and down-scaling issues

MBR technology has become a more attractive solution for medium sized plants. However, in small and decentralized systems, MBR suffer from 'down-scale' design approach of large MBRs. On the other hand, MBfRs are tackling the scale-up of the technology issues.

Fouling & biofilm control aspects

Fouling remains the main operational drawback of MBRs, increasing the operating expenses due to the need for membrane scouring and chemical cleaning. Hence, the life span of the membrane itself decreases. The specific challenges related to fouling control are the following:

- Development of innovative antifouling membranes
- Stable flux production for long term operation

The irreversible and irrecoverable fouling is hampering long term operation of membrane processes gradually reducing production rates and/or intervals between membrane cleanings.

- Effective membrane cleaning procedures

Novel fouling mitigation methods providing effective and/or low-energy membrane cleaning methods are still needed.

- Identification of appropriate pre-treatment protocols for mitigating fouling problem (tailored pretreatment)
- Maintaining of an optimum biofilm thickness, in cases when biofilm development is required e.g. MBfR.

Environmental aspects

- Development of a holistic approach for the environmental impact of full-scale MBRs and MBfRs integrating LCA, model analysis, water quality indicators, impact categories.
- Biogas up-grading and removal of methane dissolved in the effluent of anaerobic reactors treating raw sewage (AnMBR).

7.4 REFERENCES

Achilli A., Cath T. Y., Marchand E. A. and Childress A. E. (2009). The forward osmosis membrane bioreactor: a low fouling alternative to MBR processes. *Desalination*, **239**(1–3), 10–21.

Achilli A., Marchand E. A. and Childress A. E. (2011). A performance evaluation of three membrane bioreactor systems: aerobic, anaerobic, and attached-growth. *Water Science and Technology*, **63**(12), 2999–3005.

Adham S., Gillogly T., Lehman G., Rittmann B. and Nerenberg R. (2004). Membrane Biofilm Reactor Process for Nitrate and Perchlorate Removal. AWWA Research Foundation, Denver, Colorado, USA.

Akamatsu K., Lu W., Sugawara T. and Nakao S.-i. (2010). Development of a novel fouling suppression system in membrane bioreactors using an intermittent electric field. *Water Research*, **44**(3), 825–830.

Aybar M., Pizarro G., Boltz J. P., Downing L. and Nerenberg R. (2014). Energy-efficient wastewater treatment via the air-based, hybrid membrane biofilm reactor (hybrid MBfR). *Water Science and Technology*, **69**(8), 1735–1741.

BioGill, (a) Decentralised sewage. Available at: http://www.biogill.com/uploads/74946/ufiles/BioGill_Decentralised_Sewage_email.pdf. 26 May 2016.

BioGill technology comparison. Available at: www.biogill.com/uploads/74946/ufiles/BioGill_Technology_Comparison.pdf. 26 May 2016.

BioGill, (b) BioGill technology comparison. Available at: www.biogill.com/uploads/74946/ufiles/BioGill_Technology_Comparison.pdf. 26 May 2016.

Baek S. H. and Pagilla K. R. (2006). Aerobic and anaerobic membrane bioreactors for municipal wastewater treatment. *Water Environment Research*, **78**, 133–140.

Bani-Melhem K. and Elektorowicz M. (2011). Performance of the submerged membrane electro-bioreactor (SMEBR) with iron electrodes for wastewater treatment and fouling reduction. *Journal of Membrane Science*, **379**(1–2), 434–439.

Barillon B., Ruel S. M., Langlais C. and Lazarova V. (2013). Energy efficiency in membrane bioreactors. *Water Science and Technology*, **67**(12), 2685–2691.

Batstone D. J., Puyol D., Flores-Alsina X. and Rodríguez J. (2015). Mathematical modelling of anaerobic digestion processes: applications and future needs. *Reviews in Environmental Science and Bio/Technolog*, **14**, 595–613.

Bérubé P. R., Hall E. R. and Sutton P. M. (2006). Parameters governing permeate flux in an anaerobic membrane bioreactor treating low-strength municipal wastewaters: a literature review. *Water Environment Research*, **78**, 887–896.

Bilad M. R., Arafat H. A. and Vankelecom I. F. (2014). Membrane technology in microalgae cultivation and harvesting: a review. *Biotechnol Advances*, **32**(7), 1283–1300.

BioGill (b) BioGill technology comparison.

Boyle-Gotla A., Jensen P. D., Yap S. D., Pidou M., Wang Y. and Batstone D. J. (2014). Dynamic multidimensional modelling of submerged membrane bioreactor fouling. *Journal of Membrane Science*, **467**, 153–161.

Brepols C. (2010). Operating Large Scale Membrane Bioreactors for Municipal Wastewater Treatment. IWA Publishing, London, UK.

Brindle K., Stephenson T. and Semmens M. J. (1999). Pilot-plant treatment of a high-strength brewery wastewater using a membrane-aeration bioreactor. *Water Environment Research*, **71**(6), 1197–1204.

Buck R. P. (2014). Ion-selective Membranes and Electrodes. AccessScience, McGraw-Hill Education. http://accessscience.com/content/351810

Casey E., Syron E., Shanahan J. W. and Semmens M. J. (2008). Comparative Economic Analysis of Full Scale MABR Configurations. IWA North American Membrane Research Conference, University of Massachusetts, Amherst, USA.

Chae K. J., Choi M., Ajayi F. F., Park W., Chang I. S. and Kim I. S. (2007). Mass transport through a proton exchange membrane (Nafion) in microbial fuel cells. *Fuels*, **22**, 169–176.

Chen X., Liang P., Wei Z., Zhang X. and Huang X. (2012). Sustainable water desalination and electricity generation in a separator coupled stacked microbial desalination cell with buffer free electrolyte circulation. *Bioresource Technology*, **119**, 88–93.

Christian S., Grant S., Wilson D., McCarthy P., Mills D. and Kolakowski M. (2011). The first two years of full-scale anaerobic membrane bioreactor (AnMBR) operation treating a high strength industrial wastewater at Kens Foods Inc. *Water Pract Technol*, **6**, 4019–4033.

Chu L. B., Yang F. L. and Zhang X. W. (2005). Anaerobic treatment of domestic wastewater in a membrane-coupled expended granular sludge bed (EGSB) reactor under moderate to low temperature. *Process Biochemistry*, **40**, 1063–1070.

Cornelissen E. R., Harmsen D., de Korte K. F., Ruiken C. J., Qin J.-J., Oo H. and Wessels L. P. (2008). Membrane fouling and process performance of forward osmosis membranes on activated sludge. *Journal of Membrane Science*, **319**(1–2), 158–168.

Coté P., Masini M. and Mourato D. (2004). Comparison of membrane options for water reuse and reclamation. *Desalination*, **167**, 1–11.

Côté P., Peeters J., Adams N., Hong Y., Long Z. and Ireland J. (2015). A New Membrane-Aerated Biofilm Reactor for Low Energy Wastewater Treatment: Pilot Results. WEFTEC 2015, Chicago, USA.

Cuevas-Rodriguez G., Cervantes-Aviles P., Torres-Chavez I. and Bernal-Martinez A. (2015). Evaluation of different configurations of hybrid membrane bioreactors for treatment of domestic wastewater. *Water Science and Technology*, **71**(3), 338–346.

De Wever H., Brannock M., Leslie G. and Brepols C. (2008). Inside or Outside Submerged MBR: Which one is Better? Membranes in Drinking Water Production and Wastewater Treatment. IWA conference, Toulouse, France.

Dong Q. (2015). Characterization of Anaerobic Membrane Bioreactors (AnMBR) Treating Municipal Wastewater. University of Waterloo, Waterloo, Ontario, Canada.

Downing L. S. and Nerenberg R. (2008). Total nitrogen removal in a hybrid, membrane-aerated activated sludge process. *Water Research*, **42**, 3697–3708.

Drexler I. L. C. and Yeh D. H. (2014). Membrane applications for microalgae cultivation and harvesting: a review. *Reviews in Environmental Science and Bio/Technolog*, **13**(4), 487–504.

EMEFCY (2011). Spiral aerobic biofilm reactor low energy wastewater treatment. Available at: http://www.emefcy.com/files/files/Emefcy%20-%20SABRE%2011–2011.pdf. 01 July 2016.

EMEFCY (2016). MABR-Membrane Aerated Biofilm Reactor. A revolution in the economics of wastewater treatment. Available at: http://www.emefcy.com/files/files/emefcy%20brochure%202016–1.pdf. 01 July 2016.

Engelhard N. and Lindner W. (2006). Experiences with the world's largest municipal waste water treatment plant using membrane technology. *Water Practice and Technology*, **1**(4). http://wpt.iwaponline.com/content/1/4/wpt2006058

Fan Y., Hu H. and Liu H. (2007). Enhanced coulombic efficiency and power density of air-cathode microbial fuel cells with an improved cell configuration. *Journal of Power Sources*, **171**(2), 348–354.

Fenu A., Roels J., Wambecq T., De Gussem K., Thoeye C., De Gueldre G. and Van De Steene B. (2010). Energy audit of a full scale MBR system. *Desalination*, **262**, 121–128.

Ferrer J., Pretel R., Durán F., Giménez J. B., Robles A., Ruano M. V., Serralta J., Ribes J. and Seco A. (2015). Design methodology for submerged anaerobic membrane bioreactors (AnMBR): a case study. *Separation and Purification Technology*, **141**, 378–386.

Gabarron S., Ferrero G., Dalmau M., Comas J. and Rodriguez-Roda I. (2014). Assessment of energy-saving strategies and operational costs in full-scale membrane bioreactors. *Journal of Environmental Management*, **134**, 8–14.

Ge Z., Dosoretz C. and He Z. (2014). Effects of numbers of cell pairs on the performance of microbial desalination cells. *Desalination*, **341**, 101–106.

Giménez J. B., Robles A., Carretero L., Durán F., Ruano M. V., Gatti M. N., Ribes J., Ferrer J. and Seco A. (2011). Experimental study of the anaerobic urban wastewater treatment in a submerged hollow-fibre membrane bioreactor at pilot scale. *Bioresource Technology*, **102**, 8799–8806.

Gouveia J., Plaza F., Garralon G., Fdz-Polanco F. and Peña M. (2015b). A novel configuration for an anaerobic submerged membrane bioreactor (AnSMBR). Long-term treatment of municipal wastewater under psychrophilic conditions. *Bioresource Technology*, **198**, 510–519.

http://www.waterworld.com/articles/2015/03/stockholm-to-upgrade-wastewater-treatment-plant-with-ge-technology.html

He Z., Zhang F. and Ge Z. (2013). Using Microbial Fuel Cells to Treat Raw Sludge and Primary Effluent for Bioelectricity Generation: Final Report.

Herrera-Robledo M. and Noyola A. (2015). The evolution of pore-blocking during the ultrafiltration of anaerobic effluent-like mixtures. *Separation and Purification Technology*,, **147**, 172–178.

Ho J. and Sung S. (2009a) Effects of solid concentrations and cross-flow hydrodynamics on microfiltration of anaerobic sludge. *Journal of Membrane Science*, **345**, 142–147.

Ho J. H. and Sung S. W. (2009b) Anaerobic Membrane Bioreactor Treatment of Synthetic Municipal Wastewater at Ambient Temperature. *Water Environment Research*, **81**, 922–928.

Huang L. and Lee D.-J. (2015). Membrane bioreactor: a mini review on recent R&D works. *Bioresource Technology*, **194**, 383–388.

Ieropoulos I., Greenman J. and Melhuish C. (2009). Improved energy output levels from small-scale microbial fuel cells. *Bioelectrochem*, **78**, 44–50.

Itokawa H., Tsuji K., Yamashita K. and Hashimoto T. (2014). Design and operating experiences of full-scale municipal membrane bioreactors in Japan. *Water Science and Technology*, **69**(5), 1088–1093.

Jacobson K. S., Drew D. M. and He Z. (2011). Use of a Liter-Scale Microbial Desalination Cell As a Platform to Study Bioelectrochemical Desalination with Salt Solution or Artificial Seawater.

Jacobson K. S., Drew D. M. and He Z. (2011a) Efficient salt removal in a continuously operated upflow microbial desalination cell with an air cathode. *Bioresource Technology*, **102**, 376–380.

Jacobson K. S., Drew D. M. and He Z. (2011b) Use of a liter-scale microbial desalination cell as a platform to study bioelectrochemical desalination with salt solution or artificial seawater. *Environmental Science & Technology*, **45**, 4652–4657.

Jeison D. and van Lier J. B. (2006). On–line cake–layer management by trans–membrane pressure steady state assessment in anaerobic membrane bioreactors for wastewater treatment. *Biochemical Engineering Journal*, **29**, 204–209.

Jiang T., Zhang H., Gao D., Dong F., Gao J. and Yang F. (2012). Fouling characteristics of a novel rotating tubular membrane bioreactor. Chemical Engineering and Processing. *Process Intensification*, **62**, 39–46.

Judd S. and Judd C. (2011). The MBR Book, 2nd edn., Elsevier, pp. 536. https://www.elsevier.com/books/the-mbr-book/judd/978-0-08-096682-3

Kim J. R., Cheng S., Oh S.-E. and Logan B. E. (2007). Power generation using different cation, anion and ultrafiltration membranes in microbial fuel cells. *Environmental Science & Technology*, **41**, 1004–1009.

Kim Y. and Logan B. E. (2011). Series assembly of microbial desalination cells containing stacked electrodialysis cells for partial or complete seawater desalination. *Environmental Science & Technology*, **45**, 5840–5845.

Kim Y. and Logan B. E. (2013). Microbial desalination cells for energy production and desalination. *Desalination*, **308**, 122–130.

Krzeminski P. (2013). Activated Sludge Filterability and Full-Scale Membrane Bioreactor Operation. Delft University of Technology, Delft.

Krzeminski P., van der Graaf J. H. J. M. and van Lier J. B. (2012). Specific energy consumption of membrane bioreactor (MBR) for sewage treatment. *Water Science and Technology*, **65**(2), 380–394.

Lackner S. (2009). Towards Rational Design of Redox-Stratified Biofilms: A Novel Ppproach for Developing Robust Biotechnologies for Nutrient Removal from Wastewaters. Technical University of Denmark, Lyngby, Denmark.

Laera G., Chong M. N., Jin B. and Lopez A. (2011). An integrated MBR–TiO2 photocatalysis process for the removal of Carbamazepine from simulated pharmaceutical industrial effluent. *Bioresource Technology*, **102**(13), 7012–7015.

Landes N. C., Jackson W. A. and Morse A. N. (2011). Limitations encountered for the treatment of a low C:N waste using a modified membrane-aerated biofilm reactor. *Water Environment Research*, **83**(2), 128–139.

Leong J. X., Wan R., Wan D., Mostafa G., Kien B. L. and Manal I. (2013). Ion exchange membranes as separators in microbial fuel cells for bioenergy conversion. *Renewable and Sustainable Energy Reviews*, **28**, 575–587.

Lesjean B., Gnirrs R., Buisson H., Keller S., Tazi-Pain A. and Luck F. (2005). Outcome of a 2-year investigation on enhanced biological nutrients removal and trace organic elimination in membrane bioreactor (MBR). *Water Science and Technology*, **52**, 453–460.

Lin H., Peng W., Zhang M., Chen J., Hong H. and Zhang Y. (2013). A review on anaerobic membrane bioreactors: applications, membrane fouling and future perspectives. *Desalination*, **314**, 169–188.

Liu L., Liu J., Gao B. and Yang F. (2012b) Minute electric field reduced membrane fouling and improved performance of membrane bioreactor. *Separation and Purification Technology*, **86**, 106–112.

Luo H., Jenkins P. E. and Ren Z. (2010). Concurrent desalination and hydrogen generation using microbial electrolysis and desalination cells. *Environmental Science & Technology*, **45**, 340–344.

Luo H., Xu P. and Ren Z. (2012a) Long-term performance and characterization of microbial desalination cells in treating domestic wastewater. *Bioresource Technology*, **120**, 187–193.

Luo H., Xu P., Roane T. M., Jenkins P. E. and Ren Z. (2012b) Microbial desalination cells for improved performance in wastewater treatment, electricity production, and desalination. *Bioresource Technology*, **105**, 60–66.

Martin K. J. and Nerenberg R. (2012). The membrane biofilm reactor (MBfR) for water and wastewater treatment: principles, applications, and recent developments. *Bioresource Technology*, **122**, 83–94.

Meda A. and Cornel P. (2010). Energy and Water: Relationships and Recovery Potential. IWA Water and Energy conference, 10–12 November 2010, Amsterdam, The Netherlands.

Meng F., Chae S.-R., Shin H.-S., Yang F. and Zhou Z. (2011). Recent Advances in Membrane Bioreactors: configuration Development, Pollutant Elimination, and Sludge Reduction. *Environmental Engineering Science*, **29**(3), 139–160.

Onishi, H., Numazawa, R. and Takeda, H. (1980). Process and apparatus for waste water treatment. U.S. Patent (ed), U.S.

O'Flaherty V., Collins G. and Mahony T. (2006). The microbiology and biochemistry of anaerobic bioreactors with relevance to domestic sewage treatment. *Reviews in Environmental Science and Bio/Technol*, **5**, 39–55.

Padmasiri S. I., Zhang J., Fitch M., Norddahl B., Morgenroth E. and Raskin L. (2007). Methanogenic population dynamics and performance of an anaerobic membrane bioreactor (AnMBR) treating swine manure under high shear conditions. *Water Research*, **41**, 134–144.

Palmowski L., Veltmann K. and Pinnekamp J. (2010). Energy Optimization of Large-Scale Membrane Bioreactors – Importance of the Design Flux. IWA MTWR Conference, Istanbul, Turkey.

Pant D., Bogaert G. V., Smet M. D., Diels L. and Vanbroekhoven K. (2010). Use of novel permeable membrane and air cathodes in acetate microbial fuel cells. *Electrochimica Acta*, **55**(26), 7710–7716.

Perez-Calleja P., Aybar M., Picioreanu C., Esteban-Garcia A. L. and Nerenberg R. (2015 Enhanced Operation of Membrane-Biofilm Reactor (MBfR) for Waste Water Treatment by Mitigation of Gas Back-Diffusion Effect. WEFTEC 2015, Chicago, USA.

Ping Q., Cohen B., Dosoretz C. and He Z. (2013). Long-term investigation of fouling of cation and anion exchange membranes in microbial desalination cells. *Desalination*, **325**, 48–55.

Ping Q., Huang Z., Dosoretz C. and He Z. (2015a) Integrated experimental investigation and mathematical modeling of brackish water desalination and wastewater treatment in microbial desalination cells. *Water Research*, **77**, 13–23.

Pollice A., Laera G., Cassano D., Diomede S., Pinto A., Lopez A. and Mascolo G. (2012). Removal of nalidixic acid and its degradation products by an integrated MBR-ozonation system. *Journal of Hazardous Materials*, **203–204**, 46–52.

Pretel R., Robles A., Ruano M. V., Seco A. and Ferrer J. (2013). Environmental impact of submerged anaerobic MBR (SAnMBR) technology used to treat urban wastewater at different temperatures. *Bioresource Technology*, **149**, 532–540.

Pretel R., Durán F., Robles A., Ruano M. V., Ribes J., Serralta J. and Ferrer J. (2015). Designing an AnMBR-based WWTP for energy recovery from urban wastewater: the role of primary settling and anaerobic digestion. *Separation and Purification Technology*, **156**, 132–139.

Pretel R., Robles A., Ruano M. V., Seco A. and Ferrer J. (2016). Economic and environmental sustainability of submerged anaerobic MBR-based (AnMBR-based) technology as compared to aerobic-based technologies for moderate-/high-loaded urban wastewater treatment. *Journal of Environmental Management*, **166**, 45–54.

Rafiei B., Naeimpoor F. and Mohammadi T. (2014). Bio-film and bio-entrapped hybrid membrane bioreactors in wastewater treatment: comparison of membrane fouling and removal efficiency. *Desalination*, **337**, 16–22.

Robles A., Ruano M. V., Ribes J. and Ferrer J. (2012). Sub-critical long-term operation of industrial scale hollow-fibre membranes in a submerged anaerobic MBR (HF-AnMBR) system. *Separation and Purification Technology,* **100**, 88–96.

Robles A., Ruano M. V., Ribes J., Seco A. and Ferrer J. (2014). Model-based automatic tuning of a filtration control system for submerged anaerobic membrane bioreactors (AnMBR). *Journal of Membrane Science,* **465**, 14–26.

Robles A., Durán F., Ruano M. V., Ribes J., Rosado A., Seco A. and Ferrer J. (2015). Instrumentation, control, and automation for submerged anaerobic membrane bioreactors. *Environmental Technology*, **36**, 1795–1806.

Rozendal R. A., Hamelers H. V. V. and Buisman C. J. N. (2006). Effects of membrane cation transport on pH and microbial fuel cell performance. *Environmental Science & Technology*, **40**, 5206–5211.

Rozendal R. A., Sleutels T., Hamelers H. V. M. and Buisman C. J. N. (2008). Effect of the type of ion exchange membrane on performance, ion transport, and pH in biocatalyzed electrolysis of wastewater. *Water Science and Technology*, **57**, 1757–1762.

Saito T., Yan J., Hickner M., Cao X., Huang X. and Logan B. E. (2010). Using microbial desalination cells to reduce water salinity prior to reverse osmosis. *Environmental Science Biotechnolgy*, **3**, 1114–1120.

Semmens M. J. (2005). Membrane technology: pilot studies of membrane-aerated bioreactors.

Shechter R. (2015). Energy Efficiency and Performance of a Full Scale Membrane Aerated Biofilm Reactor. Proceedings of the 88th Annual Water Environment Federation Technical Exposition & Conference [CD-ROM]; Chicago, Illinois, September 26–30; Water Environment Federation: Alexandria, Virginia, pp. 5224–5228.

Shin C., McCarty P. L., Kim J. and Bae J. (2014). Pilot-scale temperate-climate treatment of domestic wastewater with a staged anaerobic fluidized membrane bioreactor (SAF-MBR). *Bioresource Technology*, **159**, 95–103.

Skouteris G., Saroj D., Melidis P., Hai F. I. and Ouki S. (2015). The effect of activated carbon addition on membrane bioreactor processes for wastewater treatment and reclamation – A critical review. *Bioresource Technology*, **185**, 399–410.

Smith A. L., Stadler L. B. and Love N. G., Skerlos S. J. and Raskin L. (2012). Perspectives on anaerobic membrane bioreactor treatment of domestic wastewater: a critical review. *Bioresource Technology*, **122**, 149–159.

Smith A. L., Stadler L. B., Cao L., Love N. G., Raskin L. and Skerlos S. J. (2014). Navigating wastewater energy recovery 1 strategies: a life cycle comparison of anaerobic membrane bioreactor and conventional treatment systems with anaerobic digestion. *Environmental Science & Technology*, **48**, 5972–5981.

Stenstrom M. K. and Rosso D. (2008). Biological Wastewater Treatment: Principles, Modeling and Design, IWA Publishing, London, UK.

Sun F., Sun B., Hu J., He Y. and Wu W. (2015). Organics and nitrogen removal from textile auxiliaries wastewater with A2O-MBR in a pilot-scale. *Journal of Hazardous Materials*, **286**, 416–424.

Syron E., Vale P. and Casey E. (2014). Where did the Bubbles Go? How to Reduce the Energy Requirements for Municipal Wastewater Treatment. IWA LET, Abu Dhabi.

Tao G., Kekre K., Oo M. H., Viswanath B., Yusof A. M. and Seah H. (2010). Energy Reduction and Optimisation in Membrane Bioreactor Systems. *Water Practice & Technology*, **5**(4), 1–7.

Terada A., Hibiya K., Nagal J., Tsuneda S. and Hirata A. (2003). Nitrogen removal characteristics and biofilm analysis of a membrane-aerated biofilm reactor applicable to high strength nitrogenous wastewater treatment. *Journal of Bioscience and Bioengineering*, **95**(2), 170–178.

Thanh B. X., Visvanathan C. and Ben Aim R. (2013). Fouling characterization and nitrogen removal in a batch granulation membrane bioreactor. *International Biodeterioration & Biodegradation*, **85**, 491–498.

Timberlake D. L., Strand S. E. and Williamson K. J. (1988). Combined aerobic heterotrophic oxidation, nitrification and denitrification in a permeable support biofilm. *Water Research*, **22**(12), 1513–1517.

Trad Z., Vial C., Fontaine J.-P. and Larroche C. (2015). Modelling of hydrodynamics and mixing in a submerged membrane bioreactor. *Chemical Engineering Journal*, **282**, 77–90.

van Haandel A., Kato M. T., Cavalcanti P. F. F. and Florencio L. (2006). Anaerobic reactor design concepts for the treatment of domestic wastewater. *Reviews in Environmental Science and Bio/Technolog,* **5**, 21–38.

Xiao K., Xu Y., Liang S., Lei T., Sun J., Wen X., Zhang H., Chen C. and Huang X. (2014). Engineering application of membrane bioreactor for wastewater treatment in China: current state and future prospect. *Frontiers of Environmental Science & Engineering*, **8**(6), 805–819.

Yang J., Vedantam S., Spanjers H., Nopens I. and van Lier J. B. (2012b). Analysis of mass transfer characteristics in a tubular membrane using CFD modelling. *Water Research*, **46**, 4705–4712.

Yang S., Hai F. I., Nghiem L. D., Nguyen L. N., Roddick F. and Price W. E. (2013). Removal of bisphenol A and diclofenac by a novel fungal membrane bioreactor operated under non-sterile conditions. *International Biodeterioration & Biodegradation*, **85**, 483–490.

Zhang J., Satti A., Chen X., Xiao K., Sun J., Yan X., Liang P., Zhang X. and Huang X. (2015). Low-voltage electric field applied into MBR for fouling suppression: performance and mechanisms. *Chemical Engineering Journal*, **273**, 223–230.

Zhang Q., Jie Y. W., Loong W. L. C., Zhang J., Fane A. G., Kjelleberg S., Rice S. A. and McDougald D. (2014). Characterization of biofouling in a lab-scale forward osmosis membrane bioreactor (FOMBR). *Water Research*, **58**, 141–151.

Chapter 8

Enhanced primary treatment

Dang Ho, Zachary Scott, Siva Sarathy and Domenico Santoro

8.1 INTRODUCTION

Enhanced primary treatment and carbon diversion is a key step being included in most of the emerging wastewater treatment schemes where energy optimization, resilience and capacity of the existing infrastructure are of concern. Indeed, it is well accepted that removal of particulate and colloidal fractions will lead to considerable operational savings in the downstream aerobic biological processes while allowing the recovery of energy in the form of methane via anaerobic sludge treatment processes.

Moreover, carbon management plays a very important role for biological nutrient removal processes where certain carbon fractions and species are preferred for optimal performance without the addition of external electron donors. For example, soluble and readily biodegradable chemical oxygen demand (COD) is required in denitrification schemes while a VFA-rich wastewater is desirable for efficient biological phosphorus removal.

Similarly, controlling particle content prior to membrane-based technologies is essential for smooth membrane bioreactor operations. In this case, the consequence of having excessive particulate reaching the membrane modules may have profound implications in terms of operational costs, as the system would experience excessive clogging as a result of the accumulation of fibers and debris.

Finally, particle management and carbon fractionation is an enabling step for a number of emerging mainline processes such as upflow anaerobic sludge blanket, moving bed biofilm reactor, and short-cut deammonification (such as Anammox) where particle hydrolysis or excessive carbon content in the influent may prevent stable operations.

In light of all the above, the decision on whether to use primary treatment technology and the performance required from the primary treatment step is a complex one, and appropriate metrics such as footprint, required wastewater quality of the primary effluent, plant capacity, greenhouse emission, etc. should all simultaneously be considered.

In this chapter, an overview of the most promising primary treatment technologies for enhanced particle removal are presented and discussed against to the conventional option of primary clarification. The technologies are discussed not only in terms of total suspended solids (TSS) removal, but also considering other metrics, often neglected, which can affect the operation of the entire plant such as nutrient diversion, footprint, cost, sludge production, greenhouse emission.

8.2 ENHANCED, HIGH-RATE PRIMARY TREATMENT

The classical approach for primary treatment is based on the use of the gravity-driven separation of solids in large sedimentation basins, commonly referred to as clarifiers. Clarifiers are designed to achieve complete removal of settleable suspended solids during average daily flow and their performance is typically measured by removal percentage of TSS (WEF 2008). A properly designed primary clarifier, operating at average flow, typically removes 50% to 70% of TSS. While removal of settleable TSS is the goal of primary sedimentation, it is the removal of the biochemical oxygen demand (BOD) component in the TSS that has a major impact on downstream biological processes. Metcalf and Eddy (2014) reports primary clarifiers typically achieve 25 to 40% removal of BOD. Conventional primary sedimentation is neither high rate, i.e., hydraulic residence times (HRTs) usually range from 1.5 to 2.5 hours at average flow, nor does it have a small footprint. Enhancements to sedimentation basins include processes such as ballasted flocculation, chemically enhanced primary treatment, and inclined plate settlers.

High-rate clarification (HRC) is distinguished from conventional clarification by the use of a physical device, alone or in combination with chemical treatment, to aid in the settling process thus increasing the hydraulic loading rate and particle removal efficiency. HRC can involve, but is not limited to: microscreens, inclined surface settlers, vortex separators and ballasted flocculation systems.

Traditionally, TSS removal has been measured to evaluate the performance of primary sedimentation. However, when evaluating primary sedimentation alternatives (e.g., sieving, filtration, vortexing) it is important to consider other metrics such as the removal efficiency of COD and/or BOD as well. For a nutrient removal plant, the ability of the primary treatment system to fractionate and remove particulate nitrogen and phosphorus and the overall removal of total nitrogen (TN) and total phosphorus (TP) should also be considered. By identifying a correlation between the liquid and solid fractions of the BOD, TN, and TP versus TSS, the latter can be used as surrogate to estimate BOD, TN, and TP removal (with the generally verified assumption that soluble materials are not removed during primary treatment).

8.2.1 Chemically enhanced primary treatment

The removal of TSS and colloidal fractions from wastewater can be increased by the addition of chemicals with a practice known as chemically enhanced primary treatment (CEPT). With CEPT, it is possible to achieve 80 to 90% TSS removal and 50 to 80% BOD removal (Metcalf & Eddy, 2014). Chemicals used in CEPT typically include, alone or in combination, ferrous sulfate, ferric sulfate, ferric chloride, alum, aluminum chloride, lime, and polymers (Table 8.1). CEPT, combined with disinfection, has been proposed to be a cost-effective alternative to conventional primary plus activated sludge for urban wastewater treatment in the developing world (Harleman & Murcott, 2001). Parker *et al.* (2001) provided a comprehensive review of the development of CEPT, from the early work done in the Great Lakes region of North America to recent implementation in Southeast Asia and Latin America.

The practice of CEPT has been extended beyond the removal of suspended solids and BOD to the chemical precipitation of phosphorous. Here, iron or alum salts are dosed to achieve 0.5 to 1 molar ratio of metals salts to phosphorous, causing soluble orthophosphate to precipitate as metal phosphates (Metcalf & Eddy, 2014). While this variant of the CEPT process can achieve over 80% removal of phosphate (Harleman & Murcott, 2001), processes for releasing and recovering phosphorous from CEPT sludge have not been adequately developed yet (Wilfert *et al.* 2015).

Table 8.1 Most commonly used coagulants and flocculants in CEPT to enhance clarification process.

Chemicals	Unit cost (€/ton)	Average dose (g/m³)	Cost (€/m³)
Alum	175–200	15–45	0.003–0.010
Polyaluminum chloride (PACl)	250–400	15–40	0.004–0.016
Calcium hydroxide (lime)	50–80	100–250	0.005–0.008
Ferric chloride	250–350	30–60	0.007–0.021
Ferric sulfate	200–280	45–60	0.010–0.017
Polymer (Polyelectrolyte)			
–cationic	2200–3080	0.1–2.0	0.001–0.006
–anionic	2000–2300	0.5–3.0	0.001–0.007

Source: CH2MHill (2009), Metcalf and Eddy (2014), U.S. EPA (2000).

8.2.2 Microscreen-based technologies

Screening has been extensively used in preliminary treatment for the removal of particulate materials that may create operation and maintenance problems in downstream processes. Typical opening sizes for screens vary from 1.0 up to 6 mm. Recently, microscreens, with openings ranging from 50 to 500 microns, have been employed to increase the capture of suspended solid to levels that are closer to those achieved by primary sedimentation. Microscreen filters are compact units which occupy a fraction of, approximately one-tenth, the footprint of a primary clarifier (Franchi & Santoro, 2015). Moreover, they combine the functions of particle removal, thickening and dewatering in one unit process. For this reason, microscreen based technologies are emerging as a competitive alternative to primary clarifiers when space is limited.

Microscreens, also known as microstrainers, remove suspended solids based solely on mechanical filtration or straining through well-defined apertures. Filter media can be made of metal screens, cloth fabrics or synthetic materials of different weaves (Bourgeous *et al.* 2003). Performance of microscreen filters depends upon the size distribution of influent solids as well as the pore size of the mesh selected for the filter screen, which typically ranges from 50 to 500 μm. Particles larger than the selected pore openings are effectively separated. Smaller particles can also be retained as the effective pore size is reduced by retained material built up on the screen during filtration, which is referred to as filter cake. This phenomenon is known as cake filtration (Tien, 2012). While cake formation enhances the filtration process, it also contributes to the increase in filtering resistance and decrease in permeability and flow through the filter. Therefore, maintaining an optimal cake thickness and pressure differential is critical for the operation of microscreen based filtration processes.

Microscreen filters are commercially available in a variety of configurations, including rotating drum filters, rotating disc filters, and rotating belt filters. Some commercially available microscreen based technologies include the Trojan Technologies Salsnes Filter, the Veolia Hydrotech drumfilter and Hydrotech disc filter, the Huber Technologies RoMesh®, and the Blue Water Technologies Eco MAT®.

8.2.2.1 Rotating belt filters

A rotating belt filter (RBF) consists of a removable mesh screen attached to an inclined moving belt of wire cloth (Figure 8.1). The system is designed to simultaneously filter wastewater and dewater the solids, for ease of solids handling and disposal. Wastewater flows into an inlet compartment and then passes through an inclined, continuously rotating belt upon which suspended solids are retained. Filtered water passing through the mesh is conveyed by gravity to an outlet pipe. The speed of the rotating belt can be automatically adjusted based on the level of water on the influent side of the belt, serving to modulate the cake thickness in order to obtain a target water height while avoiding overflow. An automatic self-cleaning system (air knife, water jet, or mechanical device) is installed at the upper end of the belt filter to continuously clean the filter mesh and remove any sludge accumulated on the mesh. Sludge drops into the collection area from the thickening process at 3–8% dry solids and is conveyed across the unit by an auger. The enclosed system also contains an optional integrated dewatering unit which can produce a cake with solids contents of up to 15 to 30% ready for disposal, transportation to a central sludge processing facility or subsequent on-site treatments (Rusten & Odegaard, 2006). This technology can provide potential savings in solids handling costs such as sludge dewatering and polymer costs. In addition, recent sludge characterization studies have observed a higher volatile solids fraction and cellulose content in sludge generated by rotating belt filters, compared to sludge produced by gravity based settling in primary clarifiers (Ruiken *et al.* 2013; Paulsrud *et al.* 2014; Sarathy *et al.* 2015).

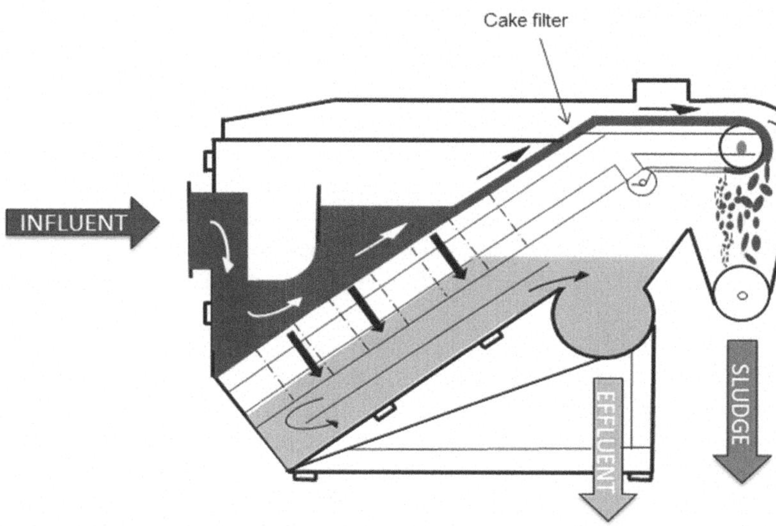

Figure 8.1 Schematic diagram of an inclined belt filter (www.salsnes.com, last accessed 12 December 2016).

8.2.2.2 Rotating drum filters

A drum filter consists of a rotating drum covered with filter cloth with pore openings ranging from 10 to 500 μm. Wastewater enters by gravity through the upstream end of the drum and flows radially outwards through the screen, leaving the solid particles retained on the filter mesh. The water level inside the chamber slowly increases due to progressive screen clogging and eventually triggers a level senor to start the backwashing jets mounted across the top of the drum. Filtration is continuous and not interrupted by backwashing. Subsequently, solids are rinsed off the surface of the screen and collected in a trough. Sludge produced by a drum filter has typical solids contents of 0.4% (CH2MHill, 2009). The filtration area for one unit varies from 0.35 to 22.5 m^2, and is capable of handling a flow range from 2 to 1000 l/s. The drum filters can be supplied in two versions: a self-contained tank version or a frame version for installation in a concrete chamber.

8.2.2.3 Rotating disc filters

The configuration and operation of a disc filter is similar to that of a drum filter. The rotating chamber consists of vertically mounted discs (5–30 discs per unit) with filter cloth on each side. Wastewater flows from the central drum into the discs and solids are retained on the inner side of the filter panels. As the water level inside the discs rises a level sensor triggers the discs to start rotating and a backwash cycle begins. Nozzles spray filtered water from the outside of the discs, removing particles for discharge via a solids collection trough. The rotation with backwash cycle stops after the water level drops below the level sensor. Typically, the backwash requires 1–3% of the total flow. Compared with drum filters on an equipment footprint basis, disc filters have significantly larger filtration area, as high as 230 m^2/unit.

Performance of microscreens depends on wastewater characteristics, belt mesh size, and flow rate. Microscreen filters can achieve TSS removal between 30% and 70% and BOD removal between 20 and 50% (Rusten & Odegaard, 2006; Franchi *et al.* 2012; McElroy, 2012). Filter mesh sizes between 150 to 850 μm are commonly used for municipal wastewater applications. While it is expected that smaller pore openings would be needed to achieve greater solids removal, the overall performance of a microscreen can be significantly improved by the filter mat that facilitates cake filtration in conjunction with mechanical sieving. When a sufficiently thick filter mat is allowed to develop, a large pore size mesh can be as effective, in terms of both TSS removal and filtration rate, as a smaller pore size mesh without a filter mat (Rusten & Odegaard, 2006; Franchi & Santoro, 2015). This indicates that larger mesh sizes (350 to 850 μm) are adequate for wastewater with high TSS concentrations and larger particles; however, if the influent wastewater consists primarily of small particles a smaller mesh size would be necessary to assist in the formation of a filter mat. Use of smaller openings (40–100 μm) is not practical for primary treatment due to the problems of clogging from oil and grease. Also, the drop in hydraulic capacity, as a result of smaller pore size, makes the system less economically competitive as more equipment is needed to treat a given flow.

Microscreens can also be operated with upstream in-line chemical addition. The pretreatment involves either a single stage polymer addition for flocculation or a two-stage process including both coagulation and flocculation. The use of chemical addition is beneficial in increasing TSS reduction up to 80–90% and COD reduction to 50–60% (Franchi & Santoro, 2015; Vaananen *et al.* 2016).

8.2.3 Vortex-based technologies

Vortex separators, also known as hydrodynamic vortex separators (HDVS), are compact devices that employ the rotary flow mechanism to remove suspended solids from the influent flow. Unlike the conventional clarifiers that rely only on gravity settling, HDVS utilize both gravity and centrifugal forces to achieve higher rates of solids-liquid separation, and thus provide the performance equivalent to conventional systems in a considerably smaller footprint (Andoh *et al.* 2001). The technology was first applied in England in the 1960's for the removal of grit and other large particles. As reviewed in 2002, over 1,500 HDVSs have been installed worldwide for stormwater, combined sewage, municipal and industrial wastewater treatment (Andoh & Saul, 2003). Three main designs that are in common use and described intensively in the literature are EPA Swirl Concentrator (non-proprietary), FluidSep™ Vortex Separator (Umwelt-und Fluid-Technik (UFT)), Storm King (Hydro International US). Recent developments in HDVS such as Downstream Defender® (Hydro International), and Vortechs™ Stormwater Treatment Systems (Contech) incorporate self-cleansing screening systems to capture oil and floatables in combined sewage overflows (Andoh & Saul, 2000; Vortechnics™, 2004).

Typically, the influent flow enters tangent to a cylindrical chamber and generates a rotational flow regime which creates a spiral flow path that a particle takes from entrance to outlet. The extended flow path provides greater time for gravity, aided by other inertial forces induced by the kinetic energy of the flow, to act on the particles and thus improve solids separation efficiency of a HDVS as compared to a conventional sedimentation basin of similar volume (Andoh *et al.* 2001). Settled particles are swept towards the central collection hopper at the bottom of the chamber and removed. In practice, the effectiveness of a given type of HDVS depends on the design of the internal components which helps to control flow patterns and ensure that captured solids are not subsequently re-entrained. Most vortex separators are relatively low-energy rotary flow devices and can operate at higher hydraulic loading rates than conventional clarifiers. Combined with the fact that they have no moving parts, and thus minimal maintenance requirements, operating costs of these HDVS systems tend to be low.

While it is important to have the internal components properly sized and configured, the solids removal performance of a HDVS is also dependent on the nature and characteristics of the influent wastewater itself. These characteristics include particle size and density that influence the particle settling properties (Andoh & Smisson, 1994; Guo, 2005). Research work has demonstrated that the removal efficiencies decrease with decreasing particle size, due to reduced particle settling velocity, and the limit of suspended particles that can be removed in a HDVS is with a settling velocity of 0.1–0.14 cm/sec, corresponding to a particle size of >50 μm (Field & O'Connor, 1996; Brueske, 2000). For influent containing solids with poor settling characteristics, a lower loading rate; i.e. a longer residence time, may be required to achieve the desired levels of solids removals. Adjusting the hydraulic loading rate impacts the balance of the forces acting on the particles, and this in turn determines which sizes and densities of particles can be separated.

Like other clarification processes, HDVS can be operated in conjunction with settlement aids such as coagulants and flocculants to further improve the removal efficiency. The observed TSS and BOD removals of HDVS average at 50% and 20% respectively without chemical addition and increase up to 70% and 50% respectively with chemical addition (Table 8.2) (Field & O'Connor, 1996; Boner, 2003). With respect to footprint, vortex separators are generally 50 to 70% the size of a conventional sedimentation tank (Weiβ, 1997), and with chemical dosage, these systems can have a smaller footprint than lamella plate clarifiers (Landon *et al.* 2002).

Table 8.2 Typical design parameters for a hydrodynamic vortex separator with and without chemical addition.

Technology	Chemical		Mixing Tank (min)	Flocculation Tank (min)	Overflow loading rate (m/h)	TSS removal (%)	Sludge Concentration (mg/L)
	Metal (mg/L)	Polymer (mg/L)					
Vortex/Swirl	–	–	–	–	10–20	40–60	30,000
Vortex/Swirl	60	1.5	1–2	4–6	10–20	60–70	40,000

Source: Menezes *et al.* (1996); CH2MHILL (2007).

8.2.4 Inclined-surface settlers

Inclined-surface settlers are settling devices consisting of stacks of plates or tubes that are installed near the tops of sedimentation basins to reduce the distance that particles need to travel before they reach the bottom of the basins. The rapid settling effect enables to maximize the effective settling area in a significantly smaller footprint (Figure 8.2). This technology was developed based on the theory proposed by Hazen (1904) that the proportion of particles removed in a sedimentation basin is primarily a function of the surface area of the basin and is dependent on the detention time. Inclined plate settlers were first introduced into the market in 1970 under the trade names of Lamella Separator^{TM} (Parkson Corporation) and GEWE^{TM} Lamella Sedimentation System (Purac Coporation) whereas tube settlers were manufactured by Neptune Microfloc, Inc. during the mid-1960's. Plate and tube settlers have been used predominantly in water treatment applications and later introduced into wastewater treatment for primary, secondary and tertiary sedimentation.

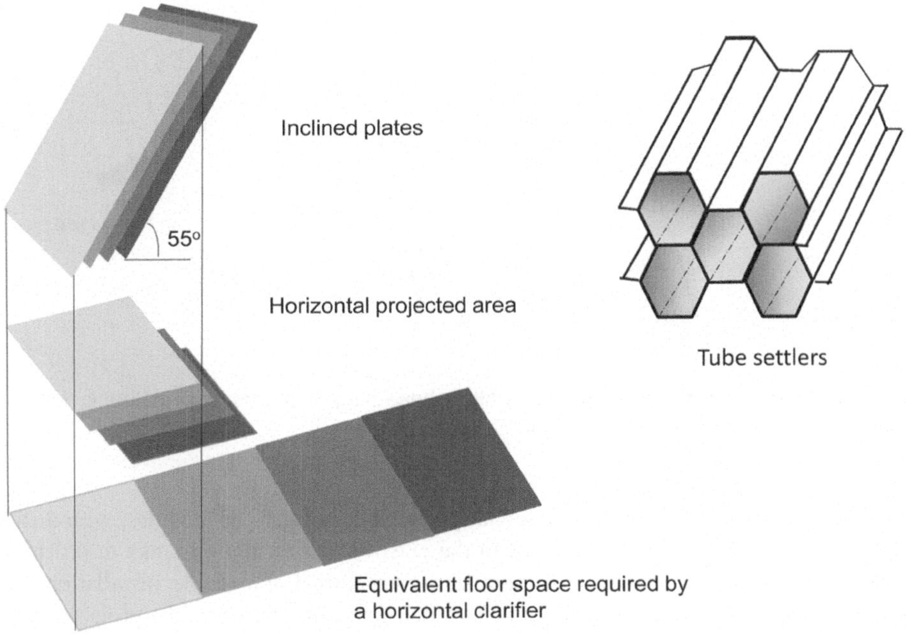

Figure 8.2 Effective settling area (projected plate area) of an inclined plate settler and tube settler.

The removal efficiency of the inclined surface settlers is dependent on the geometry of the inclined surfaces which is described by the distance between plates or tubes, their length and width, and the angle of inclination. Plate spacing depends upon the concentration of the suspended solids and varies from 50 to 102 mm with typical dimensions of 1 to 2 m long and 1.2 m wide (Hendricks, 2011). For tubes, four common cross-sectional shapes are used including circular, rectangular, square and hexagonal tubes with the dimensions of 610–1040 mm high, 1.8 to 2.4 m long, and 305 to 610 mm wide. The angle of inclination directly affects the projected settling area and ensures the mass of accumulated solids slides down the plates or tubes. A typical inclination of 55° is adopted for plates and 60° for tubes (Hendricks, 2011), with resultant increases in the effective settling surface area of the sedimentation basin by a factor of 6 to 12, thus allowing a higher peak flow to be treated in a given surface area. When the angle exceeds 60° the settling efficiency decreases. In contrast, decreasing the inclination angle increases the total settling area. However, when the angle is too shallow (<45°) the solids do not slide down the surface of the plates or tubes (Ross *et al.* 1999), and thus periodic flushing is required to remove the accumulated solids and prevent biological growths on the inclined surfaces with resulting odor issues.

With respect to the direction of water flow in relation to the direction of particle settlement, three arrangements are introduced (Figure 8.3): (1) co-current in which the influent is fed on top of the plate or tube module, and both water and the accumulated solids flow downward; (2) counter-current in which the influent is fed under the module and flow is upwards while the solids slide down the plate or tube; and (3) cross-flow, applied to plate settlers only, in which the water moves horizontally between the plate while the solids again flow downward. Counter-current designs are the most widely used as they are less expensive to install and operate (Dudley *et al.* 1994). Co-current designs are suitable for the removal of particles with high settling velocities to avoid sludge resuspension whereas crossflow pattern is normally used for the removal of both floating and settling material (Dudley *et al.* 1994).

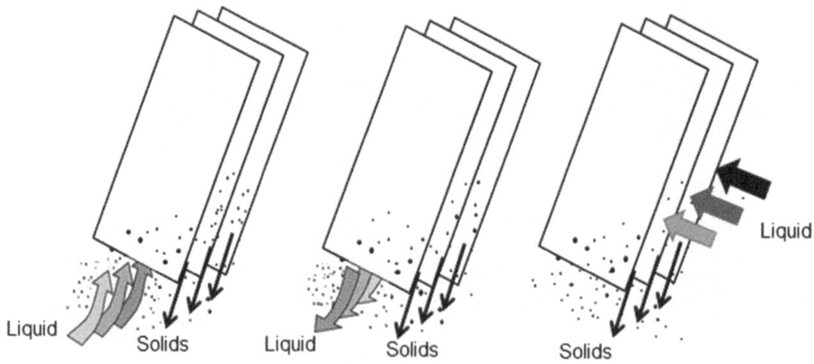

Figure 8.3 Classification of inclined-surface settlers based on flow direction including counter-current (Left), co-current (Middle) and crossflow (Right).

Inclined plates and tube settlers are commercially available in packs or modules, with different sizes and geometries, which can be easily retrofitted to the existing sedimentation tanks in order to enhance their efficiency to handle peak flow without expanding the structure. For example, installation of a Lamella plate module with a 55° of inclination and 50 to 100 mm of spacing increases the projected surface area relative to the floor area by approximately 10 to 16 times, allowing for significant increases in the hydraulic loading while maintaining similar overflow quality to that of the existing settling tanks. Therefore, this technology is ideally suited to address the need for expanding the capacity of the existing plants or where there are space and economic limitations. Since there are few moving parts, their operating and maintenance costs are minimal.

When greater removal efficiency is desired, chemical coagulation and flocculation can be added to the Lamella settlers or tube settlers. The combination of these two results in substantial reductions in the settling area required and significant improvements in the quality of treated effluent. Additional modifications including the use of either recycled sludge as in the dense sludge process, or floc-weighting agents as in ballasted flocculation, were later introduced to the systems to increase the settling velocities and mechanical strengths of the flocs, and as a result achieving additional improvement in hydraulic overflow rate and performance. These two processes are marketing under the proprietary names of DensaDeg (Infilco Degremont) and ACTIFLO (Veolia Water Technologies).

Figure 8.4 presents typical dense sludge and ballasted flocculation processes, respectively, which consist of three stages: coagulation, flocculation and enhanced sedimentation with Lamella plates. In the first stage, coagulation is accomplished in a similar manner as CEPT, which involves adding a chemical coagulant, e.g. alum or ferric salts, for destabilization and allowing the flocs to form. Subsequently, a ballasting agent, either recirculated sludge or micro-sand particles (20 to 200 μm in diameter), are added together with a flocculant-aid polymer to promote the formation of larger and more stable flocs with very high settling velocities. The fully formed ballasted flocs then settle rapidly when entering the settling zone equipped with Lamella plates or inclined tubes and the ballasts could be removed and recycled. Both processes can achieve TSS removals in the range of 85% to 95% at very high rates and low footprints (Delporte *et al.* 1995; U.S. EPA, 2003; Imasuen *et al.* 2004).

The primary advantages of the dense sludge and ballasted flocculation processes are attributed to the use of high mixing energy (150 to 400 s^{-1}), shorter detention times for flocculation (less than 10 min), floc settling velocities 20 to 60 times greater than conventional flocculation and sedimentation (Blumenschein *et al.* 2006; Sigmund *et al.* 2006) due to the introduction of ballasting agent and the use of inclined plates or tube modules to accelerate particle removal. As a result, the footprints of the high-rate clarification (HRC) processes are 5 times smaller than traditional Lamella clarifiers and up to 20 times smaller than conventional sedimentation (Blumenschein *et al.* 2006). Other reported advantages of the HRC processes are their robustness, quick start-up requirement and low sensitivity to fluctuating concentrations of raw influent, and thus they are suitable for treatment of highly varying flow rates and inlet conditions.

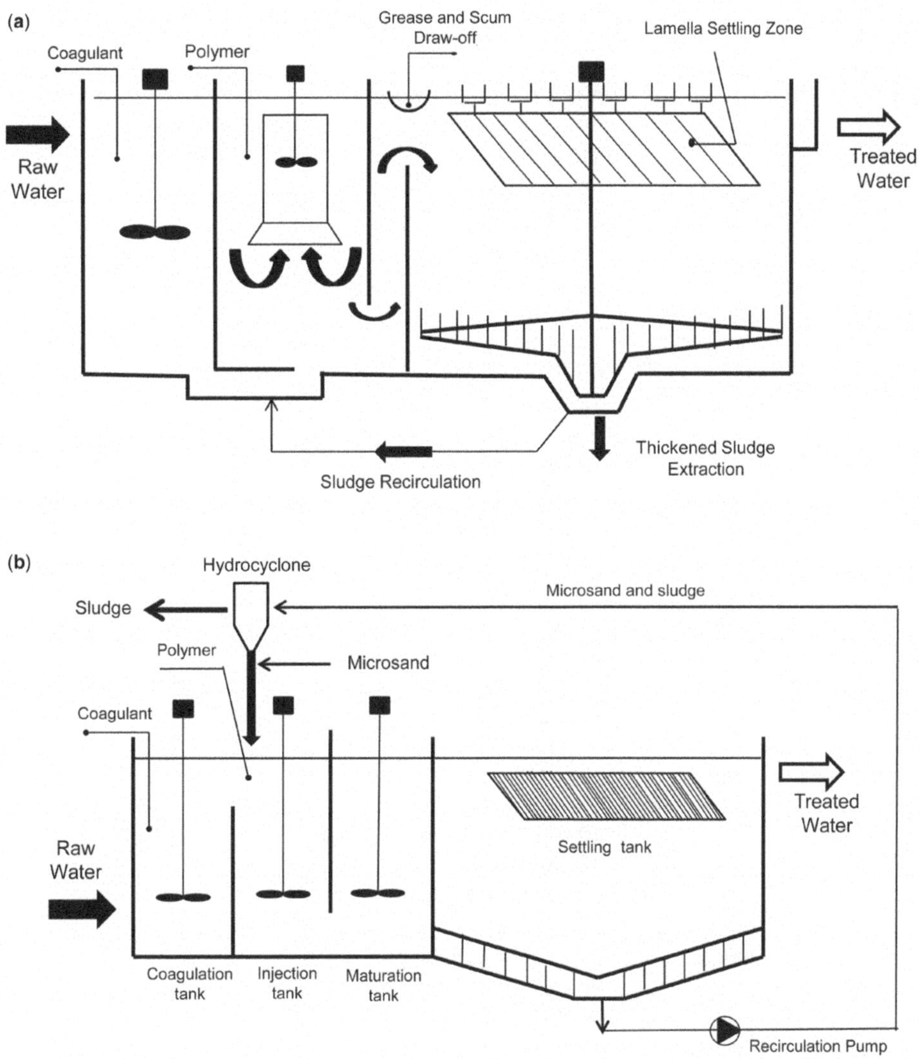

Figure 8.4 High-rate clarification processes: (a) dense sludge and (b) ballasted flocculation.

In terms of operation, HRC processes require increased doses of coagulant and polymer (Table 8.3), resulting in higher annual chemical cost (Gasperi *et al.* 2012; Keller *et al.* 2001; Landon *et al.* 2006;

Table 8.3 Case studies of pilot tests.

Location	Type	Overflow Rate and Chemical Usage	Removal Efficiency (%)			
			TSS	BOD	TP	TkN
Pine Road Eastside CSO Treatment Facility, Bremerton, Washington, US	Ballasted flocculation	45 g/m^3 PACl <1 g/m^3 anionic polymer	90–95	80	85–90	
Bayview WWTP, Toledo, Ohio, US	Ballasted flocculation	80–110 m/h 0.8–1.0 g/m^3 polymer				
		35–90 g/m^3 $FeCl_3$	82–94	50–84[a]		
		110 g/m^3 Alum	65–92	36–54[a]		
	Dense sludge	50–110 m/h 0.8–1.0 g/m^3				
		35–90 g/m^3 $FeCl_3$	74–91	36–56		
		110 g/m^3 Alum	63–86	37–79		
Southeast Plant, San Francisco, California, US	Ballasted flocculation	147 m/h 1.5 g/m^3 polymer 60–80 g/m^3 $FeCl_3$	70	70[b]	90–95	20–30
	Dense sludge	147 m/h 2 g/m^3 polymer 70–90 g/m^3 $FeCl_3$	70	60[b]	90–95	10–20
Forth Worth, Texas, US	Ballasted flocculation	97–170 m/h 0.75–1 g/m^3 anionic	70–90	35–65	90–95	25–30
	Dense sludge	polymer 70–125 g/m^3 $FeCl_3$	80–90	38–62	88–95	27–40
King County, Washington, US	Ballasted flocculation	147 m/h 0.95 g/m^3 polymer				
		110 g/m^3 Alum	93	74–93	81	
		200 g/m^3 PACl	93	75–87	92	
		110 g/m^3 $FeCl_3$	94	78–96	92	
Southerly Wastewater Treatment Plant, Ohio, US	Ballasted flocculation	120–144 m/h 1.4 g/m^3 polymer				
		40–65 g/m^3 Alum	85			
		17–25 g/m^3 PACl	<85			
		45–60 g/m^3 $FeCl_3$	85			
	Dense sludge	72–120 m/h 2–2.5 g/m^3 polymer				
		80–120 g/m^3 Alum	85			
		80–120 g/m^3 $FeCl_3$	77			
Seine Aval WWTP, Paris, France	Ballasted flocculation	40–90 g/m^3 $FeCl_3$ 0.5–1 g/m^3 polymer	80–87	42–69	90–93	13–35

PACl = Polyaluminium hydroxychloride.
[a]CBOD; [b]COD.
Sources: Gasperi *et al.* (2012); Landon *et al.* (2006); Metcalf and Eddy *et al.* (2007).

Metcalf & Eddy, 2007; Moffa *et al.* 2000). In addition, since the processes depend heavily on the mechanical equipment and a short processing time they may experience higher wear rates for pumps and piping moving sludge and sand. Often, pre-treatment with fine screens is required prior to the processes. Furthermore, with the presence of recycled sludge and sand in the reactor basins, care must be taken to avoid septic conditions which may lead to odors and corrosion problems (Keller *et al.* 2001).

8.3 PLANT-WIDE IMPACT OF ENHANCED PRIMARY PROCESSES

The primary treatment processes presented in this chapter perform solid-liquid separation methods by either gravity driven settling or physical separation by filtration. Both of these mechanisms can be enhanced by chemical additives. Depending on the separation method employed, operational settings, the influent wastewater characteristics, and the extent of chemical addition, there is large variation in the total mass and fractionation of BOD removed by primary treatment. The maximum BOD removal by primary processes is constrained by the particulate BOD fraction, near 55% for municipal wastewater (STOWA 1996). Many factors affect the expected performance of primary treatment; thus it is difficult to generalize quantitative assessments of the plant-wide impacts from implementing enhanced primary processes. Given some reasonable assumptions about primary performance under typical conditions, one can qualitatively generalize how downstream processes can be impacted by enhanced primary processes.

8.3.1 Impact on secondary stage aeration demand

The presence of primary treatment will result in the diversion of some fraction of BOD away from downstream processes. This capture of BOD by a primary process significantly affects aeration demand in downstream, mainline biological process and energy recovery in the sludge anaerobic digestion process. Reduced loading of BOD to aerobic biological processes results in a reduced aeration energy demand. Besides the mass of BOD captured in primary treatment, characterization is also important such as the fractionation between soluble and particulate forms and biodegradability (i.e., inert, slowly, and readily biodegradable fractions). Speciation and biodegradability affect the oxygen uptake rate, and available substrate to drive biological nutrient removal processes.

Gori *et al.* (2013) investigated the effect of primary sedimentation on BOD (measured as COD) and solids fractionation and the resulting effects on the wastewater plant's carbon and energy footprint. Similarly, Scott *et al.* (2015) compared gravity settling to a Rotating Belt Filter (RBF), considering aeration demand and energy recovery potential with anaerobic digestion of sludge. Calculations from both studies suggest that the addition of a primary treatment step results in a reduction in energy demand for secondary treatment, an increase in energy recovery via biogas production, and a reduction in the overall carbon footprint (Gori *et al.* 2013; Scott *et al.* 2015), Gori *et al.* (2013) also calculated that impacts on energy demand and recovery are highly dependent on the fractionation of COD in a given wastewater; the extent of benefit from applying primary sedimentation to reduce energy demand and increase energy recovery depends on the soluble and particulate fractionation of COD (Gori *et al.* 2013). As the fraction of soluble COD increases, the efficacy and benefit of applying primary treatment diminishes.

8.3.2 Impact on production, properties, and anaerobic degradability of sludge

The primary treatment technology chosen will impact the total sludge production of the plant. The typical trend is that with greater removal of TSS as primary sludge, the total sludge production as a combination of primary and secondary sludge is higher. If the produced sludge is sent to a properly operating digester, then the overall digested sludge production is very similar regardless of the efficiency of the primary treatment, as the total biodegradable organics that entered with the plant influent doesn't change with the mode of primary treatment. In the case of greater BOD removal in primary, more BOD is converted to CH_4 and CO_2 in the digester. In the case of less BOD removal in primary, more BOD is converted to CO_2 in the aeration basin. So when there is greater BOD removal in primary, then there is less energy consumed oxidizing carbon into CO_2, and more potential energy created by converting complex organics into CH_4. When an enhanced primary treatment removes less BOD than a conventional primary clarifier, its downstream aeration costs are higher, and downstream energy production is lower. However, without anaerobic digestion, the overall sludge production is expected to be lower, and thus might have lower sludge handling costs. The scenario analysis is summarized in Table 8.4.

Table 8.4 Primary removal rate effect on aeration demand, sludge production and energy recovery.

Scenario	Primary Sludge Production	Aeration Demand	Total Raw Sludge	Energy Recovery	Digested Sludge
High Primary BOD Removal, Optimal Digestion	High	Low	High	High	Low
Low Primary BOD Removal, Optimal Digestion	Low	High	Medium	Medium	Low
High Primary BOD Removal, Poor Digestion	High	Low	High	Low	Medium
Low Primary BOD Removal, Poor Digestion	Low	High	Medium	Low	Low

Mechanical screening separates solids primarily by size exclusion, in comparison to the gravity settling process that is constrained by particle settling velocities and the particles hydraulic residence time (HRT) in the settling basin. The mechanical screen will remove particles as constrained by the mesh size, and the matt of materials that develops on the screen between cleanings. Conventional clarifiers will typically be able to remove particles finer than the 350 μm particle sizes common to mechanical filters. Though the HRT in a primary clarifier will vary throughout the day and may be short for small periods, on average they are long. Depending on the wastewater source and influent characteristics, this difference in separation mechanisms may result in the primary clarifier producing sludge with either higher or lower energy content. The difference in sludge quality depends on the qualities of these finer particles, and how different they are from the larger particles which are preferentially captured by the mesh filtration. If the fine particles are mostly inert or non-biodegradable, the primary clarifier sludge will likely contain less energy/mass collected. If the fine particles are readily degradable, the primary clarifier sludge will have more energy/mass than the mechanically filtered sludge.

The retention time of the primary solids captured in a mechanical filter is likely to be significantly lower than the retention time of primary solids in a conventional primary clarifier. Solids are often held in a primary clarifier for many hours, while the solids are retained in a mechanical filter closer to the scale of minutes. Because primary sludge is energy rich and biodegradable, this usually means the primary sludge exiting the conventional process is significantly fermented, has a lower pH, and a significant concentration of volatile fatty acids (VFAs) Depending on the downstream processes following primary treatment, sludge fermentation could be advantageous or problematic. Fermented sludge is typically more difficult to dewater, but can be a good source of readily biodegradable carbon for nutrient removal, or a rapidly biodegradable feedstock for anaerobic digestion. Rapid feeding of highly degradable waste in a digester can lead to foaming problems.

The biodegradability of secondary sludge is known to be lower than primary sludge, and this difference increases with the mean cell residence time (MCRT) of the biomass in the aeration basin. Primary sludge can have a volatile solids reduction (VSR) near 65%, while secondary sludge VSR often ranges from 50% to less than 35% with extended aeration (Gosset & Belsner 1982). A plant removing less BOD in primary will produce more secondary sludge, and this secondary sludge will have a lower VSR in the digester. A digester fed with a higher ratio of secondary sludge/ primary sludge produces less biogas and energy per mass of combined sludge, as a large fraction of the BOD that came into the plant has left as CO_2. On the other hand, for an equivalent digester solids retention time (SRT), the digester fed with more secondary sludge is receiving less overall sludge, and thus the heating, mixing, and volume requirements of this digester are lower. Also, most primary sludge are thickened to less than 4% TS, while secondary sludge is often thickened to higher concentrations, thus additionally lowering the volumetric requirements of the digester fed with more secondary sludge. The optimal removal of primary sludge in terms of sludge handling operating costs is difficult to assess because it involves total sludge production at both primary and secondary steps, sludge thickening assumptions at both steps, digester SRT, the mode of dewatering after digestion, and finally the costs of sludge disposal. For most plants, especially smaller plants with no means to capture its energy, the value of biogas is likely less than the sludge handling costs. Increasing primary removal efficiency, strangely, may not be an enhancement in regards to overall plant operating costs.

The solids content and chemical composition varies significantly between type and operation of primary separation. Paulsrud *et al.* (2014) sampled 19 full-scale water resource recovery facilities (WRRFs) using RBFs as primary process and 10 WRRFs where primary treatment was carried out using primary clarifiers. They reported that the range of total solids (TS) for 19 samples obtained from RBFs, after screw pressing, was 13.6–36.9%, while 10 sludge samples from primary clarifiers ranged from 0.5–6.6% TS. Also, on average the volatile solids (VS) fraction of sieved sludge was significantly higher for sieved vs. clarified sludge, with mean values of 0.92 and 0.81 respectively. Ruiken *et al.* (2013) investigated the constituents of the RBF sludge, indicating that cellulose comprised 79% of the total mass and 84% of the organic mass.

Sarathy *et al.* (2015) investigated the physical and chemical characteristics of primary solids sequestered from municipal wastewater by RBF and a primary clarifier (PC). The core hypothesis were that engineered fractionation of primary solids could be applied to (i) improve performance of BNR by preferentially allowing readily biodegradable carbon to pass through primary treatment, (ii) increase biogas/energy yield (volumetric basis) in anaerobic digestion, and (iii) recover cellulose. The results from this study showed that sieving solids from municipal wastewater using RBF allows for simultaneous separation and thickening of primary sludge, and this sludge is easily dewatered. The reduced water content of the primary sludge, in comparison to sludge settled in PC, could facilitate increased digester loading rates and capacity. Also, as sieving was able to target specific size fractions of solids, it could preferentially capture certain fractions, such as cellulose, from wastewater. On the other hand, preferential capture of cellulose was not achievable with sedimentation. Finally, the overall methane yields of primary sludge captured by RBF and sedimentation were the same on a mass of volatile solids basis. Primary sedimentation was also found capable of removing fine particles fractions with very high methane yields, these fractions could not be removed by sieving. Because of this fine particle removal, primary sedimentation removed a small fraction of slowly degradable carbon that sieving did not, and thus sieving could be preferred to sedimentation where the carbon:nitrogen ratio is limiting. Neither primary process removes significant levels of soluble carbon (Sarathy *et al.* 2015), and thus neither is expected to negatively impact downstream biological phosphorus removal.

8.3.3 Impact on nutrient removal

Microscreen enhanced with polymer, or chemical addition may remove finer particles than a conventional clarifier, potentially even capture a portion of semi-soluble and colloidal carbon. Microscreen, especially above 350 μm, is not expected to remove non-particulate BOD, or what is considered readily biodegradable carbon. When polymer is added alongside mechanical filtration, a small portion of the colloidal or semi-soluble BOD can be captured, likely adsorbed or adhered to the larger particles, as its particles are still much smaller than even the finest filter mesh sizes. Microscreen coupled with polymer addition is unlikely to capture any of the soluble BOD (Trojan Technologies 2015). This suggests that microscreen technologies, even when coupled with polymer, should allow most of the readily degradable carbon to pass undisturbed from primary treatment to secondary, and thus not interfere with the suggested C:P ratios required for biological phosphorus removal. Biological phosphorus removal is enhanced by the availability of VFA's, which are very soluble, and not removed by filtration. Nitrogen removal processes, alternatively, are more impacted by the carbon removed by primary processes. A primary process that removes more particulate BOD may actually be a detriment to the BNR process, because it disturbs the needed C:N ratio. The soluble carbon/total carbon ratio of influent municipal wastewater is low, while the soluble nitrogen/total nitrogen ratio is high, and thus primary processes removing a large fraction of particulates remove more carbon than nitrogen, and thus disturb the C:N ratio needed for nitrogen removal (Sarathy *et al.* 2015). If the C:N ratio of primary effluent is insufficient for biological nitrogen removal, then costly chemical addition of carbon may be needed. This suggests that a microscreen process, though lower in particulate removal than a conventional clarifier, may have lower operating costs for nutrient removal when influent C:N ratios are low and less particulate removal also means lower requirements for carbon supplementation.

Rusten *et al.* (2015) reported that RBF treatment with a 33 μm mesh resulted in no negative effect on denitrification as no fermentable solids present in the raw wastewater were captured in the sludge. Moreover, the denitrification rates of raw wastewater and RBF effluent were almost identical at 0.96 ± 0.18 and 0.91 ± 0.35 g NO_3-Neq/m^2-d, respectively (Rusten *et al.* 2015).

8.3.4 Impact on power consumption and greenhouse gas emissions

Primary treatment, conventional settling or otherwise, is expected to reduce downstream aeration demand and divert BOD from aerobic respiration to anaerobic conversion with generation of chemical energy as methane Accordingly, one would expect that inclusion of primary processes in wastewater treatment leads to a reduction in greenhouse gas (GHG) production. A primary processes could be credited for the reduction in electricity demand in the downstream aeration system, and this electrical energy could be matched with a CO_2 equivalent. A fraction of the BOD diverted from aerobic respiration is biodegradable via anaerobic digestion, and a majority of the biodegradable fraction is converted to methane gas. This methane would eventually be oxidized to CO_2 after combustion, but still can provide an energy credit as it can offset demands for natural gas, which can be assigned a CO_2 equivalent. The extent of the energy credit from the conversion of BOD to methane, and then CO_2, instead of via biological respiration to CO_2, is dependent on the extent of anaerobic digestion. Incomplete digestion of the extra primary solids could lead to fugitive methane emissions as sludge degrading over time via land disposal. The oxidation of wastewater BOD eventually produces near the same amount of CO_2 from the wastewater itself, but a plant without primary treatment is likely to require more input energy to operate and thus has greater GHG production. The mechanical energy requirements of the primary treatment (skimmers and sludge pumps) are about 10 times smaller than energy requirements for aeration, so including a primary process in a wastewater plant is likely to reduce its GHG emissions.

How the inclusion of a primary process will affect the energy requirements for anaerobic digestion is debatable, as plants with primary treatment will likely generate more sludge, and thus need larger digesters, with greater heating and mixing requirements. A primary process usually feeds the digester near 4% TS, while thickened waste activated sludge (TWAS) is often fed near 7% TS. This means that a digester fed with more primary sludge will need to process a larger volume of wastewater in its effluent, and thus should have higher energy use for dewatering its effluent. Though they have become more efficient in recent years, the power and energy requirements for dewatering with a centrifuge are very high, so an increase in dewatering requirements after digestion can have a significant impact on the plants overall energy balance. To accurately quantify the overall impact of the primary process on GHG production, the impact of sludge production on digester size, heating, mixing, and final dewatering should be considered.

Primary processes have the potential to both reduce the plants net generation of GHG and also remove nitrogen and phosphorus. The cost of wastewater treatment, and energy requirements will increase with increasing nutrient removal requirements, so when nutrient removal is required it is useful to have an understanding of how the primary process could help meet nutrient removal requirements without adding additional GHG production. The removal of nutrients by primary treatment is generally limited by the fractionation of phosphorus and nitrogen between particulate and soluble forms. Municipal wastewater nitrogen is usually mostly soluble, so primary treatment has little potential to remove this even if the solids removal is maximal. The particulate fraction of phosphorus is slightly higher, so its potential removal by primary treatment is slightly higher. Further, chemical addition can precipitate most of the soluble phosphorus so its removal by primary processes can be almost complete. Given assumptions about the influent water quality and the TSS removal of the primary process, one can estimate the nitrogen and phosphorus content of the primary sludge. Given more assumptions about the energy requirement of the primary process, one can estimate the energy required to remove a given mass of nitrogen or phosphorus from the primary process.

Given the multiple assumptions listed below Table 8.5, the impact of primary treatment on GHG production and nutrient removal has been estimated for a primary process removing between 20% and 90% of the TSS, a range which is expected depending on the various available technologies and operating variables. The table suggests greater energy savings with increasing TSS removal, with a larger impact on GHG reduction calculated for energy recovered in anaerobic digestion than for aeration savings.

8.3.4.1 Calculation assumptions:

- GHG emission of CO_2 was calculated as reduction from BOD diverted from aeration. The oxygen transferred to oxygen supplied ratio was assumed to be 20%, accounting for an assumed basin depth of 4 m, and alpha factor of 0.7, head losses of 1 m of H_2O, a blower efficiency of 75%, and blower inlet air temperature of 25°C. The electricity required to power the aeration system was assumed to be provided from burning coal, producing 2 kg CO_2/kWh (U.S. EPA 2015).
- The BOD diverted from aeration was assumed to be 65% degradable in a digester, probably a conservative estimate because the VSS/BOD, and COD/BOD is >1, and primary solids have a typical VSR or COD removal of 65%, 95% of the BOD is converted to methane with 5% to anaerobic metabolism. The methane value for CO_2 reduction is equivalent to 1.15 kg CO_2/kWh, as avoided natural gas consumption.
- The greenhouse savings from digestion are overestimated because the primary sludge capture will increase the overall sludge production and increase the digester heating and mixing requirements.

Table 8.5 Impact of primary treatment on energy savings and Greenhouse Gas (GHG) emissions.

Performance Efficiency	TSS Removal (%)	BOD Removal (%)	N Removal (%)	P Removal (%)	Energy Savings from Aeration (10^3 kWh/m^3)	CO_2 Reduction from Aeration (ton CO_2/m^3)	CO_2 Reduction from Anaerobic Digestion (ton CO_2/m^3)	Total CO_2 Reduction (ton CO_2/m^3)	Net GHG (Net ton CO_2 reduced/m^3)
Low	20–30	<10	<5	<10	5–10	5–10	20–35	25–45	20–30
Medium	40–60	20–40	10–20	15–30	10–20	10–20	40–65	50–85	35–60
High	70–90	40–65	25–30	80–90	20–30	20–30	75–100	95–130	60–85

Table 8.6 High-rate primary treatment technology summary.

Parameters	Conventional primary clarifier	CEPT with conventional clarifier	Rotary belt filter	Rotary drum and disc filter	Vortex/ Swirl	Inclined plate/tube settlers	CEPT with Lamella Clarifier	Ballasted flocculation process	Dense sludge process
Process Performance									
1. TSS removal efficiencies	40–60	70–80	40–60	40–60	60–65	50–70	60–90	40–90	70–90
2. BOD removal efficiencies	25–40	40–60	15–30	15–30	30–40	30–50	40–60	35–60	35–60
3. TN removal efficiencies	<10	10–20	<10			15–30	15–20	25–30	25–35
4. TP removal efficiencies	16–20	60–80	15–20		15–20	30–40	60–75	92–96	88–95
Operational parameters									
5. Surface overflow rate ($m^3/m^2.h$)	2–5	2–5	17–70	2–150	4–10	10–30	30–40	60–200	75–100
6. Chemical addition	No	Yes (20–60 g/m^3 coagulant + 0.5–2.0 g/m^3 polymer)	Optional (0.5–4.0 g/m^3 polymer)	Optional	Optional	No	Yes (20–60 g/m^3 coagulant + 0.5–2.0 g/m^3 polymer)	Yes (40–80 g/m^3 coagulant + 0.5–1.5 g/m^3 polymer)	Yes (60–120 g/m^3 coagulant + 1.5–2.5 g/m^3 polymer)
7. Sludge concentrations (g/m^3)	10,000–25,000	20,000–30,000	30,000–200,000	600–1,400	30,000–40,000	15,000–25,000	20,000–35,000	10,000–15,000	25,000–40,000
Economic analysis									
8. Relative footprint	Large	Large	Small 20–70 m^2/unit. Up to 90% less of clarifiers	Small Up to 90% less of conventional clarifiers	Small Up to 70% less of conventional clarifiers	Medium Up to 70% less of conventional clarifiers	Medium Up to 80% less of conventional clarifiers	Small Up to 50% less of conventional clarifiers	Small Up to 50% less of conventional clarifiers
9. Capital Cost (€/m^3)	High 50–100 €/m^3 assuming a 90,000 m^3/d site	High	Medium 25–70 €/m^3	Medium	Low 0.5–8 €/m^3 unit alone, not including design and construction costs	Medium 25–40 €/m^3 assuming a 90,000 m^3/d site	Medium	High 80–120 €/m^3 assuming a 90,000 m^3/d site	High

(Continued)

Table 8.6 High-rate primary treatment technology summary (*Continued*).

	Parameters	Conventional primary clarifier	CEPT with conventional clarifier	Rotary belt filter	Rotary drum and disc filter	Vortex/ Swirl	Inclined plate/tube settlers	CEPT with Lamella Clarifier	Ballasted flocculation process	Dense sludge process
10.	Chemical cost (€/kg TSS removed)	n/a	0.02–0.05 (alum) 0.08–0.15 ferric salts)	0.04–0.10 (polymer)		0.07–0.13 (metals + polymer)	n/a	0.02–0.05 (alum) 0.08–0.15 ferric salts)	0.07–0.16	0.12–0.20
11.	kWh per kg of TSS removed	0.05–0.09	0.02–0.06	0.15–0.20			0.05–0.07	0.04–0.07	0.07-.0.20	0.07–0.20
12.	kWh per kg of BOD removed	0.11–0.16	0.05–0.12	0.35–0.50			0.08–0.14	0.06–0.16	0.20–0.35	0.20–0.35
13.	GHG emission, assuming 40,000 m^3/d plant	240 ton CO_2/ year	360 ton CO_2/year assuming 30 g/m^3 of metal salts						450 ton CO_2/year assuming 60 g/m^3 of metal salts and 0.2 g/m^3 of polymer	450 ton CO_2/year assuming 60 g/m^3 of metal salts and 0.2 g/m^3 of polymer
14.	Maintenance	Low	Low	Medium Backwash regularly to clean mesh filters	Medium Periodic maintenance to clean out the grit collection zone	Low. Periodic maintenance to clean out the grit collection zone ($300-$450 per cleaning)	Low	Medium	High Require operator judgement and frequent inspection	High Need sludge inventory, longer startup times, susceptible to upset with temperature change
15.	Problem Frequency	Rare	Rare	Low	Low	Medium. Unreliable performance at high flow	Low	Low	Medium. Require preliminary fine screening	Medium Foaming can be an issue but can be mitigated

16.	Safety and Health Risk	None	Health and safety risks with chemical storage	Health and safety risks with chemical storage	Health and safety risks with chemical storage	Health and safety risks with chemical storage	None	Health and safety risks with chemical storage	Health and safety risks with chemical storage	Health and safety risks with chemical storage
17.	Robustness	High	High	High	High	High	High	High	High	High
18.	Flexibility			Can be used for satellite treatment at overflow sites. Intermittent operation possible	Can be used for satellite treatment at overflow sites. Intermittent operation possible	Can be used for satellite treatment at overflow sites. Intermittent operation possible			Adapt to flows 25–150% its nominal design capacity with minimal effect on effluent quality	
19.	Stability	High	High	High	High	Medium Unreliable performance at variable flows	High	High	High Short startup time, fast response to influent variations	High. Longer startup time, fast response to influent variations
20.	State of development (emergent, adaptive use, innovative, established)	Established	Established	Adaptive Use (>500 installations worldwide)	Adaptive Use (>50 installations in UK)	Adaptive Use	Established (>150 US installations worldwide)	Established	Adaptive Use (>300 installations worldwide)	Adaptive Use (>200 installations)
21.	Compatibility with other technologies	Yes	Yes	Yes. Enhanced primary sludge can be used for resource recovery	Yes Can be used in secondary or tertiary treatment	Yes	Yes	Yes	Yes Can be used in tertiary and Filter Backwash treatment	Yes Can be used in tertiary and Filter Backwash treatment

Sources: CH2MHILL 2007; Crum & West 2000; Falk *et al.* 2013; Metcalf & Eddy 2014; U.S. EPA 2007; Vaananen *et al.* 2016; Wang *et al.* 2009.

Diversion of BOD of primary away from secondary will reduce the energy requirements for dewatering secondary sludge before digestion, but may slightly increase the energy requirements for dewatering digested sludges after digestion. Calculating a meaningful general value for the GHG reductions of primary BOD diversion is difficult due to the variety of specific assumptions need to quantify the net benefit.

- Medium and High Removal rates require the addition of chemicals as polymer and metal salts to remove colloidal materials, phosphorus, and nitrogen.

8.4 MINI-ASSESSMENT

Finally, the technical, economic and environmental sustainability of each technology was assessed using the indicators used in The Water_2020 Environmental Decision Support System and is summarized in Table 8.6.

8.5 REFERENCES

Andoh R. Y. G. and Saul A. J. (2000). Field Evaluation of Novel Wet-Weather Screening System. In: Proc. of Collection Systems Wet Weather Pollution Control: Looking into Public, Private and Industrial Issues, 7–10 May, New York.

Andoh R. Y. G. and Saul A. J. (2003). The use of hydrodynamic vortex separators and screening systems to improve water quality. *Water Science and Technology*, **47**(4), 175–183.

Andoh R. Y. G. and Smisson R. P. M. (1994). High-rate Sedimentation in Hydrodynamic Separators. In: Proc. of the 2nd International Conference on Hydraulic Modelling, 14–16 June, Stratford.

Andoh R. Y. G., Faram M. G., Stephenson A. G. and Kane A. (2001). A Novel Integrated System for Stormwater Management. In: Proc. of the 4th International Conference on Innovative Technologies in Urban Storm Drainage, 25–27 May, Lyon-Villeurbanne.

Blumenschein C. D., Latker E. and Banerjee K. (2006). Sand Ballasted High-rate Clarification Process for Treatment of Process Water. In: Proceedings of the 67th Annual International Water Conference, 22–26 October, Pittsburg.

Boner M. C. (2003). Peer Review: Wet Weather Demonstration Project in Columbus, Georgia (98-WWR-1P). Water Environ. Res. Found. Alexandria, Virginia.

Bourgeous K. N., Riess J., Tchobanoglous G. and Darby J. L. (2003). Performance evaluation of a Cloth-Media disk filter for wastewater reclamation. *Water Environment Research*, **75**(6), 532–538.

Brueske C. C. (2000). Technology Review: Ultra-Urban Stormwater Treatment Technologies. Department of Civil and Environmental Engineering, University of Washington, Settle, Washington.

CH2MHill. (2009). Final Tech Memo: Wet Weather Treatment Alternatives at Northeast, Southeast and Southwest WPCPs, Philadelphia.

Crumb S. F. and West R. (2000). After the rain. *Water Environment & Technology*, **12**(4), 41–45.

Delporte C., Pujol R. and Vion P. (1995). Optimzed lamellae settling for urban stormawater waste. *Water Science and Technology*, **32**(1), 127–136.

Dudley J., Mulhall E. and Solomon S. (1994). Primary Treatment Processes to Reduce BOD and Suspended Solids - Final Report. Swindon, Wiltshire, United Kingdom.

Falk M. W., Reardon D. J., Neethling J. B., Clark D. L. and Pramanik A. (2013). Striking the balance between nutrient removal, greenhouse gas emissions, receiving water quality and costs. *Water Environment Research*, **85**(12), 2307–2316.

Field R. and O'Connor T. P. (1996). Swirl Technology: Proper Design, Application and Evaluation. In: Proc. of WEFTEC, the 68th Annual Conference and Exposition. Miami, Florida.

Field R., Averill D., O'Connor P. and Steel P. (1997). Vortex separation technology. *Water Quality Research Journal of Canada*, **32**(1), 185–214.

Franchi A. and Santoro D. (2015). Current status of the rotating belt filtration (RBF) technology for municipal wastewater treatment. *Water Practice & Technology*, **10**(2), 319–327.

Franchi A., Stedman K. and Mitchell C. (2012). Demonstration-Scale of Primary Solids Harvesting for Energy Recovery at the UC Davis Wastewater Treatment Plant. In: Proc. of the 85th Annual Water Environment Federation Technical Exhibition and Conference. Water Environment Federation, 29 September –3 October, New Orleans.

Gasperi J., Laborie B. and Rocher V. (2012). Treatment of combined sewer overflows by ballasted flocculation: removal study of a large broad spectrum of pollutants. *Chemical Engineering Journal*, **211–212**, 293–301.

Gori R., Giaccherini F., Jiang L. M., Sobhani R. and Rosso D. (2013). Role of primary sedimentation on plant-wide energy recovery and carbon footprint, *Water Science and Technology*, **68**(4), 870.

Gossett J. M. and Belser R. L. (1982). Anaerobic-digestion of waste activated sludge. *Journal of the Environmental Engineering Division ASCE*, **108**(6), 1101–1120.

Guo Q. (2005) Development of Adjustment and Scaling Factors for Measured Suspended Solids Removal Performance of Stormwater Hydrodynamic Treatment Devices. In: Proc. of the World Water & Environmental Resources Congress. *ASCE-EWRI*, 15–19 May, Anchorage, Alaska.

Harleman D. and Murcott S. (2001). An innovative approach to urban wastewater treatment in the developing world. *Water 21*, **45**, 44–48.

Hazen A. (1904). On sedimentation. *Transactions of the American Society of Civil Engineers*, **53**, 45–88.

Hendricks D. (2010). Fundamentals of Water Treatment Unit Processes – Physical, Chemical and Biological. CRC Press: Taylor & Francis Group, New York.

Hendricks D. (2011). Sedimentation. In: Fundamentals of Water Treatment Unit Processes: Physical, Chemical, and Biological. Boca Raton: CRC Press: Taylor & Francis Group, New York, pp. 122–128.

Imasuen, E., Judd, S. and Sauvignet, P. (2004). High-rate clarification of municipal wastewaters: A brief appraisal. *Journal of Chemical Technology and Biotechnology*, **79**, 914–917.

Keller, J. (2005). Actiflo: A Year's Worth of Operating Experience from the Largest SSO System in the US. In: Proc. of 78th Annual Water Environment Federation Technical Exposition and Conference (WEFTEC). 29 October – 2 November, Washington, D.C.

Keller, J., Schultze, M. and Wagner, D. (2001). Lawrence Kansas: detailed design issues for a ballasted flocculation system. In: Proc. of the 74th Annual Water Environment Federation Technical Exposition and Conference. Water Environment Federation, Atlanta, Georgia, 13–17 October.

Landon D., Dee T., Dudley J. and Wood J. (2002). Intermittent Discharges- Environmental Review, Treatment and Solutions. WRcPLC, WaPUG Spring Meeting 2002, www.wapug.org.uk

Landon, S., Donahue, C., Jeyanayagam, S. and Cruden, D. (2006). Rain check. *Water Environment and Technology*, **18**, 30–35.

Ljunggren, M. (2006). Micro screening in wastewater treatment – an overview. *Vatten*, **62**, 171–177.

Mels, A. (2001). Physical-chemical pretreatment as an option for increased sustainability of municipal wastewater treatment plants. PhD thesis. Wageningen University, Wageningen, The Netherlands.

Menezes, F. M., Amal, R. and Luketina, D. (1996). Vortex with flocculation. *Powder Technology*, **88**(1), 27–31.

Metcalf & Eddy, Greeley and Hansen LLC and Limno Tech, Inc. (2007). *Blue Plains Total Nitrogen Removal/Wet Weather Plan – Long Term Control Plan Supplement No. 1.* District of Columbia – Water and Sewer Authority.

Metcalf and Eddy, Tchobanoglous, G., Burton, F. L., Stensel, H.D. (2014). Wastewater Engineering: Treatment and Resource Recovery, 5th edn, McGraw-Hill, New York.

Moffa, P. E., Goegel, H. M., Davis, D. P., LaGorga, J. J. and Pitt, R. (2000). Retrofitting Control Facilities for Wet-Weather Flow Treatment. U.S. EPA, Washington, D.C., EPA/600/R-00/020.

Parker, D. S., Barnard, J., Daigger, G. T., Tekippe, R. J. and Wahlberg, E. J. (2001). The future of chemically enhanced primary treatment: evolution not revolution, *Water*, **21**, 49.

Reardon, D. J. (1995). Turning down the power. *Civil Engineering*, **65**(8), 54–56.

Ross, G., Zabel, T. F. and Edzwald, J. K. (1999). Sedimentation and Flotation, In: Water Quality and Treatment, 5th edn, McGraw-Hill, New York.

Ruiken, C. J., Breuer, G., Klaversma, E., Santiago, T. and van Loosdrecht, M. C. M. (2013). Sieving wastewater - Cellulose recovery, economic and energy evaluation. *Water Research*, **47**(1), 43–48.

Rusten, B. and Ødegaard, H. (2006). Evaluation and testing of fine mesh sieve technologies for primary treatment of municipal wastewater. *Water Science and Technology*, **54**(10), 31–38.

Rusten, B, Razafimanantsoa, V. A., Andriamiarinjaka, M. A., Otis, C. L., Sahu, A. K. and Bilstad, T. (2015). Impact of Fine Mesh Sieve Primary Treatment on Nitrogen Removal in Moving Bed Biofilm Reactors. In Proc. of IWA Nutrient Removal and Recovery Conference, 17–21 May, Gdansk, Poland.

Sarathy, S., Ho, D., Murray, A., Batstone, D. and Santoro, D. (2015). Engineered Fractionation of Primary Solids – A Comparison of Primary Treatments Using Rotating Belt Filters and Primary Clarifiers. In: Proc. of the 88th Annual Water Environment Federation Technical Exhibition and Conference (WEFTEC), 26–30 September, Chicago.

Scott, Z., Sarathy, S., Franchi, A., Ho, D., Santoro, D., Batstone, D. and Rosso, D. (2015). Primary Wastewater Treatment by Rotating Belt Filters: Energy Assessment. In: Proc. of IWA Nutrient Removal and Recovery Conference, 17–21 May, Gdansk, Poland.

Sigmund, T., Siczka, J., Elliott, T. and Awad, J. (2006). Operating Chemically Enhanced Clarification for Optimum Disinfection Performance. In: Proc. of the 79th Annual Water Environment Federation Technical Exhibition and Conference (WEFTEC), 22–25 October, Dallas, Texas.

STOWA (1996). Variants on presettling - a feasibility study (in Dutch); report nr 96-15, Dutch Foundation for Applied Research on Water Management, Utrecht.

Tien, C. (2012). Principles of Filtration. Elsevier, Oxford, U.K.

TROJAN Technologies (2015). Unpublished internal studies of removal of BOD,TSS,COD,N,P removal by mesh filtration enhanced by polymer addition, UC Irvine WEX Center.

U.S. EPA (2000). Wastewater Technology Fact Sheet: Chemical Precipitation. EPA 832-F-00-018.

U.S. EPA. (2003). Wastewater Technology Fact Sheet: Ballasted Flocculation. EPA 832-F-03-010.

U.S. EPA. (2007). Wastewater Management Fact Sheet In-Plant Wet Weather Peak Flow Management. EPA 832-F-07-016.

U.S. EPA (2015). www.epa.gov.

Vaananen, J., Cimbritz, M. and la Cour Jansen, J. (2016). Microscreening in primary treatment: effect of chemical dosing. *Water Science and Technology*, **74**(2), 438–447.

Weiß, G. J. (1997). Vortex separator: proposal of a dimensioning method. *Water Science and Technology*, **36**(8–9), 201–206.

Wilfert, P., Kumar, P. S., Korving, L., Witkamp, G. J. and van Loosdrecht, M. C. M. (2015) The relevance of phosphorus and iron chemistry to the recovery of phosphorus from wastewater: a review. *Environmental Science & Techonology*, **49**(16), 9400.

Water Environment Federation (1998). Design of Municipal Wastewater Treatment Plants: WEF Manual of Practice No. 8. Water Environment Federation, Alexandria, Virginia.

Water Environment Federation (2005). Clarifier Design, 2nd edn., Water Environment Federation, Alexandria, Virginia.

Part 1

Reducing Requirements and Impacts

Part 1c: *Reducing Impacts*

Chapter 9

Innovative primary and secondary sewage treatment technologies for organic micropollutants abatement

Marta Carballa, Teresa Alvarino, Gianluigi Buttiglieri, Jean-Marc Choubert and Marie-Noëlle Pons

9.1 INTRODUCTION

Effluents discharged by wastewater treatment plants (WWTPs) have been singled out as one of the main sources of chemicals release into the aquatic environment: flame retardants, plasticizers, detergents, urban pesticides, pharmaceuticals and personal care products (Carballa *et al.* 2004; Lishman *et al.* 2006). As several of these chemicals are suspected to have a potential ecotoxic impact on the aquatic organisms, on-going regulation in Europe reinforces the requirements towards surface water quality (Water Framework Directive). The latter presently includes 45 priority substances to be reduced (EC, 2013) and 17 additional emerging substances to be closely scrutinized (EC, 2015). A broad strand of research has highlighted how WWTPs remove some micropollutants from wastewater, particularly using primary and secondary (i.e. biological) processes, even though they were not originally designed for this purpose (Luo *et al.* 2014a).

The specific load of micropollutants accounts for about 0.4 to 0.8 g/PE/day (Coquery *et al.* 2011), which is about up to 100 times lower than the BOD load (i.e. 60 g/PE/day). Organic micropollutants account for around 20% of the total load of micropollutants received by a WWTP (metals accounts for the majority of the load of micropollutants received by WWTPs); pharmaceuticals and personal care products (PPCPs) represent 15% which confirms the potential relevance of these substances in terms of impact on receiving bodies (if not treated); phtalates, alkylphenols and volatiles organic compounds (VOC) represent respectively 2%, 0.3% and 0.1% of the total load, confirming that these priority organics are of high concern with respect to Water Framework Directive objectives (Figure 9.1). About 20% of the received load of micropollutants is usually discharged to the environment by treated effluent in conventional WWTPs facilities.

Different reviews have reported the occurrence of micropollutants in raw and treated effluents, with various quantification frequencies and concentrations levels (Table 9.1). Mean concentrations of micropollutants can be extremely different from one WWTP to another (relative standard deviations higher than 100%), caused by different sources of micropollutants depending on the period of the year (summer/winter), the location (rural/urban), the type of connected activities to sewers (hospitals, industries) and also medicinal prescriptions.

Some substances used in the industry are usually quantified at higher concentrations in urban sewers (compared to rural ones): alkylphenols (except 4-NP1EC), VOCs (dichloromethane, trichloromethane trichloroethylene and tetrachloroethylene), chloroalcanes, dichlorophenol and bisphenol A. The concentration of alkylphenol polyethoxylates (additives of detergents in textile industries, additives in paper industries) is usually 2 to 3 times higher in urban WWTPs; these compounds are responsible for the release of alkylphenols (nonylphenol and octylphenol) by biodegradation. Glyphosate is more frequently used as herbicide in urban environments. A thorough study dealing with 117 micropollutants and 15 different WWTPs in France has reported that raw domestic wastewaters contain around twelve priority substances and around sixty other substances at concentrations higher than 0.1 μg/L (Martin Ruel *et al.* 2012).

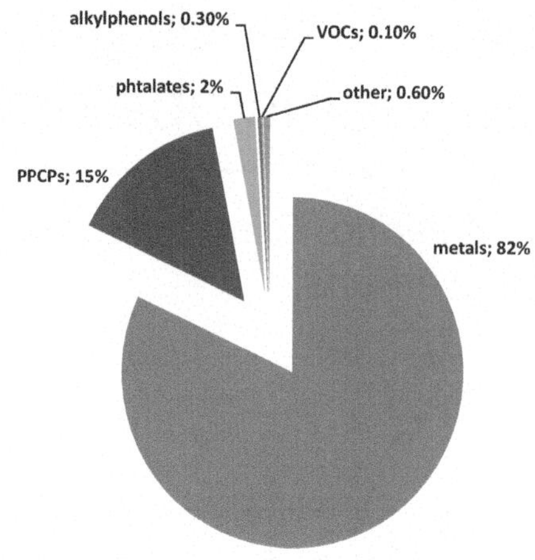

Figure 9.1 Proportion of micropollutants in domestic wastewaters (Coquery *et al.* 2011).

Diethylhexylphtalate (DEHP), alkylphenols and VOCs were very frequently quantified (>80%) and present at high concentrations (67 μg/L for DEHP; 4.3 to 9.7 μg/L for octylphenol and nonylphenol; 1.4 to 2.9 μg/L for di-, trichloromethane and tetrachloroethylene). Light polycyclic aromatic hydrocarbons (PAHs, naphthalene and fluoranthene) were frequently found (>50%) with a mean concentration higher than 0.1 μg/L. Among pesticides, diuron was the most frequently quantified (80%) with a mean concentration of 0.25 μg/L, whereas atrazine and simazine were more seldom found (0.02 μg/L). Alkylphenol ethoxylates and carboxylates (4-NP1EO, 4-NP2EO and 4-NP1EC) were systematically quantified in raw wastewaters at mean concentrations between 2.1 and 6.1 μg/L, which is the same level as priority substances nonyl- and octylphenol. Some other organic micropollutants were less frequently quantified (30–70%) but at (very) high mean concentrations in urban wastewaters (up to 49 μg/L for triclosan, 5.5 μg/L for C_{10}-C_{13} chloroalcanes, 1.6 to 2.6 μg/L for flame retardants deca- and tri-bromodiphenylether) or with concentrations between 0.1 and 1 μg/L (trichlorobenzenze, pentachlorophenol, tert-butylphenol, dichlorophenol, benzothiazole, tributylphosphate and AMPA). Some priority substances were never quantified, either because their use is now prohibited in Europe (pesticides alachlore, aldrine, dichlorodiphenyltrichloroethane (DDT), endrine, chlorfenvinphos, trifluraline) or because their use is very specific (e.g., hexachlorobutadiene, pentachlorobenzene). Estrone and 17β-estradiol are systematically quantified in raw wastewater with mean concentrations lower than 0.1 μg/L, while mean concentration of estriol reached 0.34 μg/L; a majority of pharmaceuticals are very frequently quantified in raw wastewater (> 80%): paracetamol and aspirin present a very high mean concentration (> 100 μg/L); acebutolol, atenolol, sotalol, sulfamethoxazole, roxithromycin, cafeine, theophylline, carbamazepine, ibuprofen, ketoprofen, naproxene, diclofenac and gemfibrozil present mean concentrations between 0.1 and 15 μg/L; other pharmaceuticals are never quantified above 1 μg/L.

Table 9.1 Concentration range of organic micropollutants in raw and secondary treated wastewater.

	Raw domestic wastewaters (μg/L) *Mean range (Max.)*	**Secondaty treated effluents (μg/L)** *Mean range (Max.)*
Alkylphenols		
Nonylphenol, octylphenol	4.3–9.7 (<39)	0.2–0.8 (<4.4)
Alkylphenol ethoxylates (4-NP1EO, 4-NP2EO)	2.5–6.1 (<60)	< 2 (<10)
Alkylphenol carboxylates (NP1EC)	2.1 (<14)	2.3 (<22)
Volatile organic compounds (VOCs)		
Di-, trichloromethane, tetrachloroethylene	1.4–2.9 (<24)	0.3–0.9 (<3.9)
Polycyclic aromatic hydrocarbons (PAHs) *(dry weather conditions)*		
Naphthalene, fluoranthene, benzo(b) fluoranthene, benzo(g,h,i)perylene, benzo(k) fluoranthene, indeno(1,2,3-cd)pyrene	<1	<0.4
Pesticides		
Diuron	0.25 (<2.4)	0.2 (<1.5)
Atrazine/simazine	0.02 (<0.1)	0.02 (<0.1)
Alachlore, aldrine, Dichlorodiphenyltrichloroethane (DDT), endrine, chlorfenvinphos, trifluraline, hexachlorobutadiene, pentachlorobenzene	not quantified	not quantified
Glyphosate	0.7 (<3)	0.5 (<2)
Other chemicals		
Diethylhexylphtalate (DEHP)	67 (<640)	4.6 (<23)
Triclosan	49 (<350)	0.1 (<0.5)
C_{10}-C_{13} chloroalcanes	5.5 (<22)	<0.1
Deca- and tri-bromodiphenylether	1.6–2.6 (<8)	<0.1
Trichlorobenzenze, pentachlorophenol, tert-butylphenol, dichlorophenol, benzothiazole, tributylphosphate	0.1–1	
Phenoxyacetic acid (AMPA)	1 (<2.7)	3 (<9)
Hormones		
Estrone, 17β-estradiol	<0.1	<0.1
Estriol	0.34 (<0.8)	0.04
Pharmaceuticals		
Paracetamol, aspirin	120 (<394)	0.1–0.2 (<2)
Acebutolol, atenolol, sotalol, sulfamethoxazole, roxithromycine, cafeine, theophylline, ibuprofen, ketoprofen, naproxene	0.1–15 (<25)	0.1–75 (<3)
Carbamazepine, diclofenac, gemfibrozil	0.7–1 (<3.4)	0.25–0.75 (<1.7)

Source: Martin Ruel *et al.* (2012).

Many organic micropollutants are partially removed from wastewater during conventional primary and secondary treatment processes, even though they are not originally designed for this objective (Martin Ruel *et al.* 2012; Luo *et al.* 2014a). The fate of micropollutants in WWTPs strongly depends on the physicochemical properties of the organic chemicals. The dissolved fraction in raw wastewater (i.e. dissolved concentration to total concentration ratio) indicates the propensity to remain in effluent or to be retained in sludge. Hydrophobic substances (ex. PAH with high molecular mass, polybrominated diphenyl ethers (PBDE), chloroalcanes) are only quantified in particulate phase; DEHP, nonylphenol and octylphenol are equally distributed between dissolved and particular phases, meaning that they can be found in effluent and in sludge. The other micropollutants are mainly present in dissolved phase (>60%): transformation product 4-NP1EC, most hydrophylic pesticides (log Kow <3), hormones, and most of the pharmaceuticals (ex: paracetamol, carbamazepine) since most of them are hydrophilic (log Kow between −0.5 and 3), except amitriptyline, doxepine and fluoxetine that are equally distributed between dissolved and particulate fractions (log Kow between 4 and 5).

The fate of organic micropollutants within conventional wastewater treatment systems is associated to three main mechanisms: (1) Volatilisation to the atmosphere, transferring micropollutants from liquid to gas phases; it occurs at the surface of biological reactors and mainly while forced aeration is applied. It concerns micropollutants for which Henry's law constant is higher than 100 $Pa \cdot m^3 \cdot mol^{-1}$ like VOCs, light PAHs and other organics like solvents acetone or phenol. Pharmaceutical compounds and hormones are not concerned by volatilisation (Urase & Kikuta, 2005; Plósz *et al.* 2010). (2) Sorption to sludge occurs in biological reactors, transferring micropollutants from liquid to solid phases. It concerns a large number of micropollutants, those for which a partition coefficient (K_d) is higher than 1 L/g MLSS like PAHs (pyrene, phenanthrene), nonyl- and octylphenol, or some pharmaceuticals (amitriptyline, fluoxetine. It is a rapid process influenced by the physico-chemical properties of chemicals, by the sludge quality (e.g., organic matter content) and the colloidal fraction (Barret *et al.* 2010), and by local WWTP conditions (SRT, pH, temperature). Desorption is assumed to be significantly slower than sorption. (3) Biological transformation (biodegradation) occurs in biological reactors, converting initial chemical to other one(s); it concerns several micropollutants, those for which a biodegradation constant (k_{biol}) is higher than 1 L/g MLSS/day, like analgesics (ibuprofene and paracetamol), betablockers (atenolol, bisoprolol, betaxolol, acebutolol, nadolol) or even some alkyphenol (nonyl- and octylphenol) and pesticides (isoproturon, simazine). Most of these micropollutants are not submitted to direct biodegradation (i.e. consumed by microorganisms for growth), but are submitted to cometabolism (i.e. indirect pathway) caused by the capacity of the enzymes to degrade many chemicals while organic matter is degraded.

Discharged WWTP effluents to the environment (even from nitrifying-denitrifying facilities) still contain several micropollutants due to high initial concentrations in raw wastewater, or low affinity with suspended solids (SS)/non-volatile property, or poorly biodegradable chemical structure, so as they have insufficient elimination in WWTPs. As a consequence, some micropollutants are usually measured between 1 and 10 μg/L in effluent, e.g. DEHP, carbamazepine, diclofenac and gemfibrozil; other chemicals are measured at concentrations between 0.1 and 1 μg/L (Table 9.1), e.g. sulfamethoxazole, roxithromycin, ibuprofen, propranolol, acebutolol, atenolol, sotalol, paracetamol, ketoprofen, diuron, atrazine, simazine, glyphosate. Other are present as they partly result from the transformation of other micropollutants, like phenoxyacetic acid (AMPA), alkylphenol carboxylates (NP1EC) and estrone. Several priority substances may be a problem regarding the Environmental Quality Standards of the WFD, like for pesticides (diuron, isoproturon, atrazine, simazine, chlorpyrifos, endosulfan, hexachlorocyclohexane, tributyltin), DEHP, alkylphenols, COVs (chloroform, trichlorobenzene, dichloromethane and tetrachloroethylene), PAHs (naphthalene, fluoranthene, anthracene, benzo(b)fluoranthene, benzo(k)fluoranthene), pentabromodiphenylether, pentachlorophenol.

The next future of WWTPs will strongly involve research, engineering and stakeholders to decrease the concentrations of organic micropollutants in the effluents discharged by WWTPs to the environment. Besides the source-separation and the resource recovery approaches, optimizing existing facilities and implementing polishing treatment processes are promising tracks.

9.2 ENHANCEMENT OF PRIMARY AND SECONDARY SEWAGE TREATMENT FOR ORGANIC MICROPOLLUTANTS ELIMINATION

Water quality has seen dramatic improvements over recent decades as a result of the construction of wastewater treatment plants. However, pressures on watercourses, such as growing development density and lifestyle, are continuously increasing. Various research projects and publications have investigated the organic micropollutants problem and have indicated the need to improve wastewater treatment, since for most of the substances, the elimination is only partial. This section evaluates technical processes which might enable wastewater treatment infrastructure and/or operational conditions to be upgraded so that adequate protection of ecosystems and drinking water resources against organic micropollutants can be guaranteed.

9.2.1 Enhanced primary clarification

Primary treatment encompasses grit removal, sand and FOG (fats-oils-grease) removal and primary settling (sometimes enhanced by the addition of chemicals). The fate of micropollutants in the primary treatment is a combined result of the type of processes involved and the physico-chemical properties of the micropollutants. As the processes are mainly mechanical, the most important properties are related to volatilization and sorption. For example VOCs, mostly associated with the liquid phase, will be eliminated by volatilization favored by gas stripping. PAHs, which are mostly associated with solid particles such as street particles, will be eliminated by settling (Tian *et al.* 2012). Concerning PPCPs, volatilization is totally negligible for pharmaceuticals and estrogens but is variable for musk fragrances (nearly negligible for tonalide or galaxolide, but not for celestolide).

There are few data related to the fate of micropollutants along the primary treatment. When they exist, they are global, although several steps are usually applied with different operation conditions, which can affect differently the micropollutant fate.

A FOG separator is necessary in most WWTPs as the removal of FOG avoids foaming in the secondary treatment (especially in aeration tanks) and the excessive development of filamentous bacteria. Most of the collected FOG is incinerated but in-situ aerobic treatment is also possible (Wakelin & Forster, 1997), with a recycle of the treated FOG to the wastewater treatment line. There is no information on the fate of micropollutants in such a treatment.

Grit corresponds to solid materials that are heavier than the organic biodegradable solids present in the wastewater. Their removal helps to increase the lifespan of mechanical devices such as pumps and to limit the decrease of volume that can result from their settling further downstream (biological reactor, anaerobic digester). Several technologies can be used to facilitate the separation of the grit from the liquid phase. Aerated systems, in which the apparent density of the liquid phase is decreased to increase grit settling, can promote the volatilization of micropollutants depending upon their Henry's coefficient. The mixing can also favor the detachment of some micropollutants adsorbed onto particles. In case of grit washing, the cleaning water is returned to the wastewater treatment line. There is no information on the fate of micropollutants in such treatments.

Primary settling is the final step: it consists in separating from the liquid particles able to aggregate together, forming larger particles able to settle faster. The aggregation can occur naturally or by the help of chemical flocculants. The fate of organic micropollutants during primary treatment is often assessed through the amount trapped into the primary sludge. In order to avoid the selection of specific micropollutants and to limit the cost of analyses, Gianico *et al.* (2013) have investigated the fate of groups of substances, namely extractable organic halogens (EOX), total hydrocarbons and methylene blue active substances (MBAS). Linear alkylbenzene sulphonates (LAS), which are widely used anionic surfactants present for example in detergents, represent 70 to 80% of the MBAS according to Di Corcia *et al.* (1994). The primary sludge is far more polluted by these three groups of substances than the secondary sludge itself. PAHs and heavy metals which are mostly connected to particles are removed from the wastewater treatment line either in the grit chamber or the primary settler. These micropollutants are therefore trapped into the primary sludge and higher hydraulic residence time in the primary treatment increases the removal yield for some metals (zinc, copper and lead). Such findings favor to separate the treatment of primary and secondary sludges to avoid the transfer of organic micropollutants to digested sludge, which can be used as fertilizer in agriculture.

The fate of other organic micropollutants is less documented. DEHP (Bis(2-ethylhexyl) phthalate) is mostly adsorbed to primary sludge, and a reduction of about 70% has been observed between the inlet and the outlet of a primary settler and of 35% between the raw wastewater and the outlet of the primary settler (Marttinen *et al.* 2003). The remaining DEHP is then mostly found in the secondary sludge. The fate of some polybrominated diphenyl ethers (PBDEs) have been investigated by Clarke *et al.* (2010). The fate of some congeners (BDE17, 47, 99, 100, 153, 154, 183 and 209) which account for >90% of the PBDEs are not or very slightly removed from the water phase in primary treatment.

Neither steroid estrogens nor nonylphenols were removed during the primary treatment in the pilot-plant tests run by Petrie *et al.* (2014) in UK. Samaras *et al.* (2013) have drawn different conclusions in their study of full-scale wastewater treatment plants in Greece: nonylphenols, triclosan and bisphenol A were mostly bound to particles and removed in the primary sludge.

In a survey of 19 French WWTPs Choubert *et al.* (2011) concluded that about 35% of the 93 substances tested (20 priority substances and 73 other substances including pharmaceuticals, hormones, metals and pesticides) were removed to some extent in the primary treatment (with or without chemical enhancement). But the removal efficiency was higher than 70% only for a handful set of substances (PBDEs, C_{10}-C_{13} chloroalkanes, PAHs and some metals). The removal of bisphenol A was investigated by Guerra *et al.* (2015) in 25 Canadian WWTPs. Some employed enhanced primary treatment. The efficiency was mixed, ranging from 0% to 90%. Operation conditions (short hydraulic residence times decrease the efficiency) and design (lamellar settlers improve the settleability) strongly influence the outcome.

A key property for removal of non-volatile organic micropollutants in the primary treatment is the hydrophobicity (Joss *et al.* 2005). Therefore, high sorption potential is expected for substances with high log K_{OW}. However, this rule does not work for all PPCPs and the solid-liquid partition coefficient (K_d), the ratio between the concentrations of a substance in the solid and in the aqueous phase at equilibrium conditions, is often preferred. In enhanced coagulation-flocculation tests run with $FeCl_3$, $Al_2(SO_4)_3$ or PAX, lipophilic musks (galaxolide and tonalide which have a log $K_d \approx 3.5$) or acidic substances (diclofenac), are removed from a urban wastewater matrix with efficiencies between 50% and 70% (Carballa *et al.* 2005). For the same substances, Suarez *et al.* (2009), applying similar coagulation-flocculation conditions but with a hospital wastewater matrix, the removal of both musks was increased but the removal of diclofenac was strongly decreased. The efficiency decreases as the K_d decreases. The same authors tested flotation and the removal was similar (highest for musks, lowest for anti-inflammatories)

Although the data are scarce, it seems that the primary treatment, with or without enhancement, has a limited effect on micropollutants. The key phenomena driving their removal are volatilization and adsorption. But operation conditions, such as hydraulic residence time, aeration (for sand removal), as well as design or nature of the wastewater, play a role (Luo *et al.* 2014a). Research on new organic flocculants should not only focus on the removal of suspended solids and colloids, but also on the fate of micropollutants. Attention should be paid to the reject water from the sludge treatment that can recycle the organic micropollutants as well as their metabolites.

9.2.2 Role of nitrifiers on organic micropollutants biotransformation

Since stringent nitrogen limits were in force a few decades ago, nitrification has become widespread in WWTPs. Nitrification is the biological oxidation of ammonium (NH_4^+) to nitrite (NO_2^-) and subsequently to nitrate (NO_3^-), being part of the biogeochemical nitrogen cycle. Ammonia-oxidizing bacteria (AOB) and ammonia-oxidizing archaea (AOA) are the main contributors to the first step of nitrification in WWTP and are usually referred to as ammonia-oxidizing microorganisms (AOM), while nitrite oxidizing bacteria (NOB) are responsible for the nitratation step. Several studies concluded that nutrient removal schemes, operated at longer SRT to allow the growth of nitrifiers, in activated sludge reactors of WWTPs enabled a better removal of different emerging OMPs (Clara *et al.* 2005a). However, the role of nitrifying microorganisms in the enhanced degradation is not fully clarified yet. It is clear that heterotrophic bacteria possess ability for the metabolic and also cometabolic degradation of many OMPs, but in the case of autotrophic bacteria, previous studies with pure AOB cultures (*Nitrosomonas europaea*) have shown that ammonia monooxygenase (AMO), the enzyme responsible for the first step of nitrification, is not able to provoke the breakage of aromatic rings, leading to an incomplete transformation of micropollutants.

However, ammonia oxidizing bacteria (AOB) are capable of biotransforming aliphatic hydrocarbons and the enhanced removal of some pharmaceuticals has been attributed to the role of nitrifiers. Cometabolic oxidation by the ammonium monooxygenase (AMO) enzyme is probably initiating the biotransformation of these substances, as demonstrated with the synthetic hormone 17α-ethinylestradiol (EE2) by Yi and Harper (2007) employing AMO extracted from the enriched nitrifying biomass grown in a nitrifying activated sludge (NAS) bioreactor. Khunjar *et al.* (2011) found that, although AOBs were able to biotransform EE2 more rapidly, the complete mineralization could only be accomplished by heterotrophs, which possess ring cleaving catechol dioxygenases; while TMP was only biotransformed by heterotrophs. Other authors employed the AMO inhibitor allylthiourea (ATU) to distinguish the effect of AMO from heterotrophic microorganisms. Roh *et al.* (2009) obtained degradation of triclosan and bisphenol A only in the absence of ATU in NAS and observed competitive inhibition of these OMPs on AMO. Tran *et al.* (2009) observed significant decreases in removal efficiency of naproxen (NPX) and also the recalcitrant compounds carbamazepine (CBZ) and diclofenac (DCF) when ATU was added to an enriched nitrifying culture. These results contradict those obtained by Suárez *et al.* (2010) in a NAS reactor where CBZ and DCF were not significantly removed.

Fernandez-Fontaina *et al.* (2012) studied the influence of nitrifying conditions on the biodegradation and sorption of emerging micropollutants. They observed high biodegradation efficiencies of different emerging micropollutants with NAS working at high nitrifying activities (> 1 g $N\text{-}NH_4^+$/g VSS · d). Cometabolic biotransformation seemed to be responsible for the removal of most compounds, since working at constant nitrogen loading rate (0.11 g $N\text{-}NH_4^+$/L · d) and temperature (25°C), a linear relationship ($r^2 > 0.90$) between specific micropollutant degradation rate and the specific nitrogen removal rate was found for all compounds (eg. ibuprofen (IBP) and fluoxetine (FLX), Figure 9.2). This means that the higher the rate of primary substrate consumed by bacteria, the higher the rate of micropollutant cometabolized. Therefore, an enhanced biotransformation of micropollutants could be expected in reactors operation at higher nitrogen loading rates. However, NAS showed a different affinity for each compound (94.2 µg IBP/g N v. 35.4 µg

FLX/g N, Figure 9.2), probably due to steric hindrance, activation energy limitations or the presence of specific functional groups. Increasing loading rates of micropollutants were removed at shorter hydraulic retention times (Figure 9.3), although the biodegradation efficiencies of compounds with slow/intermediate kinetics, such as fluoxetine, erythromycin, roxithromycin and trimethoprim, diminished due to kinetic and/or stoichiometric limitations. In another work, the same authors applied a cometabolic Monod-type kinetics model, linking biotransformation of micropollutants with primary substrate degradation, to a NAS reactor and they observed that such model with two parameters (the micropollutant transformation capacity (T_C) and the half-saturation constant (K_{SC})) predicts more accurately the biotransformation of several organic micropollutants (IBP, NPX, erythromycin (ERY) and roxithromycin (ROX)) in a NAS reactor (Fernandez-Fontaina *et al.* 2014). The cometabolic term (T_C) takes into account the effect of the nitrification rate on the biotransformation of these micropllutants and is the most relevant for improving the prediction accuracy of the biotransformation model. The Monod kinetics (K_{SC}) does not improve accuracy when compared to the commonly applied pseudo-first order kinetics, but could allow a higher flexibility if the reaction order changes. The estimated parameters (T_C and K_{SC}) at 25°C were successfully employed to predict the operational conditions of a NAS reactor required to achieve 90% biotransformation efficiency (Figure 9.4).

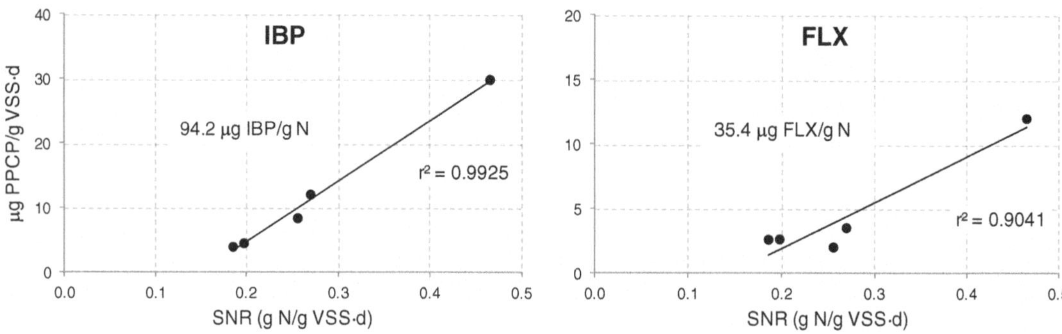

Figure 9.2 Specific micropollutant biodegradation rate (µg/g VSS · d) vs. specific nitrification rate (mg N/g VSS · d) operating at 25°C and at a nitrogen loading rate of 0.11 g N/L · d. (Fernandez-Fontaina *et al.* 2012).

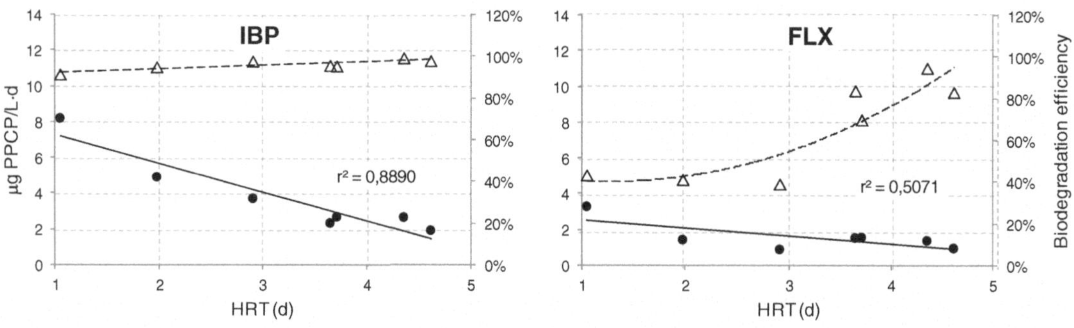

Figure 9.3 Micropollutant degradation efficiency (%, Δ) and micropollutant biodegradation rate (µg/L · d, •) vs. hydraulic retention time (HRT). (Fernandez-Fontaina *et al.* 2012).

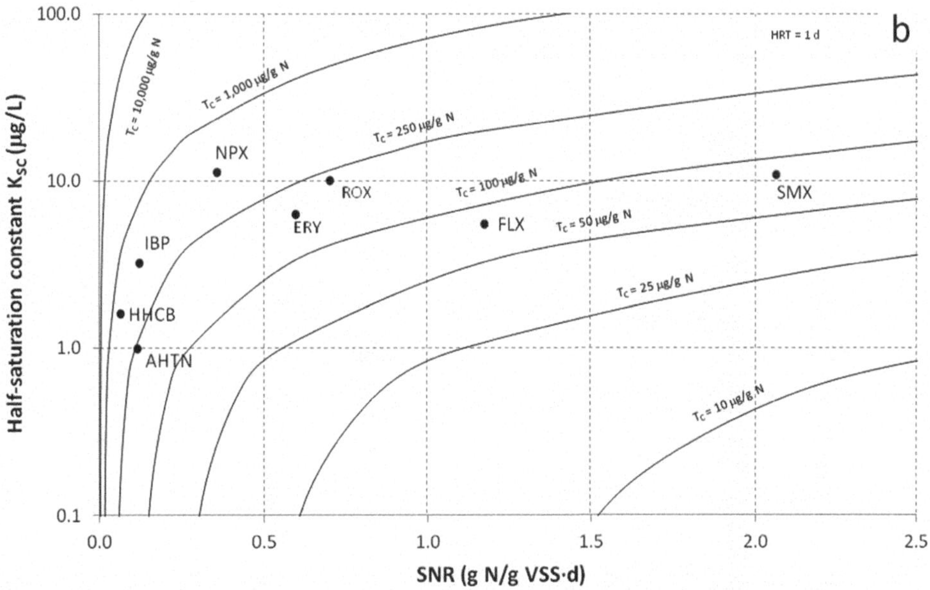

Figure 9.4 Influence of transformation capacity (T_C) and half-saturation constant (K_{SC}) of a micropollutant (K_d: 50 L/kg TSS; inlet concentration: 10 µg/L) on the minimum specific nitrogen removal rate (SNR) required to achieve 90% biodegradation efficiency in a NAS reactor modelled as a continuous stirred tank reactor (T: 25°C; TSS: 0.85 g/L; VSS: 0.65 g/L; SNR: 0.15 g N/g VSS · d; HRT: 1d). The selected organic micropollutants are located in the graph according to their experimental kinetic coefficients. (Fernandez-Fontaina *et al.* 2014.)

9.2.3 Membrane bioreactors

In the last years, some studies have presented membrane bioreactor (MBR) technology as holding a promise for the removal of several OMPs, compared to conventional activated sludge system (CAS) (Radjenovic *et al.* 2009). The use of MBRs instead of conventional activated sludge systems should be mainly more effective on biodegradation, transformation and sorption of micropollutants.

MBR effluents are free of suspended solids and the membranes (ultrafiltration or microfiltration) are able to retain big particles such as colloids inside the bioreactor, increasing the adsorption surface and therefore, micropollutants that present a tendency to sorb to the biomass will more likely adsorb onto suspended solids. In CAS systems, if the OMPs adsorb on a non-settleable solid, they leave the CAS through the effluent. On the other hand, the main advantage of the MBR process is its independence regarding the settling conditions. Therefore, most of the non-settleable solids should be retained by the membrane, leading to a higher removal rate. Radjenovic *et al.* (2009) found higher concentrations in MBR sludge rather than conventional activated sludge for hydrochlorothiazide, azithromycin, carbamazepine and ketoprofen (Figure 9.5).

Although sorption of most pharmaceutical micropollutants to activated sludge in MBR is negligible at neutral pH, enhanced sorption was observed for some acidic pharmaceuticals at lower pH. Urase and Kikuta (2005) reported that the removal rate of acidic pharmaceuticals (e.g. gemfibrozil, ibuprofen, ketoprofen, and diclofenac) in MBR was much higher at pH of 4.3–5.0 than that at pH of 6.8–7.6 and 7.5–8.0. On the other hand, the removal of neutral pharmaceutical carbamazepine is not usually affected by pH. Nevertheless, in general, the substances attached to the sludge do not accumulate in the reactor and are likely to be biodegraded (Radjenovic *et al.* 2009).

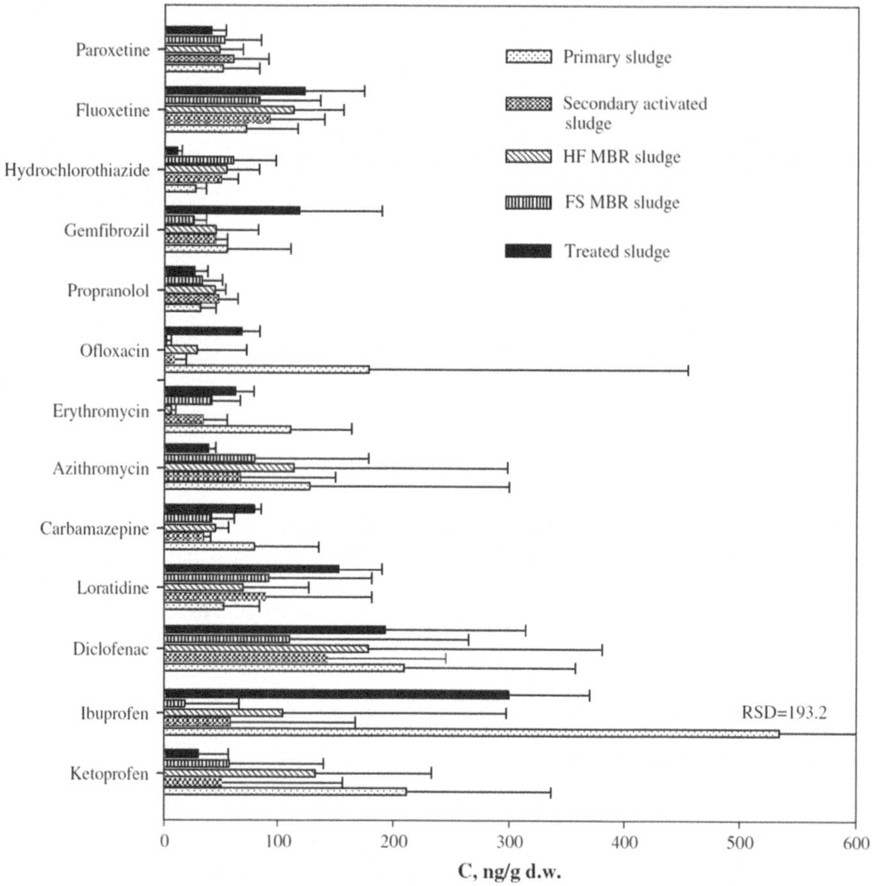

Figure 9.5 Mean concentrations of the OMPs encountered in the primary, secondary activated, advanced MBR and treated sludge in WWTP Terrassa, with their standard deviations. (Adapted from Radjenovic *et al.* 2009.)

Biodegradation processes are strictly correlated to the characteristics of the biomass, the compounds characteristics, the plant configuration and operation parameters. The biodegradation kinetics can be described by the Michaelis-Menten model, broadly used in activated sludge systems, and due to the low concentrations of OMPs, the kinetic tends to the first order. As a consequence, the lower the concentration is, the harder the biodegradation becomes, as the rate decreases. Longer retention times (independently from HRT) and greater amounts of biomass could be useful to overcome these concerns. Apart from the final liquid/sludge separation stage, providing a competitive advantage for organisms able to degrade persistent compounds by eliminating bacterial washout, MBR systems are mainly distinguished by their higher sludge ages, higher sludge concentration and with faster start-up phases without affecting substantially the separation step (Bouju *et al.* 2008). Sludge ages are generally longer for MBR (15–80 d) with respect to CAS (7–20 d), as well as by their biomass concentration, generally higher in the MBR than in the CAS (8–10 kg/m^3 in MBRs and 3–5 kg/m^3 in CAS). This gives the possibility to the slowly growing biomass to develop, and therefore the biomass presents a larger biodiversity and broader physiological capabilities and, potentially, more likely capable of degrading some of the specific present OMPs (Figure 9.6, Clara *et al.* 2005a).

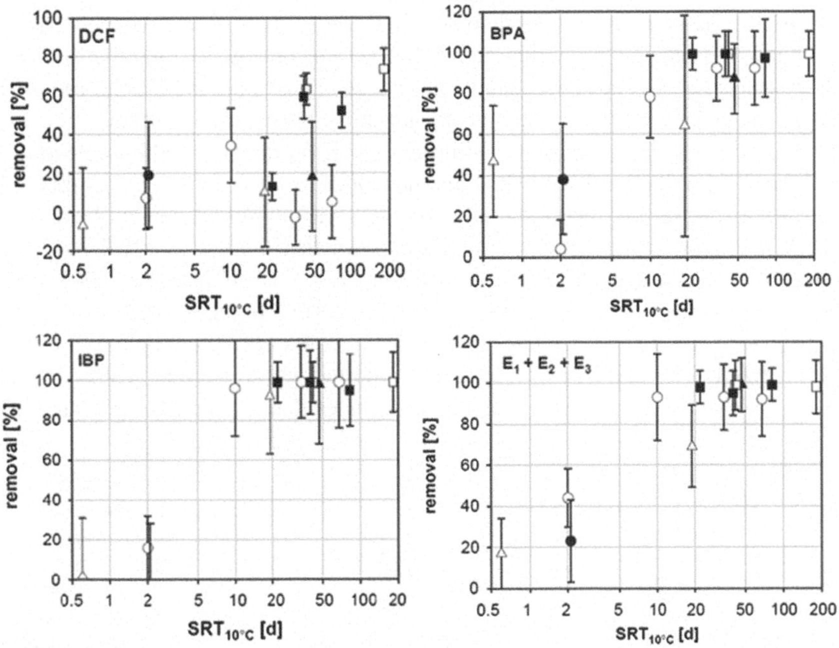

Figure 9.6 Calculated removal efficiencies (%) of diclofenac (DCF), bisphenol A (BPA), ibuprofen (IBP) and estrone (E1) + 17B-estradiol (E2) + estriol (E3) at 10°C in relation to the SRT in the different treatment facilities. (Adapted from Clara *et al.*, 2005a.)

Although high sludge age does not lead necessarily to complete biodegradation of all the target micropollutants (Figure 9.6), it has been observed that in general the biological degradation was enhanced. Therefore, MBRs, which allow a much higher sludge age, in a reasonable footprint area, should be able to lead to a greater removal of OMPs. High sludge ages have also beneficial effects on the removal of micropollutants that tend to accumulate in the sludge flocs (i.e. tetracycline, ciprofloxacin, ofloxacin, norfloxacin). Clara *et al.* (2005a) found that to get high removal of some biodegradable pharmaceutical compounds (e.g. hormones, bezafibrate and ibuprofen) a SRT >10 d is needed. Adversely, in other studies no clear correlation was observed between percentage elimination and SRT, in particular for beta-blockers, carbamazepine and the antibiotics ciprofloxacin, ofloxacin and norfloxacin (Joss *et al.* 2005).

In addition, the elevated biomass concentrations may encourage interactions between microorganisms, and therefore lead to an intensification of the biological processes. The higher biomass concentration results in a lower food to mass ratio, thus, limit the substrate availability. High sludge ages combined with reduced F/M ratios may result in an increased biodiversity and this could give a more complete biodegradation of micropollutants, since microorganisms would have no other option than degrading recalcitrant pollutants present at lower concentrations (Clara *et al.* 2005b). Moreover, MBR viable fraction is usually higher than in conventional systems (Cicek *et al.* 1999) with an improved mass transfer due to the presence of smaller flocs (10–100 μm in MBR against 100–500 μm in CAS). These factors may enhance enzymatic activities and the contact between microorganisms and pollutants and stimulate their biodegradation. Nonetheless, for most compounds, the observed biodegradation kinetic constants and the solid-liquid partitioning coefficients in MBR system were similar to literature data referring to conventional biomass (Verlicchi *et al.* 2012).

Data pertaining to full scale MBRs are quite limited (Verlicchi *et al.* 2012). As a matter of fact, in several studies, MBR were operated at unrealistically high sludge ages, or fed with synthetic feed. Hence, comparison of both conventional processes and MBR system operated in parallel and in comparable conditions are missing. For the most frequently investigated pharmaceuticals, the removal efficiencies variability ranges are generally quite wide, but, in general, higher removal efficiencies are reported in MBRs (Figure 9.7), with respect to conventional processes (Kimura *et al.* 2005; Radjenovic *et al.* 2009). Some micropollutants are well eliminated by membrane bioreactors (ex. fluoxetine, gemfibrozil, diclofenac, AMPA, diuron, sulfamethoxazole, ibuprofen, alprazolam, amitriptyline, & several betablockers, Martin-Ruel *et al.* 2011). Adversely, other authors have also shown that the use of MBR did not significantly improve the removal of all pharmaceutical substances (Clara *et al.* 2005b; Joss *et al.* 2005) and the different behaviours for each substance makes it difficult to establish a general trend. No significant differences in the removal efficiency of several pharmaceuticals (e.g. diclofenac, ibuprofen, carbamazepine, bezafibrate and ethinylestradiol) between CAS and MBR systems was reported when operated at similar sludge retention times, which suggests that the reactor type is of less importance than the sludge age (Clara *et al.* 2004).

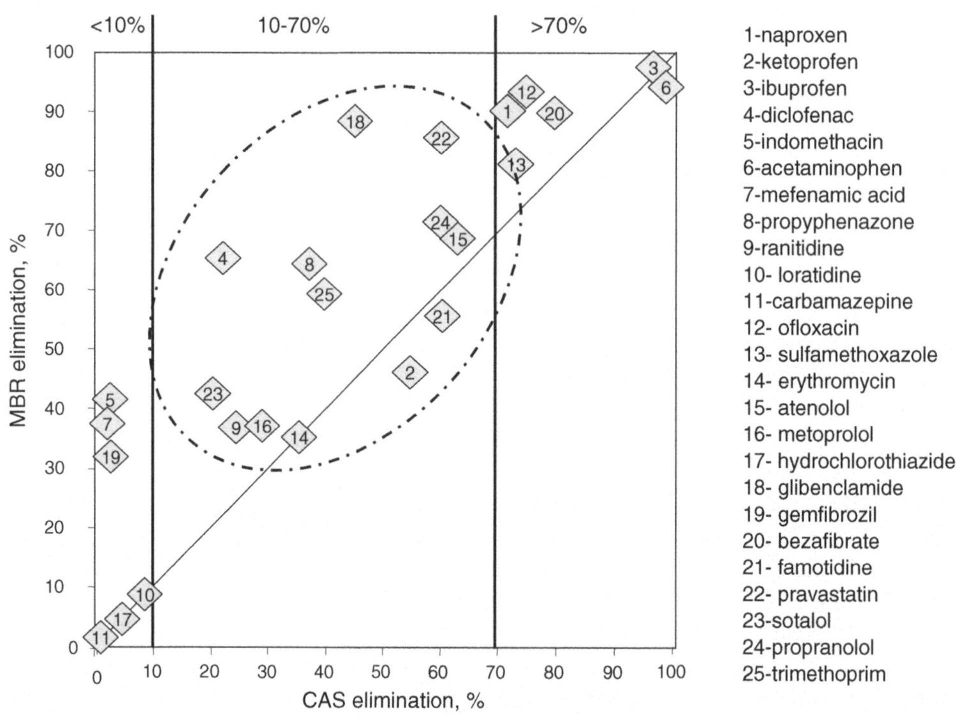

Figure 9.7 Comparison of the mean removals of encountered pharmaceuticals in full-scale CAS and pilot-scale MBRs. (Radjenovic *et al.* 2009.)

In terms of compounds characteristics, it has been observed that micropollutants with simple structures such as ibuprofen are removed easily in both conventional systems and MBRs, while those with a more complex structure are removed with a greater efficiency in MBRs. Nevertheless, it was shown (Kimura *et al.* 2005) that specific structures, like the presence of chlorine atoms or double ring in the chemical structure (e.g. ciprofloxacin, clarithromycin, furosemide and diclofenac) decreases the biodegradation potential of micropollutants in both cases and no significant difference has been observed in removal efficiency between conventional systems and MBRs. Instead, pharmaceutical substances poorly biodegraded in conventional systems but without chlorine atoms were better removed in MBRs.

It can be concluded that MBRs do not impact the removal neither of recalcitrant micropollutants, nor of readily biodegradable ones (Kimura *et al.* 2005; Radjenovic *et al.* 2009) while it can be more effective on intermediate biodegradable compounds. These results tend to show that the tertiary treatment is a key step in micropollutants removal. The higher and more stable effluent quality of MBRs is advantageous compared to a CAS effluent, and therefore, it should result in a better efficiency of post-treatments.

9.2.4 Granular sludge reactors

Aerobic granular sludge offers an interesting alternative for conventional activated sludge systems due to its excellent physical characteristics as a biofilm composed of microbial self-immobilized cells which perform different and specific roles in biodegradation of pollutants during wastewater treatment. In comparison to activated sludge, aerobic granular sludge has a denser and stronger microbial aggregate structure, a higher biomass concentration, a better settling capacity and the ability to withstand shock loads. The stratification of conversion processes and redox zones within the granules provide aerobic and anaerobic/anoxic layers, which allow the simultaneous removal of carbon, nitrogen, and phosphorous. Moreover, the systems based on aerobic granular biomass are known to perform better in front of inhibitory or toxic compounds compared to activated sludge system, because granule architecture causes diffusion gradients protecting sensitive bacteria. The development of aerobic granules has been extensively reported using sequencing batch reactor (SBR) systems, using an anaerobic feeding period, followed by an aeration period in which simultaneous nitrification/denitrification takes place. In the anaerobic stage, most or all chemical oxygen demand (COD) is taken up by microorganisms, such as polyphosphate accumulating organisms (PAOs) and glycogen-accumulating organisms (GAOs), which store COD as intracellular polymers. In the aerated stage, nitrification occurs at the oxygen-containing outer part of the granules while PAOs and GAOs oxidize their storage polymers in order to grow. This is done with oxygen in the outer part of the granules and with nitrate or nitrite in the inner part of the granules. The operational flexibility, simplicity and small footprint of the SBR (ability to decrease settling time, initial reactor volume, etc.) played a key role to promote formation and maintenance of a compact granular biomass. The system has been used to treat different municipal wastewater and wastewater containing specific pollutants (Duque *et al.* 2011). But few studies focused on the application of granular sludge for OMPs removal.

Amorim *et al.* (2014) evaluated the removal of fluoroquinolones (FQs), namely ofloxacin, norfloxacin and ciprofloxacin, at concentrations ranging between 3 and 11.5 ppm, using aerobic granular sludge and their effect on the performance and microbial community in the reactor. They found no evidence of FQ biodegradation, but they were adsorbed to the aerobic granular sludge, being gradually released into the medium in successive cycles after stopping the FQ feeding. Overall, COD removal and ammonium and nitrite oxidation were not inhibited by the presence of FQ, but nitrate accumulation was observed suggesting that denitrification was inhibited. Also the activity of phosphate accumulating organisms was affected. Moreover, exposure to FQs promoted disintregration of the granules leading to an increase of the effluent solid content. However, granules and effluent solids recovered normal levels after removing

FQs in the feed. In another work (Moreira *et al.* 2015), the same authors investigated the fate and effect of fluoxetine (FLX), at concentrations of 0.9 and 1.2 ppm, on aerobic granular SBR performance and on microbial population. Again, COD, ammonium (after 20-day adaptation period) and nitrite were not inhibited, but denitrification was. In this case, phosphate removal was affected at the beginning, but it recovered later. FLX was removed by 8–69%, depending on the operational conditions (Table 9.2), with sorption to granules as the main removal mechanism, since no intermediate metabolites or fluoride release were detected in the reactor outlet. FLX was sorbed onto the biomass till its sorption capacity was exhausted and then desorption occurred, being the latter more evident in the periods of FLX absence in the feed.

Table 9.2 Summary of the SBR performance for FLX removal.

Phase	FLX mass balance (mmol)	
	FLX in the inlet	**FLX in the effluent**
I	–	–
II	0.37	0.11
III	0.07	0.07
IV	0.31	0.44
V	0.15	0.14
VI	0.00	0.06
VII	0.38	0.29
VIII	0.00	0.08
Total	1.30	1.20

Source: Moreira *et al.* (2015).

Zhao *et al.* (2015) investigated the removal of five OMPs (ibuprofen, naproxen, prednisolone, norfloxacin and sulfamethoxazole) during urban wastewater treatment using an aerobic granular SBR. Initially, OMPs addition affected negatively granular sludge quality, but after microbial adaptation, the system recovered and effectively removed four of the five spiked substances (Figure 9.8). In another study (Xia *et al.* 2015), the same authors used a membrane bioreactor seeded with aerobic granular sludge to study the removal of the same five OMPs. They also analysed the effects of different sludge retention and hydraulic retention times (SRT and HRT, respectively) and influent organic loading rates on the OMP removal efficiencies. They observed that, in general, longer SRT and HRT improved OMP removal, while the influent organic loading rate only affected the elimination of sulfamethoxazole and ibuprofen: the higher the organic loading, the higher the removal efficiency.

The differences in the removal efficiencies can be explained by the adsorption and degradation patterns of the different substances. OMPs in the liquid phase are only in direct contact with the surface of the granule, while a slow intramolecular diffusion is necessary to retain the OMP inside the granule (Shi *et al.* 2010). The granule diameter size is inverse correlated to the specific surface. Other parameter that influences the sorption onto the granular biomass is the contact time. Alvarino *et al.* (2015) showed a continuous increase with the time of the relative amount of musk fragrances sorbed, which indicates that the solid-liquid equilibrium was not achieved.

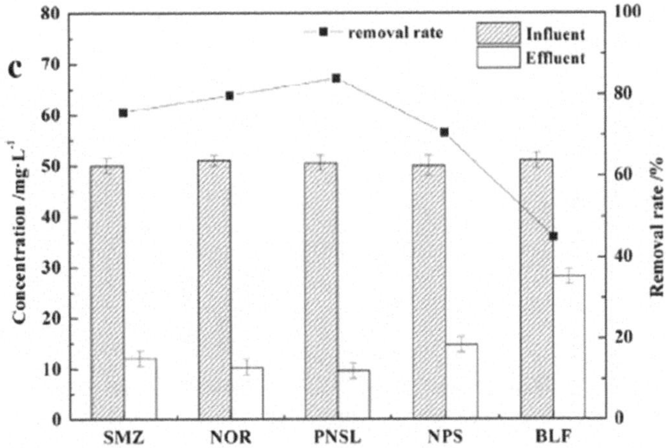

Figure 9.8 Removal efficiencies of sulfamethoxazole (SMZ), norfloxacin (NOR), prednisolone (PNSL), naproxen (NPS) and ibuprofen (BLF) in an aerobic granular sludge sequencing bioreactor. (Zhao *et al.* 2015.)

9.2.5 Partial nitritation – Anammox process

An alternative to the conventional nitrification-denitrification process is the combination of the nitritation and the anaerobic oxidation of ammonia (anammox). The advantages of the anammox process compared to the conventional nitrification-denitrification are the lower energy consumption, the reduction in the sludge production and greenhouse gas emissions and the no necessity of an external carbon source (Kartal *et al.* 2008). Additionally, the nitritation-anammox process can be carried out in a single reactor with the biomass being developed in a biofilm or granules, so different redox potentials are present in a single reactor and an enhancement in the microbial diversity is achieved. The presence of this process in WWTPs is continuously increasing in Europe and several plants at real scale have been implemented mainly in The Netherlands and Austria. However, the information about the removal of organic micropollutants in this new technology is scarce compared to other processes.

Alvarino *et al.* (2015) studied the removal of several organic micropollutants in a nitritation-anammox process treating the supernatant of an anaerobic digester (Figure 9.9). Ibuprofen was a readily biodegradable compound (removal efficiency of 98%), as previously observed by de Graaff *et al.* (2011), who studied the treatment of black water by a nitritation-annamox process after anaerobic pretreatment (Table 9.3). These authors stated that ibuprofen removal occurred mainly in the anammox reactor (77%), which is inconsistent with the results achieved by Falas *et al.* (2012), who achieved a limited IBP removal with the anammox bacteria. The same limitation was observed in the removal of other acidic pharmaceuticals, such as naproxen (NPX), ketoprofen (KTP) or diclofenac (DCF) (Falas *et al.* 2012). Nevertheless, Alvarino *et al.* (2015) classified the acidic pharmaceuticals NPX as readily biodegradable compound with a removal efficiency above 75%, which probably occurred in the aerobic phase. Similar removal efficiency was shown by Fernandez-Fontaina *et al.* (2012) in an enriched nitrifying culture (90%) and Suarez *et al.* (2010) in a nitrification-denitrification process (80%). In fact, Suarez *et al.* (2010) confirmed the removal of NPX only under aerobic conditions, while the removal was negligible under anoxic conditions. The low elimination of DCF observed by Falas *et al.* (2012) is in agreement with the results of other authors (Suarez *et al.* 2010; de Graaff *et al.* 2011; Alvarino *et al.* 2015). As in the case of the conventional nitrification-denitrification, the main removal mechanism of OMPs in the nitritation-anammox process is biotransformation, while sorption is only significant for the lipophilic compounds, such as musk fragrances and fluoxetine (Suarez *et al.* 2010; Alvarino *et al.* 2015).

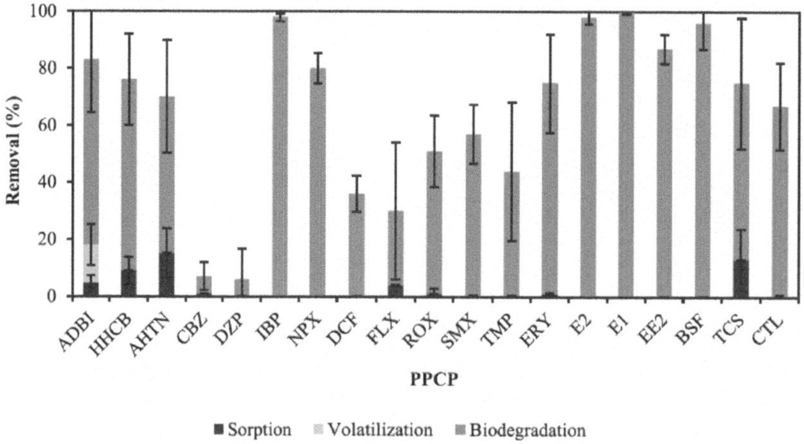

Figure 9.9 Removal efficiencies of 19 organic micropollutants in a nitritation-anammox processes treating the supernatant of an anaerobic digester. (Alvarino *et al.* 2015.)

Table 9.3 Comparison of the removal efficiency (%) of organic micropollutants in conventional nitrification-denitrification and nitritation-anammox processes.

	Anammox	Nitritation - Anammox	Nitrification - Denitrification
Ibuprofen	77	98	82
Diclofenac	0	36	36
Naproxen	N.A.	80	68
Carbamazepine	N.A.	7	10
Diazepam	N.A.	6	10

Note: N.A. = not analyzed.
Source: Suarez *et al.* (2005); de Graaf *et al.* (2011); Alvarino *et al.* (2015).

Alvarino *et al.* (2015) classified the three hormones (El, E2, EE2), BPA and ADBI as readily biodegradable compounds (removal efficiency above 80%) in a nitritation-anammox process, whereas a medium-high removal was obtained in the case of HHCB, AHTN, ROX, ERY, TCS or CTL. In the case of the hormones, similar results were obtained by Suarez *et al.* (2010) under nitrifying conditions, while the removal of EE2 was below 20% during hetrotrophic denitrification. CBZ, DZP, SMX and TMP are persistent compounds in conventional nitrification-denitrification processes (Suarez *et al.* 2005; Suarez *et al.* 2010). In contrast, the antibiotics SMX and TMP were eliminated (57% and 45%, respectively) in the nitritation-anammox process (Alvarino *et al.* 2015). The recalcitrant behaviour of CBZ and DZP is clear in both autotrophic and heterotrophic biological nitrogen removal processes.

A correlation between the specific anammox activity and the removal of several OMPs in a nitritation-anammox process was observed by Alvarino *et al.* (2015). For instance, the removal rate of ERY was linear correlated to the specific anammox activity and an enhancement in the removal efficiency was obtained from 20 to 73%, when the anammox activity increased from 50 to 150 mg N/g VSS · d (Figure 9.10).

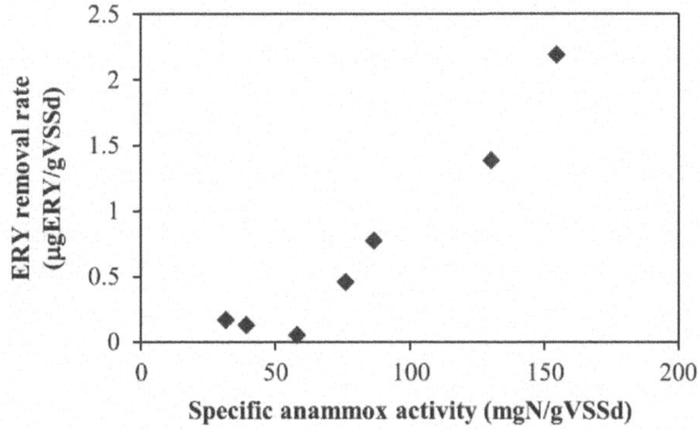

Figure 9.10 Influence of specific anammox activity in the removal of erythromycin (ERY). (Alvarino *et al.* 2015.)

To summarize, OMPs removal efficiencies obtained in the nitritation-anammox process are similar to those obtained under enriched nitrifying conditions (Suarez *et al.* 2010; Fernandez-Fontaina *et al.* 2012). Although this technology has many advantages related to the reduction in the operational costs, the persistent organic micropollutants are not suitable to be removed by the nitritation-anammox process, so a physical or chemical postreatment is necessary to reduce their release into the environment.

9.2.6 Anaerobic treatment

The main advantages of the anaerobic process are the low production of excess sludge and the reduction on the energy consumption by the conversion of organic matter into biogas (Buntner *et al.* 2011). However, anaerobic digestion in WWTP is mainly applied for the treatment of the excess of sludge and it is not a common treatment in the water line. Additionally, the information about the removal of OMPs under anaerobic conditions is focussed mainly in the sludge line, but in the water line, the information is scarce and dependant on the technology used (Table 9.4).

Ibuprofen and naproxen are two readily biodegradable pharmaceuticals in conventional activated sludge, with removal efficiencies above 90%. However, their behaviour under anaerobic conditions is completely the opposite (Table 9.4). Ibuprofen has a recalcitrant behaviour under negative redox potentials, with removal efficiencies below 50% (Musson *et al.* 2010; Alvarino *et al.* 2014; Dutta *et al.* 2014). This low removal might be related to interference of the branched substitutions on para position of the aromatic ring. In contrast, the removal efficiency of naproxen is similar and high (above 70%) under aerobic and anaerobic conditions (Alvarino *et al.* 2014). The behaviour of ketoprofen under anaerobic conditions is not clear. Dutta *et al.* (2014) classified it as a medium-high biodegradable compound in a two-stage anaerobic fluidized membrane bioreactor, while a removal below 15% was observed by Monsalvo *et al.* (2014) in an anaerobic membrane bioreactor.

Macrolide antibiotics, such as roxithromycin and erythromycin, are poorly biodegraded under anaerobic conditions, while trimethoprim and sulfamethoxazole are readily biodegradable (Table 9.4). The high biodegradation of these antibiotics under anaerobic conditions is related to the chemical structure. The unsubstituted heterocyclics are difficult to be degraded under anaerobic conditions, whereas the substituted heterocyclic compounds are prone to be biodegraded, except when halogen groups are present (Adrian *et al.* 1994). Therefore, the presence of a substituted pyridine group in the structure of the trimethoprim explains its removal, as well as the presence of the 3-amino-5-methyl-isoxazole ring in the case of the sulfamethoxazole. Additionally, electron withdrawing groups, like the sulfonyl group in sulfamethoxazole are readily converted by reductive biotransformations in anaerobic environments, while they are difficult to be degraded under aerobic conditions (Field, 2002).

Table 9.4 OMPs removal under anaerobic conditions in the water line.

	UASB reactor[a]	Anaerobic fluidized bed reactor[b]	Anaerobic MBR two stage[b]	Anaerobic MBR[c]
17a-ethinylestradiol	50	N.A.	N.A.	15
17B-estradiol	62	N.A.	N.A.	60.2
Bisphenol A	N.A.	N.A.	N.A.	31.5
Caffeine	N.A.	86.7	98.4	76.9
Carbamazepine	5	73	96.4	4.8
Diclofenac	2	31.3	78.2	<1
Estrone	38	N.A.	N.A.	<1
Ibuprofen	12.7	53.8	90.9	<1
Ketoprofen	N.A.	73.7	100	14.9
Naproxen	92	77.8	96.4	70.3
Paracetamol	N.A.	87.8	100	58.1
Sulfamethoxazole	88	35.5	89.1	95.2
Trimethoprim	98	87.4	100	35.4
Erythromycin	20	58.7	86.3	N.A.

N.A. = not analyzed.
[a]Alvarino *et al.* (2014).
[b]Dutta *et al.* (2014).
[c]Monsalvo *et al.* (2014).

High removal efficiencies (greater than 90%) were also achieved under anaerobic conditions for other compounds such as triclosan, metropolol, nonylphenol, androsterone, testosterone or acetylsalicylic acid (Musson *et al.* 2010; Monsalvo *et al.* 2014). Although the substitutions upon the aromatic ring can affect the removal of OMPs, COOH and OH functional groups promote the hydroxylation (Musson *et al.* 2010). Metoprolol contains a substituted aromatic ring that can be readily biodegraded under anaerobic conditions, as well as an OH functional group which enhances the hydroxylation. Therefore, in spite of the presence of an amine group that is recalcitrant under anaerobic conditions and readily biodegradable under positive redox conditions, metoprolol is a compound with medium-high biodegradability under anaerobic conditions (Musson *et al.* 2010).

Under anaerobic conditions, the biodegradation of the molecules can be by hydrogenation or hydroxylation. Although the hydroxylation of carbon associated to the aromatic ring allows the ring cleavage, the hydroxylation can be interfered by the substitutions upon the ring. On example is the low biodegradation of acetaminophen under anaerobic conditions due to a branched chain substitution in the aromatic ring and the stability of the amide (Musson *et al.* 2010). Aliphatic compounds are more biodegradable than cyclic compounds under anaerobic conditions, while the polycyclic are less biodegradable than cyclic compounds. This fact explains the resistance to be biodegraded of the progesterone that is a large polycyclic compound (Musson *et al.* 2010). Also, heterocyclic compounds, such as uracils or pyridines, or the presence of halogens hamper the biodegradation under anaerobic conditions (Adrian *et al.* 1994). The latter explains the persistence under anaerobic conditions of other compounds, such as diclofenac, carbamazepine, diazepam, DEET, primidone, meprobamate and atrazine (Dutta *et al.* 2014; Monsalvo *et al.* 2014; Alvarino *et al.* 2015).

There are few studies comparing the removal of OMPs under anaerobic and aerobic conditions. Alvarino *et al.* (2014) compared the removal of 16 OMPs in a conventional activated sludge (CAS) and an anaerobic UASB reactor (Figure 9.11) and most OMPs were more easily removed under aerobic conditions, except sulfamethoxazole, trimethoprim and naproxen. Monsalvo *et al.* (2014) also observed higher removal efficiencies for trimethoprim, sulfamethoxazole and ketoprofen in an anaerobic MBR compared to a CAS process. Joss *et al.* (2004) compared the removal of three estrogens (E1, E2, EE2) under different redox potential (anoxic, aerobic and anaerobic). Although the the three estrogens were eliminated under all the different redox conditions, the biological kinetic constants were significantly different and higher under aerobic conditions compared to the negative redox potentials. Under anaerobic conditions, E2 was the most biodegradable hormone and El was reduced to E2, while EE2 was the most recalcitrant. The lower biotransformation of EE2 compared to the natural hormone E2 is due to an esteric impediment in the EE2 molecule related to the presence of the ethinyl group, which does not allow the formation of a ketone, as in the case of E2 (Czajka & Londry, 2006).

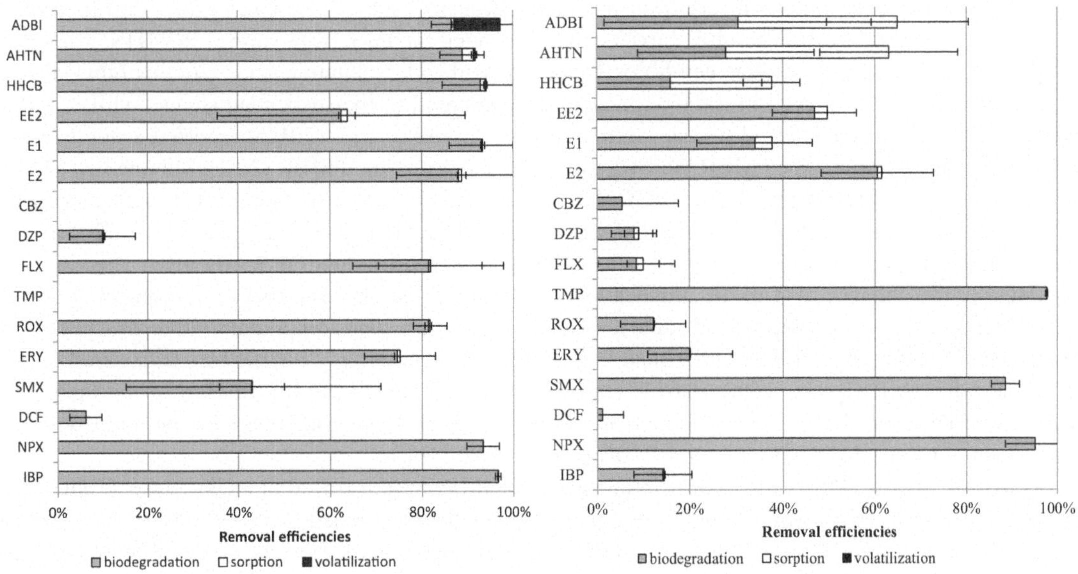

Figure 9.11 Comparison of the removal efficiencies of organic micropollutants under aerobic (a) and anaerobic (b) conditions with a HRT of 1 d. (Alvarino *et al.* 2014.)

The removal efficiencies of OMPs observed in the anaerobic wastewater treatment are clearly lower than those obtained during the anaerobic digestion of sewage sludge, especially for low-medium biodegradable compounds. For instance, Carballa *et al.* (2007) showed removal efficiencies of 15–60%, 15–80% and 20–60% for diazepam, diclofenac and ibuprofen, respectively, during the anaerobic digestion of sewage sludge, while these three compounds were recalcitrant during the anaerobic treatment of wastewater (Alvarino *et al.* 2014). These differences might indicate a positive effect of the HRT and the temperature on the removal of OMPs since the anaerobic digestion of sludge is carried out under mesophilic and thermophilic conditions and with HRT above 6 days.

Overall, the removal of OMPs during anaerobic treatment is quite limited. Yet, the combination of an anaerobic and aerobic treatment can be a good alternative in order to combine the positive effect of both redox conditions and increase the microbial diversity.

9.2.7 Hybrid systems

Hybrid bioreactors have been increasingly employed in the treatment of wastewater due to the advantageous combination of suspended and biofilm biomasses as well as the use of diferent redox potentials The use of supports in biological reactors enhances the transfer of oxygen and the nitrification rate, as well as giving robustness to the process (Di Trapani *et al.* 2014). Additionally, the presence of supports in hybrid systems allows the coexistence of different redox potentials (anoxic, aerobic and anaerobic conditions) in the same reactor, which facilities the biotransformation of OMPs (Luo *et al.* 2014b; de la Torre *et al.* 2015).

The use of supports exerts a positive influence on the elimination of OMPs in biological process. Alvarino *et al.* (2016) compared the removal of OMPs in an aerobic MBR with and without carriers in order to determine the influence of the presence of attached biomass. They observed an enhancement of the nitrification efficiency from 25 to 60%, which was related to the improvement on the removal of several OMPs, such as ibuprofen, erythromycin and estrogens (Table 9.5). Falas *et al.* (2012) studied the removal of seven OMPs in biofilm carriers and suspended biomass from several WWTPs. For some compounds (diclofenac, ketoprofen, gemfibrozil, clofibric acid and mefenamic acid), a higher removal rate per unit biomass was observed in the biofilm biomass compared to the suspended sludge, while similar removal efficiencies were showed for naproxen and ibuprofen. Luo *et al.* (2014b) reported a significantly higher removal in a moving bed MBR (46%) compared to a conventional MBR (20%). Moving bed MBRs (MBMBR) are effective for the removal of OMPs with a strong electron donating functional group, like hydroxyl groups. For instance, removal efficiencies above 80% were observed by Luo *et al.* (2014b) for this group of compounds (ibuprofen, naproxen, salicylic acid or primidone) in a MBMBR. This fact can be related to the enhancement in the nitrification rate with the use of supports (Di Trapani *et al.* 2014) because ammonia oxidizing bacteria are able to oxidize by cometabolism compounds with hydroxyl groups (Fernandez-Fontaina *et al.* 2012). Other readily removed compounds in MBMBRs are triclosan, E1, E2, EE2, nonylphenol, fenoprop, metronidazole and fluoxetine (Luo *et al.* 2014b). In spite of the lipophilic character of triclosan, E1, E2, acetaminophen and nonylphenol, biodegradation was the main removal mechanism of these OMPs in the moving MBR (Luo *et al.* 2014b).

De la Torre *et al.* (2015) compared the removal of 29 OMPs in four different technologies (Table 9.6): a MBR, a MBMBR, a IFAS-MBR which combines suspended and attached biomass and a CAS unit. The highest elimination was observed in the IFAS-MBR which combines the positive effects of both types of biomass: a high concentration of biomass due to the retention by a membrane, the presence of a biofilm and a high sludge age. Removal efficiencies above 65% were achieved for all the studied compounds in the IFAS-MBR, except for nonylphenols and di-(2)-ethylhexyl)phthalate that were recalcitrant in the three different technologies studied. The worst behaviour was observed in the MBMBR, which operated only with attached biomass and lower HRT and biomass concentration. Escola Casas *et al.* (2015) developed a four stage hybrid biofilm and activated sludge system to treat hospital wastewaters. The OMP removal efficiencies obtained in this system were similar to the conventional MBR, except for the X-ray iodinated contrast media. This group of compounds has a recalcitrant character in conventional wastewater treatments, while removals of 60–80% were obtained for most of them in the hybrid system. In these hybrids systems, OMP sorption onto the supports has to be taken into account as well.

Luo *et al.* (2014b) compared the removal of OMPs in sponge-based carriers with and without attached biomass in batch tests and observed a considerable elimination of bisphenol A, E1, E2, EE2, nonylphenol and triclosan (above 80%) in both supports, independently of the presence of biomass, so that the sorption onto the support (polyurethane) played a significant role in the removal of these OMPs. The removal of other compounds, such as ketoprofen, ibuprofen, diclofenac, naproxen, acetaminophen, gemfibrozil and salicylic acid, improved considerably in the sponge with attached biomass. The sorption of OMPs in the support depends on two main parameters: the material of the support and the presence of polar and non-polar groups in the OMP structure.

Table 9.5 Effect of the use of Kaldness rings as support on the biotransformation kinetic constants of OMPs in the aerobic chamber of an anaerobic hybrid MBR.

k_{biol}	*Without carriers*	*With carriers*
Ibuprofen	1.2 ± 0.4	8 ± 3
Naproxen	0.0	0.5 ± 0.2
Diclofenac	3 ± 1	3 ± 1
Sulfamethoxazole	2 ± 1	2 ± 1
Erythromycin	0.05 ± 0.1	0.1 ± 0.1
Roxithromycin	0.4 ± 0.2	0.5 ± 0.1
Trimethoprim	0.7 ± 0.2	0.7 ± 0.2
Carbamazepine	0.08 ± 0.09	0.08 ± 0.09
Diazepam	0.7 ± 0.5	0.7 ± 0.5
Fluoxetine	0.7 ± 0.2	0.7 ± 0.2
Estrone	3.8 ± 3	14 ± 3
17B-estradiol	7.8 ± 3	25 ± 3
17a-ethinylestradiol	0.03 ± 0.01	0.3 ± 0.1
Galaxolide	0.2 ± 0.3	0.2 ± 0.3
Tonalide	0.3 ± 0.2	0.3 ± 0.2
Celestolide	0.09 ± 0.1	0.09 ± 0.1

Source: Alvarino *et al.* (2016).

Table 9.6 Comparison of the organic micropollutant removal efficiencies (%) in different wastewater treatments.

	IFAS-MBR	MBMBR	MBR	CAS
Pharmaceutical compounds	82.8	64.2	82.4	69.9
Hormones	100	66.7	75.4	93.3
Nonylphenols	19.4	13	29.8	36.3

Source: Adapted from de la Torre *et al.* (2015).

The application of different redox potentials enhances the microbial diversity, as well as the removal of COD and nutrients (Luo *et al.* 2014b). Therefore, it is expected a positive effect of the combination of different redox potentials in the spectrum of OMPs removal in wastewater treatments. In general, OMPs are more biodegradable under aerobic than under negative redox potentials (Table 9.7). Burke *et al.*

(2015) studied the removal of 27 OMPs under aerobic and anoxic conditions and six of these compounds (phenazone, p-TSA, propyphenazone, doxycycline, 1,5-dimethyl-1,2-dehydro-3-pyrazolone and 1-Acetyl-1-methyl-2-dimethyl-oxamoyl-2-phenylhydrazide) were only removed efficiently under aerobic conditions, whereas roxithromycin, clarithromycin and clindamycin were readily biodegradable compounds under anoxic conditions and recalcitrant under positive redox potentials. Zeng *et al.* (2013) studied the removal of hormones in an anaerobic-anoxic-aerobic system and observed that EE2 was removed by sorption in the anaerobic stage, while the removal mechanism in the anoxic and aerobic chambers was the biodegradation. Wang *et al.* (2014) studied the removal of six pharmaceuticals in a WWTP which combines the anaerobic, anoxic and aerobic processes. Although most of the pharmaceuticals exhibited a poor removal (below 40%) in all the biological treatment stages, caffeine and N, N-diethyl-meta-toluamide (DEET) were biodegraded in the anaerobic and aerobic stages. Under anoxic conditions, the concentration of most of the selected OMPs remained stable or even negative due to the deconjugation of conjugated metabolites during the treatment process and desorption from the sludge (Wang *et al.* 2014). The same negative effect of the application of anoxic conditions was observed for DEET, metropolol and E2 by Xue *et al.* (2010) in an anaerobic-anoxic-aerobic process treating municipal wastewater. These authors studied the removal of 19 OMPs in an anaerobic, anoxic, aerobic process combined with a MBR and observed that the elimination of most OMPs occured mainly in the anaerobic compartment. This fact might be due to several aspects: internal recirculations that dilute the concentrations in the anaerobic stage, the readily biodegradation of some OMPs under anaerobic conditions and the adsorption onto the sludge. Sorption of OMPs is a rapid process compared to the hydraulic retention time applied in biological treatments, so that in a combined anaerobic-anoxic-aerobic process, the equilibrium is reached in the first anaerobic stage. Phan *et al.* (2014) studied the effect of the internal recirculation (IR) between the aerobic and the anoxic stages in an anaerobic-anoxic-aerobic MBR process on the removal of 30 OMPs. The internal recirculation influenced the supply of nitrate from the aerobic chamber, so that no IR corresponded to anaerobic conditions. For nine OMPs (triclosan, pirimidone, o-nonylphenol, octocrylene, amitriptyline, metronidazole, benzophenone, 17b-estradiol-17-acetate, oxybenzone), moderate to high removal were attained under negative redox potentials (50–90%) with no influence of the IR rate in their removal. However, for 11 OMPs, such as ketoprofen, ibuprofen, salicylic acid or formononetin, a negligible removal was observed under anaerobic conditions, while a moderate-high biodegradation was achieved under anoxic conditions when an IR was applied.

Alvarino *et al.* (2016) determined the removal of 16 OMPs in a UASB reactor coupled to a hybrid MBR pilot plant. According to their behaviour, OMPs were classified in four categories: OMPs removed only in the UASB reactor (tonalide, celestolide, naproxen, sulfamethoxazole, trimethoprim), recalcitrant under anaerobic conditions and readily biodegraded in the aerobic stage (ibuprofen, E1 and E2), OMPs removed partially in both chambers (galaxolide, roxithromycin) and the poorly removed in the overall system (carbamazepine, diazepam, diclofenac, fluoxetine and EE2). Although the removal efficiencies observed in the aerobic chamber were higher than under anaerobic conditions, the anaerobic pretreatment was crucial to achieve high removal efficiencies those OMPs not biodegraded under aerobic conditions. The removal efficiencies obtained in the aerobic stage were significantly lower than those observed in a lab-scale CAS reactor (Alvarino *et al.* 2014). The differences were related to limitations in terms of organic matter and nitrogen, as well as the kinetic limitations due to the low HRT and the low OMP concentration in the aerobic stage of the hybrid system.

The removal of OMPs in hybrid bioreactors depends on several operational parameters. For instance, the elimination can be improved by enhancing the filling ratio of supports, since it is related to the concentration of attached biomass. Luo *et al.* (2014b) showed an enhancement in the removal efficiency of carbamazepine, diclofenac, ketoprofen and naproxen by increasing the filling ratio from 10 to 20%.

De la torre *et al.* (2015) operated an IFAS-MBR at different SRT (10 and 20 d). A reduction on the SRT and the biomass concentration decreased the removal efficiencies. Surprinsingly, a removal above 80% of carbamazepine in the IFAS-MBR operated at a SRT of 20 d was achieved, which can be related to the higher microbial diversity and the presence of low F/M ratios which induces the microorganism to degrade the persistent compounds with a poor biodegradability (Verlicchi *et al.* 2012). Nevertheless, low or even negligible removal of carbamazepine was reported in other hybrid systems, like moving bed or hybrid MBR (Luo *et al.* 2014b; de la Torre *et al.* 2015; Alvarino *et al.* 2016), so the effect of the attached biomass in the removal of this substance is still not clear.

Table 9.7 Removal efficiencies of organic micropollutants in biological reactors under anaerobic, anoxic and aerobic conditions.

	Anaerobic	Anoxic	Aerobic
Ibuprofen	––	+/–	++
Naproxen	––/++	–/––	++
Estrone	––/++	++	++
17β-estradiol	––/++	–/++	++
17α-ethinylestradiol	–	–/––	–/+/++
Doxycycline	n.a.	––	++
Caffeine	++	––	+
Erythromycin	––/++	––	++
Trimethoprim	++	n.a.	–/––
Sulfamethoxazole	++	n.a.	–/––
Roxithromycin	––	––/++	––/++

Source: Alvarino *et al.* (2014); Phan *et al.* (2014); Wang *et al.* (2014); Zeng *et al.* (2013); Burke *et al.* (2015); Xue *et al.* (2010); Suarez *et al.* (2010).
n.a. = Not available.
++ = 75–100%, + = 50–75%, – = 25–50%, –– = 0–25%.

To summarize, the use of supports in hybrid systems improves the removal of most OMPs due to an enhancement in the microbial diversity and the nitrification capacity. Additionally, the removal of some OMPs is function of the redox potential and the presence of a biofilm formed in the supports allows the coexistence of different redox potentials in the same bioreactor.

9.3 FATE OF TRANSFORMATION PRODUCTS DURING SEWAGE TREATMENT

OMPs fate and removal in WWTPs has been presented in the previous paragraphs. It is now documented that the disappearance of the original form of micropollutants (= parent compound) may be achieved by a variety of processes. Whatever the case may be, to fully understand the fate of organic micropollutants, it is vital to consider also micropollutant metabolites, transformation products and conjugates. In general, it is recommended to only call "metabolites" those molecules that result from changes in human and treated animals (i.e. to be applied in particular for pharmaceutical active compounds), and to call "transformation products" (TPs) those generated from biotic (biodegradation by bacteria and fungi)/non biotic (hydrolysis or photolysis) processes in environment. In some cases, human metabolites can also be generated by microbial biodegradation (e.g. ibuprofen, Ferrando *et al.* 2012) but it is not the general case and attention needs to be paid in this context. The metabolites that can be categorized as phase I metabolites (from in vivo biochemical oxidation, reduction, and hydrolysis reactions increasing aqueous solubility and facilitating the elimination from the body) and phase II metabolites (usually named as conjugated metabolites) are the result of biochemical reactions that add a molecule (i.e. glucuronic acid) to the parent compound (Evgenidou *et al.* 2015). Phase II metabolites may deconjugate back to the parent compound in wastewater. The main critical issue in this context is the often insufficient database regarding the metabolism in the human body and the excretion ratio of the pharmaceutical and the corresponding metabolites in urine and faeces. In the case of pharmaceutical active compounds, a medical substance will be excreted as a mixture of metabolites, as unchanged substance, or conjugated with an inactivating compound attached to the molecule.

When they enter a wastewater treatment plant, OMPs can be metabolized, if the parent compound is degradable, to a more hydrophilic form (and usually of a lower molecular weight) and/or transformed into conjugates that can be hydrolysed later and release the parent compound. In the latter way, conjugates are acting as reservoirs of drugs from which the target compounds can then be released to the environment. If persistent, TPs pass through WWTP and end up in the receiving waters (Buttiglieri *et al.* 2008). The importance of the issue is reflected by the fact that national and/or international research strategy, priorities or even international legally binding tools have being developed to assess the health risks associated with exposure to trace concentrations of multiple TPs (Evgenidou *et al.* 2015).

TPs and parent compound can behave differently in the sewage treatment process with different pattern. TPs can be less biodegradable, more toxic and inhibitory compared to the parent compound. However only a minority of studies include work on the identification of the biotic and abiotic TPs formed and/or on the assessment of the biological potency of the treated solution/effluent produced. In just few cases, the complete (bio)degradation pathway has been clarified and verified in matrixes other than laboratory controlled conditions and at much higher concentration than the environmental one (e.g. mg/L and not µg/L or ng/L levels). In some cases, TPs activity has been confirmed, like for example sulfamethoxazole TPs transformed in the para-position that still exhibit antibiotic activity (Majewsky *et al.* 2015).

No specific pattern is observed in the occurrence of TPs. Generally speaking, TPs are found at lower concentration than the parent compound they derive from but it is not always the case. As shown in Table 9.8, some compounds can be present at higher concentrations in the inflow and the outflow of WWTP indicating their formation from the parent compound and/or their release from conjugates during the treatment process.

Table 9.8 Target compounds concentration (µg/L) in influent and effluent from an urban WWTP at different sampling campaigns.

Compound	Sampling A		Sampling B		Sampling C		Sampling D	
	In	Out	In	Out	In	Out	In	Out
Ibuprofen (IBP)	12.21	0.48	13.74	1.22	8.04	1.90	7.00	1.48
1-OH IBP	2.64	0.95	2.96	1.41	5.00	0.92	5.78	0.81
2-OH IBP	38.70	3.67	44.04	5.87	93.65	3.15	93.98	5.56
CBX IBP	36.23	0.72	38.40	10.65	19.88	5.37	18.38	9.40

Source: Modified from Ferrando *et al.* (2012).

A decrease in parent compound effluent concentrations compared to the influent is the most frequent scenario, but not the only one. A higher mass load in the effluent than in the influent can be detected if the TP is more recalcitrant than the parent compound (e.g. metoprolol acid vs. metoprolol). It is rather well known that alkylphenol carboxylates are produced during aerobic biological oxidation of alkylphenol ethoxylates (NP1EC) and aminomethyl phosphonic acid (AMPA) is a degradation product of both glyphosate and phosphonic acids contained in detergents. Both chemicals are usually measured with higher concentrations in the effluent of WWTP compared to the raw influent. Estrone is the hormone the most frequently quantified in effluent because it results partly from the biotransformation of other hormones.

Antibiotics are usually poorly absorbed by humans or animals and consequently 25 to 75% are excreted unaltered through urine and faeces. Vast bibliography exists concerning the occurrence of antibiotics in wastewaters while very few information are usually available of antibiotics TPs in wastewaters (Table 9.9, Evgenidou *et al.* 2015). Up to 20 sulphonamides TPs (directly originating from photolysis or biodegradation, Majewsky *et al.* 2015) have been predicted. The observed removal variability of sulfamethoxazole, combined with the extreme shortage of data on production and removal of its TPs, which can present similar or higher ecotoxicological effects (Majewsky *et al.* 2015), highlights the necessity of more in depth exploration. It has been observed, for example, that sulfamethoxazole could be temporarily and reversibly affected by denitrifying conditions.

Many articles have been reporting about ibuprofen degradation and fate in the environment and in water treatment processes, but there is still a lack of information about the presence and fate of its TPs in the environment (Table 9.8). A large proportion of the active compound is excreted as the parent compound together with its known human metabolites, hydroxyl ibuprofen (OH-IBP) and carboxyl ibuprofen (CBX-IBP). Only 15% of ibuprofen is, in fact, excreted unaltered whereas 2-OH-IBP and CBX-IBP account for 26% and 43% of total ibuprofen ingested respectively (Lishman *et al.* 2006). However, the human metabolites of ibuprofen are identical to its TPs (Figure 9.12i). Unlike foreseen based on excretion rates, 2-OH IBP was found at higher levels than CBX IBP in influent wastewater samples, which points out the contribution of biological degradation and consequent formation of ibuprofen metabolites before the entrance to WWTPs. Regarding their removal rates in WWTP, it can be observed that ibuprofen certainly exhibits removal rates above 90%, whereas CBX IBP, 2-OH IBP and 1-OH IBP were eliminated approximately at 72–100%, 77–100% and 58–100%, respectively (Collado *et al.* 2012).

Metoprolol (MTP) is a β-blocker of high consumption and its metabolism in mammals leads to the formation of three main metabolites, metoprolol acid (MTPA), a-hydroxymetoprolol (a-HMTP) and O-desmethylmetoprolol (O-DMTP) which account for the 85% of the urinary excretion (Figure 9.12ii). MTPA is known to be the major human metabolite of MTP and it can also be formed from its parent compound as a biodegradation product. However, MTPA is also known as a biodegradation product also of another β-blocker, atenolol. Accordingly, the closing of the mass balance of these specific TPs needs to be calculated with more accuracy. Rubirola *et al.* (2014) found that MTPA was the major transformation product (up to 40% of initial MTP concentration after 96 h treatment) and with high persistence through the treatment and also two new TPs were identified.

Estrone, estradiol and estriol are the three main estrogens produced by the human body but they are also interconnected. In some cases, in fact, it can be observed that estradiol was almost quantitatively oxidized to estrone as well an increase (up to 74 and 360%) of both hormone sulfate conjugates (Schüsener & Bester, 2008). The latter phenomenon can be explained by the assumption that other conjugates not measured (like disulfates and sulfate-glucuronides) are transformed to b-estradiol 3-sulfate during the wastewater treatment. A transformation of estrone 3-sulfate and other hormone sulfates to b-estradiol 3-sulfate is also possible as well as the increase of the steroid hormones may be based on the transformation of 16a-hydroxyestrone to estrone and estradiol (Schüsener & Bester, 2008).

Table 9.9 Maximum concentrations of some of the most frequently TPs of OMPs in WWTP influents and effluents.

(Parent compound/(excretion unchanged (%))	Max influent (ng/L)	Max effluents (ng/L)
Antibiotics		
ERY-H_2O[a,b] (erythromycin/(5–25))	1,200	7,840
OH-METR[b] (metronidazole/(20))	145	11,344
CLF acid[a,b] (clofibrate/(6))	740	258
FEN acid[b] (fenofibrate/(little))	117	349
Analgesics — NSAIDs		
SA[a,b] (acetylsalycilic acid/(little))	89,135	6,825
OH-IBP[b] (ibuprofen/(10))	6,840	1,130
CBX-IBP	23,000	1,270
4-OH-DCF[b] (diclofenac/(5–10))	237	860
4′-OH-ACF[b] (aceclofenac/(<5))	82	1,600
Antidepressants		
NOR-FLX[a] (fluoxetine/(11))	11	2.4
DES-Me-SER[a] (sertraline/(trace))	31	11
N-DES-Me-VNF[a] (venlafaxin/(5))	259	–
O-DES-Me-VNF[a]	2,602	–
DES-Me-VNF[a]	5	–
DES-Me-CT[a] (citalopram/(10))	427	301
Psychoactive drugs		
CBZ-Ep[a] (carbamazepine/(3))	75,500	3,581
DiOH-CBZ[a,b]	3,700	3,600
2-OH-CBZ[a]	163,800	48
3-OH-CBZ[a]	95	–
β-blockers		
MTPA[a] (metoprolol/(10–30))	298	2506
a-HMTP[a]	36	–
Statins (antilipidemics)		
p-OH-ATV[a] (atorvastatin/(b2))	280	–
o-OH-ATV[a]	196	–
SMV-OH acid[a] (simvastatin/(little))	10	–
Antidiabetics		
GUA[a] (metformin/(high))	400	28
Anesthetics		
NK[a,b] (ketamine/(90))	330	–
Estrogens		
4-OHE1[a] (estrone/(–))	14	82
16α-OHE1[a]	72	–

[a]Biotic transformation.
[b]Abiotic transformation.

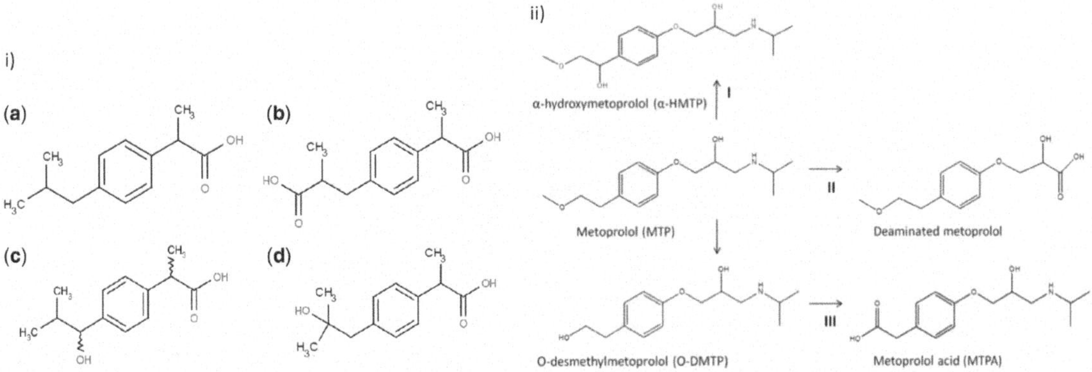

Figure 9.12 i) Structures of: (a) ibuprofen; (b) carboxyl ibuprofen (IBP-CBX); (c) 1-hydroxyl-ibuprofen (IBP-1OH); (d) 2-hydroxyl-ibuprofen (IBP-2OH) (Collado *et al.* 2012). ii) metabolic pathway of metropolol in the human organism (Rubirola *et al.* 2014).

Carbamazepine (CBZ) is heavily metabolized in the human body: 72% via urine, with only 1% excreted as parent compound, and 28% via feces, with 13% of unabsorbed CBZ (Evgenidou *et al.* 2015). Several of the metabolites of CBZ are difficult to distinguish by means of mass spectrometry (especially the constitutional isomers 1/2/3/4-OH-CBZ, EP-CBZ & Ox-CBZ, Balmann *et al.* 2014) but more than 30 metabolites have been identified so far and three key metabolic pathways have been reported. CBZ is usually very persistent with little to no degradation during conventional wastewater treatment. Several studies even reported higher CBZ concentrations, and in some cases of its TPs 2-OH-CBZ and DiOH-CBZ, after wastewater treatment. This behavior can be explained by the cleavage of the glucuronic acid moiety in the sewage network (especially when the residence time is long) and in WWTPs, thus leading to the release of parent compounds and metabolites in free form (Evgenidou *et al.* 2015). In some cases, CBZ is exceeded by CBZ metabolites (e.g. DiOH-CBZ up to 4000 ng/L; Bahlmann *et al.* 2014). EP-CBZ is probably excreted at much lower concentrations and also transformed into DiOH-CBZ in sewage. On the contrary, 10-OHCBZ is not a significant human metabolite of CBZ and it is routinely found at concentrations higher than CBZ, and detected more frequently than other CBZ metabolites. 4-OH-CBZ was identified in wastewater influents and effluents of WWTPs, although not quantified, suggesting a low removal rate (Bahlmann *et al.* 2014).

Based on the information presented in this chapter, it is apparent that more information is now available on occurrence and fate of TPs of micropollutants in WWTPs. However, knowledge in this field is in its formative stage and there are still many knowledge gaps on various aspects of TPs and metabolites in the fields of occurrence, fate, exposure, toxicity and risks.

9.4 MODELLING MICROPOLLUTANTS FATE DURING SEWAGE TREATMENT

Modelling the fate of micropollutants in a classical wastewater treatment requires taking into account the different processes susceptible to take place in the different sections of the plant (Figure 9.13) (Pomiès *et al.* 2013, 2014).

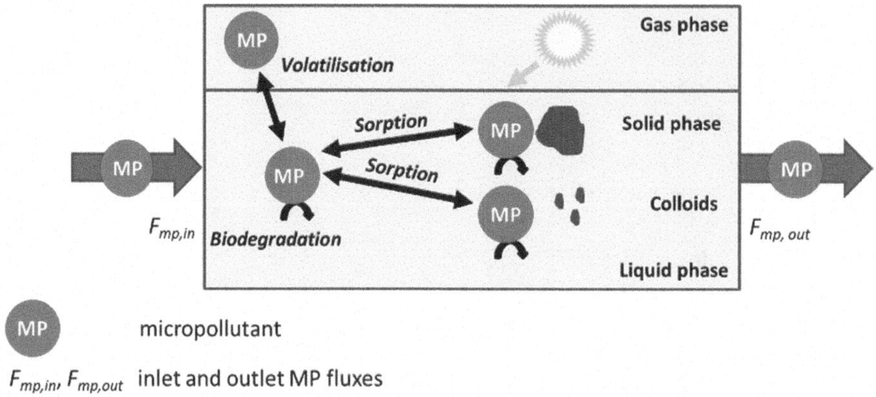

Figure 9.13 Processes governing the fate of organic micropollutants (adapted from Pomiès *et al.* 2013).

The general mass balance that summarizes the fate of organic micropollutants in the dissolved compartment can be written as:

$$\frac{dS_{mp}}{dt} = \frac{F_{smp,in}}{V} - \frac{F_{smp,out}}{V} + \left(\frac{dS_{mp}}{dt}\right)_{volatilization} + \left(\frac{dS_{mp}}{dt}\right)_{sorption} + \left(\frac{dS_{mp}}{dt}\right)_{degradation}$$

With V the biological reactor volume, $F_{smp,in}$ and $F_{smp,out}$ the dissolved micropollutant fluxes and S_{mp} the dissolved micropollutant concentration.

Part of the micropollutants is sorbed on wastewater suspended solids, colloids and sludge. The general mass balance for these sorbed micropollutants (X_{mp}) is written as:

$$\frac{dX_{mp}}{dt} = \frac{F_{xmp,in}}{V} - \frac{F_{xmp,out}}{V} + \left(\frac{dX_{mp}}{dt}\right)_{desorption}$$

With $F_{xmp,in}$ and $F_{xmp,out}$ the fluxes of micropollant sorbed on solids

Volatilization depends upon the operating conditions (i.e. aeration, mixing, temperature, pressure) and of the Henry's law constant (H) of the micropollutant. Two types of volatilization should be considered:

- The stripping, which is mostly driven by the gas flowrate:

$$\left(\frac{dS_{mp}}{dt}\right)_{stripping} = \frac{Q_{air} \cdot H \cdot S_{mp}(t)}{V \cdot R \cdot T}$$

  With Q_{air} the air flowrate, R the gas constant and T the temperature.

- The surface volatilization, rarely taken into account: its key factor is the interfacial area between the wastewater and the atmosphere:

$$\left(\frac{dS_{mp}}{dt}\right)_{surface\ volatilization} = -K_v \cdot (S_{mp} - S^*_{mp})$$

Where S_{mp}* is the equilibrium concentration of the micropollutant in water and K_V the overall mass transfer coefficient (Oskuie *et al.* 2008):

$$K_V = \frac{1}{K_{Lv}} + \frac{1}{K_{GV} \cdot H/(R \cdot T)}$$

With K_{Lv} and K_{Gv} the liquid and gas phase mass transfer coefficient, respectively. The volatilization in the primary settler and final clarifier can also be considered but have been omitted here for sake of simplicity.

Sorption and desorption are the two phenomena driving the distribution of the micropollutant between the dissolved and the solid phases. Desorption is usually much slower than sorption. Sorption isotherms can be represented by different models (e.g. Langmuir, Freundlich) but to keep things simple, a linear model is generally assumed for organic micropollutants, as their concentration is low (Limousin *et al.* 2007). The general models for sorption and desorption are written as:

$$\left(\frac{dS_{mp}}{dt}\right)_{sorption} = -k_{sor} \cdot S_{mp} \cdot SS + k_{desor} \cdot X_{mp}$$

$$\left(\frac{dX_{mp}}{dt}\right)_{desorption} = k_{sor} \cdot S_{mp} \cdot SS - k_{desor} \cdot X_{mp}$$

Where k_{sor} and k_{desor} are the sorption and desorption kinetic constants, respectively, and SS the suspended solids concentration.

The partition coefficient (K_d) is the ratio of the kinetic constants:

$$K_d = \frac{k_{sor}}{k_{desor}} = \frac{X_{mp}/SS}{S_{mp}}$$

Degradation can be abiotic or biotic. Photolysis is an abiotic transformation that can occur at the surface of an open settler, an open storm tank or lagoons. First-order kinetics is usually assumed (Fabbri *et al.* 2015):

$$\left(\frac{dS_{mp}}{dt}\right)_{photolysis} = -k_{phot} \cdot S_{mp}$$

$$k_{phot} = k_{phot,dir} + k_{phot,indir}$$

With $k_{phot,dir}$ reflecting the part of photolysis due to direct solar irradiation and $k_{phot,indir}$ the part due to the action of radicals (Bao *et al.* 2005) formed in the water phase. The time-dependency of the environmental conditions such as temperature and solar irradiation at the WWTP location should be taken into account.

Biotic degradation (= biodegradation) is the combination of two routes as described in Figure 9.14. It can occur aerobically or anaerobically (Barret *et al.* 2010). In the metabolic route, specific microorganisms are required to degrade the molecules. Most models assumed a biodegradation of the dissolved micropollutant only. The recent models describing this route are based on a Monod's formulation, taking into account dissolved oxygen (S_o) (in the case of an aerobic pathway) and/or substrate limitations (Plósz *et al.* 2010):

$$\left(\frac{dS_{mp}}{dt}\right)_{direct\ biodegradation} = -\frac{1}{Y_{mp}} \cdot \mu_{\max,mp}\left(\frac{S_O(t)}{K_{O,mp} + S_O(t)}\right) \cdot \left(\frac{S_{mp}(t)}{K_{mp} + S_{mp}(t)}\right) \cdot X_{active}$$

With Y_{mp} the conversion yield, $\mu_{max,mp}$ the maximum growth rate, $K_{O,mp}$ the oxygen half saturation coefficient, K_{mp} the micropollutant half saturation coefficient and X_{active} the active biomass. The active biomass is not measurable experimentally and in practice is chosen as a fraction (α) of the total active biomass (X):

$$X_{active} = \alpha \cdot X$$

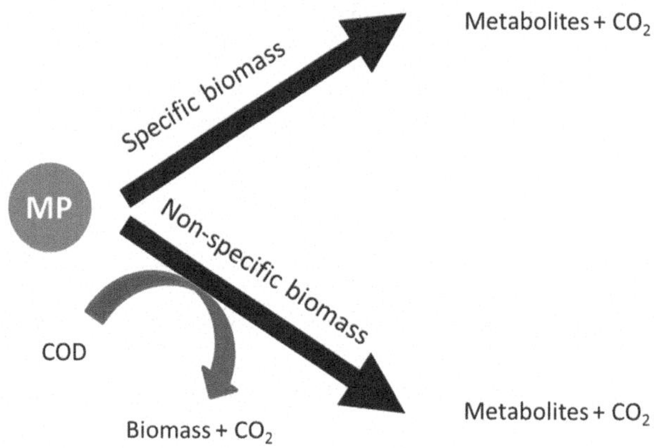

Figure 9.14 Schematic representation of micropollutant biodegradation (adapted from Pomiès *et al.* 2013).

In the cometabolic route, the biomass (X) is non-specific but a co-substrate (S_s) is needed for the degradation. The growth of the biomass is described classically by a Monod function:

$$\frac{dX}{dt} = \mu \cdot X = \left(\frac{\mu_{\max} \cdot S_S}{K_S + S_S}\right) \cdot X$$

With μ the growth rate, μ_{max} the maximum growth rate and K_S the co-substrate half saturation coefficient.

Taking the example of the degradation of PAHs proposed by Delgadillo-Mirquez *et al.* (2011), the degradation of the organic micropollutant can be written as:

$$\left(\frac{dS_{mp}}{dt}\right)_{cometabolism} = \left(T_c\frac{\mu}{Y} + k_c\right) \cdot \left(\frac{S_{mp}}{K_{SC} + S_{mp}}\right) \cdot X$$

With T_c the micropollutant transformation capacity, k_c the maximum specific rate when S_s is null, K_{SC} the half-saturation coefficient for the micropollutant and Y the yield.

Many micropollutants, especially pharmaceuticals, are present in the waste water as parent molecules (e.g. human metabolites, conjugated forms). Furthermore by-products, formed during the degradation, can remain toxic (Mitchell *et al.* 2015). Their quantification is still difficult, and little is known about their fate and their modelling (Plósz *et al.* 2010).

Micropollutants such as antibiotics and biocides can have an impact on the biomass growth and the degradation of other pollutants. Inhibition can be acute (Pala-Ozkok *et al.* 2014) or chronic (Pala-Ozlok *et al.* 2013). Inhibition can be non-competitive (by modification of μ_{max}):

$$\mu = \left(\frac{\mu_{\max} \cdot S}{K_S}\right) \cdot \left(\frac{K_I}{K_I + S_{mp}}\right)$$

Or competitive (by modification of K_S) (Bertolazzi, 2005):

$$\mu = \frac{\mu_{\max} \cdot S}{K_S\left(1 + \dfrac{S_{mp}}{K_I}\right) + S}$$

Where S is the growth substrate (co-substrate or else) concentration and K_I the inhibition constant.

The quantification of the model parameters for any micropollutant is difficult and time-consuming. Tools based on the development of relation between the structure of molecules and their properties or activities are under development (Lee & von Gunten, 2012).

9.5 CONCLUSION

One important challenge that WWTPs are facing or must face in the near future is the abatement of organic micropollutants. This chapter highlights that this is not trivial since organic micropollutants removal is affected by several factors.

The biotransformation efficiency of OMPs is strongly dependant on the redox potential, which to some extent is related to the chemical structure of each compound. Although higher removal efficiencies are attained under aerobic conditions for most compounds, anaerobic processes enhance the elimination of some substances that are not easily biotransformed during aerobic treatment. Consequently, the integration of different redox potentials in a hybrid biological system has shown to be a good alternative to enhance the elimination of a wide variety of micropollutants.

Biomass activities (heterotrophic, nitrifying, denitrifying, methanogenic) affect the biotransformation of OMPs. It has been shown that, regardless the low proportion of nitrifiers present in activated sludge reactors of WWTPs (usually below 5% of the total active biomass), this type of microorganisms exerts a huge influence on the removal of some compounds. Therefore, promoting high nitrifying activities will improve the elimination of OMPs.

Among operational parameters, the most relevant effect on the biotransformation of organic micropollutants was exerted by the retention time of the OMP in the reactor and by the sludge retention time (SRT). An increase in the retention time of the OMP in the reactor (for example, by increasing the hydraulic retention time (HRT) or by adding some sorbent that retains the OMP in the reactor) would improve the removal efficiency of compounds with slow and intermediate kinetics. A higher SRT not only enables the development of nitrifiers, but also increases the microbial diversity in the reactor, which could result in the enrichment of microorganisms with better capabilities for the biotransformation of OMPs.

However, there are still some knowledge gaps which future research should focus on. For example, little is known about the fate of OMPs during the first steps of the WWTP (fat removal and primary treatment) in order to elucidate how elimination can be fostered. Parameter estimation from OMP chemical structure is a key issue for modeling OMPs fate in WWTP. In addition, complete mineralization is normally not achieved, but OMPs are biotransformed to intermediate compounds (transformation products). Few studies include the identification of transformation products, and more importantly, their fate during wastewater treatment, since they may behave very differently to the parent compound and be even more toxic. Finally, understanding the biotransformation mechanisms, i.e. linking enzymatic activities to OMP biotransformation, is required to identify process conditions favouring cometabolic OMP biotransformation, which will help to optimize the operational conditions in WWTPs for OMP abatement.

9.6 REFERENCES

Adrian N. R. and Suflita Ü. M. (1994) Anaerobic biodegradation of halogenated and non-halogenated N-, S-, and O-heterocyclic compounds in aquifer slurries. *Environmental Toxicology and Chemistry*, **13**, 1551–1557.

Alvarino T., Suarez S., Lema J. M., Omil F. (2014) Understanding the removal mechanisms of PPCPs and the influence of main technological parameters in anaerobic UASB and aerobic CAS reactors. *Journal of Hazardous Materials*, **278**, 506–513.

Alvarino T., Suarez S., Katsou E., Vazquez-Padin J., Lema J. M., Omil F. (2015) Removal of PPCPs from the sludge supernatant in a one stage nitritation/anammox process. *Water Research*, **68**, 701–709.

Alvarino T., Suarez S., Garrido M., Lema J. M. and Omil, F. (2016) A UASB reactor coupled to a hybrid aerobic MBR as innovative plant configuration to enhance the removal of organic micropollutants. *Chemosphere*, **144**, 452–458.

Amorim C. L., Maia A. S., Mesquita R. B. R., Rangel A. O. S. S., van Loosdrecht M. C. M., Tiritan M. E. and Castro, P. M. L. (2014) Performance of aerobic granular sludge in a sequencing batch bioreactor exposed to ofloxacin, norfloxacin and ciprofloxacin. Water Research, **50**, 101–113.

Bahlmann A., Brack W., Schneider R. J. and Krauss M. (2014). Carbamazepine and its metabolites in wastewater: analytical pitfalls and occurrence in Germany and Portugal. *Water research,* **57**, 104–114.

Bao Y. and Niu, J. (2015) Photochemical transformation of tetrabromobisphenol A under simulated sunlight irradiation: kinetics, mechanism and influencing factors. *Chemosphere*, **134**, 550–556.

Barret M., Carrère H., Delgadillo L. and Patureau, D. (2010) PAH fate during the anaerobic digestion of contaminated sludge: do bioavailability and/or cometabolism limit their biodegradation? *Water Research*, **44**, 3797–3806.

Bertolazzi E. (2005) A combination formula of Michaelis-Menten-Monod type. *Computers and Mathematics with Applications*, **50**, 201–215.

Bouju H., Buttiglieri G. and Malpei F. (2008) Perspectives of persistent organic pollutants (POPS) removal in an MBR pilot plant. *Desalination*, **224**, 1–6.

Buntner D., Sánchez Sánchez A. and Garrido J. M. (2011) Three stages MBR (methanogenic, aerobic biofilm and membrane filtration) for the treatment of low-strength wastewaters. *Water Science and Technology*, **64**(2), 397–402.

Burke V., Richter D., Hass U., Duennbier U., Greskowiak J. and Massmann G. (2015) Redox-dependent removal of 27 organic trace pollutants: compilation of results from tank aeration experiments. *Environ Earth Sci*, **71**(8), 3685–3695.

Buttiglieri G. and Knepper T. P. (2008) Removal of emerging contaminants in wastewater treatment: conventional activated sludge treatment. In: Chapter in the Book "The Handbook of Environmental Chemistry, volume: Emerging Contaminants from Industrial and Municipal Wastewaters", D. Barceló and M. Petrovic (eds.) Springer-Verlag Berlin Heidelberg. Vol **5S/2**, 1–35.

Carballa M., Omil F., Lema J. M., Llompart M., García-Jares C., Rodríguez I., Gómez M. and Ternes T. (2004) Behavior of pharmaceuticals, cosmetics and hormones in a sewage treatment plant. *Water Research*, **38**(12), 918–926.

Carballa M., Omil F. and Lema J. M. (2005) Removal of cosmetic ingredients and pharmaceuticals in sewage primary treatment. *Water Research*, **39**, 4790–4796.

Carballa M., Omil F., Ternes T. and Lema J. M. (2007). Fate of Pharmaceutical and Personal Care Products (PPCPs) during anaerobic digestion of sewage sludge. *Water Research*, **41**(10), 2139–2150.

Choubert J. M., Martin Ruel S., Esperanza M., Budzinski H., Miege C., Lagarrigue C. and Coquery, M. (2011) Limiting the emissions of micro-pollutants: what efficiency can we expect from wastewater treatment plants? *Water Science and Technology*, **63**(1), 57–65.

Cicek N., Franco J. P., Suidan M. T., Vincent U. and Manem J. (1999). Characterization and comparison of a membrane bioreactor and a conventional activated sludge system in the treatment of wastewater containing high-molecular weight compounds. *Water Environment Research*, 71(1), 64–70.

Clara M., Strenn B., Ausserleitner M. and Kreuzinger N. (2004) Comparison of the behavior of selected micropollutants in a membrane bioreactor and a conventional wastewater treatment plant. *Water Science and Technology*, **50**, 29–36.

Clara M., Kreuzinger N., Strenn B., Gans O. and Kroiss H. (2005a). The solids retention time—a suitable design parameter to evaluate the capacity of wastewater treatment plants to remove micropollutants. *Water Research*, **39**, 97–106.

Clara M., Strenn B., Gans O., Martinez E., Kreuzinger N. and Kroiss H. (2005b). Removal of selected pharmaceuticals, fragrances and endocrine disrupting compounds in a membrane bioreactor and conventional wastewater treatment plants. *Water Research*, **39**, 4797–4807.

Clarke B. O., Porter N. A., Symons R. K., Marriott P. J., Stevenson G. J. and Blackbeard J. R. (2010) Investigating the distribution of polybrominated diphenyl ethers through an Australian wastewater treatment plant. *Science of the Total Environment*, **408**, 1604–1611.

Collado N., Buttiglieri G., Ferrando-Climent L., Rodriguez-Mozaz S., Barceló D., Comas J. and Rodriguez-Roda I. (2012) Removal of Ibuprofen and its Transformation Products: experimental and Simulation Studies. *Science of the Total Environment*, **433**, 296–301.

Coquery M., Pomiès M., Martin Ruel S., Budzinski H., Miège C., Esperanza M., Soulier C. and Choubert J. M. (2011) Concentrations et flux de micropolluants dans les eaux usées brutes et les rejets de stations d'épuration: méthodologie et principaux résultats du projet AMPERES. *Techniques Sciences et Méthodes*, **1/2**, 25–43.

Czajka C. P. and Londry K. L. (2006) Anaerobic biotransformation of estrogens. *Science of the Total Environment*, **367**(2–3), 932–994.

de Graaff M. S., Temmink H., Zeeman G., van Loosdrecht M. C. and Buisman C. J. (2011) Autotrophic nitrogen removal from black water: calcium addition as a requirement for settleability. *Water Research*, **45**(1), 63–74.

de la Torre T., Alonso E., Santos J. L., Rodríguez C., Gómez M. A. and Malfeito J. J. (2015) Trace organics removal using three membrane bioreactor configurations: MBR, IFAS-MBR and MBMBR. *Water Science and Technology*, **71**(5), 761–768.

di Trapani D., di Bella G., Mannina G., Torregrossa M. and Viviani G. (2014) Comparison between moving bed-membrane bioreactor (MB-MBR) and membrane bioreactor (MBR) systems: influence of wastewater salinity variation. *Bioresource Technology*, **162**, 60–69.

Delgadillo-Mirquez L., Lardon L., Steyer J. P. and Patureau D. (2011) A new dynamic model for bioavailability and cometabolism of micropollutants during anaerobic digestion. *Water Research*, **45**, 4511–4521.

Di Corcia A, Samperi R, Bellioni A, Marcomini A, Zanette M, Lemr K. and Cavalli L (1994) LAS pilot study at the Roma-Nord sewage treatment plant and in the Tiber river. *Rivista Italiana Delle Sostanze Grasse*, **71**(9), 467–475.

Duque A. F., Bessa V. S., Carvalho M. F., De Kreuk M. K., van Loosdrecht M. C. M. and Castro P. M. L. (2011) 2-Fluorophenol degradation by aerobic granular sludge in a sequencing batch reactor. *Water Research*, **45**(20), 6745–6752.

Dutta K., Lee M. Y., Lai W. W., Lee C. H., Lin A. Y., Lin C. F. and Lin J. G. (2014) Removal of pharmaceuticals and organic matter from municipal wastewater using two-stage anaerobic fluidized membrane bioreactor. *Bioresource Technology*, **165**, 42–49.

Escolà Casas M., Chhetri R. K., Ooi G., Hansen K. M., Litty K., Christensson M., Kragelund C., Andersen H. R. and Bester K. (2015) Biodegradation of pharmaceuticals in hospital wastewater by a hybrid biofilm and activated sludge system (Hybas). *Science of the Total Environment*, **530–531**, 383–392.

European Community (2013) Directive 2013/39/EU of the European Parliament and of the Council of 12 August 2013 amending Directives 2000/60/EC and 2008/105/EC as regards priority substances in the field of water policy. *Official Journal of the European Union*, **L226**, 1–17.

European Community (2015) Commission Implementing Decision (EU) 2015/495 of 20 March 2015 establishing a watch list of substances for Union-wide monitoring in the field of water policy pursuant to Directive 2008/105/EC of the European Parliament and of the Council (notified under document C(2015) 1756. 3 p.

Evgenidou E., Konstantinou I. and Lambropoulou D. (2015) Occurrence and removal of transformation products of PPCPs and illicitdrugs in wastewaters: a review. *Science of the Total Environment*, **505**, 905–992.

Fabbri D., Minella M., Maurino V., Minero C. and Vione D. (2015) A model assessment of the importance of direct photolysis in the photo-fate of cephalosporins in surface waters: possible formation of toxic intermediates. *Chemosphere*, **134**, 452–458.

Falas P, Baillon-Dhumez A., Andersen H. R., Ledin A. and la Cour Jansen J. (2012) Suspended biofilm carrier and activated sludge removal of acidic pharmaceuticals. *Water Research*, **46**(4), 1167–1175.

Fernandez-Fontaina E., Omil F., Lema J. M. and Carballa M. (2012). Influence of nitrifying conditions on the biodegradation and sorption of emerging micropollutants. *Water Research*, **46**(16), 5434–5444.

Fernandez-Fontaina E., Carballa M., Omil F. and Lema J. M. (2014). Modelling cometabolic biotransformation of organic micropollutants in nitrifying reactors. *Water Research*, **65**, 371–383.

Ferrando-Climent L., Collado N., Buttiglieri G., Gros M., Rodriguez-Roda I., Rodriguez-Mozaz S. and Barcelo D. (2012) Comprehensive study of ibuprofen and its metabolites in activated sludge batch experiments and aquatic environment. *Science of the Total Environment*, **438**, 404–13.

Field J. A. (2002) Anaerobic Degradation of Xenobiotic Compounds. VII Latin America Workshop and Symposium on Anaerobic Digestion. 22–25 October, Mérida (Mexico), p. 343.

Gianico A., Braguglia C. M., Mascolo G. and Mininni G. (2013) Partitioning of nutrients and micropollutants along the sludge treatment line: a case study. *Environ Sci Pollut Res*, **20**(8), 6256–6265.

Joss A., Andersen H., Ternes T., Richle P. R. and Siegrist H. (2004) Removal of estrogens in municipal wastewater treatment under aerobic and anaerobic conditions: consequences for plant optimization. *Environmental Science & Technology*, **38**(11), 3047–3055.

Joss A., Keller E., Alder A. C., Gobel A., McArdell C. S., Ternes T. and Siegrist H. (2005) Removal of pharmaceuticals and fragrances in biological wastewater treatment. *Water Research*, **39**(14), 3139–3152.

Kartal B., van Niftrik L., Rattray J., van de Vossenberg J. L. C. M., Schmid M. C., Sinninghe Damsté J., Jetten M. S. M., Strous M., de Vossenberg J. and Damste J. S. S. (2008) Candidatus "Brocadia fulgida": an autofluorescent anaerobic ammonium oxidizing bacterium. *FEMS Microbiology Ecology*, **63**, 46–55.

Khunjar W. O., Mackintosh S. A., Skotnicka-Pitak J., Baik S., Aga D. S. and Love N. G. (2011). Elucidating the relative roles of ammonia oxidizing and heterotrophic bacteria during the biotransformation of 17aethinylestradiol and trimethoprim. *Environmental Science and Technology*, **45**(8), 3605–3612.

Kimura K., Hara H. and Watanabe Y. (2005) Removal of pharmaceutical compounds by submerged membrane bioreactors (MBRs). *Desalination*, **178**, 135–140.

Lee Y. and von Gunten U. (2012) Quantitative structure-activity relationships (QSARs) for the transformation of organic micropollutants during oxidative water treatment. *Water Research*, **46**, 6177–6195.

Limousin G., Gaudet J. P., Charlet L., Szenknect S., Barthès V. and Krimissa M. (2007) Sorption isotherms: a review on physical bases, modeling and measurement. *Applied Geochemistry*, **22**(2), 249–275.

Lishman L., Smyth S. A., Sarafin K., Kleywegt S., Toito J., Peart T., Lee B., Servos M., Beland M. and Seto P. (2006) Occurrence and reductions of pharmaceuticals and personal care products and estrogens by municipal wastewater treatment plants in Ontario, Canada. *Science of the Total Environment*, **367**(2–3), 544–558.

Luo Y., Guo W., Ngo H. H., Nghiem L. D., Hai F. I., Zhang J., Liang S. and Wang S. L. (2014a) A review on the occurrence of micropollutants in the aquatic environment and their fate and removal during wastewater treatment. *Science of the Total Environment*, **473–474**, 619–641.

Luo Y., Guo W., Ngo H. H., Nghiem L. D., Hai F. I., Kang J., Xia S., Zhang Z. and Price W. E. (2014b) Removal and fate of micropollutants in a sponge-based moving bed bioreactor. *Bioresource Technology*, **159**, 311–319.

Majewsky M., Glauner T. and Horn H. (2015) Systematic suspect screening and identification of sulphonamide antibiotic transformation products in the aquatic environment. *Analytical and Bioanalytical Chemistry*, **407**, 5707–5717.

Martin Ruel S., Choubert J. M., Esperanza M., Miège C., Navalon Madrigal P., Budzinski H., Le Ménach K., Lazavora V. and Coquery M. (2011) On-site evaluation of the removal of 100 micro-pollutants through advanced wastewater treatment processes of reuse applications. *Water Science & Technology*, **63**(11), 2486–2497.

Martin Ruel S., Choubert J. M., Budzinski H., Miège C., Esperanza M. and Coquery M. (2012) Occurrence and fate of relevant substances in wastewater treatment plants regarding Water Framework Directive and future legislations. *Water Science & Technology*, **65**(7), 1179–1189.

Marttinen S. K., Kettunen R. H., Sormunen K. M. and Tintala J. A. (2003) Removal of bis(2-ethylhexyl) phthalate at a sewage treatment plant. *Water Research*, **37**, 1385–1393.

Mitchell S. M., Ullman J. L., Teel A. L. and Watts R. J. (2015) Hydrolysis of amphenicol and macrolide antibiotics: chloramphenicol, florfenicol, spiramycin, and tylosin. *Chemosphere*, **134**, 504–511.

Monsalvo V. M., McDonald J. A., Khan S. J. and Le-Clech P. (2014) Removal of trace organics by anaerobic membrane bioreactors. *Water Research*, **49**, 103–112.

Moreira I. S., Amorim C. L., Ribeiro A. R., Mesquita R. B. R., Rangel A. O. S. S., van Loosdrecht M. C. M., Tiritan M. E. and Castro P. M. L. (2015) Removal of fluoxetine and its effects in the performance of an aerobic granular sludge sequential batch reactor. *Journal of Hazardous Materials*, **287**, 93–101.

Musson S. E., Campo P., Tolaymat T., Suidan M. and Townsend T. G. (2010) Assessment of the anaerobic degradation of six active pharmaceutical ingredients. *Science of the Total Environment*, **408**(9), 2068–2074.

Oskouie A. K., Lordi D. T., Granato T. C. and Kollias L. (2008) Plant-specific correlations to predict the total VOC emissions from wastewater treatment plants. *Atmospheric Environment*, **42**, 4530–4539.

Pala-Ozkok I. and Orhan D. (2013) Chronic effect of erythromycin on substrate biodegradation kinetics of activated sludge. *Biochemical Engineering Journal*, **81**, 29–39.

Pala-Ozkok I., Kor-Bicakci G., Ural A., Katipoglu-Yazan T., Yagci N., Ubay-Cokgora E. and Orhon D. (2014) Modeling acute impact of sulfamethoxazole on the utilization of simple and complex substrates by fast growing microbial culture. *Journal of Chemical Technology and Biotechnology*, **89**, 603–615.

Petrie B., McAdam E. J., Lester J. N., Cartmell E. (2014) Assessing potential modifications to the activated sludge process to improve simultaneous removal of a diverse range of micropollutants. *Water Research*, **62**, 180–192.

Phan H. V., Hai F. I., Kang J., Damd H. K., Zhang R., Price W. E., Broeckmann A. and Nghiem L. D. (2014) Simultaneous nitrification/denitrification and trace organic contaminant (TrOC) removal by an anoxic–aerobic membrane bioreactor (MBR). *Bioresource Technology*, **165**, 96–104.

Plósz B. G., Leknes H. and Thomas K. V. (2010) Impacts of competitive inhibition, parent compound formation and partitioning behavior on the removal of antibiotics in municipal wastewater treatment. *Environmental Science and Technology*, **44**, 734–742.

Pomiès M., Choubert J. M., Wisniewski C. and Coquery M. (2013) Modelling of micropollutant removal in biological wastewater treatments: a review. *Science of the Total Environment*, **443**, 733–748.

Pomiès M, Choubert J. M., Wisniewski C. and Coquery M. (2014) Corrigendum to Modelling of micropollutant removal in biological wastewater treatments: a review. *Science of the Total Environment*, **466–467**, 1115.

Radjenovic J., Petrovic M. and Barcelo D. (2009) Fate and distribution of pharmaceuticals in wastewater and sewage sludge of the conventional activated sludge (CAS) and advanced membrane bioreactor (MBR) treatment. *Water Research* **43**, 831–841.

Roh H., Subramanya N., Zhao F., Yu C.-P., Sandt J., Chu K.-H. (2009) Biodegradation potential of wastewater micropollutants by ammoniaoxidizing bacteria. *Chemosphere*, **77**(8), 1084–1089.

Rubirola A., Llorca M., Rodriguez-Mozaz S., Casas N., Rodriguez-Roda I., Barceló D. and Buttiglieri G. (2014) Characterization of metoprolol biodegradation and its transformation products generated in activated sludge batch experiments and in full scale WWTPs. *Water Research*, **63**, 21–32.

Samaras V. G., Stasinakis A. S., Thomaidis N. S., Mamais D. and Lekkas T. C. (2013) Fate of selected pharmaceuticals and synthetic endocrine disrupting compounds during wastewater treatment and sludge anaerobic digestion. *Journal of Hazardous Materials*, **244–245**, 259–267.

Schüsener M. P. and Bester K. (2008). Behavior of steroid hormones and conjugates during wastewater treatment – a comparison of three sewage treatment plants. *Clean*, **36**(1), 25–33.

Shi Y. J., Wang X. H., Qui Z., Diao M. H., Gao M. M., Xing S. F., Wang S. G. and Zhao X. C. (2010) Sorption and biodegradation of tetracycline by nitrifying granules and the toxicity of tetracycline on granules. *Journal of Hazardous Materials*, **191**(1–3), 103–109.

Suarez S., Ramil M., Omil F. and Lema J. M. (2005) Removal of pharmaceutically active compounds in nitrifying–denitrifying plants. *Water Science and Technology*, **52**(8), 9–14.

Suárez S., Lema J. M. and Omil F. (2009) Pretreatment of hospital wastewater by coagulation-flocculation and flotation. *Bioresource Technology*, **100**(7), 2138–46.

Suárez S., Lema J. M. and Omil F. (2010) Removal of pharmaceutical and personal care products (PPCPs) under nitrifying and denitrifying conditions. *Water Research*, **44**, 3214–24.

Tian W., Bai J., Liu K., Sun H., Zhao Y. (2012) Occurrence and removal of polycyclic aromatic hydrocarbons in the wastewater treatment process. *Ecotoxicology and Environmental Safety*, **82**, 1–7.

Tran H. T., Urase T., Kusakabe O. (2009). The characteristics of enriched nitrifier culture in the degradation of selected pharmaceutically active compounds. *Journal of Hazardous Materials*, **171**(1–3), 1051–1057.

Urase T. and Kikuta T. (2005) Separate estimation of adsorption and degradation of pharmaceutical substances and estrogens in the activated sludge process. *Water Research*, **39**(7), 1289–1300.

Verlicchi P., Al Aukidy M. and Zambello E. (2012) Occurrence of pharmaceutical compounds in urban wastewater: removal, mass load and environmental risk after a secondary treatment - A review. *Science of the Total Environment*, **429**, 123–155.

Wakelin N. G. and Forster C. F. (1997) An investigation into microbial removal of fats, oils and greases. *Bioresource Technology*, **59**, 37–43.

Wang D., Sui Q., Lu S. G., Zhao W. T., Qiu Z. F., Miao Z. W. and Yu G. (2014) Occurrence and removal of six pharmaceuticals and personal care products in a wastewater treatment plant employing anaerobic/anoxic/aerobic and UV processes in Shanghai, China. *Environmental Science and Pollution Research International*, **21**(6), 4276–4285.

Xia Z., Xiao-chun, W., Zhong-lin, C., Hao X. and Qing-fang Z. (2015) Microbial community structure and pharmaceuticals and personal care products removal in a membrane bioreactor seeded with aerobic granular sludge. *Applied Microbiology and Biotechnology*, **99**, 425–433.

Xue W., Wu C., Xiao K., Huang X., Zhou H., Tsuno H. and Tanaka H. (2010) Elimination and fate of selected micro-organic pollutants in a full-scale anaerobic/anoxic/aerobic process combined with membrane bioreactor for municipal wastewater reclamation. *Water Research*, **44**(20), 5999–6010.

Yi T., Harper W. F. Jr (2007). The link between nitrification and biotransformation of 17a-ethinylestradiol. *Environmental Science and Technology*, **41**(12), 4311–4316.

Zeng, Q., Li, Y., Yang, S. (2013) Sludge retention time as a suitable operational parameter to remove both estrogen and nutrients in an anaerobic-Anoxic-Aerobic activated sludge system. *Environmental Engineering Science*, **30**(4), 161–169.

Zhao, X., Chen, Z., Wang, X., Li, J., Shen, J., Xu, H. (2015) Remediation of pharmaceuticals and personal care products using an aerobic granular sludge sequencing bioreactor and microbial community profiling using Solexa sequencing technology analysis. *Bioresource Technology*, **179**, 104–112.

Chapter 10

Post-treatment for micropollutants removal

Brigitte Helmreich and Steffen Metzger

10.1 INTRODUCTION

In both surface and ground water, effluents of municipal wastewater treatment plants (WWTPs) are the major source of organic micropollutants (e.g. pharmaceuticals, pesticides, personal care products, endocrine disruptors and other industrial chemicals) (e.g. Zuccato *et al.* 2006; Kasprzyk-Hordern *et al.* 2008; Stamm *et al.* 2015). Well-assessed treatment systems currently exist for the classical issues in WWTPs, such as removal of biodegradable organic substances and nutrients (phosphorus and nitrogen). However, polar and semi-polar micropollutants are not, or only incompletely, removed by these classical technologies. These micropollutants can be characterized as follows:

- Non volatile
- Not adsorbable to sludge matrix
- Non or hardly biodegradable, due to:
  - their chemical properties
  - their low substrate concentration for a target degradation
  - the absence of co-substrate in the biological treatment step

In accordance with the Water Framework Directive, European legislation defines the goal of achieving good ecological and chemical status of water bodies (Directive, 2013). Hence, the growing number of micropollutants identified in surface and ground water has led to political decisions in some European countries (e.g. Switzerland), to mitigate the risk for aquatic systems by upgrading WWTPs. Such decisions are based on many years of scientific research which examined removal efficiencies, energy demand, costs, and the feasibility of implementing the recommended technologies at existing WWTPs (Stamm *et al.* 2015).

In Europe, two main technologies have been identified currently with a potential for large-scale application concerning efficiency, energy requirements, and costs: Adsorption onto activated carbon, and oxidation with ozone (O_3) (Abegglen & Siegrist, 2012; Joss *et al.* 2008). Germany and Switzerland already have some experience with these techniques on pilot or technical scales. In 2014, the Swiss parliament agreed to upgrade 100 selected WWTPs with an additional step (ozonation or activated carbon) to remove micropollutants (Stamm *et al.* 2015). Switzerland is also considering separation using both nanofiltration (NF) and reverse osmosis (RO), but there is no pilot or technical plant under operation yet. In Germany, two states are focusing on improving the WWTPs effluents by adding a post-treatment process. Baden-Wuerttemberg is using adsorption onto activated carbon, while North Rhine-Westphalia is trying both adsorption onto activated carbon and ozonation (Maus *et al.* 2014).

10.2 CHEMICAL METHODS

10.2.1 Ozonation

Oxidation with ozone has been shown to have a high potential to decompose organic micropollutants in WWTPs effluents (e.g. Huber *et al.* 2005; Hollender *et al.* 2009; Antoniou *et al.* 2013). Organic micropollutants are degraded by the direct reaction with ozone and the target molecule, or with the secondarily formed hydroxyl radicals (e.g. Hübner *et al.* 2015). While ozone selectively reacts with electron rich groups of organic micropollutants, such as alkenes and activated aromatic rings, hydroxyl radicals are very reactive but non-selective for most organic molecules. Due to their electrophilic nature, hydroxyl radicals oxidize almost all electron-rich organic substances. Additionally, hydroxyl radicals can be scavenged by dissolved organic carbon (DOC), carbonate/bicarbonate, and nitrite (Lee & von Gunten, 2010), which could also be present in WWTPs effluents. Normally ozonation only partially oxidizes the organic micropollutants and does not accomplish complete mineralisation (Huber *et al.* 2005). The specific activity of micropollutants, such as endocrine disruption or antibiotic activity, is often removed with single transformation steps (Huber *et al.* 2004; Dodd *et al.* 2009) and some studies confirm that the transformation of micropollutants was sufficient to significantly reduce toxicity (e.g. Hollender *et al.* 2009; Reungoat *et al.* 2012; Margot *et al.* 2013). However, there is also evidence of the formation of toxic transformation products (TPs) (Stalter *et al.* 2010; Magdeburg *et al.* 2014). In general, ozonation results in the formation of more biodegradable TPs (e.g. Hübner *et al.* 2012; Prasse *et al.* 2012), and of low molecular weight compounds such as oxalate, which cannot be further oxidized by ozone but are easily biodegradable (Ramseier *et al.* 2011). Increased biodegradability can be documented using biochemical oxygen demand (BOD), the relationship of BOD to chemical oxygen demand (COD) (e.g. Yavich *et al.* 2004), or by increased DOC removal in subsequent biological steps (e.g. Hübner *et al.* 2012). Therefore, an additional biological treatment step, e.g. biofiltration, is generally recommended to remove TPs and potentially toxic effects (Hollender *et al.* 2009; Stalter *et al.* 2010; Magdeburg *et al.* 2014; Luo *et al.* 2014).

Because ozone is not in aqueous solutions, it will be produced on-site via silent electrical discharge of oxygen (O_2) molecules from clean air or pure O_2 (storage tank). Solubility and stability in the water matrix strongly depends on:

- Ozone partial pressure and ozone concentration in feed gas
- Water temperature (O_3 solubility decreases with increasing temperature)
- pH value (more stable at lower pH)
- Concentration of DOC and scavengers

Two main parameters influence removal efficiency of organic micropollutants: ozone dose and contact time. Because ozone production is an energy intensive process, it is very important to use optimum ozone doses to achieve effective micropollutant degradation while maintaining low operational costs (e.g. Bahr *et al.* 2007; Antoniou *et al.* 2013). The ozone dosage to treat WWTPs effluent strongly depends on the water matrix. Ozone doses in bench-scale experiments ranged from 2–15 mg/L, which led to the complete disappearance of a wide range of organic micropollutants (e.g. Huber *et al.* 2005; Margot *et al.* 2013; Antoniou *et al.* 2013). Nevertheless, organic micropollutants exhibit different susceptibilities to ozone degradation, which can vary by up to 10 orders of magnitude (Hollender *et al.* 2009; Huber *et al.* 2003). Hollender *et al.* (2009) classified micropollutants depending on rate constants with ozone as: fast reacting ($k_{O_3} > 10^4\ M^{-1}s^{-1}$), moderately reacting ($10^2\ M^{-1}s^{-1} < k_{O_3} < 10^4\ M^{-1}s^{-1}$), and persistent ($k_{O_3} < 10^2\ M^{-1}s^{-1}$). Additionally, organic micropollutants are competing with other DOC in the wastewater matrix, which vary in the quality of WWTPs effluents depending on the precursory treatment steps (Hollender *et al.* 2009). Without taking the concentration of DOC into consideration, it is difficult to predict the ozone doses in g/m^3 (Antoniou *et al.* 2013). Therefore, ozone dosage is often given in the specific dose of ozone, which is the ratio of dissolved ozone to DOC ($gO_3/gDOC$).

Effluents of WWTPs containing electron-rich organic substances contribute a specific ozone dose of 0.2–0.6 $gO_3/gDOC$ and 3.4 gO_3/gNO_2-N, respectively (Wert *et al.* 2009, 2011). Thus, the ozone dosage should be regulated to ensure enough residual ozone for the oxidation of organic micropollutants and a sufficient and constant ozone exposure (Margot *et al.* 2013). In a pilot-scale study in Lausanne, Margot *et al.* (2013) found that when the ozone dose was normalized by the concentration of scavenger equivalent, a weighted sum of DOC and NO_2-N concentrations (0.38 DOC + 3.4 NO_2-N), then the higher ozone doses lead to higher removal rates for most micropollutants (average reduction of 80% of 65 studied micropollutants). Based on pilot-scale and technical-scale experiments and a contact time of 10 to 30 min, Abegglen *et al.* (2009) and Barjenbruch *et al.* (2014) recommend specific ozone doses between 0.6 $gO_3/gDOC$ and 0.8 $gO_3/gDOC$. This means an ozone dose of 5–15 g/m^3 for typical WWTPs effluents.

For the ozonation process (Figure 10.1) of WWTPs effluents, it is necessary to have a reaction chamber, an ozone generator, an ozone injection device, and an ozone destructor (off-gas). A downstream (bio) filter is recommended and has been implemented (e.g. a sandfilter, granular activated carbon (GAC) or biological activated carbon (BAC)) in technical scale facilities.

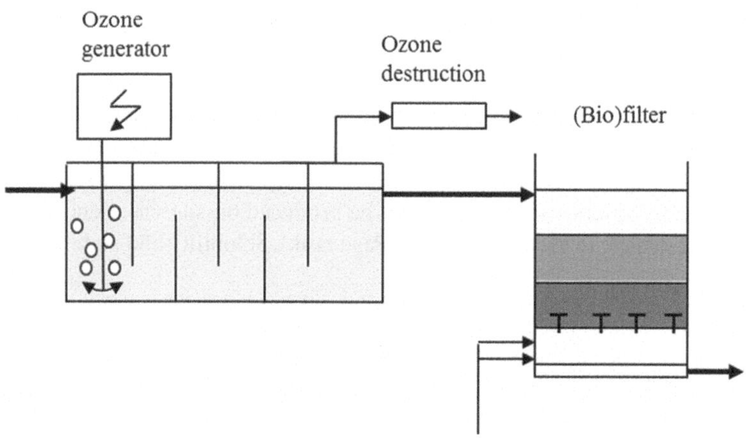

Figure 10.1 Principle of ozonation process for WWTPs effluent treatment and additional (bio) filter.

During the last few years, some WWTPs, especially in Germany and Switzerland, have been upgraded with an ozonation process. The results are often only available as reports or presentations at national conferences (in German) and are not internationally published. Table 10.1 gives an overview of different studies for pilot and full scale.

Table 10.1 Different pilot- and full-scale studies on ozonation of WWTPs effluents.

WWTP (Country)	Plant Capacity (pe)	PS/ FS	Q_{max} (m^3/h)	Q_{DW} (m^3/h)	O_3 Dose ($gO_3/gDOC$)	Contact Time (min)	(Bio) filter	References
Pilot scale								
Kloten-Opfikon (Switzerland)	54,000	PS	2	n.a.	0.5–5 mg/L	4.2	n.a.	Huber *et al.* 2005
Berlin-Ruhleben (Germany)	1.2 Mio	PS	2	n.a.	0.2–1.4	12	n.a.	Bahr *et al.* 2007b
Waldbröl (Germany)	Hospital 800	PS	1.5	n.a.	1.0	15	n.a.	Pinnekamp *et al.* 2009
Wien (Austria)	3 Mio.	PS	36.5	n.a.	0.6–1.1 (opt: 0.7)	16–20	n.a.	Schaar *et al.* 2011
Lausanne (Switzerland)	220,000	PS	360	n.a.	0.6–1	20–60	SF	Margot *et al.* 2013
Baden (Switzerland)	Hospital, 346 beds	PS	0.012–0.023	n.a.	0.64–1.08	12–23	n.a.	Kovalova *et al.* 2013
Full sale								
Regensdorf (Switzerland)	30,500	FS	900	430	0–1.2	3–15	SF	Abbeglen *et al.* 2009
Neugut, Dübendorf (Switzerland)	105,000	FS	1,833	854	0.8–1.0	17–34	SF	Wittmer *et al.* 2012
Caboolture (Australia)	40,000	FS	n.a.	334	0.6–0.8	n.a.	BAC	Reungoat *et al.* 2012
Landsborough (Australia)	10,000	FS	n.a.	84	0.2–0.3	n.a.	BAC	Reungoat *et al.* 2012
Gerringong (Australia)	11,000	FS	n.a.	38	0.4–0.5	n.a.	BAC	Reungoat *et al.* 2012
Schwerte (Germany)	50,000	FS	2,304	1,152	n.a. (10 g/m^3)	12	n.a.	Türk *et al.* 2013
Duisburg-Vierlinden (Germany)	30,000	FS	800	400	0.4 –1.0	30	FB	Maus *et al.* 2014
Bad Sassendorf (Germany)	12,000 + hospital: 1200 beds	FS	650	300	0.5	6–13	MB	Maus *et al.* 2014

Notes: PS: Partial stream; FS: Full stream; Q_{max}: Maximum flow rate; Q_{DW}: Dry weather flow; BAC: Biological active carbon; SF: sand filter; FB: fluid bed; MB: Maturation bond.
n.a. not available.

For a mixture of different organic micropollutants, the overall elimination efficiency is often described as over 80% because the treatment steps are often designed for the elimination of 80% of target organic micropollutants, as there is currently no legal demand for cleaning performance. The optimization of the ozone dosage is often based on the 80% removal of selected organic micropollutants as indicator substances; e.g. Margot *et al.* (2013) describe a removal of organic micropollutants over 80% on average by using an ozone dose of 5.7 gO_3/m^3. However, every treatment facility uses different indicator substances with different chemical properties, and therefore it is not possible to compare the different studies or give maximum elimination efficiencies. Even within the individual studies different values are given. For example, Yang *et al.* (2011) described a removal of caffeine of 100%, carbamazepine of 80%, sulfamethoxazole from 67 to 94%, and diethyltoluamide of lower than 40%. Kovalova *et al.* (2013) described a removal of 50% of iodinated contrast media, 90% of pharmaceuticals, and 98% of industrial chemicals when using 1.08 gO_3/gDOC. As a sum of all analyzed micropollutants, Kovalova *et al.* (2013) give a value of 52% elimination.

Jekel *et al.* (2015) proposed a number of organic micropollutants that could serve as indicator substances in the future. The proposed substances are: the artificial sweetener acesulfame, the antiflammatory drug ibuprofen, the anticonvulsant carbamazepine, the corrosion inhibitor benzotriazole, and the herbizide mecoprop.

Besides micropollutant degradation, ozonation also provides disinfection – which is especially relevant for hospital wastewater. Disinfection is necessary in order to prevent pathogens from entering the environment (Kovalova *et al.* 2013).

The assessment of the ozonation process (e.g. contact time, ozone concentration) is normally done for dry weather flow (Q_{DW}). It has to be assumed that the contact time is shorter in wet weather situations. Treatment could be performed until maximum flow rate (Q_{max}) is achieved. Specific ozone doses and contact times are based on first-hand experience in pilot or full scale. Further investigations have to be done to assess the reliability. For the optimal operation of ozone reactors, it is required to provide staff training as well as specific safety measures due to the toxicity of ozone gas. Therefore, ozonation is not suitable for small WWTPs which do not have permanent staff. Suitable qualified staff is also required in terms of energy and resource requirements.

There is also the need for an optimal control strategy for ozone doses requiring minimum maintenance, which gives in-time insights into process behaviour in parallel with micropollutants reduction (Wittmer *et al.* 2015). Both of these require online measurement in order to continuously monitor. Ozone reacts with various light-adsorbing unsaturated organic substances within the UV and visible spectra, which leads to a decrease of absorbance. Therefore, UV absorbance could be used as a feedback control to achieve optimal ozone dosage in wastewater (Wittmer *et al.* 2015).

10.2.2 Advanced Oxidation Processes

Advanced oxidation processes (AOPs) are considered to be very effective in the oxidation of numerous organic micropollutants. AOPs are based on the generation of free radicals, mainly the hydroxyl radical, with high oxidizing power (after fluorine hydroxyl radical, it is the strongest oxidant), which can successfully attack most organic molecules with elevated reaction constants ranging from 10^6 to 10^{10} $M^{-1}s^{-1}$ (e.g. Huber *et al.* 2003; Rivera-Utrilla *et al.* 2013). Regarding the method to generate hydroxyl radicals, AOPs are traditionally divided into four groups: chemically (e.g. O_3/H_2O_2; $H_2O_2/Fe^{2+/3+}$ (Fenton)), electro-chemically (e.g. boron doped diamond electrodes); sono-chemically (e.g. H_2O_2/ultrasonic); and photo-chemically (e.g. H_2O_2/UV; H_2O_2/UV/ $Fe^{2+/3+}$ (Photo-Fenton)) (e.g. Babuponnusami & Muthukumar, 2014; Ribeiro *et al.* 2015; Oturan & Aaron, 2014; Rajab *et al.* 2013).

Additionally, a complete mineralization of organic micropollutants using AOPs is also not always possible which results in TPs, which may be toxic for aquatic life. Another shortcoming is that hydroxyl radicals react with DOC and inorganic substances (e.g. sulphide, bromide, nitrite, and carbonate/bicarbonate). Since WWTPs effluents contain some of these scavengers, they must be taken into account during the optimization of AOPs in wastewater. Similar to the ozonation process, AOPs result in a higher biodegradability (Ribeiro *et al.* 2015). Therefore, a biological post-treatment after oxidation is recommended.

There is a lot of literature available concerning the degradation of organic micropollutants in WWTPs effluents using AOPs (see reviews: e.g. Luo *et al.* 2013; Oturan & Aaron, 2014; Riveiro *et al.* 2015; Schröder *et al.* 2016). But most studies are performed at laboratory scale and for individual substances at concentrations higher than those found in typical WWTPs effluents. These methods are emerging AOPs. Some AOPs have been established in pilot-scale investigations (Table 10.2) but no full-scale investigations are available.

Table 10.2 Pilot-scale studies of AOP application on WWTPs effluents.

WWTP (Country)	WWTP Effluent	Q_{max} (L/h)	AOP	Contact Time (min)	References
Reno, Nevada, (USA)	Municipal	0.6	3.5 g/m³ H_2O_2 and 0.8 gO_3/gDOC	5	Gerrity *et al.* 2011
Baden (Switzerland)	Hospital	12–23	2.5 g/m³ H_2O_2 and 0.8 gO_3/gDOC	12–23	Kovalova *et al.* 2013

Gerrity *et al.* (2011) reported high removal efficiency of >99% for most investigated micropollutants in a pilot-scale treatment plant of WWTPs effluent with O_3/H_2O_2 and a contact time of only 5 min. At another site, pilot-scale investigations with a combination of O_3 and H_2O_2 for the post-treatment of hospital wastewater effluent, showed no significant improvement of micropollutant degradation compared to ozone alone (±10%) (see values chapter 10.2.1) (Kovalova *et al.* 2013).

Full-scale experiments with realistic concentrations have yet to be applied. Other significant limitations associated with chemicals (e.g. H_2O_2) are additional costs, complexities associated with storage of chemicals, handling, exact dosage and safety measures (Gerrity *et al.* 2011).

10.3 PHYSICAL METHODS

10.3.1 Adsorption to activated carbon

Adsorption onto activated carbon (adsorbent) has been considered as one of the most effective methods to eliminate a broad spectrum of organic micropollutants from WWTPs effluents with a general preference to hydrophobic compounds. The affinity of a target compound for its adsorbent is quantified by the specific sorption coefficient, representing the ratio of adsorbed and dissolved concentrations of a target compound in equilibrium (Silva *et al.* 2012). The adsorption efficiency can be affected by the properties of both adsorbate (e.g. hydrophobicity, molecular size, aromaticity versus aliphaticity, presence of functional groups) and adsorbent (e.g. surface area, pore size, surface chemistry).

The following different mechanisms have been proposed for adsorption to activated carbon: (a) Disperse interactions between π electrons of the aromatic ring of an organic molecule and π electrons of the graphene planes of the activated carbon surface; (b) Formation of a donor-acceptor complex involving carbonyl surface groups, acting as donors, and the aromatic ring of organic molecules, which acts as the acceptor; and (c) Electrostatic interactions and the formation of hydrogen bridge bonds (e.g. Bautista-Toledo *et al.* 2005; Rivera-Utrilla *et al.* 2013).

One major benefit to use activated carbon to eliminate organic micropollutants from WWTPs effluents is that it does not generate toxic or pharmacologically active by-products. One drawback is that the adsorption capacity for organic micropollutants can be reduced through competition with residual DOC in WWTPs effluents (Rivera-Utrilla *et al.* 2013). Besides competing for adsorption sites, DOC clogs the outer pores of the activated carbon and thus restricts access to the inner micropores. Therefore, if the DOC concentration is high, larger amounts of activated carbon are required (Margot *et al.* 2013).

In Europe, the most commonly applied activated carbons are powdered activated carbon (PAC, 5–50 μm diameters) and granular activated carbon (GAC, 100–2,400 μm diameters). PAC dosage into the activated sludge tank or post-treatment configuration is a major application of PAC in pilot- and full-scale experiments (see Figure 10.2). GAC is mostly used as a filter. For monitoring the removal of organic micropollutants by activated carbon in practice, Sperlich *et al.* (2014) developed a quick and cost effective test method. They recommend the spectral absorption coefficient with a wave length of 254 nm (SAK_{254}) as a suitable surrogate parameter for the control of the elimination. As for ozonation, a maximum elimination efficiency for the different technologies in PAC and GAC application is difficult to specify. Since there is no legal requirement, the systems were often designed to remove 80% of the organic micropollutants, considering indicator substances (different substances with different polarities were used in the studies).

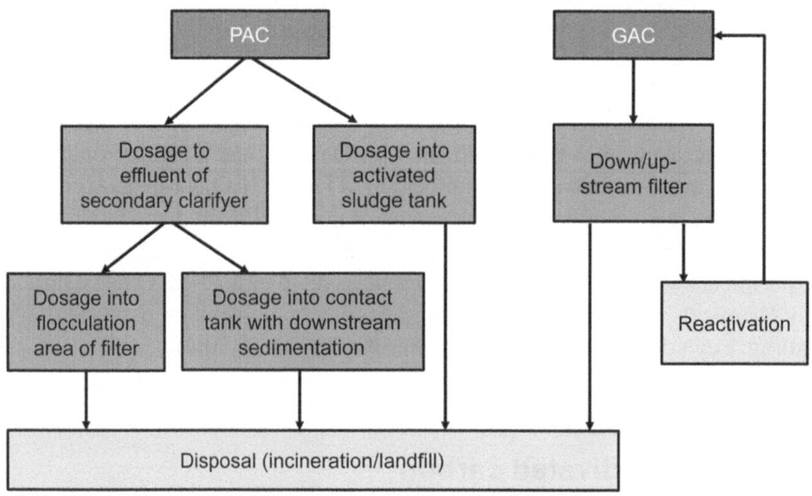

Figure 10.2 Process variations for treatment of wastewater with activated carbon.

10.3.1.1 PAC

In general, the applications of PAC differ according to the number of activated carbon loaded contacting stages (one or more stages) and the adsorbent direction in relation to the flow of the treated waste water. In a one-step system the PAC is completely loaded in the continuous flow and separated afterwards (e.g. by sedimentation or filtration). For this purpose, a contact time of more than 24 hours is necessary (Sperlich *et al.* 2014). This requires large basins, for which space is not normally available in wastewater treatment plants. To make a better use of the adsorption capacity, it requires a decoupling of the residence time of the PAC from the residence time of the wastewater in the adsorptive stage. This can be achieved by recirculation of the PAC to the reaction space of the adsorptive stage after separation.

The highest efficiency can be achieved when the PAC is dosed after biological treatment (post-treatment). The dosage and mixing of PAC to effluent can be implemented quite easily. It is possible to dose

PAC into a contact reactor between the biological treatment step and secondary clarification (Figure 10.3) or directly into the WWTPs effluent prior to downstream filtration (Figure 10.4). PAC could also be added into the activated sludge tank (Figure 10.5). Contrary to the addition of PAC, efficiently separating the PAC prior to the final discharge of treated effluent is a technical challenge. For separation, sedimentation and filtration (e.g. sand filter or multilayer filter) are the most common methods (Margot *et al.* 2013). The addition of flocculants (F) and flocculation aids (FA) (e.g. iron(III) salts) supports both the sedimentation and the filtration process. Ultrafiltration (UF) is another possible technique which offers complete PAC and bacteria retention, high virus removal and requires less space (Löwenberg *et al.* 2014).

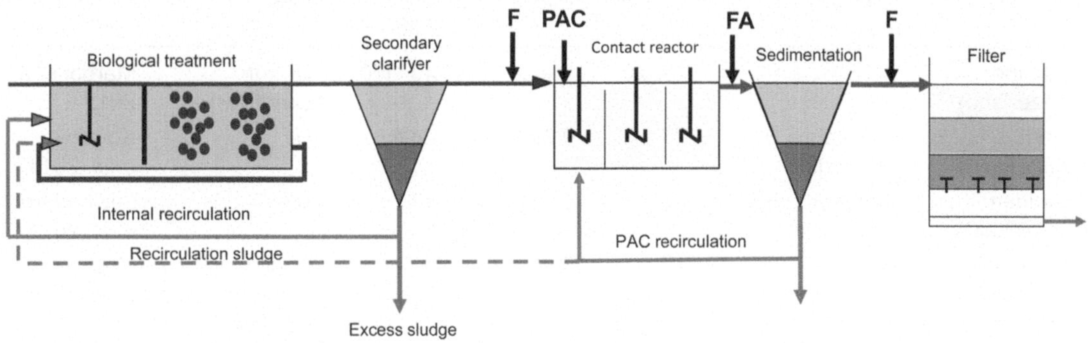

Figure 10.3 Process diagram of PAC dosage with downstream contact reactor, sedimentation and filtration. PAC: powdered activated carbon; F: flocculants, FA: flocculation aids.

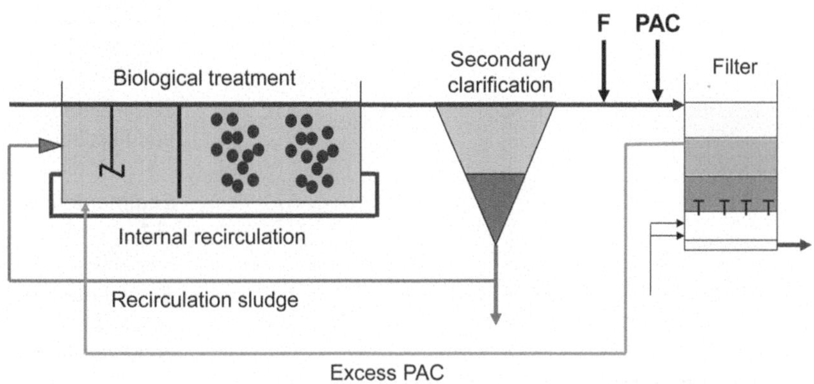

Figure 10.4 Process diagram of PAC dosage directly before downstream filtration. PAC: powdered activated carbon; F: flocculants.

Since some years, several pilot- and full-scale applications with contact reactors are available, especially in Germany and Switzerland (Table 10.3). Meanwhile, the WWTP Albstadt-Ebingen, Germany, has almost 20 years of experience (Vogel *et al.* 2014). Based on the available pilot- and full-scale experiments, the following design values are recommended for PAC dosage into contact reactors in Germany (Barjenbruch *et al.* 2014):

- *Minimum retention time in contact reactor*: 0.5 h
- *PAC dosage (dependent on DOC concentration)*: 10–20 g/m^3

- *Retention time in sedimentation basin*: min. 2 h, surface feeding: 2 m/h
- *Maximum filtration velocity filter (spatial filter)*: 12 m/h
- *Filter design*: 0.75 m Hydro anthracite layer (1.4–2.5 mm) and 0.75 m sand layer (0.71–1.25 mm)

Table 10.3 Different pilot and full-scale studies on treatment of WWTPs effluents with PAC downstream contact reactor, sedimentation or filtration.

WWTP (Country)	Plant Capacity (pe)	FS/PS	Dosage (g/m^3)	Q_{max} (m^3/h)	Contact Time (min)	Separation of PAC	References
Lausanne (Switzerland)	13,000	PS	10–20	36–54	40–170	UF (or sandfilter)	Margot *et al.* 2013
Baden (Switzerland)	Hospital, 346 beds	PS	23	0.008	24 h	UF	Kovalova *et al.* 2013
Birsfelden (Switzerland)	0.03 m^3/d	PS	20	0.001	120	UF	Löwenberg *et al.* 2014
Kressbronn-Langenargen (Germany)	24,000	FS	10	900	35	Sedimentation (2 h) and sand filtration	Metzger *et al.* 2014
Stockacher Aach (Germany)	43,000	PS	10	900	57	Sedimentation (1 h) and sand filtration	Metzger *et al.* 2014
Lahr (Germany)	100,000	PS	10	1,260	49	Sedimentation (2.8 h) and cloth sand filtration	Metzger *et al.* 2014
Albstadt-Ebingen (Germany)	125,000	FS	5–45	3,528	n.a.	Sedimentation and filtration	Vogel *et al.* 2014
Langwiese (Germany)	184,000	FS	10	3,960	57	Sedimentation (2.1 h) and sand filtration	Metzger *et al.* 2014
Mannheim (Germany)	725,000	FS	10	1,080	30	Sedimentation (2.2 h) and sand filtration	Metzger *et al.* 2014
Paris (France)	900,000	PS	5–20	Reactor 5 m × 4 m^2, 7 m/h fluidized bed)	10–20	n.a.	Mailler *et al.* 2015
Böblingen-Sindelfingen (Germany)	250,000	PS	10	3,600	30	Sedimentation (2 h) and sand filtration	Schwentner, 2015
Ulm-Steinhäusle	440,000	FS	10	5,760	40–75	Sedimentation and filtration	Metzger *et al.* 2015

Notes: PS: Partial stream; FS: Full stream; UF: Ultrafiltration.

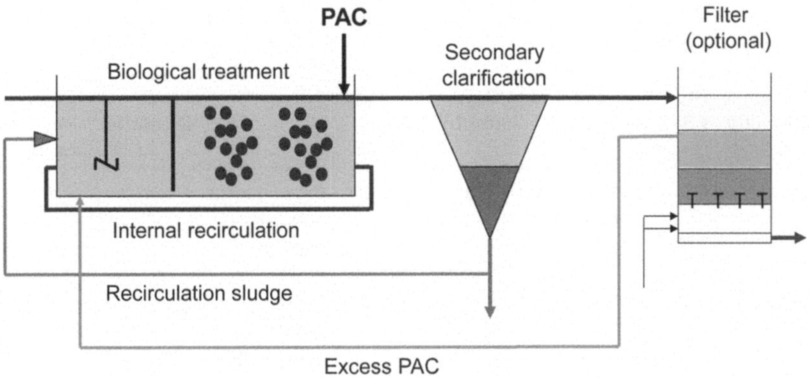

Figure 10.5 Process diagram of PAC dosage directly into biological treatment step. PAC: powdered activated carbon.

Flocculants are normally added at the beginning of the contact reactor, and flocculation aid is added at the end of the reactor. An increase of contact time could improve the adsorption efficiency for micropollutants, but it could also lead to desorption displacement effects from other substances (Barjenbruch *et al.* 2014). The countercurrent use of re-circulated activated carbon from the sedimentation step in the contact reactor increases contact time. The recommended dosage of PAC also depends on different applications. In some examples, only 10 mg/L is used, but it is possible to dose up to or over 20 mg/L. Kovalova *et al.* (2013) described a removal of 61% of iodinated contrast media, 86% of pharmaceuticals, and 91% of industrial chemicals when using 23 gPAC/m^3.

Practical operation shows that the PAC-dosage is kept low, as yet there are no legal requirements for cleaning performance. For the example at the WWTP Albstadt-Ebingen (Germany), which treats municipal and industrial wastewater from the textile industry, the PAC dose could be increased up to 45 g/m^3. In this case, the cleaning requirement depends on the color of the WWTPs effluent at wavelength of 436 nm, 525 nm and 620 nm and varies from 5 and 45 g/m^3 with a yearly mean value of 6 g/m^3 (Vogel *et al.* 2014). Pollutants and residual PAC from filter backwashing will be returned to activated sludge tank, where the PAC improves the activated sludge sedimentation properties (Vogel *et al.* 2014).

The direct dosing of PAC upstream of a filter gives the possibility to use the PAC directly after the biological treatment without a contact reactor. For this so-called flocculation filtration, dosage of PAC in addition with flocculants is made prior to filtration following process scheme Figure 10.4. The adsorbent is applied counter-currently. Loading is continuous in the downstream adsorptive stage through the enrichment of the adsorbent in the separation apparatus. The following design values are recommended in Germany (Barjenbruch *et al.* 2014):

- *Retention time in the flocculation area*: 13–28 min (filtration 30–47 min)
- PAC dosage of 15 g/m^3, dependent on DOC concentration
- *Prior retention time in sedimentation basin*: min. 2 h, surface feeding: 2 m/h
- *Maximum filtration velocity*: 15 m/h
- *Filter design*: quartz sand and expanded shale (together 1.6 m)

The recommended design values are based on only a few experiences in pilot-scale (Böhler & Siegrist, 2012; Barjenbruch, 2014). To underpin this, additional full-scale investigations still need to be done. Table 10.4 gives two examples from Switzerland and Germany.

Table 10.4 Different pilot and full-scale studies on treatment of WWTPs effluents with PAC dosage directly before downstream filtration.

WWTP (Country)	Plant Capacity (pe)	PS/ FS	PAC Dosage (g/m^3)	Q_{max} (m^3/h)	Contact Time (min)	Filter	Flow Direction	Filter Velocity (m/h]	References
Kloten/ Opfikon (Switzerland)	54,500	PS	15 PAC + 0.4 g gFe/gPAC	990	15–30	0.4 m quartz sand, 1.2 m expanded shale	Down-flow	7.5	Böhler *et al.* 2011
Buchenhofen (Germany)	600,000	PS	20 PAC + 0.1 gFe/gPAC	150–650	12.5–50	0.4 m quartz sand, 1.4 m hydro anthracite	Down-flow	5–11	Bornemann *et al.* 2012

Notes: PAC: powdered activated carbon; PS: Partial stream.

When there is less space available, the PAC is dosed directly into the activated sludge process (Figure 10.5) (Obrecht *et al.* 2015). Because PAC also adsorbs other organic substances, a higher PAC concentration (30–40 g/m^3) is required to achieve similar results if compared to post-treatment (Luo *et al.* 2014). This result in an increase of sludge concentration in the activated sludge reactor and an increase of excess sludge (loaded with activated carbon). The sedimentation process could be enhanced by adding PAC into activated sludge process (Luo *et al.* 2014). If sedimentation in the secondary clarifier is insufficient, a downstream filtration step (e.g. sand filter, cloth filter) is recommended to separate the PAC from the WWTPs effluent before final discharge (Metzger *et al.* 2014). The countercurrent use of PAC by recycling waste PAC from the post-treatment tank to the biological treatment tank could enhance micropollutants removal by 10 to 50% (Luo *et al.* 2014). Also, only a few examples are available (see Table 10.5) for this particular design proposal.

Table 10.5 Different pilot-scale studies on treatment of WWTPs effluents with PAC dosage directly into biological treatment step.

WWTP	Plant Capacity (pe)	PS/FS	PAC Dosage (g/m^3)	Q_{max} (m^3/h)	Contact Time (min)	Separation	References
Flos Wetzikon (Switzerland)	n.a.	PS	11–18	485	n.a.	Sedimentation and sand filtration	Obrecht *et al.* 2015
Düsseldorf-Holthausen	n.a.	PS	10	1.3	n.a.	Sedimentation and sand filtration	Clausen *et al.* 2014

Notes: PAC: powdered activated carbon; PS: Partial stream.
n.a. not available.

10.3.1.2 GAC

Normally, granular activated carbon is used after biological treatment (Figure 10.6). Due to the residual contamination in the effluent of the biological treatment step, a biofilm forms on GAC. The solutes adsorb on the surface of the activated carbon during their way through the spatial filter, but they can also be biodegraded. There are a number of variants of GAC adsorption that cannot all be enumerated in detail. Fundamentally, they differ with respect to the operating pressure (gravity or pressure), the flow direction (upflow or downflow), and the number of cells (single-stage or multi-stage). Depending on the solids loading of the influent, a pre-filtration may be necessary (e.g. sandfilter). When using GAC filters, a backwash is recommended.

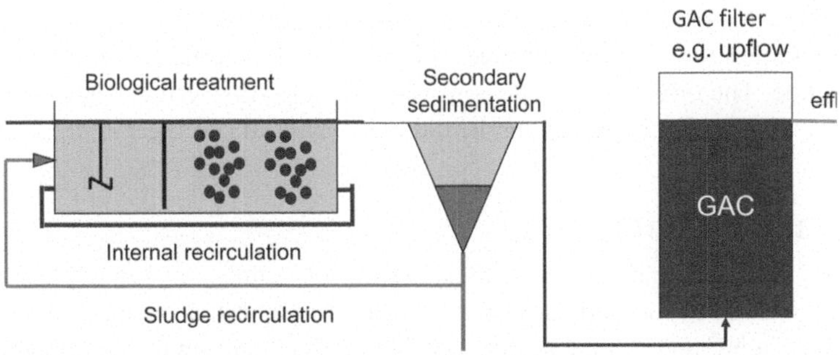

Figure 10.6 Process diagram of GAC filtration, e.g. upflow direction.

A great advantage of GAC filtration is that the activated carbon can be regenerated (with slight losses) and therefore less fresh coal is consumed (Aktaş & Çeçen, 2007). GAC filters were designed with a contact time (EBCT = empty bed contact time) of 5–10 minutes, a filter velocity of 5–15 m/h (Metcalf & Eddy, 2004), and the recommended filter bed height is between 2–4 m. If the contact time is too short, it is likely to lead to significantly lowered adsorption efficiency (Luo *et al.* 2014). The efficiency of GAC-based removal technology will decrease over time due to saturation of the adsorption sites. Therefore, the reactor should be carefully operated (Luo *et al.* 2014). Because the competition for adsorption sites and/or pore blocking can reduce the removal efficiency of activated carbon, GAC tends to perform poorly if the WWTPs effluent is highly contaminated. Snyder *et al.* (2007) suggested that a stream-treated GAC could be employed to overcome the drawbacks of GAC due to the greater adsorption capacity. GAC regeneration and contact time play important roles in the efficient removal of micropollutants. Yang *et al.* (2011) reported the elimination of 15–100% of organic micropollutants in a full-scale fixed bed GAC filter depending on the chemical moieties of the target molecule. Table 10.6 shows some GAC applications in pilot and full scale.

Table 10.6 Different pilot-scale studies on treatment of WWTPs effluents with GAC filtration.

WWTP	Plant Capacity (pe)	FS/PS	Q_{max} (m^3/h)	Filter	Flow Direction	Filter Velocity (m/h)	Filter Pack Height (m)	References
Waldbröl (Switzerland)	Hospital 800 beds	PS	485	Fixed-bed	Down-flow	10	3.0	Pinnekamp *et al.* 2009
Gütersloh-Prtzhagen (Germany)	150,600	PS	421	Fixed-bed	Upflow	10	3.0	Metzger *et al.* 2015
Obere Lutter (Germany)	38,000	PS	2401	Fixed-bed	Upflow	8	2.5	Metzger *et al.* 2015
Rietberg (Germany)	46,500	PS	360	Fluidized bed	Upflow	6	3.5	Metzger *et al.* 2015
Westerheim[a] (Germany)	5500	PS	80	Fixed-bed	Down-flow	7.5	3.5	Metzger *et al.* 2015
Gwinnett County (USA)	n.a.	PS	9460	Fixed-bed	Down-flow	n.a.	n.a.	Yang *et al.* 2011

Notes: PS: Partial stream;
[a]Upstream sand filtration.

The filter lifetime is the most important factor for the operation and the cost. This parameter depends on the desired degradation target. However, due to lack of field experience, no information is available regarding filter lifetime. The investment costs, however, have a positive effect when the GAC material can be installed into an existing filter by replacing the old filter material (Metzger *et al.* 2015).

10.3.2 Membrane filtration

Nanofiltration (NF: 1–10 nm) and reverse osmosis (RO: 0.1–1 nm) can remove a broad spectrum of organic micropollutants. Both methods are based on the principle that water is forced under pressure through a membrane. In NF and RO membrane processes, the rejection of organic micropollutants can generally be achieved by size exclusion/steric hindrance, adsorption onto membrane, and/or charge repulsion (Bellona *et al.* 2008; Xu *et al.* 2006). The removal efficiency is dependent on the membrane properties and the properties of the target compound (e.g. molecular weight (MW), molecular diameter (MWd), pk_a hydrophobicity/hydrophobicity (log K_{ow}), and diffusion coefficient). Key membrane properties affecting rejection are pore size, molecular weight cut-off (MWCO), surface charge (measured as zeta potential), hydrophobicity/hydrophobicity, and surface morphology (measured as surface roughness). Additionally, the rejection of organic micropollutants is influenced by operational conditions such as pH value, hardness, ionic strength, the presence of organic matter, and membrane fouling (e.g. Bellona *et al.* 2008; Xu *et al.* 2006). Membrane operation conditions as well as hydrodynamic conditions, such as feedwater recovery and velocity, and concentration polarization have been found to influence the rejection of organic micropollutants. Lab-scale investigations with WWTPs effluents showed removal rates for organic micropollutants up to 100% (e.g. Drewes *et al.* 2005; Röhricht *et al.* 2010; Schäfer *et al.* 2011).

NF and RO have already found their use in the treatment of drinking water, in seawater desalination, and in production of ultrapure water. However, they are less suitable for the application in municipal wastewater treatment plants (Abbeglen *et al.* 2012). One drawback of NF and RO is membrane fouling. This may influence the performance of the process by causing a noticeable decrease in the rejection of organic micropollutants (Ng & Elimelech, 2004).

For the treatment and disposal of the retained substances, treatment methods must still be developed. Furthermore, the anticipated operating costs are very high. A positive aspect of the use of NF and RO would be the additional removal of bacteria and germs.

10.4 COSTS

Currently, there is little reliable data for costs of post-treatment steps. Most cost estimates are the result of the first operation experience or feasibility studies in Germany and Switzerland. In the collection and comparison of costs for organic micropollutants elimination it should be noted that the processes differ in the number of necessary constructions and the post-treatment requirements.

Regardless of the chosen method, the magnitude of the costs for the elimination of organic micropollutants in a wastewater treatment plant is determined by several factors. The construction dimensions affect all procedures depending on the hydraulic load. Therefore, the investment costs, which consist of the cost of structural engineering, mechanical engineering, EMSR-technology, and additional building costs, are primarily determined by the selected hydraulic treatment flow. Other factors influencing the investment costs are the soil characteristics, the use of already existing structures as well as the redundancy. A study commissioned by the Association of Swiss Water Pollution Control Association (VSA) showed that the treatment plant size influences the annual cost since the operating costs and the design capacity of a wastewater treatment plant increase simultanesouly (Holinger & Hunziker Betatech AG, 2015).

The operating costs for the elimination of organic micropollutants are divided into the operating part and the consumption-bound part. The operating part is comprised of the cost of the employed staff, the servicing/maintenance, and the additional analysis. The consumption-bound part depends on the method, the cost of the consumption of electricity, activated carbon or pure oxygen, and whether conditioning agents are included. For the PAC application method, further compliance costs for sludge disposal must be considered. The consumption bound fraction is significantly influenced by the required removal efficiency as well as the existing wastewater matrix. Rule of thumb: With increasingly higher DOC of WWTPs effluent, a higher amount of activated carbon or ozone is needed for the adequate removal of organic micropollutants.

As part of a study by Biebersdorf *et al.* (2014), the cost-effectiveness of three different methods for the targeted elimination of organic micropollutants was examined. They determined and compared the annual cost of the use of PAC in a separate adsorption with internal carbon recirculation and downstream cloth filter, filtration through GAC, and ozonation with downstream sand filters for model treatment plants expanding sizes (population equivalents (pe)) of 10,000 pe, 50,000 pe and 100,000 pe. All of the variants were designed as part of the current treatment (90% of Q_{DW}). The consumption has been calculated for the following resources depending on the size and expansion of a plant-specific present DOC content:

- *PAC*: 10–12 g/m^3
- *GAC*: 7,000–9,000 bed volumina
- *Ozone*: 5.6–8.4 gO_3/m^3

The study showed that for the selected expansion sizes, the most economical method was almost always ozonation with a downstream sand filter. However, for smaller wastewater treatment plants, a GAC filter proved to be the cheapest option. The use of PAC is competitive to ozonation from a design capacity of about 50,000 pe. However, the annual cost of the methods under consideration were so close together that when choosing the procedure for the extension of a water treatment plant, the deciding factor will ultimately be the individual boundary conditions as well as non-monetary aspects such as positive spin-offs.

As part of a project funded by the German Federal Environment Agency, the annual sum of money required for a nationwide expansion of the German wastewater treatment plants of size class 3 (5,000–10,000 pe), 4 (10,000–100,000 pe), and 5 (>100,000 pe) has been calculated from a total of 82 cost data from 17 sources (Hillenbrand *et al.* 2014). The data used is based on information from 28 studies with ozonation data, from 17 GAC filters, and 37 on the cost of a PAC application. Data coming from the references to the cost components of different investments have been adjusted for comparability to a uniform evaluation scheme. The reference year for costs was defined as 2012. The study showed that for one step of organic micropollutants elimination, the costs were on average 0.124 €/m^3 of treated wastewater. For treatment plants serving more than 1 million inhabitants, the cost of micropollutants elimination 0.051 €/m^3 of treated wastewater, is to be expected (see Figure 10.7). The cost estimate for an expansion of all 3013 German sewage treatment plants is approximately 1.3 billion € (net). This considers the specific costs of treatment, such as for the construction and the operation of a filter, for size class 3 to 5 and targeted for trace substance elimination.

In Baden-Wuerttemberg, several sewage treatment plants of different sizes (24,000 pe–725,000 pe) have already been upgraded by an adsorption stage consisting of a contact and a sedimentation tank with internal carbon recirculation. Evaluating the production and operating costs for these post-treatment stages on a uniform standard indicates that the improved wastewater treatment costs between 2 and 8 € per year for the citizens. It is important to note that the construction and operation of the necessary post-treatment step after the actual adsorption was not taken into account in the calculation of costs, since five of the considered treatment plants already possessed a sand filter before expanding. Furthermore, cost-reducing factors such as subsidies or lower wastewater charges due to the reduction of chemical oxygen demand (COD) and total phosphorus (P_{tot}) in the effluent were taken into consideration. Without any government funding, the cost for the citizens would be between 2.50 and 13 € per year (Metzger *et al.* 2014).

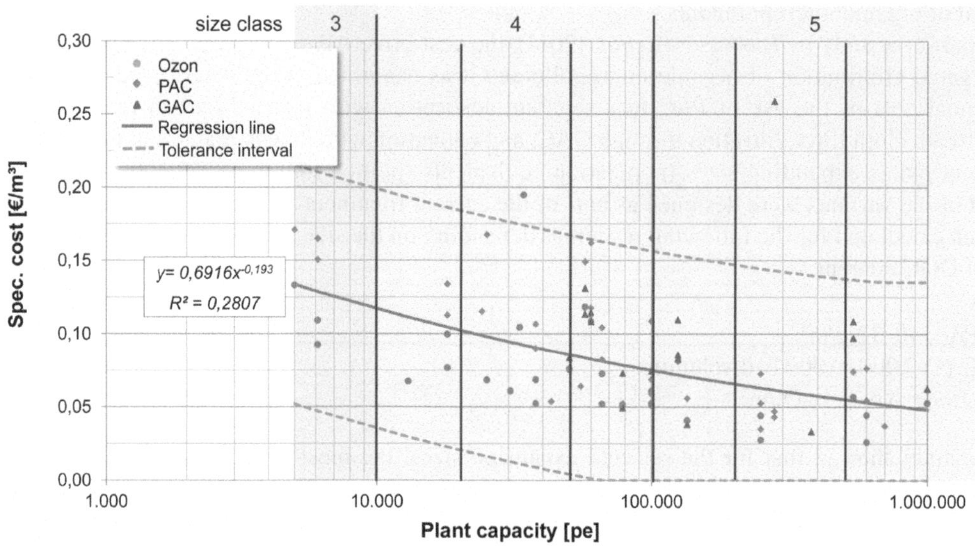

Figure 10.7 Specific costs for post-treatment steps depending on wastewater treatment plant capacity (Hillenbrand *et al.* 2014).

10.5 CONCLUSIONS

In conclusion, many years of experience are missing to make reliable statements on the costs, operation, and applicability of different post-treatment technologies. At this time, for two main post-treatment technologies, ozonation and adsorption, data for pilot- and full-scale applications are available. In principle, a wide range of organic micropollutants can be eliminated from wastewater with the use of ozone as well as with the use of activated carbon. Deciding which of these methods should be used depends on several factors. Overall, it must be said that none of the technologies can eliminate 100% of organic micropollutants. The optimization of post-treatment technologies in full scale is often based on the 80% removal, because yet there are no legal requirements for cleaning performances. Therefore, it is necessary to meet legal regulations. Additionally, indicator substances must be established to determine the efficiency of technologies.

A part of the decision process which technique should be considered is to check whether specific compounds are dissolved in the WWTPs effluent, which may exclude the application of ozone. High bromide concentrations favor for example bromate formation during the treatment with ozone. During the operation of adsorptive methods, the substances to be removed are attached to the activated carbon, so that there is no formation of critical byproducts. In the case of activated carbon, the coal has to be burned (often together with sludge) or regenerated. It should be noted that as the DOC content increases, the required level of ozone or activated carbon also increases.

According to current knowledge, both ozone and activated carbon treatments require a follow-up treatment. In the case of ozone, this is for the retention of transformation products which are formed during ozone treatment, and in the case of PAC treatment, the removal of particles is especially essential. The use of an post-treatment step can also bring additional positive effects. For example, the application of ozone has an additional disinfecting effect. When using activated carbon, the residual organic concentration generally decreases in WWTPs effluents. The addition of precipitation agents in the downstream process may favor an additional phosphorus removal. Nevertheless, to date, only a few experiences on the technical scale are available and are essentially limited to the application of ozone and activated carbon. Many variants are currently being tested on a laboratory scale, particularly the use of AOPs.

The first experiences in Germany and Switzerland with ozonation and activated carbon application demonstrates that for the selected expansion sizes, the most economical method was almost always ozonation with a downstream sand filter. However, for smaller wastewater treatment plants, a GAC filter proved to be the cheapest option.

10.6 REFERENCES

Abegglen C. and Siegrist H. (2012). Micropollutants dans les eau usées urbaines. Etapes de traitement supplémentaire dans les stations d`épuration. Swiss federal office for the environment. *Connaissance de l'environment*, **1214**, 87.

Abegglen C., Escher B., Hollender J., Koepke S., Ort C., Peter A., Siegrist H., von Gunten U., Zimmermann S., Koch P., Schärer M., Braun R., Gälli R., Junghans M., Brocker S., Moser R. and Rensch D. (2009). Ozonation of treated wastewater (Ozonierung von behandeltem Abwasser). Report pilot experiments Regensdorf, Dübendorf, Swizzerland, 16 June 2009 (in German).

Aktaş Ö. and Çeçen F. (2007). Bioregeneration of activated carbon: a review. *International Biodeterioration & Biodegradation*, **59**(4), 257–272.

Antoniou M. G., Hey G., Rodriguez Vega S., Spiliotopoulou A., Fick J., Tysklind M., Jansen J. and Andersen H. R. (2013). Required ozone doses for removing pharmaceuticals from wastewater effluents. *Science of the Total Environment*, **456–457**, 42–49.

Babuponnusami A. and Muthukumar K. (2014). A review on Fenton and improvements to the Fenton process for wastewater treatment. *Journal of Environmental Chemical Engineering*, **2**(1), 557–572.

Barjenbruch M., Firk W. and Fröhlich A. P. (2014). Possibilities for the elimination of trace organic elements in municipal wastewater treatment plants (Möglichkeiten der Elimination von organischen Spurenstoffen in kommunalen Kläranlagen). *KA Abwasser, Abfall*, **61**(10), 861–875 (in German).

Bahr C., Schumacher J., Ernst M., Luck F., Heinzmann B. and Jekel M. (2007). SUVA as control parameter for the effective ozonation of organic pollutants in secondary effluent. *Water Science and Technology*, **55**(12), 267–274.

Bahr C., Ernst M. and Jekel M. (2007b). Pilot studies on the combined oxidative and biological treatment of sewage works effluents for the removal of trace organics and active ingredients and for disinfection (Pilotuntersuchungen zur kombinierten oxidative-biologischen Behandlung von Klärwerksabläufen für die Entfernung von organischen Spuren- und Wirkstoffen und zur Desinfektion). In: Publication Series Kompetenzzentrum Wasser Berlin, B. Weigert, Band 5, Berlin (in German).

Bautista-Toledo I., Ferro-García M. A., Rivera-Utrilla J., Moreno-Castilla C. and Vegas Fernández F. J. (2005). Bisphenol a removal from water by activated carbon. Effects of carbon characteristics and solution chemistry. *Environmental Science & Technology*, **39**(16), 6246–6250.

Bellona C., Drewes J. E., Xu P. and Amy G. (2004). Factors affecting the rejection of organic solutes during NF/RO treatment – A literature review. *Water Research*, **38**(12), 2795–2809.

Biebersdorf N., Kaub J. M., Urueta F. and Schröder M. (2014). Economic employment of powdered activated carbon (PAC), granular activated carbon (GAC) and ozonation for small and medium sized model wastewater treatment plants (Wirtschaftlicher Einsatz von Pulveraktivkohle (PAK), granulierter Aktivkohle (GAK) und Ozonung für kleine und mittelgroße Midellkläranlagen). *KA Abwasser, Abfall*, **61**(10), 894–901 (in German).

Böhler M. and Siegrist H. (2012). PAC Dosage to Influent of Sand Filtration at Wastewater Treatment Plant Kloten/Optfikon – full-scale Application for MicrPoll Project. In: Proc. of DWA seminar Advanced wastewater treatment, 2012, Bonn.

Bornemann C., Erbe V., Hachenberg M., Kolisch G., Osthoff T. and Taudien Y. (2012). Application of Powdered Activated Carbon in Existing Flocculation Filtration Process at the Example WWTP Buchenhof (Einsatz von Pulveraktivkohle am Beispiel der Kläranlage Buchenhofen). In: Proc. of 45. Essener Tagung, Essen, 14–16 March. ISBN 978-3-938996-36-2 (in German).

Clausen K., Lübken M., Pehl B., Bendt T. and Wichern M. (2014). Employment of reactivated active carbon from waterworks for the elimination of trace elements in municipal wastewater treatment plants using Düsseldorf as an example (Einsatz reaktivierter Aktivkohle von Wasserwerken zur Spurenstoffelimination in kommunalen Kläranlagen am Beispiel Düsseldorf. *KA Abwasser, Abfall*, **61**(11), 1007–1012 (in German).

Directive (2013). Directive 2013/39/EU of the European Parliament and of the Council of 12 August 2013 amending Directives 2000/60/EC and 2008/105/EC as regards priory substances in the field of water policy. *Official Journal of the European Union*, **L226**, 1–17.

Dodd M. C., Kohler H. P. E. and von Gunten U. (2009). Oxidation of antibacterial compounds by ozone and hydroxyl radical: elimination of biological activity during aqueous ozonation processes. *Environmental Science & Technology*, **43**(7), 2498–2504.

Drewes J. E., Bellona C., Oedekoven M., Xu P., Kim T. U. and Amy G. (2005). Rejection of wastewater-derived micropollutants in high-pressure membrane applications leading to indirect porTable 10.reuse. *Environmental Progress*, **24**(4), 400–409.

Gerrity D., Gamage S., Holady J. C., Mawhinney D. B., Quiñones O., Trenholm R. A. and Snyder S. A. (2011). Pilot-scale evaluation of ozone and biological activated carbon for trace organic contaminant mitigation and disinfection. *Water Research*, **45**(5), 2155–2165.

Hillenbrand T., Tettenborn F., Menger-Krug E., Fuchs S., Toshovski, Kittlaus S. S., Metzger S., Tjoeng I., Wermter P., Kersting M. and Abegglen C. (2014). Measures to reduce the entry of micro-pollutants into waters (Maßnahmen zur Verminderung des Eintrages von Mikroschadstoffen in die Gewässer). Umweltbundesamt Texte 85/2014 https://www.umweltbundesamt.de/sites/default/files/medien/378/publikationen/texte_85_2014_massnahmen_zur_verminderung_des_eintrages_von_mikroschadstoffen_in_die_gewaesser.pdf (accessed 21 December 2015) (in German).

Huber M. M., Canonica S., Park G. Y. and von Gunten U. (2003). Oxidation of pharmaceuticals during ozonation and advanced oxidation processes. *Environmental Science & Technology*, **37**(5), 1016–1024.

Huber M. M., Ternes T. A. and von Gunten U. (2004). Removal of estrogenic activity and formation of oxidation products during oxidation of 17α-ethinylestradiol. *Environmental Science & Technology*, **38**(19), 5177–5186.

Huber M. M., Göbel A., Joss A., Hermann N., Löffler D., Mcardel C. S., Ried A., Siegrist H., Ternes T. A. and von Gunten U. (2005). Oxidation of pharmaceuticals during ozonation of municipal wastewater effluents: a pilot study. *Environmental Science & Technology*, **39**(11), 4290–4299.

Hübner U., Miehe U. and Jekel M. (2012). Optimized removal of dissolved organic carbon and trace organic contaminants during combined ozonation and artificial groundwater recharge. *Water Research*, **46**(18), 6059–6068.

Hübner U., von Gunten U. and Jekel M. (2015). Evaluation of the persistence of transformation products from ozonation of trace organic compounds – A critical review. *Water Research*, **68**(1), 150–170.

Hollender J., Zimmermann S. G., Koepke S., Krauss M., McArdell C. S., Ort C., Singer H., von Gunten U. and Siegrist H. (2009). Elimination of organic micropollutants in a municipal wastewater treatment plant upgraded with a full-scale post-ozonation followed by a sand filtration. *Environmental Science & Technology*, **38**(19), 5177–5186.

Holinger AG und Hunziker Betatech AG (2015). Dimensioning water amount and redundancies of steps for the elimination of micropollutants (Dimensionierungswassermenge und Redundanzen von Stufen zur Elimination von Mikroverunreinigungen). Study for VSA, not published (in German).

https://www.micropoll.ch/fileadmin/user_upload/Redaktion/Dokumente/01_Berichte/03_Projektierungshilfen/v4_d_Technischer_Bericht_VSA_MikroPoll.pdf (accessed 06 December 15).

Jekel M., Dott W., Bergmann A., Dünnbier U., Gnirß R., Haist-Gulde B., Hamscher G., Letzel M., Licha T., Lyko S., Miehe U., Sacher F., Scheurer M., Schmidt K. C., Reemtsma T. and Ruhl A. S. (2015). Selection of organic process and source indicator substances for the anthropogenically influenced water cycle. *Chemosphere*, **125**(4), 155–167.

Joss A., Siegrist H. and Ternes T. A. (2008). Are we about to upgrade wastewater treatment for removing organic micropollutants? *Water Science and Technology*, **57**(2), 251–255.

Kasprzyk-Hordern B., Dinsdale R. M. and Guwy A. J. (2009). The removal of pharmaceutical, personal care products, endocrine disruptors and illicit drugs during wastewater treatment and its inpact on the quality of receiving waters. *Water Research*, **43**(2), 363–368.

Kovalova L., Siegrist H., von Gunten U., Eugster J., Hagenbuch M., Wittmer A., Moder R. and McArdell C. S. (2013). Elimination of micropollutants during post-treatment of hospital wastewater with powdered activated carbon, ozone and UV. *Environmental Science & Technology*, **47**(11), 7899–7908.

Lee C. O., Howe K. J. and Thomson B. M. (2012). Ozone and biofiltration as an alternative to reverse osmosis for removing PPCPs and micropollutants from treated wastewater. *Water Research*, **46**(4), 1005–1014.

Lee Y. and von Gunten U. (2010). Oxidative transformation of micropollutants during municipal wastewater treatment: comparison of kinetic aspects of selective (chlorine, chlorine dioxide, ferrate^{VI}, and ozone) and non-selective oxidants (hydroxyl radical). *Water Research*, **44**(2), 555–566.

Luo Y., Guo W., Ngo H. H., Nghiem L. D., Hai F. I., Zhang J., Liang S. and Wang X. C. (2014). A review on the occurrence of micropollutants in the aquatic environment and their fate and removal during wastewater treatment. *Science of the Total Environment*, **473–474**, 619–641.

Löwenberg J., Zenker A., Baggenstos M., Koch G., Kazner C. and Wintgens T. (2014). Comparison of two PAC/UF processes for the removal of micropollutants from wastewater treatment plant effluent: process performance and removal efficiency. *Water Research*, **56**(6), 26–36.

Magdeburg A., Stalter D., Schlüsener M., Ternes T. A. and Oehlmann J. (2014). Evaluating the efficiency of advanced wastewater treatment: target analysis of organic contaminants and (geno-)toxicity assessment tell a different story. *Water Research*, **50**(1), 35–47.

Mailler R., Gasperi J., Coquet Y., Deshayes S., Zedek S., Cren-Olivé C., Cartiser N., Eudes V., Bressy A., Caupos E., Moilleron R., Chebbo G. and Rocher V. (2015). Study of a large scale powdered activated carbon pilot: removals of a wide range of emerging and priority micropollutants from wastewater treatment plant effluents. *Water Research*, **72**(4), 315–330.

Margot J., Kienle C., Magnet A., Weil M., Rossi L., de Alencastro L. F., Abegglen C., Thonney D., Chèvre N., Schärer M. and Barry D. A. (2013). Treatment of micropollutants in municipal wastewater: ozone or powdered activated carbon? *Science of the Total Environment*, **461–462**, 480–498.

Maus C., Herbst H., Ante S., Becker H. P., Glathe W., Börgers A. and Türk J. (2014). Information on configuration and design of ozone facilities for micropollutant elimination (Hinweise zur Auslegung und Design von Ozonanlagen zur Mikroschadstoffelimination). *KA Abwasser, Abfall*, **61**(11), 998–1006 (in German).

Metcalf and Eddy (2004). Wastewater Engineering, Treatment and Reuse, 4th edn., (International), McGraw-Hill, New York, USA, ISBN 007-124140-X.

Metzger S., Tjoeng I. O., Rößler A., Schwentner G. and Rölle R. (2014). Costs of the powdered activated carbon application for the elimination of trace elements using the example of implemented plants and those under consideration (Kosten der Pulveraktivkohleanwendung zur Spurenstoffelimination am Beispiel ausgeführter und in Bau befindlicher Anlagen). *KA Abwasser, Abfall*, **61**(11), 1029–1037 (in German).

Metzger S., Rößler A., Türk J., Antakyali D., Schulz J., Wunderlin P. and Meier A. (2015). Status quo of the expansion of wastewater treatment plants by one level for the targeted micropollutant removal (Status quo der Erweiterung von Kläranlagen um eine Stufe zur gezielten Spurenstoffelimination). Modernisierungsreport 2015/2016. wwt 14–19 (in German).

Ng H. Y. and Elimelech M. (2004). Influence of colloidal fouling on rejection of trace organic contaminants by reverse osmosis. *Journal of Membrane Science*, **244**(1–2), 215–226.

Obrecht J., Thomann M., Holinger M., Stoll J. M., Frank K. and Sobaszkiewskz M. (2015). PAC dosage in activated sludge process (PAK Dosierung ins Belebungverfahren), *Aqua & Gas*, **2**, 20–32.

Oturan M. A. and Aaron J.-J. (2014). Advanced oxidation processes in water/wastewater treatment: principles and applications. A Review *Critical Reviews in Environmental Science & Technology*, **44**(23), 2577–2641.

Pinnekamp J., Beier S., Cramer C., Schröder H. F., Mauer C. and Selke D. (2009). Elimination von Spurenstoffen mit Membrantechnik und weitergehenden Behandlungsverfahren – Pilotprojekt Krankenhaus Wäldbröl (Report for Ministry of the Environvent and Conservation, Agriculture and Consumer Protection of the State of North Rhine-Westphalia, IV-9-042 1B4 0020) (in German).

Prasse C., Wagner M., Schulz R. and Ternes T. A. (2012). Oxidation of the antiviral drug acyclovir and its biodegradation product carboxy-acyclovir with ozone: kinetics and identification of oxidation products. *Environmental Science & Technology*, **46**(4), 2169–2178.

Ramseier M. K., Peter A., Traber J. and von Gunten U. (2011). Formation of assimilable organic carbon during oxidation of natural waters with ozone, chlorine dioxide, chlorine, permanganate, and ferrate. *Water Research*, **45**(5), 2002–2010.

Rajab M., Heim C., Greco G., Helmreich B. and Letzel T. (2013). Removal of sulfamethoxazole from wastewater treatment plant effluents by a boron-doped diamond electrode, *International Journal of Environment and Pollution*, **1**(3), 88–97.

Reungoat J., Escher B. I., Macova M., Argaud F. X., Gernjak W. and Keller J. (2012). Ozonation and biological activated carbon filtration of wastewater treatment plant effluents. *Water Research*, **46**(3), 863–872.

Ribeiro A. R., Nunes O. C., Pereira M. F. R. and Silva A. M. T. (2015). An overview on the advanced oxidation processes applied for the treatment of water pollutants defined in the recently launched Directrive 2013/39/EU. *Environment International*, **75**(1), 33–51.

Rivera-Utrilla J., Sánchez-Polo M., Ferro-García M. A., Prados-Joya G. and Ocampo-Pérez R. (2013). Pharmaceuticals as emerging contaminants and their removal from water. A review. *Chemosphere*, **93**(7), 1268–1287.

Röhricht M., Krisam J., Weise U., Kraus U. R. and Dürin R. A. (2010). Elimination of pharmaceuticals from wastewater by submerged nanofiltration plate modules. *Desalination*, **250**(3), 1025–1026.

Schaar H. and Kreuzinger N. (2011). Technical realization and implementation of an ozone stage for purified according to the prior art, urban waste water - Pre-accession to the starte-of-the-art (Technische Umsetzung und Implementierung einer Ozonstufe für nach dem Stand der Technik gereinigtes kommunales Abwasser – Heranführung an den Stand der Technik). Report Lebensministerium, Wien (in German).

Schäfer A. I., Akanyeti I. and Semião A. J. C. (2011). Micropollutant sorption to membrane polymers: a review of mechanisms for estrogens. *Advances in Colloid and Interface Science*, **164**(1–2), 100–117.

Schröder P., Helmreich B., Škrbić B., Carballa M., Papa M., Pastore C., Emre Z., Oehmen A., Langenhoff A., Molinos M., Dvarioniene J., Huber C., Tsagarakis K. P., Martinez-Lopez E., Pagano S. M., Vogelsang C. and Mascolo G. (2016). Status of hormones and pain killers in wastewater effluents across several European states – Consideration for the EU watch list considering estradiol and diclofenac. *Environmental Science and Pollution Research*, DOI: 10.1007/s11356-016-6503-x, in press.

Schwendtner G. (2015). Operating Experience end Results with Tertiary Treatment at WWTP Böblingen-Sindelfingen. In: Proc. of 48. Essener Tagung, 15–17 March, Essen, ISBN 978-3-938996-42-3 (in German).

Silva C. P., Otero M. and Esteves V. (2012). Process for the elimination of estrogenic steroid hormones from water: a review. *Environmental Pollution*, **165**(6), 38–58.

Snyder S. A., Adham S., Redding A. M., Cannon F. S., DeCarolis J., Oppenheimer J., Wert E. C. and Yoon Y. (2007). Role of membranes and activated carbon in the removal of endocrine disruptors and pharmaceuticals. *Desalination*, **202**(1–3), 156–181.

Sperlich A., Altmann J., Zietzschmann F., Meinel F., Ruhl A. S., Gnirss R. and Jekel M. (2014). Qick and cost effective test procedure for activated carbon for the monitoring of trace elements (Schnelle und kostengünstige Testverfahren für Aktivkohle zur Kontrolle der Spurenstoffelimination in der Praxis). *KA Wasser Abwasser*, **61**(11), 989–997 (in German).

Stalter D., Magdeburg A. and Oehlmann J. (2010). Comparative toxicity assessment of ozone and activated carbon treated sewage efflluents using an in vivo test battery. *Water Research*, **44**(8), 2610–2620.

Stamm C., Eggen R. I. L., Hering J. G., Hollender J., Joss A. and Schärer M. (2015). Micropollutant removal from wastewater: facts and decision making despite uncertainties. *Environmental Science & Technology*, **49**(11), 6374–6375.

Türk J., Dazio M., Dinkel F., Ebben T., Herbst H., Hochstrat R., Madzielewski V., Matheja A., Montag D., Remmler F., Schaefer S., Schramm E., Vogt M., Werbeck N., Wermter P. and Wintgens T. (2013). Final report on the research project Economic benefits of upgrading waste water treatment plants for the elimination of organic trace materials, pharmaceuticals, industrial chemicals, bacteriological relevant germs and viruses (Volkswirtschaftlicher Nutzen der Ertüchtigung kommunaler Kläranlagen zur Elimination von organischen Spurenstoffen, Arzneimitteln, Industriechemikalien, bakteriologisch relevanten Keimen und Viren) (TP9), The Ministry for Climate Protection, Environment, Agriculture, Conservation and Consumer Protection of the State of North Rhine-Westphalia (MKULNV), AZ IV-7-042600 001, 08/0581 (in German).

Vogel H. J., Baur S., Triebskorn R., Rößler A. and Metzger S. (2014). The Albstadt-Ebingen wastewater treatment plant: 20 years employment of powdered activated carbon in full flow (Die Kläranlage Ablstadt-Ebingen: 20 Jahre Pulveraktivkohleeinsatz im Vollstrom). *KA Abwasser, Abfall*, **61**(10), 902–907 (in German).

Wert E. C., Rosario-Ortiz F. L. and Snyder S. A. (2009). Effect of ozone exposure on the oxidation of trace organic contaminants in wastewater. *Water Research*, **43**(4), 1005–1014.

Wert E. C., Gonzalez S., Dong M. M. and Rosario-Ortiz F. L. (2011). Evaluation of enhanced coagulation pre-treatment to improve ozone oxidation efficiency in wastewater. *Water Research*, **45**(16), 5191–5199.

Wittmer A., Thomann M., Blunschi M., Abegglen C., Schachtler M. and Siegrist H. (2012). Elimination of Micro-Pollutants, Germs and Bacteria in Municipal Wastewater Treatment plants – Waste Water Treatment Plant Concept Neugut, Dübendorf (Elimination von Mikroschadstoffen, Keimen und Bakterien in kommunalen Kläranlagen – Konzept Kläranlage Neugut, Dübendorf). 2nd Symposium on Micropollutants, 21 June 2012, Düsseldorf, Germany (in German).

Wittmer A., Heisele A., McArdell C. S., Böhler M. and Longree P. (2015). Decrease UV absorbance an indicator of micropollutant removal efficiency in wastewater treated with ozone. *Water Science and Technology*, **71**(7), 980–985.

Xu P., Drewes J. E., Kim T. Bellona, C. and Amy G. (2006). Effect of membrane fouling on transport of emerging organic contaminants in NF/RO membrane applications. *Journal of Membrane Science*, **279**(1-2nnn), 165–175.

Yavich A. A., Lee K. H., Chen, K. C., Pape, L. and Masten, S. J. (2004). Evaluation of biodegradability of NOM after ozonation. *Water Research*, **38**(12), 2839–2846.

Zuccato E., Castiglioni S., Fanelli R., Reitano G., Bagnati R., Chiabrando C., Pomati F., Rossetti C. and Calamari D. (2006). Pharmaceuticals in the environment in Italy: causes, occurrence effects and control. *Environment Science Pollution Research*, **13**(1), 15–21.

Chapter 11

Technologies limiting gas and odour emissions

David Gabriel, Xavier Gamisans, Raúl Muñoz and Javier Lafuente

11.1 INTRODUCTION

Gaseous emissions from WWTP are made of complex mixtures of compounds produced in different parts of the facility, as well as in the sewer network nearby the plant. Composition and emission rates mainly depend on plant configuration and operation (ventilation rates of headworks, dewatering, thickeners…), thus being site specific (Vincent, 2001). Compounds emitted can be either non-odorants such as most Green House Gases (GHG) or odorants such as most Volatile Organic Compounds (VOCs) and Volatile Inorganic Compounds (VICs) with a wide range of odour detection thresholds. Their concentration in the liquid phase as well as their physical-chemical characteristics, such as solubility in water, will have an impact on the emission rates and in the selection of the most appropriate technology for their treatment (Kennes & Veiga, 2001). The later will also be influenced by the biodegradability characteristics of each compound. In addition to their complexity, gaseous emissions from wastewater treatment plants (WWTPs) are highly variable and depend on external factors such as temperature fluctuations or the wastewater loading rate. Thus, large seasonal effects are often found in most facilities (Gabriel & Deshusses, 2003). Design of containment, ventilation and treatment systems must consider such complexity to avoid emissions impact.

11.2 PHYSICAL-CHEMICAL TECHNOLOGIES

11.2.1 Absorption

In gas absorption, a gaseous solute is transferred from an inert carrier gas into a liquid where the solute is soluble (McCabe *et al.* 2007). The mass transfer process is enhanced by high gas-liquid interfacial areas and also with the occurrence of a chemical reaction between the solute transferred and different chemical species from the liquid phase. Absorption processes, also known as chemical scrubbing, are currently the most extended technology for odour abatement in WWTP (Tchobanoglous *et al.* 2003). Selection of the scrubbing solution is largely dependent on the composition of the odorous gaseous stream. Ammonia, hydrogen sulfide and organic sulfur compounds, organic nitrogen compounds (such as amines) and organic acids can be removed using water as liquid phase. Generally, water scrubbing can be a suitable option for odorous compounds with Henry's Law constant lower than 0.07 atm/M (Estrada *et al.* 2013).

In this sense, greenhouse gases, such as CH_4 and N_2O are not advisable compounds to be treated with this technology (Van der Heyden *et al.* 2015). Only the addition of titrant agents (either NaOH or H_2SO_4) is required for appropriate pH maintenance of the aqueous phase. Hydrophobic odorous compounds, such as terpenes or hydrocarbons, with Henry's Law constants above 20 atm/M, need the addition of chemical oxidants to ensure high removal efficiencies. In this sense, NaOCl, $KMnO_4$ and H_2O_2 are commonly used to oxidize both, organic and inorganic (e.g. H_2S) compounds (Card, 2001; EPA, 1985).

Scrubber designs are generally vertical cylinders, operated at countercurrent mode (Figure 11.1).

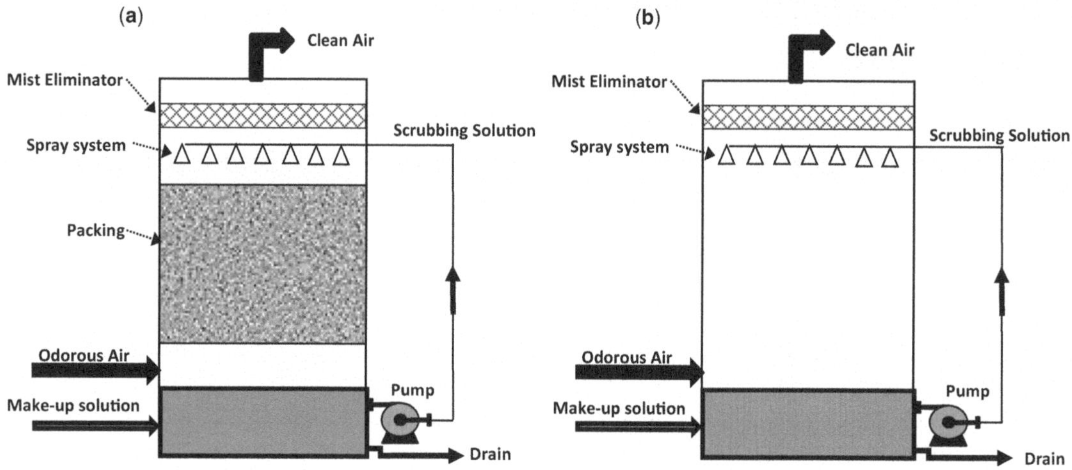

Figure 11.1 Typical wet scrubber configurations: countercurrent packed tower (a) and spray chamber absorber (b).

Horizontal oriented, cross-flow designs are also used. In both cases, foul air is blown into the scrubber and passed through a bed of packing material (Tchobanoglous *et al.* 2003). Typical packing materials include plastic, ceramic, metal or graphite with different shapes (rings, spheres, saddles). A media support plenum, a scrubbing solution distribution (spray nozzles) and pumping system, a mist eliminator and an exhaust fan completes the usual system configuration. Mist scrubbers (without a packed bed) are a suitable configuration when low pressure drops are required (Card, 2001). Venturi scrubbers can also be used in those applications where simultaneous dust and odorous compounds abatement is required (Gamisans *et al.* 2002). Instrumentation and monitoring devices are required mainly for pressure drop control and to ensure the appropriate quality of the scrubbing solution (pH and ORP measurement). Construction materials depend on the composition and characteristics of the odorous effluent and usually must be corrosion resistant.

Due to the complex nature of odorous emissions in WWTP, typical absorption system configurations include three columns: an acidic step for nitrogen-based compounds, two steps with diluted caustic/oxidant reagents for hydrogen sulfide and organic compounds (Prado *et al.* 2009). Typical gas contact times (empty bed residence time, EBRT) are below 4 sec (Gabriel & Deshusses, 2004). However, this parameter depends on the flooding velocity which, in turn, depends on the physical properties of the gas and the packing material characteristics. From these low contact time requirements, the scrubber vessels themselves have the smallest footprint requirements of any odour control system. However, additional footprint is needed for the chemical storage tanks, pumps, controls and other ancillaries.

Chemical scrubbers can be constructed in a wide range of sizes, from small 10,000 m^3/h to massive 10,000,000 m^3/h units (EPA, 1985). Chemical scrubbers are one of the most economic alternatives in terms of investment costs, due to the reduced contact times needed. A recent compilation of costs for industrial-scale chemical scrubbers allowed to establish the following relationship between investment costs and the odorous gas flowrate to be treated (Estrada *et al.* 2011):

$$I_C = 509 \times F^{-0.4}$$

Chemicals, power and maintenance can be expensive, and large amounts of water are needed. Also, the spent chemicals must be properly disposed, and water softening is sometimes required. Annual reagents consumption reported fluctuate between 1–3 $kg/(m^3/h)_{air\ treated}$, besides water consumptions of $3.3 \times 10^{-2}\ l/(m^3/h)_{air\ treated}$ (Estrada *et al.* 2011). Therefore, total annual operating costs, including electrical consumption, are considerably high (3.6 $€/(m^3/h)_{air\ treated}$).

11.2.2 Adsorption

Generally, adsorption can be defined as a superficial phenomenon, which implies the transfer of one or more solutes (sorbate) from a fluid (either gas or liquid) to a solid phase (Volesky, 2003). Basically, two possible mechanisms are involved in adsorption processes: physical adsorption and chemical adsorption (also called chemisorption). Physical adsorption is based on weak interaction forces (such as van der Waals and electrostatic attraction), while chemisorption involves the formation of chemical bonds between the odorous compounds and the adsorbent surface (Kennes *et al.* 2001). The type of interaction between the adsorbent and the sorbate (odorous compounds) strongly influences the possible regeneration of the former. Therefore, physical adsorption is basically an exothermic and reversible process while chemisorption entails more difficulties in regeneration when the sorbent is exhausted. From a constructive point of view, adsorption systems for odour control in WWTPs consist of fixed beds of different granular materials in vertical cylindrical columns (Turk & Bandosz, 2001). Foul air is passed through the bed and odorants are adsorbed due to the porous nature of the granular materials (Figure 11.2). The adsorbent is considered to be exhausted when the odorous substances concentration at the exit of the column reaches an arbitrary value (breakthrough condition) (Clark, 1987).

Several granular materials have been used for adsorption systems. These include activated aluminas, silica gel and molecular sieves. However, activated carbon is the most popular material for odour treatment (European Comission, 2003). Activated carbon is obtained from several carbonaceous materials such as peat, coal, wood, nutshells, coconut husk and petroleum pitch (Kennes *et al.* 2001). Typical activated carbon production involves pyrolysis of the carbonaceous source followed by physical and chemical activation. Activation of the raw materials increases their porosity and specific surface area, besides providing functional groups (carboxylic acids, phenols, hydroxyls and carbonyls among others) able to interact with the odorous compounds. In order to improve and/or to make more selective adsorbent materials, activated carbons can also be impregnated with chemical substances (NaOH, KOH, $KMnO_4$, $CuSO_4$, $Pb(C_2H_3O_2)_2$) (Turk & Bandosz, 2001). The physical characteristics of virgin (non-impregnated) activated carbon are shown in Table 11.1 (Kennes *et al.* 2001).

Non-impregnated activated carbon has a greater capacity for removing VOCs than impregnated carbon. Virgin carbon can also remove H_2S, despite the fact that its adsorption capacity for this compound is much lower than that of impregnated carbon. Removal efficiency obtained with this technology is high (>98%) for moderate to low odorous compounds concentrations. Greenhouse gases has been successfully treated using modified zeolites (Centi *et al.* 2000; Saha *et al.* 2010) and both, thermal and pressure swing adsorption (Kumar *et al.* 2003). Moisture content of the influent to the adsorption column can hinder the system performance since water molecules can compete for the sorption sites (Dorado *et al.* 2012). In this sense, relative humidity below 70% is advised. The technology is robust, simple and do not requires complex instrumentation (only breakthrough conditions measurement). Adsorption capacity of most activated carbons can reach values between 5 and 40 percent of its weight, despite the fact that a typical value of 10% (0.1 g absorbed compound/g activated carbon) is used for design criteria. Factors affecting the sorption capacity of a specific activated carbon include: characteristics of the sorbent material (surface area, pore volume availability, presence of functional groups-polarity), characteristics of the sorbates (composition, concentration, molecular weight, boiling point, solubility, size, polarity) and characteristics of the odorous effluent to be treated (temperature, pressure, flowrate, humidity). An expression to calculate the time to breakthrough (or service life) is (EPA, 1985):

$$t = 6.7 \times 10^6 \frac{S.W}{E.Q.M.C}$$

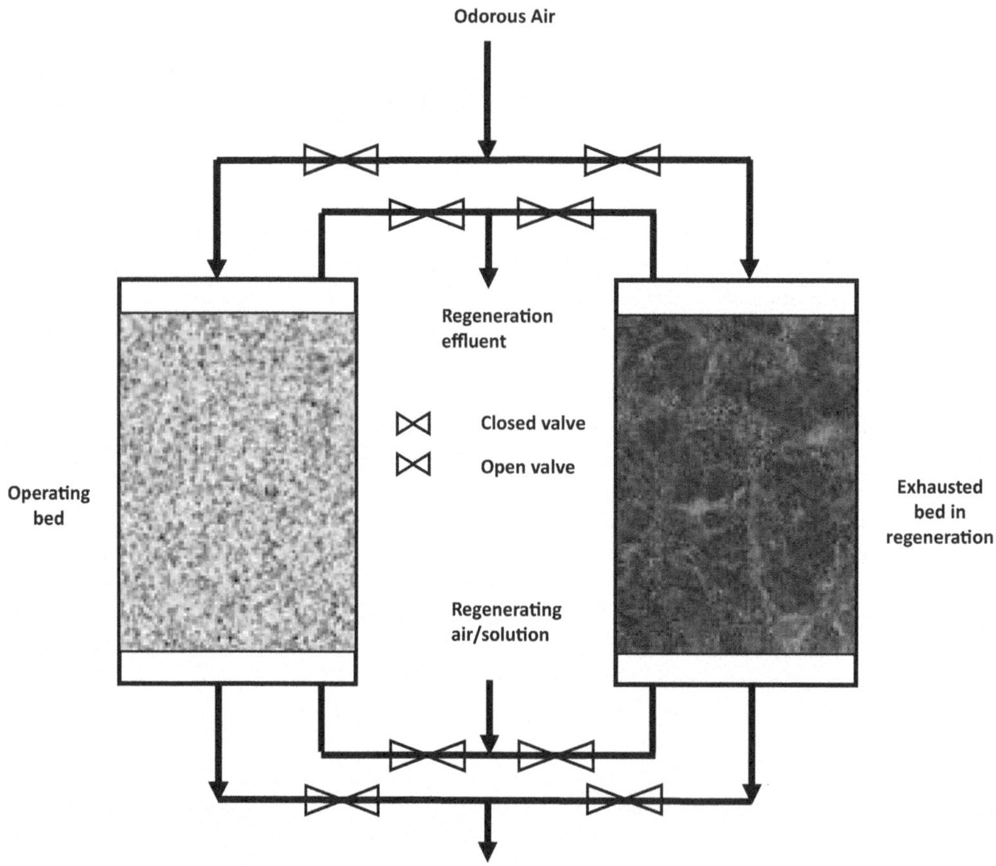

Figure 11.2 Schematic of an activated carbon adsorption column with regeneration system.

Table 11.1 Physical characteristics of activated carbon.

Parameter	Value
Surface area (m^2/g)	500–1800
Bulk porosity (–)	0.25–0.8
Mean Pore diameter (Å)	10–100
Bulk density (kg/m^3)	150–550
Thermal stability (max. T, °C)	150

Where S is the bed adsorption capacity at breakthrough (dimensionless), W is the weight of adsorbent (kg), E is the desired adsorption efficiency (dimensionless), Q is the odorous source flowrate (l/s), M is the average molecular weight of the odorous substances (g/mol) and C is the odorous substances concentration (ppm_v). Typical values for S and E are close to 0.1 and 1, respectively (Estrada *et al.* 2013).

Adsorption-based technologies have been applied for odour removal for 50 years. Therefore, it is a well-stablished technology, with low requirements in terms of footprint due to the small contact times required. Typical adsorption columns contain up to 10,000 kg activated carbon with bed heights up to 1 m and linear velocities from 15 to 25 m/min. The cost of adsorption-based systems is quite variable, depending on the gas flowrate to be treated. A recent compilation (Estrada *et al.* 2011) of investment costs allowed to formulate the following equation:

$$I_C = 113.2 \times F^{-0.33}$$

Operating annual costs are highly variable, usually from 7.2 to 200 (€/(m^3/h), depending on the adsorbent regeneration/replacement cost. An average value for energy requirements is 19 MJ/(m^3/h).

11.2.3 Incineration

Incineration is a combustion process where atmospheric oxygen and heat are utilized to oxidize odorous compounds to the corresponding non-odorous oxides. The incineration unit, also called thermal oxidizer, provides the energy required for the complete oxidation of malodorous compounds. The thermal energy is supplied by an external fuel source, typically natural gas. Since the oxidation is produced in the gas phase, no mass transfer limitations are found. In order to optimize fuel consumption, several alternatives can be currently found (EPA, 1985). In this sense, regenerative thermal oxidizers (RTO) recover a portion of the enthalpy remaining in the exhaust gas and uses it to preheat the incoming odorous gas stream (Figure 11.3). The use of ceramic materials for heat exchanging allows recovering 90–95% of the energy from the exhaust gases. Operation temperatures can be reduced by using catalysts in the oxidation chamber.

Main parameters to be considered in designing thermal/catalytic oxidation systems include residence time, temperature and mixing (Wang & Chou, 2000). Representative residence times required for odorous compounds oxidation range from 0.25 to 0.6 s. In this sense, this technology requires small footprint for its implementation. Typical temperature operation values for these systems are between 800–1000°C (Estrada *et al.* 2013). The presence of catalyst (usually finely divided Pt, Ni, Cu, Cr and Mn) permits the oxidation of malodorous compounds at 300–500°C. Mixing into the oxidation chamber is one of the most important parameters to avoid the partial oxidation of malodorous compounds. Partial oxidation might cause the emission of pollutant substances such as organic acids, aldehydes, ketones and carbon monoxide among others (Kennes *et al.* 2001). Besides, complete oxidation of halogenated and sulphur-compounds can potentially produce harmful substances like HCl, phosgene and SO_2. Proper mixing conditions are related to the appropriate combined design of burners, combustion chamber and reactor configuration.

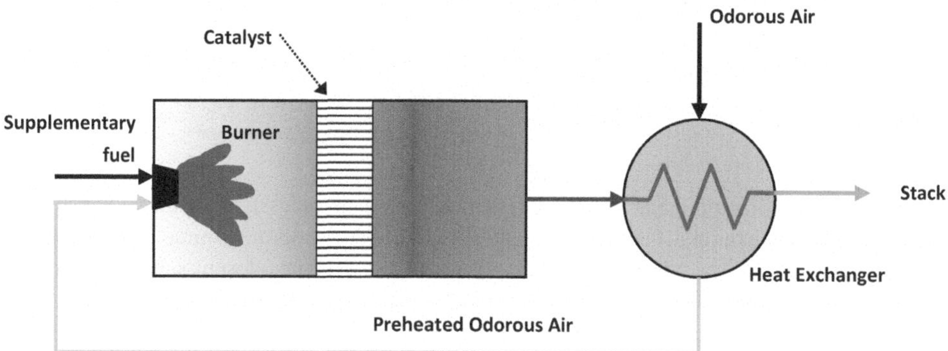

Figure 11.3 Schematic of a regenerative thermal oxidizer (RTO).

Regarding to economic aspects, incineration is the most costly available technology for odour treatment. In terms of energy consumption, regenerative thermal oxidizers are between 70 to 160 times more expensive than any other (either physical-chemical or biological) technology. Also, this technology presents very high investment costs, ranging from 8–52 €/(m^3/h) $_{\text{air treated}}$ (European Comission, 2003), with operating costs higher than 8 €/(m^3/h) $_{\text{air treated}}$. Apart from the economic issues, there are also environmental issues associated to combustion-based technologies. In this sense, the use of an external fuel source implies the emission of additional greenhouse gases (mainly CO_2). Due to these major drawbacks, there are relatively few applications of RTOs used specifically for odour control (Burguess *et al.* 2001). Only when facing recalcitrant, difficult-to-treat air streams, combustion-based technologies are considered.

11.2.4 Advantages and drawbacks of physical-chemical techniques

Absorption systems are a proven and reliable technology highly effective for H_2S and NH_3 abatement even at high concentrations. Wet scrubbers are simple and stable on its operation and maintenance besides being able to treat large volumes of malodorous emissions (containing both odorous compounds and particles) economically. Single towers require moderate footprint. Main drawbacks appear when treating organic based odour-causing compounds due to limited effectiveness. Storage and handling of potentially hazardous chemicals is required. Conversely, activated carbon adsorbers do not require chemicals storage and recirculation systems. These are the most simple systems in operation and maintenance, besides being able to remove a wider range of compounds than wet scrubbers. They are also suitable for VOC removal. However, carbon bed replacement is expensive in terms of fresh carbon and the labor necessary to change out beds. In addition, spent carbon becomes a waste solid that must be disposed properly (based on its hazardous nature). Regarding to regenerative thermal oxidizers, they are considered neither practical nor cost effective to treat odorous emissions when compared to sorption-based technologies. However, it is still used in landfill and industrial applications due to its high removal efficiency regardless the characteristics (hydrophobicity, molecular weight, concentration, etc.) of the target compounds.

11.3 MATURE BIOLOGICAL TECHNOLOGIES

11.3.1 Biofilters

Odour impact of WWTP emissions can be significantly diminished with proper design and operation of sewers and plant units (Vincent, 2001), even if collection and treatment of waste gases is usually needed, particularly when WWTP sites are close to residential areas. Biofilters are probably the most mature biological technology (MBT). Biofilters have become the most popular biological treatment technology for odour emissions emanating from WWTP because of their simplicity and low construction and operating costs.

In general terms, a conventional biofilter consists of a reactor packed with a packing material (usually organic) that serves as a support of a microbial biofilm that degrades the pollutants previously transferred from the gas to the biofilm phase (Figure 11.4). For the biological treatment to become effective, pollutants of interest should be somewhat biodegradable, non-toxic and partly water soluble. To sustain biofilms, either the waste gas to be treated is previously humidified in a humidification column or a discontinuous water flow is regularly supplied. Thus, biofilters are particularly suitable for the treatment of relatively poor water-soluble compounds, with a dimensionless gas-liquid partition coefficient above 1 (Kennes *et al.* 2009). Often, a leachate is collected from the bottom of the biofilter where oxidation products of VOCs, H_2S and NH_3 are collected as carbonates, sulfate, nitrate and nitrite, respectively. However, improper design, operation and, particularly, poor maintenance of biofilters (not uncommon at WWTP), leads to the accumulation of less oxidized species in the packed bed and biofilter leachates. The lack of maintenance of biofilters is often the principal issue reported in biofilters malfunctioning. A list of parameters and operating ranges are recommended in literature (Devinny *et al.* 1999). Water content of the packing material, packing material characteristics and nutrients availability, temperature and pH and alkalinity are the most important to avoid potential gas channelling and sub-optimal growth of microbial cultures (Kennes & Veiga, 2001). From a construction perspective, open-bed biofilters are simpler than enclosed biofilters, but are more exposed to changing meteorological conditions. Thus, the proper alternative is highly dependent on the location of the biofilter.

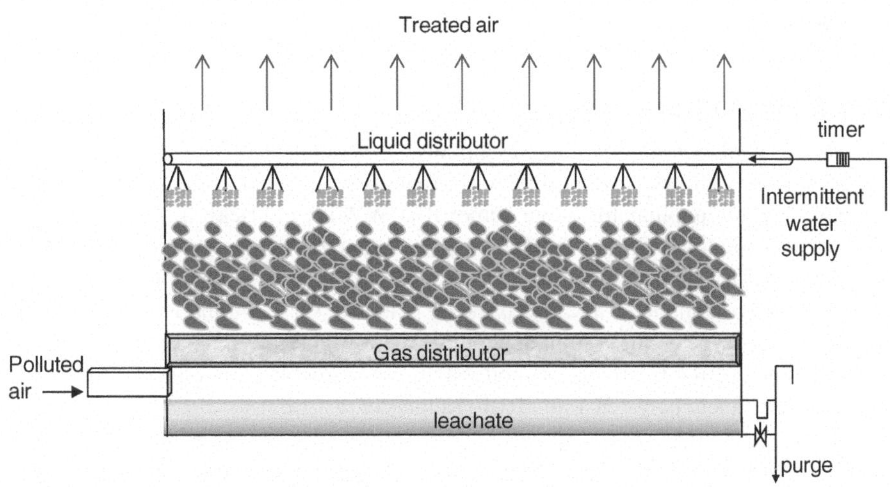

Figure 11.4 Schematic of an open-bed biofilter.

Waste gases biofiltration at WWTP has been demonstrated through years as an adequate technology for dealing with complex emissions from WWTP. However, excessive loads of inorganic volatile inorganic compounds such as H_2S and NH_3 may affect process performance due to inhibition of microbial cultures. An excessive H_2S load (above 10–15 g H_2S m^{-3} h^{-1}) may lead to biofilter acidification due to proton production (Yang & Allen, 1994) while NH_3 loads above 5–10 g NH_3 m^{-3} h^{-1} may lead to partial to nil nitrification due to accumulation of free ammonia and free nitrous acid (Baquerizo *et al.* 2009). Despite NH_3 is rarely found to produce inhibition in WWTP biofilters due to low emissions, H_2S commonly found in the pretreatment area of WWTPs needs a previous treatment to avoid biofilter upsets. In terms of VOCs, an excessive carbon supply (loads above 350 g C m^{-3} h^{-1}) would quickly produce an overgrowth of heterotrophic biomass that would eventually lead to packed bed clogging (Dorado *et al.* 2012). However, typical total VOC loads at WWTPs are much below clogging limits. The main disadvantage of biofilters for VOC removal at WWTPs is the large reactor volumes needed to cope with large flowrates of largely diluted waste gases. Such problem is even worse to reach acceptable removals of poorly soluble odorants such as some aldehydes (nonanal, decanal), sulphides (dimethyl disulfide, dimethyl sulfide) or aromatic and aliphatic hydrocarbons (toluene, dodecane). Even in some cases, biofilters have been reported to produce odorants as partial intermediates of degradation of more complex pollutants (Dorado *et al.* 2014). Dimensionless gas–liquid Henry coefficients for N_2O, NO, CH_4 and CO of 1.7, 21.5, 29.2 and 43.1 (Sander, 1999), respectively, indicate that GHG are sparingly soluble in water. Thus, biofilters are the only suitable biological technique for their treatment. However, few references exist about biofiltration of GHG. While CH_4, CO and NO can be treated to a certain extent in conventional biofilters, N_2O has been shown to be generated rather than removed in biofiltration systems (Sánchez *et al.* 2015).

Despite biofilters limitations, biofilters are the most economical technology in terms of operating and installation costs for odour treatment, which is a must in facilities producing low-added value products such as WWTP. Based on the comprehensive cost assessment protocol developed by Devinny *et al.* (1999), Prado *et al.* (2009a) performed an economical analysis of several biofilter configurations showing that investment costs were deeply influenced by the packing material selected. Installation costs ranged between 3.6 and 19.0 € m^{-3} h for pig manure plus sawdust and Biosorbens™, an inorganic commercial packing, respectively, which were referenced to a 400 m^3 biofilter operated at a gas contact time of 60 sec. Total annualized costs ranged between 3.4 and 4.1 € m^{-3} h, which were strongly influenced by the packing material replacement costs, necessary because of packing materials degradation along time.

11.3.2 Biotrickling filters

Probably, H_2S is likely the most annoying compound generated at WWTPs, which is produced in anaerobic environments such as sewers, anaerobic digesters and sludge thickeners. Despite the fact that biofilters have become popular for odour removal at WWTP, biotrickling filters have gained much attention in the last decade for H_2S removal. Biotrickling filters work in a very similar manner compared to biofilters (Figure 11.5a). Their main differences are related to (1) a continuous water phase trickling over a packed bed, which is made of an inert packing material, (2) a continuous water make-up supply, and purge and (3) a tower-type configuration due to lower footprint requirements. Thus, biotrickling filters are particularly suitable for the treatment of moderately water-soluble compounds, with a dimensionless gas-liquid partition coefficient between 1 and 0.01 (Kennes *et al.* 2009).

Several works have demonstrated the capability of biotrickling filters to treat moderate-to-high pollutant loads of H_2S (Deshusses & Gabriel, 2005; Montebello *et al.* 2013). Headworks and sludge thickening off-gases containing H_2S are produced at concentrations usually below 100 ppm_v but at high flowrates due to ventilation requirements, while H_2S in biogas produced from the anaerobic digestion of WWTP sludge is found at concentrations usually above 3000 ppm_v but at much lower flowrates. In both cases, H_2S loads to be treated in biotrickling filters might easily be above 60–70 g m^{-3} h^{-1}. Such H_2S loads would hinder a biofilter operation due to protons accumulation in the packed bed, thus chemical scrubbers have been traditionally used in most WWTP for H_2S removal. In the last years, several works have demonstrated that it is possible to operate biotrickling filters for the treatment of low concentrations of H_2S at gas contact times (EBRT) equivalent to these of chemical scrubbers (Gabriel & Deshusses, 2003; Prado *et al.* 2009b; Santos *et al.* 2015) to reach Elimination Capacities (ECs) as high as 110–120 g H_2S m^{-3} h^{-1}. Because of G-L mass transfer issues, the packing material plays a key role in ensuring reaching such performance limits. The main consequences of such studies are that chemical scrubbers can be converted to biotrickling filters with a relatively simple and economical protocol (Gabriel *et al.* 2004), thus saving costs on chemicals usage, or that new odour removal systems can be built based solely on biological processes (Dillon & Kyi, 2013). Despite the fact that H_2S from WWTP headworks can be easily removed in biotrickling filters, the operating conditions of biotrickling filters (EBRT < 2 sec and pH < 2 to favor autotrophic sulfide-oxidizing biomass – SOB-activity) do not favor simultaneous removal of organic compounds, particularly these low-soluble odorants found in WWTP together with H_2S. Operation of a series of converted chemical scrubbers to biotrickling filters operated at near neutral pH and EBRTs around 5 sec has demonstrated that larger REs of VOCs are reached simultaneously to H_2S removal at expenses of a larger water consumption to maintain the reactor pH (Santos *et al.* 2015). However, complete odour removal from WWTP off-gases in biotrickling filter is still challenging. Thus, biofiltration and biotrickling filtration must be seen as complementary treatment techniques to be installed at WWTP for proper odor removal, acting biotrickling filters as a pretreatment for H_2S removal prior to a second-stage biofilter.

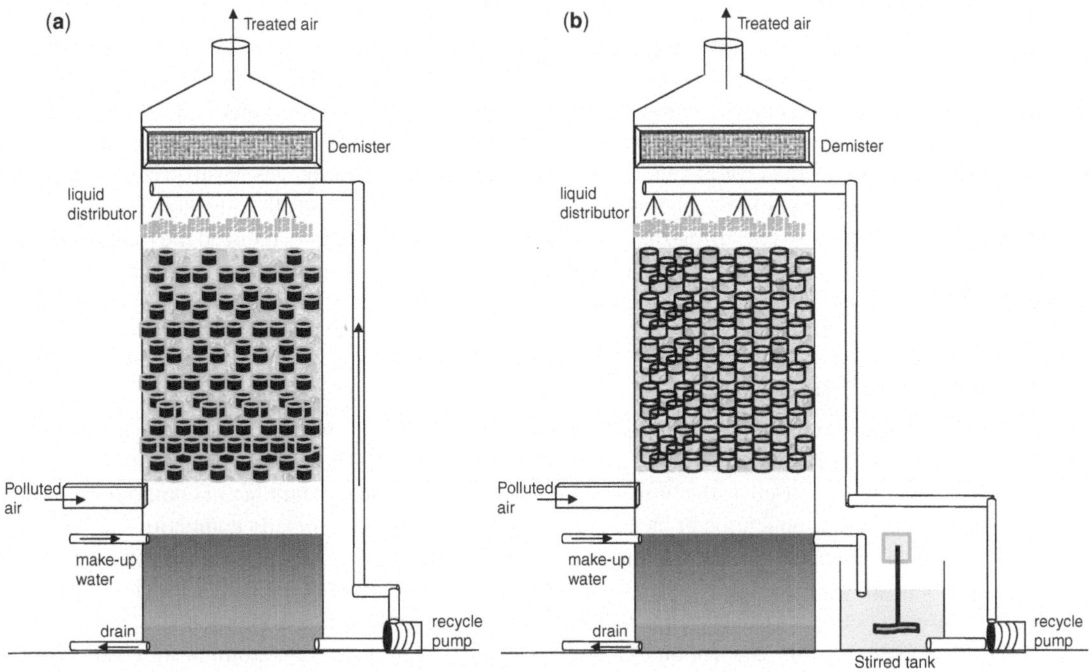

Figure 11.5 Schematic of (a) a biotrickling biofilter and (b) a bioscrubber.

Similarly, biotrickling filters have also been shown as a suitable technology to treat similar, and even higher loads of H_2S as these contained in biogas produced in anaerobic digesters (Fortuny *et al.* 2008; Montebello *et al.* 2013; Montebello *et al.* 2014). Biogas desulfurization is needed in order to ensure proper functioning of downstream equipments such as cogeneration engines for biogas processing for heat or electricity production at WWTP. Despite the fact that similar H_2S loads to these of odour causing off-gases can be found in WWTP biogas, concentrations of H_2S above 2000–3000 ppm_v require larger EBRTs (above 1 min) to ensure H_2S transfer from the gas to the liquid phase. In addition, an electron acceptor must be provided to the biotrickling filter for biological activity of SOB. Nitrate and oxygen are the two alternatives available that, provided in the proper amount, allow the complete oxidation of sulfide to sulfate to occur. If a limited amount of any electron acceptor is provided, then elemental sulfur is produced as an oxidation intermediate, which may hinder biotrickling filter performance due to solids build up in the packed bed. However, robust and reliable processes for autotrophic denitrification (Soreanu *et al.* 2008; Fernández *et al.* 2013) as well as for aerobic (Chaiprapat *et al.* 2011; Fortuny *et al.* 2011) biogas desulfurization have been developed so far to treat biogas flowrates with inlet H_2S concentrations of around 7000–8000 ppm_v. Main advantages of the anoxic process are that 1) no biogas dilution is produced, which is necessary if final biogas usage is other than heat or electricity production, and 2) no mass transfer limitations exist in the supply of the electron acceptor. However, a nitrate source must be found, which must provide a nitrate excess to ensure no elemental sulfur production. Nitrification of a fraction of the reject water produced at WWTPs could be an alternative for providing an oxidized source of nitrogen. Aerobic desulfurization of H_2S has the main advantage of much reduced operating costs due to air usage. Biogas dilution with air is the main drawback identified in this process. However, a reduced amount of oxygen in biogas after desulfurization (2–3%) is not an issue in heat and cogeneration engines, the most common way to recover energy at WWTP (Gabriel *et al.* 2013).

Cost-benefit analyses published so far demonstrate that conversion of chemical scrubbers to biotrickling filters for odor removal are very attractive from an economical perspective even if estimates are highly dependent on site conditions. Savings on chemicals were as high as 50 \$/kg H_2S removed for a 10,000 $m^3\ h^{-1}$ biotrickling filter directly compared with a chemical scrubber (Gabriel & Deshusses, 2004). Conversion payback times can be easily below 2 years per scrubber. Conversely, no in-depth economical analysis has been performed to evaluate the cost-benefit of using biotrickling filters for biogas desulfurization. Gabriel *et al.* (2013) compared the investment, operating costs and payback time of a full-scale biotrickling filter in front of adding $FeCl_3$ for biogas desulfurization at a WWTP. Biotrickling filters reduced the treatment costs to half of these of $FeCl_3$ addition demonstrating also that the biological treatment is a rather economical alternative.

11.3.3 Bioscrubbers

Despite the fact that bioscrubbers have been applied at industrial scale for the treatment of several off-gases, their application in WWTP is mostly limited to biogas desulfurization. The complex nature of odorant off-gases produced at WWTP as well as the limited solubility in water of a large number of compounds in such emissions clearly limit the application of bioscrubbers, which are an economically competitive technology for the treatment of pollutants with moderate-to-high solubilities (gas-liquid partition coefficient below 0.01). Bioscrubbers are very similar to biotrickling filters from a configuration point of view (Figure 11.5b). The main difference between both systems is that the absorption and the reaction processes are separated in a bioscrubber. Generally, the absorption unit is a non-colonized packed bed column in which absorption of pollutants occurs. The bottom liquid is directed to a bioreactor, commonly operated with a suspended culture, which is placed as an intermediate unit prior the recirculation of the outlet liquid of the bioreactor to the absorption unit. Such configuration increases operational flexibility and improves the controllability of the process. Also, treatment of off-gases that would produce a large amount of solids, either biomass due to VOCs degradation or salts, is favored in a bioscrubber since packed bed clogging is avoided. Conversely, investment costs are higher than in biofilters and biotrickling filters.

The main advantage of bioscrubbers for biogas desulfurization lies in the possibility of elemental sulfur recovery in the reaction unit without clogging problems in the absorption unit. In the case that limited supply of nitrate or oxygen may occur in biotrickling filters, which has often been related to the treatment of loads of H_2S above 150–200 g m^{-3} h^{-1} (Fortuny *et al.* 2008; Montebello *et al.* 2012), bioscrubbing is a more competitive process compared to biotrickling filtration. This process, patented by Paques BV (Thiopaq®), is driven by SOB that partially oxidize sulfide to elemental sulfur by limiting the amount of oxygen fed to a continuous stirred tank reactor coupled to a sedimentation unit for solids separation. In any case, sulfate can also be produced as end-product if excess air is supplied in the bioreactor, which does not lead to biogas dilution. In this case, a larger proton production leads to a larger caustic consumption to maintain bioreactor pH, which increases operational costs of the bioscrubbing process. Consequently, in terms of economics, viability of the bioscrubbing process must be balanced according to the feasibility of obtaining a benefit out of the elemental sulfur produced. In small plants or plants treating medium loads of H_2S, sulfate must be preferred as end-product over elemental sulfur due to the difficulty to market low amounts of elemental sulfur. In this case, operating costs of the bioscrubber are significantly higher than those of an aerobic biotrickling filter.

11.3.4 Advantages and drawbacks of mature biological technologies

MBTs discussed have been implemented in full-scale WWTP to face odour problems as well as to improve performance of energy recovery systems from biogas produced in WWTP, showing proper performance results and positive cost-benefit outcomes. However, chemical scrubbing is a mature, consolidated technology still more widespread than bioreactors in WWTPs despite of its higher operating costs. Biofilters and biotrickling filters are achieving satisfactory odour removal efficiencies, thus gaining recognition from plant designers and operators. However, bioreactors for off-gases treatment are still seen as emerging technologies. For a full implementation and confidence in the technology some technical issues must be overcame such as larger start-ups and larger recovering periods after process upsets in comparison to chemical scrubbing. In addition, technical limitations often claimed by plant operators such as incomplete odour removal or unsuccessful performance, which are commonly related with a lack of proper maintenance of the bioreactors as well as limited process knowledge, must be properly addressed through personnel training. Despite the fact that cost-effectiveness, robustness, reliability and performance of biological reactors for waste gases treatment at WWTPs have been extensively demonstrated; permeation into the market is still limited.

11.4 EMERGING BIOLOGICAL TECHNOLOGIES

11.4.1 Two-phase partitioning bioreactors

Despite the advantages of MBTs in terms of low operating costs and environmental impacts, a poor abatement performance is still recorded in biological systems during the treatment of hydrophobic gas pollutants (e.g. alkanes, nitrous oxides, terpenes, etc.) or high loading rates of moderately soluble toxic VOCs (Muñoz *et al.* 2007). This requires the development of innovative biotechnologies, namely emerging biological technologies (EBTs), capable of both increasing the gas-liquid concentration gradient of hydrophobic pollutants and buffering the microbial community against surges in the loading of moderately soluble VOCs. In this regard, two-phase partitioning bioreactors (TPPBs) have recently emerged as a novel biotechnology capable of overcoming the above limitations using conventional bioreactor configurations as a platform. TPPBs are based on the addition into the biological process of a non-aqueous phase (NAP) with a high affinity for the target gas pollutant, which entails a higher VOC absorption capacity and driving forces for pollutant mass transfer to the microbial community (Figure 11.6b) (Kraakman *et al.* 2011). For instance, a bioreactor using silicone oil as a NAP can enhance the biodegradation performance of hexane by a factor of 17 compared to a conventional aqueous-phase bioreactor (Hernandez *et al.* 2012). The microbial community in the TPPB can be present both in the aqueous phase (promoting a gas-NAP-aqueous phase mass transport pathway) and/or in the NAP (promoting a gas-NAP mass transport pathway and a direct pollutant uptake), the latter supporting the highest mass transfer enhancement since the NAP-aqueous phase pathway is often less efficient than the gas-NAP pathway (Figure 11.6c) (Muñoz *et al.* 2012). Recent studies have also shown that the presence of a NAP into the bioreactor can increase both the gas/water and the gas/NAP interfacial areas, which ultimately enhances overall gas pollutant mass transfer rates (Quijano *et al.* 2010). The presence of a NAP can also protect process microbiology against VOC loading surges and starvation periods by temporarily decreasing the VOC concentration in the aqueous phase and by acting as a VOC reservoir, respectively (Kraakman *et al.* 2011).

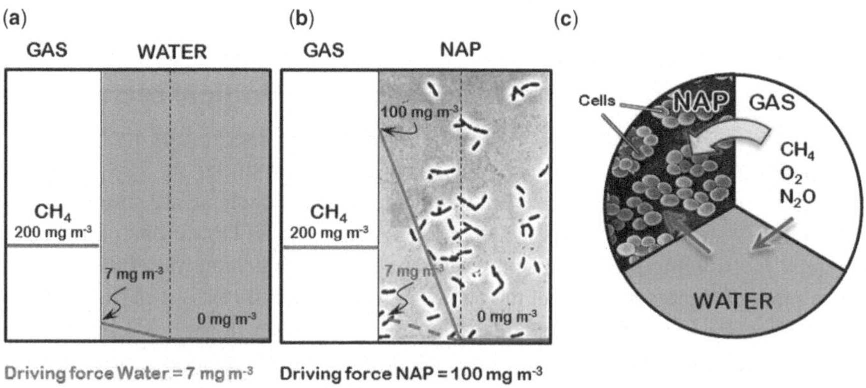

Figure 11.6 Gas-liquid concentration gradients during the treatment of CH_4 in a conventional bioreactor (a) and in a TPPB constructed with silicone oil (b). Mass transfer pathway in a TPPB operated with hydrophobic biomass growing inside the NAP (c).

The performance of TPPBs is governed by the type of NAP used and the target gas pollutants to be treated, the characteristics of the microbial community (which determine the mechanisms of pollutant uptake) and the bioreactor configuration (which determines the hydrodynamics of the NAP). Thus, the selection of the optimum NAP in a TPPB must consider the following criteria (Muñoz *et al.* 2012):

- *Biocompatibility*: The NAP must be non-toxic to the microbial community supporting pollutant biodegradation.
- *Non-biodegradability*: Undesirable effects such as NAP losses and microbial competition with the target gas pollutant must be avoided.
- *High affinity for the target gas pollutant*: This feature, quantified via the gas-NAP partition coefficient, determines the pollutant concentration gradient available for mass transport.

- *Low-emulsion forming tendency and difference in density with water*: These characteristics will facilitate the separation of the NAP and prevent losses by striping.
- *Availability in bulk quantities and at a low cost*: The process must be cost-efficient at a large scale and the cost of the NAP might limit the implementation of this biotechnology.
- *Immiscibility in water and low vapour pressure*: This will prevent NAP losses and facilitate phase separation.
- *Good hydrodynamic characteristics*: A proper NAP dispersion is crucial to benefit from the increased concentration gradients mediated by the presence of the NAP in the bioprocess.

Silicone oil at fractions of 10–30% (v/v) has been the most commonly used NAP in the past decade based on the fulfilment of all the above cited criteria. Solid inert polymers based on copolymers of polyurethane, vinyl acetate and ethylene have been also evaluated as NAPs, gaining acceptance due to their low cost, facilitated separation and recycling during VOC treatment (Hernandez *et al.* 2011). However, the use of solid polymers in TPPBs did not support superior elimination capacities under steady-state operation compared to conventional single-phase systems (likely due to a limited pollutant diffusion). On the other hand, most TPPBs for the treatment of off-gas emissions have been operated with mixed or pure cultures growing in the aqueous phase, which results in a mass transfer mechanism based on the transfer of the gas pollutant from the gas to the NAP prior to its transfer to the aqueous phase where biodegradation occurs. However, recent studies confirmed that hydrophobic microorganisms can grow immersed in silicone oil (since both nutrients and water can diffuse into NAPs) and uptake both O_2 and the gas pollutant in the NAP (Hernandez *et al.* 2012; Cantera *et al.* 2015). A TPPB based on the confinement of the biomass inside the NAP can result in an enhancement in pollutant abatement of one order of magnitude due to the fact that microbial cells growing inside the NAP actually use the entire potential of the gas/NAP mass transfer pathway. Finally, TPPBs devoted to the abatement of gas pollutants have been implemented in stirred tank reactors, airlift, biotrickling filters, biofilters and even bubble columns (Muñoz *et al.* 2012). Stirred tank reactors are by far the most common TPPB configuration, but are also the most energy-demanding system (Littlejohns & Daugulis, 2009). For instance, the power requirements per bioreactor volume in TPPBs implemented in stirred tank reactors range from 0.2 to 15 kW $m^{-3}_{reactor}$, while the specific power consumption decreases to 0.02 to 0.06 kW $m^{-3}_{reactor}$ in TPPB-airlift reactors and to 0.01 kW $m^{-3}_{reactor}$ in two-phase biotrickling filters (Muñoz *et al.* 2012). In the particular case of odour and greenhouse gas treatment in WWTPs, TPPBs can cope with the abatement of the hydrophobic fraction of malodorous emissions, CH_4 and N_2O (Cantera *et al.* 2015). The removal efficiencies of TPPBs when treating moderately soluble VOCs such as benzene, toluene, ethylbenzene and xylene (BTEX), styrene remained above 95%, while these removals decreased to 70–90% during the treatment of highly hydrophobic pollutants such as methane or hexane. In this context, while the removal of CH_4 has been successfully tested in TPPBs, no report is available (to the best of our knowledge) on the performance of TPPB for N_2O abatement. No competitive advantage of TPPBs compared to conventional biotechnologies has been reported for the treatment of highly soluble compounds such as H_2S or oxygenated VOCs.

11.4.2 Activated sludge diffusion

In activated sludge diffusion systems, the malodorous emissions generated in the WWTP are introduced in the aeration tank as a part of the air needed to satisfy the biological oxygen demand of the wastewater (Figure 11.7) (Hardy, 2001). Malodorous and GHGs emissions are diluted emissions of VOCs and VICs with an O_2 content of ≈21%, which can make up the aeration air or even substitute it. The malodorous compounds and CH_4 diffuse from the pressurized emission (introduced at the bottom of the tank via fine or coarse bubble diffusers) into the mixed liquor together with the O_2, odorants and CH_4 being subsequently oxidized by the microbial community present in the aeration tank (Burgess *et al.* 2001). The biogenic nature of most odorants, along with the high metabolic versatility of activated sludge, guarantee the biodegradation of most malodorous compounds diffused into the mixed liquor. Malodorous emissions in properly designed and operated WWTPs represent between 20 and 100% of the total aeration requirements in the plant, although the use of high-level covers might generate larger malodorous emissions and therefore the need for implementation of additional end-of-the pipe units. Activated sludge diffusion exhibits the merits from mature biological technologies (low operating costs and low environmental impacts), while overcoming their main limitations (packing media drying, channelling, structural problems, pH control, accumulation of toxic metabolites, high investment costs). Apart from the obvious decrease in investment costs derived from the use of the already existing aeration tanks, the implementation of activated sludge diffusion can provide a competitive advantage in plants with severe land limitation problems. From an operational view point, the use of activated sludge diffusion only entails the cost of conveying the malodorous emissions to the aeration compressor and the frequent substitution of moisture traps.

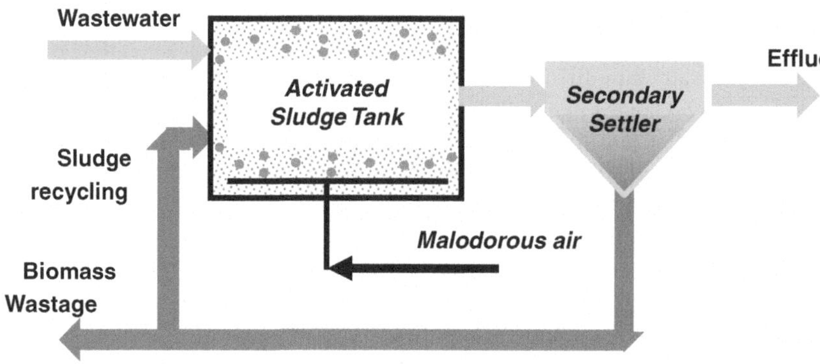

Figure 11.7 Biological treatment of malodorous emissions in an activated sludge diffusion system.

Activated sludge diffusion systems provide very high gas-liquid interfacial areas (as a result of the small air bubbles), which supports high volumetric mass transfer rates and consequently high odour removal efficiencies. Indeed, odour removal efficiencies higher than 99% have been reported in activated sludge systems treating malodorous emissions from nearby composting facilities. Likewise, Bowker (2000) reported odour removal efficiencies of 99.9% at H_2S concentrations of up to 120 ppm_v, which confirmed the high abatement potential of this biotechnology. The performance of this technology for the removal of the greenhouse gases CH_4 and N_2O has not been evaluated yet. On the other hand, operational problems derived from H_2S-mediated corrosion of pipelines, compressors and filters might be minimized by using corrosion resistant materials (PVC, glass fiber, stainless steel) and moisture traps (Burgess *et al.* 2001). In addition, there is still controversy on the impact of H_2S addition into the activated sludge process, since H_2S can promote the development of filamentous bacteria such as *Thiotrix spp*, which might ultimately hinder the settleability of the sludge. Finally, recent studies have also shown the high robustness of this technology against H_2S concentration surges, electricity shutdown, pH control failures or temperature fluctuations (Lebrero *et al.* 2011).

11.4.3 Membrane bioreactors

Membrane bioreactors for off-gas treatment were also devised to overcome the main limitations of conventional packed bed biotechnologies such as the limited mass transfer of hydrophobic gas pollutants, packing media deterioration and microbial inhibition by acidification or metabolite accumulation (Lebrero *et al.* 2013). This technology is based on a membrane-based selective separation of the target pollutants (via differential affinity or size exclusion) from the off-gas, followed by their biocatalytic oxidation (since O_2 diffuses concomitantly with the gas pollutants) by a microbial biofilm attached on the other side of the membrane or by a microbial community in suspension (Figure 11.8). Indeed, the membrane is used as an interphase between the off-gas emission and the microbial community responsible of pollutant biodegradation. In this system, while the membrane facilitates the selective separation of the gas pollutants, the recirculating aqueous solution on the other side of the membrane provides the nutrients required for microbial growth and allows for an easy removal of inhibitory metabolites and pH control (Lebrero *et al.* 2014). On the main advantages of this technology is its compact nature, which supports interfacial areas of up to 30000 m^2 m^{-3}. Three main types of membranes have been so far tested for the abatement of gas pollutants in bioreactors:

- Porous membranes, which can be classified into microporous (pore diameter < 2 nm), mesoporous (2 nm < pore diameter < 50 nm) and macroporous (pore diameter > 50 nm). Their porous nature allows for a facilitated pollutant diffusion across the membrane, which increases the pollutant mass transfer rates. Unfortunately, the blockage of the pores by cell debris, water molecules or salt precipitates gradually increases the mass transfer resistance to the transfer of gas pollutants, which ultimately results in a reduced bioreactor performance. They are preferentially manufactured in teflon and polypropylene (Kumar *et al.* 2008).
- Dense membranes have no macroscopic pores and the selective separation of gas pollutants is based on solubility and diffusivity across the membrane Their non-porous nature makes them more resistant to biofouling, being manufactured in silicone, latex, polyethylene and polypropylene (Kumar *et al.* 2008).
- Composite membranes are comprised of a thin dense top layer (in direct contact with the biofilm or the aqueous phase) mounted on a porous membrane layer (in direct contact with the polluted gas emission). This type of membranes allows for an enhanced mass transfer through the pores while being protected from membrane blockage by the thin dense membrane. Polydimethylsiloxane, polyvinylidene and polyacrylamide are the preferred materials for their manufacture (Kumar *et al.* 2008).

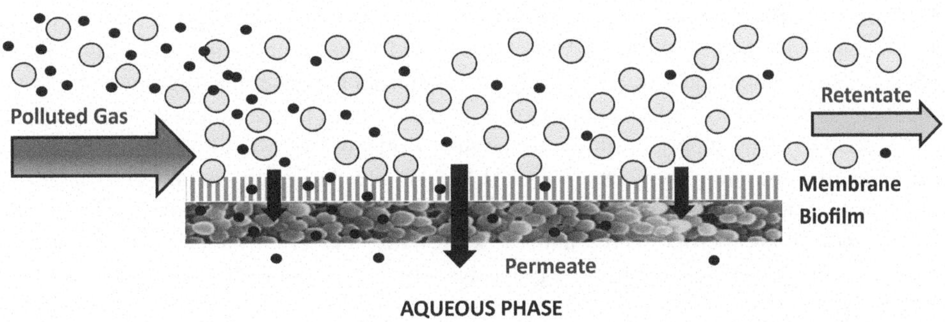

Figure 11.8 Mechanism of selective permeation of gas pollutants in off-gas treatment membrane bioreactors.

A case-specific balance between permeability and selectivity for the target gas pollutants and the long-term mechanical strength, determines the type and material of membrane selected.

This type of bioreactor has been implemented in plate and frame, tubular, capillary and hollow fibre membrane modules. However, while plate and frame membrane bioreactors exhibit an easy membrane replacement, their interfacial area is relatively low (100–400 m^2 m^{-3}) compared to capillary membranes (600–1200 m^2 m^{-3}) or hollow fibre (<30000 m^2 m^{-3}) bioreactors. This technology has been successfully evaluated for the treatment of dimethyl sulfide, dichloroethene, ethyl acetate, benzene, toluene and nitrous oxides using both pure and mixed bacterial cultures. This technology has not been tested yet for the removal of CH_4. Gas residence times of 2–30 s are often required to support an almost complete pollutant removal in membrane bioreactors at inlet loadings comparable to those typically applied in biofilters or biotrickling filters (Lebrero *et al.* 2013).

11.4.4 Activated sludge and oxidized ammonium recycling

Activated Sludge Recycling (ASR) has emerged in the past decade as an innovative technology for odour control in WWTPs. This biotechnology, which in fact can be classified as an odour prevention strategy rather than as an *end-of-pipe* technology, is based on the recirculation of activated bacterial sludge from the secondary clarifier or from the aerobic tank into the head works of the WWTPs (Estrada *et al.* 2015) (Figure 11.9). Hence, the recycled aerobic and anoxic heterotrophs present in the activated sludge are capable of biodegrading any inorganic or organic odorous compounds present in the wastewater, thus preventing their emission. The potentially recycled sludge flowrates can decrease by a factor of 2–3 when settled activated sludge (6–8 g TSS L^{-1}) rather than mixed liquor (2.5–3.5 g TSS L^{-1}) is recycled. The low investment and operating costs represent the main advantages of this biotechnology since there is no need for covers or gas pipelines (only a liquid pump and conventional liquid pipelines) and the operating costs are limited to the pumping of the activated sludge, which has flow rates and pressure drops that are lower than those often required during off-gas treatment. The amount of fundamental information on the microbial mechanisms of odorant removal is rather limited, but preliminary studies at pilot and full scale have been recently conducted in WWTPs with promising results. In this context, Schmidt *et al.* (2014) reported 87–98% reductions in the H_2S gas concentration and a 60% odour reduction by recirculating activated sludge to the inlet works at a ratio of 0.92 (recycled sludge:raw wastewater). Likewise, a 50% reduction in the odour emission rates was reported by Kiesewetter *et al.* (2012) by recycling activated sludge to the inlet works at ratios as low as 0.12–16.

Similarly, the recycling of residual process streams with oxidized NH_4^+ has been also tested as an alternative or complementary strategy to ASR. Both mixed liquor (containing NO_3^- concentrations of 6–10 mg N L^{-1}) and nitrified reject water (namely centrate) from the anaerobic digestion of mixed sludge (containing NO_3^- concentrations of 400–1000 mg N L^{-1}) have been tested as low cost source of electron acceptors (Figure 11.9). The supply of nitrate to the inlet works of the WWTP prevents the formation of septic conditions, thus avoiding the formation of H_2S, CH_4 or volatile fatty acids (Estrada *et al.* 2015). In addition, any odorant or the CH_4 already present in the receiving wastewater at the inlet works can be oxidized by the indigenous microflora present in the wastewater or externally supplied in the case of mixed liquor recycling. This odour prevention strategy has been mainly tested at full scale (all studies reported in the USA) with reductions in H_2S gas concentrations of 70% when recycling partially nitrified centrate (Husband *et al.* 2010) or reductions in the H_2S gas concentrations down to 0.1–0.5 ppm_v when recycling mixed liquor from the aeration tanks (at a ratio of 0.5) was applied, with associated reductions in the operating costs of the already existing scrubbers of ≈60%.

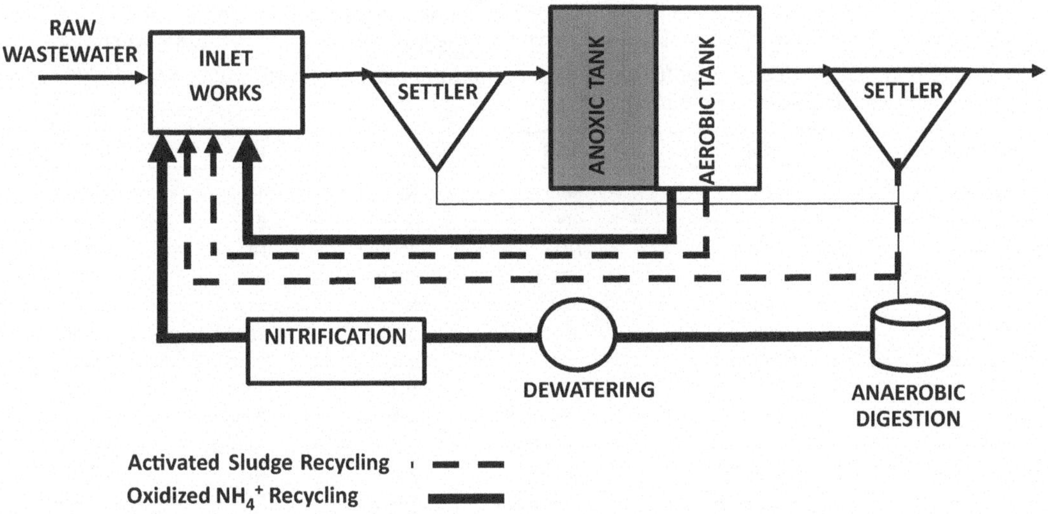

Figure 11.9 Schematic representation of a WWTP with activated sludge recycling and oxidized ammonium recycling.

11.4.5 Advantages and drawbacks of emerging biological technologies

Among the four EBTs discussed, only activated sludge diffusion, ASR and oxidized ammonium recycling have been successfully tested at full scale with satisfactory odour removal efficiencies. Their low operating and investment cost, since they are based on units or residual streams present in the plant at zero cost, makes them very attractive as odour control technologies to be considered during plant design. However, the lack of clear criteria for design and operation still hinder the uptake of these promising technologies by WWTP operators. In addition, the performance of these EBTs for the abatement of CH_4 has yet to be assessed. On the other hand, TPPBs and membrane bioreactors have been mainly operated at lab scale for the treatment of high concentrations of VOCs, and little experience exists nowadays for the treatment of malodorous emissions. The enhancement in the mass transport of gas pollutants can reach one order of magnitude compared to conventional aqueous phase systems during the treatment of VOCs, although more modest enhancements (40–80%) have been recorded during the treatment of CH_4. The high cost of the NAP and the technical difficulties to completely recover it rank among the main limitations of TPPBs, while biomass clogging and the high cost of the membrane modules are often pointed out as the main bottlenecks of membrane bioreactors despite their good performance at low gas residence times.

11.5 CONCLUSIONS

Characteristics, advantages and drawbacks of physical-chemical technologies and mature and emerging biological technologies to limit gas and odour emissions have been outlines in this chapter. A list of indicators for each of the technologies discussed in this chapter is provided in Table 11.2.

Table 11.2 Table for mini–assessment of the following technologies: adsorption (ADS), absorption (ABS), incineration (INC), biofilter (BF), biotrickling filter (BTFs), bioscrubber (BS), two–phase partitioning bioreactor (TPPB), activated sludge diffusion (ASD), membrane bioreactor (MB), activated sludge recycling (ASR) and oxidized sludge recycling (OAR).

	Parameter	ADS	ABS	INC	BF	BTF
1	H_2S LR[a]	375	550	ND	18	125
2	VOC LR[b]	3000	900	4500	8	62
3	GHG LR	3000	ND	4500	ND	ND
4	NH_3 LR	250	375	1000	7	75
5	H_2S RE	>99	>99	ND	>97	>97
6	VOC RE	>99	50–90	90–99	70–80	20–30
7	GHG RE	80–90	<5*	80–95	20–30*	<5*
8	NH_3 RE	>99	>99	ND	100	100
9	H_2S conc.[c]	150–13000	150–13000	ND	7–90	50–4500
10	VOC conc.	10–2000	50–500	<25% LEL	14–40	18–54
11	GHG conc.	2000	ND	2000	ND	ND
12	NH_3 conc.	40–2400	200–6000	50–5000	8–65	60–1800
13	H_2S EBRT[d]	2–8	1–3	1.5–10	2–25	2–180
14	VOC EBRT	2–8	4–8	1.5–10	25–70	4–12
15	GHG EBRT	2–8	4–8	1.5–10	>300	>180
16	NH_3 EBRT	2–8	1–3	1.5–10	3–25	2–60
	Flow rate able to be treated (m^3/h)	100000	500000	90000	350000	50000
17	Capital cost (€/m^3)[e]	10000	15000	16000	520–2800	15000
18	Energy cost (€/kg) **	<1	<1	3–8	<1	<1
Operating conditions						
19	pH	–	2–10	–	5–8	2–10
20	Reagents consumption	Adsorbant Regeneration solution	$3\cdot10^{-5}$ m^3 water/ m^3 air treated	Supplementary Fuel	$0.25\cdot m^3$ water/ m^2/h	$3\cdot10^{-5}$ m^3 water/ m^3 air treated
21	Pressure drop (mm wc)	200–500	40–100	100–500	2–10	20–80
Qualitative operational data						
22	Control needs	Adsorbant Saturation	pH ORP	T	Water content	pH
23	State of development (emergent, adaptive use, innovative, established)	Established	Established	Established	Established	Innovative
24	Robustness	High	High	High	moderate	Moderate
25	Stability	High	High	Moderate	moderate	High
26	Major drawback	Regeneration requirements	Reagents consumption	Fuel consumption	Packing degradation	Potential solids accumulation in the long–run
27	Compatibility with other technologies	Stand–alone or before BTF–BS	Stand–alone or before BS	Stand–alone	Stand–alone or after ADS–ABS–BTF–BS	Stand–alone
28	Problem frequency	Rare	Rare	Low	Low	Rare
29	Need of specialized staff	No	No	No	Yes	Yes

[a]LR: maximum volumetric loading rate (g X m^{-3} h^{-1}, where X indicates S, C or N) to avoid inhibition/clogging problems.

[b]Average molecular weight of VOCs of 94 g/mol at T = 25°C.

[c]Range of maximum inlet concentration (ppm_v) corresponding to the maximum recommended LR.

[d]Range of minimum EBRT (s) corresponding to the maximum inlet concentration range and maximum recommended LR.

[e]A standard deviation of ±30% is assumed.

*Based on CH_4 removal. Production of N_2O observed in bioreactors for NH_3 removal.

**Only those related to the reactor unit. Electrical costs of air/gas supply not considered.

BS	TPPB	ASD	MB	ASR	OAR
375	–	2.6	–	–	–
175	60–1200	0.8	0.15–320	–	–
ND	65–190	–	–	–	–
250	–	–	–	–	–
>97	–	99.9	–	<98	<70
20–30	30–99	>95	12–77	–	–
<5*	20–60	–	–	–	–
100	–	–	–	–	–
150–13,000	–	85	–	–	–
50–380	500–14,000	0.007	0.005–2.6	–	–
ND	550–20,000	–	–	–	–
200–6000	–	–	–	–	–
2–180	–	94	–	–	–
4–30	30–120	94	20–30 s	–	–
>180	240–600	–	–	–	–
2–60	–	–	–	–	–
50000	Systems limited to lab scale tests	Plant specific–Limited to aeration needs	Systems limited to lab scale tests	–	–
17,000	–	–	–	–	–
<1	–	–	–	–	
2–10	7	7	7	7	7
$3 \cdot 10^{-5}$ m3 water/ m^3 air treated	–	–	–	–	–
20–80	–	5000	<200	–	U to few meters for water pumping
pH	pH	D.O	Pressure drop	–	ORP
Innovative	Emergent	Emergent	Emergent	Emergent	Emergent
High	High	High	High	High	High
High	High	High	moderate	moderate	moderate
Limited RE for hydrophobic compounds	High cost of organic phases	Risk of sludge bulking	High investment costs expected	Poor know–how	Poor know–how
Stand–alone	Stand–alone	Stand–alone	Stand–alone	Complementary–Reduces odour treatmenent needs	Complementary–Reduces odour treatmenent needs
Rare	Rare	Rare	Low	Rate	Rare
Yes	Yes	No	Yes	No	No

11.6 REFERENCES

Baquerizo G., Maestre J. P., Machado V. C., Gamisans X. and Gabriel D. (2009). Long-term ammonia removal in a coconut fiber-packed biofilter: analysis of N fractionation and reactor performance under steady-state and transient conditions. *Water Research*, **43**, 2293–2301.

Bowker R. P. (2000). Biological odour control by diffusion into activated sludge basins. *Water Science and Technology*, **41**(6), 127–132.

Burgess J. E., Parsons S. A. and Stuetz R. (2001). Developments in odour control and waste gas treatment biotechnology: a review. *Biotechnology Advances*, **19**, 35–63.

Cantera S., Estrada J. M., Lebrero R., García-Encina P., Muñoz R. (2015). Comparative performance evaluation of conventional and two-phase hydrophobic stirred tank reactors for methane abatement: Mass Transfer and Biological Considerations. *Biotechnology and Bioengineering*, **113**(6), 1203–1212.

Card T. (2001). Chemical odour scrubbing systems. In: Odour in Wastewater Treatment: Measurement, Modelling and Control, R. Stuetz and F.B. Frechen (eds), IWA Publishing, London, pp. 308–344.

Centi G., Genertali P. dall'Olio, L. and Perathoner S. (2000). Removal of N_2O from industrial gaseous streams by selective adsorption over metal-exchanged zeolites. *Ind. Eng. Chemical Research*, **39**, 131–137.

Chaiprapat S., Mardthing R., Kantachote D., Karnchanawong S. (2011). Removal of hydrogen sulfide by complete aerobic oxidation in acidic biofiltration. *Processes Biochemical*, **46**, 344–352.

Clark R. M. (1987). Evaluating the cost and performance of field-scale granular activated carbon, *Environmental Science and Technology*, **21**, 573–580.

Deshusses M. A. and Gabriel D. (2005). Biotrickling filter technology. In: Biotechnology for Odor and Air Pollution Control, Shareefdeen Z. and Singh A. (eds), Springer, Berlin, pp. 147–167.

Devinny J.S., Deshusses M. A. and Webster T. S. (1999). Biofiltration for air Pollution Control. Lewis Publishers, Boca Raton. Florida, EEUU.

Dillon C. D. and Kyi M. (2013). Performance and stress testing of a large-scale biotrickling scrubbers. Proceedings of the 5th IWA Conference on Odors and Air Emissions + 10th Conference on Biofiltration for Air Pollution Control, San Francisco, CA.

Dorado A. D., Lafuente J., Gabriel D. and Gamisans X. (2010). The role of water in the performance of biofilters: parametrization of pressure drop and sorption capacities for common packing materials. *Journal of Hazardous Materials*, **180**, 693–702.

Dorado A. D., Lafuente J., Gabriel D. and Gamisans X. (2012). Biomass accumulation in a biofilter treating toluene at high loads – Part 1: Experimental performance from inoculation to clogging. *Chemical Engineering Journal*, **209**, 661–669.

Dorado A. D., Husni S., Pascual G., Puigdellivol C. and Gabriel D. (2014). Inventory and treatment of compost maturation emissions in a municipal solid waste treatment facility. *Waste Management*, **34**, 344–351.

Estrada J. M., Kraakman N. J. R. B., Muñoz R. L. and Lebrero R. (2011). A comparative analysis of odour treatment technologies in wastewater treatment plants. *Environmental Science and Technology*, **45**(3), 1100–1106.

Estrada J. M., Lebrero R., Quijano G., Kraakman N. J. R. and Muñoz R. (2013). Strategies for odour control. In: Odour Impact Assessment Handbook, V. Belgiorno, V. Naddeo and T. Zarra (eds.), John Wiley & Sons, Ltd., West Sussex, UK, pp. 85–124.

Estrada J. M., Kraakman N. J. R., Lebrero R. and Muñoz R. (2015). Integral Approaches to Water Resource Recovery Facilities Upgrading for Odor Prevention: Activated Sludge and Oxidized Ammonium Recycling. *Bioresource Technology*, **196**, 685–693.

European Comission (ed.) (2003). EC, Integrated Pollution Prevention and Control. Reference Document on Best Available Techniques in Common Waste Water and Gas Treatment/Management Systems on the Chemical sector.

Fernandez M., Ramirez M., Perez R. M., Gomez J. M. and Cantero D. (2013). Hydrogen sulphide removal from biogas by an anoxic biotrickling filter packed with Pall rings. *Chemical Engineering Journal*, **225**, 456–463.

Fortuny M., Baeza J. A., Deshusses M. A., Gamisans X., Casas C., Lafuente J. and Gabriel, D. (2008). Biological sweetening of energy gases mimics in biotrickling filters. *Chemosphere*, **71**, 10–17.

Fortuny M., Gamisans X., Deshusses M. A., Lafuente J., Casas C. and Gabriel, D. (2011). Operational aspects of the desulfurization process of energy gases mimics in biotrickling filters. *Water Research*, **45**, 5665–5674.

Gabriel D., Deshusses M. A. (2003). Retrofitting existing chemical scrubbers to biotrickling filters for H2S emission control. *Proceedings of the National Academy of Sciences of the USA*, **100**, 6308–6312.

Gabriel D. and Deshusses M. A. (2004). Technical and economical analysis of the conversion of a full-scale scrubber to a biotrickling filter for odor control. *Water Science Technolology*, **50**(4), 309–318.

Gabriel D., Cox H. H. J. and Deshusses M. A. (2004). Conversion of full-scale wet scrubbers to biotrickling filters for H2S control at publicly owned treatment works. *Journal Environmental Eng-ASCE*, **13**, 1110–1117.

Gabriel D., Deshusses M. A. and Gamisans, X. (2013). Desulfurization of biogas in biotrickling filters. In: Air Pollution Prevention and Control, C. Kennes and M. C. Veiga (eds), John Wiley & Sons, Ltd., West Sussex, UK, pp. 513–524.

Gamisans X., Sarrà M. and lafuente F. J. (2002). Gas pollutants removal in a single- and two-stage ejector-venturi scrubber. *Journal of Hazardous Materials*, **90**(3), 251–266.

Hardy P. B. (2001). Simultaneous activated sludge wastewater treatment and odour control. *Water Science Technolology*, **44**(9), 189–196.

Hernandez M., Muñoz R. and Daugulis A. (2011). Biodegradation of VOC mixtures of different hydrophobicities in two-phase partitioning bioreactors containing tailored polymer mixtures. *Journal Chemical. Technolology Biotechnology*, **86**, 138–144.

Hernandez M., Quijano G. and Muñoz R. (2012). The key role of microbial characteristics on the performance of VOC biodegradation in two-liquid phase bioreactors. *Environmental Science Technology*, **46**, 4059–4066.

Husband J. A., Phillips J., Coughenour J. R., Walz T. and Blatchford G. (2010). Innovative approach to centrate nitrification accomplishes multiple goals: nitrogen removal and odour control. *Water Science Technology*, **61**(5), 1097–1103.

Kennes C. and Veiga M. C. (2001). Bioreactors for Waste Gas Treatment. Kluwer Academic Publishers, The Netherlands.

Kennes C., Rene E. R. and Veiga M. C. (2009). Bioprocesses for Air Pollution Control. *Journal Chemical Technology Biotechnology*, **84**, 1419–1436.

Kraakman N. J. R., Rocha-Rios J. and van Loosdrecht M. C. M. (2011). Review of mass transfer aspects for biological gas treatment. *Appl Microbiol Biotechnology*, **91**, 873–886.

Kiesewetter J., Kraakman B., Cesca J., Trainor S. and Witherspoon J. (2012). Expanding the use of activated sludge at biological waste water treatment plants for odor control. WEF Odors and Air Pollutants Conference, Lousiville, Kentucky.

Kumar A., Dewulf J. and Van Langenhove H. (2008). Membrane-based biological waste gas treatment. *Chemical Engineering Journal*, **136**, 82–91.

Kumar R., Huggahalli M., Deng S. and Andrecovich M. (2003). Trace Impurity Removal from Air. *Adsorption*, **9**, 243–250.

Lebrero R., Rodriguez E., Martín M., García-Encina P. and Muñoz R. (2010). H_2S and VOCs abatement robustness in biofilters and air diffusion bioreactors: a comparative study. *Water Research*, **44**, 3905–3914.

Lebrero R., Muñoz R., Kumar A. and Van Lagenhove H. (2013). Membrane bioreactors for waste gas treatment. In: Air Pollution Prevention and Control - Bioreactors and Bioenergy, C. Kennes and M. C. Veiga (eds), Wiley-Blackwell, pp. 155–184.

Lebrero R., Gondim A. C., Perez R., García-Encina P. and Muñoz R. (2014). Comparative assessment of a biofilter, a biotrickling filter and a hollow fiber membrane bioreactor for odour treatment in wastewater treatment plants. *Water Research*, **49**, 339–350.

Littlejohns J. V. and Daugulis A. J. (2009). A two-phase partitioning airlift bioreactor for the treatment of BTEX contaminated gases. *Biotechnology and Bioengineering*, **103**, 1077–1086.

McCabe W. L., Smith J. C. and Harriott P. (2007). Unit Operations of Chemical Engineering, 7th ed., McGraw-Hill, New York.

Montebello A. M., Fernandez M., Almenglo F., Ramirez M., Cantero D., Baeza M. and Gabriel D. (2012). Simultaneous methylmercaptan and hydrogen sulfide removal in the desulfurization of biogas in aerobic and anoxic biotrickling filters. *Chemical. Engineering Journal*, **200**, 237–246.

Montebello A. M., Bezerra T., Rovira R., Rago L., Lafuente J., Gamisans X., Campoy S., Baeza M. and Gabriel D. (2013). Operational aspects, pH transition and microbial shifts of a H_2S desulfurizing biotrickling filter with random packing material. *Chemosphere*, **93**(11), 2675–2682.

Montebello A. M., Mora M., López L. R., Bezerra T., Gamisans X., Lafuente J., Baeza M. and Gabriel D. (2014). Aerobic desulfurization of biogas by acidic biotrickling filtration in a randomly packed reactor. *Jouranl Hazardous Materials*, **280**, 200–208.

Muñoz R., Villaverde S., Guieysse B. and Revah S. (2007). Two phase partitioning bioreactors for the treatment of volatile organic compounds. *Biotechnology Advances*, **25**(4), 410–422.

Muñoz R., Daugulis A., Hernandez M. and Quijano G. (2012). Recent advances in two-phase partitioning bioreactors for the treatment of volatile organic compounds. *Biotechnology Advances*, **30**, 1707–1720.

Prado O. J., Gabriel D. and Lafuente J. (2009a). Economical assessment of the design, construction and operation of open-bed biofilters for waste gas treatment. *Journal Environmental Management*, **90**, 2515–2523.

Prado O. J., Redondo R. M., Lafuente J. and Gabriel D. (2009b). Retrofitting of an industrial chemical scrubber into biotrickling filter: performance at a gas contact time below 1 s. *Journal Environmental Eng-ASCE*, **135**, 359–366.

Quijano G., Rocha Rios J., Hernandez M., Villaverde S., Revah S., Muñoz R. and Thalasso, F. (2010). Determining the effect of solid and liquid vectors on the gaseous interfacial area and oxygen transfer rates in two-phase partitioning bioreactors. *Journal Hazardous Materials*, **175**, 1085–1089.

Saha D., Zongbi B., Jia F. and Deng S. (2010). Adsorption of CO_2, CH_4, N_2O and N_2 on MOF-5, MOF-177 and Zeolite 5ª. *Environmental Science Technology*, **44**, 1820–1826.

Sánchez A., Artola A., Font X., Gea T., Barrena R., Gabriel D., Sánchez-Monedero D., Roig, A., Cayuela M. L. and Mondini C. (2015). Greenhouse gas emissions from organic waste composting. *Environmental Chemical Letters*, **13**, 223–238.

Santos A., Guimerà X., Dorado A. D., Gamisans X. and Gabriel D. (2015). Conversion of chemical scrubbers to biotrickling filters for VOCs and H2S treatment at low contact times. *Applied Microbiology and Biotechnology*, **99**, 67–76.

Schmidt D., Karkeskint J., Porter J., Parker D. and Hoskins K. (2014). Mixing RAS with sewage to reduce odor emissions while avoiding adverse impacts to processes. Proceedings of the WEF Odors and Air Pollutants Conference, 2014, Miami, Florida.

Soreanu G., Beland M., Falletta P., Edmonson K. and Seto P. (2008). Laboratory pilot scale study for H_2S removal from biogas in an anoxic biotrickling filter. *Water Science Technology*, **57**(2), 201–207.

Tchobanoglous G., Burton F. L. and Stensel H. D. (2003). Wastewater Engineering: Treatment and Reuse. McGraw-Hill, New York.

Turk A. and Bandosz T. J. (2001). Adsorption systems for odour treatment. In: Odour in Wastewater Treatment: Measurement, Modelling and Control, R. Stuetz and F. B. Frechen (eds), IWA Publishing, London, pp. 345–364.

US EPA. (1985). Design Manual for Odor and Corrosion Control in Sanitary Sewerage Systems and Treatment Plants. EPA/625/1-85/018. US Environmental Protection Agency.

Van der Heyden C., Demeyer P. and Volcke E. I. P. (2015). Mitigating emissions from pig and poultry housing facilities through air scrubbers and biofilters: State-of-the-art and perspectives. *Biosystems Engineering*, **134**, 74–93.

Vincent A. J. (2001). Sources of odours in wastewater treatment. In: Odours in Wastewater Treatment: Measuring, Modelling and Control, R. Stuetz and F. B. Frechen (eds), IWA Publishing, London, pp. 69–92.

Volesky B. (2003). Sorption and Biosorption. B. V. Sorbex (ed.), Hoben, NJ, USA.

Wang J. B. and Chou M. S. (2000). Kinetics of catalytic oxidation of benzene, n-hexane, and emission gas from a refinery oil/water separator over a chromium oxide catalyst. *Journal of the Air & Waste Management*, **50**, 227–233.

Yang Y. H., Allen E. R. (1994). Biofiltration Control of Hydrogen Sulfide 1: Design and Operational Parameters. *Journal of the Air & Waste Management*, **44**, 863–868.

Chapter 12

Reducing the impact of sludge

Sara Isabel Pérez-Elvira, Fernando Fdz-Polanco, Maria Concetta Tomei, Giorgio Bertanza, Camilla Maria Braguglia, Andrea Gianico and Hélène Carrère

12.1 INTRODUCTION

For several reasons, such as reliability, simplicity and yield, the activated sludge process is the most widely used for biological wastewater treatment in the world, but it results in the generation of a considerable amount of excess sludge that has to be disposed of. This sludge contains high fractions of volatile solids (VS) and retain large amounts of water (>95% by weight), resulting in extremely large volumes of residual solids produced, and significant disposal costs. In fact, treatment and disposal of excess sludge from WWTPs account for 25–65% of the total plant operation cost. Thereby, the conventional method converts a water pollution problem into a solid waste disposal problem.

This problem is becoming more and more pressing both in developing and industrial countries. In the latter, the disposal of excess sludge is one of the most serious challenges in biological wastewater treatment, for two main reasons:

(1) New wastewater treatment regulations are causing a rise in the number of plants. This increase in the number of WWTPs is translated into a higher production of sewage sludge.
(2) Sludge disposal routes such as incineration or landfilling are being subjected to more stringent environmental quality requirements. Land application is considered as the most sustainable alternative. However, legislation concerning land application of sludge is being tightened in order to prevent health risks to man and livestock due to the potentially toxic elements in the sewage sludge, i.e. heavy metals, pathogens and persistant organic pollutants.

Therefore, the current legal constraints, the rising costs and public sensitivity of sewage sludge disposal necessitate the development of strategies for reduction and minimisation of excess sludge production. Reducing sludge production in the wastewater treatment instead of post-treating the sludge produced appears to be an ideal solution to this issue, because the problem would be treated at its roots. Several strategies are currently being developed for minimisation of sludge production on biological wastewater treatment plants. In this paper we will give an overview of processes aimed to reducing the impact of sludge.

The existing alternatives for sludge reduction can be classified according to the place of the plant where the minimisation takes place (Pérez-Elvira *et al.* 2006). Figure 12.1 shows potential locations for sludge disintegration units in a classical urban wastewater treatment plant.

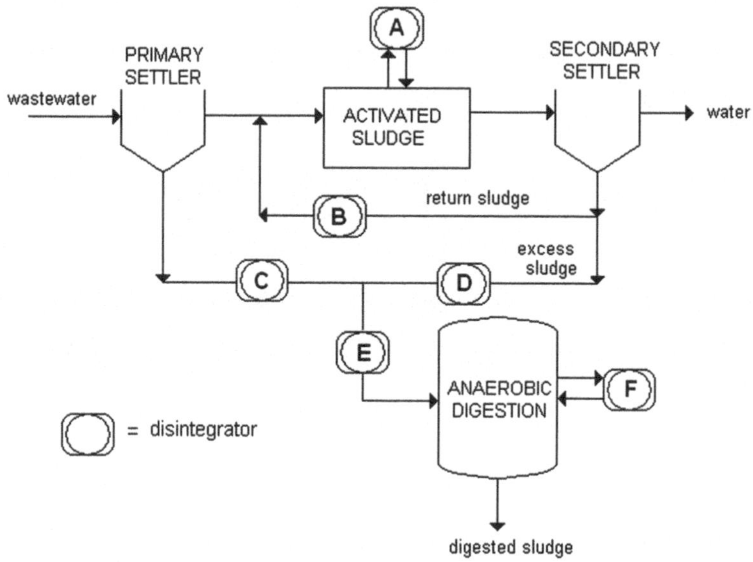

Figure 12.1 Possible location of sludge disintegration processes in a plant's layout.

Four main strategies are identified (Table 12.1): processes in the water line, pre-treatment processes in the sludge line, enhanced digestion or final sludge removal alternatives.

- *Processes in the water line*: The idea is to reduce sludge production in the wastewater treatment rather than the post-treatment of sludge after generation, by introducing a treatment unit in the water line. This unit can be located either directly in the aeration tank (A) or in the sludge recirculation loop, after thickening (B).
- *Pre-treatment processes in the sludge line*: Anaerobic digestion shows certain limitations in the first hydrolytic step, leading to slow degradation of the organic matter and too high retention times. In order to improve the kinetics of anaerobic biodegradation of sludge, a pre-treatment process can accelerate the hydrolysis step enhancing biogas productivity as well as improve the characteristics of the digested sludge. This pre-treatment unit preceding sludge anaerobic digestion, may be applied either to primary sludge (C), to excess waste activated sludge (WAS) (D) or to the mix of primary and WAS (E). Another option is to place the treatment unit in the recirculation loop of the digester to treat digested sludge (F).
- *Enhanced sludge stabilization*: The conventional mesophilic anaerobic digestion of sludge can also be enhanced by modifying the digestion temperature or combining digestion phases, at different temperatures (G).
- *Final sludge removal*: A combination of the wet oxidation with anaerobic digestion of the liquid residue (H) is reported as an example of a promising sustainable alternative to mitigate the drawback of energy consumption.

Table 12.1 Classification of existing processes to reduce sludge impact.

Location in the WWTP	Principle	Process
Water line (A, B)		
	Lysis cryptic growth	Chemical oxidation
		Enzymatic reactions
		Mechanical treatment
	Maintenance metabolism	Extended aeration process
		Membrane bioreactor
	Uncoupling metabolism	Chemical uncoupler
		Side stream anaerobic reactor
	Predation on bacteria	Oligochaetes (worms)
Sludge line		
Enhanced Hydrolysis (C,D,E, F)	Physical pre-treatments	High Pressure homogenizers
		Ultrasonic treatment
		Thermal hydrolysis
		Focused-pulse technology
		Grinding-Stirred ball mills
		Lysis centrifugation
	Chemical pre-treatments	Oxidation
		Alkaline hydrolysis
	Biological pre-treatments	Enzymatic hydrolysis
	Treatment unit in the digester recirculation loop	Physical, chemical, biological
Enhanced Stabilization (G)	Thermophilic digestion with pre-treatment	Pre-treatment – Thermophilic anaerobic digestion
	Temperature phased anaerobic digestion	Multi-step anaerobic digestion at different temperatures
	Sequential anaerobic-aerobic digestion	Combination of different reaction environment aerobic and anaerobic at differerent temperature
Final sludge removal (H)	Wet oxidation coupled with anaerobic digestion of residue	

In the discussion that follows, processes are briefly reviewed in order to show a comparison among them. The purpose of this evaluation is to outline consolidated and emerging technologies, and to provide a general dialog for categorising processes. Information was obtained from personal research activities and reported experience.

12.2 PROCESSES IN THE WATER LINE (A,B)

As presented in Table 12.1, there are several processes to minimise the excess sludge production during secondary WWT rather than the post-treatment of sludge generated. As shown in Figure 12.1, the different processes can be located either directly in the aeration tank (A) or in the sludge recirculation loop, after thickening (B). As will be latter discussed, although both configurations have scarce real applicability, location B is preferred to A in order to minimize the treatment cost.

12.2.1 Lysis-cryptic growth

When certain external forces are applied, microbial cells undergo lysis or death during which cell contents (substrates and nutrients) are released into the medium, providing an autochthonous substrate that is used in microbial metabolism (Mason, 1986). The biomass growth due to this substrate is termed as cryptic growth (Mason, 1987).

There are two stages in lysis-cryptic growth: lysis (which is the rate-limiting step) and biodegradation (Müller, 2000). The approach to optimize the lysis step in the aeration tank by an external disintegration treatment can be performed with different mechanisms (mechanical, thermal, chemical or biological), below described.

12.2.1.1 Chemical oxidation

Chemical or thermo-chemical treatments are based on the addition of alkaline or acid substances. The working principle is as follows: a fraction of activated sludge in the aeration tank is treated with chemicals in a reactor (where most activated sludge microorganisms would be killed and oxidized to organic substances), and then, these organic substances produced from the sludge chemical treatment will be degraded in the subsequent biological treatment.

The most promising results referenced at laboratory scale correspond to alkaline reagents (NaOH), chlorination and ozonation. Chlorine is a lower cost alternative to ozone. With *NaOH*, optimal conditions obtained were pH > 10, temperature >50–60°C, contact time less than 1 h, since longer time did not improve solubilisation effectively. *Chlorination* treatment of excess sludge resulted in a 60% reduction of excess sludge (Chen *et al.* 2001; Saby *et al.* 2002), at the chlorine dose of 0,066 g Cl_2/g MLSS. For ozonation several references (Yasui & Shibata, 1994; Sakai *et al.* 1997; Kamiya & Hirotsuki 1998; Egemen *et al.* 1999; Egemen *et al.* 2001; Ahn *et al.* 2002; Böhler & Siegrist 2003) report that the excess sludge production was reduced by 50% at an ozone dose of 10 mg/g MLSS in the aeration tank per day, while no excess sludge was produced when the ozone dose was kept as high as 20–50 mg/g MLSS.

Regarding full-scale, *Biolysis O* is an ozonation process developed by Ondeo-Degrémont to reduce sludge generation. The recommended ozone dosage is in the range 0.03–0.05 gO_3/gTSS produced (much smaller than lab-scale), resulting in sludge reductions of between 30–80%. At the moment, sludge ozonation is economically sustainable for WWTPs with large capacity or in the areas where sludge disposal costs are very high, or in the case of operational problems such as sludge foaming and bulking.

Positive (✓) and negative (×) aspects of the ozonation technology are:

- ✓ Sludge settleability improvement (in terms of SVI).
- ✓ Successful full-scale experience.
- × High costs involved in chemical treatment.
- × Consumption of chemicals in the degradation of other possible organic materials that may be present.
- × Significant increase of soluble chemical oxygen demand in the effluent.
- × More research needed, regarding dosage and reactor configuration.

12.2.1.2 Enzymatic reactions

Enzymatic reactions are biological processes based on hydrolytic enzymes which adsorb to the sludge and attack the polymeric substances leading to solid solubilisation and biodegradation enhancement. Considering the high presence of proteins, carbohydrates and lipids in the composition of excess sludge, the addition of enzymes such as protease, lipase, cellulase, hemicellulase and amylase could be advisable. The enzymatic treatment can be achieved with added commercial enzymes (generally patented and confidential) or by promoting the enzyme activity (instead of buying them) in the activated sludge system.

This second option is commercialized: *process S-TE* (Shiota *et al.* 2002; Sakai *et al.* 2000) and *Biolysis E* (by Ondeo-Degrémont). Both technologies combine an activated sludge system with a thermophilic aerobic sludge digester for biomass solubilisation. A fraction of return sludge from the wastewater treatment is injected into a thermophilic aerobic sludge digester (50–60°C), in which the sludge is solubilised by a particular type of microbe (*Bacillus stearothermophillus*) that produces enzymes. The solubilised sludge is returned to the aeration tank for its further degradation. Pilot scale facilities showed up to 93% reduction in the overall excess sludge production, and a full-scale plant treating domestic sewage was operated for three years, showing from 30 to 80% reduction of overall excess sludge production (depending on the quantity of the sludge sent daily to the reactor).

Positive (✓) and negative (×) aspects of the enzymatic reaction process are:

- ✓ Full-scale experience.
- ✓ Decrease in the growth of filamentous organisms.
- ✓ Average cost (investment and operation) similar to or lower than that of classic treatment systems.
- × Small increases of the effluent SS and COD concentrations.
- × Variable results, not predictable neither reproducible.
- × High enzyme dosages needed.
- × Full-scale feasibility not yet completely demonstrated.

12.2.1.3 Mechanical treatment

The mechanical disintegration of sludge is aimed to enhance sludge solubilisation as a consequence of the bacteria cell disintegration and the disaggregation of biological flocs. The process is based on the placement of a device in which energy is supplied as pressure or rotational/translation movement. Sludge is disintegrated and the lysate obtained is recirculated into the activated sludge reactors. In general, at low applied energy only floc disintegration is observed, while high energy is required to damage microbial cells.

The systems proposed for mechanical disintegration can be milling, high pressure and cavitation (Vaxelaire *et al.* 2008). The different options differ widely with regards to configuration, operational conditions, and energy consumption. All of them are highly energy consuming, and therefore no full-scale application is reported.

Positive (✓) and negative (×) aspects of the mechanical treatment are are:

- × Huge operation cost (treatment of large quantities of sludge).
- × Sludge concentration as a key technical and economic issue.
- × Increases of SS and COD concentrations in the final effluent.
- × Worsening of sludge settleability.

12.2.2 Maintenance metabolism

Maintenance metabolism refers to the fraction of energy source of microorganisms used for maintaining living functions (turnover of cell materials, active transport, motility). Thus, the sludge production should be inversely related to the activity of maintenance metabolism (Chang *et al.* 1993). A significant reduction of sludge production can be achieved by maximising the energy used for maintenance requirements rather than for cellular synthesis (Martinage & Paul, 2000).

By increasing the biomass concentration (controlling sludge retention time or sludge loading rate) it would be theoretically possible to reach a situation in which the amount of energy provided equals the maintenance demand. This can be achieved by extended aeration processes. Canales *et al.* (1994), demonstrated that higher sludge ages increased the biomass viability. However, these processes have a very high footprint and energy demand. The only possibility to achieve full oxidation in a more compact plant is to be able to retain higher quantity of sludge per unit volume, or to increase the oxidation. While the latter can be obtained adopting pure oxygen processes, the high sludge age strategy can be implemented by adopting membrane bioreactors. In a membrane reactor, solids retention time (SRT) can be controlled independently from hydraulic retention time (HRT), which will result in a higher sludge concentration (typically 15–20 g/L), and subsequently in a lower sludge loading rate. When this sludge loading rate becomes low enough, little or no excess sludge is produced (Ghyooy & Verstraete, 2000; Wagner & Rosenwinkel, 2000; Rosenberger *et al.* 2002), but this option is quite expensive in terms of energy requirements. Full scale results are reported by Churchouse and Wildgoose 1999.

Positive (✓) and negative (×) aspects of membrane bioreactors are:

- ✓ Small footprint.
- ✓ Flexibility of operation.
- × Sludge settling and dewatering becomes worse.
- × *Poor oxygenation*: increased aeration cost.
- × Membrane fouling, which requires frequent cleaning and replacement (high cost).
- × Not feasible to operate membrane bioreactors with complete sludge retention in practice.
- × Energy requirements.

12.2.3 Uncoupling metabolism

In an environmental engineering sense, the concept of energy uncoupling can be extended to the phenomenon in which the rate of substrate consumption is higher than that required for growth and maintenance. As a result, under energy uncoupling conditions the observed growth yield of activated sludge would be reduced markedly (Tsai & Lee, 1990; Mayhew & Stephenson, 1998; Liu, 2000). Uncoupled metabolism is observed under some conditions, such as: existence of inhibitory compounds or heavy metals, abnormal temperature, excess energy source, limitation of nutrients, and alternative aerobic-anaerobic cycle.

12.2.3.1 Chemical uncoupler

Organic protonophores carry protons through cells' intracellular cytoplasm membrane such as 2,4-dinitrophenol (dNP), para-nitrophenol (pNP), pentachlorophenol (PCP) and 3,3', 4',5-tetrachlorosalicylanilide (TCS) which result in the oxidation of organic substrate to carbon dioxide rather than used for biosynthesis. As a result, the growth efficiency can be lowered.

However, although research with organic protonophores has shown that the dissipation of energy through oxidative phosphorylation can directly reduce biomass production (Okey & Stensel, 1993; Low & Chase 1998; Mayhew & Stephenson 1998; Low *et al.* 2000; Chen *et al.* 2000), the current use of organic protonophores is impractical for several reasons, which include the inherent toxicity of protonophores. Because of this, the removal of the additives is required prior to discharge. Further experimentation to establish alternative methods of uncoupling metabolism is desirable.

Positive (✓) and negative (×) aspects of chemical uncouplers are:

- ✓ It only needs to add a defined uncoupler dosing.

× Little is known about the uncoupling mechanisms and the connections between chemical uncouplers impact on sludge yield and process conditions.
× Most of the organic protonophores are xenobiotic and potentially harmful to the environment.
× Unexpected increase in the O_2 requirement (obtained in full-scale application).
× Acclimation problems for the microorganisms.

12.2.3.2 Side stream anaerobic reactor

The idea is to integrate an anaerobic reactor (at ambient temperature) in an activated sludge process. Three technologies are identified: The Oxic-Settling-Anaerobic process *(OSA®)*, The ANaerobic ANoxic-OXic process (*ANANOX®*), ad the *Cannibal®* process.

The three systems have so far been thoroughly tested on laboratory scale pilot prototypes (Chudoba, 1991; Chudoba *et al.* 1992a, b; Ghiglizza *et al.* 1996; Copp & Dold, 1998; Chen *et al.* 2003; Coma *et al.* 2013). However, in view of industrial scale application, further investigation on the carbon balance and microbial examination population is needed to understand the process.

Positive (✓) and negative (×) aspects of the side stream anaerobic reactor are:

✓ Easy to introduce the anaerobic zone to the conventional activated sludge process.
✓ Control of the growth of filamentous organisms.
✓ No physical or chemical forces are needed.
✓ Improved COD removal and settle ability of activated sludge.
× Further research is needed to understand the process.
× Optimum operational conditions not defined.
× Odour generation.
× Not recommended for very low sewage temperatures.

12.2.4 Predation on bacteria

Considering a biological wastewater treatment process as an artificial ecosystem (habitat for bacteria and other organisms), sludge production could be reduced by bacteriovory. Both living and death bacteria can be utilised as a food source by higher bacteriovoric microorganisms, such as protozoa (ciliates, flagellates, amoeba and heliozoa) and metazoa (rotifera and nematoda), that predate on the bacteria. Protozoa are considered to be the most common predators of bacteria, making up around 5% of the total dry weight of a wastewater biomass.

Two stage systems and Oligochaetes have been investigated. Main research on the predation on bacteria can be found in Welander and Lee, 1994; Lee and Welander, 1996a, b; Rensik and Rulkens, 1997; Luxmy *et al.* 2001; Ratsak *et al.* (1993). From 12 to 80% reduction in the overall biomass production have been obtained in laboratory tests with a mixed microbial culture.

The use of predatory activity to reduce the overall biomass production requires some caution. Despite efforts to control the growth and reproduction of predators in the biological systems, the conclusion is that it is very difficult to manage predators directly within activated sludge. The feasibility at full-scale has not been fully tested in municipal wastewater with nutrient removal.

Positive (✓) and negative (×) aspects of predation on bacteria are:

✓ Protozoa and metazoa are already present in the activated sludge process.
× The worms growth is still uncontrollable, specially in the full-scale application.
× High capital and operation costs.

12.3 PRE-TREATMENT PROCESSES IN THE SLUDGE LINE (C,D,E,F)

The introduction of a pre-treatment process prior to anaerobic digestion is nowadays one of the most interesting option to optimize the sludge digestion and subsequent wastewater treatment plant energy balance (Carrère *et al.* 2010).

By means of an efficient pre-treatment, the substrate can be made better accessible to the anaerobic bacteria, accelerating the digestion process (enhance the solubility of sludge solids), increasing the degree of degradation (increment of methane production), and consequently decreasing the amount of sludge to be disposed of. Other benefits can be, depending on the pre-treatment, the improvement of sludge dewatering, the reduction of pathogens, or the suppression of foaming.

These sludge disintegration processes are based on physical, thermal, chemical, biological techniques, or a combination of them, and can treat either primary sludge (C), excess WAS (D) or the mix of primary and WAS (E). Since primary sludge is already readily degradable, pretreatment may be less effective (Ge *et al.* 2010). WAS however, has relatively low degradability, especially at long sludge ages (Gosset & Belser, 1982). In addition, WAS is generally hydrolysis limited. Thus, activated sludge pretreatment (D) is mostly used in preference to primary sludge pretreatment (C). Pretreatment of mixed sludge (E) may be useful when the treatment also leads to sludge sanitation (for example thermal treatment). Finally, a hydrolysis unit can be implemented in the recirculation loop of the digester (F). This is where degradability of inert or slowly degradable material is to be enhanced, as the digester has already removed readily biodegradable material.

12.3.1 Physical pre-treatments

The disintegration of solid particles present in the sludge releases cell compounds and creates new surface where biodegradation takes place. Various disintegration machines are investigated in research projects. All of them are presented here.

12.3.1.1 High pressure homogeneizers

These units consist of a multistep high-pressure-pump and a homogenising valve. The pump compresses the suspension to pressures up to several hundred bars. When passing through the homogenising valve, the pressure drops below the vapour pressure of the fluid, and the velocity increases up to 300 m/h. The cavitation bubbles formed implode, inducing into the fluid temperatures of several hundred °C, which disrupts the cell membranes.

This process has been developed at full-scale, named: **Crown® process** (Biogest, 2010), **Cellruptor** or Rapid non-equilibrium decompression, RnD® process (Ecosolids, 2010) and **Microsludge® process** (Paradigm Environmental Technologie Inc.), and claim to nearly double biogas production and solids removal.

Positive (✓) and negative (×) aspects of high pressure homogenizers are:

✓ Compact equipment.
✓ Better dewaterability of the final sludge.
× Low reduction of pathogens.
× Clogging problems caused by coarse and fibrous particles.
× High maintenance cost.
× High tensions and erosion in the pump and homogenising valve.
× High operation cost.

12.3.1.2 Ultrasonic treatment

The ultrasonic disintegration treatment consists of an ultrasound generator operating at frequencies of 20–40 kHz to transmit mechanical impulses to the bulk liquid through a sonotrode. Pressure waves lead to cavitation bubbles forming in the liquid phase, which grow and then implode releasing localised high energy (local heating and high pressure), which cause sludge disintegration and, at high energy, the rupture of microbial cells. Since the most important mechanism of ultrasonic disintegration is ultrasonic cavitation, it is advantageous to apply ultrasounds at low frequencies (20–40 kHz) and at high energy levels.

Ultrasonic treatment acts to mechanically disrupt the cell structure and floc matrix. The mechanical phenomena of sludge sonication leads to sludge floc disintegration and microorganisms lysis, according to the treatment time and power, equating to specific energy applied. A threshold for specific energy is often reported for sludge solubilisation (Chu *et al.* 2002; Li *et al.* 2009), ranging from 1,000 to 16,000 kJ kg^{-1} TS and depends on sludge TS concentration. Indeed, the higher the sludge concentrations, the lower the specific threshold energy, since cavitation bubbles have higher probabilities of contacting sludge particles. However, if the solids concentration is too high, increased viscosity hinders cavitation bubble formation. For a given specific energy, power input is more effective than retention time (Gronroos *et al.* 2005). Taken as a whole, biogas enhancement ranges from 24% to 140% in batch systems and from 10% to 45% in continuous or semi-continuous systems. Sludge dewaterability and settling may be improved by sonication (Kim & Kim, 2003; Feng *et al.* 2009). Another advantage of sonication is the mitigation of sludge bulking problems (Wunsch *et al.* 2002; Neis *et al.* 2008) and potential digester foaming.

There are some commercial devices, such as Sonix (Sonico, UK), Biosonator (Ultrawaves, Germany), Smart DMS (Weber Ultrasonics) and Sonolyzer (Ovivo), reporting an energy consumption impressively smaller compared to those employed in laboratory experiences.

Positive (✓) and negative (×) aspects of ultrasonic homogenizers are:

- ✓ Reliability of operation (high degree of research and development).
- ✓ No clogging problems.
- ✓ Simple management and compactness.
- × Erosion in the sonotrode.
- × High maintenance cost.
- × Negative energy balance due to the high energy consumption of the equipment.

12.3.1.3 Grinding – Stirred ball mills

This technology consist of a cylindrical grinding chamber almost completely filled with grinding beads. An agitator forces the beads into a rotational movement. The micro-organisms are disintegrated in between the beads by shear- and pressure- forces. Although batch laboratory studies showed potential volatile solids destruction and enhanced biogas production (Baier & Schmidheiny, 1997; Lehne *et al.* 2001; Müller, 2001; Winter, 2002; Müller *et al.* 2004), at full scale the results were poor. The technology was developed by Kady International (Bio-Lysis® Sludge reduction process), consisting of high-speed "rotary mills" that shear the activated sludge. The Bio-Lysis did not have an effect on the sludge yield in a full-scale demonstration at the Plum Island WWTP near Charleston, S.C.

Positive (✓) and negative (×) aspects of stirred ball mills are:

- ✓ Equipment robustness.
- × Huge erosion in the grinding chamber.
- × High energy friction losses.

× Clogging problems.
× The degree of disintegration of the sludge is lower compared to other techniques.

12.3.1.4 Lysis centrifugation

The centrifugal forces created in the Lysat-thickening centrifuge are deliberately applied to cell destruction. This disruption takes place using a special ring which is integrated into the centrifugal thickener and which dissipates the kinetic energy provided by the centrifuge. This device has been implemented in several WWTPs as a pretreatment for anaerobic digestion: Liberec (100,000 PE, Czech Republic), Furstenfeldbruck (70,000 PE) and Aachen-Soers (650,000 PE) in Germany. The increase of biogas production is 15–26%.

Positive (✓) and negative (×) aspects of the Lysat centrifuge are:

✓ Moderate energetic consumption.
× Low degree of sludge disintegration.
× High operation and maintenance cost.
× Wear on the equipment plates.

12.3.1.5 Focused-pulse technology

This technology consists of an electro-hydraulic method. The idea is to treat the sludge by a rapidly pulsing, high voltage electric field (20–30 kV), in very rapid pulse periods. The shockwaves created in the sludge induce sudden disruption of cellular membranes, and polymeric structures (Weise & Jung, 1998; Weise & Jung, 2001; Müller, 2001).

Commercial technologies are *OpenCel®* (Trojan Technologies) and *BioCrack®* (Vogelsang). Both combine mechanical maceration with an electrokinetic process. According to the commercializing companies, the cyclic exposure to positive and negative charges weakens the cell wall, and eventually the cyclic forces cause cell rupture and release of internal contents.

Positive (✓) and negative (×) aspects of the high performance pulse technique are:

✓ Compact equipment.
× Erosion in the electrodes.
× Low research and development.
× Contradictory results.

12.3.1.6 Thermal hydrolysis

Thermal hydrolysis leads to partial solubilisation of sludge, which enhances anaerobic digestion, as can be seen in numerous studies on thermal hydrolysis for pretreatment of anaerobic digestion (Haug *et al.* 1978; Tanaka *et al.* 1997; Bougrier *et al.* 2008). Most studies report an optimal temperature in the range of 160–180°C and treatment times from 30 to 60 min. However, treatment time is often shown to have little effect at this temperature range (Neyes & Baeyens, 2003). On the other hand, thermal treatments at moderate temperature (70°C) may last several days (Gavala *et al.* 2003; Ferrer *et al.* 2008), because the main mechanism in such a case is assumed to be enzymatic hydrolysis. Treatments at excessively high temperatures (higher than 170–190°C) lead to decreased sludge biodegradability in spite of achieving high solubilisation efficiencies. This is usually ascribed to the so called Maillard reactions (Dwyer *et al.* 2008), involving carbohydrates and amino acids in the formation of melanoidins, which are difficult or impossible to degrade (Bougriier *et al.* 2008). The increase of methane production depends on the initial biodegradability of the sludge, with higher impacts on hardly biodegradable sludge (Carrère *et al.* 2008) and better results on WAS than on primary sludge. Additional advantages of thermal treatments include sludge sanitation, reduction of sludge viscosity with subsequent enhancement of sludge handling, and no extra energy needs, since energy requirements can be covered by excess biogas production and energy balance is positive (Kepp *et al.* 2000). Positive (✓) and negative (×) aspects of thermal hydrolysis are:

- ✓ Most effective treatment, according to energetic considerations.
- ✓ Very good dewaterability of the final sludge.
- ✓ Best sludge disinfection (Class A).
- ✓ Sludge viscosity reduction; better digester mixing.
- ✓ Removal of digester foaming.
- × Fouling of the heat exchangers (when existing).
- × Pumps maintenance (if existing).

There are several commercial thermal pre-treatment processes nowadays, with different operation schemes: *Cambi®, Biothelys® and Exelys®, Lysotherm®, Haarslev's Hydrolysis System (HCHS), Turbotec®, Aqualysis®*, and *teCH$_4$$^+$*. The technologies differ in many aspects, such as: operation (batch or continuous), reactor configuration, operation conditions (pressure, temperature, time), energy integration scheme (key for the process economics).

In order to roughly compare the existing processes, the available information is summarized in Table 12.2.

Table 12.2 Summary of operation and performance reported for the commercial thermal pre-treatment processes.

Company	Cambi	Veolia		Lysotherm	Haarslev	Sustec	Aqualogy	teCH$_4$
Tech.	THP	Biothelys	Exelys	SH + E	HCHS	Turbotec	Aqualysis	*tH$_4$$^+$*
Operation	Batch	Batch	Cont.	Cont.	Cont.	Batch/ Cont.	Cont.	Cont.
%TS in feeding	12–16	10–15	>20	6–8	17–22	10	8–12	6–15
T (°C)	160–180	150–170	N.A.	175	150–170	140–190	165–180	220
P (bar)	6–9	7–9	N.A.	5–15	7–8	4–8	7–10	12–14
t (min)	20–30	30–60	N.A.	30–60	20	30–75	15–30	<5

12.3.1.7 Chemical oxidation

The most widely used chemical method is ozonation. Ozonation leads to partial sludge solubilisation and yield increases with ozone dose. A too high ozone dose will result in reduced apparent solubilisation due to oxidation of the solubilised components (Yeom *et al.* 2002). In addition, it is oxidative, and may therefore increase destruction at the expense of methane yield. Several studies have shown an optimal ozone dose for the enhancement of anaerobic biodegradability: 0.1 gO_3 g^{-1} COD (Weemaes *et al.* 2000), 0.2 gO_3 g^{-1} TSS (Yeom *et al.* 2002), 0.15 gO_3 g^{-1} TS (Bougrier *et al.* 2007). Positive (✓) and negative (×) aspects of ozone pre-treatment are:

- ✓ Better dewaterability of the final sludge.
- × High energy consumption.
- × Metals present in the initial sludge (Fe, Zn, Ag, Cu), are transferred to the liquid phase, which should be purified.

12.3.1.8 Alkaline hydrolysis

During the alkaline pre-treatment, the pH of the sludge is increased up to 12, maintaining this value for a period of time (normally 24 hours). This process may be used to hydrolyse and decompose lipids, hydrocarbons and proteins into smaller soluble substances such as aliphatic acids, polysaccharides and amino acids (Chiu *et al.* 1997; Mukherjee & Levine, 1992).

Alkali treatment is relatively effective in sludge solubilisation, with in order of efficacy being (NaOH > KOH > $Mg(OH)_2$ and $Ca(OH)_2$) (Kim *et al.* 2003). However, too high concentrations of Na^+ or K^+ may cause subsequent inhibition of anaerobic digestion (Mouneimne *et al.* 2003). It is normally combined with thermal treatment. Compared to thermal hydrolysis, alkali treatment temperature is normally lower, and the increase in temperature normally driven by chemical processes (from 170 to 120–130°C). Sludge solubilisation and anaerobic biodegradability increase with alkali dose and temperature, with an upper limit (Kim *et al.* 2003; Valo *et al.* 2004). However, since the addition of alkali increases mineral content of digested sludge, it reduces the interest of cotreatment on sludge reduction. In addition, sludge dewaterability may be diminished by KOH addition (Everret, 1974).

Positive (✓) and negative (×) aspects of acid or alkaline hydrolysis are:

✓ Low energetic requirements.
✓ Very good dewaterability of the final sludge.
× Modification of the sludge composition.
× Possible damage to the bacteria responsible for the microbiologic activities.
× Bad odour generation.
× Corrosion and fouling of the equipment.
× Higher COD in the final effluent due to the presence of non-biodegradable substances.

12.3.1.9 Biological pre-treatment

The biological-enzymatic pretreatments include the possibility of applying commercial enzymes, which can be selected based of the type of enzyme and quantity required (Davidson *et al.* 2007). It is also possible to use an enzyme-producing microorganism that releases hydrolytic enzymes during the hydrolysis step (Hasegawa *et al.* 2000). Finally, it is possible to use the inherent enzymatic activity of the secondary sludge, called autohydrolysis. This option is the most interesting one from a practical point of view and has been demonstrated by various groups (Mayhew *et al.* 2002; Guellil *et al.* 2001; Burgess *et al.* 2008; Yan *et al.* 2008; Carvajal *et al.* 2013).

The autohydrolysis pretreatment is a biological-enzymatic pretreatment, which involves subjecting the secondary sludge to a temperature of 55°C and a limited amount of oxygen in batch operation. These conditions allow the microorganisms in the secondary sludge to release the hydrolytic enzymes contained in their own metabolic system, which ultimately leads to the hydrolyzation of the secondary sludge. As a consequence, the product expected is a solubilized and hydrolysed organic matter, with improved anaerobic degradation, but without associated high costs. An improvement in dewaterability and pathogen have also been reported.

The full scale development of this technology corresponds to the Monsal process (GE Water & Process technologies). The GE's Monsal advanced digestion use bacteria in the absence of oxygen to break down the sludge before anaerobic digestion.

Positive (✓) and negative (×) aspects of biological pre-treatment are:

✓ Low energy consumption.
✓ No stress on the equipment.

× High operation cost if commercial enzymes are used.
× The enzymatic mechanisms are still not clear.
× Autohydrolysis: more research needed.
× The usage of enzymes to better the sludge stability is not clear.
× High generation of odours.

12.4 TECHNOLOGIES FOR ENHANCING SLUDGE STABILIZATION (G)

12.4.1 Thermophilic anaerobic digestion: effect of thermal pre-treatment

It is well known that enzymatic hydrolysis of particulate matter is the rate-limiting kinetic step in the anaerobic digestion of substrates as WAS. Effectiveness of pre-treatments to enhance sludge biodegradability (Braguglia *et al.* 2011) and potentially improve dewaterability after digestion has been highlighted in the previous paragraphs. Among these pre-treatments, thermal hydrolysis has the potential to produce high quality sludge for agricultural use. In a recent study performed on WAS provided from a full-scale WWTP, Gianico *et al.* (2013) investigated the effects of low-temperature (134°C) thermal pre-treatment on semi-continuous thermophilic WAS digestion performances at different organic loading rates (OLRs). Results are summarized in Tables 12.3 and 12.4.

Table 12.3 Effect of thermal pre-treatment on gravity and dynamic thickened WAS.

	Gravity Thickened WAS		Dynamic Thickened WAS	
	Untreated	**Pre-Treated**	**Untreated**	**Pre-Treated**
TS (g/L)	20.8 ± 0.3	20.5 ± 0.3	41.4 ± 0.5	40.2 ± 0.5
VS/TS (%)	69	69	71	71
Soluble COD (mg/L)	45 ± 6	2,834 ± 320	206 ± 15	5,908 ± 565
Soluble NH_4^+-N (mg/L)	14 ± 2	192 ± 16	32 ± 2	279 ± 23
Surface Charge (mC/gTS)	54 ± 4	1,527 ± 138	24 ± 2	1,683 ± 151

Consistent increase of soluble COD and suface charge occurred after pre-treatment, due to the release of soluble/colloidal particles, thus worsening sludge filterability. Untreated and pre-treated WAS digestion performances, in terms of organic matter removal, specific biogas production and methane production are reported in Table 12.4. The specific biogas productions with respect to VS fed did not vary significantly by increasing OLR in the range 1.0–3.7 kg VS/m^{3}d, indicating that the conversion rate was not affected by the changing load in the digester. Methane content in biogas ranged between 60–70%. Methane production rate varied between 0.15–0.80 Nm^3/m^{3}digester.d and increased by increasing the OLR from 1.0 to 3.7 kg VS/m^{3}d, for both untreated and pre-treated WAS digestion.

The effectiveness of thermophilic process (with or without pre-treatment) in enhancing the digestate hygienization is fully described in Levantesi *et al.* (2014). The investigated processes efficiently reduced *E. coli* to below detection limit, indicating that the increased temperature is the driving force bacterial inactivation. Notably the *E.coli* limits proposed for the restricted and unrestricted use of sludge, 2 log *E. coli* removal and *E. coli* <500 CFU/gTS respectively, were always achieved. Instead, the removal of viral indicators to below detection limit was observed only at high temperature during thermal pre-treatment. Somatic coliphages (SOMCPH) removal increase of only 1–1.5 log units from typical mesophilic to thermophilic digestion, while SOMCPH were always absent in digestates after thermal pre-treatment. Similar results were reported in previous studies (Guzman *et al.* 2007), confirming the higher resistance of viral particles to thermal treatments.

Table 12.4 Digestion performances at different OLRs.

	Test #1 OLR = 1.0 kg VS/m³d		Test #2 OLR = 1.8 kg VS/m³d		Test #3 OLR = 3.7 kg VS/m³d	
	Untreated	**Pre-Treated**	**Untreated**	**Pre-Treated**	**Untreated**	**Pre-Treated**
Specific biogas production ($Nm^3/kgVS_{fed}$)	0.26 ± 0.02	0.36 ± 0.04	0.31 ± 0.03	0.32 ± 0.03	0.31 ± 0.02	0.32 ± 0.03
CH_4 production rate ($Nm^3/m^3_{digester}d$)	0.15 ± 0.01	0.22 ± 0.01	0.38 ± 0.03	0.41 ± 0.03	0.74 ± 0.06	0.80 ± 0.07
Specific CH_4 production ($Nm^3/kgCOD_{fed}$)	0.11 ± 0.01	0.16 ± 0.01	0.15 ± 0.01	0.15 ± 0.01	0.15 ± 0.01	0.15 ± 0.01

Positive (✓) and negative (×) aspects of thermophilic digestion are:

- ✓ More efficient organic solids removal.
- ✓ High methane yields.
- ✓ Reduced digester volume.
- ✓ Good hygienization.
- × COD, ammonia, fatty acids accumulation.
- × Bad odours.
- × Worse dewaterability of the digested sludge.

12.4.2 Temperature-phased anaerobic digestion

A temperature-phased anaerobic digestion (TPAD) system could bring together the advantages of both mesophilic and thermophilic systems. Conventional TPAD consists of a first short thermophilic step, acting as pre-treatment, followed by a longer mesophilic methanogenic step, aimed to achieve an effective organic matter removal, degrading the soluble compounds released under thermophilic conditions. This two-stage process based on the separation between hydrolysis/acidogenesis, with fast kinetics, and methanogenesis. In a recent study, Gianico *et al.* (2014) proposed an inverse temperature-phased anaerobic digestion (iTPAD) process with the aim to enhance energy recovery, producing a hygienized final product suitable for agricultural use. In such approach, the first kinetic step, namely hydrolysis of particulate matter, was performed with ultrasounds. The acidogenic step was carried out in mesophilic reactors aimed to produce volatile fatty acids (VFAs), and the last methanogenic step was performed in thermophilic reactors to achieve higher methane yields and well-hygienized digestates. The main operating conditions are described in Gianico *et al.* (2014), the plant lay-out is shown in Fig. 12.2, and the results are summarized in Table 12.5.

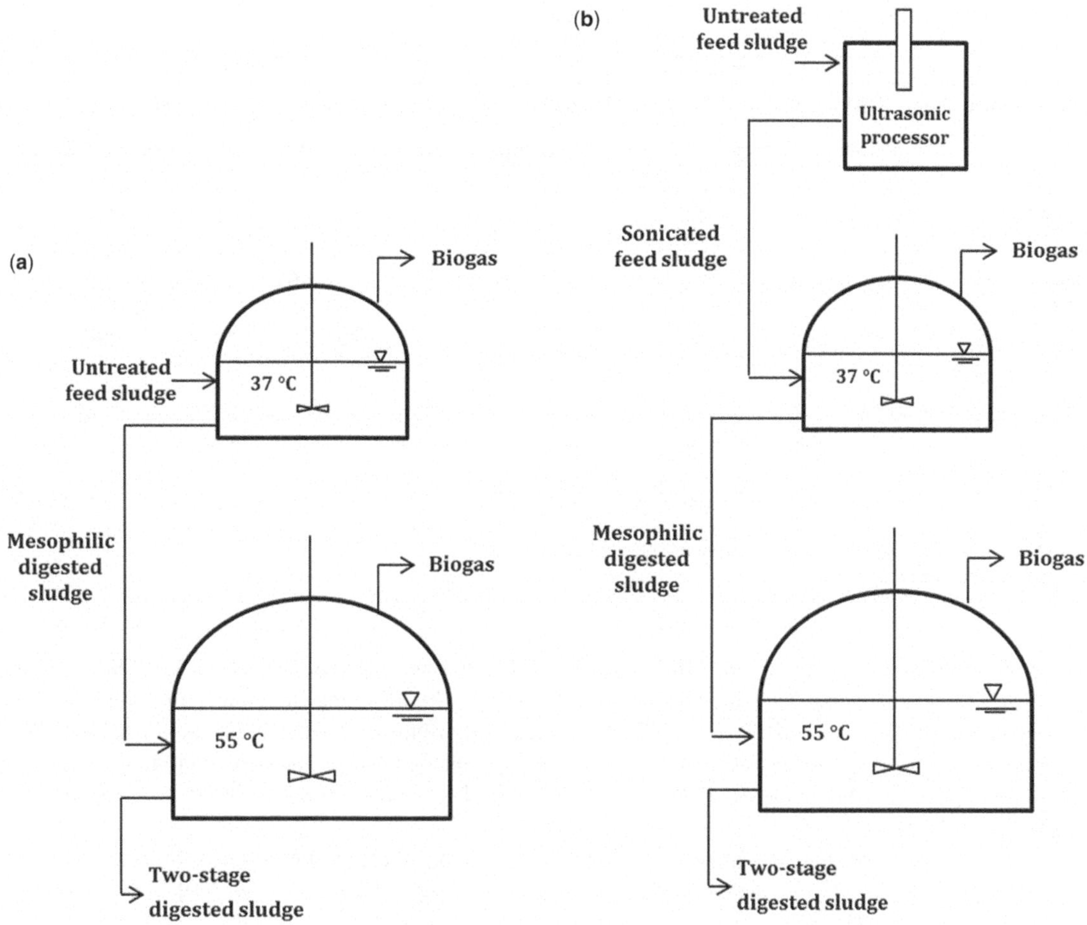

Figure 12.2 Schematic diagram of the two-stage process with untreated (a) and sonicated (b) WAS.

The production of long-chain VFAs was observed by feeding sonicated sludge (Gianico *et al.* 2014; Gagliano *et al.* 2014). Nevertheless, after the thermophilic stage, the soluble COD of both digestates arising from "untreated" and "sonicated" line resulted comparable (Table 12.5), demonstrating high organics removal in thermophilic conditions.

In the mesophilic stage of Test#1, a gain in specific biogas production (+19%) due to sonication was observed, while sonication had no effects on the following thermophilic stage. On the contrary a biogas gain due to sonication (+17%) was observed in the thermophilic stage of Test#2. Moreover, in Test#2 the thermophilic conversion of higher amounts of VFAs produced in the first stage led a +8% gain in methane content.

The mesophilic stage of Test#1 resulted long enough to establish methanogenesis, after a change in metabolic pathway from acetoclastis to acetate oxidation, under which acetate is first oxidized to H_2 and CO_2, and subsequently converted to CH_4 (Gagliano *et al.* 2014). Low concentrations of VFAs, namely acetate and propionate, have been detected only during the thermophilic stage.

Table 12.5 Performances of two-stage digestion tests.

Performance Parameters	Test #1				Test #2			
	Untreated		Sonicated		Untreated		Sonicated	
	1st Stage	2nd Stage	1st Stage	2nd Stage	1st Stage	2nd Stage	1st Stage	2nd Stage
TS (g/L)	22 ± 2.9	17.5 ± 1.6	21 ± 2.4	17.5 ± 1.9	41 ± 3.7	31 ± 1.5	39.5 ± 2.8	31 ± 1.7
VS (%TS)	53	47	53	47	62	56	61	55
pH	7.6 ± 0.1	8.2 ± 0.1	7.7 ± 0.1	8.4 ± 0.1	7.5 ± 0.2	8.5 ± 0.1	7.5 ± 0.1	8.5 ± 0.1
Biogas ($Nm^3/kgVS_{fed}$)	0.21 ± 0.02	0.10 ± 0.01	0.25 ± 0.02	0.10 ± 0.01	0.13 ± 0.01	0.18 ± 0.01	0.13 ± 0.01	0.21 ± 0.02
CH_4 (%)	58	46	58	50	49	52	50	56
Supernatant:								
COD_{sol} (mg/L)	210 ± 25	930 ± 135	250 ± 36	870 ± 99	1,520 ± 198	2,180 ± 247	2,890 ± 308	2,480 ± 267
VFA_{tot} (mg HAc/L)*	<10	150 ± 36	<10	115 ± 23	700 ± 106	240 ± 47	1,290 ± 185	220 ± 69
NH_4^+-N (mg/L)	550 ± 36	785 ± 88	615 ± 37	920 ± 38	510 ± 67	1,260 ± 136	615 ± 69	1,300 ± 62

*The VFA concentrations are expressed as acetic acid equivalents.
Source: Gianico *et al.* (2014).

On the contrary, during Test#2 the HRT reduction led to an intense production of VFAs, 700 mg/L and 1,290 mg/L for control and experimental reactor, respectively, highlighting the effect of the high load on acidogenic phase (Gianico *et al.* 2014). VFAs were efficiently removed during thermophilic stage. Thus, the amount of VFAs fed to thermophilic stage definitely changed the pathway of methanogenesis and microbial populations (Gagliano *et al.* 2014) resulting in higher methane production rates compared with Test#1 (Table 12.5).

The effectiveness of this two-stage technology in improving the digestate quality to assure its safe agricultural use was moreover investigated in Gianico *et al.* (2014), Braguglia *et al.* (2014) and Levantesi *et al.* (2014). The iTPAD technology showed good hygienization performances, all the microbial indicators were in fact reduced with a logarithmic removal >3.5, >2.3 and >0.8, respectively for *E. coli*, SOMCPH and *Salmonella*. A very limited disinfection capacity of the first mesophilic stage, and no benefits due to sonication were reported (Levantesi *et al.* 2014).

The assessment of pollution level (in terms of organic and inorganic micropollutants) and ecotoxicity of final digestate is reported in Braguglia *et al.* (2014). Overall, it can be concluded that iTPAD process provided high removals (up to 70%) of micropollutants as phthalates and pharmaceuticals. Toxicity reduction was related to efficient removal of organic micropollutants, both conventional and emerging ones.

Positive (✓) and negative (×) aspects of iTPAD including a sonication pre-treatment are:

✓ Increased biogas production depending on the OLR.
✓ High methane yelds.
✓ Good hygienization.
✓ Efficient reduction of organic and inorganic micropollutants.
× COD, ammonia, fatty acids accumulation.
× Bad odours.
× Worse dewaterability of the digested sludge.
× Increased comeplexity of the plant.

12.4.3 Sequential anaerobic-aerobic digestion of waste and mixed sludge

Sequential anaerobic aerobic digestion to enhance sludge stabilization has been proposed in previous studies (Kumar *et al.* 2006a; Kumar *et al.* 2006b; Novak *et al.* 2011): the rationale behind this approach is the availability of optimal operating conditions for the different VS sludge fractions. An additional positive effect of the post- aerobic digestion step is the ammonia nitrogen removal in the supernatant stream originated from the anaerobic digestion, which can consistently reduce the ammonia load recycled to the treatment plant. Sequential digestion, in its conventional configuration i.e. mesophilic anaerobic stage followed by post-aerobic digestion at room temperature, has been extensively investigated for mixed (Kim & Novak, 2011; Novak *et al.* 2011; Tomei *et al.* 2011), and WAS (Tomei *et al.* 2011; Tomei & Carozza, 2015). In both cases its superior performance for VS and COD removal, in comparison to a single digestion phase (anaerobic or aerobic), has been demonstrated.

Recent applications of the sequential process suggest the operation of the aerobic stage under alternating aeration cycles to achieve simultaneous nitrification-denitrification with the added advantages of nitrogen removal and energy saving for aeration (Tomei & Carozza, 2015).

Another operating alternative to improve the process performance consists in a modified version of the sequential digestion, with operation of the aerobic step under controlled mesophilic conditions (Tomei *et al.* 2016). The motivation for this approach is that the higher aerobic temperature has in principle a beneficial effect on the degradation kinetics of soluble and particulate COD and, in presence of alternating aeration conditions, on nitrification-denitrification kinetics. Mesophilic conditions in the aerobic reactor can be attained with a reduced external energy input because the influent to the aerobic digester (coming from the anaerobic-one) is at a mesophilic temperature ($T = 37°C$) and, additionally, exothermic reactions take place in the aerobic digestion step. Given these conditions, an effective insulation of the reactor avoids (or strongly minimizes) the external energy demand. The two-step anaerobic-aerobic digestion has been successfully investigated, and results reported in Tomei *et al.* (2016), and Tomei and Carozza (2015). Experiments were performed on two digesters operated in series at SRT of 15 days the anaerobic and 12 days the aerobic. Anaerobic step was operated under mesophilic conditions while the aerobic step was operated under intermittent aeration to achieve simultaneous nitrification-denitrification. The aerobic reactor was operated at room temperature when fed with mixed sludge while the temperature was varied from 20 to 37°C in the WAS digestion. A summary of the main results achieved in the above mentioned studies is below reported.

Case study 1: Mixed sludge

Average VS removal efficiencies on weekly basis are 50 ± 8 and $45 \pm 5\%$ for the anaerobic and aerobic phase respectively. The efficiency in the aerobic stage is comparable to the anaerobic one, and it is higher than the values reported in Parravicini *et al.* (2006) (16%) and Kumar *et al.* (2006a) (20%). This finding may be explained with the higher aerobic SRT in these experiments (12 days vs 3–4 days of the previous studies), which allows the degradation of slower biodegradable fractions constituting the VS matrix.

A first assessment of the dewaterability trend in sequential digestion is performed through CST measurement: a worsening of dewaterability has been observed after the anerobic phase, as demonstrated by the CST increase from 154 ± 53 to 538 ± 78 sec, followed by an improvement after the aerobic phase (CST equal to 253 ± 72 sec).

The specific biogas production (SGP) was 0.82 ± 0.15 $Nm^3/(kgVS_{destroyed})$. The value is in within the range of 0.19–1.6 $Nm^3/(kg/VS_{destroyed})$ reported in the literature (Speece, 1988; Bolzonella *et al.* 2005) for mesophilic digestion of sewage sludge. Also the methane fraction (0.67) is consistent with literature data (Boušková *et al.* 2005).

As regard as the nitrogen removal, intermittent aeration was efficient in achieving simultaneous nitrification-denitrification reaching up to 97% ammonia nitrogen removal and 70% denitrification efficiency.

Case study 2: Secondary sludge

Figures 12.3a and 12.3b show the VS removal for test T1 (T = 20°C) and T2 (T = 37°C), respectively. Anaerobic VS removal efficiencies are (~43%) in both series, while an additional VS removal of 19.6% for T1 and 32.8% for T2 was achieved in the subsequent aerobic stage. This corresponds to a total VS removal efficiency of 54% and 62% for T1 and T2, respectively. Better results for VS removal were obtained in the second series of test, thus demonstrating the beneficial effect of the increased temperature.

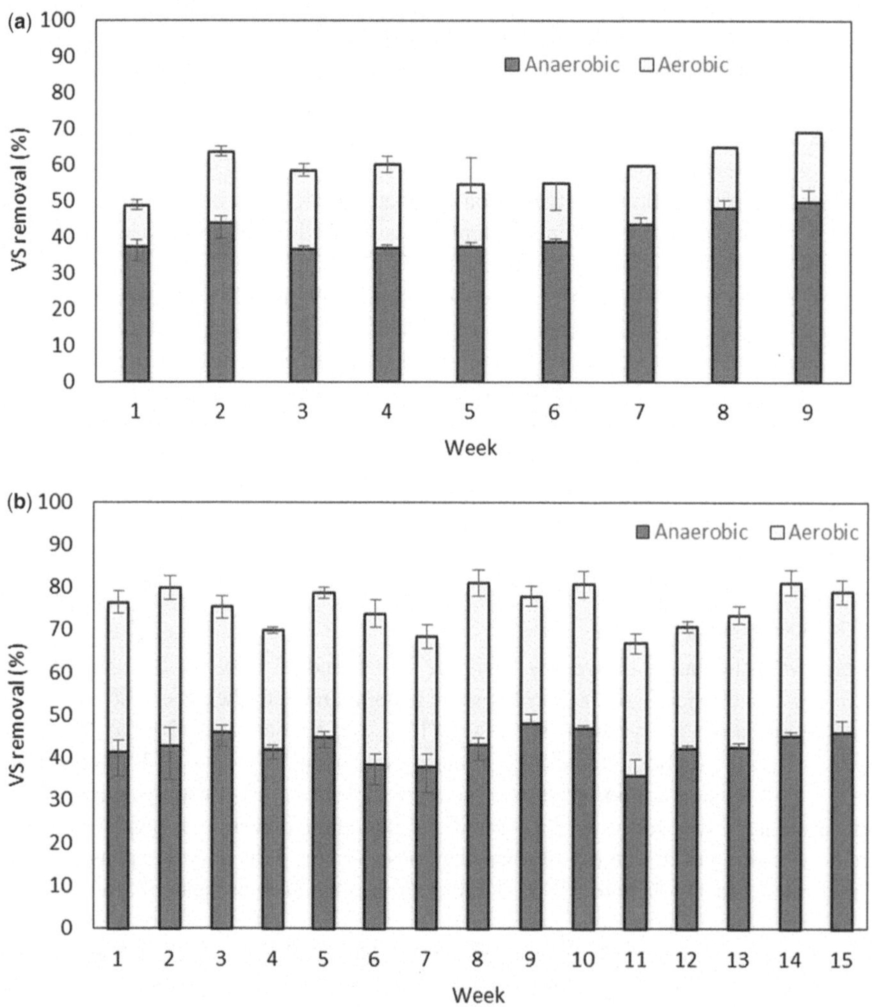

Figure 12.3 Removal efficiencies for VS in T1 (a) and T2 (b) tests as average week values and related SD.

The average specific biogas production (SGP) showed a progressive increase from the start-up phase, reaching a quite stable value of 0.95 ± 0.21 $Nm^3/(kg/VS_{destroyed}$ in T1, which was also maintained in T2. Even for WAS, SGP values are within the range of 0.19–1.6 $Nm^3/(kg/VS_{destroyed})$ reported in the scientific literature (Speece, 1988). Methane content of the biogas varied in a very limited range of values i.e. 65.6 ± 2.9% for the entire experimental period.

The aerobic phase provided effective nitrification in both runs, with efficiencies increasing from 81.5% for T1 to 83.8% for T2. The denitrification efficiency showed a marked improvement from 75.7 to 97.3%. Achieved total nitrogen removal, i.e. 57.5% in T1, and 75.3% in T2 demonstrated the positive effect of the higher aerobic temperature on nitrogen removal.

The positive effect on sludge dewaterability is confirmed also for WAS and it is more evident for the higher aerobic temperature (CST decrease of 51%).

Detailed results on this case study are reported in Tomei *et al.* (2016).

Finally, a relevant additional feature of the sequential process exploiting the presence of different reaction environments is the very efficient removal of micropollutants reported in (Braguglia *et al.* 2015) showing the very good performance in comparison to other stabilization alternatives.

Positive (✓) and negative (×) aspects of sequential digestion are:

- ✓ Very high organic solids removal increasing with aerobic temperature.
- ✓ High removal of micropollutants.
- ✓ Consistent reduction of the nitrogen load recycled to the plants.
- ✓ Easy to be implemented in the upgrading of WWTPs.
- ✓ Improvement of the dewaterability characteristics of the digested sludge.
- × Increased energy consumption for aeration.
- × Needs of post-thermal treatment for achiewing Class A biosolids.

12.5 WET OXIDATION OF SEWAGE SLUDGE COUPLED WITH ANAEROBIC DIGESTION OF LIQUID RESIDUE (H)

12.5.1 Wet oxidation and its role in sewage sludge treatment

Wet oxidation (WO), also known as wet combustion, is a process that transforms the organic and inorganic substances, in solution or suspended, through chemical oxidation (air or pure oxygen are used) and hydrolysis reactions. The insoluble organic matter is turned in simple soluble organic compounds that subsequently are oxidized and maybe converted in carbon dioxide and water. Effluent water contains fatty acids with a low molecular weight, ammonia, inorganic acids and inorganic salts, and can be biologically treated. The output gases can contain ammonia, CO and some low molecular weight compounds, in addition to nitrogen, CO_2, water steam and oxygen (Genç *et al.* 2002).

The main affecting operating parameters are temperature, pressure, contact time and, although on minor extent, pH. Operative temperatures range from 150 to 360°C and running pressures are generally kept between 30 and 250 bar. In order to operate with lower temperature and pressure, catalysts may be utilized. Practically all ions of heavy or transition metals together with peroxides (used as generators of radicals) exert catalytic activity (with different efficiency) (Gomes *et al.* 2005).

WO technology is well-known: it was proposed 100 years ago for the treatment of both high strength industrial wastewaters and sewage sludge (Strehlenert, 1911). Today, 200 wet oxidation full scale plants are operating around the world for the treatment of industrial wastewaters (Roy *et al.* 2010).

In the field of sewage sludge treatment, WO, belonging to the category of hydrothermal oxidation methods, can be considered as an alternative solution to conventional incineration. The number of recent publications confirms the renewed interest: many studies have investigated wet oxidation, at lab scale, as a technology for excess sludge treatment (see, for instance, Chung *et al.* 2009; Gielen *et al.* 2011; Baroutian *et al.* 2013; Bertanza *et al.* 2016) and conceptual and mathematical models have been proposed for both industrial wastewaters and sewage sludge (Li *et al.* 1991; López Bernal *et al.* 1999; Zhang & Chuang, 1999; Lopes *et al.* 2007; Bertanza *et al.* 2015a; Menoni & Bertanza, 2016).

The main goals of sludge WO are: reduction of TSS (Total Suspended Solids) and, in particular, of VSS (Volatile Suspended Solids), high stabilization of the solid residue (dramatic reduction of the VSS/TSS ratio), improved dewaterability of the sludge, increased biodegradability of the liquid residue. Actually, final products consist in wastewater (characterized by the presence of biodegradable compounds and ammonia and which require further treatment), gas (which doesn't contain dangerous compounds) and inorganic solids (which can be disposed of or, possibly, recovered) (Chung *et al.* 2009).

Several patents of this technology are available.

12.5.2 WO of sewage sludge: effect of process parameters

Typical process conditions for sludge treatment by WO are the following: 230–250°C temperature, ~60 atm pressure and contact time in the range 15–120 minutes. WO process typically becomes energetically self-sufficient at medium-high temperature (>200°C): Hii *et al.* (2014).

The role of temperature: Khan *et al.* (1999) and Mucha and Zarzycki (2008) report the effect of temperature on the degradation of TCOD during WO of sludge. Considering TSS concentration reduction, Chung *et al.* (2009) have shown that a temperature of 200°C represents a threshold: only slight differences were observed for higher values, especially for contact times exceeding 30 minutes. Bertanza *et al.* (2016) found that, depending on sludge origin, the increase of temperature from 200 to 250°C led to a variation of COD and VSS removal rate in the ranges 22–79% and 54–99%, respectively, the reaction time being fixed at 60 min.

The role of contact time: Beside the expected effect of reduction of total organic matter as the reaction proceeds, it is interesting to investigate the conversion pattern of soluble organics. Bernardi *et al.* (2010) have shown that soluble TOC increases rapidly during the first minutes of reaction, in accordance with the observed decrease of solid TOC concentration. This can be explained considering that, during the early treatment phase, hydrolysis prevails on oxidation. Increasing contact time, the concentration of sTOD decreases at an extent depending on operating conditions (e.g. employed catalyst). Similar results were obtained by Genç *et al.* (2002) and Mucha and Zarzycki (2008). The latter also evidenced that organic matter oxidation leads to the progressive increase of organic acids concentration, which are formed rapidly at the early stages of reaction.

Bertanza *et al.* (2016) recorded an increase of COD and VSS removal from 44 to 85% and from 71 to 99%, respectively, by increasing the reaction time from 15 min up to 120 min, at a temperature of 250°C.

The role of pressure: In order to avoid oxygen to be the limiting factor of the reaction, the required minimum pressure should be calculated, based on initial COD concentration of the sludge. For instance, the full scale plant described in Slavik *et al.* (2015) works at operating pressures in the range 50–55 bars. Once reached favourable conditions, further increase of pressure does not lead to significant effects (Chung *et al.* 2009).

12.5.3 Reaction kinetics and process modelling

Global reaction rate of a chemical oxidation process can be described as follows (Kolaczkowski *et al.* 1999):

$$r_c = k_0 e^{\left[-\frac{E}{RT}\right]} \cdot (C)^{\alpha} \cdot (O_2)^{\beta} \tag{12.1}$$

where:

k_0 = rate constant;
E = activation energy [J];
R = ideal gas constant [8.315 J/(mol · K)];
T = temperature [K];
C = organic compound concentration [mol/m^3];
O_2 = oxygen concentration in liquid phase [mol/m^3];
α = reaction order (with respect to organic compound concentration);
β = reaction order (with respect to oxygen concentration).

Literature data for α and β are available for WO of single compounds (syntetic solutions): α is generally equal to 1 and β varies in the range 0.4–1 (Debellefontaine & Foussard, 2000).

This model is not able to describe observed rate in a complex mixture of compounds, like sludge; in this case, some organics are transformed immediately to CO_2, and others are transformed into intermdiates (typically short-chain alcohols and fatty acids) that have a lower reactivity. The global velocity of reaction depends on formation rate of final products and formation and destruction rate of intermediate products.

For this reason, Generalized Lumped Kinetic Models (GLKM) have been proposed for the WO process, for describing COD or TOC reduction or N transformation. Models used for sludge treatment are for instance reported in Mucha and Zarzycki (2008), Bertanza *et al.* (2015a) and Menoni and Bertanza (2016).

12.5.4 Treatment/Disposal of residues

The liquid stream originated by WO is easily biodegradable and therefore the recirculation to the biological WWTP may be a feasible solution. However, the WO effluent has a residual organic and nitrogen content so that its pre-treatment may be required when the receiving WWTP has no surplus treatment capacity left. In Bertanza *et al.* (2015b), the innovative matching of WO with anaerobic digestion of the liquid residue is proposed. In this way, energetic valorization (biogas production) and the reduction of COD load to be recycled back to the WWTP can be simultaneously achieved. A techno-economic and environmental assessment of this solution has been performed (Bertanza *et al.* 2015c): it turned out as a sustainable alternative to conventional sludge stabilization and incineration for large WWTPs or for centralized sludge treatment plants.

As for the solid residue, it mainly consists of inorganic material: only a very small organic residue is still present after treatment, the VSS removal efficiency being very high (see data reported above). A cake (dry solid content up to 55–60%, according to Luck (1999) and Bertanza *et al.* (2015c) can be obtained by conventional dewatering. The improvement of sludge dewaterability is an important advantage of WO in view of the amount of residue to be disposed, as non-hazardous waste. Moreover, Slavik *et al.* (2015) investigated the possibility of recovering the residue as filler for bituminous materials and similar.

12.6 COMPARATIVE ANALYSIS OF THE PROCESSES

Table 12.6 summarizes the development level of the different processes, indicating if it is laboratory (LAB) or full-scaled (FULL), and the corresponding name of the process. The activity regarding research and development (R&D) is also presented, indicating those processes that are nowadays subject of mayor interest.

Table 12.6 General comparison of existing processes to reduce sludge impact.

Location in the WWTP	Process	Technology	Development	R&D Activity
Processes in the water line	Chemical oxidation	Biolysis O®	FULL	NO
	Enzymatic reactions	Biolysis E®	FULL	NO
	Mechanical treatment		LAB	NO
	Extended aeration process		FULL	NO
	Membrane bioreactor		FULL	NO
	Chemical uncoupler		LAB	NO
	Side stream anaerobic reactor	OSA®, Ananox®, Cannibal®	FULL	YES
	Oligochaetes (worms)		LAB	NO
Enhanced hydrolysis	High Pressure homogenizers	Crown®, Cellruptor®, Microsludge®	FULL	YES
	Ultrasonic treatment	Sonix®, Biosonator®, Sonolyzer	FULL	YES
	Thermal hydrolysis	Cambi®, Biothelys®, Exelys®, Lysotherm® Haarslev, Turbotec®, Aqualysis®, tH_4^+	FULL	YES
	Focused-pulse technology	OpenCEL®, BioCrack®	FULL	YES
	Grinding-Stirred ball mills	Bio-Lysis®	FULL	YES
	The lysis centrifuge		FULL	NO
	Oxidation		LAB	NO
	Alkaline hydrolysis		LAB	NO
	Enzymatic hydrolysis	Monsal®	FULL	YES
Enhanced stabilization	Thermophilic digestion with pre-treatment		FULL/LAB	YES
	Temperature phased anaerobic digestion		LAB	YES
	Sequential anaerobic-aerobic digestion		FULL/LAB	YES
	Wet oxidation coupled with anaerobic digestion of residue	For WO (examples) DUAL TOP®, Athos®	FULL/LAB	YES*

*The WO technology is established; research is ongoing on the coupled solution WO + anaerobic digestion.
LAB: proven at laboratory scale; FULL: full-scaled.

From Table 12.6 it can be first observed that although there are several processes developed full scale in the water line, there is no current interest on their research, except for those involving anaerobic digestion units (OSA®, Ananox® y Cannibal®). The reason is related to the key principle related to the general energy balance in a wastewater treatment plant: the aerobic process is highly energy consuming, while in contrast the anaerobic digestion of the sludge allows recovering the energy present in the wastewater.

Therefore, those configurations based on the water line propose to minimize sludge production by promoting aerobic degradation, increasing the removal of organic matter in the aeration basin (not only wastewater pollution, but also sludge). This approach clearly leads to additional O_2 consumption, CO_2 emissions, and energy costs. As a consequence, the interest and application of these processes is negligible, and will not be further discussed in the present section. On the contrary, attention has to be paid to those configurations based on optimizing the anerobic digestion of sludge and enhancing sludge stabilization, which are potentially able to consistently reduce the amount and improve the "quality" of the produced sludge interms of micropollutant and pathogen contents.

12.6.1 Enhanced hydrolysis. Processes in the sludge line

The configurations based on the sludge line by optimizing the anaerobic digestion with a pre-hydrolysis step, are however more profitable, and subject of research and development. The factors that make sludge pre-treatments specially interesting are: (i) increased energy production and positive balance in WWTPs, (ii) increased final handling costs (especially for dewaterability and transport), and (iii) increased legislative requirements for stabilisation performance and pathogen removal.

Although knowledge and experience is being gained when operating full-scale installations, and there is still limited knowledge concerning routine operation of some of the technologies, a general analysis can be done based on the current state-of-the-art.

Table 12.7 presents some features regarding applicability of the pre-treatment processes in the sludge line.

Table 12.7 Main features of the pre-treatment processes in the sludge line.

	Increased Biogas	Dewaterability Improvement	Biosolids Hygienization	Maintenance Requirements	Operation Costs
High Pressure homogenizers	++	+	–	—	—
Ultrasonic treatment	++	–	+	—	+
Thermal hydrolysis	++	++	++	+/++	++
Focused-pulse technology	–	–	–	—	—
Stirred ball mills	–	–	–	—	—
Lysis centrifuge	+	–	–	–	–
Oxidation	+	–	–	–	–
Alkaline hydrolysis	+	–	–	–	–
Enzymatic hydrolysis	+	+	+	+	+

First, although lab and full-scale studies generally indicate the increase in biogas production, the best results corresponds to high pressure homogeneizers, ultrasonic treatment and thermal hydrolysis. Worse results are presented for technologies such as chemical or enzymatic hydrolysis and even contradictory for processes like focused-pulse or grinding. When analysing the characteristics of the biosolids, regarding dewaterability and hygienization, thermal hydrolysis appears as the only technology that guarantees good results in both items. Sonication normally worsens sludge dewaterability, while high pressure does not assure sanitization. Enzymatic hydrolysis also presents good performance in both aspects. Finally, no energy assessments neither maintenance nor operation considerations are usually considered in scientific reports. From the point of view of the maintenance requirements, all the alternatives except thermal and enzymatic demand continuous replacement of parts (generally technologically expensive for those mechanic technologies) or reagents consumption (for chemical treatment).

Regarding operation costs (energy) Cano *et al.* (2016) present an energy feasibility study of the main sludge pre-treatment technologies. By making a simple evaluation of the energy consumed by the pretreatments, they stated that unfortunately not all the pretreatment technologies have an energy self-sufficiency to be implemented in a WWTP, requiring many times a continuous energy investment. Generally, pretreatments consuming electricity do not satisfy its energy demands from the biogas production in the same process, although high solubilization or biogas production increases are reached. Just ultrasounds applied in full-scale plants, with commercial technologies such as Sonix or Biosonator, provide an energetically self-sufficient pretreatment. In the case of thermal pretreatments, the potential to be implemented with full energy integration is much higher, since they can recover heat from the biogas engine as well as electrical energy in the same extent as for electric pretreatments. This way, full energy integration can be achieved in thermal hydrolysis plants with a complete energy recovery and self-sufficiency, being sludge concentration the main key factor to assure energy self-sufficiency.

12.6.2 Enhanced sludge stabilization processes

Regardless of which sludge disposal alternative is employed, all can take advantage of more effective stabilization processes, and this justifies the increased attention paid to sludge stabilization processes aimed at increasing their efficiency and reducing costs.

The proposed alternatives for enhancing sludge stabilization are aimed to improve the removal of organic solids and to improve the "quality" of the stabilized sludge in terms of micropollutant and pathogen content so achieving biosolids suitable for agricultural reuse.

In Table 12.8 a qualitative comparison among the different technologies and the conventional anaerobic digestion is reported.

Table 12.8 Comparison of technologies for enhancing sludge stabilization.

Parameter	AD	SAA	TT	iTAD	WO-AD
Energy requirement and associated costs	✓	✓	×	×	✓
Complexity of the installation and operation	✓	×	×	×	×
Solid reduction	×	✓	✓	✓	✓
Micropollutant removal	×	✓	×	✓*	✓
Hygienization potential	×	✓**	✓	✓	✓

*suitable only for specific groups of micropollutants.
**increased with respect to AD but lower than the level achieved with thermophilic processes.
Note: ✓ = positive; × = negative; AD = anaerobic digestion; SAA = Sequential Anaerobic-Aerobic digestion; TT = Thermophilic digestion with Thermal pre-treatment; iTPAD = inverse Temperature-Phased Anaerobic Digestion; WO-AD = Wet Oxidation and Anaerobic Digestion of the liquid residue.

Data reported in Table 12.8 show that the proposed technologies are generally characterized by higher complexity of the installations and energy demand in comparison to the conventional anaerobic digestion, but these drawbacks can be compensated by the better performance detected for solid reduction and, in case of agricultural application, for efficient micropollutant removal and increased hygienization.

WO and TT are the solutions characterized by highest energy demand, but this negative aspect can be mitigated in both cases with opportune strategies for the energy recovery. In the case of SAA the additional energy required for mixing and air supply of the aerobic digester is compensated by the energy saving achieved by operating the aerobic step at intermittent aeration, so reducing the nitrogen load recycled to the WWTP and the oxygen demand in the aeration basin. It is also worth noting that the availability of the different reaction environments in SAA (i.e. anaerobic and aerobic) is effective for the removal of micropollutants. Among the analysed alternatives, SAA is easier to install and, at the same time, characterized by lower complexity, and very good performance, so it is suitable also for medium potentialities WWTPs. When higher hygienization level are required, it is mandatory to employ higher temperature digestion processes, and, given the higher energy demand and complexity of installation and operation, the high plant potentiality is a key-factor for their applicability.

12.7 REFERENCES

Ahn K. H., Park K. Y., Mueng S. K., Hwang J. H., Lee J. W., Song K. G. and Choi S. (2002). Ozonation of wastewater and ozonation for recycling. *Water Science and Technology*, **46**(10), 71–77.

Baier U. and Schmidheiny P. (1997). Enhanced anaerobic degradation of mechanically disintegrated biosolids. IAQW International Conference on waste water sludge, Cestochowa, Poland, Part 1, 106–112.

Böhler M. and Siegrist H. (2003). Partial ozonation of activated sludge to reduce excess sludge, improve denitrification and control scumming and bulking. Proceedings IWA Specialised Conference BIOSOLIDS 2003: Wastewater as a Resource. Throndheim, Norway, 23–25 June, 2003.

Bertanza G., Galessi R., Menoni L. and Zanaboni S. (2016). Wet oxidation of sewage sludge from municipal and industrial WWTPs. *Desalination and Water Treatment*, **57**(6), 2422–2427.

Bertanza G., Galessi R., Menoni L., Salvetti R., Slavik E. and Zanaboni S. (2015a). Wet oxidation of sewage sludge: process modeling and full scale experience. *Environmental Science and Pollution Research*, **22**, 7306–7316.

Bertanza G., Galessi R., Menoni L., Pedrazzani R., Salvetti R. and Zanaboni S. (2015b). Anaerobic treatability of liquid residue from wet oxidation of sewage sludge. *Environmental Science and Pollution Research*, **22**, 7317–7326.

Bertanza G., Canato M., Heimersson S., Laera G., Salvetti R., Slavik E. and Svanström M. (2015c). Techno-economic and environmental assessment of sewage sludge wet oxidation. *Environmental Science and Pollution Research*, **22**, 7327–7338.

Bolzonella D., Pavan P., Battistoni P. and Cecchi F. (2005). Mesophilic anaerobic digestion of waste activated sludge: influence of the solid retention time in the wastewater treatment process. *Processes Biochemistry*, **40**, 1453–1460.

Braguglia C. M., Gianico A. and Mininni G. (2011). Laboratory-scale ultrasound pre-treated digestion of sludge: Heat and energy balance. *Bioresource Technology*, **102**, 7567–7573.

Braguglia C. M., Coors A., Gallipoli A., Gianico A., Guillon E., Kunkel U., Mascolo G., Richte E., Ternes T. A., Tomei M. C. and Mininni G. (2015). Quality assessment of digested sludges produced by advanced stabilization processes. *Environmental Science and Pollution Research*, **22**(10), 7216–7235.

Canales A., Pareilleux A., Rolls J.L., Goma G. and Huyard A. (1994). Decreased sludge production strategy for domestic wastewater treatment. *Water Science and Technology*, **30**, 96–106.

Chang J., Chudoba P. and Capdeville B. (1993). Determination of the maintenance requirement of activated sludge. *Water Science and Technology*, **28**, 139–142.

Chen G. H., Mo H. K., Saby S., Yip W. K. and Liu Y. (2000). Minimization of activated sludge production by chemically stimulated energy spilling. *Water Science and Technology*, **42**(12), 189–200.

Chen G. H., Saby S., Djaer M. and Mo H. K. (2001). New approaches to minimize excess sludge in activated sludge systems. *Water Science and Technology*, **44**(10), 203–208.

Chen G. H., An K. J., Saby S., Brois E. and Djafer M. (2003). Possible cause of excess reduction in an oxic-settling-anaerobic activated sludge process (OSA process). *Water Research*, **37**(16), 3855–3866.

Chiu Y. C., Chang C. N., Lin L. G. and Huang S. J. (1997). Alkaline and ultrasonic pre-treatment of sludge before anaerobic digestion. *Water Science and Technology*, **36**(11), 155–162.

Chudoba B., Morel A. and Capdeville B. (1992b). The case of both energetic uncoupling and metabolic selection of microorganisms in the OSA activated sludge system. *Environmental Technology*, **13**, 761–770.

Chudoba P., Chudoba J. and Capdeville B. (1992a). The aspect of energetic uncoupling of microbial growth in the activated sludge process: OSA system. *Water Science and Technology*, **26**(9–11), 2477–2480.

Chudoba P. (1991). Etude et intérêt du découplage énergetique dans les processus d'epuration des eaux par voie biologique procédé OSA. Ph.D. thesis, Inst Natl des Sci Appl, Toulouse, France, 1991.

Copp J. B. and Dold P. L. (1998). Comparing sludge production under aerobic and anoxic conditions. *Water Science and Technology*, **38**, 285–294.

Egemen E., Corpening J., Padilla J., Brennan R. and Nirmalakhandan N. (1999). Evaluation of ozonation and cryptic growth for biosolids management in wastewater treatment. *Water Science and Technology*, **39**(10–11), 155–158.

Egemen E., Corpening J. and Nirmalakhandan N. (2001). Evaluation of an ozonation system for reduced waste sludge generation. *Water Science and Technology*, **44**(2–3), 445–452.

Gagliano M. C., Braguglia C. M., Gallipoli A., Gianico A. and Rossetti S. (2014). Microbial diversity in innovative mesophilic/thermophilic temperature-phased anaerobic digestion of sludge. *Environmental Science and Pollution Research*, **22**(10), 7339–48.

Garuti G., Dohanyos M. and Tilche A. (1992a). Anaerobic-Aerobic combined process for the treatment of sewage wit nutrient removal: the ANANOX process. *Water Science and Technology*, **25**(7), 383–394.

Garuti G., Dohanyos M. and Tilche A. (1992b). Anaerobic-aerobic wastewater treatment system suitable for variable population in coastal areas: the ANANOX process. *Water Science and Technology*, **25**(12), 185–195.

Ghiglizza R., Lodi A., Converti A., Nicolella C. and Rovarti M. (1996). Influence of the ratio of the initial substrate concentration to biomass concentration on the performance of a sequencing batch reactor. *Bioprocess Engineering*, **14**, 131–137.

Ghyooy W. and Verstraete W. (2000). Reduced sludge production in a two-stage membrane-assisted bioreactor. *Water Research*, **34**, 205–215.

Gianico A., Braguglia C., Cesarini R. and Mininni G. (2013). Reduced temperature hydrolysis at 134°C before thermophilic anaerobic digestion of waste activated sludge at increasing organic load. *Bioresource Technology*, **143**, 96–103.

Gianico A., Braguglia C. M., Gallipoli A. and Mininni G. (2014). Innovative two-stage mesophilic/thermophilic anaerobic degradation of sonicated sludge: performances and energy balance. *Environmental Science and Pollution Research*, **22**(10), 7248–7256.

Guzmán C., Jofre J., Montemayor M. and Lucena F. (2007). Occurrence and levels of indicators and selected pathogens in different sludges and biosolids. *Journal of Applied Microbiology*, **103**(6), 2420–2429.

Kamiya T. and Hirotsuki J. (1998). New combined system of biological process and intermittent ozonation for advanced wastewater treatment. *Water Science and Technology*, **38**(8–9), 145–153.

Kepp U., Machenbach I., Weisz N. and Solheim O. E. (1999). Enhanced stabilisation of sewage sludge through thermal hydrolysis – 3 years of experience with full scale plants. *Water Science and Technology*, **42**(9), 89–96.

Kim J. and Novak J.T. (2011). Combined anaerobic/aerobic digestion: effect of aerobic retention time on nitrogen and solids removal. *Water Environmental Research*, **83**, 802–806.

Kumar N., Novak J. T. and Murthy S. (2006a). Effect of secondary aerobic digestion on properties of anaerobic digested biosolids. Water Environmental Federation 79th Annual Technical Exibition and Conference, Dallas, 6806–6829.

Kumar N., Novak J.T. and Murthy S. (2006b). Sequential Anaerobic-Aerobic Digestion for Enhanced Volatile Solids Reduction and Nitrogen Removal. Water Environmental Federation Residuals and Biosolids Management Conference, Cincinnati, OH, March 12–14.

Kunz P. and Wagner S. (1994). Ergebnisse und Perspektiven aus Untersuchungen zur Klärschlammdesintegration. *awt Abwassertechnik*, **44**(2), 33–40.

Lee N. M. and Welander T. (1996a). Reducing sludge production in aerobic wastewater treatment through manipulation of the ecosystem. *Water Research*, **30**(8), 1781–1790.

Lee N. M. and Welander T. (1996b). Use of protozoa and metazoan for decreasing sludge production in aerobic wastewater treatment. *Biotechnology Letters*, **18**(4), 429–434.

Lehne G., Müller A. and Schwedes J. (2001). Mechanical disintegration of sewage sludge. *Water Science and Technology*, **43**(1), 19–26.

Levantesi C., Beimfohr C., Blanch A. R., Carducci A., Gianico A., Lucena F., Tomei M. C. and Mininni G. (2014). Hygienization performances of innovative sludge treatment solutions to assure safe land spreading. *Environmental Science and Pollution Research*, **22**(10), 7237–7247.

Liu Y. (2000). Effect of chemical uncoupler on the observed growth yield in batch culture of activated sludge. *Water Research*, **34**, 2025–2030.

Low E. W. and Chase H. A. (1998). The use of chemical uncouplers for reducing biomass production during biodegradation. *Water Science and Technology*, **37**(4–5), 399–402.

Low E. W., Chase H. A., Milner M. G., Curtis T. P. (2000). Uncoupling of metabolism to reduce biomass production in the activated sludge process. *Water Research*, **34**(12), 3204–3212.

Luxmy B. S., Kubo T. and Yamamoto K. (2001). Sludge reduction potential of metazoan in membrane bioreactors. *Water Science and Technology*, **44**(10), 197–202.

Martinage V. and Paul E. (2000). Effect of environmental parameters on autotrophic decay rate. *Environmental Technology*, **21**, 31–41.

Mason C. A., Hamer G. and Bryers D. (1986). The death and lysis of microorganisms in environmental processes. *FEMS Microbiology Reviews*, **39**, 373–401.

Mayhew M. and Stephenson T. (1998). Biomass yield reduction: Is biochemical manipulation possible without affecting activated sludge process efficiency? *Water Science and Technology*, **38**(8–9), 137–144.

Menoni L. and Bertanza G. (2016). Wet Oxidation of sewage sludge: a mathematical model for estimating the performance based on the VSS/TSS ratio. *Chemical Engineering Journal*, **306**, 685–692.

Mukherjee S. R. and Levine A. D. (1992). Chemical solubilization of particulate organics as a pre-treatment approach. *Water Science and Technology*, **26**(9–11), 2289–2292.

Müller J. (1996). Mechanischer Klärschlammaufschluβ, Ph.D. thesis, Technical University of Braunschweig, ISBN 3-82-2053.

Müller J. (2000). Disientegration as a key-step in sewage sludge treatment. *Water Science and Technology*, **41**(8), 123–130.

Müller J. A. (2001). Prospects and problems of sludge pre-treatment processes. *Water Science and Technology*, **44**(10), 121–128.

Müller J. A., Winter A. and Strünkmann G. (2004). Investigation and assessment of sludge pre-treatment processes. *Water Science and Technology*, **49**(10), 97–104.

Novak J. T., Banjade S., Murthy S. N. (2011). Combined anaerobic-aerobic digestion for increased solids reduction and nitrogen removal. *Water Research*, **45**, 618–624.

Okey R. W. and Stensel D. H. (1993). Uncouplers and activated sludge – the impact on synthesis and respiration. *Toxicology Environmental Chemical*, **40**, 235–54.

Parravicini V., Svarda K., Hornek R. and Kroiss H. (2008). Aeration of anaerobically digested sewage sludge for COD and nitrogen removal: optimization at largescale. *Water Science and Technology*, **57**, 257–264.

Ratsak Ch, Kooijman S. A. L. M. and Kooi B. W. (1993). Modelling the growth of an oligochaete on activated sludge. *Water Research*, **27**(5), 739–747.

Rensik J. H. and Rulkens W. H. (1997). Using metazoan to reduce sludge production. *Water Science and Technology*, **36**(11), 171–179.

Rosenberger S., Kruger U., Witxig R., Manz W., Szewzyk U. and Kraume M. (2002). Performance of a biorreactor with submerged membranes for aerobic treatment of municipal wastewater. *Water Research*, **36**(2), 413–420.

Saby S., Djafer M., Chen G. H. (2002). Feasibility of using a chlorination step to reduce excess sludge in activated sludge process. *Water Research*, **36**(3), 656–666.

Sakai Y., Aoyagi T., Shiota N., Akashi A. and Hasegawa S. (2000). Complete decomposition of biological waste sludge by thermophilic aerobic bacteria. *Water Science and Technology*, **42**(9), 81–88.

Sakay Y., Fukase T., Yasui H. and Shibata M. (1997). An activated sludge process without sludge production. *Water Science and Technology*, **36**(11), 163–170.

Shiota N., Akashi A. and Hasegawa S. (2002). A strategy in wastewater treatment process for significant reduction of excess sludge production. *Water Science and Technology*, **45**(12), 127–134.

Speece R. E. (1988). A survey of municipal sludge digesters and diagnostic activity assays. *Water Research*, **22**, 365–372.

Tomei M. C. and Carozza N. A. (2015). Sequential anaerobic/anaerobic digestion for enhanced sludge stabilization: comparison of the process performance for mixed and waste sludge. *Environmental Science and Pollution Research*, **22**, 7271–7279.

Tomei M. C., Rita S. and Mininni G. (2011). Performance of sequential anaerobic/aerobic digestion applied to municipal sewage sludge. *Journal of Environmental Management*, **92**, 1867–1873.

Tomei M. C., Mosca Angelucci D. and Levantesi C. (2016). Two-stage anaerobic and post-aerobic mesophilic digestion of sewage sludge: Analysis of process performance and hygienization potential. *Science of the Total Environment*, **545–546**, 453–464.

Tsai S. P. and Lee Y. H. (1990). A model for energy-sufficient culture growth. *Biotechnology Bioengineering*, **35**, 138–145.

Wagner J., Rosenwinkel K. H. (2000). Sludge production in membrane bioreactors under different conditions. *Water Science and Technology*, **41**(10–11), 251–258.

Weise Th H. G. G. and Jung M. (1998). Klärschlammbehandlung mit der Hochleistungspulstechnik. *Veröffentlichung des Instituts für Siedlungswasserwirtschaft der TU Braunschweig, Heft*, **61**, 75–82.

Weise Th H. G. G. and Jung M. (2001). Sludge disintegration by electrical shockwaves. International Conference on Pulsed Power Applications. Gelsenkirchen, March 2001.

Welander T. and Lee N. M. (1994). Minimization of Sludge Production in Anaerobic Treatment by use of Predators. The Second International Symposium on Environmental Biotechnology, Brigton, UK, 4–6 July 1994.

Westgarth W. C., Sulzer F. T. and Okun D. A. (1964). Anaerobiosis in the activated sludge process. Proceedings of the Second IAWPRC Conference. Tokyo, 1964. pp. 43–55.

Winter A. (2002). Minimisation of costs by using disintegration at a full-scale anaerobic digestion plant. *Water Science and Technology*, **46**(4–5), 405–412.

Yamamoto K., Hiasa M., Mahmood T. and Matsuo T. (1989). Direct solid-liquid separation using hollow fibre membrane in an activated sludge aeration tank. *Water Science and Technology*, **21**, 43–54.

Yasui H. and Shibata M. (1994). An innovative approach to reduce excess sludge production in the activated sludge process. *Water Science and Technology*, **30**(9), 11–20.

Part 2

Re-using Water and Sludge

Chapter 13

Producing high-quality recycled water

Jörg E. Drewes, Nils Horstmeyer, Philipp Michel and Stuart Khan

13.1 INTRODUCTION

An increasing number of regions worldwide have suffered from water scarcity in the recent past. This has resulted in a growing interest in utilizing unconventional water resources to augment drinking water supplies (Drewes & Khan, 2015; Drewes & Horstmeyer, 2016). Water reuse is increasingly being recognized as an effort to re-use water as part of a 3R concept in contemporary urban water systems design (Hering *et al.* 2013). This chapter will discuss options to produce high-quality recycled water for potable reuse applications through reuse of municipal wastewater effluents. In addition, the role of an environmental buffer will be discussed in particular in lieu of recent developments to transition from indirect to direct potable reuse. The key water quality issues will be addressed as well as current and potentially future regulatory requirements, which will determine treatment scheme design. Different treatment scheme approaches to potable water reuse will be compared regarding water quality but also regarding their specific energy requirements. The chapter will also touch upon recent developments to combine water reuse and energy recovery concepts driving innovation in contemporary design of water reclamation and reuse schemes.

13.2 WATER QUALITY CONSTITUENTS OF CONCERN AND REGULATORY REQUIREMENTS

Health risks in potable water reuse applications are primarily associated with microbial and chemical contaminants arising from many sources that can have adverse effects on human health (Drewes & Khan, 2011). In addition, an important consideration for public acceptance of potable water reuse projects are aesthetic issues related to taste and odor that need to be addressed through treatment (Agus *et al.* 2013). While conventional wastewater treatment in many locations provides an effluent quality that is suitable to be discharged to streams or oceans, treated effluents are still composed of a wide range of naturally occurring and anthropogenic trace organic and inorganic contaminants, residual nutrients, total dissolved solids, residual heavy metals, and pathogens (Drewes & Khan, 2011). Microbial contaminants including bacteria, viruses and protozoan parasites are acknowledged as the most critical constituent in reclaimed water due to potential acute human health impacts in public water supplies. Chemical contaminants, of which a large number can still be present in reclaimed water, can be of concern due to potential adverse acute and chronic health effects (NRC, 2012).

In order to quantify the potential for human health effects as a result of exposure to microbial and chemical contaminants, regulatory agencies have adopted the concept of a 'tolerable level of risk' to assist in setting water quality guidelines or standards. In the regulatory realm, *de minimis* risk, which is defined as a level of risk characterized by the risk being virtually non-existent to describe risks that are 'below regulatory concerns'. Traditionally, for drinking water supplies, *de minimis* risk levels are related to public health criteria (i.e. toxicity of the constituent; characteristics of the population; exposure). Different risk levels are commonly used, depending on the specific situation and type of contaminant. The United States Environmental Protection Agency (EPA), Office of Drinking Water, uses a 'regulatory window' of 10^{-6} to 10^{-4} risk per person per year for evaluation of risk where 10^{-4} is the baseline risk for all regulations and 10^{-6} is the *de minimis* risk level (EPA, 1992). Performance goals for potable water reuse projects for instance in California have been proposed that are based on a low tolerable or *de minimis* risk level of 10^{-4} annual risk of infection and occurrence data of pathogens in raw wastewater (NWRI, 2013). These tolerable risk levels refer to the final drinking water quality and have resulted in proposals of performance goals for viruses and protozoa (Table 13.1).

Table 13.1 Performance goals for microbial contaminants for potable water reuse projects in California.

	Enteric Virus	*Giardia*	*Cryptosporidium*
Required log reduction	12	10	10

Source: NWRI, 2013.

Performance goals for chemical contaminants for a proposed potable water reuse scheme should include contaminants of recognized health concern that have published guideline values or standards (regulated and unregulated contaminants) as well as surrogate measures for bulk water characteristics that can provide assurance for proper removal of chemical contaminants in potable reuse schemes (Figure 13.1). Regulated contaminants include chemicals with an acceptable health risk specified, for example, as primary or secondary maximum contaminant levels (MCLs), and as notification levels commonly used in California. For unregulated contaminants with available toxicological information, a *de minimis* risk approach can be used, where a reference dose (RfD), acceptable daily dose (ADD), or predicted no-effect concentration (PNEC) that expresses their toxicological relevance can be adopted (Schwab *et al.* 2005; Snyder *et al.* 2008; Bull *et al.* 2011; Khan, 2013).

Performance validation and verification of treatment trains for potable reuse projects can be obtained through direct measurements of certain performance-based indicator contaminants that correlate with the core removal mechanisms (i.e., biotransformation, adsorption, size exclusion, chemical oxidation) of individual unit processes (Drewes *et al.* 2008; Dickenson *et al.* 2009, 2011). The following factors need to be considered for the selection of performance-based indicator contaminants for the assessment of treatment efficacy in potable water reuse treatment schemes (Drewes & Horstmeyer, 2016):

- Target contaminants chosen to assess treatment performance must occur frequently enough and at concentrations significantly above their analytical method detection limit (preferably the ratio between measured environmental concentration and method detection limit should exceed 10).
- Appropriate and commercially available analytical methods must exist to quantify the target contaminants in reclaimed water.

- The performance-based indicator contaminants used for monitoring should broadly represent the range of physicochemical and biological properties affecting their removal by the various unit processes within a potable reuse treatment train.

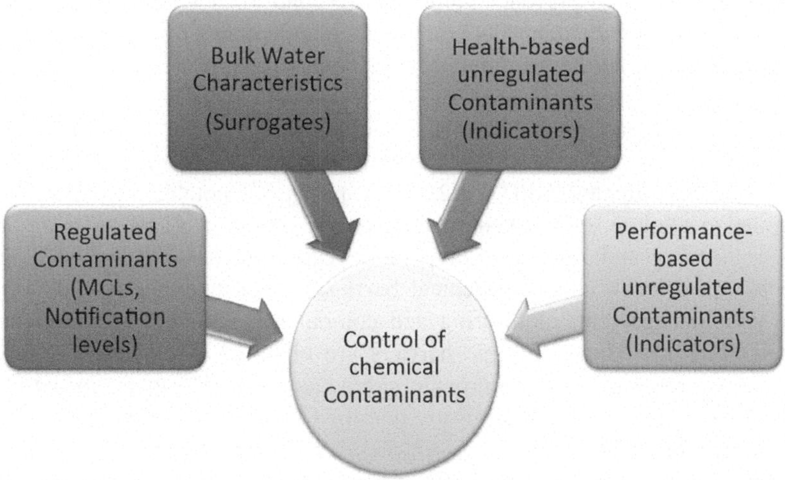

Figure 13.1 Classes of chemical contaminants that should be considered in monitoring programs of potable reuse projects.

In addition, specific performance-based bulk measurements that can be monitored continuously (e.g. electrical conductivity, UV absorbance) can serve as surrogate parameters for contaminants of interest whose removal is similar. Such surrogate parameters can also indicate out-of-specification performance or treatment process failure (Drewes *et al.* 2008, 2010; Wert *et al.* 2009). Such approaches have the advantage that they can be established as real-time monitoring strategies where high-resolution system performance control is desired.

13.3 TREATMENT SCHEMES FOR POTABLE WATER REUSE

Drinking water augmentation using reclaimed water is practiced in the form of indirect potable reuse (IPR) for more than 50 years in the USA, Singapore, Europe and Australia (Drewes & Khan, 2015; Drewes & Horstmeyer, 2016). Thus, a long track record exists worldwide that planned potable reuse can be practiced safely without compromising public health (NRC, 2012; Khan, 2013). A key element of an indirect potable reuse system is its reliance on an environmental buffer (e.g. groundwater aquifer, reservoir). While some environmental buffers might offer opportunities for further treatment (Drewes & Khan, 2011), the core functions of the environmental buffer are to provide – through storage – some level of water quality equalization and time to respond to any process failures or out-of-compliance water quality monitoring results. While eliminating the environmental barrier in direct potable reuse (DPR), where reclaimed water is directly piped into the raw water supply of a drinking water facility or into the drinking water distribution system, these core functions need to be replaced (Leverenz *et al.* 2011). Options to replace these functions exist by considering additional treatment barriers, engineered storage, and real-time process control.

The main goal of potable water reuse while producing high-quality recycled water is to provide a high degree of consistent public health protection (reliability). *Reliability* of a potable reuse system is defined as the ability to verifiably deliver a water quality that consistently exceeds the public health protection expected of conventional drinking water supplies. The goal of reliably exceeding public health protection is accomplished by concepts that prevent failures and respond to those that occur. The prevention of failures is achieved by adding redundancy and robustness to a treatment train. *Redundancy* advocates for the addition of measures beyond the minimum requirements to ensure treatment goals are reliably met and performance targets are consistently achieved if not exceeded. *Robustness* is the ability of a potable reuse system to address a broad variety of contaminants and resist catastrophic failures. In particular considering the broad variety of chemical contaminants, robustness is the use of a diversity of barriers to control a variety of contaminants (NRC, 2012). The capacity of a potable reuse system to successfully adapt and/or respond to a failure is described as *resilience*.

Redundancy, robustness and resilience in potable reuse schemes are provided through use of multiple technical, operational, and management barriers, which together provide opportunities for risk prevention, risk management, monitoring and compliance, and individual action (NRC, 2012; Tchobanoglous *et al.* 2015). Here, multiple barriers are obtained through use of several different treatment processes operated in series to provide redundancy and robustness in the removal of both pathogens and unwanted chemicals, and to insure that the failure of a single process does not result in the failure of the entire treatment system.

The multiple barrier concept is comprised of five levels: (a) source control program; (b) conventional wastewater treatment; (c) advanced water treatment; (d) management of the environmental (engineered) buffer; and (e) drinking water treatment including management of the distribution system (Drewes & Khan, 2011). Among the technical barriers are engineered physical, chemical, and biological processes, each of which removes multiple classes of contaminants. The concept of multiple barriers is a core design principle in producing high-quality recycled water and in particular in potable water reuse.

13.4 ENERGY EFFICIENCY OF POTABLE WATER REUSE SCHEMES

Potential conventional process combinations of water reuse schemes are presented in Figure 13.2. These treatment combinations usually utilize secondary or tertiary treated effluents from conventional biological wastewater treatment. Advanced water treatment (AWT) is applied worldwide in different scenarios to generate high-quality effluent for potable reuse. For example, common treatment combinations for AWT are: (1) microfiltration (MF), reverse osmosis (RO) and advanced oxidation processes (UV/H_2O_2) (Orange County, USA); (2) ultrafiltration (UF), biologically-active activated carbon filtration (BAC), ozone, UV disinfection (Gwinnett County, USA); or (3) ozone, BAC, chlorine, UV disinfection (Melbourne, Australia). The AWT processes commonly utilized in potable reuse schemes do increase the energy footprint significantly. The total energy footprint of potable water reuse schemes using AWT can vary between 0.9 and 1.8 kWh/m^3. There are opportunities to reduce this significant energy demand by considering elements of the 3R concept (see section *Design Requirements of Potable Water Reuse Schemes/Energy Potential*).

The energy requirements and associated carbon footprint of selected treatment processes for conventional drinking water facilities, conventional biological wastewater treatment, and advanced treatment processes representing barriers to pathogens and chemical constituents are listed in Table 13.2.

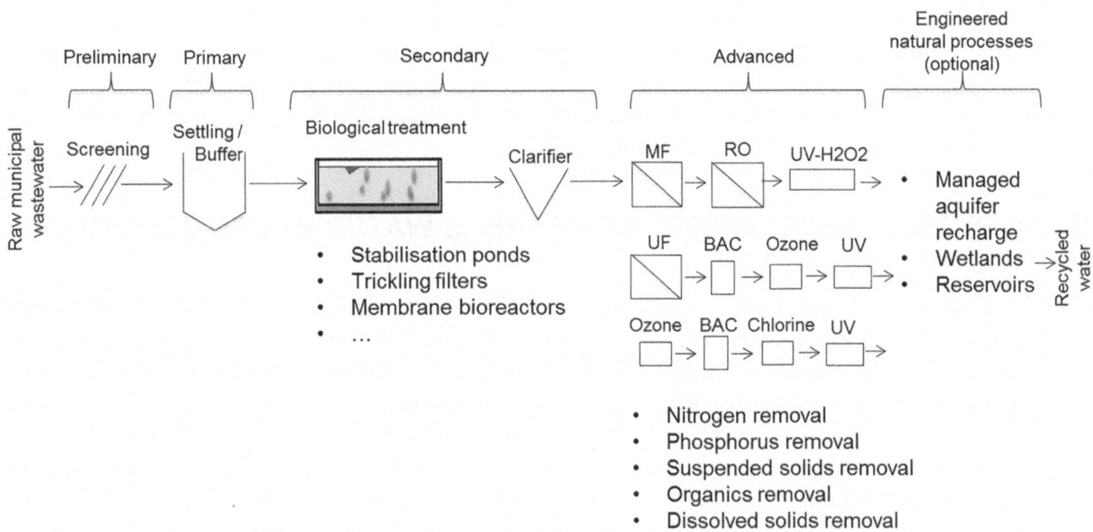

Figure 13.2 Process combinations of water reuse schemes using advanced treatment process to produce high-quality water.

Table 13.2 Average energy requirements and CO_2-Footprint of conventional drinking water augmentation, wastewater treatment, and water reuse processes.

Process	Energy Demand		CO_2-Footprint ($kg\ CO_2/m^3$)
	Range (kWh/m^3)	Typical (kWh/m^3)	
Conventional drinking water treatment	0.1–1.13	0.1	0.05
Conventional biological wastewater treatment without nutrient removal	0.37–1.0	0.33	0.17
Conventional biological wastewater treatment without nutrient removal and subsequent filtration	0.29–1.22	0.49	0.25
MF – RO – UV/H_2O_2 – Stabilization – Chlorination	0.86–1.06	0.95	0.48
O_3	0.04–0.17	0.05	0.06–0.13
PAK	0.03–0.04	0.03	0.15–0.24
MF, UF	0.06–0.1	0.07	–
NF	0.16–0.25	0.2	–
Brackish water desalination	0.82–1.64	1.55	0.6
Seawater desalination	2.51–3.9	3.17	1.59

MF = Microfiltration, RO = Reverse osmosis, UV = Ultraviolet light treatment, H_2O_2 = Hydrogen peroxide, O_3 = Ozonation, PAC = Powdered activated carbon, UF = Ultrafiltration, NF = Nanofiltration.
Source: Tchobanoglous *et al.* 2015; Lazarova *et al.* 2012; Bolle and Pinnekamp, 2015.

Energy optimizations and additional savings of individual unit operations and processes employed in water reuse schemes have either already been implemented or have limited potential. However, integrating energy and heat recovery concepts into water reuse schemes might have a significantly larger savings potential. This could be accomplished by implementing anaerobic biological processes and subsequent energy utilization with combined heat and power (CHP) units.

13.5 DESIGN REQUIREMENTS OF POTABLE WATER REUSE SCHEMES/ ENERGY POTENTIAL

The design of potable water reuse schemes delivering high-quality recycled water requires a thorough understanding of (i) source water characteristics (source control, flow equalization); (ii) the performance of reliable treatment systems (redundancy, robustness, resilience, emergency facilities); and (iii) storage and blending (interactions with distribution systems, mineral balance, disinfection residual, blending with other supplies) Drewes and Horstmeyer (2016). Any contemporary potable water reuse scheme aimed to be more efficient with respect to energy and nutrient recovery should be designed as an integrated concept while considering the site-specific conditions, plant capacity, alternative available freshwater sources, and water pollution control. In considering these aspects, water reuse schemes can be extended by energy-recovery, nutrient recovery, and heat recovery concepts while focusing on core processes and process combinations with high-energy demand and considering potentially more energy-efficient alternatives with similar performance.

By identifying high-energy unit processes and process combinations, it is possible to potentially consider alternative treatment steps. The substitution of energy-intensive biological treatment in particular aeration during conventional activated sludge treatment by coupling biological, chemical and physical treatment processes could reduce the overall energy demand while providing more organic material present in primary effluent for increased biogas production. Juby (2013) proposed an alternative wastewater treatment train compromised of primary clarification (sedimentation) followed directly by MF (or UF), RO (or NF), and UV/H_2O_2. In this scheme, primary solids are diverted towards anaerobic digestion for subsequent energy recovery. The soluble organic matter is able to pass the membranes (MF/UF) and is still available for later energy recovery. Solids rejected by the MF or UF treatment are also diverted to the anaerobic digestion process utilizing primary solids for energy recovery.

The chemical energy content present in wastewater is the energy contained in organic matter which could be released by chemical reactions (Tchobanoglous *et al.* 2014). The concentration of organic molecules in wastewater is commonly expressed by the chemical oxygen demand (COD), a parameter which sums up all organic constituents in the wastewater (Tchobanoglous *et al.* 2014). Alternative treatment schemes could consist of an improved physical separation of the particulate organic matter by using a microsieve (100 μm) resulting in increased primary sludge removal and downstream membrane filtration. Municipal raw wastewater contains high concentrations of organic matter (around 500–1000 mg/L COD). The energy content per gram of oxidizable material represents approximatelty 13.9 kJ/g COD (=3.86 kWh/kg COD) (Heidrich *et al.* 2011). This results in a theoretical energy content of 169 kWh/person, year or 1.93–3.86 kWh/m³ in raw wastewater (assuming 120 g COD/person, day or 500–1000 mg/L COD). The increased primary sludge removal can result in an improved biogas generation of approximately 50% more compared to conventional pre-treatment. In addition, this treatment approach results in a reduced organic load on subsequent purification steps (up to 70–80% of the total COD of the raw sewage influent are removed).

McCarty *et al.* (2011) quantified the energy potential in municipal wastewater treatment by recovering energy from organic matter with approximately 1.93 kWh/m³ (assuming low COD concentrations of 500 mg/L in raw sewage). Anaerobic treatment with an anaerobic fluidized membrane bioreactor (AFMBR) utilizing a granular activated carbon (GAC) fouling control strategy exhibited significant lower energy demand (total energy demand 0.058 kWh/m^3) compared to conventional membrane bioreactor designs treating municipal wastewater (500 mg/L COD) (McCarty *et al.* 2011; Kim *et al.* 2011). However, an important issue that needs to be addressed is the management of nutrients and in particular removal of ammonia, which is not converted in this treatment train.

Alturki *et al.* (2010) investigated combining membrane bioreactors (MBR) with nanofiltration (NF) or RO. Results revealed that the MBR effectively removes hydrophobic and biodegradable trace organic compounds. The remaining organic constituents, mostly hydrophilic trace organic compounds, were effectively removed by NF or RO treatment. Other studies have also demonstrated that alternative integrated membrane systems consisting of UF/NF are more energy-efficient than UF/RO while providing a similar permeate quality for many key water quality constituents (Bellona *et al.* 2012; Shahmansouri & Bellona, 2015). Contemporary (potable) water reuse schemes following the '3R' concept are characterized by (i) recovery of energy and nutrients; (ii) reuse of (potable) water; and (iii) reduction the energy footprint. By following the '3R' concept the alternative supply becomes also a constant and drought-independent available freshwater source by closing local and regional water cycles, a more tailored treatment design to meet local water quality requirements (*'fit for purpose'*), and a more flexible treatment system that can cope with demographic changes, load variations, and changing water requirements.

13.6 STATE-OF-THE-ART WATER QUALITY MONITORING APPROACHES FOR HIGH-QUALITY RECYCLED WATER

Currently, regulatory strategies for monitoring programs where high-quality product water is desired offer different approaches to ensure a consistent water quality. The most common approach is to define threshold values of the target contaminants in the final product water. These threshold values are commonly based on health-effect studies, which estimate the acute and chronic risk to human health (or the environment) providing an exposure risk. These defined thresholds are embedded into a monitoring and control concept of treatment schemes designed to provide high-quality water. While these schemes can differ among countries, the threshold values for certain contaminants, since they are based on scientific evidence, usually are similar.

Water quality management of projects delivering high-quality product water depends on many factors and variables and represents a complex dynamic system. The major elements of that dynamic system are the source water (i.e. dynamic flow, dilution processes, industrial and commercial discharger), the treatment system (i.e. dynamic contaminant load and inherent system variability), and other holistic parameter, which can influence the natural degradation of several contaminants in the environmental buffer (e.g. variably sun irradiation triggering photolytic degradation processes). In traditional water quality schemes these dynamic changes are rarely considered and the different forms of dynamics are compensated by buffer values or safety factors added to the threshold. This is assumed to address the variability of complex dynamic systems, but there is no anticipation when these systems will occur and what the overall variability might be. Generally, anticipation of water quality changes needs four core elements:

- Continuous monitoring and a reliable continuous simulator prediction
- Dynamic system operation
- Scientific model
- A cost-benefit approach

The last point might not be of primary focus in potable reuse projects where protection of public health is concerned, but it has been demonstrated that systems operating by a cost-benefit incentive are more likely to be implemented in today's world of tight economic budgets. For traditional contaminants (e.g. nitrate, phosphorous), advanced dynamic water quality scheme might not be needed due to cost intense monitoring. However, for contaminants with acute and chronic risk (e.g. pathogens, certain trace organic chemicals) monitoring of water quality variability might be more appropriate. Based on a literature review, three typical monitoring approaches dealing with system dynamics have been identified:

- Static regulatory standards but dynamic treatment approach to meet the target concentrations.
- Static regulatory standards, but extended with integrated seasonal dilution factor which is then a form of variability.
- Regulatory authorities accept a dynamic treatment goal (e.g. different seasons, different reuse applications) and adapt the regulatory thresholds, which could be in different dynamic resolutions (e.g. minutes, hours day months or seasons). This can be again combined with treatment strategies offering dynamic characteristics.

In this context the usage of advanced stochastically simulation and predictions should be emphasized. Although research is able to provide a diverse range of such predictive programs to simulate contaminant faith, no real world example can be found where these simulations are accepted and integrated into a regulatory real-time water quality scheme. However, a dynamic real-time water quality scheme strongly increased monitoring complexity and eventually increased cost. Conventional treatment targets in the past have been met using static thresholds, where the main purpose was to discharge to the aquatic environment. The requirements for treatment efficiency in these conventional treatment plants did not justify a new thinking in terms of advanced anticipation to different system dynamics. However, in providing high-quality recycled water significantly different approaches are needed when dealing with complex contaminants and mixtures including emerging trace organic chemicals or pathogens. Being able to adapt to those challenges will make the difference in term of cost efficiency and more appropriately optimizing removal of critical contaminants.

Advanced wastewater treatment technologies (e.g. ozonation, UV/hydrogen peroxide, activated carbon adsorption, and membrane technologies) have been proposed to reduce contaminant in potable reuse schemes. However, these advanced treatment technologies are energy-intensive, operationally complex and expensive. As such, the investment of implementing advanced treatment should be balanced against reliably meeting public health goals versus environmental impacts and financial costs. The relationships between these benefits and costs will vary between locations and also over time, for instance as a function of seasonal or even daily variations. In balancing the benefits and costs of advanced water treatment, there are potential opportunities to establish more flexible modes of operation, for instance by incorporating a dynamic load-dependent dosing regime (e.g. for ozonation) or a dynamic hydraulic flux adjustment (e.g. for membrane systems). However, the ability to apply these flexible and potentially more ideal operational modes will depend upon our ability to dynamically assess and respond to a range of diverse factors affecting treatment process performance and its desired outcome. These aspects may include considering variable concentrations of contaminants in raw wastewater, the inherent treatment process performance, environmental dilution, natural attenuation capability, variable exposure factors as well as energy costs and energy sources.

As mentioned previously, static thresholds respectively standards have been stated as the central elements regulatory authorities use for water quality management. Stepping into responsibility for the upcoming treatment challenges this won't be sufficient. In a general scheme the treatment situation can be visualized in Figure 13.3.

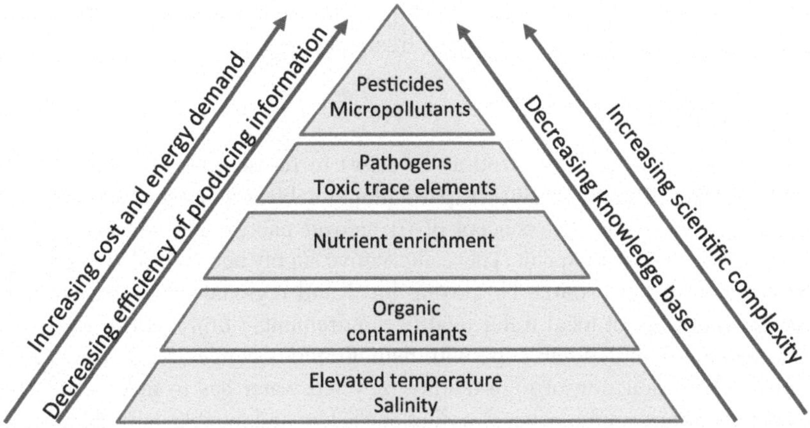

Figure 13.3 Hierarchical complexity of wastewater treatment in producing high-quality water.

Figure 13.3 illustrates that advanced treatment in particular of emerging contaminants achieves limits in the following 4 criteria.

- Scientific complexity
- Limited knowledge base
- Highest treatment cost
- Most limited information about the contaminant

This position raises the question whether the traditionally used static threshold approach is comprehensive enough to fulfill the various requirements for treatment process design. For emerging micropollutants the scientific complexity limits direct or strongly simplified monitoring concepts. Therefore, only very advanced simulations techniques appear to be capable to make realistic predictions about real-time contaminant concentrations. These predictive models make use of advanced stochastic simulations and allow probabilistic predictions. Once a threshold based on probabilities is accepted it is the next step to optimize the prediction by knowing as much as possible about surrogate and system parameters. This knowledge is likely to be triggered by holistic and treatment plant specific parameters. Eventually the system and schematic real-time knowledge will be increasingly advanced. This allows a dynamic operation of the treatment plant to reduce the high treatment costs associated with the removal of these contaminants as illustrated Figure 13.3.

Nevertheless, this is not the only way to optimize the cost-benefit ratio between treatment cost and target removal. There are natural processes of the environmental buffer of indirect potable reuse schemes, which decrease the contaminant concentration immediately or in an acceptable time scale. This might include dilution and natural attenuation processes in a reservoir or groundwater aquifer. They can cause a tremendous reduction of the contaminant concentration and are provided at low or no cost. The only price is the real-time knowledge and management of those additional processes. Those advanced dynamic treatment schemes and scenarios require a dynamic threshold management and it is necessary that this can be transparently displayed. There will be a fundamental change in designing monitoring programs for facilities producing high-quality recycled water. Benchmarking approaches of water reclamation facilities will be combined with regulatory targets. This will accelerate the demand for smart monitoring technologies and push shared data and also big data processing in a reuse scheme designed to deliver high-quality water. The traditional static way of dealing with contaminations is not impossible or unlikely, but cost pressure will favor intelligent and dynamic treatment approach.

13.7 CONCLUSIONS

Water reuse is increasingly being recognized as an effort to re-use water as part of a '3R' concept in contemporary urban water systems design. In particular potable water reuse schemes providing high-quality recycled water follow the '3R' concept of (i) recover energy and nutrients; (ii) reuse (potable) water; and (iii) reduce the energy footprint. These alternative supply becomes also a constant and drought-independent available freshwater source by closing local and regional water cycles, providing a more tailored treatment design to meet local water quality requirements ('*fit for purpose*'), and establishing a more flexible treatment system that can cope with demographic changes, load variations, and changing water requirements. The production of high-quality recycled water has to meet stringent water quality requirements. Novel monitoring approaches to assure consistent and reliable delivery of high-quality water are needed to appropriately address source water and inherent treatment process train variability rather than just meeting static final product water threshold levels.

13.8 REFERENCES

Agus E., Lim M. H., Zhang L. and Sedlak D. L. (2011). Odorous compounds in municipal wastewater effluent and potable water reuse systems. *Environmental Science & Technology*, **45**(21), 9347–9355.

Alturki A. A., Tadkaew N., McDonald J. A., Khan S. J., Price W. E. and Nghiem L. D. (2010). Combining MBR and NF/RO membrane filtration for the removal of trace organics in indirect potable reuse applications. *The Journal of Membrance Science*, **365**, 206–215.

Bellona C., Heil D., Yu C., Fu P. and Drewes J. E. (2012). The pros and cons of using nanofiltration in lieu of reverse osmosis for indirect potable reuse applications. *Separation and Purification Technology*, **85**, 69–76.

Bolle F. and Pinnekamp J. (2015). Energiebedarf von Verfahren zur Elimination von organischen Spurenstoffen – Phase II. Abschlußbericht Ministerium für Klimaschutz, Umwelt, Landwirtschaft, Natur- und Verbraucherschutz des Landes Nordrhein-Westfalen Aktenzeichen IV-7-042 600 003 J.

Bull R., Crook J., Whittaker M. and Cotruvo J. (2011). Therapeutic dose as the point of departure in assessing potential health hazards from drugs in recycled municipal wastewater. *Regulatory Toxicology and Pharmacology*, **60**(1), 1–19.

Dickenson E. R. V., Drewes J. E., Sedlak D. L., Wert E. and Snyder S. A. (2009). Applying surrogates and indicators to assess removal efficiency of trace organic chemicals during chemical oxidation of wastewater. *Environmental Science and Pollution Research*, **43**, 6242–6247.

Dickenson E., Drewes J. E., Snyder S. A. and Sedlak D. L. (2011). Indicator compounds: an approach for using monitoring data to quantify the occurrence and fate of wastewater-derived contaminants in surface waters. *Water Research*, **45**, 1199–1212.

Drewes J. E. and Khan S. (2011). Water reuse for drinking water augmentation. In: Water Quality and Treatment, J. Edzwald (ed.) 6th Edition, American Water Works Association, Denver, Colorado, 16.1–16.48.

Drewes J. E. and Khan S. (2015). Contemporary design, operation and monitoring of potable reuse systems. *Jouranl Water Reuse and Desalination*, **5**(1), 2–7.

Drewes J. E. and Horstmeyer N. (2016). Recent Developments in Potable Reuse. D. Fatta-Kassinos *et al.* (eds.), Advanced Treatment Technologies for Urban Wastewater Reuse, Springer International Publishing, Switzerland. *Handbook of Environmental Chemistry*, **45**, 269–290.

Drewes J. E., Sedlak D., Snyder S. and Dickenson E. (2008). Development of Indicators and Surrogates for Chemical Contaminant Removal during Wastewater Treatment and Reclamation. Final Report. WateReuse Foundation, Alexandria, VA.

Drewes J. E., Khan S. J., McDonald J. A., Trang T. T. T. and Storey M. V. (2010). Chemical monitoring strategy for the assessment of advanced water treatment plant performance. *Water Science and Technology: Water Supply*, **10**(6), 961–968.

Environmental Protection Agency (1992). Guidelines for Exposure Assessment. EPA/600/Z-92/001. *Federal Register* 57(104), 22888–22938. Washington, D.C.

Heidrich E. S., Curtis T. P. and Dolfing J. (2011). Determination of the internal chemical energy of wastewater. *Environmental Science & Technology*, **45**, 827–832.

Hering J. G., Waite T. D., Luthy R. G., Drewes J. E. and Sedlak D. L. (2013). A changing framework for urban water systems. *Environmental Science and Pollution Research*, **47**(19), 10721–10726.

Juby G. (2013). Starting over: Moving closer to energy independence in water reclamation. *Water Environment and Technology*, **25**, 52–56.

Khan S. (2013). Drinking water through recycling. A Report of a Study by the Australian Academy of Technological Sciences and Engineering (ATSE). Melbourne.

Lazarova V., Choo K. and Cornel, P. (2012). Water-Energy Interactions in Water Reuse. IWA Publishing. London, UK.

Leverenz H. L., Tchobanoglous G. and Asano T. (2011). Direct potable reuse: a future imperative. *J Water Reuse Desalination*, **1**(1), 2–10.

McCarty P. L., Bae J. and Kim J. (2011). Domestic Wastewater Treatment as a Net Energy Producer – Can This be Achieved. *Environmental Science & Technology*, **45**, 7100–7106.

National Research Council (NRC). (2012). Water Reuse – Potential for Expanding the Nation's Water Supply Through Reuse of Municipal Wastewater. *The National Academies Press*, Washington, DC.

National Water Research Institute (NWRI). (2013). Final Report Examining the Criteria for Direct Potable Reuse Recommendations of an NWRI Independent Advisory Panel. Prepared for WateReuse Research Foundation Project WRRF-11-02, Fountain Valley, CA.

Schwab B. W., Hayes E. P., Fiori J. M., Mastrocco F. J., Roden N. M., Cragin D., Meyerhoff R. D., D'Aco V. J. and Anderson P. D. (2005). Human pharmaceuticals in US surface waters: A human health risk assessment. *Regulatory Toxicology and Pharmacology*, **42**(3), 296–312.

Shahmansouri and Bellona. (2015). Nanofiltration technology in water treatment and reuse: applications and costs. *Water Science and Technology*, **71**, 309–319.

Snyder S. A., Trenholm R., Snyder E. M., Bruce G. M., Pleus, R. C. and Hemming J. D. (2008). Toxicological relevance of EDCs and Pharmaceuticals in Drinking Water. Awwa Research Foundation Report. Denver, CO: Water Research Foundation.

Tchobanoglous G., Stensel H. D., Tsuchihashi R., Burton F. L., Bowden G., Abu-Orf M. and Pfrang W. (2014). Wastewater Engineering: Treatment and Resource Recovery, 5th edition. McGraw-Hill Education, New York.

Tchobanoglous G., Cotruvo J., Crook J., McDonald E., Olivieri A., Salveson A. and Trussell R. S. (2015). Framework for Direct Potable Reuse. WateReuse Association, Alexandria, VA.

Wert E., Rosario-Ortiz F., Snyder S. (2009). Using UV absorbance and color to assess the oxidation of pharmaceuticals during the ozonation of wastewater. *Environmental Science & Technology*, **43**(13), 4858–4863.

Chapter 14

Producing sludge for agricultural applications

Joan Colón, Manuel Alarcón, Mark Gerard Healy, Ayten Namli, F. Dilek Sanin, Carlota Tayà and Sergio Ponsá

14.1 INTRODUCTION

The progressive implementation of the Urban Waste Water Treatment Directive 91/271/EEC (amended by 98/15/EC) in all Member States is increasing the quantities of sewage sludge requiring disposal. When treating municipal wastewater, the disposal of sludge is a problem of growing importance, representing up to 50% of the current operating costs of a wastewater treatment plant. Based on sewage sludge production data from 2002 until 2007, an increase from 5.5 million tons to an annual EU-27 sewage sludge production of 10 million tons is reported (European Commission, 2014). This increase is mainly due to the practical implementation of the Directive as well as the slow but constant rise in the number of households connected to sewers and the increase in the level of treatment.

During the last decades there has been a major change in the ways sludge is disposed. Prior to 1998, municipal sludge was primarily disposed at seawaters or was either used as a fertilizer on agricultural land (Ødegaard *et al.* 2002); alternatives were sludge incineration or simply landfilling. Since 1998 onwards, European legislation prohibits the sea disposal of sewage sludge, in order to protect the marine environment. Moreover, the European Union published in 1999 the Landfill Directive (1999/31/EC), which requires the member states to reduce the amount of biodegradable waste being dumped by promoting the adoption of measures to increase and improve sorting activities at the origin, recovery and recycling. The main article of the Directive 91/271/EEC dealing with sludge is Article 14, which stipulates that 'sludge arising from wastewater treatment shall be re-used whenever appropriate'. This is a clear priority given to the use of sludge in agriculture, when this use is appropriate considering in particular the quality of the sludge. To emphasize the nutritional value of sewage sludge, the term biosolids is normally used when sewage sludge is applied for agricultural purposes. Accordingly, the agricultural use has become the principal disposal method for biosolids; 37% of the sludge produced is being utilized in agriculture, 11% is being incinerated, 40% is landfilled, while 12% is used in some other areas such as forestry, silviculture and land reclamation (Fytili and Zabaniotou 2008). Last Eurostat reported data shows that 54% of the sludge produced was used in agriculture or composted (Eurostat 2016). The latest trends in the field of sludge management, i.e. wet oxidation, pyrolysis, gasification and co-combustion of sewage sludge with other materials for further use as energy source, have generated significant scientific interest (Fytili and Zabaniotou, 2008).

Since 1986 the utilization of sewage sludge has been subject to provisions stipulated in the EU Directive (86/278/EEC). The Directive sets out requirements with respect to the quality of sludge, the soil on which it is to be used, the loading rate, and the crops that may be grown on treated land (European Commission, 2001a). The Directive seeks to encourage the use of sewage sludge in agriculture. At the same time it regulates its use in such a way that any potential harmful effect on soil, vegetation, animals and human beings is prevented. According to the above principle, the use of untreated sludge in agriculture is prohibited, unless it is injected or incorporated in the soil. To provide protection against potential health risks from residual pathogens, sludge must not be applied to soil in which fruit and vegetable crops are growing or grown, or less than ten months before fruit and vegetable crops are to be harvested. Grazing animals must not be allowed access to grassland or forage land less than three weeks after the application of sludge. The Directive also requires that sludge should be used in such a way that account is taken of the nutrient requirements of plants and that the quality of the soil and of the surface and groundwater is not impaired. Moreover, the term 'treated sludge' is defined in the Directive as the sewage sludge which 'has undergone biological, chemical or heat treatment, long-term storage or any other appropriate process so as significantly to reduce its fermentability and the health hazards resulting from its use' (Directive 86/278/EEC).

All the EU member states have transposed the European limits of Directive 86/278/EEC for sludge use in agriculture into their own regulations. Since its adoption, several Member States have enacted and implemented stricter limit values for heavy metals. The member states imposing more stringent limits than those of the sludge directive are Austria, Belgium, Czech Republic, Denmark (with respect to Zn), Finland, Germany, Netherlands, Slovenia and Sweden. On the contrary, the member states that still have the limits close to those of the sludge directive are Cyprus, Estonia, France, Greece, Hungary, Luxembourg, Ireland, Italy, Latvia and Spain (Mininni *et al.* 2015). The perspective of the revision of Directive 86/278/EEC, which could lead to the implementation of more stringent limit values for heavy metals in sludge, could therefore have an impact in the latter countries, at least on the provisions to be set by national regulations (average heavy metals content in sludge is in most cases well below regulatory requirements) (European Commission, 2001b).

The European Commission is currently assessing whether the current Directive should be reviewed – and if so, the extent of this review. For example, Directive 86/278/EEC sets limit values for only six heavy metals, but some countries have already incorporated limits for other metals (e.g. Se, Mo, As). Table 14.1 shows the limit values of heavy metals for sludge intended to be used in agriculture. Also Table 14.2 shows the maximum heavy metal concentration in soils and the maximum amount of each heavy metal that can be added annually in agricultural land (Directive 86/278/EEC).

The Directive does not have limit values for organic and emerging micropollutants in sewage sludge, which could contaminate terrestrial and aquatic environment when the sludge is used in agriculture. Although there is no uniform approach to set limits for micropollutants, several countries such as: Austria, Denmark, France, Germany and Sweden have established limits concentrations in sludge for:

(i) *Polycyclic aromatic hydrocarbons (PAH)*: 1–6 mg kg^{-1} DS
(ii) *Polychlorinated biphenyls (PCB)*: 0.1–1 mg kg^{-1} DS
(iii) *PCDD/F*: 30–100 mg kg^{-1} DS
(iv) *Absorbable organic halogens (AOX)*: 400–500 mg kg^{-1} DS
(v) *Linear alkylbenzene sulphonates (LAS)*: 1,300–5,000 mg kg^{-1} DS
(vi) *Nonylphenol and –ethoxylates (NPE)*: 10–450 mg kg^{-1} DS
(vii) *Di(2-ethylhexyl)phthalates (DEHP)*: 50–100 mg kg^{-1} DS

Table 14.1 Limits of Cd, Cu, Hg, Ni, Pb and Zn for sludge in agriculture (mg/kg DS of sewage sludge).

State	Cd (mg kg^{-1} DS)	Cu (mg kg^{-1} DS)	Hg (mg kg^{-1} DS)	Ni (mg kg^{-1} DS)	Pb (mg kg^{-1} DS)	Zn (mg kg^{-1} DS)	Cr (mg kg^{-1} DS)
Directive 86/278/EEC	20–40	1,000–1,750	16–25	300–400	750–1,200	2,500–4,000	–
Austria	2–10	300–500	2–10	25–100	100–400	1,500–2,000	50–500
Belgium	6–10	375–600	5–10	50–100	300–500	900–2,000	250–500
Bulgaria	30	1,600	16	350	800	3,000	500
Cyprus	20–40	1,000–1,750	16–25	300–400	750–1,200	2,500–4,000	–
Czech republic	5	500	4	100	200	2,500	200
Denmark	0.8	1,000	0.8	30	120	4,000	100
Estonia	20	1,200	20	400	900	2,500	1,200
Finland	3	600	2	100	150	1,500	300
France	20	1,000	10	200	800	3,000	1,000
Germany	2*–10	600*–800	1.4*–8	60*–200	100*–900	1,500*–2,500	80*–900
Greece	40	1,750	25	400	1,200	4,000	500
Hungary	10	1,000	10	200	750	2,500	1,000
Ireland	20	1,000	16	300	750	2,500	–
Italy	20	1,000	10	300	750	2,500	–
Latvia	10	800	10	200	500	2,500	600
Lithuania	PTE regulated throuhg limits in soil						
Luxembourg	20–40	1,000–1,750	16–25	300–400	750–1,200	2,500–4,000	1,000–1,750
Malta	5	800	5	200	500	2,000	800
Netherlands	1.25	75	0.75	30	100	300	75

(Continued)

Table 14.1 Limits of Cd, Cu, Hg, Ni, Pb and Zn for sludge in agriculture (mg/kg DS of sewage sludge) (*Continued*).

State	Cd (mg kg^{-1} DS)	Cu (mg kg^{-1} DS)	Hg (mg kg^{-1} DS)	Ni (mg kg^{-1} DS)	Pb (mg kg^{-1} DS)	Zn (mg kg^{-1} DS)	Cr (mg kg^{-1} DS)
Poland	10	800	5	100	500	2,500	500
Portugal	20	1,000	16	300	750	2,500	1,000
Romania	10	500	5	100	300	2,000	500
Slovakia	10	1,000	10	300	750	2,500	1,000
Slovenia	2	300	2	70	100	1,20	150
Spain	40	1,750	25	400	1,200	4,000	1,500
Sweden	2	600	2.5	50	100	800	100
UK	PTE regulated through limits in soil						
Brazil	39	1,500	17	40	300	2,800	1,000
China	5–20	800–1,500	5–15	100–200	300–1,000	2,000–3,000	–
Japan	5	–	2	300	100	–	500
Jordania	40	1,500	17	300	300	2,800	900
Russia	15	750	7.5	200	250	1,750	500
USA	39–85	1,500–4,300	17–57	420	300–840	2,800–7,500	–
Range in Europe	0.5–40	75–1,750	0.2–25	30–400	40–1200	100–4,000	75–1,750

* Proposed new limits

PTE (Potential toxic elements)

Sources: Adapted from Mininni *et al.* 2015; Healy *et al.* 2016; LeBlanc *et al.* 2008.

Although in the EU a common norm on the maximum allowed values of pathogenic microorganisms or indicators in fertilizing products does not exist, *Salmonella* and *Escherichia coli* has been proposed as marker microorganisms, in such a way that the sludge produced must not contain *Salmonella* in 50 g (fresh matter), and the treatment must induce a concentration reduction of *Escherichia coli* of 6 log10 or the concentration be less than 5 × 102 CFU/g of final product. Some European countries have already set limits for pathogens such as *salmonella spp., fecal streptococci, enterovirus, helminthes eggs, Escherichia coli* and enterobacteria. More detailed information about organic micropollutants and pathogens limits in sewage sludge can be found in Mininni *et al.* (2015).

Table 14.2 Limit values for concentrations of (i) heavy metals in soil and (ii) amounts of heavy metals which may be added annually to agricultural land, based on a 10 year average (Directive 86/278/EEC).

Parameters	Heavy Metals in Soil (mg kg^{-1} DS)	Heavy Metals Added Annually (kg ha^{-1} y^{-1})
Cd	1–3	0.15
Cu	50–140	12
Ni	30–75	3
Pb	50–300	15
Zn	150–300	30
Hg	1–1.5	0.1
Cr	–	–

In general, EU legislation on sewage sludge is based on the pre-cautionary scheme and the limits set for its agricultural use are in general stricter than the USEPA's (Environmental Protection Agency, USA). Furthermore, sewage sludge falls under numerous restrictions and it cannot be included in ecological bio-products (as compost), in organic farming fertilizers, etc. Although the EU's policy towards a sustainable use of phosphorus is currently promoted, P recovered from sewage sludge is not yet identified as a possible raw product, due to its 'waste' origin. Sewage sludge cannot be regarded solely as 'waste'; it is a renewable resource for energy and material recovery. From this perspective, legislation on sewage sludge management tends to incorporate issues related to environmental protection, public health, climate change impacts and socio-economic benefits.

Future trends on sludge management are mainly dependent on future alignment of legislation. It does not seem that a new sludge directive is pending (Mininni *et al.* 2015). In fact, the European Union developed the draft of a 'Working document on sludge' (European Commission 2000) to promote the use of sewage sludge in agriculture while improving the safety and harmonize quality standards but the draft was finally withdrawn. Moreover, works in progress have not evidenced a health and environment impact due to sludge agricultural use although some attention is already paid by many member states on organic pollutants and pathogens. Finally, it is expected that stabilized sludge will be used in agriculture in large quantities in future in many member states such as France, Germany, Italy, Spain and the UK (Mininni *et al.* 2015).

14.2 SLUDGE PRODUCTION PROCESSES

The main objective of wastewater treatment is to reduce the pollution load on receiving waters. However, the treatment processes concentrate most of the impurities and the microbial excess biomass in the sludge. Sludge, originating from the treatment process of wastewater, is the residue generated during the primary (physical and/or chemical), the secondary (biological) and the tertiary (additional to secondary, often nutrient removal) treatment (Fytili & Zabaniotou, 2008). The treatment and disposal of sludge should be considered as an integral part of the treatment process; therefore, wastewater treatment should be regarded as a low-solids stream (treated water effluent) and a high-solids stream (sludge).

Table 14.3 shows the main physico-chemical properties of primary and biological sludge and Table 14.4 shows the main physico-chemical properties of mixed sludge.

Table 14.3 Typical chemical composition and properties of primary and activated sludge.

Item/Sludge	Primary Sludge	Activated Sludge
Total dry solids (DS), %	2–8	0.83–1.16
Volatile solids (% of DS)	60–80	50–88
Grease and fats (% of DS)		
Ether soluble	6–30	–
Ether extract	12966	5–12
Protein (% of DS)	20–30	32–41
Nitrogen (N, % of DS)	1.5–4	2.4–5
Phosphorus (P_2O_5, % of DS)	0.8–2.8	2.8–11
Potash (K_2O, % of DS)	0–1	0.5–0.7
Cellulose (% of DS)	8–15	–
Iron (not as sulfide)	2–4	–
Silica (SiO_2, % of DS)	15–20	–
Alkalinity (mg/L as $CaCO_3$)	500–1500	580–1,100
Organic acids (mg/L as Hac)	200–2000	1,100–1,700
Energy content	10,000–12,500	8,000–10,000
pH	5–8	6.5–8.0

Source: Fytili and Zabaniotou, 2008

14.2.1 Sludge production

14.2.1.1 Primary sludge production

Primary sludge is drawn from the primary sedimentation tanks. It contains all the readily sedimentable matter from the wastewater; plus another 1% collected as scum; it has a high organic content (mainly fecal matter and food scraps) and is thus highly putrescible. In its fresh state, raw sludge is grey in color with a heavy fecal odor.

Primary sludge accounts for 50–60% of initial suspended solids in the wastewater inlet stream. Typical solids concentrations in raw primary sludge from settling municipal wastewater are 6%–8% and the portion of volatile solids varies from 60% to 80%. Primary precipitates can be dewatered readily after chemical conditioning because of their fibrous and coarse nature.

14.2.1.2 Biological sludge production

Activated sludge (AS) is the most common secondary biological treatment used to treat sewage and industrial wastewater, and was developed around 1912–1914. There is a large variety of designs; however, in principle all AS consist of three main components: (i) an aeration tank, which serves as bioreactor, (ii) a settling tank

Table 14.4 Chemical composition and properties of (i) untreated sludge, (ii) Dewatered anaerobically digested sludge (iii) composted sludge, (iv) thermally dried sludge, (v) biochar and (vi) sludge ash.

Properties	Units	Mixed Sewage Sludge[a]	Anaerobically Digested Sludge[b]	Compost[c]	Dried Sludge[d]	Biochar[e]	Sludge Ash[f]
VS content (db%)	%	43–80	56–76	55–67	64–69	25	<1
Ash content	%	20–57	44–24	32–45	31–36	75	>99
pH		4.5–8.3	8.4	6.5–7.8	6.9–7.2	6.7–9.5	–
EC	mS cm^{-1}	1.1–11.9	2.3	1.2–8.5	1.5–6.2	0.6–1.9	–
CEC	cmol kg^{-1}	9.2	–	–	–	2.3–35	–
RI	mg O_2 g^{-1} VS h^{-1}	2.5–9.5	2–5	1.3–2.1	–	–	–
TOC	g kg^{-1}	360–412	340–412	181	296	179	–
C/N ratio		42562	42560	7.5–13	8.3	9	–
Total N	g kg^{-1}	15–62	39–59	22–39	36–61	22525	–
Total P	g kg^{-1}	15432	34	13–28	13–29	20–42	23–93
S	g kg^{-1}	8.9	8–15	–	–	0.4–1	6.6–8.9
Ca	g kg^{-1}	10–38	19–50	–	–	2–15	66–163 Mg
Mg	g kg^{-1}	4–26	0.3–19.2	–	–	3–15	16–35
Na	g kg^{-1}	0.7–1.5	–	2.5	–	1–2.7	3.6–32 K
K	g kg^{-1}	1.9–6.5	2.3	2.8–5.0	2.2–4.3	1.2–16	42669
Al	g kg^{-1}	8	–	–	–	–	37–67
Cu	mg kg^{-1}	151–800	993	139–743	645–823	400–2100	553–4775
Co	mg kg^{-1}	30	–	15	–	21	42–553
Cr	mg kg^{-1}	54–500	54	30–345	30–217	230	114–1402
Ni	mg kg^{-1}	17–80	64	19–105	42–85	35–740	63–369
Cd	mg kg^{-1}	0.6–3.6	3.2	<0.5–4.4	<0.5–3.6	1.8–9.8	1.7–15.6
Zn	mg kg^{-1}	588–1700	998	600–1385	800–1346	900–3300	384–4303
Pb	mg kg^{-1}	28–3940	78	67–1196	75–3747	130–750	122–999
Mn	mg kg^{-1}	188–395	–	173–241	–	253–667	470–2510
Hg	mg kg^{-1}	0.4–8	–	2.4–2.8	2.7	–	1.1
NPE	mg kg^{-1}	489–2556	513–981	24–363	14–3150	–	–
PCBs	mg kg^{-1}	0.01–0.35	0.023	0.01–0.06	0.01–0.06	–	–
PHAs	mg kg^{-1}	0.1–5.3	1.1	<0.01–16	0.2–7.4	1–100	–
DEHP	mg kg^{-1}	2–164	143	2–120	589138	–	–
LAS	mg kg^{-1}	816–3240	3240	214–2879	331–5572	–	–
PCDD/Fs	ng TEQ kg^{-1}	7–15	7.7	11–55	12–77	–	–

[a]Herzel *et al.* 2016; Hossain *et al.* 2015; Manara and Zabaniotou, 2012; Ponsá *et al.* 2010; Ponsá *et al.* 2009; Ramírez *et al.* 2008; Tarrasón *et al.* 2008.
[b]Ramírez *et al.* 2008; Ponsá *et al.* 2010; Epstein, 2002.
[c]Domene *et al.* 2009; Fernández *et al.* 2007; Ponsá *et al.* 2009; Ramírez *et al.* 2008; Tarrasón *et al.* 2008.
[d]Domene *et al.* 2009; Domene *et al.* 2010; Fernández *et al.* 2007; Ramírez *et al.* 2008; Tarrasón *et al.* 2008.
[e]Hossain *et al.* 2010; Hossain *et al.* 2015; Mayer *et al.* 2016; Cai *et al.* 2013.
[f]Herzel *et al.* 2016; Mattenberger *et al.* 2008; Zhang *et al.* 2002.

('final clarifier') for separation of AS solids and treated waste water and (iii) a return activated sludge (RAS) equipment to transfer settled AS from the clarifier to the influent of the aeration tank.

In a biological treatment processes, biomass growth occurs concurrent with the oxidation of organic or inorganic compounds. The ratio of the amount of biomass-produced respect to the amount of substrate consumed is defined as the biomass yield. In aerobic conditions the growth yield can reach 0.60–0.70, which means that 60–70% of organic biodegradable matter removed in the biological treatment is converted into new cellular biomass (Foladori *et al.* 2010). Due to both biological growth and accumulation of partially degraded solids present in the raw wastewater, excess sludge eventually accumulates beyond the desired mixed liquor suspended solids (MLSS) concentration in the aeration tank. This amount of secondary sludge (called Waste Activated Sludge) is removed from the treatment process to keep the ratio of biomass to food supplied in balance. Typical solids concentrations in secondary sludge from an activated sludge processes are 1–2% and the portion of volatile solids varies from 50% to 85%. Additionally, nutrients from wastewater should be treated before discharging. There is a minor production of sludge produced after biological nutrient removal (BNR) processes.

14.2.2 Characteristics of sewage sludge

The characteristics of sludge play an important role when considering the ultimate disposal of the processed sludge, especially in their use for land application. Sludge characteristics can be broken down in three categories: (i) physical, (ii) chemical and (iii) biological.

The important physical characteristics are the solid content and the organic matter content. The total solids content affects the method of land application. Liquid or low-solids sewage sludge will generally be injected into soil to prevent vectors and provide better aesthetics. Dewatered or semisolid biosolids are usually spread on the surface and subsequently plowed into the soil (Epstein, 2002). The organic matter is an important constituent of biosolids and its use for land application enhances the organic content of soils. In sandy soils the organic matter increases the water-holding capacity, soil aggregation and other soil physical properties. It reduces the soil bulk density and increases the cation exchange capacity (a very important property for supplying plant nutrients). The positive effect of organic matter on the soil physical properties enhances the plant root environment. Therefore, plants are better able to withstand drought conditions, extract water, and utilize nutrients (Epstein, 2002).

Chemical properties affect plant growth as well as the soil's chemical and physical properties. The important chemical characteristics are: (i) pH, (ii) soluble salts, (iii) plant nutrients (macro and micro), (iv) essential and non-essential trace elements to humans and animals and (v) organic chemicals. A detailed list of heavy metals, trace elements, priority pollutants and organic chemicals can be found in Epstein (2002). The pH of most biosolids (whether liquid, semisolid, or solid) is generally in the range of 7–8, unless lime is added during the wastewater treatment process. Plant nutrients are among the most important chemical characteristics of biosolids; the major plant nutrients are nitrogen (N), phosphorus (P) and potassium (K). Other macronutrients are calcium (Ca), magnesium (Mg), and sulfur (S). The micronutrients essential to plant growth are boron (B), copper (Cu), iron (Fe), manganese (Mn), molybdenum (Mo), nickel (Ni) and zinc (Zn). It has been recognized for centuries that sewage sludge contain plant nutrients. Table 14.3 and Table 14.4 show nutrient typical values found in raw and treated sludge.

Regarding biological properties, pathogens are the most important biological property of biosolids for land application. Since pathogens survive the wastewater treatment processes (primary and secondary treatment), land application of sewage sludge directly from these processes needs to be avoided or restricted to land management systems. Further treatment, such as digestion, composting, alkaline stabilization, or heat drying, increases the opportunities for land application (Epstein, 2002). The presence of pathogenic microorganisms such as viruses, bacteria and fungi involve a potential risk that may affect soil organisms or plants and produce changes in the microbial community structure and soil properties. A more detailed description of health risks involved in application of sludge in agriculture due to pathogens is found in Section 2.2.6.2 of this chapter. Application of sewage sludge in agriculture enhances microbial population, which increases the rate of organic matter decomposition in soils. As a result, there is a significant change in the soil physical properties. This produces a marked improvement in the plant-root environment and better plant growth (Epstien, 2002).

14.3 SLUDGE PRE-TREATMENT PROCESSES

14.3.1 Sludge pre-treatment technologies

Pretreatment, which aims to reduce the eventual amount of sludge production, is typically done by the application of external forces and agents in order to destruct sludge solids (Müller et al. 2004). Pretreatment is mostly used for biological sludge, even though it also has applications for mixed sludge. Applied forces lead to rupture of the cell membrane of bacteria in biological sludge resulting in release of organic substances outside the cell (Wang *et al.* 1999). Hence, sludge disintegration achieves solubilization and conversion of slowly biodegradable, particulate organic materials to low molecular weight, readily biodegradable compounds ending up producing much less sludge after stabilization such as digestion.

There are different kinds of pretreatment methods, which are conducted by the application of mechanical (ultrasound, homogenizer, mill, and others), physical (thermal treatment, microwave), chemical (use of ozone, acids, alkali and other chemicals) and biological (with or without enzyme addition) means (Müller, 2001). These methods can be applied individually or one method can be combined with another (such as thermo-chemical) to disintegrate sludge more effectively. It is known that sludge pretreatment if applied before anaerobic digestion, increases the stabilization and biogas production, decreases the sludge to be disposed, solves bulking and foaming, improves dewatering and disinfects sludge. Below is brief discussion of mechanisms of different pretreatment methods.

Satisfactory results obtained in lab scale tests encouraged many companies to commercialize thermal pretreatment methods. Some of these methods have become a part of the sludge treatment systems in a number of WWTPs. Examples to these are the patented thermal pretreatment systems such as Cambi, Biothelys and Zimpro Processes. There are many full-scale applications of thermal pretreatments processes in Ireland, Denmark, Norway, USA and Sweden. Full-scale application of ultrasonication is not as widespread as thermal treatments and one example at full scale is found in Ulu Pandan Water Reclamation Plant in Singapore. Ozonation pretreatment are not widespread at full scale but some full applications can be found in industrial WWTPs. An example of ozone application in municipalities located in the southern part of the Marche region in Italy.

These Pre-treatment technologies have been extensively described in Chapter 12, therefore in the present Chapter only the physico-chemical and biological changes and its implication for soil agricultural application are considered.

14.3.2 Effects of pretreatment on the agricultural use and value of sludge

The nutritional and beneficial value of sludge for land application is highly affected by the process that sludge goes through; more specifically, whether the sludge is pretreated or not; has it gone through a digestion process, or what is the sequence of treatment units for sludge. As mentioned above, sludge pretreatment processes break up the flocs either by physical or chemical means, or by different combinations of these. During these processes, floc components are solubilized from the solid phase and introduced into the liquid medium. As expected, this affects the quality of solid sludge (the typical form of sludge that is used in land application).

14.3.2.1 Organic Matter Reduction

As previously mentioned, the function of organic matter in land-applied sludge is to enrich the soil and enhance the soil properties such as aggregation and water holding capacity. On the other hand, the main purpose of pretreatment is to advance digestability of sludge and enhance biogas production. Therefore, pretreatment is most commonly applied prior to anaerobic digestion. For systems applying pretreatment and further anaerobic digestion, most solubilized organics are converted to biogas (Braguglia *et al.* 2015). Thus, digested sludge for land application contains much less organic matter. If no digestion exists, the solubilized organics are lost with the liquid fraction obtained in dewatering operations, which also end up producing lower organic content sludge. Despite the fact that pretreatment reduces organic matter in solid sludge, there is still sufficient organics remaining following pretreatment due to the fact that only partial removal of organics is achievable during these processes.

14.3.2.2 Nutrients Solubilization

Most nutrients such as nitrogen and phosphorus have the potential to be solubilized along with other floc components during the pretreatment process. Once they are soluble, they are either uptaken during the digestion process, or released with the liquid fraction discharged from the digesters or obtained in dewatering processes. Zhang *et al.* (2015) showed that about 39% more ammonia nitrogen and about 82% more nitrate nitrogen are lost with centrate when thermal pretreatment and anaerobic digestion are employed together in a full scale WWTP. Srinivasan *et al.* (2015) compared ozone, peroxide, radiofrequency heating and combinations of these methods for their effect on sludge properties. Their results showed that very significant amounts of nutrients are solubilized from sludge solids into the liquid phase. According to their data, solubilized Total Kjeldahl Nitrogen (TKN), NH_4^+, orthophosphate, calcium, magnesium and potassium reach up to 97%, 95%, 96%, 94%, 100%, and 85%, respectively (Srinivasan *et al.* 2015). Dogan and Sanin (2009) report that about 30% more ammonia nitrogen is present in digester effluent when alkali-microwave pretreatment and anaerobic digestion is applied together in a lab scale digester. The study observed no release of orthophosphate phosphorus during these treatments. These findings indicate some nutrients are clear to be lost from sludge during pretreatment due to solubilization effect. Therefore, pretreated sludge is expected to have typically lower nutrient contents when compared to untreated sludge.

14.3.2.3 Pathogen and Indicator Reductions

One remarkable effect of pretreatment methods is their achievement of better microbial quality of sludge (pathogen or indicator content reduction). For example, after thermal hydrolysis sludge of higher microbiological quality and cleaner sludge is obtained. Levantesi *et al.* (2015) found that thermal hydrolysis (135°C, 20 min) caused over a 3.2 logs removal of *E. coli*, almost 4 logs removal of smatic coliphages, more than 2.5 logs removal in spores and higher than 0.9 logs removal in *Salmonella*. Among a number of pretreatment methods tested, thermal hydrolysis at 130°C provided the highest removal of microorganisms tested, reducing their concentration to non-detectable levels in almost all analyzed samples (Levantesi *et al.* 2015). In the study of Foladori *et al.* (2007), the mechanism of ultrasound on microbial decay was explained by an initial disaggregation of cells clumped in aggregates of different sizes with no observation of death. With increasing ultrasound energy, both permeabilisation and cell disruption start. They found that activated sludge, *E. coli* and *E. feacalis* showed differences in their inactivation by sonication. *E. coli* underwent cell disintegration at lower levels of ultrasonic energy. On the other hand, a complete disaggregation of activated sludge flocs required ultrasonic energy around 80 kJ L^{-1}, while for the damage and death of the released free cells, higher levels of energy need to be applied. Ozonation is also a means of effective disinfection. In the study of Park *et al.* (2008) the fecal coliform concentration was below the limits of detection when the ozone dose was above 0.3 g/g DS. At this dose, the impact of ozone on the inactivation of *Streptococcus* and *Salmonella* was also significant. At an ozone dose of 0.4 g/g DS complete reduction of these organisms was observed. In the study, an ozone dose of 0.3 g/gDS was suggested to fulfill the criteria for the disinfection for class A biosolids in USA.

14.3.2.4 Trace Organic Contaminants Removal

Concerns have been raised due to the presence of numerous trace organic contaminants such as pharmaceuticals, antibiotics, hormones, pesticides, etc. in sludge. From this perspective, pretreatment may bring some relief, since a number of methods used are able to oxidize refractory compounds. Removal of these persistent organics depends on the method applied and the chemical's structure; so the results and success vary from one system to another. Since some of these pretreatment methods have oxidative properties, the toxic organics have the potential to be degraded during these treatments, although if the oxidation is not complete, there is a risk of forming by-products which are even more toxic than the initial molecule. Methods such as ultrasonication, ozone application and thermal treatment are considered among the advanced methods that are able to break some bonds of the trace organic contaminants. In one study, the ultrasound treatment of sludge was tested on the removal of pesticides. The treatment resulted in a significant reduction in the sludge pesticide content (90% of the total pesticide mass was removed). Investigation into the sono-degradation of three characteristic pesticides (thiabendazole, acetamiprid and imazalil) revealed the formation of transformation products already reported in studies on the degradation of these compounds by advanced oxidation processes, thus confirming that ultrasonication involves hydroxyl radical reactions (Rivas Ibañez *et al.* 2015). The fate of pharmaceutical residues in WWTP sludge was evaluated during mesophilic anaerobic digestion and six treatment technologies (pasteurization, thermal hydrolysis, advanced oxidation processes using Fenton's reaction, ammonia treatment, thermophilic dry digestion, and thermophilic anaerobic digestion) were compared. Advanced oxidation processes using Fenton's reaction affected several compounds, including substances showing general stability over the range of treatments such as carbamazepine, propranolol, and sertraline. Pasteurization, ammonia treatment, and thermophilic dry digestion exhibited relatively modest reductions. Interestingly, only thermal hydrolysis efficiently removed the ecotoxicologically potent estrogenic compounds from the sludge (Malmborg & Magner, 2015). Ak *et al.* (2013) demonstrated that anaerobic digestion of waste activated sludge when coupled with mild ozone treatment (e.g. 1.33 mg O_3/g-VSS), affects enhanced removal of endocrine disrupting compounds (acetaminophen, estrone, benzyl butyl phthalate, progesterone, diltiazem and carbamazepine) sorbed onto the sludge. Anaerobic reactors receiving return activated sludge feed ozonated at different ozone doses indicated substantial pollutant removals as compared to the control. Fate of nonylphenol compounds (NPEs) were studied in thermally hydrolyzed and anaerobically digested (15 day SRT) sludges. In this study even though the transformation between the target compounds occurred, the total concentrations of NPE did not change between influent and effluent for thermally pretreated and anaerobically digested sludges (Manara & Zabaniotou, 2012).

14.3.2.5 Heavy Metals

Heavy metals, which constitute a historical concern for sludge land application, are conservative and accumulative pollutants. Their fate in pretreated and digested sludge has been an interest. Most of the studies in literature shows much less can be done on heavy metals concerning their removal during pretreatment processes. One mechanism of removal is by solubilization from sludge solids to liquid and therefore they may be removed by physico-chemical technologies. In one study, sonication time and power density greatly affected the heavy metals solubilization degree. Soluble heavy metals increased almost linearly with sonication time within the first 15 min and then stabilized. A minimum power density of 0.8 W/mL was required for heavy metal solubilization. The study showed that the effect of sonication time on heavy metal release was higher than that of power density. Besides, each heavy metal behaved differently during the ultrasonic treatment. Arsenic and nickel release were easier and the solubilization degree reached 58.4% and 34.9% after 30 min of sonication, respectively. On the other hand, solubilization degree of copper was low. Cadmium was stable and could not be released by sonication. Other studies indicated no solubilization of metals from sludge. Braguglia *et al.* (2015) reported that due to typical weight loss during anaerobic digestion, and conversion of biodegradable matter to biogas, the heavy metal concentrations in the digested samples are expected to be higher with respect to the feed. They observed no removal during the investigated processes (thermal treatment and sonication), and because of the mass loss during the treatments, the effective heavy metal concentration increased at the end. In the study of Yan *et al.* (2015), the effect of hydrothermal treatment at various temperatures (120–200°C) on the properties of sewage sludge derived solid fuel was investigated. Similar heavy metal enrichment in solid particles was found after hydrothermal treatment. These results indicate that the possibility of conservative pollutants such as heavy metals to enrich in sludge can pose risks for land application.

14.4 SLUDGE TREATMENT PROCESSES

The solids resulting from wastewater treatment must undergo further treatment prior to land application. Land application of biosolids requires the disinfection and stabilization of biosolids. The objective is to reduce the level of pathogens, reduce vector attraction and produce a stabilized product - that is, a product that would not decompose very rapidly and produce offensive odors (Epstein, 2002). Studies carried out during the last years, showed that raw sewage sludge in the conditions of its land application can be a significant source of undesirable substances in the soil and plants. The main contaminants of sewage sludge are heavy metals, organic pollutants, pharmaceutical residues and pathogens (Dichtl *et al.* 2007).

Changes in legal requirements for sewage sludge application are planned that will set lower limit values for hazardous substances and higher quality requirements in general. Although direct application of raw sewage sludge to agricultural land is the current most commonly applied management technique in Europe, several technologies focused on minimizing the negative impacts of direct soil application of sewage sludge are used at industrial level or being under development.

An overview of most used processes applied to sewage sludge including biological, drying, thermal and chemical processes and the implications of the application to agricultural soils of its final products (e.g. compost, dried sludge, biochar, ashes) are explained in the following sections:

- *Biological processes*: (i) anaerobic digestion, (ii) composting, (iii) vermicomposting and (iv) bioleaching
- *Drying processes*: (i) thermal drying, (ii) biodrying and (iii) solar drying
- *Thermal processes*: (i) incineration, (ii) pyrolysis (iii) gasification
- *Chemical processe*: (i) lime addition

14.4.1 Biological processes

14.4.1.1 Anaerobic digestion

Anaerobic digestion used to treat primary and secondary sludge resulting from the aerobic treatment of municipal wastewater is a standard technology around the world. The technology is used in thousands of installations as part of modern treatment systems of municipal wastewaters. Anaerobic digestion is defined as a biological process in which the biodegradable matter is degraded or decomposed in the absence of oxygen using specific microorganisms that produce biogas composed mainly of methane and carbon dioxide. Overall, the process converts about 40% to 60% of the organic solids to methane (CH_4) and carbon dioxide (CO_2), as thereby it also reduces the amount of final sludge solids for disposal whilst limiting odor problems associated with residual putrescible matter. The chemical composition of the gas is 60–65% methane, 30–35% carbon dioxide, plus small quantities of H_2, N_2, H_2S and H_2O. Of these, methane is the most valuable because it is a hydrocarbon fuel (giving 36.5 MJ/m^3 in combustion).

In general, mesophilic anaerobic digestion of sewage sludge is more widely used compared to thermophilic digestion. In mesophilic anaerobic treatment, Gantzer *et al.* (2001) reported that pathogens (*Salmonella* and viable pathogen nematode eggs) were still present at concentrations above the sanitation requirements (under the provision of French Decree N. 97–1133). On the contrary, the enhanced hygienization effect of the thermophilic process complies with the EU policy for elimination of pathogens and it has been reported that thermophilic anaerobic digestion of sewage sludge can lead to EPAs class A biosolids, which are suitable for subsequent land application. More information about pathogenic disinfection during anaerobic digestion processes can be found in Epstein (2002).

Epstein (2002) reported data on the nutrient content in 250 sewage sludge samples from 150 wastewater treatment plants. Nitrogen, P, Ca, and S are present in relatively large amounts, whereas K and Mg are found in much smaller amounts. Anaerobically digested sludge showed median and average concentrations of total nitrogen of 4.2 and 5% respectively. Similar values were obtained for aerobic sludge. On the contrary the median concentration of N-NH_4 was four times higher in anaerobic digested sludge compared to aerobically treated sludge.

Although anaerobic treatment itself is very effective in removing biodegradable organic compounds, leaving mineralized compounds like NH_4^+, PO_4^{3-}, S^{2-} in the solution, several organic compounds such as pharmaceutical and personal care products (PPCPs) can persist after the process. Carballa *et al.* (2007) showed removal efficiencies of PPCPs higher than 60% for antibiotics, natural estrogens, musks and naproxen. For the other compounds (e.g. ibuprofen, diazepam, etc.), the values ranged between 20% and 60%, except for Carbamazepine, which showed no elimination.

To be applied in soils, sufficiently stabilized sewage sludge should be used in order to avoid negative effects on plant growth. Ramirez *et al.* (2008) showed a reduction of toxicity of anaerobically digested sludge compared to raw sludge (from two to five times less toxicity) in *B. Rappa, L. perenne* and *T. pratense*. On the contrary digested sludge showed higher ecotoxicity compared to composted sludge (much more stabilized). Anaerobic sludge usually undergoes an aerobic post-treatment (e.g. composting) to improve stability and to decrease its final moisture, facilitating its storage and transport.

14.4.1.2 Composting

Composting is defined as the biological decomposition and stabilization of organic substrates, under conditions that allow development of thermophilic temperatures as a result of biologically produced heat, to produce a final product that is stable, free of pathogens and plant seeds and that can be beneficially applied to land (Haug, 1993). It is the main biological process applied to sewage sludge in Europe and is a generally accepted and highly beneficial method of stabilizing its organic matter (Oleszczuk, 2008). In fact, according to the last Eurostat available data, composting was used to treat 14% of sludge produced in Europe in 2013.

Composting reduces the volume of sludge and its transporting costs, eliminates the risk of disseminating pathogens and removes mal-odorous compounds. Moreover, the addition of compost to agricultural soils has the following positive effects: (i) lead to a slow release of nutrients (95% of N is in organic form),

(ii) has a high binding capacity for organic and inorganic elements (contaminants or nutrients), (iii) improves water storage and soil water content due to its increased water-holding capacity and (iv) aids in the creation of soil agglomerates that can facilitate aeration of plant roots and improve water infiltration into the soil (Sánchez *et al.* 2015). It also prevents soil erosion and runoff. Gantzer *et al.* (2001) among many other studies (Haug, 1993) showed that composting processes achieving thermophilic temperatures (>45°C) were able to fulfil the sanitation requirements of pathogen micro-organism (*Salmonella* and viable pathogen nematode eggs among others).

When composted materials are used as organic amendments in soils, it is of great importance that the material is sufficiently stabilized in order to avoid negative growth effects due to N mineralization, oxygen depletion or the presence of phytotoxic compounds. Ramírez *et al.* (2008) showed that composting is an effective way to reduce phytotoxicity of sewage sludge before being applied to agricultural soils: a strong positive correlation was found between higher values of half maximal effective concentration (EC50) (less toxicity) and the stability degree of their organic matter. In addition, negative correlations between EC50 and total nitrogen, hydrolysable nitrogen or ammonium content were found. On the contrary, no ecotoxicity correlations were found with heavy metals or organic pollutant content comparing raw and composted sewage sludge. In similar studies, Domene *et al.* (2011) showed that mortality and reproduction of soil microinvertebrates were clearly explained by the stability of wastes (the higher the stability the higher the LC50 and EC50), which was probably related to releases of secondary metabolites, mainly ammonium, during the decomposition in soil of unstable raw sewage sludge.

14.4.1.3 Vermicomposting

Another method used in some countries such as India is vermicomposting. Vermicomposting has been widely used as a method of sludge stabilization because of simple technology. Vermicomposting is a complex mechanical and biochemical transformation of sludge achieved through the action of earthworms. Earthworms have potential both to increase the rate of aerobic decomposition and composting of organic matter and also to stabilize the organic residues and removing the harmful pathogens and heavy metals in the sludge (Sinha *et al.* 2010). Earthworm metallothioneins (MTs) proteins have a very high capacity to bind metals. Numerous studies have documented earthworm's bioaccumulation capability of Cd, Pb, Cu, Ca, Mg, Fe, Zn. Ireland (1983) determined up to 100 mg kg^{-1} Cd and 7600 mg g^{-1} dry weight Pb in tissue of earthworm after vermicomposting of biosolid. Basja *et al.* (2003) suggested that earthworm may not be able to remove toxic substances completely, but at least it changes the 'chemical make-up' of the sludge to rendering it harmless to the soil. They found that vermicomposting complies with 'grade A' standards for biosolid stabilization. Tiger Worm (*E. foetida*), Red Tiger Worm (*Eisenia andrei*), the Indian Blue Worm (*Perionyx excavatus*), the African Night Crawler (*Eudrilus euginae*), and the Red Worm (*Lumbricus rubellus*) are most appropriate for vermicomposting of biosolid under all climatic conditions (Sinha *et al.* 2010).

14.4.1.4 Bioleaching

Bioleaching (bio-acidification process) is the most common metal leaching method and is based on the oxidation of sulfur or iron by chemiolithotrophic bacteria. The most widely used microorganisms in metal leaching are *Acidithiobacillus thiooxidans* (formerly known as *Thiobacillus ferrooxidans*) and *Acidithiobacillus ferrooxidans* (Cho *et al.* 1999). Bioleaching of biosolid before land application can be used to remove a significant fraction of the heavy metals content of the agricultural product (Shanableh & Ginige, 1999). Ghavidel *et al.* (2010) reported that bioleaching is an efficient and powerful tool for removal of heavy metals from biosolid. Researchers were able to remove 24.73% of Fe,

83.96% of Cu, 81.46% of Ni and 38.96% of Pb from biosolid using the bioleaching method. Wen *et al.* (2009) reported that the removal efficiencies of Cr, Cu, Pb, and Zn from biosolids were 43.6%, 96.2%, 41.6%, and 96.5%, respectively. However, Shanableh and Ginige (1999) found that the bioleaching process also reduces the nutrient content of the biosolid.

14.4.2 Drying processes

Drying is a relatively simple technological operation in which thermal energy is provided to sludge to evaporate water. Although usually the ultimate goal of dried sludge is its energy valorization (e.g. via incineration or pyrolysis), it is frequently used directly as soil amendment; therefore, drying processes can be used as standalone sludge post-treatments or be applied prior to further thermal treatment. The most common sludge drying technology is thermal drying. The application of this conventional drying technology can be technically and economically challenging because of the use of high amounts of external energy (e.g. natural gas). Recently bio-drying, extensively described in Chapter 15, has been presented as an economical and energy-saving emerging technology to reduce the sludge content and to evaporate bound water by biologically produced heat (Dufour, 2006). Solar drying could be an economic alternative to conventional drying systems, especially in areas with proper climatic conditions (Dichtl *et al.* 2007).

The process of drying sludge reduces volume of the product, making its storage, transportation, packaging and retail easier. Sludge drying also inactivates pathogens and volatile chemicals and leads to a sanitized final product in pellets in relatively short time, with low odours and good handling characteristics (Fernández *et al.* 2007). Gantzer *et al.* (2001) showed that a drying process carried out during 10 h at 108°C was able to fulfil the sanitation requirements of two categories of pathogen micro-organism (*Salmonella* and viable pathogen nematode eggs).

Tarrasón *et al.* (2008) showed that thermal-drying of sewage sludge modifies its behaviour as a source of nitrogen when applied to soil. As consequence, mineral nitrogen concentrations ($N\text{-}NH_4$ and $N\text{-}NO_3$) of soil treated with thermally dried sludge can be high at short time after amendment far before growth of vegetation, increasing the risk of nitrate leaching. Soil amended with thermally dried sludge shows a greater degree of carbon mineralization because the organic matter is not stabilized yet (on the contrary than in a composting process) and, as highlighted before, it is of great importance that the organic ammendment applied to soil is sufficiently stabilized in order to avoid negative growth effects. Several authors showed higher ecotoxicity, both in plants and microinvertebrates, of thermally dried sludge compared to composted sludge. For example, in germination tests of *Brassica rapa*, EC50 for composted sludge were 10 times higher (less toxicity) than for thermally dried sludge. Moreover there was no statistical difference between thermally dried sludge and fresh sludge in terms of its ecotoxicty.

14.4.3 Thermal processes

The direct use of sewage sludge in agriculture is controversial because it acts as a sink for pollutants in wastewater treatment plants and it is often contaminated with heavy metals and organic contaminants. Lately, various modern thermal technologies have been introduced, offering an alternative trend to the sewage sludge disposal, especially with the decreasing availability and the increasing price of land for landfilling. The principal goal of thermal processing of sewage sludge is the utilization of the stored energy in sludge and the minimization of environmental impacts at the same time, in order to meet the increasingly stringent standards (Fytili & Zabaniotou, 2008). Processing of raw sewage sludge before thermal treatment, usually by means of a drying process, is often necessary from a technological and economical point of view. Thermal processes generate by-products that can be potentially used in agriculture, if they comply with the regulations, such as sludge ashes produced in incineration processes and biochar produced in pyrolysis processes.

14.4.3.1 Incineration

Incineration is the most popular thermal treatment used for the processing and management of sewage sludge. In fact, according to the last Eurostat available data, 23% of sludge treated in Europe in 2013 was treated by incineration. During incineration, organic matter is combusted to CO_2 and other trace gases, with water removed as vapour. The process cannot be considered as a complete disposal option because significant quantities of inorganic incinerated sewage sludge ash remain.

Application of waste ash to agricultural land presents an opportunity for the recovery of essential plant nutrients (Zhang *et al.* 2002). However, the amount of ashes that may be applied to agricultural land is restricted by their heavy metal contents, because concentration of Ni, Cu, Zn, Cd, Sn, Pb are much higher in sludge ashes than they are in soils. From the aspect of liming effect and plant nutrients, the waste ashes can be used as liming agents on acid soil and may also bring agronomic benefits (Zhang *et al.* 2002).

Sludge ash is rich in phosphorus content, ranging between 4% and 9%, and contains amounts of phosphorus comparable to commercial superphosphate. Sludge ash could replace phosphate rock-based products and reduce EU dependence on phosphorus imports (Herzel *et al.* 2016), but its direct utilization is usually not possible because of its content of heavy metals. Therefore, the focus is on alternative sludge ash treatment technologies that gain the most economic and ecological benefit from the sludge's valuables. Different technologies such as (i) BioCon-Process, (ii) SEPHOS-Process, (iii) ASH DEC Umwelt AG or (iv) RuePa-Process have been recently developed for phosphorus recovery from sludge ash (Dichtl *et al.* 2007). Compared to co-incineration, only during mono-incineration phosphorus recovery from sludge ash is possible. The phosphorus concentration in ashes resulting from co-incineration processes is too low so that the recovery of nutrients is uneconomic (Dichtl *et al.* 2007).

14.4.3.2 Pyrolysis and Gasification

Pyrolysis is the process through which, organic substances are thermally decomposed in an oxygen-free atmosphere, at temperatures varying in the range of 300 and 900°C. Gasification is the thermal process during which carbonaceous content of sewage sludge is converted to combustible gas and ash in a net reducing atmosphere. A by-product of these thermochemical processes of sewage sludge is a solid product containing char (mostly carbon) and ash called biochar.

This biochar can be combusted for heat and power, gasified, activated for adsorption applications, or applied to soils as a soil amendment. Biochar has received much attention in the context of carbon sequestration, climate change mitigation, and soil improvement (Mayer *et al.* 2016). For instance, biochar contributes to carbon sequestration when land-applied because the carbon does not readily degrade; the mean residence time of carbon in biochars made at 550°C was estimated to be over 1000 years (Singh *et al.* 2012). Additionally, biochar may act as a soil conditioner, enhancing plant growth and crop yields by supplying and, more importantly, retaining nutrients (reducing nutrient runoff from land via adsorbtion) and by providing other services such as improving soil physical and biological properties. Specifically, the incorporation of the biochar can influence the structure, texture, porosity, particle size distribution and density of the soil, and in this way it potentially alters the air oxygen content, water storage capacity and microbial and nutritional status of the soil within the plant rooting zone (Amonette & Joseph, 2012). Biochar can also neutralize the pH conditions of acidic soils as it has a positive liming effect (Hossain *et al.* 2010). Agrafioti *et al.* (2013) indicated that there is no environmental risk using biosolid biochars as a soil amendment.

Recent studies showed that pyrolysis could also contribute to the removal of organic micropollutants of wastewater sludge. Pyrolysis carried out at 450°C removed polychlorinated biphenyls (PCBs) by 75% from industrial sewage sludge, and pyrolysis of contaminated sediment at 800°C removed greater than 99% of dioxins and PCBs (Ross *et al.* 2016).

Biochar from sewage sludge also has a high heavy metal content and the amount of char that can be intended for agricultural utilization is therefore also restricted. Liu *et al.* (2013) investigated the biosolid biochar's capability of remedying contaminated soils. They found that pyrolysis increased heavy metals (Cu, Pb, Zn, Cd, and Cr) contents of biosolid biochar, but heavy metal availability of them were lower than those of air-dried biosolid and the plant availability of heavy metals was reduced in polluted soil. Conversely, Van Wesenbeeck *et al.* (2014) found that heavy metals were retained in the biochar during carbonization, whereas Hg, As, Cd, and Se were released and thereby depleted in the biochar. Zhang *et al.* (2015) investigated the immobilization of As (III) of biosolid biochar. According to their results, biosolid pyrolyzed at a higher temperature showed a lower As (III) sorption capacity and sorption of As (III) was faster than that of Cr (VI) but slower than that of Pb (II). Biosolid biochar reduced plant productivity because of increased electrical conductivity associated with the biochar amendment.

As this biochar still contains organic matter, combustion of the biochar is suggested for the full exploitation of sewage sludge. Ash from biochar combustion and gasification is enriched in P, facilitating P recovery. Therefore, the combination of thermochemical processes, such as pyrolysis followed by char combustion or gasification, combined with phosphorus recovery leads to value added products, energy and nutrients, all contributing to a greater use of this waste (Atienza-Martínez *et al.* 2014).

14.4.4 Chemical processes

Sewage sludge tends to increase acidity of the soils as a result of proton release from organic matter decomposition and mineralization of $N\text{-}NH_4$. Increased soil acidity could cause greater solubility of metals and consequently their enhanced plant availability and leaching potential, particularly in soils with poor buffering capacity. Increased attention is paid to the sludge stabilization process aiming to minimize the mobility of heavy metals by using various additives due to compliance to more stringent regulations (Samaras *et al.* 2008). Lime is considered as one of the most common amendment materials for sewage sludge stabilization, as it plays significant role in reducing the microbial content of sludge (pathogens), as well as the availability of heavy metals, enhancing the agricultural benefits and lowering the respective environmental risks (Wong & Selvam, 2006). However, the application of lime for the stabilization of sewage sludge depends upon a number of parameters, such as the availability of lime, the associated costs, the required period for stabilization, etc.; thus, alternative materials other than lime such as fly ash should be considered for sludge stabilization.

Samaras *et al.* (2008) showed that sewage sludge amended with stabilizing agents (lime and fly ash) initially provoked strong phytotoxic effects on three examined plant species. On the contrary, samples stabilized for an extended time (35 d) presented negligible seed germination inhibition. For liming treatments, various studies have demonstrated the necessity of a stable pH between 12 and 12.6 for 20–60 days for the elimination of *Salmonella* and viable nematode eggs. On the contrary, when these sanitation conditions are not fulfilled, as in the study carried out by Gantzer *et al.* (2001) (using quick lime 25% and a retention time of 1 day), the samples analysed contained viable nematode eggs and/or *Salmonella*. In the same study with lime concentrations up to 62%, the sludge was sanitized, six months storage at pH not less than 11.5 were necessary to produce sanitized sludge.

14.5 GENERAL EFFECTS OF BIOSOLIDS ON AGRICULTURE

14.5.1 Effect on agricultural productivity and soil fertility

The reuse of biosolids in agriculture provides the necessary nutrients and micronutrients necessary for plant and crop growth. They may be used as a soil conditioner, improving its physical and chemical properties and reducing the possibility of soil erosion. Their use also addresses EU policy on sustainability and reuse of resources. Numerous studies have documented their efficacy in increasing crop yields and their use in biofuel cropping systems, and in general, biosolids application to land have been found to have a statistically significant impact on crop yields (Latare *et al.* 2014) and soil phosphorus (Shu *et al.* 2016), while having negligible adverse ecological impacts (Adair *et al.* 2014). A selection of recent studies that report impacts of biosolids application on crop growth, soil fertility, water holding capacity, and soil pH (a lowered pH upon biosolids application is known to enhance the uptake of most metals; Carvalho *et al.* 2013) is shown in Table 14.5.

Table 14.5 Impacts of biosolids application on soil fertility and plant productivity.

Country	Area of Study Focus	Biosolids Application Rate	% Increase of Parameter Measured Versus no Treatment (Zero Biosolids Addition)					References
			Biomass Yield	Mehlich P	Organic Matter	Water Holding Capacity	pH	
USA	Switchgrass growth	0 kg N ha^{-1}	0					Liu *et al.* 2015
		153 kg N ha^{-1}	25					
		306 kg N ha^{-1}	37					
		459 kg N ha^{-1}	46					
Turkey	Wheat growth	0 kg N ha^{-1}	0					Sanin *et al.* 2013
		80 kg N ha^{-1}	30					
		160 kg N ha^{-1}	10					
Canada	Soil test phosphorus	0 t ha^{-1}		0				Shu *et al.* 2016
		28 t ha^{-1}		30				
S. Africa	Organic matter, water holding capacity, pH	0 t ha^{-1}			0	0	0	Cele and Maboeta, 2016
		25 t ha^{-1}			157	3	−12	
		100 t ha^{-1}			576	5	−8	

For example, Mantovi *et al.* (2005) in a study carried out during 12 years, showed that biosolids gave crop yields similar to the highest mineral fertiliser dressing. Applied at a normal rate (5 tons DS ha^{-1} y^{-1}), they can completely surrogate mineral fertilisers, giving crop yields similar to that by mineral dressing. However, with a higher sludge (liquid or dewatered) application rates up to 10 ton DS ha^{-1} y^{-1}, excessive N supply was harmful, leading to wheat lodging and poor quality of crops such as sugar beet or wheat. On the contrary sludge compost could be applied at these higher rates without causing negative effects on yield and quality of crops. These results highlights the suitability of compost as a treatment alternative for sewage sludge.

14.5.2 Health risks involved in application of sludge in agriculture

There are several issues associated with the reuse of municipal sewage sludge in agriculture. While many of these are issues of perception, there is considerable concern, which is scientifically based, over the presence of persistent and emerging contaminants in biosolids (Clarke & Cummins, 2014), the risk of contamination of soil and water (Fu *et al.* 2016), the presence of toxic metals and pharmaceuticals in the sludge, which may build up in the soil and enter the food chain following continuous applications to land (Latare *et al.* 2014; García-Santiago *et al.* 2016), and the risk of emission and transport of bioaerosols containing pathogens following land application of biosolids (Jahne *et al.* 2015). The potential impact of land application of biosolids may also be very long lasting: for example, micro-plastics, which have been found in high concentrations in sewage sludge and have been detected on soils 15 years post-application (Magnusson & Norén, 2014).

The risk of indirect exposure to humans can occur through several pathways (consumption of food-crops, animal up-take to meat or milk or drinking water). Risk assessment approaches have been adopted to assess the environmental fate of contaminants in biosolids, with Quantitative Structure Activity Relationships (QSAR) model approaches dominating (Clarke & Cummins, 2015). Studies that have made links between biological effects and individual compounds in field trials are extremely rare (Zhang *et al.* 2015). While most commentators have stated that the risk to human health following dietary intake of organic contaminants from crops grown on biosolids-amended lands is minimal (Verslycke *et al.* 2016), they acknowledge that a certain amount of uncertainty still exists (Oun *et al.* 2014).

As shown in Table 14.1, there are considerable differences in national legislation regarding the reuse of biosolids in agriculture related to health risk policies and perception. In some countries, such as Belgium (Brussels and Flanders), Switzerland and Romania, the reuse of biosolids in agriculture is prohibited (Milieu *et al.* 2013), whereas in other countries, such as the Republic of Ireland, restrictions govern their reuse in agriculture (Bord Bia, 2013). Moreover, there are differences governing the application rates of biosolids to land. In Europe, the application of biosolids is based on the nutrient and metal content of the biosolids whereas in the majority of states of the USA, biosolids are applied to land based on the nitrogen requirement of the crop being grown and not on a soil-based test. This means that excessive metal accumulation may build up in soil and plants (Antoniadis *et al.* 2008), or may be lost to surface waters (Oun *et al.* 2014) or groundwater. In the EU, the rate of application of six metals (Cu, Ni, Pb, Zn, Cd and Hg) are currently regulated, but the possibility exists that other potentially harmful, unregulated metals, such as arsenic, selenium and antimony, for which no international standards exist for reuse in agriculture, may accumulate in the soil upon repeated application. In a study of the sludge from a range of wastewater treatment plants in Ireland, Healy *et al.* (2016) measured antimony concentrations from 17 to 20 mg kg^{-1}, which were appreciably higher than recorded in wastewater treatment plant sludge elsewhere (<0.01 to 0.06 mg kg^{-1}; LeBlanc *et al.* 2008) and in non-polluted soils (0.53 mg kg^{-1}; Fay *et al.* 2007).

Losses to surface and subsurface waters may occur in two ways: as short-term (incidental losses) whereby losses occur in a rainfall event immediately following land application of biosolids, or as long-term (chronic losses), which occurs when there is a build-up of contaminants in the soil. Surface and groundwater losses of nitrogen and phosphorus species following land application of biosolids have mainly been reported in the literature (Peyton *et al.* 2016). Research has also focused on the presence of human enteric pathogens (Peyton *et al.* 2016), persistent organic pollutants (POPs), endocrine disrupting compounds (EDCs) and pharmaceutical and personal care products (PPCPs) in biosolids (García-Santiago *et al.* 2016) and their potential for loss in rainfall events. Prior to land application, sludge is treated using techniques such as thermal drying, composting, anaerobic digestion and pasteurisation, but complete inactivation of pathogens is difficult to achieve and even though reductions in wastewater treatment may reduce the densities of pathogens in sludge by a number of orders of magnitude. Depending on factors such as pH, soil texture, temperature, moisture content and competition with other microorganisms, may actually regrow following land application (Erickson *et al.* 2014). As the survival time of pathogens, following land application, may be up to four months (Brennan *et al.* 2012), there is a very high possibility that they may be transported to surface and groundwater in incidental rainfall events after land application. For example, Peyton *et al.* (2016) measured total coliform concentrations of up to 1.0×10^6 MPN (Most Probable Number) per 100 ml in surface runoff 15 days after land applications of three common types of biosolids (thermally dried, lime stabilised and anaerobically digested sludge). Alternatively, it is possible that viable pathogens could be present on the crop surface following biosolid application, or may become internalised within the crop tissue where they are protected from conventional sanitization (Solomon *et al.* 2002). In this case, a person may become infected if they consume the contaminated products. To prevent this risk stric application policies are stablished in current European Legislation (Directive 86/278/EC). However, at the time of writing, there has been no documented case of outbreaks or illnesses that have occurred from exposure to pathogens arising from the landspreading of biosolids.

According to Erbardt and Prüeb (European Commission, 2001a), organic contaminants are not expected to pose major health problems to the human population when sludge is re-used for agricultural purposes. Furthermore, many organic compounds will be biodegraded in the soil, and because of their size, organic compounds are generally not taken up by plant roots and translocated to the above-ground edible crop (Epstein, 2002). The presence of organic environmental pollutants, like dioxins and PCBs in agricultural crops is more the result of atmospheric deposition than direct absorption from contaminated soil. On the other hand, there are environmental reasons for monitoring sludge for detergents like LAS and nonylphenols because they are high volume chemicals with an extensive household and industrial use. They are also more water soluble than the organics previously discussed and therefore more mobile and bioavailable in soils. The impact on human health is low because of a low transfer from soil to human consumers (European Commission, 2001a). The ecotoxicological impacts of some of these organic compounds have been studied: as an example, a low ecotoxicologial risk might be expected for plants and soil invertebrates considering the usual levels of NPE in soils receiving polluted sludge (Domene *et al.* 2010; Domene *et al.* 2009). The environmental impact, however, could be significant through leaks to surface waters. Many detergents are clearly toxic and harmful to aquatic organisms and detergents have been indicated as responsible for changes in aquatic populations (European Commission, 2001a). PAHs have become one of the primary pollutants in sludge: it is essential to reduce their contents before the sludge can be used in agriculture through proper treatment. Paraiba *et al.* (2011) investigated the presence of PAHs in biosolids and in soil with biosolids applied as agricultural fertilizer and simulated a long-term risk of soil contamination by PAH. Their results evidenced that PAH concentration levels found in biosolids might raise potential contamination risks to the soil. It is important to perform a close monitoring of PAHs contents and to conduct a more detailed study of the PAHs migration mechanisms in order to obtain data to make changes in existing legislation to ensure full safety of the procedure of agricultural biosolid use. According to Baran and Oleszczuk (2002), biosolid below 5% with a PAH content up to 6000 $\mu g\ kg^{-1}$, should not disturb to natural soil conditions.

The risk to soil fertility of organic contaminants in biosolid spread on farmland has been designated as 'possible' in Table 14.6 by Smith (2009).

Table 14.6 Assessment of risks to health and the environment from recycling sewage sludge to agricultural land.

Environmental Parameter	Risk Attributed
Human health	P[a]
Crop yields	L[b]
Animal health	L
Groundwater quality	L
Surface water quality	L
Air quality	L
Soil fertility	P[c,d]

(L, low risk[a]; P, possible risk[b])
[a]Risk is designated as 'possible' (P) where there is some reported evidence that current operational practice may result in a potential impact on the environment on the basis that one or more of the following conditions apply.
[b]Risk is designated as 'low' (L) where environmental effects are minimized by current operational practice.
[c]There is uncertainty about the environmental implications of particular sludge components.
[d]Effects may occur under certain extreme 'worst-case' conditions, given the current regulations and codes of practice.
Source: Smith, 2009.

14.6 CASE STUDIES ON AGRICULTURAL APPLICATION OF SLUDGE

Legislation governing the land application of sludge in agriculture is designed to minimize the risk of danger to the public, either through the contamination of soil, surface and groundwater, or through the risk to public health. On account of this, regulations governing the reuse of sewage sludge in agriculture are frequently conservative, overly reactive to issues of public perception and local custom, and are discriminatory between the reuse of human sludge and potentially more dangerous, but more socially acceptable, wastes (e.g. animal wastes such as dairy cattle slurry) in agriculture. Therefore, quantifying the environmental persistence and fate of organic and inorganic contaminants following land application of biosolids is necessary, as it provides a sound scientific basis for management practices governing their use in agriculture and, moreover, it allows the potential risks associated with its reuse to be evaluated against other wastes that are commonly applied to agricultural land. The potential benefits of reuse of treated municipal sludge on land are well known and detailed in Table 14.5, and while the potential risks associated with its reuse, which ultimately govern legislation and practice, are also well examined, the quantification of those risks relative to the reuse of other wastes on land are less frequently examined in the literature. Therefore, for the purposes of this review, issues of concern investigated in case studies will be (i) surface runoff of contaminants relative to equivalent applications of dairy slurry, arguably the common agricultural waste applied to land, and (ii) potential bioaccumulation in the food chain through uptake by crops.

Very few studies have compared surface runoff of contaminants arising from the land application of biosolids to equivalent applications of dairy cattle slurry, but of those studies that have, it has been found that, in general, biosolid applications to land to not pose any greater threat to surface runoff than dairy slurries (or their derivatives, e.g. compost). It is important to note that these studies did not investigate the surface runoff of new and emerging contaminants, which may be present in runoff waters. Peyton *et al.*

(2016) applied three types of biosolids (anaerobically digested, thermally dried and lime stabilized slugde sludge) and dairy cattle slurry to small grassland plots at the same rate (40 kg P ha^{-1}) and measured the surface runoff of nutrients, microbial matter and metals, over three successive rainfall events that occurred within 15 days of application. Soil types, on which the study was conducted, ranged from sandy silt to sandy loam, and did not impact the results. The study found that with the exception of total and faecal coliforms and some metals (Ni, Cu), the greatest losses were from the dairy cattle slurry-amended plots. The study concluded that when compared with slurry treatments, biosolids generally do not pose a greater risk (in terms of losses of the parameters measured) along the runoff pathway. Mamedov *et al.* (2016) also examined the relative impacts of land applications of anaerobic digested biosolids to dairy waste (applied as a compost) on surface runoff in a laboratory-based runoff box study. When they were applied at the same rate (50 t ha^{-1}) to three types of soil (loamy sand, loam, clay), surface runoff of suspended solids from the biosolids-amended runoff boxes was the same as or significantly lower than the composted manure amended-runoff boxes, depending on the soil type (no other organic or inorganic contaminants were measured in that study).

Metal bioavailability and uptake by plants is affected by contamination levels and several soil properties such as pH, organic matter and clay content, element speciation in the soil, absorption of the element onto the root, and translocation into the plant (McGrath & Fleming, 2006). Alkaline soil conditions reduce metal bio-availability, but metal cations are more active under acid conditions, with increases of Mn, Zn, Cu, Ni and Cd content in ryegrass being reported when soil pH is reduced (to around 4) following biosolids application (Smith, 1994). In addition to modifying soil pH, the rate at which biosolids is applied also potentially impacts metal uptake by plants (Antoniadis *et al.* 2008) to a point at which plant phytotoxicity may be likely. Antoniadis *et al.* (2008) measured appreciable differences in Cd, Ni, Pb and Zn concentrations in ryegrass between AD biosolids-amended and un-amended plots. Of the studies that have investigated bioaccumulation of various elements by crops, no field-based study could be found that compared uptake rates arising from the landspreading of biosolids with dairy cattle slurry or indeed any other type of animal waste. The heavy metal transfer factor (TF) is the ratio of heavy metal concentration in the plant to that in the soil and is the slope of the proportional line between plant and soil heavy metal. Sanin *et al.* (2013) found that TFs of Cd, Pb, and Ni for wheat plant have increased due to increasing doses of biosolid and wheat germs had higher Cd, Pb, and Ni concentrations than corn-stalk.

It is important to determine the cumulative and residual effects of repeated applications of biosolid on agricultural land. Sigua *et al.* (2005) indicated that successive land application of biosolid for at least three years followed by no sewage sludge application for at least two years may well be a good practice economically because it will boost and/or maintain sustainable forage productivity and at the same time minimize probable accumulation of nutrients, especially trace metals. Land-receiving of biosolid should be periodically monitored to ensure that heavy metal levels in the soil and plants remain within acceptable limits and to assess acceptable biosolids doses and maximum application.

The use of the combinations of biosolid and chemical fertilizers may be more effective than alone application of biosolid (Sanin *et al.* 2013). Erdal *et al.* (2000) found that the combinations of biosolid and triple super phosphate fertilizer significantly increased P content of corn when compared with control. On the other hand, the more amount of biosolid among combinations increased, the less content of P was found for the plant. However, this lower amount among combinations (until 80 mg kg^{-1} biosolid treatment) was not significant. According to Erdal *et al.* (2000) the biosolid can be used for supplying some part of phosphorus needs of plant. Li *et al.* (2005) concluded that combined application of sewage sludge and chemical fertilizer could help quickly establishing a self-maintaining vegetation system in the primary process of nutrient demand. Kahiluoto *et al.* (2015) compared biosolid with chemical fertiliser and they suggested that phosphorus was more plant-available from biosolid than from chemical fertiliser. Iron coagulants are sometimes added to sewage to prevent phosphorus from entering waterways and causing eutrophication. However, adding iron brings a risk: iron-bound phosphorus may not be as usable by plants as non-iron bound forms of phosphorus. However, increasing the amount of sludge used reduced the proportion of phosphorus taken up by plants, even though there was a greater amount of potentially available phosphorus. High levels of iron binding were found to prevent take-up of phosphorus.

14.7 CONCLUSIONS

Reuse in agriculture has become the principal disposal method for sewage sludge in Europe. Agricultural use accounts for 37% of the total sludge produced and it is expected that stabilized sludge will be used in agriculture in large quantities in the next years in many EU member states. The sludge utilization in agriculture is subject to provisions stipulated in the EU Directive 86/278/EEC.

The reuse of treated municipal sewage sludge (biosolids) in agriculture provides the nutrients and micronutrients (such as N, P, K but also Fe, B, Cu and Ni among others) necessary for plant and crop growth. The use of sludge in agriculture also enhances the organic content of soils, increases the water-holding capacity, the soil aggregation, reduces the soil bulk density, increases the cation exchange capacity, and enhances the plant root environment. Therefore, plants are better able to withstand drought conditions, extract water, and utilize nutrients. However, there are several issues associated with the reuse of municipal sewage sludge in agriculture. The presence of persistent and emerging contaminants in biosolids, the risk of contamination of soil and water, the presence of toxic metals and pharmaceuticals in the sludge, and the risk of emission and transport of bioaerosols containing pathogens following land of biosolids, are among the main concerns.

Therefore, sludge treatment aiming to minimize the negative impacts of sludge direct soil application is a key step of sludge management schemes. Several biologic (e.g. anaerobic digestion, composting), thermal (e.g. drying, incineration, pyrolysis) or chemical (e.g. lime addition) treatments are widely applied to sludge aiming to: (i) reduce or even completely eliminate the presence of pathogens (specially thermal and chemical treatments but also biological treatments reaching thermophilic temperatures), (ii) stabilize the organic matter producing products that would not decompose very rapidly (iii) minimize the offensive odours generation (iv) reduce the moisture content and therefore improving its storage capacity and reducing its volume and transportation costs and (v) partially eliminate several organic pollutants and emerging contaminants. However, much less can be done on heavy metals concerning their removal during treatment processes.

Numerous studies have shown that biosolids application to agricultural land has a statistically significant impact on crop yields and soil phosphorus, while having negligible adverse ecological impacts. To be applied in soils, sufficiently stabilized sewage sludge should be used in order to avoid limitations on the plants' growth. In that sense, composted sludge have shown less toxicity in both plants and soil biota than raw, anaerobically digested and thermally dried sludge. In addition, composted sludge can be applied at higher rates than raw sludge without causing negative effects on yield and quality of crops.

14.8 REFERENCES

Adair K. L., Wratten S., Barnes A. M., Waterhouse B. R., Smith M., Lear G., Weber P., Pizey M. and Boyer S. (2014). Effects of biosolids on biodiesel crop yield and belowground communities. *Ecology Engineering*, **68**, 270–278.

Agrafioti E., Bouras G., Kalderis D. and Diamadopoulos E. (2013). Biochar production by sewage sludge pyrolysis. *The Journal of Analytical Applied Pyrolysis*, **101**, 72–78.

Ak M. S., Muz M., Komesli O. T. and Gökçay C. F. (2013). Enhancement of bio-gas production and xenobiotics degradation during anaerobic sludge digestion by ozone treated feed sludge. *Chemical Engineering Journal*, **15**, 499–505.

Amonette J. E. and Joseph S. (2012). Biochar for Environmental Management: Science and Technology. Earhtscan publications Ltd. London, UK.

Antoniadis V., Robinson J. S. and Alloway B. J. (2008). Effects of short-term pH fluctuations on cadmium, nickel, lead, and zinc availability to ryegrass in a sewage sludge-amended field. *Chemosphere*, **71**, 759–764.

Atienza-Martínez M., Gea G., Arauzo J., Kersten S. R. A. and Kootstra A. M. J. (2014). Phosphorus recovery from sewage sludge char ash. *Biomass Bioenergetics*, **65**, 42–50.

Bajsa O., Nair J., Mathew K. and Ho G. E. (2003). Vermiculture as a tool for domestic wastewater management. *Water Science and Technology*, **48**, 125–132.

Baran S. and Oleszczuk P. (2002). Chromatographic determination of polycyclic aromatic hydrocarbons (PAH) in sewage sludge, soil, and sewage sludge-amended soils. *Polish Journal of Environmental Studies*, **11**, 609–615.

Bord Bia. (2013). Sustainable dairy assurance scheme - Producer Standard. http://www.bordbia.ie/industry/farmers/quality/SDQASDocuments/SustainableDairyAssuranceScheme-ProducerStandard.pdf (accessed January 2016).

Braguglia C. M., Coors A., Gallipoli A., Gianico A., Guillon E., Kunkel U., Mascolo G., Richter E., Ternes T. A., Tomei M. C. and Mininni G. (2015). Quality assessment of digested sludges produced by advanced stabilization processes. *Environment Science Pollution Research*, **22**(10), 7216–7235.

Brennan R. B., Healy M. G., Grant J., Ibrahim T. G. and Fenton O. (2012). Incidental phosphorus and nitrogen loss from grassland plots receiving chemically amended dairy cattle slurry. *Science of the Total Environment*, **441**, 132–140.

Cai L., Chen T. B., Gao D., Zheng G. D., Liu H. T. and Pan T. H. (2013). Influence of forced air volume on water evaporation during sewage sludge bio-drying. *Water Research*, **47**, 4767–4773.

Carballa M., Omila F., Ternesh T. and Lema, J. M. (2007). Fate of pharmaceutical and personal care products (PPCPs) during anaerobic digestion of sewage sludge. *Water Research*, **41**(10), 2139–2150.

Carvalho A., Nabais C., Roiloa S. R. and Rodriguez-Echeverria S. (2013). Revegetation of abandoned copper mines: the role of seed banks and soil amendments. *Web Ecology*, **13**, 69–77.

Cele E. N. and Maboeta M. (2016). A greenhouse trial to investigate the ameliorative properties of biosolids and plants on physiochemical conditions of iron ore tailings: implications for an iron ore mine site remediation. *Journal of Environmental Management*, **165**, 167–174.

Cho K. S., Ryu H. W. and Moon H. S. (1999). Effects of sludge solid and S^0 amount on the bioleaching of heavy metals from sewage sludge using sulfur-oxidizing bacteria. *Journal of the Korean Society of Environmental Engineers*, **21**, 433–442.

Clarke R. M. and Cummins E. (2014). Evaluation of 'classic' and emerging contaminants resulting from the application of biosolids to agricultural lands: a review. *Human and Ecological Risk Assessment*, **21**, 492–513.

Dichtl N., Rogge S. and Bauerfeld K. (2007). Novel strategies in sewage sludge treatment. *Clean - Soil, Air, Water*, **35**, 473–479.

Directive 86/278/EEC. (1986). Sewage Sludge Directive 86/278/EEC. Official Journal of the European Community. http://eur-lex.europa.eu/legal-content/EN/TXT/?uri = CELEX%3A31986L0278 (accessed June 2016).

Directive 91/271/EEC. (1991). Urban Waste Water Treatment Directive 91/271/EEC. Official Journal of the European Community. http://eur-lex.europa.eu/legal-content/EN/TXT/?uri=CELEX:31991L0271 (accessed June 2016).

Directive 99/31/EC. (1991). Landfill of waste. Official Journal of the European Community. http://eur-lex.europa.eu/legal-content/EN/TXT/PDF/?uri=CELEX:31999L0031&from=EN (accessed June 2016).

Directive 98/15/EC. (1998). Directive 98/15/EEC amending Directive 91/271/EEC. Official Journal of the European Community. http://eur-lex.europa.eu/legal-content/EN/TXT/?uri=CELEX:31998L0015 (accessed June 2016).

Dogan I. and Sanin F. D. (2009). Alkaline solubilization and microwave irradiation as a combined sludge disintegration and minimization method. *Water Research*, **43**, 2139–2148.

Domene X., Mattana S., Ramírez W., Colón J., Jiménez P., Balanyà T., Alcañiz J. M. and Bonmatí M. (2009). Biossays prove the suitability of mining debris mixed with sludge for land reclamation purposes. *Journal Soils Sediments*, **10**, 30–44.

Domene X., Colón J., Uras M. V., Izquierdo R., Àvila A. and Alcañiz J. M. (2010). Role of soil properties in sewage sludge toxicity to soil collembolans. *Soil Biology and Biochemistry*, **42**, 1982–1990.

Domene X., Solà L., Ramírez W., Alcañiz J. M. and Andrés P. (2011). Soil bioassays as tools for sludge compost quality assessment. *Waste Management*, **31**, 512–522.

Dufour P. (2006). Control Engineering in Drying Technology: Review and Trends. *Drying Technology: an International Journal*, **26**, 889–904.

Epstein E. (2002). Land Application of Sewage Sludge and Biosolids. Lewis Publisher, London, ISBN 1-56670-624-6.

Erdal I., Bozkurt M. A., Cimrin K. M., Karaca S. and Saglam M. (2000). Effects of humic acid and phosphorus fertilizer application on growth and phosphorus uptake of maize (Zea mays L.) grown on a calcareous soil. *Turkish Journal of Agriculture and Forestry*, **24**, 663–668.

Erickson M. C., Habteselassie M. Y., Liao J., Webb C. C., Mantripragada V., Davey L. E. and Doyle M. P. (2014). Examination of factors for use as potential predictors of human enteric pathogen survival in soil. *Journal of Applied Microbiology*, **116**, 335–349.

Eurostat "Sewage sludge production and disposal" last update: 16/02/16. http://ec.europa.eu/eurostat/web/products-datasets/-/env_ww_spd (accessed February 2017).

European Commission. (2000). Working document on sludge 3rd draft. http://ec.europa.eu/environment/waste/sludge/pdf/sludge/_en.pdf (accessed January 2016).

European Commission. (2001a). Organic contaminants in sewage sludge for agricultural use. http://ec.europa.eu/environment/archives/waste/sludge/pdf/organics_in_sludge.pdf (accessed January 2016).

European Commission. (2001b). Disposal and recycling routes for sewage sludge. Part 2 – Regulatory report. http://ec.europa.eu/environment/archives/waste/sludge/pdf/sludge_disposal2.pdf (accessed January 2016).

European Commission. (2014). End-of-waste criteria for biodegradable waste subjected to biological treatment (compost & digestate). http://ipts.jrc.ec.europa.eu/publications/pub.cfm?id=6869 (accessed January 2016).

Fernández J. M., Plaza C., Hernández D. and Polo A. (2007). Carbon mineralization in an arid soil amended with thermally-dried and composted sewage sludges. *Geoderma*, **137**, 497–503.

Fay D., McGrath D., Zhang C., Carrigg C., O'Flaherty V., Kramers G., Carton O. T. and Grennan E. (2007). National soils database. End of Project Report – RMIS 5192. Teagasc, Johnstown Castle, Co.Wexford.

Foladori P., Laurab B., Giannia A. and Giulianoa Z. (2007). Effects of sonication on bacteria viability in wastewater treatment plants evaluated by flow cytometry-Fecal indicators, wastewater and activated sludge. *Water Research*, **41**, 235–243.

Foladori P., Bruni L., Tamburini S. and Ziglio G. (2010). Direct quantification of bacterial biomass in influent, effluent ans activated sludge of wastewater treatment plants by using flow cytometry. *Water***Research*, **44**(13), 3807–3818.

Fu Q., Sanganyado E., Ye Q. and Gan J. (2016). Meta-analysis of biosolid effects on persistence of triclosan and triclocarban in soil. *Environmental Pollution*, **210**, 137–144.

Fytili D. and Zabaniotou A. (2008). Utilization of sewage sludge in EU application of old and new methods-A review. *Renew. Renewable and Sustainable Energy Reviews*, **12**, 116–140.

Gantzer C., Gaspard P., Galvez L., Huyard A., Dumouthier N. and Schwartzbrod J. (2001). Monitoring of bacterial and parasitological contamination during various treatment of sludge. *Water Research*, **35**(16), 3763–3770.

García-Santiago X., Franco-Uría A., Omil F. and Lema J. M. (2016). Risk assessment of persistent pharmaceuticals in biosolids: dealing with uncertainty. *Journal Hazard. Material*, **302**, 72–81.

Ghavindel A., Alikha H. A. and Savaghebi G. (2010). Bioremediation of sewage sludge for land application as a fertilizer using bioleaching. *Journal of Agricultural Science*, **25**, 170–174.

Haug R. T. (1993). The Practical Handbook of Compost Engineering. CRC Press, Boca Ratón, USA.

Healy M. G., Fenton O., Forrestal P. J., Danaher M., Brennan R. B. and Morrison O. (2016). Metal concentrations in lime stabilised, thermally dried and anaerobically digested sewage sludges. *Waste Management*, **48**, 404–408.

Herzel H., Krüger O., Hermann L. and Adam C. (2016). Sewage sludge ash - A promising secondary phosphorus source for fertilizer production. *The Science of the Total Environment*, **542**, 1136–1143.

Hossain M. K., Strezov V., Yin Chan K. and Nelson P. (2010). Agronomic properties of wastewater sludge biochar and bioavailability of metals in production of cherry tomato (Lycopersicon esculentum). *Chemosphere*, **78**, 1167–1171.

Hossain M. K., Strezov V. and Nelson P. F. (2015). Comparative assessment of the effect of wastewater sludge biochar on growth, yield and metal bioaccumulation of cherry tomato. *Pedosphere*, **25**, 680–685.

Ireland M. P. (1983). Heavy Metals Uptake in Earthworms. In: Earth-worm Ecology. Chapman & Hall, London.

Jahne M. A., Rogers S. W., Holsen T. M., Grimberg S. J. and Ramler I. P. (2015). Emission and dispersion of bioaerosols from dairy manure application sites: human health risk assessment. *Environment Science Technology*, **49**, 9842–9849.

Kahiluoto H., Kuisma M., Ketoja E., Salo T. and Heikkinen J. (2015). Phosphorus in manure and sewage sludge more recyclable than in soluble inorganic fertilizer. *Environment Science Technology*, **49**(4), 2115–2122.

Latare A. M., Kumar O., Singh S. K. and Gupta A. (2014). Direct and residual effect of sewage sludge on yield, heavy metals content and soil fertility under rice-wheat system. *Journal of Ecological Engineering*, **69**, 17–24.

LeBlanc R. J., Matthews P. and Richard R. P. (2008). Global atlas of excreta, wastewater sludge, and biosolids management: moving forward the sustainable and welcome uses of a global resource. United Nations Human Settlements Programme (UN-HABITAT), Kenya. http://esa.un.org/iys/docs/san_lib_docs/habitat2008.pdf (Accessed January 2016).

Levantesi C., Beimfohr C., Blanch A. R., Carducci A., Gianico A., Lucena F., Tomei M. C. and Mininni G. (2015). Hygienization performances of innovative sludge treatment solutions to assure safe land spreading. *Environmental Science and Pollution Research*, **22**, 7237–7247.

Li S., Wu D. and Zhang J. (2005). Effects of vegetation and fertilization on weathered particles of coal gob in Shanxi mining areas. *China. Journal of Hazardous Materials*, **124**(1–3), 209–216.

Liu X. A., Fike J. H., Galbraith J. M., Fike W. B., Parrish D. J., Evanylo G. K. and Strahm B. D. (2013). Effects of harvest frequency and biosolids application on switchgrass yield, feedstock quality, and theoretical ethanol yield. *Global Change Biology*, **7**, 112–121.

Liu X. A., Fike J. H., Galbraith J. M., Fike W. B., Parrish D. J., Evanylo G. K. and Strahm B. D. (2015). Effects of harvest frequency and biosolids application on switchgrass yield, feedstock quality, and theoretical ethanol yield. *Bioenergy*, **7**, 112–121.

Magnusson K. and Noren F. (2014). Screening of microplastic particles in and downstream a wastewater treatment plant. Report to the Swedish Environmental Research Institute, C55.

Malmborg J. and Magner J. (2015). Pharmaceutical residues in sewage sludge: Effect of sanitization and anaerobic digestion. *Journal of Environmental Management*, **153**, 1–10.

Mamedov A. I., Bar-Yosef B., Levkovitch I., Rosenberg R., Silber A., Fine, P. and Levy G. J. (2016). Amending soil with sludge, manure, humic acid, orthophosphate and phytic acid: effects on infiltration, runoff and sediment loss. *Land Degradation and Development*, **27**, 1629–1639.

Manara P. and Zabaniotou Z. (2012). Towards sewage sludge based biofuels via thermochemical conversion – A review. *Renewable and Sustainable Energy Reviews*, **16**, 2566–2582.

Mantovi P., Baldoni G. and Toderi G. (2005). Reuse of liquid, dewatered and composted sewage sludge on agricultural land: effects of long-term application on soil and crop. *Water Research*, **39**, 289–296.

Mattenberger H., Fraissler G., Brunner T., Herk P., Hermann L. and Obernberger I. (2008). Sewage sludge ash to phosphorus fertiliser: Variables influencing heavy metal removal during thermochemical treatment. *Waste Management*, **28**, 2709–2722.

Mayer P., Hilber I. and Hale, S. (2016). How to Determine the Environmental Exposure of PAHs Originating from Biochar. *Environment Science Technology*, **50**(4), 1941–1948.

McGrath D. and Fleming G. A. (2006), Trace elements and heavy metals in Irish soils. Teagasc, Co. Wexford. http://www.teagasc.ie/publications/2011/823/Trace_Elements.pdf

Milieu WRC, RPA (2013). Environmental, economic and social impacts of the use of sewage sludge on land. Final Report – Part I, Part II and Part III.

Mininni G., Blanch A. R., Lucena F. and Berselli S. (2015). EU policy on sewage sludge utilization and perspectives on new approaches of sludge management. *Environmental Science and Pollution Research*, **22**, 7361–7374.

Müller J. A. (2001). Prospects and problems of sludge pre-treatment processes. *Water Science and Technology*, **44**, 121–128.

Müller J. A., Winter A. and Strünkmann G. (2004). Investigation and assessment of sludge pre-treatment processes. *Water Science and Technology*, **49**, 97–104.

Ødegaard H., Paulsrud B. and Karlsson I. (2002). Wastewater sludge as a resource: Sludge disposal strategies and corresponding treatment technologies aimed at sustainable handling of wastewater sludge. *Water Science and Technology*, **46**, 295–303.

Oun A., Kumar A., Harrigan T., Angelakis A. and Xagoraraki I. (2014). Effects of biosolids and manure application on microbial water quality in rural areas in the US. *Water*, **6**, 3701–3723.

Paraíba L. C., Queiroz S. C. N., de Souza D. R. C. and Saito M. L. (2011). Risk simulation of soil contamination by polycyclic aromatic hydrocarbons from sewage sludge used as fertilizers. *Journal of the Brazilian Chemical Society*, **22**(6), 1156–1163.

Park K. Y., Maeng S. K., Song K. G. and Ahn K. H. (2008). Ozone treatment of wastewater sludge for reduction and stabilization, *Journal Environment Science and Health*, **43**, 1546–1550.

Peyton D. P., Healy M. G., Fleming G. T. A., Grant J., Wall D., Morrison L., Cormican M. and Fenton O. (2016). Nutrient, metal and microbial loss in surface runoff following treated sludge and dairy cattle slurry application to an Irish grassland soil. *The Science of the Total Environment*, **541**, 218–229.

Ponsá S., Pagans E. and Sáncehz A. (2009). Composting of dewatered wastewater sludge with various ratios of pruning waste used as a bulking agent and monitored by respirometer. *Biosystems Engineering*, **102**, 433–443.

Ponsá S., Gea T. and Sánchez A. (2010). Different indices to express biodegradability in organic solid wastes. *Journal of Environmental Quality*, **39**, 706–712.

Quilbé R., Serreau C., Wicherek S., Bernard C., Thomas Y. and Oudinet J. P. (2005). Nutrient transfer by runoff from sewage sludge amended soil under simulated rainfall. *Environmental Monitoring and Assessment*, **100**, 177–190.

Ramírez W. A., Domene X., Ortiz O. and Alcañiz J. M. (2008). Toxic effects of digested, composted and thermally-dried sewage sludge on three plants. *Bioresource Technology*, **99**, 7168–7175.

Rivas Ibánez G. R., Esteban B., Ponce-Robles L., Casas López L. J., Agüera A. and Sánchez Pérez J. A. (2015). Fate of micropollutants during sewage sludge disintegration by low-frequency ultrasound. *Chemical Engineering Journal*, **280**, 575–587.

Ross J. J., Zitomer D. H., Miller T. R., Weirich C. A. and McNamara P. J. (2016). Emerging investigators series: pyrolysis removes common microconstituents triclocarban, triclosan, and nonylphenol from biosolids. *Environment Science: Water Research Technology*, **2**, 282–289.

Samaras P., Papadimitriou C. A., Haritou I. and Zouboulis A. I. (2008). Investigation of sewage sludge stabilization potential by the addition of fly ash and lime. *Journal of Hazardous Materials*, **154**, 1052–1059.

Sanin D., Namli A., Farasat S. and Akca M. O. (2013). Management of Domestic-Urban Wastewater Sludge, 108G190 TÜBİTAK-KAMAG, Project Expert, 2010–2013, Ankara, Turkey.

Sánchez A., Gabarrell X., Artola A., Barrena R., Colón J., Font X. and Komilis, D. (2015). Composting of wastes. In: Resource Recovery to Approach Zero Municipal Waste. CRC Press, Boca Ratón, USA.

Shanableh A. and Ginige P. (1999). Acidic bioleaching of nitrogen and phosphorus from sewage sludge. *Environment Technology*, **20**(5), 459–468.

Shu W., Price G. W., Sharifi M. and Cade-Menum B. J. (2016). Impact of annual and single application of alkaline treated biosolids on soil extractable phosphorus and total phosphorus. *Agriculture, Ecosystems & Environment*, **219**, 111–118.

Sigua G. C., Adjei M. B. and Rechcigl J. E. (2005). Cumulative and residual effects of repeated sewage sludge applications: forage productivity and soil quality implications in South Florida, USA. *International Journal of Environment and Pollution Research*, **12**(2), 80–8.

Singh B. P., Cowie A. L. and Smernik R. J. (2012). Biochar carbon stability in a clayey soil as a function of feedstock and pyrolysis temperature. *Environment Science Technology*, **46**, 11770–11778.

Sinha R. K., Herat S., Bharambe G. and Brahambhatt A. (2010). Vermistabilization of sewage sludge (biosolids) by earthworms: Converting a potential biohazard destined for landfill disposal into a pathogen-free, nutritive and safe biofertilizer for farms. *Waste Management Research*, **28**(10), 872–881.

Smith S. R. (1994). Effect of soil pH on availability to crops of metals in sewage sludge-treated soils. I. Nickel, copper and zinc uptake and toxicity to ryegrass. *Environmental Pollution*, **85**, 321–327.

Smith S. R. (2009). Organic contaminants in sewage sludge (biosolids) and their significance for agricultural recycling. *Philosophical Transactions of the Royal Society*, **367**, 4005–4041.

Srinivasan A., Young C., Liao P. H. and Lo K. V. (2015). Radiofrequency-oxidation treatment of sewage sludge. *Chemosphere*, **141**, 212–218.

Tarrasón D., Ojeda G., Ortiz O. and Alcañiz J. M. (2008). Differences on nitrogen availability in a soil amended with fresh, composted and thermally-dried sewage sludge. *Bioresource Technology*, **99**, 252–259.

Van Wesenbeeck S., Prins W., Ronsse F. and Antal M. J. (2014). Sewage sludge carbonization for biochar applications. *Fate of Heavy Metals. Energy & Fuels*, **28**(8), 5318–5326.

Verslycke T., Mayfield D. B., Tabony J. A., Capdevielle M. and Slezak B. (2016). Human health risk assessment of triclosan in land-applied biosolids. *Environmental Toxicology and Chemistry*, **35**(9), 2358–2367.

Wen J. K., Yao G. C., Chen B. W., Wu B., Liu X. and Huang S. T. (2009). Effect of pH on activities of mineral-bioleaching microorganisms and bioleaching rate. *Chinese Journal of Rare Metals*, **33**(1), 80–83.

Wang Q., Kuninobu M., Kakimoto K., Ogawa H. I. and Kato Y. (1999). Upgrading of anaerobic digestion of waste activated sludge by ultrasonic pretreatment. *Bioresource Technology*, **68**, 309–319.

Wong J. W. C. and Selvam A. (2006). Speciation of heavy metals during co-composting of sewage sludge with lime. *Chemosphere*, **63**, 980–986.

Yan M., Prabowo B., Fang Z., Wei C., Zhiqiang J. and Yanjun H. (2015). Effect of hydrothermal treatment temperature on the properties of sewage sludge derived solid fuel. *International Journal of Renewable Energy Development-IJRED*, **4**(3), 163–169.

Zhang F. S., Yamasaki S. and Kimura K. (2002). Waste ashes for use in agricultural production: II. Contents of minor and trace metals. *The Science of the Total Environment*, **286**, 111–118.

Zhang Z., Le Velly M., Rhind S. M., Kyle, C. E., Hough R. L., Duff E. I. and McKenzie C. (2015). A study on temporal trends and estimates of fate of bisphenol A in agricultural soils after sewage sludge amendment. *The Science of the Total Environment*, **515–516**, 1–11.

Part 3

Recovering Resource: Energy and Chemicals

Chapter 15

Recovering energy from sludge

Sergio Ponsa, David Bolzonella, Joan Colon, Marc A. Deshusses, Isabel Fonts, Noemí Gil-Lalaguna, Dimitrios Komilis, Gerasimos Lyberatos, Sara Isabel Pérez-Elvira and José Luis Sánchez

15.1 INTRODUCTION

Economic and social growth are strongly accompanied by increasing amounts of waste, causing unnecessary losses of materials and energy, environmental damage and negative effects on health and quality of life. This is becoming a worldwide problem and all societies should be concerned about the consequences of non-controlled industrial and urban design and societal growth. Waste generation and management is one of the most serious problems in modern societies, and consequently nowadays stringent policies on waste issues are increasingly implemented in developed countries.

As a result, fossil fuel resources are depleting rapidly and global energy consumption is growing. In this context, renewables energies will play an important role in the future worldwide.

The EU's Renewable Energy Directive (Directive, 2009/28/EC) sets a binding target of 20% final energy consumption from renewable sources by 2020, while through the 2030 EU Energy Strategy (EU COM/2014/015) that defines the policy framework for climate and energy in the period from 2020 to 2030, the EU countries have already agreed on a new renewable energy target of at least 27% of final energy consumption in the EU as a whole by 2030. To achieve this, EU countries have committed to reaching their own national renewables targets which includes having at least 10% of their transport fuels come from renewable sources by 2020.

Nowadays, there are new technologies that enable production of renewable energy in an economical and environmentally sustainable way. Apart from the classic renewable energy technologies such as solar, wind and hydropower, organic wastes are considered an excellent resource for renewable energy and materials. Many innovative technologies are successfully being developed and applied to produce renewable energy from organic wastes sometimes with production of other valuable materials (e.g. organic soil amendments).

Organic wastes comprise a wide range of wastes mainly classified as municipal organic waste and industrial organic waste that mainly differ in the management responsibility. Industrial waste falls under the "producer pays principle" while municipalities or public utilities are mainly responsible for the management of the municipal wastes. Among these wastes, the organic (biodegradable) wastes receive special attention in international legislations (e.g. Directive 2008/98/CE, Article 22) since they can be considered as a source of renewable energy and materials, and they have important negative environmental impacts when they are not treated appropriately. In Figure 15.1, the main classification of biodegradable wastes is shown.

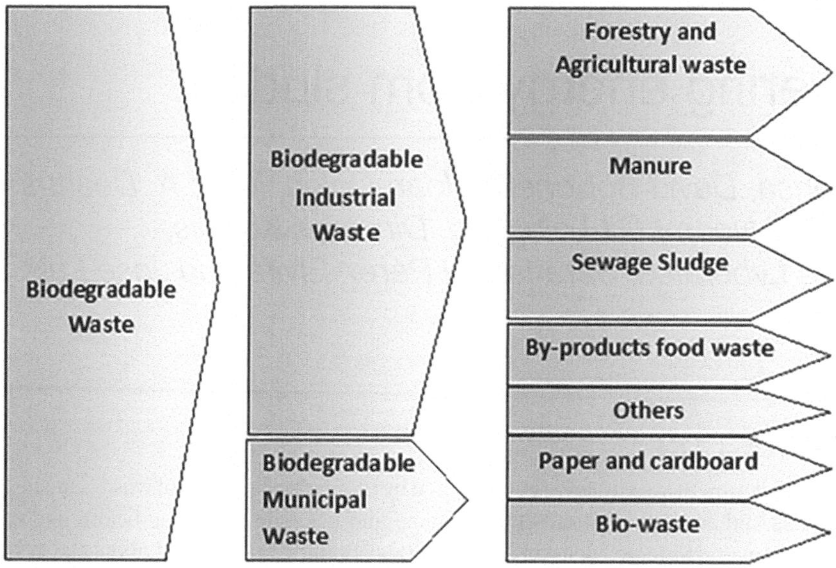

Figure 15.1 Biodegradable waste classification.

Among all biodegradable wastes considered, the generation of municipal sewage sludge (SS) is increasing lately worldwide due to increasingly stringent limits set for the treated effluent, due to population increase and due to the fact that an increasing number of households are being connected to wastewater treatment facilities. For example, in Europe it is estimated that in 2020 13 million tons of dry solids (DS) will be generated from sewerage, compared to 9.8 million tons of DS that were generated in 2005 (Kelessidis & Stasinakis, 2012).

Well established technologies for the treatment of sewage sludge exist since the 1950's. The principal ones are direct land application, anaerobic digestion (AD), which produces methane, composting, and thermal treatment (mainly incineration). Landfilling, a common and cheap disposal technique of the past is now prevented by several legislations in the world, unless sludge is stabilized and relatively dry.

15.1.1 Sewage sludge definition and production

According to the European Waste Catalogue, municipal sewage sludge is a non-hazardous waste that appears under Chapters 19 and 20 of that catalogue. Table 15.1 summarizes the sewage sludge production in the EU.

Table 15.1 Sewage sludge production in the EU.

	Sewage Sludge Production (×1000 ton DS/year)	Specific Sewage Sludge Production (kg/(p.e. × year))	Specific Sewage Sludge Production (kg/(p.e.s. × year))
EU-15	9806	21.9	25.4
EU-12	1151	11.5	21.1
EU-27	10957	17.7	24.3

p.e.: population equivalent; p.e.s.: population equivalent served by a wastewater treatment system.
Source: Kelessidis & Stasinakis, 2012.

Based on the specific sludge production, Austria (2006 data) is the largest producer (with 30.8 kg/(p.e. × year)) whilst Malta produces the lowest amount, namely 0.1 kg/(p.e. × year). More detailed information on the amounts of SS produced in the EU can be found in Kelessidis and Stasinakis (2012).

15.1.2 Legislation issues applied to SS and current status

In the EU, the two key directives on sewage sludge management are the 86/278/EEC (Sewage Sludge Directive) and the 91/271/EEC (Urban Wastewater Treatment Directive) (Fytili & Zabaniotou, 2008). The former legislation encouraged the application of sewage sludge on soil as long as it is treated (e.g. composted) or incorporated or injected in a safe manner into the soil. In this directive, limits for heavy metals concentrations in soils and sludge are set. These limits have either been adopted by some EU members, while more stringent values have been adopted by others. In some national legislations, limits for additional parameters not mentioned in 86/278/EEC, such as total Cr, pathogens and organic micropollutants have been also applied (Kelessidis & Stasinakis, 2012).

The directive 91/271/EEC (as amended by 98/15/EC) sets more stringent regulations for effluent disposal and requires that municipalities as small as with 2000 p.e. should have wastewater treatment facilities by 2005. For countries that joined the EU after 2014, this deadline becomes 2015 or 2018 (Kelessidis & Stasinakis, 2012). Apart from the above legislation, directive 99/31/EU, that has set limits on the disposal of biodegradable wastes to landfill, has affected the landfilling of SS. In addition, sludge incineration is considered a disposal technique as long as the energy recovery is below a threshold value set by 2008/98/EC (Waste Framework Directive).

Currently, the most typical treatment technologies for SS in Europe are anaerobic and aerobic stabilization. According to Kelessidis and Stasinakis (2012), 89% of the EU-27 countries have adopted anaerobic digestion, whilst 74% of the same countries use aerobic digestion. Chemical stabilization techniques (e.g. lime addition) are applied much less frequently and could be also used in combination with biological treatment techniques (e.g. anaerobic digestion). Incineration of sludge in the EU-27 has doubled from 1992 to 2005, whilst the use of landfilling, during the same period, was reduced by almost 50% (Kelessidis & Stasinakis, 2012). Incineration of sludge can achieve a large volume reduction, odor minimization, energy recovery potential and destruction of toxic organic compounds and pathogens. Ash disposal after incineration has to be accounted for when designing such systems (Fytili & Zabaniotou, 2008). The existing technologies to treat sewage sludge are included in Table 15.2.

The co-digestion of SS with other substrates has been extensively studied in the literature (Tandukar & Pavlostathis, 2015). The objective of co-digestion has always been the enhancement of biogas production and the production of a final sludge with better dewatering characteristics. Usually high strength wastes such fat-oil-grease and food processing wastes have been used for co-digestion with municipal SS.

Table 15.2 Existing technologies applied for sewage sludge treatment.

Technique	Advantages/Benefits	Disadvantages/ Constraints
Energy recovery via biological processes (AD)	• Well established technique • Green energy • Continuous process • Reduced transport cost if on-site • Can be combined with co-substrates • Nutrient recovery • Organic amendment	• High installation cost • Requires good know-how during operation • Requires some dewatering • Usually requires further composting of digestate
Aerobic treatment and land disposal	• Well established technique • Reduced transport cost if on site • Nutrient recovery • Organic amendment • Low capital cost	• Requires dewatering • Requires bulking agent • Area requirements • Odour nuisance
Thermal treatment	• Green energy • Continuous process • Reduced transport cost if on site • Ash reuse depending on situation	• Requires dewatering • Negative public perception • Low in waste hierarchy • Emissions • Ash disposal • High cost
Liming	• Reduction of microbial content (pathogens) and heavy metals availability • Well established technique • Easy technology • Nutrient recovery	• Lime availability and lime cost • Requires dewatering
Direct land application	• Nutrient recovery • Organic amendment • Low cost	• Large area requirements • Strict legislative limits • No pathogen removal • No organic matter stabilisation • Environmental impacts • Negative public perception • Not directly in line with 86/278/EEC
Landfill disposal	• Low treatment cost • Easy technology	• Against targets of 99/31/EU • Resource loss • High disposal cost due to landfill tax • Requires dewatering • Contribution in biogas emissions and leachate generation

15.1.3 Legislative constraints and policy goals

The factors that are expected to affect the current and future sewage sludge management options are:

- *Waste hierarchy:* According to the existing waste hierarchy, emphasis should be given on recovery of nutrients, organic matter and green energy via AD. On the other hand, energy recovery via thermal treatment processes is found low in the hierarchy pyramid.
- *Landfilling:* Landfilling of sludge (raw or treated) is not desirable based on the requirements of 99/31/EU that necessitates the reduction of biodegradable organics to landfill.
- *Organic matter preservation and supply of nutrients:* The low amounts of soil organic matter, especially in the South of Europe, makes the ultimate disposal of carbon sources (such as biologically treated SS) to arid land very attractive in those places. This can supply significant amounts of nutrients (N, P, K) to arable land.
- *Economics and economy of scale:* The use of decentralized or centralized units to process sludge is highly affected by site-specific situations. High transportation costs might hinder the use of centralized units to treat sludge. On the other hand, the use of several decentralized units might result in higher costs per dry ton of sludge treated. Also, traceability of sludge in case of centralized units might be problematic, especially if the treated sludge is spread on land. A careful financial analysis should be conducted in each case. Still, the principle of proximity should be accounted for when selecting the location of treatment facilities.
- *Energy recovery:* There is a tremendous need for renewable energy nowadays. The current technologies of anaerobic digestion and incineration of SS fulfil those requirements.
- *Heavy metals limits and their analytical quantification:* Those limits differ depending on the country. However, they clearly dictate the type of technology that can be applied. Furthermore, the worldwide trend of reducing these limits needs to be considered. In addition, the methodologies to quantify the environmental impact of heavy metals in sludge need improvement. The commonly used "total heavy metal content" is not necessarily the metal content that is environmentally available. Different (e.g. sequential) leaching techniques can provide more toxicologically relevant information. This, however, would mean a dramatic change in regulated limits and analytical protocols.
- *Other quality limits:* Although, traditionally, heavy metals have been the principal contaminants regulated and reported in SS, the addition of other quality parameters has made the selection of sludge disposal technologies more difficult. Organic pollutants (e.g. priority organic pollutants) and pathogen levels have been introduced in the regulations and need to be accounted for during technology selection.
- *Co-substrate (feedstock) suitability:* The use of certain co-substrates during AD is controlled or even prohibited by some regulations. The Animal By-Products (ABP) regulation (EC/1774/2002) may have an effect on the proportions of ABP mixed with SS, whilst several animal byproducts necessitate heat treatment prior to co-digestion. If SS is co-digested with biodegradable waste, it might not fall under the framework of sludge regulations. Thus, the co-digestate may not be covered by any of the approved quality protocols.
- *Public acceptance:* This parameter can highly affect the selected technology. Usually, there is a negative public perception towards certain treatment technologies, such as thermal processes. On the other hand, a positive perception exists for certain biological processes. These public perceptions might not be necessarily justified by scientific data, but they are rather controlled by the media. Still, they can determine to a high extent the final choice of a technology and the location of the treatment facilities.

The main goal of this book chapter is to present emerging and novel technologies to treat SS. The term emerging also refers to established technologies (such as AD or incineration) in which some type of game-changing modification has been included. The technologies examined in this book chapter are:

(1) Advanced thermal and high pressure pre-treatments to enhance energy recovery in AD processes
(2) Co-digestion of sewage sludge with non-sludge organic wastes (including bio-hythane production)
(3) Bio-drying
(4) Gasification
(5) Pyrolysis
(6) Supercritical water processing

The above-mentioned technologies are analysed in detail below, whereas well-established sludge treatment technologies (anaerobic digestion, composting, incineration) are not discussed in this chapter.

15.2 BIOLOGICAL BASED TECHNOLOGIES

15.2.1 Advanced thermal/high pressure pre-treatments to enhance energy recovery in AD processes

The pre-treatment of waste activated sludge before anaerobic digestion is generally accepted as one the most efficient and sustainable alternatives in response to the current bottleneck for biosolids disposal and energy recovery. The key of this alternative is the optimization of the anaerobic digestion process (biodegradability and degradation rate enhancement, digesters load increase, sludge reduction) together with the achievement of some other benefits (sanitation, dewatering enhancement, removal of emerging micropollutants). Technical and especially economical constraints of most of the technologies have limited their scale-up to field implementation Among all the existing pre-treatment alternatives (thermal, chemical, mechanical, electrical, ultrasound) the thermal/high pressure process is perhaps the most promising technology.

While the bibliography for lab-scale operation is large (Cano *et al.* 2015; Sridhar Pilli *et al.* 2015; Ariunbaatar *et al.* 2014; Carrère *et al.* 2010; Pérez-Elvira *et al.* 2006), the references for full scale operation are scarce.

15.2.1.1 General features and technology basis

A thermal/high pressure pre-treatment consists of subjecting the sludge to high temperature and high pressure in a hydrolysis reactor. Several commercial technologies have been developed that follow this common principle, but with different operation schemes. The first processes were developed for the treatment of municipal SS, aimed to enhance biodegradability in a batch process. Today, the technology has been extended to the treatment of other biosolids (food waste, agriculture biosolids, animal by-products) and new processes have been developed with different objectives (digestion improvement, sanitization, biosolids quality, energy recovery).

Regarding the process performance, in general terms, most of the processes offer the same qualitative benefits which are summarized for SS in Table 15.3:

Table 15.3 General features of thermal hydrolysis technologies.

Biogas production	Keeping digestion HRT constant, biogas yield increases by 40–60% for waste activated sludge (WAS), and by 10–25% for mixed sludge
Digested sludge (biosolids)	Reduced volumes (proportional to biogas yield increase) Lower viscosity Improved dewaterability (DS > 30%) Pasteurization: EPA Class A; no pathogen regrowth. Agricultural application No odors
Anaerobic process	Reduced energy consumption in mixing Higher DS concentration in feed (10% DS) Higher organic loading rates (4 kg SSV/$m^3 \cdot d$) No foam formation Increased elimination yield of emergent compounds Higher COD and N concentration in dewatering recycle

Regarding the design and operation scheme, there are four key aspects that differ among the processes developed. Table 15.4 describes the different options, while comparing their positive (+) or negative (–) influence on the global process.

Table 15.4 Design and operation aspects that define thermal hydrolysis processes.

Mechanism	Thermal	Combination of temperature and time 150–170°C, 20–60 min widely applied >180°C secondary reactions appear	+	Key factor, responsible for hydrolysis
	Steam explosion	Sudden decompression; flash boiling Polymer structure and cells fracture	+	Improves the global process efficiency
Heat exchange	Heat exchangers	To preheat feed and recover energy Thermal fluids (water, steam, hot oil)	–	Require maintenance Subject to scaling
	Live steam	Increase heat recovery options	+	Efficient and easy to control
Sludge flow	Pumps	Most used in commercial systems Delicate with viscous and hot sludge	–	Critical equipment Substantial maintenance
	Pressurization	Use of a pressurized vessel to feed	+	Easy, no moving parts
Operation regime	Batch	Feeding-reaction-decompression cycles Non-steady state operation	–	Parallel reactors Large footprint
	Continuous	No loading and unloading cycles Steady state operation	+	No down-time State variables; easy control

15.2.1.2 Commercial thermal pre-treatments comparison

Seven different thermal hydrolysis technologies are commercially available in 2016 for the treatment of municipal sewage sludge. Table 15.5, compares their main aspects on the basis of information available publicly (commercial literature and patents), though in some cases, it has not been possible to verify the reported data.

Regarding the operating conditions, all the technologies have two aspects in common: the operation with concentrated sludge, and no use of chemicals. While it is generally accepted to operate in the range 150–230°C for 15–30 minutes, the proposal of second-generation technologies (such as tH_4^+) aim to reduce operation time (and thus reduce footprint) by increasing hydrolysis temperature and exploiting the steam explosion effect.

The process configuration, heating mechanism and energy recovery scheme vary in the different technologies. While the origin of thermal pre-treatment was a batch process (THP-Cambi: most widely applied and considered as the bench mark), it exhibits conceptual limitations inherent to its batch, non-steady state operation. Further developments (Exelys) offer continuous operation, while lacking robustness and reliability. The most recent proposals search for better technical robustness, smaller footprint and equipment demand, and energy recovery optimization. Second-generation processes of thermal hydrolysis (such as tH_4^+) are rising with conceptual and technical advantages and cost-effective operation.

Table 15.5 Comparison of thermal/high pressure pre-treatment technologies.

Company	Technology	Heating/ Duration (°C/min)	Steam Explosion	Heat Exchangers	Pumps	Maintenance	Footprint	Energy Integration
Cambi	THP	165/30	yes	no	yes	–	–	–
Veolia	Exelys	165/30	no	yes	yes	–	–	–
SH+E	Lysotherm	165/30	no	yes	yes	–	–	–
Sustec	TurboTec	165/30	no	yes	yes	–	–	–
Haarslev	ACH	165/20	yes	no	no	+	+	–
Aqualogy	Aqualysis	170/15	yes	no	yes	–	+	–
$teCH_4^+$	tH_4^+	220/<5	yes (twice)	no	no	+	+	+

+ positive; – negative.

Finally, only scarce reporting of the disadvantages of thermal/high pressure pre-treatment have been found. Some issues to take into account are: system complexity, need for high-level operator training, shutdowns (regulatory inspections and approved maintenance), and increase in soluble inert fraction and ammonia in the recycled stream.

15.2.1.3 Economic evaluation

The economic benefit of the thermal hydrolysis technology is mainly related to the increase in biogas production and reduction in biosolids disposal, which depend on market values and the specific issues of each situation (upgrade of existing WWTPs, new WWTPs, digestion needs, treatment of mixed or secondary sludge).

The investment in a thermal pre-treatment is justified if the WWTP must increase digestion capacity, as the advanced digestion (TH + AD) allows roughly doubling the organic loading rate in the existing digesters. Furthermore, from the point of view of operation costs, the key for the thermal process profitability is the energy integration of the system, only optimized when all the biogas is used to produce energy in a gas engine and the residual heat from cogeneration is sufficient to run the thermal process (i.e., no biogas burning or heat purchase). This optimization is possible for continuous systems that use live steam heating.

An economic comparison between advanced digestion (TH + AD) and conventional digestion (AD) can be performed for the implementation of a thermal unit in a WWTP, considering the items presented in Table 15.6 and current market values (capital, disposal and energy costs). The results, converted into a final common unit (€/m^3 sludge treated) and presented in Figure 15.2, demonstrate the clear economic advantage of a thermal pre-treatment. When the capacity of the WWTP needs to be increased, the balance is clearly positive towards the thermal pre-treatment.

Table 15.6 Items quantified in the economic evaluation performed.

Steps of the Process	Items Quantified
A. Pre-treatment	• Investment for the plant • Energy consumption
B. Anaerobic digestion	• Pumping cost (mixing) • Energy from biogas in a heat and power system
C. Digestate treatment and biosolids disposal	• Dewatering energy consumption • Polyelectrolyte cost • Transport cost • Market for the product
D. Centrate treatment	• Treatment cost • Nutrients recovery possibilities

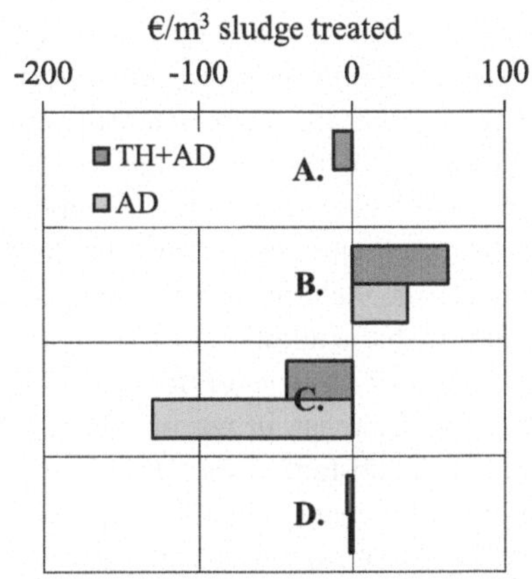

Figure 15.2 Economic comparison between advanced digestion (TH + AD) and conventional digestion (AD).

15.2.2 Co-digestion of sewage sludge with non-sludge organic wastes

Anaerobic digestion is widely applied in WWTPs to stabilize primary and biological sludge prior to their final disposal while recovering energy through biogas. When spare digester volumes are available, substrates other than sludge can be anaerobically co-digested to increase the energy recovery (and incomes). In fact, an interesting option for improving methane yields in WWTPs is Anaerobic co-Digestion (AcoD) of sludge and other organic waste like the organic fraction of municipal solid waste (OFMSW), food waste (FW), fat oil and grease, agro-waste and others (Mata-Alvarez *et al.* 2011).

Actually, the transportation cost of the co-substrate from the generation point to the AD plant is the first selection criterion: in general, costs are in the range 2–3 €/ton, when in a 30 km radius, but can go up to 5 €/ton when the transportation distance exceeds 40–50 km. Despite this fact, it is important to choose the best co-substrate and blending ratio with the aim of favoring synergisms, diluting harmful compounds and optimizing methane production, while avoiding deterioration of digestate quality.

Sewage Sludge (SS) ranks as the second main substrate for AcoD. Historically, AcoD of SS with the organic fraction of municipal solid waste (OFMSW) is the most reported co-digestion researched, but a number of different options can be found in actual practice. Even so, between 2010 and 2013, many publications can be found dealing with co-digestion of fats, oils and greases (FOG), fruit and vegetable waste (FVW) and other agro-wastes, food processing waste, glycerol, and algae. The low organic load of the SS, together with the unused capacity of many wastewater treatment plant (WWTP) digesters, frequently as much as 30%, are the main driving forces behind SS co-digestion. SS is characterized by relatively low C/N ratio and high buffer capacity. Therefore, it is suitable for mixing with co-substrates characterized by high amounts of easily biodegradable organic matter and low alkalinity values. The main aim of the approach is to improve biogas production and energy recovery up to levels similar to those of the energy demand of the WWTP (Bodik & Kubaska, 2013). While doing so, the WWTP becomes a local centre for waste disposal. Clearly, the good quality of the final digestate and its potential reuse remain a prerequisite.

There are now a considerable number of WWTPs applying the co-digestion option around Europe and worldwide and the current technology readiness level (TRL) of this technology is 9, i.e, co-digestion is fully commercial.

The most deployed option is SS co-digestion with the organic fraction of municipal solid waste: in this case the organic waste is pre-treated for inert removal and size reduction in the WWTP or in another dedicated plant and then co-treated with sludge in the anaerobic digesters. The same approach applies for FVW or food waste. The typical biogas yields for these materials are in the range 0.7–0.9 $Nm^3\ kgVS^{-1}$ (Koch *et al.* 2015), values 3–4 times higher than those of sludge. The higher biogas yield is therefore a consequence of the feeding mix. For example, Koch *et al.* (2015) reported that the addition of 1,000 m^3 of food waste to 9,000 m^3 sludge over a period of 6 months doubled the biogas production in the Garching/Alz WWTP. The corresponding specific methane yield in the two periods added up to 0.310 $Nm^3\ CH^4\ kgVS^{-1}$ and 0.393 $Nm^3\ CH_4\ kgVS^{-1}$ without and with the addition of FW, respectively.

Mattioli *et al.* (2015) found that the co-treatment of 10 tons per day of organic waste and mixed sludge in Rovereto WWTP resulted in an increase in the organic loading rate which passed from 0.73 to 1.38 kgVS $(m^3\ day)^{-1}$ and allowed for a substantial increase in biogas generation from 1321 to 2723 $Nm^3\ day^{-1}$, a significant increase in energy recovery. The daily potential power generation using turbine systems increased from 3900 kWh to 7800 kWh when shifting from simple anaerobic digestion of sludge to co-digestion. The latter value is equivalent to 85% of the total energy consumption of the wastewater treatment plant.

In general, the typical energy required for wastewater treatment is in the range 20–30 kWh per person-equivalent per year, while digestion of mixed sludge can produce up to 10–15 kWh per person-equivalent per year under the best conditions,while co-digestion can help to cover this demand offering energy independency.

Although co-digestion is largely applied in WWTPs, reports on full-scale sludge co-digestion practices are scarce in the peer-reviewed literature and they mainly focus on technical requirements and economic efficiency.

Specific constrains to the application of the AcoD regime are related to the pre-treatment step (preparation of the feeding material), to the reactor configuration (mixing, heat exchangers) and to the pollutant loading recycling determined by the reject water.

The literature contains a wide range of reports of SS and biowaste co-digestion with success often attributed to the consideration of several factors such as SS composition (primary, secondary or mixed), OLR, digester configuration, temperature range or mixing conditions.

Considering a WWTP size of 50,000 person-equivalents (p.e.), capital costs are estimated at about 200 € per p.e., while operating costs are around 40 € per p.e. per year (adapted from Pavan *et al.* 2007). Both capital and operating costs increase by 10% when considering the presence of an anaerobic digestion unit in the WWTP. On the other hand, when implementing co-digestion, operating costs can be accounted for circa 5 € per p.e. per year. These costs can be easily covered by the revenues from the tipping fees of organic waste and/or the revenues determined by tariffs for renewable energy production. As a global figure, the sum of these revenues can be valorised at 10–20 € per p.e. per year, thus covering part of the OPEX.

Righi *et al.* (2013) assessed the environmental profile of an experimental decentralized waste management system based on the anaerobic co-digestion of the organic fraction of municipal solid waste and dewatered sewage sludge in small plants (annual capacity: about 3000 t coupled at a 75 kW CHP unit) distributed locally. Two scenaria were analysed: with and without aerobic composting post-treatment. The scenaria were compared with the past and the current approaches followed in the case study area (NE Italy) and characterized by a strong centralization. Environmental performances of the different scenario were assessed using a Life Cycle Assessment methodology. The results indicated that the anaerobic co-digestion of dewatered sewage sludge and OFMSW in small plants, combined with composting post-treatment may offer an environmentally sustainable option of waste management in small communities. This was achieved by (1) a strong reduction in the distances and volumes transported by road, (2) low energy requirement for the process itself, (3) energy saving from the CHP unit and (4) energy/resources saving from the compost produced by the digested matter. Also social advantages may be gained, such as increased public acceptability of waste treatment facilities and increased awareness among citizens of waste management issues. It is worth to note that these advantages are closely linked to key assumptions such as the use of digested matter as fertilizer.

A particular application of the co-digestion approach is the two-phase AcoD of biowaste and sludge. In this approach, a two-phase AD process is applied. The first reactor, operating with high organic loading rate and short hydraulic retention time, will produce a biogas rich in H_2 while the liquid phase will be rich in volatile fatty acids (VFAs), while the second reactor will produce biogas as usual. Figure 15.3 shows a sketch of the two-phase process. The benefits of this approach are twofold: the production of a biogas rich in hydrogen and methane, called bio-hythane, and of a rich in carbon source (VFAs) liqidd stream,at the same time. The latter can be used in the wastewater treatment line to sustain the biological nutrients removal.

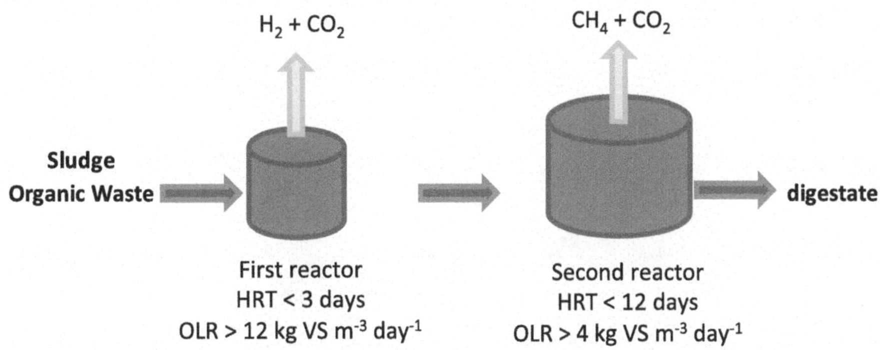

Figure 15.3 Scheme of the AD process for bio-hythane production.

Although the process could be easily implemented in existing WWTPs, at this stage there are no running plants that have adopted the two-phase AD process for the production of bio-hythane from sludge and organic waste. The actual technology readiness level (TRL), a method of estimating technology maturity ranging from 1 (basic principle observed) to 9 (actual system proven in operational environment) is therefore around 6.

An experimental test was carried out at pilot scale (Gottardo *et al.* 2015), using two stirred reactors both maintained at thermophilic temperatures and fed semi-continuously. No chemicals or recirculation were used to control the pH in the first phase. Sludge and OFMSW were fed to the AD system, each one contributing 50% of the fed VS. Sludge was the "buffer" agent and enabled a stable process operation. These experiments confirmed the possibility of obtaining a stable hydrogen production with a specific hydrogen production of 40 NL $kgVS^{-1}$ fed and a specific methane production in the second phase of 320 NL kgVS fed^{-1}. The biogas produced had a composition characterized by a stable presence of hydrogen over 5%.

The addition of small amounts (5–10%) of H_2 to a rich CH_4 biogas, although not changing significantly the energy content of the mixture, improves the quality of gas combustion while reducing NO_x, VOC and CO_2 emissions (Porpatham *et al.* 2007).

To optimize hydrogen production by fermentation, it is necessary to optimize the activity of the enzyme hydrogenase. Recent studies have shown that to maximize the yield of hydrogen production, the pH should be maintained in the range 5–6.5, with an optimum value at 5.5 (Valdez – Vazquez *et al.* 2009). The pH of the first reactor was always maintained in the optimal range for the hydrogenase enzymes operation with an average value of 5.1 ± 0.2, due to the buffer capacity of the sludge fed to the system.

In terms of bio-hythane produced, results reported in Figure 15.4 show that the percentage of methane, hydrogen and carbon dioxide were on the average 58.6%, 7.5% and 33.9%, respectively.

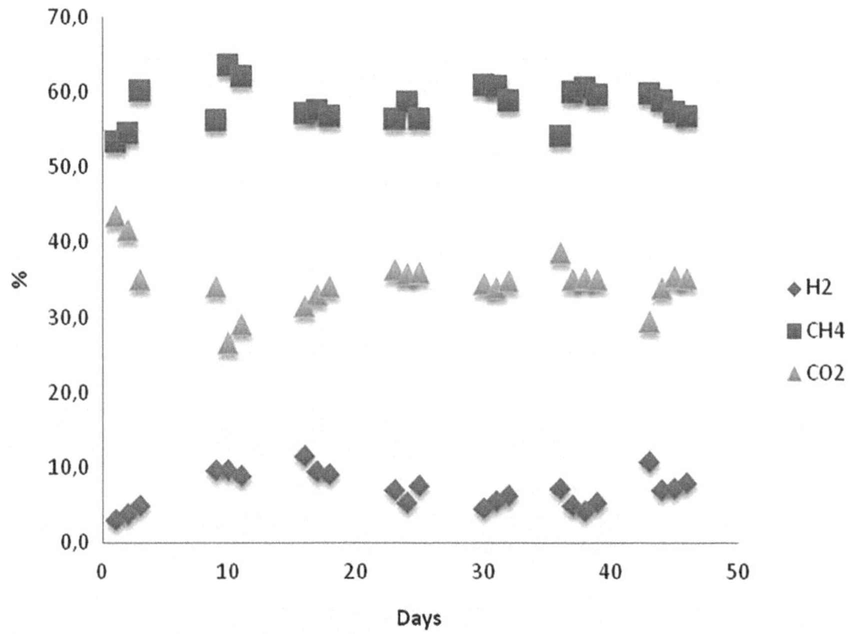

Figure 15.4 Gas composition in percentage by volume produced by this system.

The hydrogen content was never below 5% and this met the best characteristics for bio-hythane mixture. The total specific gas production obtained was 0.58 m^3 $kgVS^{-1}$ added, for a volumetric gas production rate of 1.65 m^3 per m^3 reactor per day.

15.2.3 Bio-drying of sewage sludge to produce biomass fuel

Bio-drying is a compost-like process. However, its final goal is different from the conventional composting process, as it does not aim towards a complete mineralization of the waste. Instead, the metabolic heat is used to remove water from the waste matrix at the lowest possible residence time and minimal carbon biodegradation, hence preserving most of the gross calorific value of the waste matrix (Tambone *et al.* 2011). Due to the distinct characteristics of mixed sludge such as its stickiness and relatively low heating values, the application of many conventional drying technologies can be technically and economically challenging. Bio-drying represents an economical and energy-saving emerging technology to reduce the sludge volume and to evaporate bound water by using biologically produced heat (Dufour, 2006). The bio-dried sludge has a high-energy content that can be used to produce steam and/or power if combusted, reducing fossil fuel requirements and contributing positively to the prevention of climate change. Therefore, a major benefit of bio-drying is the opportunity to incorporate the biogenic content of the input waste, a CO_2-neutral energy source, into a fuel product (Velis *et al.* 2009).

Compared with bio-drying of MSW, which is already widely applied in Europe with an overall treatment capacity of 2.000.000 tons year^{-1} (Herhof GmbH, 2008), there are some limitations associated to bio-drying of dewatered sludge. Mainly, dewatered sludge (i) contains fewer biodegradable organic substances, (ii) has higher moisture and (iii) has lower biomass matrix porosity, all of which are disadvantageous for effective aeration and energy recovery (Zhao *et al.* 2010). Consequently, there are a limited number of sludge bio-drying facilities in Europe, with one model facility present in Holland (Winkler *et al.* 2013). So far, most of the research conducted in dewatered sludge bio-drying has been developed at lab/pilot scale aiming to optimize the process and to increase the TRL from 4 to 8 in the forthcoming years.

In order to perform adequate bio-drying, several parameters must be taken into account. The initial moisture content (MC) is important because a too high initial MC limits oxygen transport and inhibits microbial activity, impeding bio-drying (Navaee-Ardeh *et al.* 2011). On the contrary, if the initial MC is too low, microbial activity will be slowed by the lack of moisture, resulting in reduced drying performance (Villegas & Huiliñir, 2013). Initial MC ranging from 50 to 70% (wet basis) has been reported as optimum MC for dewatered sludge bio-drying (Yang *et al.* 2014). To achieve the desired MC and to increase free air space and improve structural properties, dewatered sludge is usually mixed with bulking agents as well as with bio-dried sludge that serves as a well adapted inoculum for the aerobic decomposition of organic matter at thermophilic conditions (65–75°C) (Winkler *et al.* 2013). The airflow rate affects the biodrying efficiency, controlling the temperature of the process and improving the evaporation, making it the key factor influencing water loss during the process (Cai *et al.* 2013). A wide range of airflow rates, ranging from 0.5 to more than 5 L_{air} min^{-1} kg VS^{-1}, has been tested at lab-scale. As a general rule, at higher airflow rates there is increased moisture content reduction, lower temperature in the matrix and lower VS reduction. However, Huiliñir and Villegas (2014a) showed that at an airflow rate of 5.26 L_{air} min^{-1} kg VS^{-1}, there was no VS reduction, confirming that, under this condition, no bio-drying took place but only convective drying. Temperature is a crucial factor affecting both water evaporation and organic matter degradation. A temperature profile pattern similar to typical composting is expected in the bio-drying process, with four different stages: (I) warming stage, (II) thermophilic stage, (III) mesophilic stage and (IV) cooling stage. Stages II and III are the ones responsible for water evaporation.

According to Zhao *et al.* (2010), if the water content of sludge decreases from 80% to 40%, about 1,667 kJ energy is required for evaporating 0.667 kg of water from 1 kg initial sludge, as calculated from latent heat of water evaporation (2,500 kJ kg^{-1}). A fraction of this energy requirement is expected to be covered by bio-generated heat. Then, the lower heating value (LHV) of the dried sludge would increase from about 0.5 MJ kg^{-1} to 6 MJ kg^{-1} (Stasta *et al.* 2006). LHVs ranging from 7.7 to 14.6 MJ kg^{-1} biodried sludge (Winkler *et al.* 2013; Yang *et al.* 2014) have been reported, which means that if incinerated in a boiler with an efficiency of 90%, 6.9 to 13.1 MJ kg^{-1} of net energy could be recovered. The use of this CO_2-neutral energy source avoids the use of fossil fuels, minimizing the emission of CO_2 equivalent emissions. The avoided emissions will vary depending on the energy source substituted. For example if the avoided burdens for heat production are evaluated, considering a natural gas-fired condensing boiler, the avoided CO_2 equivalent emissions will range from 0.49 to 0.93 kg CO_{2eq} kg^{-1} bio-dried sludge. On the contrary, if coal is the substituted energy source for heat production, the savings increase to values ranging from 0.91 to 1.73 kg CO_{2eq} kg^{-1} bio-dried sludge. Table 15.7 summarises the main facts and results available in the literature,only considering the recent results obtained at pilot scale.

Although the energy recovered from bio-dried sludge positively contributes to climate change mitigation, there are unavoidable environmental and social concerns arising from bio-drying processes: gaseous emissions and compounds responsible for odour nuisance are the most common. Biofiltration and wet scrubbing technologies are proven technologies for the removal of ammonia and odorous compounds from composting processes (Colón *et al.* 2009) and can be easily adapted to bio-drying processes. In fact, Winkler *et al.* (2013) reported odour reductions of 80% by directing bio-drying process exhaust air to a biofilter. Moreover, nitrogen was recovered from exhaust air in a gas-scrubbing unit in the form of ammonium sulphate (48.5 kg $(NH_4)_2SO_4$ t^{-1} dewatered sludge, 40% v/v), which can be used as fertilizer.

Although bio-dried sludge is mainly intended for energy generation, it can also be used as soil amendment. Therefore, microbial and chemical quality should be taken into account. Pathogens are mostly mesophilic bacteria, which will be likely killed during the thermophilic stage of the bio-drying procedure; hence, bio-dried sludge can safely be applied as soil amendment (Navaee-Ardeh *et al.* 2010). Regarding metal content, a wide range of concentrations can be found, depending on the origin of the input wastewater (municipal, industrial, etc.). In any case, threshold concentrations of the European Union (EU 86/278) for application as fertilizer on agricultural land must be respected. It must be mentioned that every country in the EU has its own threshold values and some are more stringent than others.

From an economical point of view, Frei *et al.* (2006) performed a techno-economic study for a bio-drying process treating 100 t d^{-1} of pulp and paper wastewater sludge (26% TS) and compared the results with a traditional landfilling (16%) and incineration (84%) sludge management strategy. The results showed an estimated payback period of 2.5 years, taking into consideration capital costs of C $ 5.14M and total annual savings of C $ 2.1M. Most savings came from the reduction in fossil fuel consumption (90% estimated reduction in fossil fuel cost). However, it should be highlighted that a sensitivity analysis, changing important parameters such as the retention time, the final TS content, the sluge:bulking agent ratio, etc. showed payback times ranging from 0.6 to 16 years depending on the conditions.

Another technology for minimizing costs of the thermal drying is solar drying, in which free solar energy is used to dry the SS. In solar drying the SS is spread in thick layers, usually in the floor of a greenhouse chamber. Fans can be added in order to have homogeneous distribution of the inlet air at the surface layers of the product. Also, ventilation is used to evacuate the saturated air and replace it by fresh air. Generally, mixing the product once or several times in a day permits harmonized distribution of the dried product. Bennamoun (2012) published a review of recent SS solar drying lab- and pilot-scale experiments, showing moisture removals ranging from 20 to 55% (initial MC of ≈80%). Regarding pathogens removal, 1 log reduction of coliforms was reported in 45 days solar drying experiments. Class A pathogen requirement (<1000 CFU g^{-1}) was obtained by adding limited amounts of lime. Solar drying could be used in combination with biodrying to minimize the overall cost of this energy-consuming technology.

Table 15.7 Overview of studies reporting dewatered sludge bio-drying processes.

Substrate	Co-substrate/ Bulking Agent	Substrate Initial Dry Matter (%)	Mixture Initial Dry Matter (%)	Product Final Dry Matter (%)	VSS Reduction (%)	Residence Time (Days)	Input air Requirement	LHV (KJ $kg^{-1}_{bio\text{-}dried\ sludge}$	Scale	References
WWTP dewatered sludge	Biodried sludge + wood chips (5.6%)	25	45	67	26*	24	15613 $m^3_{air}/t_{input\ sludge}$	7700	Full-scale	Winkler *et al.* 2013
Pulp and paper mixed dewatered sludge	Biodried sludge (30% w_{db}:$w_{db;}$ bs:ss)	37	40.3	51	–	6	165873 $m^3_{air}/t_{input\ sludge}$	–	Pilot-scale	Navaee-Ardeh *et al.* 2010
Pulp and paper dewatered sludge	Rice husk (2.5:1 w:w; ss:rh)	11.75	35	35–55	0–17	10	0.51 to 5.26 L_{air} min^{-1} kg VS^{-1}	–	Pilot-scale	Huiliñir & Villegas, 2014a
WWTP dewatered sludge	Wood shavings (2:1 to 10:1 w:w; ss:ws)	11.75	59–78	52–77	2–8	8	1 to 3.5 L_{air} min^{-1} kg VS^{-1}	–	Pilot-scale	Huiliñir & Villegas, 2014b
Slaughterhouse dewatered sludge	Air-dried & biodried sludge	–	50–70	–	14–23.8	10	–	14644	Pilot-scale	Yang *et al.* 2014
WWTP dewatered sludge	Biodried sludge & saw dust (3:2:1 w:w:w; ss:bs:sd)	–	66	48.7	–	20	1265 $m^3_{air}/t_{input\ sludge}$	–	Pilot-scale	Cai *et al.* 2013
WWTP dewatered sludge	Straw and saw dust (15:1:2 w:w; ss:st:sd)	20.1	69.7	<50	31–37	20	0.75 to 1.52 L_{air} min^{-1} kg VS^{-1}	–	Pilot-scale	Zhao *et al.* 2010

*TSS reduction (%); ss: sludge, bs: biodried sludge, rh: rice husk, ws: wood shavings, ds:saw dust, st:straw.

15.3 THERMAL BASED TECHNOLOGIES

15.3.1 Gasification

Gasification is a thermo-chemical process, in which any carbon-containing material is converted into a combustible gas in the presence of a reactive atmosphere (air, steam or CO_2) and at a temperature that usually ranges between 800 and 900°C. The gasification product gas is mainly composed of a mixture of CO, CO_2, H_2, CH_4 and other light hydrocarbons, steam and N_2 if air is used as the gasification medium. Their relative proportions depend on the composition of the raw material and the gasification operating conditions. After upgrading, the producer gas can be used in a more versatile way than the original biomass, including power generation in gas engines or turbines or as a chemical feedstock to produce chemicals such as methanol and Fischer-Tropsch liquids.

Gasification is a suitable technology for sewage sludge management because it reduces waste volume, removes toxic organic compounds and fixes heavy metals in the resultant solid. The first published studies related to the gasification of sewage sludge date back to mid-1990s (Bacaicoa *et al.* 1995). Since then, further studies performed at laboratory facilities (Aznar *et al.* 2008; Gil-Lalaguna *et al.* 2014a) and even at demonstration scale (Judex *et al.* 2012; Midilli *et al.* 2011) have shown the feasibility of obtaining a fuel gas from such waste. Most of these studies focus on the gasification of anaerobically digested and thermally dried sewage sludge under an air atmosphere, using between 20 and 40% of the oxygen required for total combustion. Typical gas yields of 1.5–2.5 Nm^3 per kg of dry sewage sludge have been reported under these conditions, depending on the physicochemical properties of the feedstock as well as on the type of gasifier (Aznar *et al.* 2008; Midilli *et al.* 2001). This specific gas production is slightly lower than typical values for lignocellulosic biomass gasification, 2–3 $Nm^3 \cdot kg^{-1}$ (Knoef, 2005), because of the higher content of ash in the sewage sludge, which usually ranges between 20 and 40 wt.%. The heating value of the product gas obtained under these conditions is about 4 $MJ \cdot Nm^{-3}$, so this gas could be suitable for boiler, engine and turbine operation, but not for pipeline transportation due to its low energy density.

As thermal drying of sewage sludge involves a large consumption of energy, raising the cost of sludge disposal, researchers have also investigated the possibility of producing hydrogen from wet sewage sludge. After stabilization and mechanical dewatering of sewage sludge by filter pressing or centrifugation, the moisture content of sewage sludge can still exceed 65–70 wt.%. Hence, this natural moisture of the sludge is converted into steam at the first gasification stage, which gives rise to the partial gasification of the sludge and the reforming reactions of the organic vapors, thus contributing to the production of a H_2-rich fuel gas, with a concentration of H_2 that can exceed 50 vol.% (Domínguez *et al.* 2006). However, because of the endothermic nature of the steam-reforming reactions, an energy supply is required for the process. The simultaneous addition of air to the gasifying medium can provide the energy demand for steam gasification through the partial combustion of the solid feedstock. With the joint use of steam and enriched-air as gasification medium, the calorific value of the producer gas can be increased up to 6 MJ Nm^{-3}, also obtaining a dry gas production of around 1.1–1.3 Nm^3 kg^{-1} dry sewage sludge and a H_2/CO molar ratio in the exit gas as high as 2.5–3 (Gil-Lalaguna *et al.* 2014a). In this case, the gasification of sewage sludge with a moisture content lower than 20% could virtually be an auto-thermal process by feeding at least 23% of the oxygen required for the total combustion of the sludge (Gil-Lalaguna *et al.* 2014b). The gas product could be partially combusted, producing the heat needed for reducing the water content in the dewatered sewage sludge to the required 20 wt.%. For this purpose, about 55% of the gasification gas would be theoretically required for the drying stage, so the remaining gas could be destined to other applications. Hence, from an energy standpoint, this two-stage process involving partial thermal drying and gasification of sewage sludge appears as a self-sufficient treatment (Figure 15.5), thus reducing the high disposal cost of thermally dried sewage sludge.

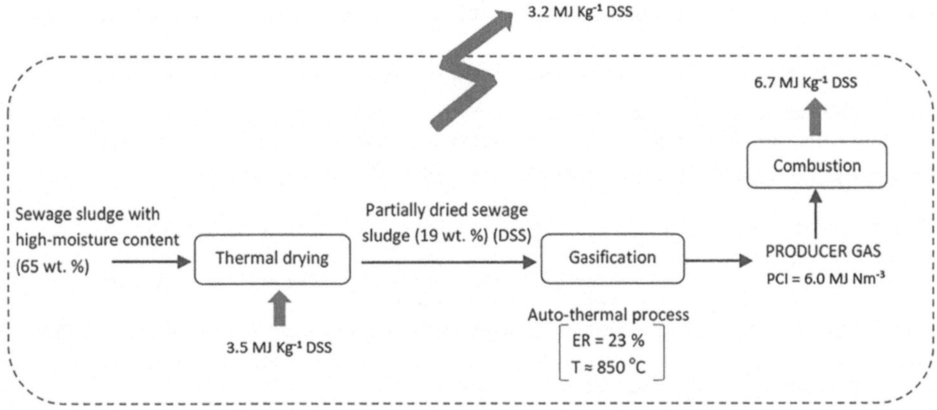

Figure 15.5 Energy balance for thermal drying and gasification of sewage sludge.

In view of the gas end-use, and as occurring in any gasification process, the cleaning stage of the gas derived from sewage sludge gasification is the bottleneck for some gas applications. One of the major issues is to deal with the tar formed during the process, which is a complex mixture of condensable organic compounds including single- to multiple-ring aromatic compounds. The presence of tar in the gasification gas leads to operational problems associated with condensation, formation of aerosols and polymerization, such as the blocking of downstream pipelines and fouling of engines and turbines. Aromatic compounds such as benzene, toluene, xylene, phenol, indene, naphthalene, anthracene, pyrene, pyridine or benzonitrile have been identified in the sewage sludge-derived tar (Adegoroye *et al.* 2004; Aznar *et al.* 2008; Gil-Lalaguna *et al.* 2014a). Furthermore, quite different tar formation rates from 5 g tar to more than 80 g tar per kg dry sewage sludge have been reported during sewage sludge gasification depending on the gasifier type and the gasification medium (Aznar *et al.* 2008; De Andrés *et al.* 2011; Gil-Lalaguna *et al.* 2014a; Midilli *et al.* 2001). The allowable tar limit in the gas is highly dependent on the gas end-use but, for example, gas engines and turbines can technically allow tar levels only as high as 50–100 mg Nm^{-3} or even lower (Anis & Zainal, 2011). As the concentration of tar in the producer gas from sewage sludge gasification is much higher than this limit, usually about 5–20 g Nm^{-3}, tar cracking during sewage sludge gasification using cheap catalysts such as alumina, dolomite and olivine has been studied, obtaining significant reductions in the concentration of tar (De Andrés *et al.* 2011).

In addition to tar formation, the particular composition of sewage sludge leads to the formation of other impurities during gasification, such as NH_3, HCl or H_2S, the latter being one the most abundant with a concentration in the gas up to 4,500 ppm (Gil-Lalaguna *et al.* 2014a). The presence of H_2S in the gasification gas entails both environmental and operational problems, related to SO_2 emissions during the combustion of a H_2S-containing gas and to corrosion in pipes, engines or turbines, respectively. Several technologies have been studied for H_2S removal, including the use of the sewage sludge ash itself as chemical adsorbent, because of its content of metal oxides, such as CaO and Fe_2O_3 being potentially reactive with H_2S (Gil-Lalaguna *et al.* 2015).

Besides the technical and environmental issues, the economic implications determine the rate of development and implementation of emerging technologies such as gasification. For example, the specific capital costs of two of the main opportunities for power generation, integrated gasification combined cycle or gasifier + engine, are around 3000 € kWe^{-1} for medium-scale plants of about 15–20 MWe of net capacity (Knoef, 2005). Regarding the operational cost, this clearly depends on the feed pre-treatment and gas clean-up requirements, but generation costs from biomass gasification plants are expected to be in the range 100–130 € MWh^{-1} (IEA report, 2007).

In summary, sewage sludge can be considered a suitable renewable energy source for gasifiers in order to produce thermal energy, while disposing of this waste. However, lab-scale facilities and pilot plants of sewage sludge gasification have shown some challenges related to the removal of organic and inorganic impurities in the gas and to the improvement of gasification efficiency.

15.3.2 Pyrolysis

Pyrolysis is a thermochemical reaction carried out in an inert atmosphere, usually N_2, and at a temperature around 500°C. Three products (char, liquid and gas) are obtained in the process. Many research works are aimed to obtain the liquid product, because it has promising physicochemical properties to be used as fuel or as source of valuable chemical products.

Pyrolysis of sewage sludge for liquid production has been studied by several authors (Fonts *et al.* 2012). The yield to sewage sludge pyrolysis liquid and its physicochemical properties depends on the treatments applied to the sewage sludge and on the operating conditions. Anaerobically digested and thermally dried sewage sludges are the ones that have been studied by a greater number of researchers. Typical yields to the pyrolysis products, obtained at 550°C, are: 51.0% to char, 33.6% to liquid and 10.4% to gas, though considerable differences in the yields and physicochemical properties of the liquid fuels have been observed. The gas product usually does not have a high enough heating value to be burned using the existing technologies. This low heating value is due to the high proportion of the fluidization agent, N_2, in the gas product. However, novel investigations in biomass pyrolysis using the own tail pyrolysis gas as fluidization agent have achieved to successfully burn the gaseous product obtained (Mullen *et al.* 2013a). If sewage sludge pyrolysis gases did not contain nitrogen, the combustion of the pyrolysis non-condensable gases would release 2.46 MJ per dry kg of sewage sludge (Ábrego *et al.* 2013). The char usually has a higher heating value (HHV) around 5.2 MJ kg^{-1} (Gil-Lalaguna *et al.* 2014), so it could be used as fuel to carry out the pyrolysis process, since it is an endothermic reaction. The enthalpy required to carry out the reaction is around 0.15 MJ kg^{-1} of dry sewage sludge, taking into account that the pyrolysis products leave the control volume at 550°C (Gil-Lalaguna *et al.* 2014). Apart from this, the main energetic challenge would be the thermal drying of the sewage sludge, since it should be taken into account that the energy required to dry the sewage sludge to a final moisture content of 5% (by weight) is about 4 to 8 MJ kg^{-1} of dry sewage sludge, depending on the initial moisture content (Gil-Lalaguna *et al.* 2014).

Sewage sludge pyrolysis liquids are heterogeneous and separate into different phases. If the pyrolysis is carried out without the use of a catalyst, three phases usually appear in the liquid: a light organic phase (LOP), a heavy organic phase (HOP) and an aqueous phase (Azuara *et al.* 2015). However, depending on the conditions, the catalytic pyrolysis of sewage sludge over γ-alumina may lead to the production of a liquid with an aqueous phase and only one organic phase (OP) (Azuara *et al.* 2015). The yield and physicochemical properties of the organic phases produced from catalytic and non-catalytic pyrolysis are shown in Table 15.8.

The presence of oxygen and nitrogen causes low storage stability of the liquid and reduces its higher heating value. Moreover, the presence of nitrogen-containing compounds would provoke the emission of atmospheric pollutants if the liquid were burned. The oxygen content of the organic phase is significantly reduced by the catalytic treatment of the vapors over γ-alumina (Azuara *et al.* 2015), because in this process, the fatty acids contained in the raw material are transformed into hydrocarbons, many of them being α-olefins. However, the liquid still has a high content of nitrogen-containing compounds, such as, aliphatic and aromatic nitriles, amides and N-heterocyclic aromatic compounds, making it difficult to use as a fuel. The reduction of the nitrogen content of pyrolysis oils by means of hydrodenitrogenation has been studied recently (Izhar *et al.* 2012; Mullen *et al.* 2013b). In most of the operating conditions studied in these two works, the reduction of the nitrogen content achieved was lower than 30% (Izhar *et al.* 2012; Mullen *et al.* 2013b). Only when part of the bio-oil was solubilized with 80% xylene was complete elimination of the nitrogen achieved (Izhar *et al.* 2012). On the other hand, the hydrogen requirement for the removal of the nitrogen from N-heterocyclic aromatic compounds at least doubles the amount needed for the reduction of the other heteroatoms present in the pyrolysis oil (Talmadge *et al.* 2014).

Table 15.8 Physicochemical properties of sewage sludge pyrolysis organic phases from non-catalytic (LOP and HOP phases) and catalytic (OP phase) pyrolysis over γ-alumina.

	LOP	HOP	OP
Yield (% over sewage sludge fed)	2	10	11
Water Content (% by mass)	0	7	2
HHV (MJ/kg)	42.1	31.2	41.2
C (% by mass)	85.9	70.5	79.5
H (% by mass)	11.8	8.8	9.8
N (% by mass)	1.8	10.0	8.4
S (% by mass)	0.2	1.0	0.8
O (% by mass)	0.3	9.7	1.5

Source: Azuara et al. (2015).

Apart from the cited chemical compounds, the qualitative studies of the composition of the sewage sludge pyrolysis oil have revealed the presence of many compounds that could be considered as added-value chemicals (Fonts *et al.* 2012). Very few works have studied quantitatively the possible production of these added–value chemicals by sewage sludge pyrolysis (Sanchez *et al.* 2009; Fonts *et al.* 2017). The most interesting compounds that could be obtained from the sewage sludge pyrolysis liquids would be α-olefins, n-paraffins, long aliphatic and aromatic nitriles, methyl-styrene, benzocyclobutene and octocrylene (Fonts *et al.* 2017). The production of added-value compounds from sewage sludge pyrolysis could be an option to valorize this waste. One of the ways forward to develop this technology would be to optimize the operating conditions that would allow to obtain the highest yields of the added-value chemicals identified, paying special attention to the nitrogen-containing compounds. The production of nitrogen-containing compounds with industrial applications is an opportunity for sewage sludge and other proteinaceous biomasses, since these compounds cannot be obtained from lignocellulosic biomasses. However, if the nitrogen-containing compounds produced have no industrial relevance, their treatment by hydrodenitrogenation is extremely costly and their combustion with the rest of the non-value-added compounds of the liquid would be problematic due to the resulting high NOx emissions. Other compounds with many industrial applications recovered from the sewage sludge pyrolysis oils are the α-olefins. Thus, it would be most relevant to develop pre-treatments to the sewage sludge and catalytic treatment of the pyrolysis vapors that would allow to maximize their yield. Lastly, other critical developments are the separation processes that are required for downstream processing of liquids. These separation processes need to overcome the technical and economic challenges posed by the large number of compounds present in the sewage sludge pyrolysis liquids.

The energy balance of the process is a key factor for the production of an oil fuel or for the production of added-value chemical compounds. Table 15.9 summarizes the energy balance of the whole process, describing separately the energy required or given off in each one of the stages, with or without the energy recovered by the combustion of the organic phases of the liquid product.

Table 15.9 Energy flows in MJ kg^{-1} (required > 0; given off < 0) in each stage of the sewage sludge pyrolysis process including drying of the sewage sludge.

Stage	Energy Requirements	
Sewage sludge thermal drying until 6.5% of moisture (MJ kg^{-1} of dry sewage sludge)	4[a] (65% initial moisture content)	8[a] (77% initial moisture content)
Sewage sludge pyrolysis (MJ kg^{-1} of dry sewage sludge)	0.15[a]	
Given off energy		
Gas LHV (MJ kg^{-1} of dry sewage sludge)	−2.46[b]	
Char Higher Heating Value (related to SS) (MJ kg^{-1} of dry sewage sludge)	−2.7[a]	
Energy balance of the process (MJ · kg^{-1} of dry sewage sludge)	**−1.01**	**2.99**
Liquid Higher Heating Value (MJ kg^{-1} of dry sewage sludge)	−4.0 (calculated from data of Table 15.1)	
Energy balance of the process taking into account liquid combustion (MJ kg^{-1} of dry sewage sludge)	**−5.01**	**−1.01**

[a]Data from reference Gil-Lalaguna *et al.* 2014.
[b]Data from reference Ábrego *et al.* 2013.

As can be observed in Table 15.9, if the wet sewage sludge contains an initial moisture content of 65% or lower, the whole process could become exothermic, if it is possible to recover all the energy given off in the combustion of the gas and the char produced in the pyrolysis reaction. If the initial moisture content of the wet sewage sludge was 77%, the whole process would be endothermic, unless the liquid product was combusted and all the heat released in its combustion was recovered. In view of these results, the production of added-value chemicals with an exothermic energy balance would be possible, depending on the initial moisture content of the sewage sludge and on the percentage of sensible heat recovered from the combustion of the gas and the char products.

Todate, only one industrial-scale sewage sludge pyrolysis plant is operated as an alternative to incineration or anaerobic digestion. This plant was developed by Professor Bayer in Germany in the mid-1980s and the technology was named Enersludge™ (Bridle & Pritchard, 2004). The construction of the plant began in 1997 in Perth (Australia); its cost was around 11 million of US $ and it was designed to treat 75,000 m^3 of wastewater per day. The aim of this process was to reduce the costs of sewage sludge management by generating power from the combustion of the pyrolysis oil and use it locally at the plant or injection to the grid. The combustion of the organic phase of the pyrolysis liquid gave off 7.7 MJ per kg of dry sewage sludge.

15.3.3 Supercritical water processing

As already eluded to, thermochemical processing of sludge has many potential advantages over biological processing. These include several orders of magnitude faster reactions, and generally complete elimination or inactivation of all pathogens, since processing temperatures exceed 100–150°C. Thermochemical processes are also rapid to start (no need to acclimate or buildup biomass) and are generally better coping with fluctuations than biological processes. On the other hand, thermochemical processes almost always require more complex equipment, they can be energy intensive, or processing may produce undesired byproducts such as NOx, SOx or other residues that can pose challenges for disposal. One promising thermochemical process for sludge treatment and energy recovery is supercritical water processing, in particular supercritical water oxidation (SCWO) and supercritical water gasification (SCWG). SCWO refers to processing in the presence of an oxidant, generally oxygen, provided to the stream undergoing treatment, either via pure oxygen or air, while SCWG is conducted in the absence of oxidant (Bermejo & Cocero, 2006). The main differences between the two approaches are highlighted in Table 15.10.

Table 15.10 Main differences between supercritical water oxidation (SCWO) and gasification (SCWG) for sludge treatment.

	SCWG	SCWO
Heat of reaction	Endothermic	Exothermic
Typical heat of reaction[a]	+8 to +10 MJ $kg^{-1}{}_{dry}$	–10 to –20 MJ $kg^{-1}{}_{dry}$
Oxidant	None	Required[b]
Reaction rate	Slower (>1 minute)	Faster (<10 sec.)
Products of reaction	CH_4, CO_2 + CO + H_2	CO_2 + heat
Fate of N	NH_4^+ in liquid effluent	N_2 gas
Corrosion potential	Moderate	Moderate-High
Risk of reactor plugging	Higher	Lower

[a]Note the difference in the sign between gasification which is endothermic and oxidation which is exothermic.
[b]Generally O_2 from air or pure oxygen is used, though H_2O_2 and $KMnO_4$ has been used in laboratory studies.

Supercritical processing relies on the unique reactivity and transport properties when slurries are brought above the critical point of water (374°C and 218 atm, see Figure 15.6). Supercritical water is a dense single phase (100–300 kg m^{-3} depending on pressure and temperature) with transport properties similar to those of a gas and solvent properties comparable to those of a non-polar solvent (Tassaing *et al.* 2002). When oxygen is present, it is fully soluble in supercritical water, resulting in extremely rapid and complete oxidation of the organics to carbon dioxide, clean water (that can be recovered), some non-leachable inorganic salts, and if engineered properly, a surplus of energy that can be converted to heat and/or electricity. Typical reaction times are in the order of seconds, resulting in SCWO systems that are quite compact compared to other technologies. The process does not generate harmful by-products such as NOx or SOx, CO, or odors (Bermejo & Cocero, 2006). As mentioned earlier, because of the stringent reaction conditions, complete destruction of all pathogens is achieved.

The many appealing benefits of supercritical water processing have stimulated engineers and entrepreneurs to invest significant efforts and resources in the development of the technology. However, even after 25 or 30 years of development, commercial success remain elusive (Marrone, 2013). Challenges that have slowed down commercial deployment of SCWO are linked to the complex nature of a high-pressure, high-temperature process. Critical issues include materials selection, reactor design and construction to withstand the corrosive nature of the reactive mass, dealing with highly exothermic reactions at high pressure and high temperature, plugging of the reactor by mineral deposits and addressing these issues within tight cost constraints.

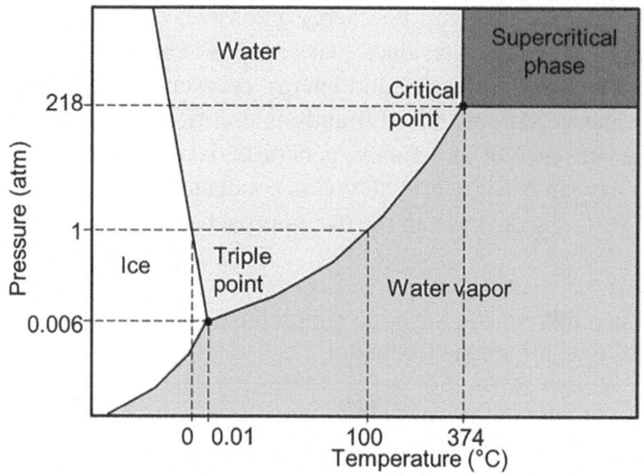

Figure 15.6 Schematic of the phase diagram of water (not to scale).

Much research has been conducted in the two most critical areas: corrosion and plugging. A complete review of this research is outside the scope of this chapter and the reader is advised to consult specific journal articles and reviews (e.g., Bermejo & Cocero, 2006; Marrone & Hong, 2009). Corrosion issues are quite complex because of the different possible corrosion mechanisms, combined with the multiple materials, feedstocks and process conditions that have been used. However, it is generally accepted that corrosion is most severe in the subcritical to critical transition zones (310–370°C). The heat up-region tends to be more susceptible to corrosion than the cool-down region because of the increased concentration of oxidant, although on the "reacted" side, concerns exist when acidic reaction products are produced. Also, corrosion is increased by high chloride concentrations and low pH. Corrosion control methods generally fall in 3 categories: (1) Preventing corrosive species from reaching reactor surfaces, (2) Utilizing a corrosion resistant material (as liner or material of construction), (3) Adjusting reactor design and/or process conditions to minimize corrosion.

Less has been published on the mitigation of plugging by solids that deposit during SCWO or SCWG. These solids can originate from the oxidation or gasification reactions (i.e. ashes) or from dissolved salts precipitating at supercritical conditions as a result of the low solubility of these salts in the non-polar supercritical water. Regarding the latter, the extent of salt precipitation will depend on the concentration of dissolved salts in the feedstock, the nature of the salts and the amount that will dissolve after cooling to subcritical conditions. Solids formed during reaction are mostly ashes. Typically sludge has an ash content of about 20–30% of its dry solids. Since it is generally admitted that dry solids concentration in the sludge feedstock must be at least 10% for economical SCWO/G of sludge, roughly 2–3% of the total throughput will be solids to be managed. Methods to manage solids have not been widely published. Even so, they fall in two categories, (a) periodic cleaning and removal of solids and (b) designs that minimize deposits. Cleaning can be accomplished during operation (which can be challenging given the process conditions) or after system shut-down, whereas designs to minimize deposits can include large diameter vertical reactors facilitating solids removal, or narrow tubular reactors with high Reynolds numbers to promote entrainment of the solids. Many solids mitigation methods are either patented or trade secrets.

In the past five years, there has been increased interest in using SCWO or SCWG for high-strength waste (including sludge) treatment. As a result, several technical-scale or full-scale systems have been built (see e.g., Marrone, 2013; SCFI-Aquacritox, 2015; Innoveox, 2015; SuperWater Solutions, 2015). These include a $600,000 pilot constructed in China, which in early trials was reported (Xu *et al.* 2012) to allow effective treatment of 3 tons/d of wet sewage sludge (or 240 kg/d dry sludge), with operating costs of \$76 per ton of dry solids, or \$950 per wet ton. The study also reports that the system was self-sufficient with respect to heat (a critical aspect for financial viability) and was producing excess heat that allowed to generate 30 tons/d of hot water (at 80°C). Unfortunately, the operation of this system has been fairly limited. Detailed data for the operation of other systems have been difficult to obtain.

At Duke University, a team of engineers has designed and built a prototype SCWO system housed in a standard 20 feet (~7 m) shipping container (Deshusses & Jacoby, 2015). This is the work of a collaborative project with engineers of the University of Missouri, funded by the Reinvent the Toilet program of the Bill & Melinda Gates Foundation. The system is designed to treat the fecal waste produced by roughly 1,000–1,500 persons daily, or about 1 m^3 of sludge at 10–20% solids (Figure 15.7). The unit has been undergoing testing at Duke since early 2015. The design includes moderate preheating of the waste slurry, followed by mixing with supercritical water (~600°C) and air (which serves as oxidant). This rapidly brings the waste undergoing treatment to supercritical conditions, thereby minimizing corrosion and risks of waste charring and plugging. Sludge is rapidly oxidized to CO_2, with the corresponding heat of combustion released in the reaction medium. The reactor has a 19 mm internal diameter and is 4 m long. After the reaction, heat recovery follows in a 39 m long heat exchanger. The system is well instrumented and its operation is controlled using a programmable logic controller. Experiments were first conducted with isopropanol, prior to treating secondary sludge. Currently, experiments are conducted with secondary sludge slurries (Figure 15.7) with a low solids content and isopropyl alcohol (IPA) as a co-fuel to minimize the risk of pumping issues that could arise with high solids slurries. The results so far show very rapid treatment of the sludge-IPA mixture with near complete removal (99.97%) of the chemical oxygen demand (COD). Interestingly, total nitrogen and phosphorus removal were both over 98%.

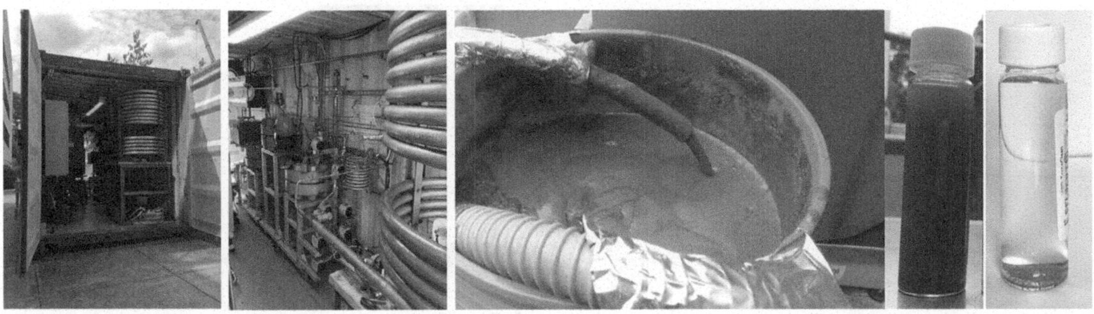

Figure 15.7 Pictures of the SCWO container unit at Duke University prior to installing thermal insulation. The system is designed to handle the fecal waste of about 1000–1500 people. On the right, pictures of a typical slurry experiment with the slurry feed (4.3% biosolids, 9% IPA (center)). Treated effluent (right), and after gravity settling (far right). The final effluent is odourless.

The development of supercritical water processing, using either oxidation or gasification, has seen unprecedented activity in the past five years, fueled by various research efforts, new corrosion resistant materials and a better understanding of the process, and new energy subsidies or concerns about sludge and hazardous waste disposal. Economically, producing energy from sludge will probably remain a challenge for some time, but the perspectives for clean, efficient and sustainable treatment of sludge using either SCWO or SCWG look excellent.

15.4 CONCLUSIONS

The generation of municipal sewage sludge has been increasing worldwide during the last decades and a production close to 13 million tons of DS of sludge is expected in Europe in 2020. Currently, the most common treatment technologies for SS in Europe are anaerobic and aerobic stabilization, as well as incineration and direct land application. On the contrary, due to stringent limits on the disposal of biodegradable wastes to landfills, direct landfilling has been strongly reduced in the recent years. According to the existing European waste management guidelines, emphasis should be placed on recovery of nutrients, organic matter and green energy. Moreover, public acceptance as well as economics and economies of scale are key factors, when deciding which SS treatment technology should be chosen.

This chapter has described some of the most relevant emerging and novel technologies for treating SS that are currently being developed worldwide. These technologies can be divided into two main groups: (i) biologically-based technologies (advanced anaerobic digestion strategies and bio-drying) and (ii) thermally- based technologies (gasification, pyrolysis and supercritical water processing). Figure 15.8 summarizes the different treatment opportunities and Table 15.11 summarizes the main advantages and limitations/bottlenecks of each described technology as well as its stage of development.

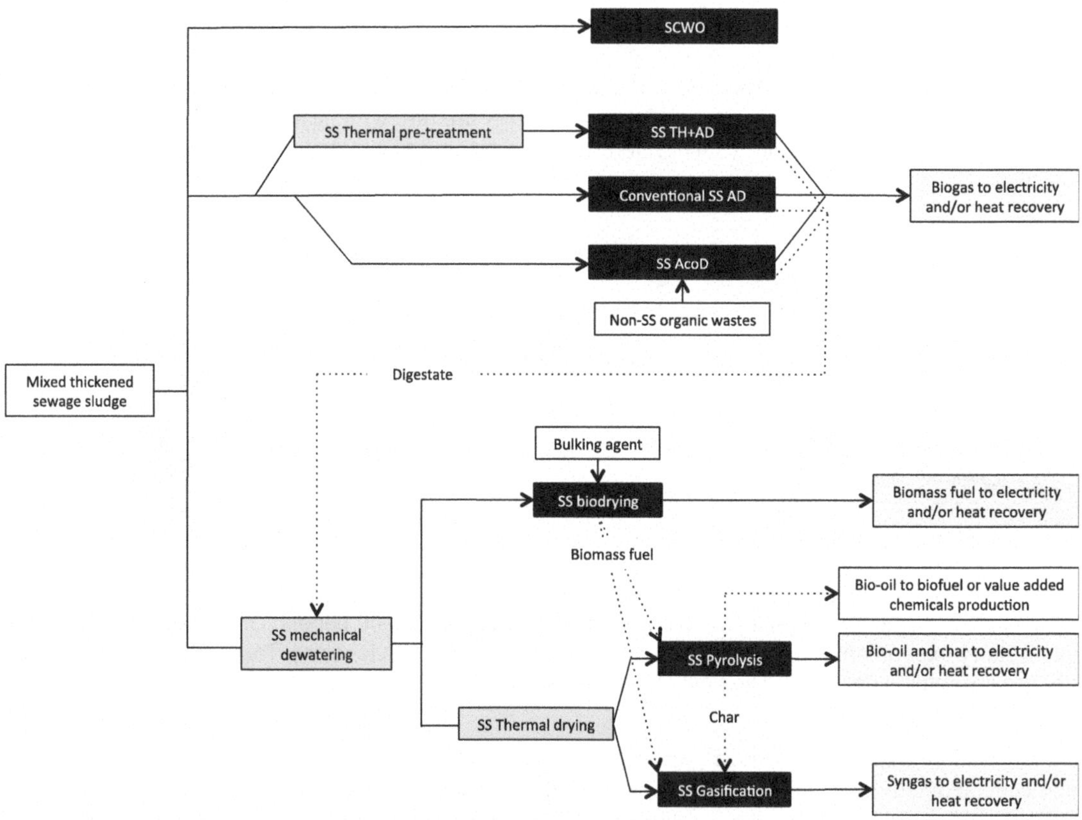

Figure 15.8 Block diagram showing the different thermal treatment options for sludge management. Dashed lines show alternative pathways for energy recovery of sludge products and by-products. Other treatment options such as composting, liming or direct land application are not included in this block diagram.

Table 15.11 Summary of the main advantages and limitations/bottlenecks of each described technology as well as stage of development.

	Parameters	Advanced Thermal and High Pressure Pre-treatments to Enhance Energy Recovery in AD Processes	Co-digestion of Sewage Sludge with non-sludge Organic Wastes	Bio-drying	Gasification	Pyrolysis	Supercritical Water Processing
1	*Energy product obtained*	Biogas	Biogas	Biomass fuel	syngas	Bio-oil, bio-Char and syngas	Heat or steam
2	*Amount produced*	400 to 450 $m^3_{biogas}\ t^{-1}$ DS (compared to ≈300 $m^3_{biogas}\ t^{-1}$ DS for conventional AD)	≈ 400 $m^3_{biogas}\ t^{-1}$ DS (mixutre of SS:OFMSW 9:1 w:w) (compared to ≈ 300 $m^3_{biogas}\ t^{-1}$ DS for conventional AD)	1.1 to 1.6 $t_{biodried\ sludge}\ t^{-1}$ DS	1.1 to 2.5 $Nm^3\ kg^{-1}\ DS_{biosolids}$	Typical yields: 51% char, 34% oil and 10% gas	1000 to 3000 kwh t^{-1} DS (as heat)
3	*LHV of energy products*	22 to 24 MJ m^{-3}_{biogas}	22 to 24 MJ m^{-3}_{biogas}	6 to 14 MJ $kg^{-1}_{biodried\ sludge}$ (depending on the final moisture content)	4 to 6 MJ Nm^{-3}_{gas}	Bio-oil: 30–45 MJ $kg^{-1}_{bio\text{-}oil}$ Bio-char: 5 MJ $kg^{-1}_{biochar}$	–
4	*Energy consumption (Electricity and/or fuels)*	≈180 kwh t^{-1} DS (50 kwh t^{-1} DS for pre-treatment and 85–135 kwh t^{-1} from AD process)	85 to 135 kwh t^{-1} DS	200 to 400 kwh t^{-1} DS (reactor aeration, other energy consumption not included)	250 to 970 kwh t^{-1} DS (mainly due to thermal drying)	1150 to 2250 kwh t^{-1} DS (mainly due to thermal drying pre-treatment)	–
5	*Electricity generation**	≈750 kwh t^{-1} DS (compared to ≈525 kwh t^{-1} DS in conventional AD)	≈ 700 kwh t^{-1} DS (compared to ≈525 kwh t^{-1} DS in conventional AD)	650 to 1250 kwh t^{-1} DS	400 to 1400 kwh t^{-1} DS (syngas used in a power generator)	925 to 1400 kwh t^{-1} DS (bio-oil for electricity production)	300 to 1000 kwh t^{-1} DS

(Continued)

Table 15.11 Summary of the main advantages and limitations/bottlenecks of each described technology as well as stage of development (*Continued*).

	Parameters	Advanced Thermal and High Pressure Pre-treatments to enhance Energy Recovery in AD Processes	Co-digestion of Sewage Sludge with Non-sludge Organic Wastes	Bio-drying	Gasification	Pyrolysis	Supercritical Water Processing
6	*Heat generation***	Heat self consumed in AD process (digestor heating)	Heat self consumed in AD process (digestor heating)	Heat self consumed in the process (sludge biodrying)	Heat self consumed in the process (sludge drying)	Heat self consumed in the process (sludge drying)	Heat self consumed in the process
7	*Sludge retention time*	21–30 days	21–30 days	7–21 days			<1 minute
8	*Reactants/ material consumption*	– Polyelectrolyte (no use of chemicals during the thermal treatment)	– OFMSW/other cosubstrates, – Polyelectrolyte	– Woodchips/ bulking agent – Sulphuric acid (acid scrubber)	– Fluidizing gas (nitrogen) for fluidized bed systems – Catalysts (usually zeolites) only in catalytic pyrolysis	– Sulphuric acid and caustic soda (scrubbers)	– Oxidant (air or pure oxygen)
9	*by-products and or waste streams*	1.4 $t_{digestate}$ t^{-1} DS (compared to ≈2.3 $t_{digestate}$ t^{-1} DS in conventional AD)	≈2.3 $t_{digestate}$ t^{-1} DS	Ashes	Ashes that can be used as cement kiln, asphalt, ceramics, etc.	biochar that can be used in agriculture, etc.	Inorganic salts and ashes
11	*Capital cost (€ t–1 treatment capacity)*	160–250 € m^{-3}	150–225 € m^{-3}	220–500 € t^{-1} DS	650 € t^{-1} DS	1200 € t^{-1} DS (only one industrial facility reported)	410 € t^{-1} DS

12	*Operational Cost*	65–105 € t^{-1} DS	65–105 € t^{-1} DS (2 to 5€ t^{-1} for transport of co-substrates)	60–140 € t^{-1} DS	30–110 € t^{-1} DS	–	60–140 € t^{-1} DS
13	*State of development (emergent, adaptive use, innovative, establisehd) and TRL*	Commercially available technologies (TRL: 9)	Commercially available technologies (TRL: 9)	1 reported full-scale facility and several lab-scale and pilot plants	Lab-scale and pilot plants	1 reported full-scale facility and several lab-scale and pilot plants	3 reported full-scale facilities, several facilities under construction and several lab-scale and pilot plants
14	*Main advantadges*	– Increase AD biodegradability – Reduced sludge production compared to traditional AD	– Improve substrate chemical properties such as C/N ratio – Explotation of non-used AD capacity of WWTP	– Reduced energy consumption compraed to thermal drying	– Potentially a self-sufficient energy treatment	– The bio-oil can be used as fuel as well as a source of chemical valuable products – Char can be used as fuelo or as soil ammedment	– Clean water can be recoverd – The process does not generate harmful by-products suchs as Nox or SOx
15	*Bottlenecks*	– Lack of robustness and reliability in continuous processes – Economical constraints limited their scale-up	– Transportation costs of co-substrates	– Necessity of co-substrates to carry out the process – Treatment of off-gases usually required	– Requires a gas cleaning stage – Dealing with tar formed during the process – Capital costs are still a limiting factor for full-scale implementation	– Endothermic process if liquid is not combusted – The liquid fraction contains N-compounds that makes it difficult to use it as a fuel	– Complex nature of a high-pressure, high-temperatrue process – Corrosion and reactor plugging are the most critical areas
16	*Compatibility with other technologies*	Digestate can be used in subsequent composting, biodrying, pyrolyis or gasification processes	Digestate can be used in subsequent composting, biodrying, pyrolyis or gasification processes	Biodried sludge can be used in subsequent, pyrolyis or gasification processes	Biochar can be used in as soil ammendment	Biochar can be used in subsequent gasification processes or as soil ammendment	Stand alone

*$CHP_{efficiency}$ = 35% for electricity and 50% for heat.
** Excess heat is considered as energy losses.
Source: Bridle & Pritchard 2004; Frei *et al.* 2006; Garrido-Baserba *et al.* 2015; Gil-Lalaguna *et al.* 2014; Hospido *et al.* 2005; Koch *et al.* 2015; Mills *et al.* 2014; Xu *et al.* 2012.

According to the proposed European waste management guidelines, when organic matter and nutrients contained in SS can be used as fertilizer or soil amendment, AD processes should be have priority, due to the fact that AD allows the simultaneous recovery of energy (biogas) and fertilizer/soil amendment (AD digestate). Modifications to the already existing AD processes have demonstrated its capacity to highly improve energy recovery. Specifically, advanced thermal/high-pressure pre-treatments have demonstrated their capacity to enhance SS biodegradability and degradation rates, while improving biogas production (up to 60%, compared to traditional SS AD processes) as well as other operational parameters (e.g. higher loading rates). Co-digestion of SS with other biodegradable materials (e.g. OFMSW, FVW or fats, oils and greases) has also proved efficient for improving the stability of the overall AD process (improving physical-chemical properties such as C/N ratio) as well for increasing both biogas production and methane content in the biogas.

Bio-drying is another bio-based technology, mainly aiming eventually at energy recovery (combustion of bio-dried SS) but can also be used as organic matter/nutrient recovery technology. The goal of bio-drying is to remove water from the waste matrix using metabolic heat of organic matter degradation, thereby highly increasing the LHV of the bio-dried sludge, which is important when the product is to be used for energy recovery purposes. At the same time, the mass reduction (water losses) achieved during the process highly reduces the transportation costs when the product is used as soil amendment/fertilized, or even when the final product is disposed to landfills (reducing both transportation and disposal costs).

When not enough agricultural land is available for safe land application of treated SS, or when the treated sludge physical-chemical properties do not comply with current legislation criteria (e.g. metal content), thermal technologies aiming to recover energy and/or high added value products (also including incineration of bio-drying end-products) should be preferred. Gasification is a thermo-chemical process, in which any carbon-containing material is converted into a combustible gas (a mixture of CO, CO_2, H_2, CH_4 and other light hydrocarbons) in the presence of a reactive atmosphere, and is suitable for use in boilers, engines and turbines. Gasification also reduces waste volume, removes toxic organic compounds and fixes heavy metals in the resultant solid.

Pyrolysis is a thermochemical process carried out in an inert atmosphere, in which 3 products (char, liquid and gas) are obtained. The liquid fraction is the most valuable one, with high calorific value and it is also a potential source of high added-value compounds (e.g. α-olefins, n-paraffins, long aliphatic and aromatic nitriles, methyl-styrene, benzocyclobutene and octocrylene). On the contrary, the gas fraction usually does not have high enough heating value to be burned using the existing technologies.

Finally, supercritical water processing (including oxidation and gasification) is a promising thermochemical process for sludge treatment and energy recovery. The main advantages of supercritical water technologies compared to other thermally-based technologies are: (i) reaction times are in the order of seconds, resulting in systems that are quite compact compared to other technologies (ii) the process does not generate harmful by-products such as NOx or SOx, CO, or odors and (iii) clean water is produced and can be recovered.

The ultimate goal of the SS treatment technologies described in this chapter is to promote a more sustainable SS management strategy, moving towards a circular economy approach, in which organic matter, nutrient and/or energy recovery are prioritized.

15.5 REFERENCES

Ábrego J., Sánchez J. L., Arauzo J., Fonts I., Gil-Lalaguna N. and Atienza-Martínez M. (2013). Technical and energetic assessment of a three-stage thermochemical treatment for sewage sludge. *Energy Fuels*, **27**(2), 1026–1034.

Adegoroye A., Paterson N., Li X., Morgan T., Herod A. A., Dugwell D. R. and Kandiyoti R. (2004). The characterisation of tars produced during the gasification of sewage sludge in a spouted bed reactor. *Fuel*, **83**, 1949–1960.

Anis S. and Zainal Z. A. (2011). Tar reduction in biomass producer gas via mechanical, catalytic and thermal methods: a review. *Renewable & Sustainable Energy Reviews*, **15**, 2355–2377.

Ariunbaatar J., Panico A., Esposito G. and Pirozzi F. (2014). Pre-treatment methods to enhance anaerobic digestion of organic solid waste. *Applied Energy*, **123**, 143–156.

Aznar M., Manyà J. J., García G., Sánchez J. L. and Murillo M. B. (2008). Influence of freeboard temperature, fluidization velocity and particle size on tar production and composition during the air gasification of sewage sludge. *Energy & Fuels*, **22**, 2840–2850.

Azuara M., Fonts I., Bimbela F., Murillo M. B. and Gea G. (2015). Catalytic post-treatment of the vapors from sewage sludge pyrolysis by means of γ-Al2O3: effect on the liquid product properties. *Fuel Processing Technology*, **130**(0), 252–262.

Bacaicoa P. G., Bilbao R. and Uson C. (1995). Sewage Sludge Gasification: First Studies. In: Proceedings of the Second Biomass Conference of the Americas: Energy, Environment and Agricultural Industry, 685–694.

Barrido-Baserba M., Molinos-Senante M., Abelleira-Pereira J. M., Fdez-Güelfo L. A., Poch M. and Hernández-Sancho F. (2015). Selecting sewage sludge treatment alternatives in modern wastewater treatment plants using environmental decision support systems. *Journal of Cleaner Production*, **107**, 410–419.

Bennamoun L. (2012). Solar drying of wastewater sludge: a review. *Renewable and Sustainable Energy Reviews*, **16**(1), 1061–1073.

Bermejo M. D. and Cocero M. J. (2006). Supercritical water oxidation: a technical review. *AIChE Journal*, **52**(11), 3933–3951.

Bianchini A., Bonfiglioli L., Pellegrini M. and Saccani C. (2015). Sewage sludge drying process integration with a waste-to-energy power plant. *Waste Management*, **42**, 159–165.

Bodik I. and Kubaska M. (2013). Energy and sustainability of operation of a wastewater treatment plant. *Environment Protection Engineering*, **39**(2), 15–24.

Boon A. and Thomas V. (1996). Resource or Rubbish?. IChemE's Research Symposium No5, ROYAUME-UNI, Belfast.

Bridle T. R. and Pritchard D. (2004). Energy and nutrient recovery from sewage sludge via pyrolysis. *Water Science and Technology*, **50**(9), 169–175.

Cano R., Pérez-Elvira S. I. and Fdz-Polanco F. (2015). Energy feasibility study of sludge pre-treatments: a review. *Applied Energy*, **149**, 176–185.

Cao Y. and Pawlowski A. (2012). Sewage sludge-to-energy approaches based on anaerobic digestion and pyrolysis: Brief overview and energy efficiency assessment. *Renewable and Sustainable Energy Reviews*, **16**, 1657–1665.

Carrère H., Dumas C., Battimelli A., Batstone D. J., Delgenès J. P., Steyer J. P. and Ferrer I. (2010). Pre-treatment methods to improve sludge anaerobic degradability: a review. *Journal of Hazardous Materials*, **183**(1–3), 1–15.

De Andrés J. M., Narros A. and Rodríguez M. E. (2011). Behaviour of dolomite, olivine and alumina as primary catalysts in air–steam gasification of sewage sludge. *Fuel*, **90**, 521–527.

Deshusses M. A. and Jacoby W. (2015). Treatment of Fecal Sludge in a Prototype Supercritical Water Oxidation Reactor. In: Proc. the 250th American Chemical Society National Meeting & Exposition, Innovation from Discovery to Application. Boston, 16–20 August.

Domínguez A., Menéndez J. A. and Pis J. J. (2006). Hydrogen rich fuel gas production from the pyrolysis of wet sewage sludge at high temperature. *Journal of Analytical and Applied Pyrolysis*, **77**, 127–132.

Fonts I., Gea G., Azuara M., Ábrego J. and Arauzo J. (2012). Sewage sludge pyrolysis for liquid production: a review. *Renewable and Sustainable Energy Reviews*, **16**(5), 2781–2805.

Fonts I., Navarro-Puyuelo A., Ruiz-Gómez N., Atienza-Martínez M., Wisniewsky A. and Gea G. (2017). Assessment of the production of value-added chemical compounds from sewage sludge Pyrolysis Liquids. *Energy Technology*, **5**(1), 151–171.

Fytili D. and Zabaniotou A. (2008). Utilization of sewage sludge in EU application of old and new methods—a review. *Renewable and Sustainable Energy Reviews*, **12**, 116–140.

Gil-Lalaguna N., Sánchez J. L., Murillo M. B., Rodríguez E., Gea G. (2014a). Air-steam gasification of sewage sludge in a fluidized bed. Influence of some operating conditions. *Chemical Engineering Journal*, **248**, 373–382.

Gil-Lalaguna N., Sánchez J. L., Murillo M. B., Atienza-Martínez M. and Gea G. (2014b). Energetic assessment of air-steam gasification of sewage sludge and of the integration of sewage sludge pyrolysis and air-steam gasification of char. *Energy*, **76**, 652–662.

Gil-Lalaguna N., Sánchez J. L., Murillo M. B. and Gea G. (2015). Use of sewage sludge combustion ash and gasification ash for high-temperature desulphurization of different gas streams. *Fuel*, **141**, 99–108.

Gottardo M., Micolucci F., Mattioli A., Faggian S., Cavinato C. and Pavan P. (2015). Hydrogen and Methane Production from Biowaste and Sewage Sludge by Two Phases Anaerobic Codigestion. *Chemical Engineering Transactions*, **43**, 379–384.

Hospido A., Moreira M. T., Marín M., Riogla M. and Feijoo G. (2005). Environmental evaluation of different treatment processes for sludge from urban wastewater treatments: anaerobic digestion versus thermal processes. *The International Journal of Life Cycle Assessment*, **10**(5), 336–345.

IEA (2007). Biomass for Power Generation and CHP. IEA Energy Technology Essentials ETE03.

Innoveox (2015). http://www.innoveox.com/ (accessed November 2015).

Izhar S., Uehara S., Yoshida N., Yamamoto Y., Morioka T. and Nagai M. (2012). Hydrodenitrogenation of fast pyrolysis bio-oil derived from sewage sludge on $NiMo/Al_2O_3$ sulfide catalyst. *Fuel Processing Technology*, **101**, 10–15.

Judex J. W., Gaiffi M. and Burgbacher H. C. (2012). Gasification of dried sewage sludge: status of the demonstration and the pilot plant. *Waste Management*, **32**, 719–723.

Kelessidis A. and Stasinakis A. (2012). Comparative study of the methods used for treatment and final disposal of sewage sludge in European countries. *Waste Management*, **32**, 1186–1195.

Knoef H. A. M. (2005). Handbook Biomass Gasification. BTG Biomass Technology Group, The Netherlands, ISBN: 90-810068-1-9.

Koch K., Plabst M., Schmidt A., Helmreich B. and Drewes J. E. (2015). Co-digestion of food waste in a municipal wastewater treatment plant: comparison of batch tests and full-scale experiences. *Waste Managagement*, **47**, 28–33.

Marrone P. A. and Hong G. T. (2007). Corrosion control methods in supercritical water oxidation and gasification processes. *Journal of Supercritical Fluids*, **51**(2), 83–103.

Marrone P. A. and Hong G. T. (2013). Supercritical water oxidation—current status of full-scale commercial activity for waste destruction. *Journal of Supercritical Fluids*, **79**, 283–288.

Mattioli A., Gatti G. B., Mattuzzi G. P., Bolzonella D. and Cecchi F. (2015). Implementation of Full Scale Codigestion of the Organic Fraction of Municipal Solid Waste and Sludge Improves the Energy Balance of the Wastewater Treatment Plants. 7th Eastern European Young Water Professionals Conference, Belgrad, Serbia, 17–19 September.

Midilli A., Dogru M., Howarth C. R., Ling M. J. and Ayhan T. (2001). Combustible gas production from sewage sludge with a downdraft gasifier. *Energy Conversion and Management*, **42**, 157–172.

Mills N., Pearce P., Farrow J., Thorpe R. B. and Kirkby N. F. (2014). Environmental & economic life cycle assessment of current & future sewage sludge to energy technologies. *Waste Management*, **34**(1), 185–195.

Mullen C. A., Boateng A. A., Goldberg N. M. (2013a). Production of deoxygenated biomass fast pyrolysis oils via product gas recycling. *Energy and Fuels*, **27**(7), 3867–3874.

Mullen C. A., Boateng A. A., Reichenbach S. E. (2013b). Hydrotreating of fast pyrolysis oils from protein-rich pennycress seed presscake. *Fuel*, **111**, 797–804.

Pavan P., Bolzonella D., Battistoni E. M. and Cecchi F. (2007). Anaerobic co-digestion of sludge with other organic wastes in small wastewater treatment plants: an economic considerations evaluation. *Water Science and Technology*, **56**(10), 45–30.

Perez-Elvira S., Nieto P. P. and Fdz-Polanco F. (2006). Sludge minimisation technologies. *Reviews in Environmental Science and Biotechnology*, **5**, 375–398.

Porpatham E., Ramesh A. and Nagalingam B. (2007). Effect of hydrogen addition on the performance of a biogas fuelled spark ignition engine. *International Journal of Hydrogen Energy*, **32**, 2057–2065.

Righi S., Oliviero L., Pedrini M., Buscaroli A. and Della Casa C. (2013). Life Cycle Assessment of management systems for sewage sludge and food waste: centralized and decentralized approaches. *Journal of Cleaner Production*, **44**, 8–17.

Sanchez M. E., Menendez J. A., Dominguez A., Pis J. J., Martinez O., Calvo L. F. and Bernad P. L. (2009). Effect of pyrolysis temperature on the composition of the oils obtained from sewage sludge. *Biomass & Bioenergy*, **33**(6–7), 933–940.

SCFI-Aquacritox (2015). http://www.scfi.eu/ (accessed November 2015)

Sridhar P., Song Yan R. D., Tyagi R. and Surampalli Y. (2015). Thermal pre-treatment of sewage sludge to enhance anaerobic digestion: a review. *Critical Reviews in Environmental Science and Technology*, **45**(6), 669–702.

SuperWater Solutions (2015). http://www.superwatersolutions.com/ (accessed November 2015)

Talmadge M. S., Baldwin R. M., Biddy M. J., McCormick R. L., Beckham G. T., Ferguson G. A., Czernik S., Magrini-Bair K. A., Foust T. D., Metelski P. D., Hetrick C. and Nimlos M. R. (2014). A perspective on oxygenated species in the refinery integration of pyrolysis oil. *Green Chemistry*, **16**(2), 407–453.

Tandukar M. and Pavlostathis S. (2015). Co-digestion of municipal sludge and external organic wastes for enhanced biogas production under realistic plant constraints. *Water Research*, **87**, 432–445.

Tassaing T., Danten Y. and Besnard M. (2002). Infrared spectroscopic study of hydrogenbonding in water at high temperature and pressure. *Journal of Molecular Liquids*, **101**(1–3), 149–158.

Tyagi V. K. and Lo S. L. (2013). Sludge: a waste or renewable source for energy and resources recovery?. *Renewable and Sustainable Energy Reviews*, **25**, 708–728.

Valdez-Vazquez I. and Poggi-Varaldo H. M. (2009). Hydrogen production by fermentative consortia. *Renewable and Sustainable Energy Reviews*, **13**, 1100–1113.

Winkler M. K. H., Bennenbroek M. H., Horstink F. H., van Loosdrecht M. C. M. and van de Pol G.-J. (2013). The biodrying concept: an innovative technology creating energy from sewage sludge. *Bioresource Technology*, **147**, 124–129.

Xu D., Wang S., Tang X., Gong Y., Guo Y., Wang Y. and Zhang J. (2012). Design of the first pilot scale plant of China for supercritical water oxidation of sewage sludge. *Chemical Engineering Research and Design*, **90**(2), 288–297.

Chapter 16

Metal recovery from sludge: Problem or opportunity

Ioannis Vyrides, Elena Anayiotou, Predrag Banković, Wim De Schepper and Xochitl Dominguez-Benetton

16.1 INTRODUCTION

The estimated production of waste sludge in Europe is 2 billion tons per year (Meulepas *et al.* 2015). Until today, this waste sludge is disposed via incineration, landfilling or ocean disposal as well as reused as soil conditioner in agriculture. Land application represents the most economical way for final disposal of residual sludge and it combines the recycling of plant nutrients and sludge disposal at the same time (Gu *et al.* 2004). The recent banning of ocean disposal and new stringent European landfilling criteria have opened new prospects for sludge management (Tyagi *et al.* 2013). There is a general agreement that the long-term goal should be to recycle the nutrients and organic matter present in sludge. Due to the physical-chemical processes involved in sludge generation however, sewage sludge tends to accumulate heavy metals as well as potentially pathogenic organisms (viruses, bacteria, etc.) and poorly biodegradable trace organic compounds. The concentration of heavy metals in sewage sludge is found to be between 0.5–2% on a dry weight basis, but may rise to as high as 6% in some cases (Pathak *et al.* 2009). Several wastewater treatment plants (WWTPs) receive influent not only from residential areas but also from industrial areas. Sludge generated at these plants contains higher heavy metal concentrations, which may vary considerably with time and mostly depend on the specific industrial activities. The application of contaminated sludge over prolonged periods may cause accumulation of heavy metals along the food chain or in the ground and surface water, resulting in negative effects on animal and human health. Removal of heavy metals prior to land application of sludge or reuse of nutrients from sludge and ashes is therefore desired.

Apart from this, rising metal prices and China's tightening grip on supplies on several metals have heightened the appeal of finding other sources of metals supply. Unlike oil, there are no bio-derived alternatives for these metals (e.g., Cobalt (Co), Zinc (Zn) and Copper (Cu)). These are unique and finite elements that are quickly dispersed throughout the environment, making their mining more costly and difficult (Carey, 2015; Dosdosn *et al.* 2012). Westerhoff *et al.* (2015) attempted to estimate the economic potential of metals in sewage sludge by calculating the monetary value of metal content in sludge from metal concentrations and spot market price of purified metals. The top 13 elements with the highest economical potential to be recovered from biosolids resulting this research are Silver (Ag), Copper (Cu), Gold (Au), Phosphorous (P), Iron (Fe), Palladium (Pd), Manganese (Mn), Zinc (Zn), Iridium (Ir), Aluminium (Al), Cadmium (Cd), Titanium (Ti), Gallium (Ga), and Chromium (Cr). The analysis was performed for a community with a population of 1,000,000 inhabitants (approx. 28,600 dry tons of biosolids per year), and the estimated value of metals in the biosolids could approach 12,000,000 euro per year (415 €/ton) with greater than 20% of the value accounted (2,300,000 €/year) for Gold (Au) and Silver (Ag). Phosphorus, which is the focus of many wastewater recovery systems, has a relatively low economic potential (51,000 €/year). Many studies (Table 16.1) point out that the metals that are usually found in relative high concentrations in sludge are Al and Fe, both resulting from coagulant addition during treatment, and to a less extent Zn, Cu, Ni, Pb and minor traces of Ag. During this study, 5 metals were pre-selected (Cu, Ni, Al, Zn, Ag) to be further analysed based on the sum of the average concentration found in dry weight sludge multiply by their average price.

Table 16.1 Concentration of heavy metals on dry weight sludge (mg/kg dry sludge).

Sludge Type	Zn	Cr	Cu	Ni	Pb	Fe	Mn	References
Anaerobically digested sludge	2823	663	255	622	57	72200	nil	Wong *et al.* (2004)
Anaerobic sewage sludge	1000	nil	300	200	180	nil	400	Lombardi *et al.* (2001)
Anaerobically digested sludge	2306	181	256	47	88	64400	nil	Wong *et al.* (2002)
Anaerobically digested sewage sludge	13301 27609	79.2 128	153 111	nil	nil	nil	nil	Gu *et al.* (2004)
Anaerobically digested sewage sludge	1690	85.6	242.6	79.6	64	29731	nil	Chan *et al.* (2003)
Sewage sludge	3756	nil	296.4	nil	351.3	nil	nil	Chen *et al.* (2005)
Tannery sludge	nil	nil	10382	nil	nil	3123	nil	Wang *et al.* (2010)
Sewage sludge	812.9	nil	171	nil	118.	nil	nil	Zhang *et al.* (2009)
Pig manure	3434	10.3	1408	nil	nil	nil	995	Zhou *et al.* (2012)
Anaerobic sludge	702	10.4	245	nil	nil	nil	nil	Zhou *et al.* (2013)
Pig slurry	2710	10.41	871	18.70	0.7	nil	510	Zhu *et al.* (2013)
Sewage sludge	1516	nil	545.6	nil	133	nil	nil	Zhu *et al.* (2013)

Therefore, an approach similar to the one used by Westerhoff *et al.* (2015) was used to estimate the economic value of these metals present in a 300,000 p.e. sludge line (~ 8,580 dry tons of sludge per year) (Table 16.2). The aim of this research was to provide a rough indication of average and maximum possible added value. Annual per capita production of biosolids is in the order of 28.6 kg D.M./person-year (Westerhoff *et al.* 2015). In addition, a sludge management overview is pointed out based on the concentration of metals in sludge and the possible technologies.

Table 16.2 Presence of some metals in EU sludge and estimation of their economic potential.

Metal	Average Metal Content Mg/kg d.m.	Maximum Metal Content Mg/kg d.m.	Sludge Ton d.m./ year	Relative Yearly Profit Potential Based on the Average Metal Content and the Average Price ($) of Metal (during, 2005–2015)	Relative Yearly Profit Potential Based on the Maximum Metal Content and the Highest Price of Metal ($) (during, 2005–2015)
Cu	285	1050	8580	18,706	108,966
Ni	29	66	8580	7962	29,446
Al	16,885	30,500	8580	314,946	2,170,740
Zn	910	2330	8580	23,540	88,761
Ag	7	108	8580	45,516	1,162,272

Values of metals was estimated from Fytili, D. and Zabaniotou, A. (2008) and EU Draft Summary Report 1 Environmental, economic and social impacts of the use of sewage sludge on land.

The economic value was evaluated based on the price of scrap metal (01/2015) and the average price for refined metal between 2005 and 2015 (London Stock Exchange). Several studies have pointed out that the recovery of those metals could be between 70–99% (Pathak *et al.* 2009), being in this case full recovery assumed. The amount of metals present in sludge for a 300,000 p.e. installation is in the order of a few tons/year (except Al) and should be considered as a minor contribution to the industrial metal cycle. For example, the EU's demand for Cu in 2014 was estimated at around 4.2 million tons whereas yearly Cu mass flow in sewage sludge for the total EU is estimated to 4854 ton or 0.1%. The maximal recoverable "scrap" metal value for the considered metals is estimated to amount to 100–200 kEuro/year for a 300,000 p.e. installation. It is therefore concluded that the main current driver for metal recovery from sludge or ashes remains environmental rather than economic (e.g. facilitating sludge disposal through land application). However, in several cases where the concentrations of metals are high and the metal's price has relatively high value (e.g. Al and Ag, Table 16.2) then the recovery of metals can be an option for metal recovery and economic profits. However, the net profit has to be estimated having also in considerations the capital and operational costs (for metal extraction and recovery) and the highest metals that can be recovered from sludge. Consequently, metals should be extracted from the sludge only if their concentrations are above certain (national) limits (Figure 16.1) and when extraction is feasible applying best available technologies. In case that the recovery of target metals has economic potential, including avoidance of disposal cost of metal contaminated fractions, and/or ecological benefit, then the effort should be on metal recovery processes and not only on metal removal (Figure 16.1). In case that the extracted metals do not have any economic or ecological benefit for recovery then, after bioleaching, they can be precipitated or adsorbed and subsequently disposed off. The leached sludge can be neutralized and be used for land application or can be directly used for incineration.

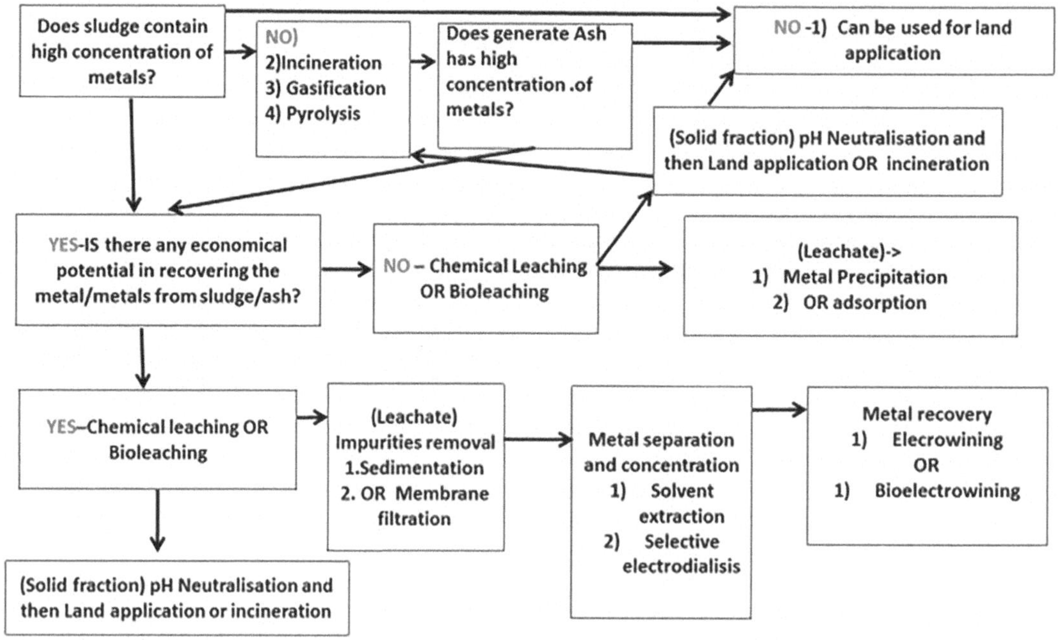

Figure 16.1 An overall sludge management diagram that points out the decision that need to be taken based on the concentration of metals in sludge and the possible technologies.

16.2 LEACHING OF METALS FROM SLUDGE

16.2.1 Chemical leaching

Heavy metals can be released from sewage sludge by chemical leaching with inorganic and organic acids Marchioretto (2003). Several acids have been tested and the most common are inorganic acids such as hydrochloric acid, nitric acid, sulfuric acid, and phosphoric acid. Organic acids like citric and oxalic acid are also applied. However, the main disadvantage of chemical leaching is the high cost due to high acid consumption. Despite this fact, it is could be useful to consider chemical leaching as a practical method for heavy metals extraction from WWTP that produce low volume of sludge with high concentration of heavy metals.

16.2.2 Bioleaching

A promising eco-friendly process for the efficient extraction of metals from waste sludge is through biological means. Biohydrometallurgical processes, which utilize microorganisms for the extraction of metals from ores/concentrates/wastes, are regarded as one of the most promising and revolutionary biotechnological options for mineral processing and metallurgy (Erust *et al.* 2013). In the bioleaching/biohydrometallurgical process, microorganisms convert insoluble metal sulphides or oxides (directly or indirectly) into water-soluble metals. Additionally, biohydrometallurgy can be conducted under mild conditions, usually without the use of any toxic chemicals whereas chemical leaching requires large volumes of chemicals and produces toxic substances. Therefore, biohydrometallurgy which has lower costs and it is more ecofriendly than chemical leaching is considered a more suitable process for metal extraction from the high volumes of sludge typically generated in WWTPs.

Common acidophilic strains used for bioleaching are Acidobacilus ferroxidans, and Leptospirilium ferooxidans. These microbes can act as a catalyst by oxidizing ferrous to ferric ion (Equation 16.1). The ferric ion oxidize the metal (Equation 16.2) thus, decreases its concentration and ferric ion is reduced back to ferrous ion. The ferrous iron resulting from this reaction is re-oxidized to ferric iron by the bacteria and as such can take part in the oxidation process again. Apart from this, acidophilic microorganisms such as Acidobacilus thioxidans can directly release metal by oxidizing sulphur (Equation 16.3). In addition, at low pH and low redox potential, several metal oxides can be leached but this will result in an increase of pH (Equation 16.4) producing a subsequent termination of the leaching process. The presence of Acidobacilus ferrooxidans and Acidobacilus thioxidans can maintain the pH at low level if an extra source of sulphur is supplemented (Equation 16.5).

$$Fe^{2+} + H^+ + O_2 \text{ (bacteria)} \rightarrow Fe^{3+} + H_2O \quad (16.1)$$

$$\text{Metal-X} + Fe^{3+} \rightarrow M^{2+} + Fe^{2+} \quad (16.2)$$

$$\text{Metal-S} + O_2 \text{ (bacteria)} \rightarrow M^{2+} + SO_2 \quad (16.3)$$

$$\text{Metal-O} + 2H^+ + SO_4^{2-} \rightarrow Me^+ + SO_4^{2-} + H_2O \quad (16.4)$$

$$S^0 + O_2 + H_2O \text{ (Bacteria)} \rightarrow H^+ + SO_4^{2-} \quad (16.5)$$

Zhu *et al.* (2013) reported that bioleaching not only has an important role in removing heavy metals, but also results in improved sludge dewaterability and elimination of pathogens. Recently a new bioleaching approach without pre-acidification was developed using indigenous iron-oxidizing bacteria, and the bioleaching time could be reduced to 2–4 days (Gu *et al.* 2004). Moreover, bioleaching using sulfur-oxidizing bacteria does not require pre-acidification, because of the production of the sulphuric acid through the sulfur-oxidizing microorganisms (Equation 16.5). The pH and ORP of the system are known to be the most important operating parameters that influence metal solubilization in bioleaching (Wong *et al.* 2002). Metal solubilization is affected by a combination of pH decrease and ORP increase, which are a result of bacterial activity. High ORP coupled with low pH have been considered as indicators of the presence of high population of Thiobacilli (Wong *et al.* 2002). For the above reasons, plus the fact that in bioleaching methods there is no need for large quantities of chemical reagents, bioleaching is estimated to be 80% cheaper than chemical methods. Several studies that use bioleaching processes for the removal of metals from sludge are summarised in Table 16.2. The solubilisation efficiency of metals such as Cu, Zn and Cr has been reported as high.

16.3 REMOVAL OF METAL FROM THE LEACHATE WITHOUT METAL RECOVERY

16.3.1 Metal precipitation

Precipitation of metals involves the addition of chemicals to alter the physical state of the dissolved or suspended metals. The chemicals react with heavy metal ions to form insoluble precipitates (Fu & Wang, 2011). The ability of a compound to precipitate/dissolve at a given temperature is determined

Table 16.3 Overview of applied process parameters in sludge bioleaching.

Type of Sludge	Solids in Sludge	Microorganisms Used	Additional Substrate	Quantity of Additional Substrate	% [ ] of Inoculated Bacteria	Experimental Conditions	Maximum Bioleaching Removal of Metals	References
Anaerobically digested sludge	2.21% solids	Acidihiobacillus ferrooxidans XL (shake flask)	Fe^{2+} (from Fe_2S & $FeSO_4 \cdot 7H_2O$)	0.52 + 0 g/L	15.0	T = 28°C pH = 3.0 16 days iron-oxidising bacteria Fe^{2+}	99% of Zn, 65% of Cr, 74% of Cu, 58% of Pb and 84% of Ni after 16 days	Wong *et al.* (2004)
Anaerobically digested sewage sludge	1% solid content	Acidithiobacillus ferrooxidans ANYL-1 (Gram-negative, rod-shaped bacterium) (shake flask)	Fe^{2+} (from $FeSO_4 \cdot 7H_2O$)	20 g/L	15.0	T = 30°C pH = 2–2.5	86% Cu and 56% Cr in 8 days	Gu *et al.* (2004)
Anaerobically digested sewage sludge	Solid (1.25%)	Iron-oxidizing bacteria & sulfur-oxidizing bacteria (shake flask)	Fe^{2+} (from $FeSO_4 \cdot 7H_2O$)	4.0 g/L	15.0	T = 30°C 16 days	Cr 52 ys8% after 12 days, 80% Cu in 2 days, 95% Zn in 4 days	Chan *et al.* (2003)
Sewage sludge	28.53 g/L (TS)	Thiobacillus (shake flask)	nil	nil	2.0	12 days pH = 2–3 T = 28°C	Cu, Pb and Zn after bioleaching accounted for 81.6, 40.2 and 75.8%,respectivelly after 12 days	Chen *et al.* (2005)
Sewage sludge	solid content, 2%;	Sulfur-oxidizing bacteria (shake flask)	Elemental sulfur	10 g/L DS	Sewage sludge adapted with sulfur oxidizing bacteria 5%	T = 28°C pH < 2 dewatering of the sludge 16 days	66% for Cu after4 64% Zn after 2 days	Zhang *et al.* (2009)
Pig manure	Solid content 3%	-A.Thioxidans TS6 - A.Ferrooxidans LX5 (shake flask)	Fe^{2+} (from $FeSO_4 \cdot 7H_2O$) S_0 from elemental sulfur	–4 g/L –2 g/L	5.0	T = 28°C pH = 2 bioleached pig manure dewatered	The removal efficiencies of Zn, Cu, and Mn after 6 days were 95.1%, 80.9%, and 87.5%, respectively	Zhou *et al.* (2012)

by the solubility product constant, Ksp, the product of the equilibrium concentrations of the ions in a saturated solution of the compound. The precipitation occurs when Ksp is exceeded. The forming precipitates can be separated from the water by sedimentation or filtration. Depending on the type of the added precipitating chemical, metals can be removed as metal oxides/hydroxides, sulfides, carbonates, and phosphates. Alkaline agents, such as NaOH, CaO, $Ca(OH)_2$, $CaCO_3$, MgO, $Mg(OH)_2$ and NH_4OH, cause the precipitation of metal ions in the form of hydroxides. Another option is the precipitation to be achieved using reagents like Na_2S, NaHS, H_2S or FeS. However, it is more expensive, and caution is necessary in the case of acidic effluents when polluting H_2S formation occurs. By the application of apatite and hydroxyapatite minerals as precipitating agents, metal ions can be separated from a solution in the form of phosphates. Phosphates in the majority of cases have lower Ksp values than the hydroxide counterparts. The removal of metal ions in the form of carbonates is caused by the addition of Na_2CO_3 and $CaCO_3$ (Blais *et al.* 2008).

In order to meet strict environmental demands, alternative precipitating agents, such as chelating agents, have been introduced. These include trimercaptotriazine, potassium/sodiumthiocarbonate and sodiumdimethyldithiocarbamate, 1,3-benzenediamidoethanethiol (BDET2-) dianion, N,N0-bis-(dithiocarboxy)piperazine (BDP) and 1,3,5-hexahydrotriazinedithiocarbamate (HTDC) potassium ethyl xanthate, and dipropyl dithiophosphate, diethyldithiocarbamate DDTC and dimethyldithiocarbamate DMTC. Lately, sulphate-reducing microorganisms have been used for the precipitation of metal ions in the form of sulphides, as well (Fu & Wang, 2011). Advantages of chemical precipitation are the simplicity, as well as the treatment at high metal ion concentrations, while the principal disadvantages are the ineffectiveness at low metal ion concentrations and the large amount of produced sludge (O'Connell *et al.* 2008).

16.3.2 Metal adsorption

Adsorption is a process that occurs when a gas or liquid substance (adsorbate) accumulates on the surface of a solid or a liquid (adsorbent) and becomes bound by physical attractive forces, ion exchange, or chemical bonds (Lakherwal, 2014). Adsorbents that are commonly used for the removal of metals are the following: activated carbon, activated alumina, biomaterials (i.e. adsorbent materials derived from low-cost agricultural wastes), ion-exchange resins, and low-cost sorbents such as bark/tannin-rich materials, dead biomass, xanthate, clay minerals, zeolites, peat moss, bone gelatin beads, leaf mould, moss, iron-oxide-coated sand, modified wool, modified cotton, lignin, chitosan, seaweed/algae/alginate, fly-ash, peat, wool, cotton, natural oxide (Gupta & Bhattacharyya, 2012). Activated carbon is rarely selected as adsorbent for the removal of heavy metals since it is expensive and shows much better performance in the adsorption of organic compounds (Gupta & Bhattacharyya, 2012). Biosorption, on the other hand, has shown to be promising for the removal of heavy metals from wastewater. It can be defined as the ability of biological materials to accumulate heavy metals from wastewater through metabolically mediated or physico-chemical pathways of uptake. Biosorption, which is the ability of certain microbial biomaterials to bind and concentrate heavy metals from even the most dilute aqueous solutions, offers a technically feasible and economically attractive alternative. Algae, bacteria and fungi and yeasts have proved to be potential metal biosorbents (Abbas *et al.* 2014). Other low cost adsorbent material can be the low-cost lignocellulosic material or agriculture waste (Salman *et al.* 2015).

Adsorption as a method for the removal of metal ions exhibits several advantages such as: a variety of available low-cost adsorbents, fast kinetics, possible selectivity depending on adsorbent, etc. Disadvantages of the method are: efficiency at relatively low metal concentrations that is dependent on the type of adsorbent, possible need for adsorbent derivatisation, etc. (Lakherwal, 2014).

16.4 METAL RECOVERY

The leachate resulting from municipal sludge is complex in composition which makes direct metal recuperation often difficult. The main technological challenges related to the reuse of metals from municipal sludge leachates are: (1) their relative low concentration, (2) the presence of other metals, inorganics and organics, (3) high ionic strength of some matrices that make the extraction more difficult. In order to facilitate the reuse of metals from leachates, it is necessary to proceed with purification and concentration followed by selective metal separation and deposition. Optimally, separation and concentration to obtain relatively pure monometallic partitions is envisaged to enhance metal recovery. The following paragraphs summarise processes that come into scope when metal recovery from sludge and ash leachates is considered, categorised as processes for removal of impurities, separation of metals and finally their recovery.

16.4.1 Removal of impurities from leach solution

Removal of suspended solids and to a lesser extent soluble organics from raw leachate is desired to improve downstream metal recuperation technologies. Either conventional physical separation technologies (e.g., sedimentation, cyclones) or membrane technologies can be applied to enhance the quality of leachate. Depending on the characteristics of the products that need to be retained, various types of membrane technology can be employed for leachate purification and concentration, including pressure driven technologies such as micro- or ultrafiltration, nanofiltration and reverse osmosis. Alternatively membrane contactors can be used for concentration by evaporation (membrane distillation). Microfiltration and ultrafiltration can be used as initial purification technology to remove suspended solids from the leachate. Gerardo *et al.* (2013) investigated a microfiltration-diafiltration process to extract metals and nutrients from digested dairy farm sludge under acid conditions in which a particle free, nutrient and metal rich leachate was obtained. Ultrafiltration (UF) can also be used to remove suspended solids from leachates. Nanofiltration (NF) is another pressure driven membrane filtration technology that can be useful in leachate purification and concentration. Nanofiltration has a cut-off between 100 and 300 Da and a relative low energy consumption when compared to reverse osmosis and evaporation technologies. In NF, monovalent ions are less retained while larger molecules and multivalent ions (including metal ions) are retained in the concentrate. Ortega *et al.* (2008) studied the application of NF for the removal of metals from leachate and found good metal ion rejection (between 62% to 100%). Divalent ions were better rejected than monovalent ions. Based on the low retention of acid and the high retention of metal ions, the possibility of acid recuperation through NF treatment is suggested (Ortega *et al.* 2008). Acid resistant NF membranes are commercially available which can withstand up to 20% sulphuric acid.

16.4.2 Metal separation

16.4.2.1 Liquid- liquid extraction

Liquid-liquid extraction (LLE) or solvent extraction is a common separation process, whereby a small quantity of an organic extractant is dissolved in a diluent to form an organic phase. During extraction the desired metal in the aqueous phase transfers into the organic phase until the equilibrium of the solute in the two immiscible phases is achieved. For each metal (e.g., copper, nickel, zink) there is a specific organic extractant. The main factors that can influence the extraction efficiency are pH and the concentration of the extractant. The second part of the LLE operation consists of a stripping, during which the desired metal is recovered from the organic phase into an aqueous phase (normal an acid phase), and the extractant can be recycled and reused. The desired metals are then removed from the organic solution by electrowinning (paragraph 16.4.3.). The LLE method has been used in mining and nuclear industries, although no study has examined the possibility of recovering metals from sludge leachate (Visser *et al.* 2001).

16.4.2.2 Electrodialysis

Electrodialysis (ED) is a faradaic process in which an electric field is applied to a stack of alternating cation (CEM) and anion exchange membranes (AEM) (Figure 16.2). Due to the electric field an ionic current is induced moving cations to the cathode and anions to the anode. Metal ions are removed from the feed stream (diluate) and concentrated in the receiving stream (concentrate). Conventional ED can be used for separation and concentration of metal salts from leachate. Additionally, various alternative stack configurations have been proposed to allow selective separation. The use of monovalent selective cation exchange membranes allows separation of mono- and multivalent ions.

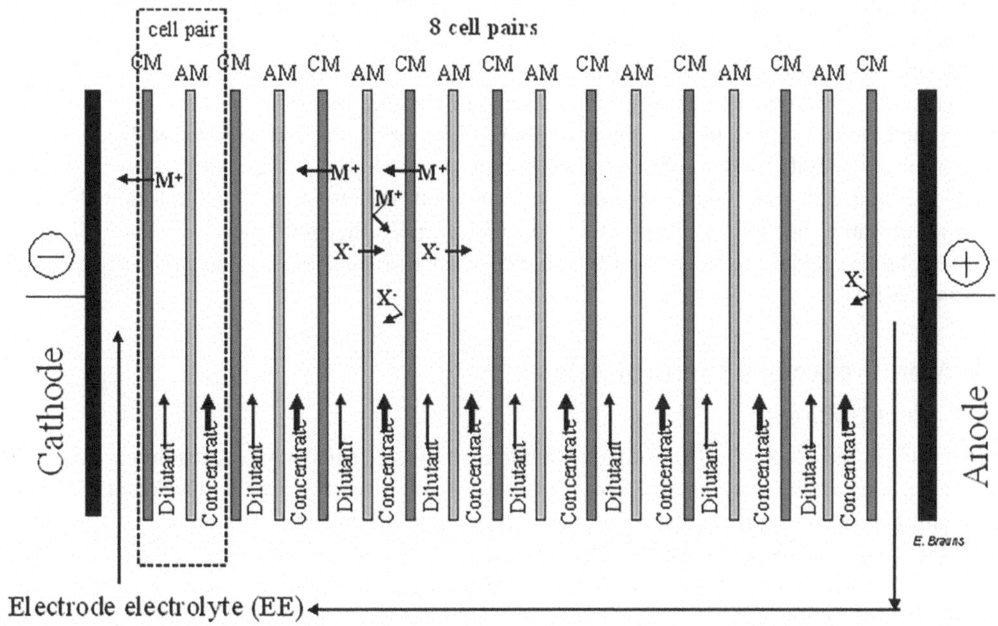

Figure 16.2 Schematic representation of a conventional ED stack, CM/AM = Cation/Anion Exchange Membrane.

Ebbers *et al.* (2015) proposed a two-compartment ED setup in which ash solution is in direct contact with the anode. Heavy metal ions are separated by migration trough a CEM towards the cathode (Ebbers *et al.* 2015). The half-reactions occurring at the anode contribute to the acidity of the stirred suspension resulting in increased dissolution of both phosphorus and heavy metals. Combinations of ED with complexation reactions have been evaluated to selectively separate metal ions from mixtures. For example, the selective separation of cobalt from nickel using a combination of conventional ED with a complexant and bipolar membrane electrodialysis for recovery of the complexing agent has been reported (Xu & Huang, 2008).

The separation of the ternary mixture Cu, Ni and Zn was also found to be feasible using malonic, oxalic or citric acid as complexant. The optimal dimensionless permeation flux ratio P is 5.3/3.6/0.7 for Zn, Ni and Cu, respectively. The combined use of a cation exchange membrane and a complexing agent is based on the difference in affinity between metal ions and ligand and is expected to be of use for various metal separations (Wang & Hsieh, 2008). Although offering improvements to the conventional electrolytic metal recovery, electrodialysis has some disadvantages such as: high operational costs due to membrane fouling and energy consumption, difficult optimal process control, membrane materials must be adequately chosen for each feed stream.

16.4.2.3 Membrane filtration

Combinations of chelation-complexation and membrane filtration have been proposed for selective metal separation. Micellar enhanced ultrafiltration (MEUF) and complexation-ultrafiltration can be applied for the retention of metals from liquid streams (Fu & Wang, 2011). In MEUF, surfactants are used that aggregate into micelles which can bind metal ions to form large metal-surfactant structures. MEUF process has been used for the removal of Cu, Cr, Zn, Ni, and Cd on laboratory scale, while no full scale installations have been reported. Still and improvement of the operational parameters of the recovery stage is required. Gerardo *et al.* (2013) estimated the cost of the process as 1.25 euro/m^3 of leachate when applied to digested anaerobic sludge from dairy farms. In complexation-ultrafiltration, a polymer is used to complex metal ions resulting in their retention by UF. The process is comparable to MEUF. Due to the different affinity of metals for complexants a selectivity can be obtained. For example, Ni(II) and Cr(VI) ions could be separated from their binary mixture using polyethyleneimine as complexant (Korus *et al.* 2014). Other chelating agents e.g. polyacrylic acid (PAA), polyethyleneimine (PEI), diethylaminoethyl cellulose and humic acid, have been used to achieve selective separation and recovery of heavy metals (Fu & Wang, 2011).

16.4.3 Metal recovery technologies

16.4.3.1 Electrowinning

Electrowinning is the electrolysis of aqueous solutions of metal salts with an insoluble anode. In the process, metal ions are reduced and deposited on the cathode while oxygen is developed at the anode. Electrowinning is usually applied following a series of purification and concentration steps to allow efficient application in terms of mass transport rate. Electrowinning/deposition can be made selective by controlling the applied potential (i.e. potential-controlled electrolysis, PC El). Variations on electrowinning include membrane electrowinning and the process can be combined with other hydro- and pyrometallurgical processes depending on feed composition and metals to be recovered. The electrochemical process for metal recovery is based on the simple mechanism of cathodic reduction (Brown, 1990; Fornari & Abbruzzese, 1999).

$$M_e^{m+}\ (\text{soluble}) + me^- \rightarrow M_e \tag{16.6}$$

Competing cathode reaction in aqueous solution is hydrogen evolution:

$$2H^+ + 2e^- \rightarrow H_2\ (g) \tag{16.7}$$

Hydrogen gas produced at the cathode creates turbulence in the system, which can enhance the mixing. The gas also serves to transport the insoluble coagulated particles to the surface of the solution. Thus, a floating layer is formed at the liquid surface consisting of both hydrogen bubbles and entrapped suspended matter. Under acidic conditions the reaction that takes place on the anode is as follows:

$$H_2O \rightarrow O_2 + 4H^+ + 4e^- \quad (16.8)$$

while in basic medium the following anodic reaction occurs:

$$4OH^- \rightarrow O_2 + H_2O + 4e^- \quad (16.9)$$

Beside parameters that govern processes in electrochemical cells, such as applied voltage, current and current density (i.e. current normalized for the electrode area), other important parameters in metal recovery by electroreduction are mass transfer, concentration, current efficiency, space-time yield, overpotential etc. Standard electrode potential and concentration of competing reactants determine which species will be deposited on the cathode. The general rule is that metals with high electrode potentials are more easily reduced (Moats & Free, 2007).

Few types of reactors have found applications in metal recovery, including quite simple tank cells, plate and frame cells, rotating cells, as well as reactor systems such as fluidized bed, packed bed cell and porous carbon packing. Tank cells are the most commonly used reactors because of their simplicity. The electrodes in tank cells can be arranged in monopolar or bipolar mode, and the number of electrodes in a stack may be in the range of 10 to 100. Capital costs associated with electrowinning include the reactor, rectifier and electrical connections, pumps and plumbing, and installation labour. Operating costs include electricity, maintenance or replacement of electrodes, and labour. The costs of anode replacement are relatively high. In spite of improvements in mass transfer achieved by increased cathode surface area (new electrode materials and novel cathode designs), stirring, or the combination of both, electrowinning may become too expensive at low wastewater concentrations because of increased ohmic resistence.

Electrowining is the final step of biohydrometallurgy in the mining sectors in order to recover metals from low grade ore. It is an established technology and can be used in case that there is a potential for metal recovery from the ore (Figure 16.1 and Table 16.1).

16.4.3.2 Bio electrochemical methods

Over the past decade, research on microbial electrochemical technologies (METs) boomed across the globe in the slipstream of which, microbial electro-metallurgy (MEM) appeared. MEM works under the overall principles of METs, but within the process metal removal and recovery are achieved. Metal recovery (Uranium) was first described in 2005 (Gregory & Lovley, 2005). Bioelectrochemical recovery of Ag, Cu, Pb, Fe, Cd, and Zn (among others) from mixed dilute solutions followed (Modin *et al.* 2012). The working principle of MEMs is the following: Microorganisms may colonize the anode (where oxidation occurs), the cathode (where reduction occurs) or both. For example, a cathode can be driven by a power supply to directly or indirectly (typically via H_2) provide reducing power to the microorganisms. The latter can use the energy gained for growth while simultaneously reducing the metallic contaminants as electron acceptor. By fine-tuning the potential at which reduction (or oxidation) occurs, it should be feasible to selectively separate metals, according to thermodynamic considerations. In the majority of MEM systems, bioanodes have been implemented. Usually mixed cultures of exoelectrogenic bacteria, inoculated from activated sludge, form biofilms on carbon anodes. These biofilms catalyse the oxidation of organic substrates and transfer of electrons to the anodes. Most studies reported have used acetate as the electron donor to drive or complement the respective cathodic reaction.

The MEM systems studied until 2012 focused on the recovery of a single type of metal in solution. However, in wastewaters and leachates, different metals often exist in mixtures. This is the case of leachate from solid waste incineration fly ash, which can contain high concentrations of Cu, Pb, Cd, and Zn. Modin *et al.* (2012) varied the control of a microbial electrochemical system to sequentially recover these metals individually from a mixture. Tao *et al.* (2014) used a MFC connected to a conventional electrolysis system to recover Cu, Pb, and Zn from a real fly ash leachate. In the MFC, 97.1% of the Cu could be removed in 36 h from an initial concentration of 52.1 $mg \cdot L^{-1}$. Recently, Zhang *et al.* (2015) reported the combination of MFC and MEC systems for recovery of Cr, Cu and Cd. In this study, MFCs using Cr^{4+} and Cu^{2+} as electron acceptors were stacked in parallel or series to drive a MEC using Cd^{2+} as the final electron acceptor. SEM and XRD confirmed the precipitation of $Cr(OH)_3$ and pure copper and cadmium on respective cathodes. Several waste streams such as fly ash leachate seem to contain a metal mix of Cu (which can be reduced in a MFC) and metals such as Pb, Cd, Zn which cannot be spontaneously reduced under typical MEC conditions. This makes the MFC a good technology to selectively extract relatively clean Cu from complex metal mixtures. If Fe^{3+} occurs in the mix, it could also be reduced in a MFC but it could potentially be precipitated as $Fe(OH)_2$ rather than deposited on the cathode.

MEMs show much promising results for metal recovery, especially from diverse dilute metal-rich aqueous streams. Succinctly, MEM can: (a) use complex solid waste or wastewaters as electron donors/acceptors, (b) reduce energy-consumption compared to traditional processing, and (c) recover and/or generate metallic commodities with lower greenhouse emissions. However, it must be emphasized that further investigation and optimization, especially for mixed metal ion systems, are required to go beyond lab and pilot scale studies.

16.5 USE OF SLUDGE AFTER CHEMICAL LEACHING OR BIOLEACHING

Sludge can be neutralised and can be used as a fertiliser. In addition, sludge can be disposed to landfills with lower risk of heavy metals leaching to surface and groundwater or uptake by plants. Sludge can be used as soil improver. Moreover, sludge can be applied with lower risk as energy source in incineration as the off-gas treatment system would be less complex than when the sludge is metal polluted.

16.6 CONCLUSIONS

The most economical metals to be recovered from sludge were (1) Al (2) Ag (3) Cu (4) Zn (5) Ni. This was found based on the highest concentration of metals in dry sludge and their highest price during the period 2005–2015. The yearly profit for Al without taking in account the operational cost and assuming full recovery was 2,170,740 $. However, when the average concentration of metals and price were taken in to account the following metals were found to be more profitable to be recovered: (1) Al (2) Ag (3) Zn (4) Cu (5) Ni. At this case, the yearly profit for Al was 314,946 $.

Based on the metal concentration in sludge and the price of metals a sludge management diagram is proposed that includes technologies of chemical leaching and bioleaching for metal extraction. The technologies of precipitation and adsorption for metal treatment from the leachate. And the technologies of membrane filtration, solvent extraction, electrowinning or bio-electrowinning for metal recovery.

Having in consideration that the (a) concentration of these metals in ore deposit will gradually decrease over time, (b) more energy will be needed for their recovery, and (c) as well as their demand could be increased due to population rise this will result in search for other alternatives for their recovery. As was shown it is possible for some metals recovery from WWTP sludge to be a viable option.

16.7 REFERENCES

Abbas S. H., Ismail I. M., Mostafa T. M. and Sulaymon A. H. (2014). Biosorption of heavy metals: a review. *Journal of Chemical Science and Technology*, **3**(4), 74–102.

Blais J. F., Djedidi Z., Cheikh R. B., Tyagi R. D. and Mercier G. (2008). Metals Precipitation from Effluents: Review. Practice Periodical of Hazardous, Toxic, and Radioactive Waste Management.

Brown C. J. (1990). Metal Recovery by Ion Exchange and Electrowinning. Proceedings of the 4th International Forum on Electrolysis in the Chemical Industry 1990, Electrosynthesis Corp., Fort Lauderdale, FL.

Carey J. B. (2015). Inner workings: finding precious metals in unlikely places. *Proceedings of the National Academy of Sciences*, **112**(29), 8803–8804.

Chan L. C., Gu X. Y. and Wong J. W. C. (2003). Comparison of bioleaching of heavy metals from sewage sludge using iron-and sulfur-oxidizing bacteria. *Advances in Environmental Research*, **7**(3), 603–607.

Chen Y. X., Hua Y. M., Zhang S. H. and Tian G. M. (2005). Transformation of heavy metal forms during sewage sludge bioleaching. *Journal of Hazardous Materials*, **123**(1), 196–202.

Dodson J. R., Hunt A. J., Parker H. L., Yang Y. and Clark J. H. (2012). Elemental sustainability: towards the total recovery of scarce metals. *Chemical Engineering and Processing: Process Intensification*, **51**, 69–78.

Ebbers B., Ottosen L. M. and Jensen P. E. (2015). Comparison of two different electrodialytic cells for separation of phosphorus and heavy metals from sewage sludge ash. *Chemosphere*, **125**, 122–129.

Erüst C., Akcil A., Gahan C. S., Tuncuk A. and Deveci H. (2013). Biohydrometallurgy of secondary metal resources: a potential alternative approach for metal recovery. *Journal of Chemical Technology and Biotechnology*, **88**(12), 2115–2132.

Environmental, economic and social impacts of the use of sewage sludge on land. Draft Summary Report 1 Assessment of Existing Knowledge http://www.svensktvatten.se/Documents/Kategorier/Avlopp%20och%20milj%C3%B6/REVAQ/Sludge%20summary%20report%201.pdf

Fornari P. and Abbruzzese C. (1999). Copper and nickel selective recovery by electrowinning from electronic and galvanic industrial solutions. *Hydrometallurgy*, **52**(3), 209–222.

Fu F. and Wang Q. (2011). Removal of heavy metal ions from wastewaters: a review. *Journal of Environmental Management*, **92**(3), 407–418.

Fytili D. and Zabaniotou A. (2008). Utilization of sewage sludge in EU application of old and new methods—a review. *Renewable and Sustainable Energy Reviews*, **12**(1), 116–140.

Gerardo M. L., Zacharof M. P. and Lovitt R. W. (2013). Strategies for the recovery of nutrients and metals from anaerobically digested dairy farm sludge using cross-flow microfiltration. *Water research*, **47**(14), 4833–4842.

Gregory K. B. and Lovley D. R. (2005). Remediation and recovery of uranium from contaminated subsurface environments with electrodes. *Environmental Science & Technology*, **39**(22), 8943–8947.

Gu X. Y. and Wong J. W. C. (2004). Characterization of an indigenous iron-oxidizing bacterium and its effectiveness in bioleaching heavy metals from anaerobically digested sewage sludge. *Environmental Technology*, **25**(8), 889–897.

Gupta S. and Bhattacharyya K. G. (2012). Adsorption of heavy metals on kaolinite and montmorillonite: a review. *Physical Chemistry Chemical Physics*, **14**, 6698–6723.

Lakherwal D. (2014). Adsorption of heavy metals: a review. *International Journal of Environmental Research and Development*, **4**, 41–48.

Lombardi A. T. and Garcia O. (2002). Biological leaching of Mn, Al, Zn, Cu and Ti in an anaerobic sewage sludge effectuated by Thiobacillus ferrooxidans and its effect on metal partitioning. *Water Research*, **36**(13), 3193–3202.

Marchioretto M. M. (2003). Heavy Metals Removal from Anaerobically Digested Sludge. Thesis Wageningen University.

Meulepas R. J., Gonzalez-Gil G., Teshager F. M., Witharana A., Saikaly P. E. and Lens P. N. (2015). Anaerobic bioleaching of metals from waste activated sludge. *Science of the Total Environment*, **514**, 60–67.

Moats M. and Free M. (2007). A bright future for copper electrowinning. *JOM*, **59**(10), 34–36.

Modin O., Wang X., Wu X., Rauch S. and Fedje K. K. (2012). Bioelectrochemical recovery of Cu, Pb, Cd, and Zn from dilute solutions. *Journal of Hazardous Materials*, **235**, 291–297.

O'Connell D. W., Birkinshaw C. and O'Dwyer T. F. (2008). Heavy metal adsorbents prepared from the modification of cellulose: a review. *Bioresource Technology*, **99**(15), 6709–6724.

Ortega L. M., Lebrun R., Blais J. F. and Hausler R. (2008). Removal of metal ions from an acidic leachate solution by nanofiltration membranes. *Desalination*, **227**(1), 204–216.

Pathak A., Dastidar M. G. and Sreekrishnan T. R. (2009). Bioleaching of heavy metals from sewage sludge: a review. *Journal of Environmental Management*, **90**(8), 2343–2353.

Salman M., Athar M. and Farooq U. (2015). Biosorption of heavy metals from aqueous solutions using indigenous and modified lignocellulosic materials. *Reviews in Environmental Science and Bio/Technology*, **14**(2), 211–228.

Tao H. C., Lei T., Shi G., Sun X. N., Wei X. Y., Zhang L. J. and Wu W. M. (2014). Removal of heavy metals from fly ash leachate using combined bioelectrochemical systems and electrolysis. *Journal of Hazardous Materials*, **264**, 1–7.

Tyagi V. K. and Lo S. L. (2013). Sludge: a waste or renewable source for energy and resources recovery? *Renewable and Sustainable Energy Reviews*, **25**, 708–728.

Visser A. E., Swatloski R. P., Griffin S. T., Hartman D. H. and Rogers R. D. (2001). Liquid/liquid extraction of metal ions in room temperature ionic liquids. *Separation Science and Technology*, **36**(5–6), 785–804.

Xu T. and Huang C. (2008). Electrodialysis-based separation technologies: a critical review. *AIChE Journal*, **54**(12), 3147–3159.

Wang J. K. and Hsieh C. S. (2008). Preferential transport behaviors of ternary system cupric–nickel–zinc ions through cation-exchange membrane with a complexing agent by dialysis. *Chemical Engineering Journal*, **138**(1), 103–110.

Westerhoff P., Lee S., Yang Y., Gordon G. W., Hristovski K., Halden R. U. and Herckes P. (2015). Characterization, recovery opportunities, and valuation of metals in municipal sludges from US wastewater treatment plants nationwide. *Environmental Science & Technology*, **49**(16), 9479–9488.

Wong J. W. C., Xiang L. and Chan L. C. (2002). pH requirement for the bioleaching of heavy metals from anaerobically digested wastewater sludge. *Water, Air, and Soil Pollution*, **138**(1–4), 25–35.

Wong J. W. C., Xiang L., Gu X. Y. and Zhou L. X. (2004). Bioleaching of heavy metals from anaerobically digested sewage sludge using FeS 2 as an energy source. *Chemosphere*, **55**(1), 101–107.

Zhang P., Zhu Y., Zhang G., Zou S., Zeng G. and Wu Z. (2009). Sewage sludge bioleaching by indigenous sulfur-oxidizing bacteria: effects of ratio of substrate dosage to solid content. *Bioresource Technology*, **100**(3), 1394–1398.

Zhu Y., Zeng G., Zhang P., Zhang C., Ren M., Zhang J. and Chen M. (2013). Feasibility of bioleaching combined with Fenton-like reaction to remove heavy metals from sewage sludge. *Bioresource technology*, **142**, 530–534.

Chapter 17

Nutrients recovery from wastewater streams

Ana Soares, Justyna Czajkowska, Jesús Colprim, Alex Gali, Sara Johansson, Alma Masic, Adrien Marchi, Andrew McLeod, Valentin Nenov, Mael Ruscalleda and Tadeusz Siwiec

17.1 INTRODUCTION

Nutrient resource recovery from wastewater and other waste streams is a challenge of the 21st century. With world population expected to increase to over 7.7 billion in 2020 there is a dramatic demand for water, energy, food and resources, especially in urban areas. These pressures have lead into a paradigm shift in wastewater treatment plants from waste treatment facilities to resource factories. Nutrients, phosphorus and ammonia, have been identified as key compounds to be recovered from wastewater and waste streams.

Phosphorus is a finite resource being used at alarming rates for fertiliser production to secure agricultural productivity and food supply. As consequence phosphorus has been named the 'disappearing nutrient' due its high demand worldwide in contrast with its finite asymmetrical distribution (Gilbert, 2019). Recent reports indicate that potential phosphate rock ore reserves might reach up to 460,000 million metric tons (IFDC, 2010) with Morocco and the Western Sahara having 60–70% of the world reserves, followed by China and USA. China is the largest exporter of phosphate ore with 45.5% of the market in 2014 but it has put in place a quote system since 2008 (Web report, 2014). This measure, to reduce exports, was first put in place by USA that adopted a national protection measure to limited phosphorus ore export and in the 1980s.

Ammonia is also a common component in fertilisers as well as in the composition of industrial products such as cleaners, pharmaceuticals and fermentation processes, just to name a few. Nitrogen is an essential nutrient for all living organism but it is strongly correlated with plant growth and development. Ammonium based fertilizers are used to enhance the agricultural productivity, increasing crop yields and high food production. According to Fertilizers Europe, in 2011 approximately 10 million tons/year of nitrogen-based fertilizers were used in the European Union (EU) (Fertilizers Europe, 2012), but these numbers will vary from region to region. On the other hand ammonia is the most common and undesirable contaminant in industrial, domestic and agricultural wastewaters. The ammonia-nitrogen concentration can vary from 5 to 1000 mg/L in industrial wastewater and from 10 to 200 mg/L in domestic wastewater (Ashrafizadeh, 2010). In the EU limits of effluent discharge concerning to nitrogen are currently <1–2 mg NH_4^+-N/L, being even stricter in sensitive water bodies (EU Council Directive, 1991).

Ammonia is mainly produced through the Haber-Bosch process from nitrogen in the air. Although the ammonia production does not have the same pressures as the phosphorus production, as it is not a finite resource (78% of the air atmosphere is N_2), the Haber-Bosch is a high-energy intensive process and hence the price of ammonia is very much linked with the energy prices (Sutton *et al.* 2013).

Nutrient recovery from wastewater and wastes is receiving much attention from the scientific, industrial communities as well regulators and public perception groups. Governments and agency all over the world are now calling for innovation to bring 'circular economy' to practice (EU, 2015). Although technological advances are occurring at a fast pace to provide opportunities for nutrient recovery, there is still the significant challenge of making these processes economically feasible as well as providing an end route and entrance of the product to a stable and welcoming supply chain or market.

The aim of this chapter is to provide an overview of the current technologies available for nutrient resource recovery from wastewater streams at various scales and to provide information on the product end-uses, bottlenecks and future perspectives.

17.2 RECOVERY OF AMMONIA BASED PRODUCTS

17.2.1 Processes

Ammonia can be found in water as un-ionized ammonia (NH_3) and ionized ammonia ($NH4^+$). Current processes used to recover nitrogen-based compounds from wastewater and other wastes often require transfer of the dissolved ionized ammonia as $NH4^+$ to ammonia gas NH_3. This can be achieved by air striping, steam stripping or membrane processes.

17.2.1.1 Air stripping

Air stripping is a process consisting on the removal of a gas dissolved in a liquid, by contacting it with the air. The removal is based on the mass-transfer process that occurs in a tank in which the liquid and the air are introduced in. The driving force for the mass transfer is the difference in ammonia partial pressure between the liquid and the stripping gas. Ammonia must be present as a gas to be stripped, which is dependent on the pH (Equation 17.1).

$$NH_4^+\ (l) + OH^-\ (l) \leftrightarrow NH_3\ (g) + H_2O\ (l) \tag{17.1}$$

As the pH of the water increases, the equilibrium shifts to the right side of the equations and the ionized ammonia converts into ammonia gas, which can be removed by gas stripping. Therefore, it is necessary to increase the pH in order to accomplish the separation. Often, lime or caustic are added to the wastewater until a pH of 10.8 to 11.5 is reached.

The equipment required for the ammonia striping process includes a tower, generally a cylindrical tank containing a high voidage random packing media, in which the liquid (e.g. wastewater) containing the ammonia to be stripped and the air used for the stripping are introduced. The packing media is used to provide a significant surface area of contact between the liquid and the air. Counter-current flow is typically used, the liquid is introduced from the top of the tank whereby the air is introduced from the bottom of the tank and released at the top, hence several flow patterns exist for the airflow (Figure 17.1).

The advantage of this process is that it is a proven technology, as it is already widely used in the industry for different applications. The process is also quite simple, as well as its design.

Figure 17.1 Ammonia gas stripping, loose fill random packing media (Left), schematic representation of a typical wet scrubber (Middle) and full-scale installation (Right). Figures reproduced with kind permission of Forbesgroup, UK (Nov, 2015).

Bottlenecks of ammonia air stripping

The drawbacks of this process are the bigger footprint, when compared with membrane processes, as generally a tank of several meters of width and height is required. Scaling and fouling are also common problems that could occur in the contact tank. The precipitation of calcium carbonate or calcium sulphate is also possible. To overcome theses problems, a periodical maintenance involving the introduction of anti-scaling chemicals, cleaning of the equipment or the replacement of the packing media should be taken into consideration.

The efficiency of the ammonia stripping varies with temperature, as this is a mass transfer regulated process. To overcome this problem, heat might be used to maintain a constant temperature.

Full-scale and cost analysis of ammonia air stripping

The capital expenditure (CAPEX) of the ammonia-air stripping process is dependant on the prices of the air stripper (the tank in which the mass transfer is occurring) which include the tank in itself and the packing media, the pump to introduce the ammonia rich stream, the blower for the air, pipes and fittings, the heater and the manometer. The operational expenditure (OPEX) is based on the electrical consumption of the pump, the blower and the heater. Potential suppliers include: Forbesgroup, UK (www.forbesgroup.co.uk), Task Environmental Engineering (www.task.be), Branch Environmental (www.branchenv.com) and Monroe Environmental (www.monroeenvironmental.com), just to name a few.

17.2.1.2 Steam stripping

Steam stripping of ammonia is similar to air stripping, with the difference that the process requires temperatures in excess of 95°C and no chemicals are needed for pH correction, which reduces operating costs. In this process, the ammonia is stripped into the steam. The ammonia-rich steam is then discharged from the stripper into a condenser to be recovered as concentrated aqueous ammonia solution (Figure 17.2). No off-gas treatment is required, since the vapours produced can be condensed into a small amount of concentrated liquid stream.

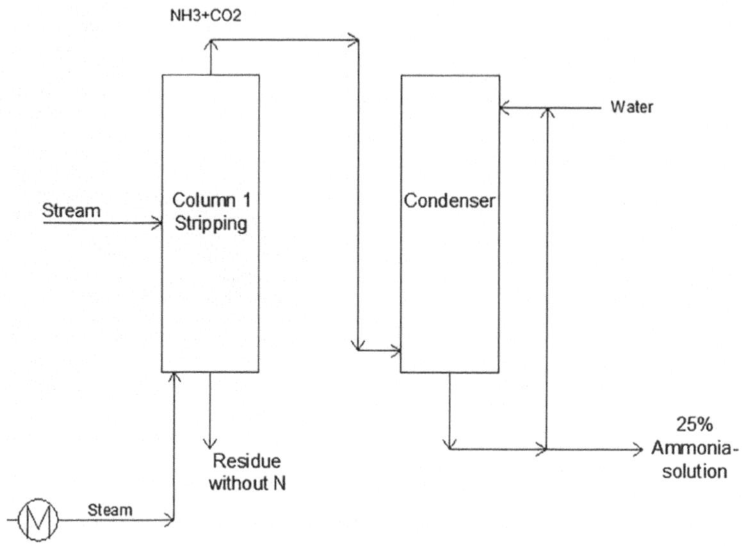

Figure 17.2 Schematic diagram of ammonia steam stripping process.

The DecaStripp©-process is a steam stripping system developed by ENVIMAC for industrial and municipal applications, mainly focused to the removal of ammonium from wastewater after fermentation processes. This process has been particularly implemented in anaerobic digestion plants and is proven in numerous full-scale applications for over 25 years ago. The key component of the DecaStripp©-process is decarbonisation step in which CO_2 is stripped out from the solution simultaneously with ammonia at ambient pressure (Maćkowiak & Górak, 2011). Recovery of the stripped ammonia can be reached by the condensation of stripping steam in a subsequent step. The DecaStripp©-process does not need chemicals and allows the production of 25%-ammonia solution, which is a valuable product on the market.

Other application of steam stripping is the Astrid® Process by Aristot Industrial Ecology Process, in which are combined different processes such as primary filtration, low pressure steam stripping and selective adsorption, to control volatile organic compounds (hydrocarbons and aromatics), minerals (NH_3, H_2S, HCN) and dust emissions. This technology can be applied to treat leachates from landfill and methanisation liquor, process effluent from dehydration, coal water in coke plant and sour water in petrochemical industries. Specifically, the mobile unit is designed to treat 3 m^3/h with the following pollutants in the effluent: SS, NH_3-NH_4, BTEX, alcohols, H_2S, THT, chloride. This configuration could achieve removal efficiency until 5 mg/L of NH_3/NH_4 in the effluent treated.

Bottlenecks of ammonia steam stripping

Stripping technology requires a high N-loaded concentration stream (>2 g NH_4-N/L) to be economically competitive. Concerning to urban WWTPs, the highest ammonium load concentration is found in the sidestream after the centrifuge step of the digestate (centrates) in the range of 500–1,500 g NH_4-N/L, far below the concentration recommended for steam stripping technologies. According to literature and providers, as a consequence of their low ammonia content, the technico-economic assessment of these technologies when applied to any of the streams generated in an urban WWTP would yield unfavourable results and they will not be cost-effective. For this reason, the current application of this technology is focused on industrial case studies, in which this ammonium values are achieved in the generated streams.

Full-scale and cost analysis of ammonia steam stripping

Some configurations of steam stripping are being marketed for ammonia removal/recovery from wastewater and other wastes, such us VALEAz process (www.alcion-env.com), ASTRID® process (www.aristot.fr), DecaStripp© process (www.envimac.de), steam stripping by RVT process equipment GMBH (www.rvt-pe.com), TSKE Tsukishima Kankyo Engineering Ltd. (www.tske.co.jp), among others. Regarding the associated costs of implementing steam technologies, no data are available or provided by suppliers, so no accurate cost analysis can be indicated at this point. However, it is expected for steam stripping be more cost effective than air stripping because of its higher market value.

17.2.1.3 Membrane processes

Hollow fibre membrane contactors (HFMC) can be used to remove and recover pollutants such as ammonia from the wastewater and has been shown to be an adequate and desirable solution for nitrogen based compounds recovery (Zhu *et al.* 2005; Tan, 2006; Ashrafizadeh, 2010; Hasanoglu, 2010; Canellas *et al.* 2014). The application of transmembrane chemisorption (liquid-gas-liquid) uses hydrophobic membranes for separation and recovery of ammonia using an acid solution as stripping phase without dispersion of one phase within another (Zheng *et al.* 2005). The liquid (e.g. wastewater) containing ammonia is pumped to the external side fluid (shell side) on the internal side of the hollow fibre (Lumen side). A stripping solution containing an acidic solution, such as sulphuric acid, flows in counter current direction. The microporous and hydrophobic nature of the membranes maximize the overall mass transfer rate creating a large interfacial area and high diffusion within the pores.

Depending on the liquid pH, the ionized ammonia (NH_4) reacts with hydroxide ion (OH^-) under elevated pHs to form free ammonia (NH_3). This chemical reaction is reversible, higher pH (pH > 10) shifts the equilibrium to the right side favouring the formation of free ammonia the volatile ammonia (Equation 17.1). passes through the gas-filled membrane micro pores and reacts with the acid stripping solution (Zhu *et al.* 2005). Due to the high solubility of ammonia in acid, the absorption of ammonia is considered to be instantaneous (Hasanoglu, 2010) and reacts to the acid to form a salt. If sulphuric acid is used, the membrane stripping process has been shown by Ulbricht *et al.* (2013) to generate ammonium sulphate solution (30% w/w) around 8% w/w of N that can be sold and be used after further treatment as a stable fertiliser in agriculture. The advantage of this process is that it is a proven technology, is easy to operate, has a low energy demand, it is a very compact technology and it has no air pollution. The drawbacks include, periodic membrane replacement, a requirement for skilled workers, requires a control system and the membrane is subject to fouling.

Bottlenecks of ammonia stripping by membrane processes

The drawbacks include, periodic membrane replacement, requires skilled workers, requires a control system and the membrane is subject to fouling.

Full-scale and cost analysis of ammonia stripping by membrane processes

Full-scale ammonia stripping using membrane processes is currently being marketed by Liqui-cel (www.liquicel.com) for a wide range of applications (Figure 17.3). The membrane contactors can be found a range of sizes, small (2.5 × 28 and 14 × 13) medium (4 × 28 and 6 × 28) and large (10 × 28 and 13 × 40). Depending on the flow, pH, temperature and ammonia concentration of the wastewater, the membrane process for ammonia stripping needs to be designed accordingly, to provide the most efficient CAPEX and OPEX.

Figure 17.3 Picture of a Liqui-Cel® Membrana module.

17.2.2 Products

17.2.2.1 Ammonium sulphate

Once a rich ammonia gas has been obtained through air striping, steam stripping or membrane processes, the ammonium sulphate production can take place. To start this process, ammonia rich gas is often contacted with a concentrated acid solution in a adsorption column. This process has been applied to recover ammonia-based fertilisers from urine, urine, wastewater, digestates, dewatering liquors as well as ion exchange regenerants. It has been reported by Morales *et al.* (2013) that in Europe ten pre-treatment plants are producing fertilizer as ammonia sulphate, using a combination of the ammonia air stripping process followed by reaction with sulphuric acid. An example is a full-scale plant in Switzerland where a new ammonia stripping method combined with CO2 pre-stripper is used to produce ammonium sulphate from digestate and urine (Morales *et al.* 2013).

Sulphuric acid has been the preferred solvent for acid stripping over nitric, phosphoric and carbonic acid partly because it is cheaper and readily available. However it is also non-volatile and non-oxidising compared to hydrochloric and nitric acids respectively. It also possesses very low pKa of –3 and 9, which maximizes availability of H+ for reaction. The low N concentration in ammonium sulphate makes it less attractive as fertiliser compared to ammonium nitrate, however the presence of sulphur increases the agronomic interest of this product. Nevertheless using sulphuric acid instead of nitric acid will be more suitable as nitric acid is more hazardous and explosive mixtures are produced when ammonium nitrate is mixed with organic matter (Evans *et al.* 2009). Diammonium phosphate (DAP) $(NH_4)_2HPO_4$ is formed when phosphoric acid is used containing 16 to 18% nitrogen and 20 to 21% phosphorus (46% P_2O_5) an excellent source of N and P. Other absorber acid that can be viable to use is carbonic acid producing ammonium bicarbonate (NH_4HCO_3) and a more stable product such as ammonium carbonate $(NH_4)_2CO_3$ (Norddahl, 2006). Ammonium bicarbonate can be used as a substrate for nitrogenous fertiliser and buffering resources in the anaerobic digestion process (Redzwan *et al.* 2010).

17.2.2.2 Ammonia water

An alternative is to recover ammonia as 25% ammonia solution in water by means of steam stripping technology. Ammonia solution, ammonium hydroxide, ammonia water, ammoniacal liquor, ammonia liquor, aqueous ammonia, or simply ammonia is a solution obtained from steam stripping processes, normally at 20–25%. Compared to ammonia sulphate (obtained from air stripping), 25%-ammonia solution has three times larger nitrogen amount per kilogram and is also utilized in numerous applications, mainly in industry and agriculture fields.

Ammonia solution is used as a reagent in SCR (Selective Catalytic Reduction) and SNCR (Selective Non-Catalytic Reduction) Denox systems (US EPA 2002). In food production it is used as acidity regulator and it is classified by the Food and Drug Administration as generally recognized as safe. Its pH control abilities make it an effective antimicrobial agent. It is used in commercial cleaning products and household cleaning applications whereby it is used for cleaning many types of surfaces. In agriculture, ammonia solution is not as popular as anhydrous ammonia as fertilizer; however, interest in ammonia solution is increasing because it is safer to use than anhydrous ammonia. Most ammonia solution is used either for direct application to the soil or the manufacture of ammoniated super phosphates.

17.2.2.3 Ammonium nitrate

Ammonium nitrate (NH_4NO_3) is a commonly used fertiliser due to its high nitrogen content. Ammonium nitrate can be produced from urine. Although its nitrogen content is lower than in urea, it is a more stable solution as it does not lose nitrogen through volatilisation. The salt can be mined in the very dry desert Atacama in Chile, but the standard way of obtaining ammonium nitrate since the early 1900s is by chemical production (Seo, 2011). The invention of the Haber-Bosch process, by which ammonia is produced by reacting nitrogen with hydrogen, enabled synthetic production of ammonium nitrate. In this process, ammonia reacts with nitric acid to form ammonium nitrate (Ahlgren *et al.* 2008). It is a rather expensive process, because the high energy demand has to be covered with fossil fuel. Moreover, its use of fossil fuels is seen as a contribution to the emission of greenhouse gases. Current research is focusing on improving the process with respect to energy demand and use of renewable sources or even on replacing the process entirely (Razon, 2015). In line with the idea that waste should be treated as a resource, a new approach has been taken with respect to recovering nitrogen from source-separated urine by production of ammonium nitrate with the aim to use it as a fertiliser (Udert & Wächter, 2012).

Production process and existing experience

Since urine is the most significant contributor of nutrients in wastewater, its separation at the source has been suggested and developed as a way to reduce the load on wastewater treatment plants as well as provide the opportunity to recover the nutrients (Larsen *et al.* 2009). Current lab- and pilot-scale experiments within the VUNA project (www.vuna.ch) have shown that a two-step process consisting of nitrification and distillation (Figure 17.4) is suitable for production of ammonium nitrate from source-separated urine (Fumasoli *et al.* 2015). Urine is collected with urine-diverting toilets and treated with nitrification in order to ensure stable long-term storage. During the nitrification process, half of the ammonia is biologically converted to nitrate by ammonia oxidizing and nitrite oxidizing bacteria before the process halts due to limited alkalinity. Thereafter, the diluted ammonium nitrate solution is processed in the distiller, from which a final concentrate is obtained. The concentrated solution also contains all other nutrients found in urine, e.g. phosphorus, potassium or sulphate. A pilot-scale system was operated for more than four years in an office building in Switzerland (Fumasoli *et al.* 2015) and a second pilot-scale system was operated in a decentralised manner with urine from 700 households in Durban, South Africa (Etter *et al.* 2015). A critical phase occurs during the start-up of the nitrification reactor and particular attention has to be given to the pH in the reactor in order to avoid accumulation of acid-tolerant bacteria, which could otherwise lead to a complete failure of the reactor (Fumasoli *et al.* 2015). Furthermore, the operation of the reactor must ensure avoidance of nitrite accumulation, as the acid of nitrite, nitrous acid, is a strong inhibitor of nitrite oxidizing bacteria, and of sudden surges of influent load. Although the biological nitrification process requires a relatively complex operation, it ensures a hygienic end-product with only distilled water and a small amount of activated sludge as by-products. Further drying of the liquid concentrate into a solid fertiliser is possible, but is not recommended due to a more complicated process and a thermally unstable end-product (Etter *et al.* 2015).

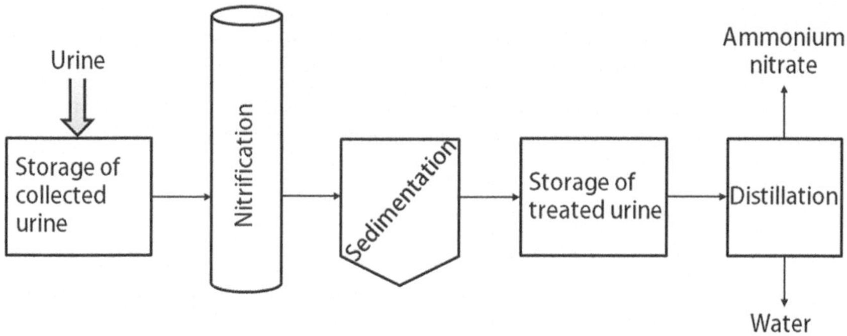

Figure 17.4 Sketch of the production process of the concentrated nutrient solution containing ammonium nitrate from source-separated urine.

Cost analysis and benefit

In the ammonium nitrate concentrate, almost 100% of the macronutrients nitrogen, phosphorus, potassium and sulphur in stored urine are recovered, along with microelements (such as zinc and boron) and a very low content of heavy metals. During production, 30 L of concentrate are obtained from 1000 L of urine (Etter *et al.* 2015). Studies on an unoptimised pilot-plant in South Africa showed that treatment of 1000 L of urine cost about 130 EUR, which could potentially be significantly decreased in a full-scale application. The current nitrification process consumes between 10 and 120 Whelectricity/L urine depending on the ammonia concentration in the influent. The electricity demand for distillation is approximately 110 Wh/L urine (Fumasoli *et al.* 2015). It has an approximately seven times higher energy demand than conventional wastewater treatment and fertiliser production; however, this is expected to be reduced by further optimisation of the process, in particular of the nitrification reactor.

Bottlenecks

Currently, no online nitrite sensors exist for high-strength wastewaters, preventing an efficient monitoring of the nitrification reactor both during the critical start-up phase as well as during normal operation. One promising approach is the use of an in-situ ultraviolet spectrophotometer with chemometric models to estimate nitrite in nitrified urine (Mašić *et al.* 2015). Another is a model-based observer, for which an activated sludge model has been set up, to be used with pH and oxygen measurements in order to estimate nitrite (Mašić *et al.* 2014). Further bottlenecks include scalability and implementation/construction of the production process (Etter *et al.* 2015).

Product end-uses

Initial studies on plant (ryegrass) uptake of phosphorus and nitrogen from an ammonium nitrate fertiliser, produced from synthetic nitrified urine, show that the nutrients were just as available to the plants as from a standard mineral fertiliser (Bonvin *et al.* 2015). The product already satisfies EU regulations regarding content of heavy metals due to the low content found in urine, however, further research is needed to determine the fate of micropollutants such as pharmaceuticals and pathogens. In Switzerland, a temporary approval has been received from the Federal Office of Agriculture in 2015 for use of the product as fertiliser for flowers, grass, and trial agricultural purposes – an approval that may become definite and may be extended to food production upon demonstration of safety with respect to micropollutants (VUNA, 2015). Studies suggest that the biggest obstacles for overall acceptance among the public lie in the handling of the urine collection system, with issues such as leakages and odours, and that the added value of fertiliser production should be used as a motivation to increase awareness and acceptance of dry sanitation (Roma *et al.* 2013).

Future perspectives

A number of challenges remain before ammonium nitrate obtained from urine can become a viable and competitive fertiliser, most importantly the fate of micropollutants, the social acceptance, and the cost of production. Nevertheless, the process is very promising with a lot of potential for improvement and with a convincing sustainability perspective.

17.3 RECOVERY OF PHOSPHORUS BASED PRODUCTS

Recovery of phosphorus products from wastewater and other wastes, including ashes and other solid materials mainly occurs through chemical processes, although biological processes are now being investigated at laboratory scale. Products that can be recovered rich in phosphorus include struvite, potassium phosphates, phosphoric acid and calcium phosphate.

17.3.1 Struvite

Magnesium ammonium phosphate (struvite, $MgNH_4PO_4 \cdot 6H_2O$) is a mineral that crystallizes as a white orthorhombic crystalline structure, which is composed of magnesium, ammonium and phosphate in equal molar concentrations. The colour can vary from completely colourless through yellow and grey to brown. The struvite forms according to equation 3.3.2 (Hanhoun *et al.* 2011).

$$Mg^{2+} + NH_4^{+} + PO_4^{3-} + 6H_2O \leftrightarrow MgNH_4PO_4 \cdot 6H_2O \tag{17.2}$$

Struvite occurs in young sediments with a high proportion of organic substances in deposits of guano, fossil bones, fertilizer mines, in numerous organic debris as well as kidney stones in humans. It can also be formed in WWTPs, especially after anaerobic reactors in EBPR processes. In the anaerobic stage of the EBPR process, phosphorus is released into the bulking liquid, causing a rapid increase in phosphorus concentration. Many anions and cations, such as magnesium, ammonium and phosphate can be present simultaneously in the liquid phase (Wu *et al.* 2010). Struvite deposits foul pumps and case pipe blockages, leading to a significant increase in the cost of sludge management operations (Uysal *et al.* 2010). Studies completed in flow-through tests, demonstrated that the weight of the deposited slurry on the inner walls of the steel pipes was the largest at values of pH 9.5 and also dependence with ammonia and phosphate concentrations (Figures 17.5 and 17.6) (Czajkowska, 2015).

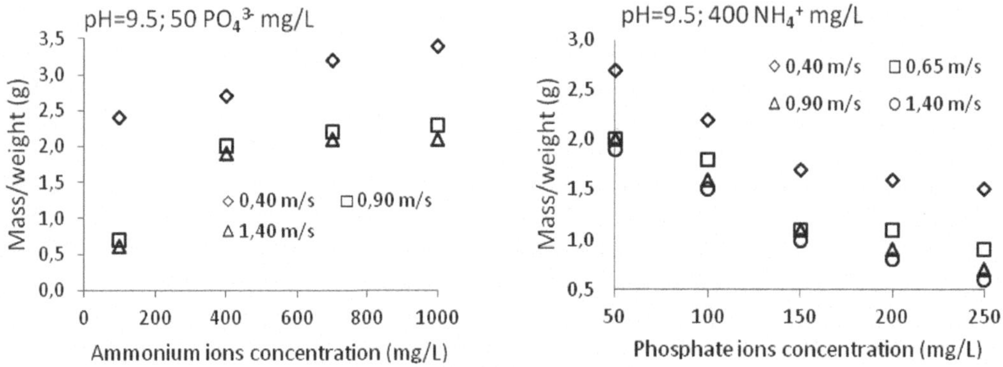

Figure 17.5 Influence of liquid velocity and ammonia concentration (Left) and liquid velocity and phosphate concentration (Right) on the on struvite mass production in pipes (Czajkowska, 2015).

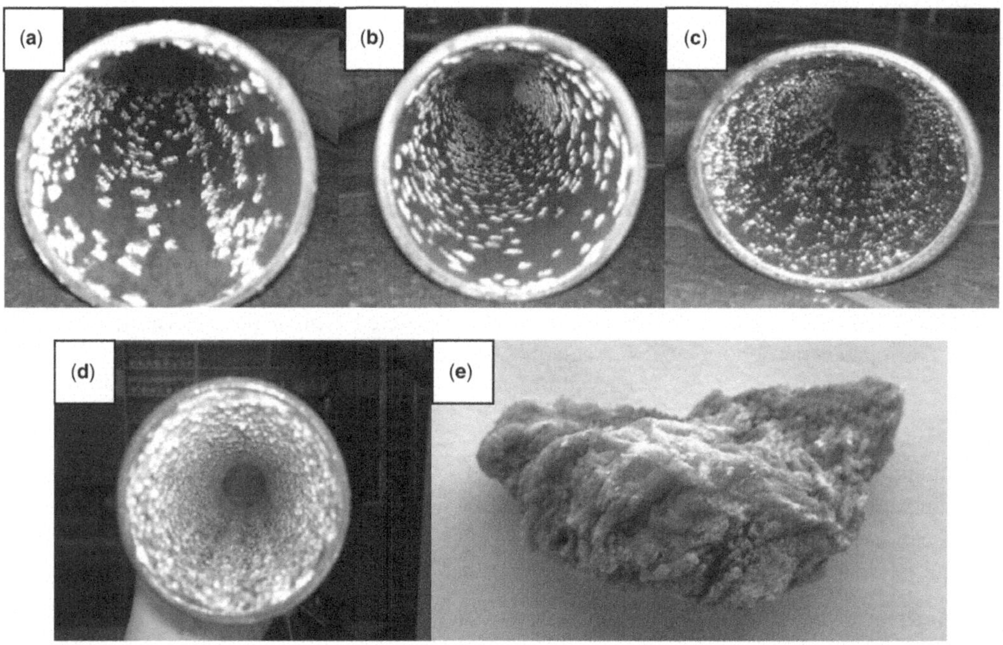

Figure 17.6 Struvite deposited on the inner walls of the steel pipes as a sludge under various flows (V, velocity wastewater in pipe) with a: V - 0.4 m/s after 70 h; b: V - 0.9 m/s after 70 h; c: V - 1.4 m/s after 70 h; d: V - 0.4 m/s after 210 h; e: struvite removed from pipe in wastewater treatment plant (Czajkowska, 2015).

17.3.1.1 Production process and existing experience

The formation of struvite is conducted by two main mechanisms: nucleation and crystal growth. During the first step, the seed or nuclei of struvite appears in the supersaturated solution. The crystal growth phase consists in the growth of the nuclei, which can be controlled to obtain the desired crystal size (Mukhlesure, 2013). After this step, the product is removed from the reactor to follow a process of filtration and drying stage.

There are several reactor types and operational strategies available for struvite nucleation and crystal growth, but the three main types include: ion exchange process followed by precipitation, stirred or air agitated reactors (where the struvite crystal grows takes place as the result of the interaction between struvite nuclei) and fluidised bed reactors (where the struvite is generated by crystallisation on seed materials) (Le Corre *et al.* 2007 & 2009) (Figure 17.7). In any case, struvite formation is initiated by adjustment of pH and chemical addition to reach the necessary ratio of Mg:N:P, which is usually 1:1:1.

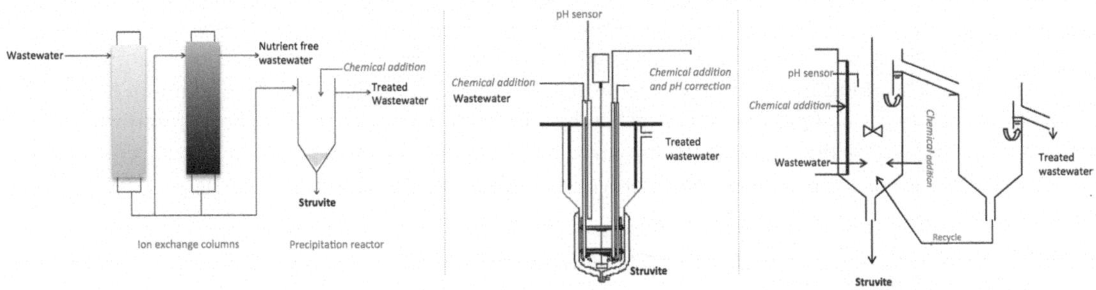

Figure 17.7 Schematic representation of struvite production processes: ion exchange process follow by precipitation (Left); stirred tank reactors (Centre and Right). Adapted from Le Corre *et al.* (2009).

pH correction and magnesium dosing

Struvite is formed in the pH range 7–11, wherein the minimum solubility occurs at pH 9–10 (Czajkowska, 2015). Although sludge dewatering liquors and digestates have pH values usually above 7.5, further pH adjustment is necessary to achieve struvite formation. Dosing of alkaline industrial-based chemicals such as calcium, sodium and magnesium hydroxides usually takes place. Municipal wastewater and several other wastewaters tend to be rich in ammonium and phosphorus, but deficient in magnesium. Hence the dosing of magnesium hydroxide, can both act as source of magnesium as well as lead to a increase in pH (Lee *et al.* 2003; Chimenos *et al.* 2003). Other strategies include the use of by-products of desalination processes (nanofiltration of sea water) that are high in magnesium concentration and high Mg/Ca-ratio, (Anne *et al.* 2001; Niewersch, 2011). Seawater (Matsumiya *et al.* 2000) or salt concentrate (Etter *et al.* 2011) present low cost alternatives and are also possible candidates.

The pH can also be elevated by air stripping, as aeration of wastewater removes CO2 and increases the pH (Battistoni *et al.* 1997). This method provides a slow pH increase and allows the excessive increase in solution saturations to be prevented. At such conditions the formation potential of other magnesium precipitates is reduced. Therefore, the utilization of CO2 stripping for pH adjustment is not only saving alkaline chemicals as NaOH, but also allows the decrease in the addition of magnesium (Young-Hui *et al.* 2011).

Struvite formation from other wastes

To increase the yield of struvite production, attention has been given to other wastes, by-product streams and natural sources that are rich in N, P and Mg, i.e. with high ionic concentration of struvite components. These include: urine, landfill leachates, semiconductor industry wastewater. In these streams, ammonia and phosphate concentration can reach values of 1000 mg N/L and 150 mgP/L, respectively, for swine wastes (Ryu, 2010) and 2500 mg N/L and 200 mg P/L, respectively, in human urine (Hug & Udert, 2013). Landfill leachates are very good sources of magnesium with concentrations 400 mg/L (Huang *et al.* 2014).

Struvite precipitation has been comprehensively studied at laboratory scale with focus on the agronomic benefits of phosphorus recycling from urine (Antonini, 2012) and swine wastewater (Ryu & Lee, 2010).

Influence of other supplements on struvite precipitation

In studies performed by Hutnik *et al.* (2008, 2009a, b) in the laboratory crystallizer (draft tube, mixed suspension mixed product removal DT MSMPR), allowed to determine the impact of certain ions on the quality of the precipitated crystals of struvite. These indicate that the presence of aluminium ions have a beneficial effect on both the crystal size and homogeneity of particle size distribution of the population. Precipitation of phosphate using magnesium and ammonium ions in the presence of aluminium ions from the solution resulting in separation of more struvite crystals, but particles of aluminium hydroxide. Similarly, in the presence of potassium ions, whereby a predominance of longer and wider uniform crystal struvite, while improving the uniformity of the structure. However, in the presence of zinc ions the average size of struvite crystals were smaller and heterogeneous (Hutnik *et al.* 2009a, b, 2008).

17.3.1.2 Struvite production in full-scale installations

Phosphorus recovery in WWTPs usually takes place in the sludge liquors line, were the concentrations of phosphate can reach >60 mg/L and the concentrations of solids are relatively low, allowing easy separation of the crystals. Table 17.1 shows the key commercial processes for struvite recovery currently applied at full-scale.

Table 17.1 Summary of commercial technologies used for struvite recovery.

Name of Company	Type of Rector/ Technology Used	Effluent Processed	Name of Commercial Product
PCS	Airlift reactor and sedimentation	Digested sludge	AirPrex
Waterschap Velt en Vecht	Aerated basin	Digested sludge	
(DHV)	Fluidised bed reactor	Sludge liquors/reject water	Crystalactor
Paques	Stirred reactor with solids separation	Sludge liquors/reject water	Phospaq
Ostara	Fluidised bed reactor	Sludge liquors/reject water	Pearl
NuReSys	Stirred reactor	Sludge liquors/reject water	
Seaborne	Stirred reactor followed by centrifuge	Sludge liquors/reject water	

The recovered struvite produced by these technologies varies in quality and size. Whilst some products are separated from the liquors and present homogenous sizes resembling chemical ferilizers (Figure 17.8) others are not separated from the solids in the effluent treated (e.g. Waterschap Velt en Vecht process).

It is difficult to estimate the number of full-scale struvite recovery plants as new ones are constantly being installed. Current figures indicate there are more than 15 struvite plants in Europe. Prices for the struvite vary from 40–60 euro/ton (2009) (Bergmans, 2011) and Ostara indicates that the sales of the recovered product are sufficient for capital cost payback periods of 3–10 years and operational costs (Bergmans, 2011).

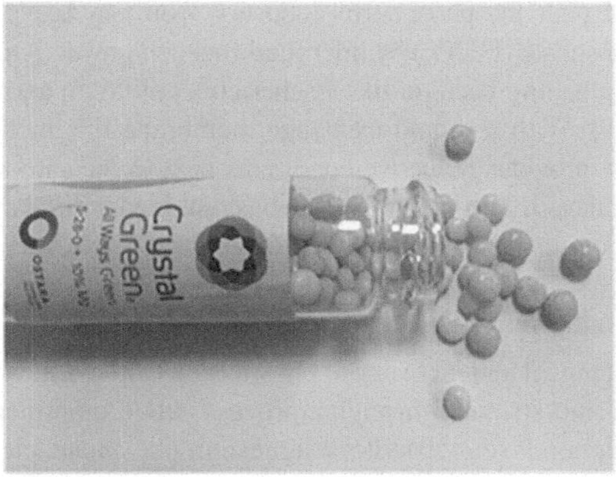

Figure 17.8 Picture of the Crystal Green struvite produced by the Pearl process developed by Ostara.

17.3.1.3 Novel processes for struvite production based on biological processes

Bio-electrochemical recovery of phosphorus

Struvite can also be crystallised at the cathode of a water electrolysis cell where the consumption of protons (via hydrogen evolution) results in a localized pH increase (Moussa *et al.* 2006). The main disadvantage of this process is the high-energy cost for producing the potential needed to split water at the anode (1.25 V in theory, or more than 1.8 V in practice). In order to make electrochemical struvite precipitation more energy efficient, production of struvite in a microbial electrolysis cell (MEC) has been investigated. In an MFC, microorganisms convert organic and inorganic matter into an electrical current at a significantly lower potential (minimum of 0.2 V when bacteria used) than that needed for splitting water (Call & Logan, 2008). In addition, no precious metals are needed for the MEC compared to a water electrolyzer using Pt catalysts on the anode. Recovery of the hydrogen gas can additionally offset energy costs. Electrical energy efficiencies in MECs have reached 400% (the ratio of the energy in the hydrogen gas produced, to the electrical energy need) (Cusick & Logan, 2012). The above-cited studies show that the increased pH at the cathode of a MEC facilitates the process of struvite precipitation (Figure 17.9).

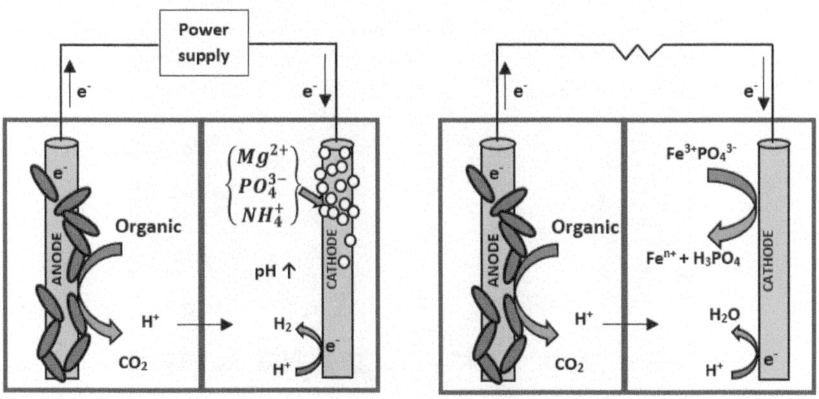

Figure 17.9 Microbial electrolysis cell for struvite precipitation (Left) and mobilization of orthophosphates from iron phosphate ($FePO_4$) by microbial fuel cell power (Right).

Another newly discovered path for phosphorus recovery from sewage sludge is the mobilization of orthophosphates from iron phosphate ($FePO_4$) by microbial fuel cell power. An electrical current is achieved in the microbial fuel cell by attaching bacteria like Escherichia coli to an anode, which the bacteria use as electron acceptor (Figure 17.9). With a proton exchange membrane the anode cell is separated from the cathode cell. The electric flow provides reductive conditions next to the cathode. Thus dissolution of up to 82% of the $FePO_4$ has been achieved. This technique can be combined with struvite crystallization by adding magnesium chloride and ammonium hydroxide in stoichiometric ratio (Fischer *et al.* 2011; Nenov *et al.* 2011).

Struvite production through biomineralisation

A new route that enables the removal and recovery of phosphorus from wastewater is the use of bacteria that can produce struvite crystals (bio-struvite) through a process called biomineralisation. Biological crystal formation of phosphorus compounds (e.g. struvite; magnesium phosphate, etc.), has been demonstrated to be a by-product of the metabolism of specific bacteria that can be found frequently in the environment. Biomineralization refers to a series of processes involving selective extraction and uptake of elements from the local micro-environment created by organisms, and the incorporation of these elements into minerals deposited under strict regulation of biological system, which distinguishes the processes from abiotic mineralization (Mann, 2001). Basic biomineralization processes include biologically controlled mineralization (BCM) and biologically induced mineralization (BIM). Beside genetic and cellular control of biomineralization, regulation of chemistry, space, structure, morphology and construction also drive the biomineralization to different extents (Mann, 2001). Recently, the possibility of mineralizing struvite from primary effluent and sludge dewatering liquors using bio-struvite producing bacteria was tested in settled wastewater (7.5 mg/L PO_4) and sludge dewatering centrifuge liquors (30.5 mg/L PO_4) showed effective phosphorus recovery of minimum 72% and 93%, respectively, together with producing final effluent containing as low as about 2 mg/L PO_4-P without chemical addition, resulting in bio-struvite yields of 200 mg/L (Soares *et al.* 2014). The studied bacteria were able to grow and producing struvite reaching 250 μm in size (Figure 17.10). Current work is progressing, to optimise the growth and production of bio-struvite using selected bacteria at lab-scale, demonstrate the feasibility of technology at pilot-scale and provide an assessment of the economic viability of the process.

Figure 17.10 Scanning electron microscope image displaying bio-struvite crystals produced by *Brevibacterium antiquum* after 10 days of incubation in sludge dewatering centrifuge liquors.

17.3.1.4 Product end-uses

Struvite recovery in WWTPs has two major benefits: to produce a product that can replace mined fertilisers (e.g. phosphate rock) and to avoid scaling issues. From the sustainability approach, struvite crystallisation is a low energy process. Many studies have suggested that struvite is an excellent fertiliser, due to its low solubility and nitrogen and phosphorus components without the danger of damaging plants (Hanhoun *et al.* 2011). In addition, the content of heavy metals is low compared to phosphate rock. According to Rittman (2011), struvite quality is comparable to standard fertilisers such as superphosphate and diammonium phosphate. Munch (2001) estimated that the sales price of struvite based on the nitrogen and phosphorus content would be 150 GBP per ton is it is going to be used in broad-scale agriculture. However, Munch (2001) said that if struvite is sold as a boutique fertiliser, a conservative estimation for the selling price would be 350 GBP per ton.

17.3.2 Potassium phosphate

Recovery and reuse of potassium has received much less attention than struvite although the EU is importing 70% of the potassium fertilizer it consumes (Fertilizers Europe, 2015). The geographical distribution of potash mines is limited, even more than the case of phosphorus mines, and the market is dominated by five countries (Canada, Russia, Belarus, China and Germany) that make up for 80% of world production (Manning, 2015). This has subjected the market to cartelization which has favoured producers rather than buyers (Ciceri *et al.* 2015; Davies & Thiemann, 2015). Potassium is not included in regulations such as the Urban Wastewater Treatment Directive (EU Council Directive, 1991) and is thus not subjected to the same legal drivers for recovery and reuse as phosphorus and nitrogen. It is rather the possibility to locally recover potassium in the form of struvite and thus cut the dependency to imported mineral fertilizers that has been the main driver.

17.3.2.1 Production process and existing experience

Phosphorus recovery by precipitation of struvite has been studied using various feedstock's and been implemented at full-scale. However, other ions can substitute for ammonium in the struvite crystal lattice and form analogues such as magnesium potassium phosphate (MPP, $MgKPO_4 \cdot 6H_2O$, K-struvite) and magnesium sodium phosphate (MSP, $MgNaPO_4 \cdot 6H_2O$, sodium struvite). Similar to ammonium struvite, potassium struvite has showed good fertilizing properties such as slow release of nutrients due to its low solubility in water (Salutsky & Steiger, 1964).

As in any crystallization process, precipitation of potassium struvite occurs when the combined concentrations of its component ions, Mg, K, and P, reaches saturation. Various attempts have been made to assess the solubility product of potassium-struvite and although results vary, they show that potassium struvite is slightly more soluble in water than ammonium struvite (Table 17.2). This means that in the presence of ammonium, ammonium struvite is more likely to precipitate than potassium struvite. This has also been experienced in batch tests on synthetic urine where removal efficiencies of K decreased when the ammonium concentration was increased while the P recovery remained constant (Xu *et al.* 2011). Due to the higher tendency of ammonium struvite to precipitate, attempts to precipitate potassium struvite are commonly targeted on feedstocks after N-removal, such as hydrolysed urine subjected to ammonia stripping or downstream of biological N-removal.

17.3.2.2 Lab-scale experience on synthetic and real urine

Urine contributes to a large amount of the nutrient load in municipal wastewaters, but adds a very small volume. Attention has therefore been focused on separately treat urine before it is diluted in the sewers.

Nitrogen in urine mainly exists in the form of urea, which can be hydrolysed to ammonium and bicarbonate. The ammonium can then either be removed by stripping or biological methods.

Table 17.2 Summary of commercial technologies used for struvite recovery.

Struvite Compound	pK_{sp}	References
Magnesium ammonium phosphate (MAP)	13.26	Ronteltap *et al.* (2010)
Magnesium potassium phosphate (MPP)	12.2	Xu *et al.* (2015)
	11.7	Luff and Reed, (1980)
	10.6	Taylor *et al.* (1963)
Magnesium sodium phosphate (MSP)	11.6	Xu *et al.* (2015)

Only a few experiments to precipitate potassium struvite have been carried out on hydrolysed real and synthetic urine. Reactor configurations include draft tube and baffle reactor and CSTR. However, results have been promising with high P-removal efficiencies either as potassium or ammonium struvite. Struvite precipitation has been suggested as a polishing step for residual ammonium downstream of biological N-removal and in the case of complete ammonium removal remaining phosphate can be precipitated in the form of potassium struvite (Wilsenach *et al.* 2007). To avoid the precipitation of sodium struvite it has been suggested that other sources for pH adjustment than NaOH are used (Dai *et al.* 2015). Co-precipitation of ammonium and potassium struvite however presents a possibility to recover a multi-nutrient product containing all three macronutrients N, P and K.

17.3.2.3 Full-scale implementation on calf manure

Whereas K-struvite precipitation still mainly has been studied in the lab-scale, one full-scale implementation has been operating in Putten, the Netherlands since 1998. The plant is treating calf manure and consists of a biological nitrification-denitrification step followed by CSTRs in series into which MgO in suspension is dosed (Schuiling & Andrade, 1999). The recovered struvite had a composition of 13, 11 and 11 weight % of P, Mg and K respectively, but also an organic carbon content of 5.5% (weight). It should be noted that the calf manure influent contains many fines, such as animal hair and organic matter that can improve crystal growth and the recovery of larger crystals.

17.3.2.4 Product end-uses

Even though recovery of potassium does not have the same legal drivers as phosphorous, the limited geographical distribution of potash mines and the effects of mining on the environment has prompted suggestions to look into locally available resources of potassium (Ciceri *et al.* 2015). Due to the limited data on potassium flows through the food system, further research is needed to estimate the potential for potassium struvite recovery from wastewaters.

17.3.2.5 Future perspectives

It is possible to recover potassium from several waste streams in the form of potassium struvite and thus reduce the demand for mined sources, but some challenges remain to be overcome such as hesitation from end-users to switch from chemical to bio-based fertilizers (ESPP European Sustainable Phosphorus Platform, 2015) and promotion of a market for bio-based fertilizers. It is vital to meet the requirements from end-users in terms of fertilizer quality and composition. Potassium struvite could be promoted in fertilizer blends for crops with high potassium need such as potatoes. Finally, legislation and policy, such as the EU Circular Economy Policy and the EU Fertilizer Regulation, will play an important part for the progress of bio-based fertilizers.

17.3.3 Calcium phosphate and hydroxyapatite

Another way of removing and recovering phosphate from ion exchange regenerants or phosphorus-rich brines is through precipitation by using calcium to create crystals of calcium carbonate or hydroxyapatite ($Ca_5(PO_4)_3OH$), according to Equation 17.3.

$$3PO_4^{3-} + 5Ca^{2+} + OH^- \rightarrow Ca_5(PO_4)_3OH \tag{17.3}$$

The technology for forming hydroxyapatite is very similar to the ones forming struvite, but the main differences come from the chemicals used and the nutrients removed. Whereas struvite formation leads to phosphate and ammonia recovery, hydroxyapatite precipitation is selective for phosphate. The process is very simple: it consists in adding calcium to the ion exchange regenerants or brines. Then, after crystallisation, the product (hydroxyapatite) is removed, and the regenerant is recovered (low concentration of phosphate). The advantages of hydroxyapatite formation are: low energy process, very good removal of P from the regenerant (up to 90%); very good characteristics of the product and small footprint. The main disadvantages are: need for addition of acid for the removal of the alkalinity, need for addition of chemicals (calcium).

17.3.3.1 Production process and existing experience

The most efficient way to create crystals of hydroxyapatite is the Crystalactor®. This technology was developed by a company named DHV in the 1970s. At first, it was used for the softening of drinking water but since 1985, it has been used for metal and nutrients recovery from wastewater. One of the main applications of the Crystalactor® is the recovery of phosphate from wastewater, especially in the food industry where it is more widely used (Giesen, 2009).

A Crystalactor® consists in a cylindrical fluidized bed reactor around 6 meters height and between 0.5 and 4 meters diameter (Stratful *et al.* 1999). The footprint of such a technology is thus very small, so it is very suitable for already built plants that want to add a phosphorus recovery process. This technology has already been tested on site. In Holland for instance, three wastewater treatment plants (Westerbork, Geestmerambacht, Waterboard Uitwaterende Sluizen) have used the Crystalactor® technology (Nieminen, 2010).

In order to have a product as pure as possible, the alkalinity must be removed. Indeed, because the formation of hydroxyapatite requires the addition of calcium, there is also formation of calcium carbonate, the calcium reacting with the alkalinity. To lower the alkalinity, the pH is lowered to 5. At that pH, the alkalinity is mostly in the form of CO_2 gas which will evaporate (Stratful *et al.* 1999). Calcium is the essential chemical that is needed in order to precipitate phosphate in hydroxyapatite. The ratio of calcium to phosphate needed is 5:3. However, since hydroxyapatite may not be the only product formed (calcium hydroxide or calcium carbonate can also be formed at this high pH), the optimum ratio Ca:P has been reported at 2:1 (Kuzawa, 2005).

17.3.3.2 Product end-uses

The products formed are crystals of hydroxyapatite, which first application is in the food industry or agriculture. Indeed, its composition is similar to apatite rock (phosphate rock), which is currently used as a fertilizer (Xuechu, 2009). Besides, hydroxyapatite can be a raw material for production of phosphoric acid or for cattle food. Nowadays, some other applications of calcium phosphate crystals are being studied. For instance, it has been shown that hydroxyapatite can be a good adsorbent when it comes to:

- *Removal of lead*: adsorption on hydroxyapatite and magnetite (Dong *et al.* 2010)
- *Removal of fluoride*: adsorption on synthetic hydroxyapatite (Badillo-Alamraz *et al.* 2007)
- Removal of proteins (Cummings *et al.* 2009)

Moreover, hydroxyapatite has multiple clinical applications, since it is already used for damaged bones replacement and repair (Sharpe *et al.* 1997). However, this solution will not be taken into consideration given the problems it causes for social acceptance. Indeed, it seems almost impossible to make people accept having their teeth or bones repaired with something coming out of wastewater. According to Giesen (2009), the purity of the product is 90% removed components, 5% water and 5% seed (sand). The removed components are mostly phosphate precipitates (hydroxyapatite and calcium phosphate $Ca_3(PO_4)_2$), but because the alkalinity did not completely disappear even with the pH lowered to 5, there is also the creation of calcium carbonate.

17.3.4 Recovery of phosphorus compounds from sludge ashes

Phosphorus present in sludge ashes exhibits low bioavailability and contains heavy metals, making it unsuitable for direct use as a fertilizer (Adam *et al.* 2009). Mono-incineration sewage sludge ashes (SSA) typically contain around 15% of phosphorus (as % P). The temperature of incineration can determine the proportion of phosphorus that gets transferred to the SSA. For example, total nitrogen and P_2O_5 volatilization increases with the incineration temperature of 400–900°C, whereas K_2O and Na_2O volatilization decrease (Huang *et al.* 2011). A number of actions during the incineration step can also contribute to improve P recovery. For instance, Han *et al.* (2008) demonstrated that adding $CaCl_2$ (5%) to the sludge during incineration could allow the separation of a proportion of heavy metals (As, Cu, Pb, etc.) from the SSA via the flue gas. This is due to the fact that heavy metal chlorides have a lower evaporation temperature than chloride combined with elements like Na, Zn, Mg, K or P.

Depending on the process and chemicals used, recovery of P from SSA could reach 80% of the total P in a WWTP. In comparison, processes for recovery of P from sludge and dewatering liquors can recover only from 20–50% of the total P in the WWTP Han *et al.* (2008). Phosphate recovery from sludge ashes appears to be one of the most promising practices in terms of recoverable quantities of P from wastewater Han *et al.* (2008). Moreover, simultaneous metal recovery like iron and aluminium can also take place.

Different processes have been developed to recover P from SSA in a form and a purity that allows its re-use by the fertilizer industry. Several pilot scale processes have been implemented over the past 10 years and give first hints of the technical potential of these processes. The economical cost-benefits, LCA of those processes, as well as the discussion about the quality of the products was kept out of the scope of this chapter. There is so far no silver bullet in this area and the most suitable technological solution will be case specific to regional circumstances.

Over the past ten years, 2 main categories of processes have emerged:

- Thermochemical processes – sewage sludge ashes (SSA) are heated at high temperatures in a furnace where phosphorus is reduced and then selectively precipitated as P-rich slag via different processes. During the thermochemical reactions, the heavy metals are usually either evaporated or removed as liquid phase.

- Wet-chemical processes – sewage sludge ashes (SSA) are contacted with a strong acid and/or alkaline chemical to solubilize the phosphorus that can then either be kept in solution as phosphoric acid product (most probably chosen option) or extracted by means of precipitation as struvite or apatite, depending on the precipitation media used and the other elements present in solution.

This section of the book describes some of the current processes used for P recovery from SSA at pilot-scale and full-scale. Table 17.3 clarifies the various P products that can be retrieved from ashes, according to the process employed.

Table 17.3 Processes for recovery of P from sludge ashes described in this book classified by the nature of the P product recovered.

	Phosphoric Acid (Technical Grade)	White Phosphorus	Struvite or Calcium Phosphate	P Rich Slags
Name of the process	EcoPhos® Tetraphos®	RecoPhos® Thermphos®	Leachphos®	ASH DEC® Mephrec®

17.3.4.1 Thermochemical processes

ASH DEC

ASH DEC is the evolution of a process developed within the SUSAN EU-FP6 project (Herzel *et al.* 2015). The basic principles used in this process are that under reductive conditions in a rotary kiln at 950°C the SSA reacts with Na_2SO_4 (the reductive agent) (Table 17.4). Heavy metals evaporate while the phosphorus present in the SSA forms a precipitate as P- rich ash, mainly $CaNaPO_4$ (Figure 17.11). This process has been demonstrated at Weimar, Germany and fed with 2 tons of SSA, for an operational period of 2 weeks (Herzel *et al.* 2015).

Mephrec

The Mephrec process was developed by the German company Ingitec for P recovery from sewage sludge and/or ash. The process consists in several steps: (i) dewatered sewage sludge (25–30% dry matter) is dried to 80 % DM; (ii) mixing of SSA with dried sludge then pressing into compact briquettes; (iii) the briquettes are thermally treated (gasification) in a shaft furnace at temperatures above 1450°C after coke addition (Table 17.4).

This furnace constitutes the core of the process where the P present in ashes and sludge is transformed into silicophosphates (comparable to 'thomas phosphate'), thus P rich slags, as output material. Heavy metal compounds are reduced under these conditions into their elemental form. The volatile toxic trace elements (As, Cd, Hg, Pb, Zn) are evaporated whereas non-volatile heavy metals (Cr, Cu, Fe, Ni) are separated from the slag in form of a liquid metal phase by density separation (Figure 17.12).

Mephrec presents the advantage of energy recovery potential (<55 kWh/kgP) thanks to a battery of off-gas treatment technology like organic rankine cycle (ORC) and combined heat and power (CHP) generator to try to reuse the thermal energy from the off-gas. It can also be complemented with a refuse-derived fuel (RDF) plant combined with a cyclone to reuse the dust from the off-gas as fuel. This process has been implemented Bergakademie Freiberg (Germany) in 2008.

Table 17.4 Overview of thermochemical processes used to recover P from SSA.

	Ash Dec	Mephrec	RecoPhos
Temperature (°C)	950°C in rotary kiln	1450 in furnace	1300°C in furnace
Additive	Na_2SO_4 and reducing agent	Coke and pre-dried sewage sludge 80%DS	Coke (carbon and SiO_2)
P transformation	P present in the SSA is transferred to a $CaNaPO_4$ phase.	P is converted into silicophosphate (comparable to 'Thomas phosphate')	P gets reduced in a thin melt film at the surface of coke particles and evaporates. Gaseous P gets condensed as white phosphorus. A facultative subsequent oxidation allows to obtain phosphoric acid;
Heavy metals elimination	Heavy metals are removed as off-gas	The volatile toxic trace elements (As, Cd, Hg, Pb, Zn) are evaporated whereas non-volatile heavy metals (Cr, Cu, Fe, Ni) are separated from the slag in form of a liquid metal phase by density separation.	Non specified
P product	P-rich slags with Ca-Na-PO_4 (15–25% P_2O_5)	P-rich slags with Ca-Si-PO_4 (10–25% P_2O_5)	White phosphorus or phosphoric acid
P recovery efficiency	98%	80% (assumed)	
Electricity demand	0.8 kWh/kgP	12 kWh/kgP electricity	
Heat demand	3.5 kWh gas/kgP	22 kWh coke/kgP 68 kWh heat/kgP	
Chemical demand	3.3 kg Na_2SO_4/kgP 1.3 kg sludge/kgP	2.7 kg coke/kgP 1.3 kg $CaCO_3$/kgP	

RecoPhos

This process was developed as part of a FP7 EU Environment Project (RecoPhos, 2013). The RecoPhos process is a thermo-chemical process, i.e. high temperature and reducing conditions are used. The P contained with the ashes is reduced by carbon and silicon dioxide to produce phosphorus. The reduction of the P contained in the sewage sludge ash takes place in a thin melt film on the surface of the coke particles. The reduced phosphorus becomes volatile under these conditions and can evaporate from the film without significantly reacting with other elements and can subsequently be retrieved either as white phosphorus or oxidized into phosphoric acid. The full-scale process RecoPhos, adapted from the triple superphosphate process, produces 1000 tons/month fertilizer from SSA.

Table 17.4 gives an overview of the principles of the thermochemical processes for P recovery from SSA and sludge as well as their recovery efficiency and some figures regarding energy and chemicals demands.

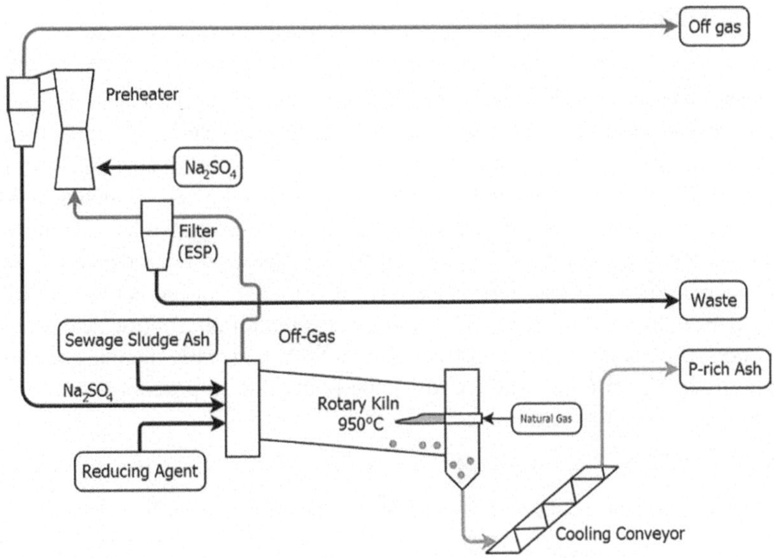

Figure 17.11 Schematic representation of the ASH DEC process (Reproduction with kind permission of Herzel *et al.* 2015).

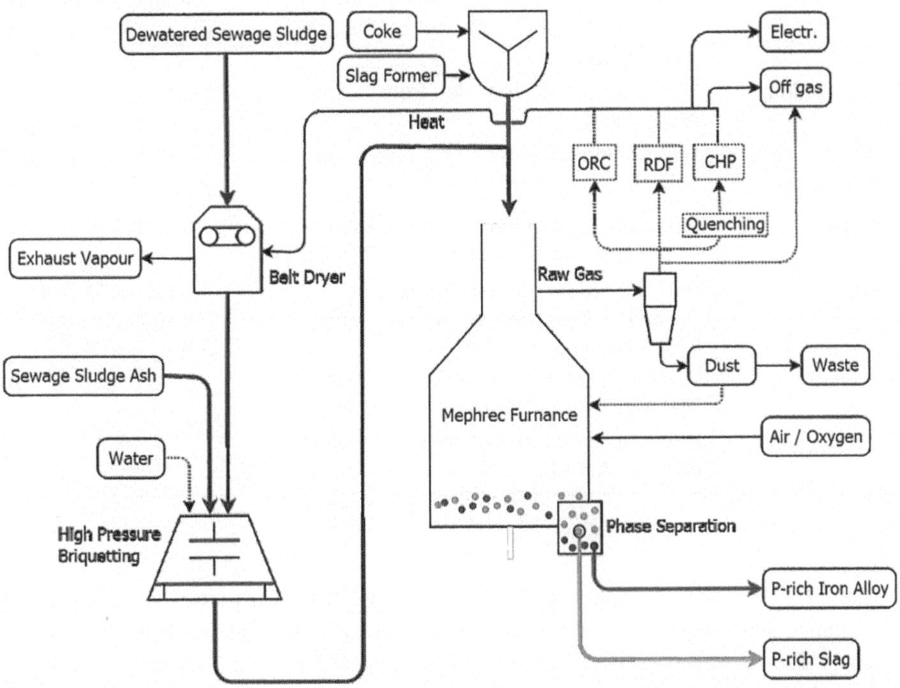

Figure 17.12 Schematic representation of the Mephrec process (Reproduction with kind permission of Herzel *et al.* 2015).

17.3.4.2 Wet-chemical processes

EcoPhos

The EcoPhos process consists on a digestion of the fly ashes with phosphoric acid followed by a filtration to remove any solid residues (Table 17.5). Then a succession of ion exchanges by means of resins allow to selectively separates Al, Fe, Mg and Ca in order to purify the phosphoric acid solution. Regeneration of the resins is done with HCl and the regeneration solutions can be valorised as such (Ca-/Mg-Cl solutions) or in wastewater treatment plants as Al-/Fe-Cl solutions (for chemical phosphorous removal for instance). A part of the phosphoric acid recovered is recycled to the digestion chamber ashes and the rest gets up-concentrated by steam heat process. The phosphoric acid thus produced is of technical grade and can be used by the fertilizer industry (Figure 17.13). Originally design for phosphoric acid production from low grade phosphorus ores, the EcoPhos can be applied also on SSA. Currently there is one plant working with low grade phosphorus in Varna (Bulgaria). A full-scale plant able to process mono-incineration sludge ashes is under construction in Dunkirk, France.

Table 17.5 Overview of wet-chemical processes used to recover P from SSA.

	EcoPhos	**LeachPhos**
Process sequence	Leaching of ashes Filtration of ashes Ion exchange targeting Mg/Ca ions Ion exchange targeting Fe/Mg ions Up-concentration of phosphoric acid by evaporation	Leaching of ashes Filtration of ashes With the filtrate of '2', precipitation of CaP Filtration of CaP precipitate
Acid used for leaching	Phosphoric acid (recycled from production)	Sulphuric acid
Other chemicals used	Hydrochloric acid (resin regeneration)	Lime slurry, NaOH
P recovery efficiency	97%	70%
Electricity demand	0.03 kWh/kg ash treated or 0.2 kWh/kgP recovered (assuming 15% P in ash)	1.6 kWh/kgP recovered
Chemical demand	0.9 L HCl (37%)/kg ash treated, or 2.7 kg HCl/kgP recovered (assuming 15% P in ash) 0.3 g resin/kg ash treated, or 2.1 g resin/kgP recovered (assuming 15% P in ash)	5.6 kg H_2SO_4/kgP 0.6 kg NaOH/kgP 3.9 kg $Ca(OH)_2$/kgP
Comments	Potential simultaneous metals recovery (Mg/Ca, Al/Fe)	

LeachPhos

The LeachPhos process has been developed and patented by BSH Umweltservice GmbH. The SSA is mixed with sulphric acid to leach the phosphorus followed by a subsequent filtration separate the SSA from the leachate. P is then recovered as CaP or MAP (Figure 17.14). A demonstration plant was operated in 2012/2013 by BSH company treating 40 tons SSA at a rate of 2 tons SSA/h.

Table 17.5 gives an overview of the two wet-chemical processes for P recovery from SSA by reporting their recovery efficiency as well as their energy and chemicals demands.

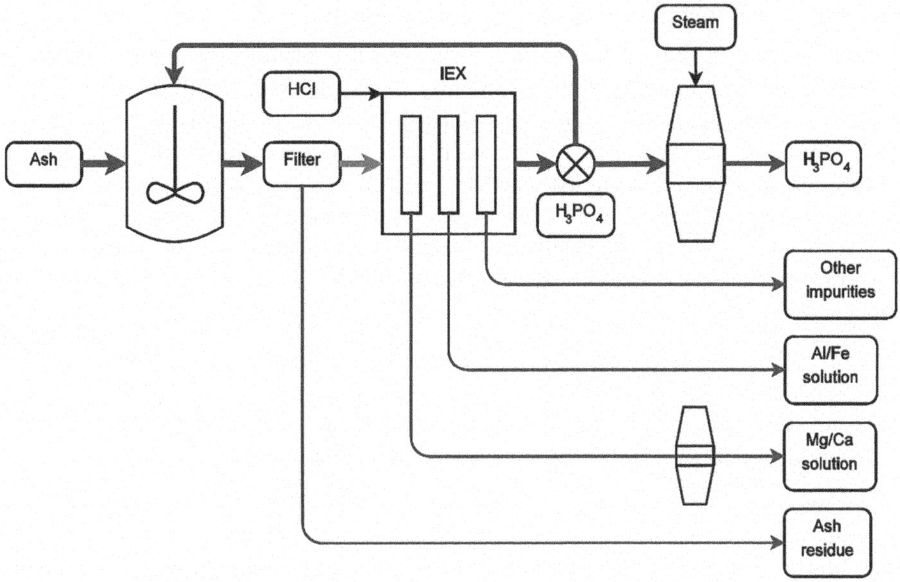

Figure 17.13 Schematic representation of the EcoPhos process (reproduction with kind permission of Remy *et al.* 2015).

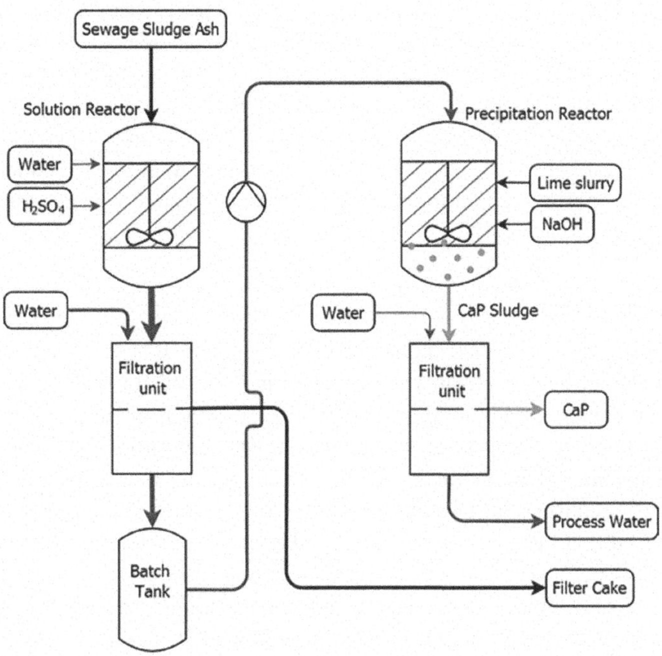

Figure 17.14 Schematic representation of the LeachPhos process (reproduction with kind permission of Herzel *et al.* 2015).

Optimal conditions for P leaching in wet-chemical processes

Lab scale studies investigating P recovery from SSA using wet-chemical processes are gaining much interest from the scientific community in the last 5 years. Donatello *et al.* (2010) studied the production of technical grade phosphoric acid. After sulphuric acid leaching, the leachate was purified and concentrated up to 80% H_3PO_4 by a cation exchange resin. The overall P recovery reached 70–90%. Different influencing parameters were investigated, including (i) reaction time; (ii) liquid-solid ratio; (iii) the nature of SSA; and (iv) acid concentration. Optimal conditions were reported to be 120 min reaction time and a liquid-solid ratio of 20. In another study completed Krupa-Zuczek *et al.* (2008) to investigate the production of phosphoric acid from ashes of meat-bones incineration indicated that the ashes contained about 16% of P as hydroxyapatite (HA). The recovery process was achieved in 2 steps:

- Dissolution of HA in phosphoric acid to obtain monocalcium phosphate. Tests showed that this dissolution happens at >37% in H_3PO_4.
- Monocalcium phosphate conversion to $CaSO_4$ with sulphuric acid at recommended temperature of 95°C. After this process a food grade phosphoric acid was obtained.

The nature of the acid using for P solubilisation also affects P recovery yields. For example, in a study published by Tarko *et al.* (2014) HNO_3 was shown to be more efficient at solubilising P than H_3PO_4. In the same line, Martinez *et al.* (2014) performed recovery tests using and two different acids, after three different combustion temperatures. When using sulphuric acid, higher P recovery yields were achieved for sludge combusted at temperatures of 900°C when compared with temperatures of 600°C and 750°C. On the opposite when oxalic acid was used (10 kg oxalic acid/kg) the combustion temperature did not impacted the recovery yields.

Another bottleneck of P leaching in wet-chemical processes is the presence of the heavy metals. Guedes *et al.* (2014) proposes to promote the separation of the leached P from heavy metals by electrodialysis. The anolyte obtained had a composition of 98% phosphorus and 2% heavy metals. Ottosen *et al.* (2014) carried out similar experiments: separation of heavy metals (positively charged) and P (negatively charged) by electrodialysis. This was possible for Fe-rich ashes, but not for Al-rich ashes. Indeed, aluminium interfered with P speciation and negatively charged P was not prevailing anymore. Considering that P recovery struggles with economic viability, it is unlikely that such an additional energy demanding process gets applied at industrial scale.

17.3.4.3 Future perspectives

Currently, mostly mono-incineration or co-incineration is implemented at full-scale for sludge stabilisation and volume reduction. Nevertheless, advanced thermal conversion technologies such as gasification and pyrolysis are currently receiving much attention and are likely to be implemented in a significant number of sites in the next decades. These processes have end products containing the P with different properties than the SSA from conventional incineration. Which of these new thermal conversion technologies is most advantageous for P recovery is a future research topic. Lately, the company AVA-CO2 Schweiz AG (Germany) announced that their new process called Hydrothermal Carbonization or 'AVA cleanphos' process would give a product more suitable to leaching (due to a more crystalline structure of phosphorus) than incineration (where phosphorus is present in amorphous phase). The HTC process does not trap phosphorus in a stable glass matrix, which is claimed to allow a lower acid use for leaching and makes a fine grinding pre-treatment useless (AVA-CO2 Schweiz Ag, 2015).

Another relevant new technology is the Euphore® process developed by EuPhoRe GmbH company; a red-ox-technique that handles dried sludge and combines pyrolysis and combustion. The Euphore carbonizes the sewage sludge in a first step under reductive conditions and in an immediately following second step; the rest carbon of the sewage sludge coal is fired spontaneously. Due to the immediate transfer from reductive conditions to oxidative conditions in the process, coinciding with high temperatures, the P compounds in the ashes became more available resulting in fertiliser with high plant P-availability. The pyrolysis gases of the reductive step and the process gases of the oxidation step can be reused thermally. Due to input of additives before the treatment, an effective heavy metal reduction in the final product is reached. Gas heavy metal compounds are separated into the flue gas and can be cleaned in the gas cleaning phase (Geerts, 2015).

In 2014, pilot experiments were carried out to investigate the viability of an end-of-sludge treatment method to be both auto-thermal (starting from dewatered sludge) and would allow for relatively easy and complete phosphorus recovery. The sewage sludge used was collected from Aquafin's sludge drying plants and processed in a rotary kiln test-reactor by IBU-tec advanced materials AG. A carbon-efficiency (the proportion of carbon removed from the ashes during the process) of more than 99% was reached for the organic matter present in the sludge (Geerts, 2015). Looking at the mass and energy balances of Aquafin sludge line cases, this efficiency would release of enough excess heat to feed the drying step, meaning having an auto-thermal sludge line starting from dewatered sludge (27% dry matter). P_2O_5 was up concentrated to around 10% of the mass. Standardized solubility tests suggested a high plant availability of the phosphorous from the product, with best results for the sludge to which additives were added.

The main strengths of this technique would be related to: (i) energy and price independence for the end-of-sludge routes; (ii) avoiding the lock down of phosphorus within cement or ashes of co-incineration, allowing for a circular use of P; (iii) no additional treatment steps for the waste water treatment company, creating a P-product that might be valorised directly in the fertilizer market.

Simultaneous P and metal recovery in wet-chemical processes

Wet-chemical P leaching processes also offer the opportunity for metal recovery such as iron and aluminium. For example, in WWTP where aluminium based coagulants are used for P-removal, the SSA are rich in P and Al offering the opportunity to recover two different products. This could be achieved by sequential acidic and alkaline treatment of SSA (Petzet *et al.* 2012). The first acidic step aims at dissolving the P fraction bound with Ca (calcium phosphate-Ca-P) that is insoluble under alkaline conditions to form Al-P instead. Al-P is then dissolved in a second alkaline step. The P can be precipitated back to Ca-P while the Al solution can be recycled in WWTP for P removal. This process claimed to have lower chemical demand than the direct acidic dissolution of all P-compounds, due to the described rearrangement of the P component from Ca-P to Al-P. On the other hand Ottosen *et al.* (2013) reports that Al-rich SSA shows a significantly higher buffering capacity and thus demands more chemicals for acid leaching. The Ecophos® claims to offer the recovery of Ca/Mg and Fe/Al, as described in section 1.1.2.1 of this chapter (Remy *et al.* 2015). Knowing whether the metal recovery can help to overcome the economic bottleneck of P recovery or not needs further study.

Valorisation of SSA after P recovery

After the P recovery processes for SSA, the question of a proper valorisation of the rest ashes, rather than landfilling, remains. The construction sector, sequestering high quantities of materials, seems to be an option. The SSA can be used for cement production (5 to 20% in composition), as replacement of sand. The result is a usable cement with good properties (density, resistance and water absorption) (Baeza-Brotons *et al.* 2014). The restriction of this valorisation route is the content of Fe and Al that should controlled, giving thus an additional push towards Fe/Al recycling from the ashes. The replacement of sand is a sustainable end-of-life option as sand is called to also become a critical scarce resource.

17.4 CONCLUSIONS

Nutrient recovery from wastewater is receiving much attention from the scientific, industrial communities as well regulators and public perception groups. Many technologies and processes are being developed for the recovery of products such as: ammonium nitrate, ammonium sulphate, ammonium water, bio-struvite, calcium phosphate, hydroxyapatite, phosphoric acid, potassium phosphate, struvite, white phosphorus, etc. Many of these processes use chemicals, reagent and practices that are typically used within the chemical industry, hence requiring specific assets and well-trained operators. The numbers of installations worldwide for nutrient recovery is increasing rapidly as technological advances are occurring at a fast pace. Nevertheless there is still the significant challenge to make these processes economically feasible as well as providing an end route and entrance of the product to a stable and welcoming supply chain or market. The attractiveness of these processes to the water industry is still very much driven by reducing problems down stream (e.g. struvite precipitation in pipes) or reducing the recirculation of nutrients within the WWTP and consequent reduce the pollutant load to the secondary treatment. Governments and agency all over the world, that are calling for innovation to bring the circular economy to practice within the wastewater industry, need to take an active role in order to open and support supply chains and markets for the recovered products.

17.5 REFERENCES

Adam C., Peplinski B., Michaelis M., Kley G. and Simon F.-G. (2009). Thermochemical treatment of sewage sludge ashes for phosphorus recovery. *Waste Management*, **29**(3), 1122–1128.

Ahlgren S., Baky A., Bernesson S., Nordberg Å., Norén O. and Hansson P.-A. (2008). Ammonium nitrate fertiliser production based on biomass – environmental effects from a life cycle perspective. *Bioresource Technology*, **99**(17), 8034–8041.

Anne C. O., Trébouet D., Jaouen P. and Quéméneur F. (2001). Nanofiltration of seawater: fractionation of mono- and multi-valent cations. *Desalination*, **140**, 67–77.

Antonini S., Alejandra Arias M., Eichert T. and Clemens J. (2012). Greenhouse evaluation and environmental impact assessment of different urine-derived struvite fertilizers as phosphorus sources for plants. *Chemosphere*, **89**, 1202–1210.

Asharafizadeh S. and Khorasani Z. (2010). Ammonia removal from aqueous solutions using hollow-fiber membrane contactor. *Chemical Engineering Journal*, **162**, 242–249.

AVA-CO_2 Schweiz Ag. (2015). AVA-CO_2 develops phosphorus recovery process. *Membrane Technology*, **2015**(1), 1–6.

Badillo-Almaraz V. E., Flores J. A., Arriola H., Lopez F. A. and Ruiz-Ramirez L. (2007). Elimination of fluoride ions in water for human consumption using hydroxyapatite as an adsorbent. *Journal of Radioanalytical and Nuclear Chemistry*, **271**(3), 741–744.

Baeza-Brotons F., Garces P., Paya J. and Saval J. M. (2014). Portland cement systems with addition of sewage sludge ash. *Application in concretes for the manufacture of blocks. Journal of Cleaner Production*, **82**, 112–124.

Battistoni P., Fava G., Pavan P., Musacco A. and Cecchi F. (1997). Phosphate removal in anaerobic liquors by struvite crystallisation without addition of chemicals. *Preliminary results, Water Research*, **31**, 2925–2929.

Bergmans B. (2011). Struvite Recovery from Digested Sludge. Thesis Master of Science in Civil Engineering. Delft University, Delft, Netherlands.

Bonvin C., Etter B., Udert K. M., Frossard E., Nanzer S., Tamburini F. and Oberson A. (2015). Plant uptake of phosphorus and nitrogen recycled from synthetic source-separated urine. *AMBIO*, **44**(Suppl. 2), S217–S227.

Call D. and Logan B. E. (2008). Hydrogen production in a single chamber microbial electrolysis cell lacking a membrane. *Environmental Science and Technology*, **42**(9), 3401–3406.

Canellas J., Cheong, S., Dias J., Fukasaku Y., Razban R. and Sebok A. (2014). Recovery of valuable products from ion exchange regenerants, MSc Water and Wastewater Engineering Project Report, Cranfield University, Cranfield, UK.

Chimenos J. M., Fernández A. I., Villalba G., Segarra M., Urruticoechea A., Artaza B. and Espiell F. (2003). Removal of ammonium and phosphates from wastewater resulting from the process of cochineal extraction using MgO-containing by-product. *Water Research*, **37**, 1601–1607.

Ciceri D., Manning D. A. C. and Allanore A. (2015). Historical and technical developments of potassium resources. *Science of The Total Environment*, **502**, 590–601.

Cusick R. D. and Logan B. E. (2012). Phosphate recovery as struvite within a single chamber microbial electrolysis cell. *Bioresource Technology*, **107**, 110–115.

Czajkowska J. (2015). Conditions of Sediments Precipitation Containing Struvite. Doctoral Thesis, Warsaw University of Life Sciences, Warsaw, Poland. (In Polish).

Czajkowska J. and Siwiec T. (2011). Crystallization of struvite from synthetic wastewater in experimental installation under flow conditions. *Scientific Review Engineering and Environmental Sciences*, **20**(2), 130–139. (In Polish).

Dai W., Xu X., Liu B. and Yang F. (2015). Toward energy-neutral wastewater treatment: a membrane combined process of anaerobic digestion and nitritation–anammox for biogas recovery and nitrogen removal. *Chemical Engineering Journal*, **279**, 725–734.

Davies J. and Thiemann A. (2015). Competition law and policy: drivers of economic growth and development. Coherence for Development. https://www.oecd.org/development/002014381_CfD_E-book_FINAL%20VERSION%20FOR%20WEB.pdf. (accessed October 2015).

Donatello S., Tong D. and Cheeseman C. R. (2010). Production of technical grade phosphoric acid from incinerator sewage sludge ash (ISSA). *Waste Management*, **30**(8–9), 1634–1642.

Dong L., Zhu Z., Qiu Y. and Zhao J. (2010). Removal of lead from aqueous solution by hydroxyapatite/magnetite composite adsorbent. *Chemical Engineering Journal*, **65**, 827–834.

Doyle J. D. and Parsons S. A. (2002). Struvite formation, control and recovery. *Water Research*, **36**(16), 3971–3978.

ESPP. (2015). European Sustainable Phosphorous Platform. http://phosphorusplatform.eu/. (accessed October 2015).

Etter B., Tilley E., Khadka R. and Udert K. M. (2011). Low-cost struvite production using source-separated urine in Nepal. *Water Research*, (45), 852–862.

Etter B., Udert K. M. and Gounden T. (2014). VUNA – Scaling up nutrient recovery from urine. In: Proceedings from the Technology for Development International Conference, Lausanne, Switzerland, 4–6 June.

Etter B., Udert K. M. and Gounden T. (eds). (2015). VUNA Final Report. EAWAG, Dübendorf, Switzerland.

EU Council Directive. (1991). Council Directive 91/271/EEC concerning urban waste water treatment. Urban waste water treatment: http://eur-lex.europa.eu/LexUriServ/site/en/consleg/1991/L/01991L0271-19980327-en.pdf. (accessed October 2009).

EU. (2015). Communication from the Commission to the European Parliament, the Council, the European Economic and Social Committee and the Committee of the Regions. Closing the loop - An EU action plan for the Circular Economy. http://eur-lex.europa.eu/legal-content/EN/TXT/?uri = CELEX:52015DC0614, (accessed Aug 2016).

Evans T. D. (2009). Recovering ammonium and struvite fertilizers from digested sludge dewatering liquors. Resource Recovery Not Wastewater Treatment Conference. Aqua-Enviro, London.

Fertilizers Europe. (2012). Forecast of food, farming and fertilizer use in the European Union 2012–2022. http://www.fertilizerseurope.com/fileadmin/user_upload/publications/agriculture_publications/Forecast_2012-final.pdf, (accessed August 2016).

Fertilizers Europe. (2015). Industry Facts and Figures. Industry facts and figures 2015. http://www.chicken.org.au/page.php?id=4. (accessed October 2015).

Fischer F., Bastian C., Happe M., Mabillard E. and Schmidt N. (2011). Microbial fuel cell enables phosphate recovery from digested sludge as struvite. *Bioresource Technology*, **102**, 5824–5830.

Fumasoli A., Etter B., Sterkele B., Morgenroth E. and Udert K. M. (2015). Operating a pilot-scale nitrification plant for complete nutrient recovery from urine. Proceedings from the IWA Specialist Conference Nutrient Removal and Recovery: moving innovation into Practice, Gdansk, Poland, 18–21 May.

Geerts S. (2015). Personal Communication. Aquafin NV, Aartselaar, Belgium.

Giesen A. (2009). P-recovery with the Crystalactor process. Proceedings from the Conference Baltic 21 on Phosphorus Recycling and Good Agricultural Practice, Berlin, Germany, 28–30 September.

Gilbert N. (2009). The disappearing nutrient. *Nature*, **461**, 716–718.

Guedes P., Couto N., Ottosen L. M. and Ribeiro A. B. (2014). Phosphorus recovery from sewage sludge ash through an electrodialytic process. *Waste Management*, **34**(5), 886–892.

Han J., Xu M., Yao H., Furuuchi M., Sakano T. and Kim H. J. (2008). Influence of calcium chloride on the thermal behavior of heavy and alkali metals in sewage sludge incineration. *Waste Management*, **28**(5), 833–839.

Hanhoun M., Montastruc L., Azzaro-Pantel C., Biscans B., Frčche M. and Pibouleau L. (2011). Temperature impact assessment on struvite solubility product: a thermodynamic modeling approach. *Chemical Engineering Journal*, **167**, 50–58.

Hasanoglu A., Romero J., Pérez B. and Plaza A. (2010). Ammonia removal from wastewater streams through membrane contactors: experimental and theoretical analysis of operation parameters and configuration. *Chemical Engineering Journal*, **160**, 530–537.

Herzel H., Adam C., Hermann L., Mallon J., Schaaf M., Eicher N. and Stemann J. (2015). Comparative review of ash processes. P-REX deliverable D4.1. www.p-rex.eu. (accessed November 2015).

Huang Y., Dong H., Shang B., Xin H. and Zhu Z. (2011). Characterization of animal manure and cornstalk ashes as affected by incineration temperature. *Applied Energy*, **88**(3), 947–952.

Huang H., Xiao D., Zhang Q. and Ding L. (2014). Removal of ammonia from landfill leachate by struvite precipitation with the use of low-cost phosphate and magnesium sources. *Journal of Environmental Management*, **145**, 191–198.

Hug A. and Udert K. M. (2013). Struvite precipitation from urine with electrochemical magnesium dosage. *Water Research*, **47**, 289–299.

Hutnik N., Wierzbowska B., Matynia A., Piotrowski K. and Gluzińska J. (2008). Influence aluminium ions on struvite crystals quality by continious struvite reaction crystallization process. *Chemik. Science - Technique – Market*, **61**, 505–508. (In Polish).

Hutnik N., Wierzbowska B., Matynia A. (2009a). Influence potassium ions on struvite crystals quality by continuous struvite reaction crystallization process. *Inż. Applied Chemistry*, **4**, 54–55. (In Polish).

Hutnik N., Wierzbowska B., Matynia A., Piotrowski K. (2009b). Influence zinc ions on struvite crystals quality in dilute aqueous solution by continuous struvite reaction crystallization process. *Inż. Applied Chemistry*, **5**, 38–39. (In Polish).

IFDC report. (2010). World phosphate rock reserves and resources.

Krupa-Zuczek K., Kowalski Z. and Wzorek Z. (2008). Manufacturing of phosphoric acid from hydroxyapatite, contained in the ashes of the incinerated meat-bone wastes. *Polish Journal of Chemical Technology*, **10**(3), 13–20.

Larry J., Cummings, Mark, A., Snyder and Brisack K. (2009). Protein chromatography on hydroxyapatite columns: Chapter 24. In: Methods in Enzymology, Burgess, R. R. and Deutscher, M. P., (eds), Academic Press, London, UK, Vol. 463, 387–404.

Larsen T. A., Alder A. C., Eggen R. I. L., Maurer M. and Lienert J. (2009). Source separation: will we see a paradigm shift in wastewater handling?. *Environment Science and Technology*, **43**(16), 6121–6125.

Le Corre K. S., Valsami-Jones E., Hobbs P., Jefferson B. and Parsons S. A. (2007). Struvite crystallisation and recovery using a stainless steel structure as a seed material. *Water Research*, **41**(11), 2449–2456.

Le Corre K. S., Valsami-Jones E., Hobbs P. and Parsons S. A. (2009). Phosphorus recovery from wastewater by struvite crystallization: a review. *Critical Reviews in Environmental Science and Technology*, **39**(6), 433–477.

Lee S. I., Weon S. Y., Lee C. W. and Koopman B. (2003). Removal of nitrogen and phosphate from wastewater by addition of bittern. *Chemosphere*, **51**, 265–271.

Luff B. B. and Reed R. B. (1980). Thermodynamic properties of magnesium potassium orthophosphate hexahydrate. *Journal of Chemical & Engineering Data*, **25**(4), 310–312.

Maćkowiak J. F. and Górak A. (2011). Experimental model validation of an integrated process for the removal of carbon dioxide from aqueous ammonia solutions. *Chemical Engineering and Research and Design*, **89**(8), 1252–1260.

Mann S. (2001). General principles of biomineralization. In: Biomineralization: Principles and Concepts in Bioinorganic Materials Chemistry, Mann, S. (ed.), 1st edn, Oxford University Press, Oxford, UK, pp. 24–37.

Manning D. A. C. (2015). How will minerals feed the world in 2050?. *Proceedings of the Geologists' Association*, **126**(1), 14–17.

Martinez M. A., Gea G., Arauzo J., Kersten S. R. A. and Kootstra A. M. J. (2014). Phosphorus recovery from sewage sludge char ash. *Biomass and Bioenergy*, **65**, 42–50.

Mašić A., Santos A. T. L., Etter B., Udert K. M. and Villez K. (2015). Estimation of nitrite in source-separated nitrified urine with UV spectrophotometry. *Water Research*, **85**(15), 244–254.

Mašić A. and Villez K. (2014). Model-based observers for monitoring of biological nitrification of urine in decentralized treatment. Proceedings from the IWA International Conference Ecotechnologies for Wastewater Treatment (EcoSTP2014), Verona, Italy, 23–25 June.

Matsumiya Y., Yamasita T. and Nawamura Y. (2000). Phosphorus removal from sidestreams by crystallisation of magnesium-ammonium-phosphate using seawater. *Water and Environmental Management*, **14**(4), 291–296.

Morales N., Boehler M. A., Buettner S., Liebi C. and Siegrist H. (2013). Recovery of N and P from urine by struvite precipitation followed by combined stripping with digester sludge liquid at full scale. *Water*, **5**, 1262–1278.

Moya-Gonzalez A. (2012). Nutrient recovery from ion exchanges regenerants. MSc thesis, Cranfield University, Cranfield, UK.

Munch E. V. et al. (2001). Making a business from struvite crystallisation for wastewater treatment: turning waste into gold. Second International Conference on Phosphate Recovery for Recycling from Sewage and Animal Wastes, Noondwijkerhout, The Netherlands, 12–13 March 2001.

Nieminen J. (2010). Phosphorus recovery and recycling from municipal wastewater sludge. MSc thesis, Aalto University, Helsinki, Finland.

Norddahl B. (2006). A membrane contactor for ammonia stripping, pilot scale experience and modeling. *Desalination*, **99**, 172–174.

Ottosen L. M., Kirkelund G. M. and Jensen P. E. (2013). Extracting phosphorus from incinerated sewage sludge ash rich in iron or aluminum. *Chemosphere*, **91**(7), 963–969.

Ottosen L. M., Jensen P. E. and Kirkelund G. M. (2014). Electrodialytic separation of phosphorus and heavy metals from two types of sewage sludge ash. *Separation Science and Technology*, **49**(12), 1910–1920.

Petzet S., Peplinski B. and Cornel P. (2012). On wet chemical phosphorus recovery from sewage sludge ash by acidic or alkaline leaching and an optimized combination of both. *Water Research*, **46**(12), 3769–3780.

Pratt C., Parsons S. A., Soares A. and Martin B. D. (2012). Biologically and chemically mediated adsorption and precipitation of phosphorus from wastewater. *Current Opinion in Biotechnology*, **23**(6), 890–896.

Rahman Md. M., Mohd M. A., Rashid S. U., Ahsan A., Hossain M. M. and Ra C. S. (2014). Production of slow release crystal fertilizer from wastewaters through struvite crystallization – a review. *Arabian Journal of Chemistry*, **7**(1), 139–155.

Razon L. F. (2015). Is nitrogen fixation (once again) "vital to the progress of civilized humanity"?. *Clean Technology and Environmental Policy*, **17**(2), 301–307.

Recohos. (2013). RecoPhos factsheet. http://www.recophos.org/c/mid,1364,Media_Centre/, (accessed November 2015).

Redzwan G. and Banks C. (2010). Supplementation of ammonium bicarbonates for the treatment of fruit cordial wastewater by anaerobic digestion process. *Anaerobe*, **16**, 34–37.

Remy C. and Jossa P. (2015). Life Cycle Assessment of selected processes for P recovery from sewage sludge, sludge liquor, or ash. P-REX deliverable D9.2. www.p-rex.eu. (accessed November 2015).

Rittmann B. E., Mayer B., Westerhoff P. and Edwards M. (2011). Capturing the lost phosphorus. *Chemosphere*, **84**, 846–853.

Roma E., Philp K., Buckley C., Xulu S. and Scott D. (2013). User perceptions of urine diversion dehydration toilets: experiences from a cross-sectional study in eThekwini Municipality. *Water SA*, **39**(2), 305–312.

Ronteltap M., Maurer M., Hausherr R. and Gujer W. (2010). Struvite precipitation from urine - Influencing factors on particle size. *Water Research*, **44**(6), 2038–2046.

Ryu H. D. and Lee S. I. (2010). Application of struvite precipitation as a pretreatment in treating swine wastewater. *Process Biochemistry*, **45**, 563–572.

Salutsky M. L. and Steiger R. P. (1964). Metal potassium phosphates. *Agricultural and Food Chemistry*, **12**, 486–491.

Schuiling R. D. and Andrade A. (1999). Recovery of struvite from calf manure. *Environmental Technology*, **20**(7), 765–768.

Sharpe J. R., Sammons R. L. and Marquis P. M. (1997). Effect of pH on protein adsorption to hydroxyapatite and tricalcium phosphate ceramics. *Biomaterials*, **18**, 471–476.

Soares A., Veesam M., Simoes F., Wood E., Parsons S. A. and Stephenson T. (2014). Bio-struvite: a new route to recover phosphorus from wastewater. *CLEAN - Soil, Air, Water*, **42**(7), 994–997.

Song Y. H., Qiu G. L., Yuan P., Cui X. Y., Peng J. F., Zeng P., Duan L., Xiang L. C. and Qian F. (2011). Nutrients removal and recovery from anaerobically digested swine wastewater by struvite crystallization without chemical additions. *Journal of Hazardous Materials*, **190**(1–3), 140–149.

Stratful I., Brett S., Scrimshaw M. B. and Lester J. N. (1999). Biological phosphorus removal, its role in phosphorus recycling. *Environmental Technology*, **20**, 681–6950.

Sutton M. A. *et al.* (2013). Our nutrient world: the challenge to produce more food and energy with less pollution. Global Overview of Nutrient Management. Centre for Ecology & Hydrology on behalf of the Global Partnership on Nutrient Management and the International Nitrogen Initiative. Centre for Ecology & Hydrology, Lancaster, UK.

Tan X., Tan S. P., Teo W. K. and Li K. (2006). Polyvinylidene (PVDF) hollow fiber membranes for ammonia removal from water. *Journal of Membrane Science*, **27**, 59–68.

Tarko B., Gorazda K., Wzorek Z., Nowak A. K., Kowalski Z., Kulczycka J. and Henclik A. (2014). Recovery of phosphorus from industrial sewage sludge ashes. *Przemysl Chemiczny*, **93**(6), 1041–1044.

Taylor A. W., Frazier A. W. and Gurney E. L. (1963). Solubility products of magnesium ammonium and magnesium potassium phosphates. *Transactions of the Faraday Society*, **59**, 1580–1584.

Udert K. M. and Wächter M. (2012). Complete nutrient recovery from source-separated urine by nitrification and distillation. *Water Research*, **46**(2), 453–464.

Ulbricht M., Schneider J., Stasiak M., Sengupta X. X. (2013). Ammonia recovery from industrial wastewater by trans-membrane-chemi-sorption. *Chemie Ingenieur Technik*, **85**(8), 1259–1262.

US EPA. (2002). Air Pollution Control Cost Manual, EPA-452/B-02-001.

Uysal A., Yilmazel Y. D. and Demirer G. N. (2010). The determination of fertilizer quality of the formed struvite from effluent of a sewage sludge anaerobic digester. *Journal of Hazardous Materials*, **181**, 248–254.

VUNA EAWAG News. (2015). http://www.eawag.ch/medien/publ/news/2015_na_01/index_EN. (accessed June 2015).

Web report. (2014). http://www.reportsnreports.com/reports/427514-china-phosphorus-industry-data-report-edition-1-.html. (accessed October 2015).

Wilsenach J. A., Schuurbiers C. A. H. and van Loosdrecht M. C. M. (2007). Phosphate and potassium recovery from source separated urine through struvite precipitation. *Water Research*, **41**(2), 458–466.

Wu Ch.-Y., Peng Y.-Z. and Wang S.-Y., Ma Y. (2010). Enhanced biological phosphorus removal by granular sludge: from macro- to micro-scale. *Water Research*, **44**, 807–814.

Xu K., Li J., Zheng M., Zhang C., Xie T. and Wang C. (2015). The precipitation of magnesium potassium phosphate hexahydrate for P and K recovery from synthetic urine. *Water Research*, **80**, 71–79.

Xu K., Wang C., Liu H. and Qian Y. (2011). Simultaneous removal of phosphorus and potassium from synthetic urine through the precipitation of magnesium potassium phosphate hexahydrate. *Chemosphere*, **84**(2), 207–212.

Xuechu C., Hainan K., Deyi W., Xinze W. and Yongyong L. (2009). Phosphate removal and recovery through crystallisation of hydroxyapatite using xonolite as seed crystal. *Journal of Environmental Sciences*, **2**, 575–580.

Zeng L., Mangan C. and Li X. (2006). Ammonia recovery from anaerobically digested cattle manure by steam stripping. *Water Science and Technology*, **54**(8), 137–145.

Zheng J., Zheng D., Wong F. and Xu Z. (2005). Shell side mass transfer in a transverse flow hollow fiber membrane contactor. *Journal of Membrane Science*, **261**, 114–120.

Zhu Z., Hao Z., Shen Z. and Chen J. (2005). Modified modeling of the effect of pH and viscosity on the mass transfer in hydrophobic hollow fiber membrane contactors. *Journal of Membrane Science*, **25**, 269–275.

Chapter 18

Recovery of organic added value products from wastewater

Anuska Mosquera-Corral, Gilda Carvalho, Andrea Fra-Vázquez, Ioanna Ntaikou, Piotr Oleskowicz-Popiel, Tania Palmeiro-Sánchez, Maria A.M. Reis and Mª Eugenia Suárez-Ojeda

18.1 INTRODUCTION

18.1.1 Potential feedstocks in wastewater treatment plants

Traditionally, sewage treatment plants (STP) operated with the objective of removing pollutants from wastewater, i.e. organic matter and nutrients, to enable discharge into surface water sources. However, the ambition of a more environmental-friendly and sustainable society has induced an extensive search for processes and innovative technologies that aim at recovering added-value products from wastewater. In this way a new perspective for achieving sustainable STPs is appearing mainly focused on the reduction of the use of resources and energy, and on the minimization of wastes production together with the implementation of processes of resource recovery. These approaches are being mostly accomplished through energy production and nutrient recycling which are outside the scope of this chapter.

The main product recovery processes from wastewater that have been proposed in the literature are: organic acids and alcohols; polyhydroxyalkanoates (PHA); methane containing biogas; and extracellular polymeric substances (EPS). Different streams inside a typical sewage treatment plant have been identified as potential feedstock to obtain these products (Figure 18.1). These streams to be used as feedstock in the STPs are the solid fraction from the pre-treatment stages, the sludge coming from the primary and secondary settling and the wastewater itself. Compounds like alginate will be directly obtained from the secondary systems based on aerobic granular biomass.

Primary and secondary sludge, or even co-fermentation of both or combined with other substrates, are susceptible of acidification to produce acids and alcohols. These produced compounds have a wide range of applications such as the production of biopolymers and/or bioenergy reducing the wastes generated (Lee *et al.* 2014).

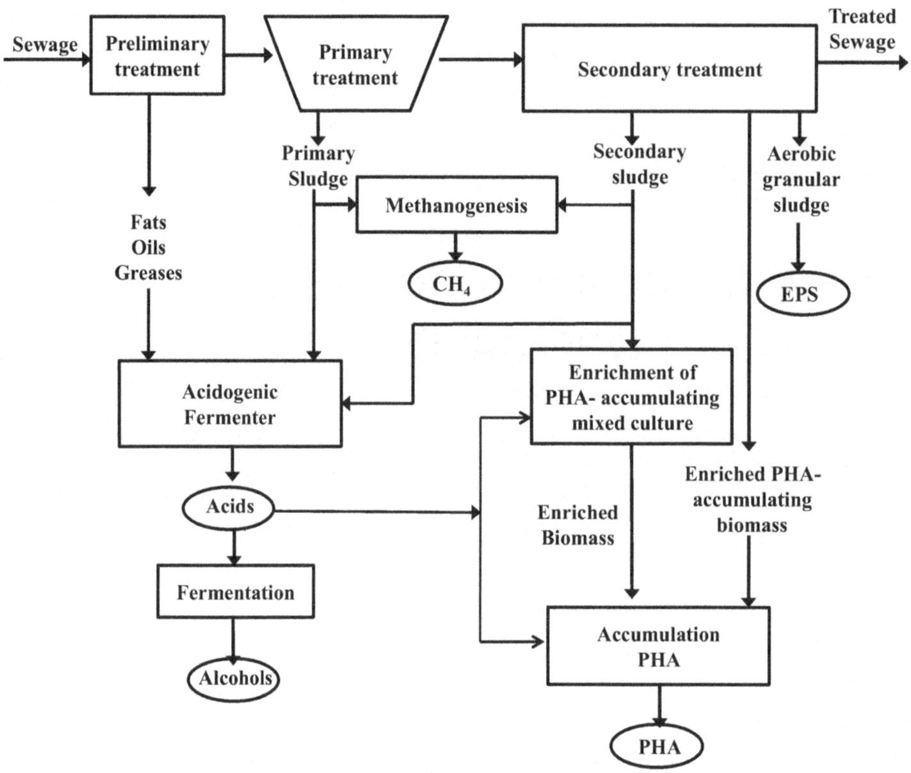

Figure 18.1 Identified streams in STPs with potential value as feedstocks for products recovery.

Both primary and secondary sludge are suitable as feedstock for acids production due to the high organic matter content in these wastes (7–40 g TCOD/L) (Ucisik & Henze, 2008). Moreover, acids production for the subsequent PHA production is currently one important research line in the resource recovery approach. On the other hand, authors like Cheung *et al.* (1997) studied the ethanol production from the cellulosic components of primary sludge which represented the 10% of the sludge mass. In these conditions alcohol production from sewage was not further studied as these authors found that economic feasibility was not possible at so low cellulose percentages.

Nowadays PHAs production is increasing in interest as these compounds are intracellular biopolymers synthesised by microorganisms with similar properties to some plastics, but with the advantage of being biodegradable and produced from renewable resources (Reis *et al.* 2003). The PHA production at industrial scale is carried out up to date by pure cultures, with large operational costs due to the requirements of the process, such as sterile conditions and use of pure substrates. The integration of the PHA production in the sewage treatment plant could reduce these costs by the use of cheap feedstocks and non-sterile atmosphere requirements. This integration can be performed using the implementation of a specific process in the STP using activated sludge as inoculum to enrich a mixed microbial culture (MMC) in PHA-accumulating microorganisms. Different wastewater compositions have been tested for the selection of MMC-PHA cultures. Chua *et al.* (2003) achieved percentage of 20% of PHA accumulated to sludge dry weight by using sewage containing 30–130 mg TOC/L. Morgan-Sagastume *et al.* (2010) investigated the production of PHA at high organic loading rates (6 g SCOD/(L · d)) and nutrients concentration (3.5 g N/L and 0.5 g P/L) using waste activated sludge (WAS) pre-treated in a high pressure thermal hydrolysis (HPTH) unit. With this pre-treatment a rich-VFA stream was obtained and used as precursor for biopolymer production, reaching up to 25% on dry biomass basis. A further research study was carried out at pilot scale in the North STP in Belgium where PHA-accumulating biomass was selected from activated sludge grown in wastewater, characterized by low COD loads (Morgan-Sagastume *et al.* 2014). Mengmeng *et al.* (2009) studied the production of PHA by activated sludge fed with VFAs generated from an alkaline fermentation of WAS, containing approximately 5 g TOC/L and achieved a maximum PHA content of 56.5% of the dry biomass.

Methane is another product commonly obtained from sludge treatment, however its production is not studied here as it is more linked to energy production which is the object of another chapter in this book.

Microorganisms are known to be able to produce EPS in response to operational or environmental conditions which have in many cases a structural function which facilitates aggregates formation. Alginate-like exopolysaccharides (ALE) are among the most studied compounds, which are recognized as essential compounds of the structure of biofilms in wastewater treatment systems. After extraction from aerobic granular sludge cultivated in a pilot plant, fed with municipal sewage containing 585 mg TCOD/L, an amount of 160 ± 4 mg/g (VSS ratio) has been measured (Lin *et al.* 2010). The implementation of a recovery process for these compounds can be easily accomplished by collecting the purge of aerobic granules from the secondary treatment as indicated in Figure 18.1.

18.1.2 Most studied processes

Among the aforementioned value-added products only a limited number of processes are under study and development to be implemented in STPs, and some of them are summarized in Table 18.1. Wastewater and primary and secondary sludge have been used as feedstock for VFA, alcohols and PHA production either in lab scale or pilot scale reactors.

18.1.2.1 Acids and alcohols

As mentioned before several streams present in STPs are eligible for acids production and the final composition of the product depends considerably on the waste and on the operating conditions of the acidogenic reactor (Ucisik & Henze, 2008). This is a very important feature because it is critical for downstream applications (production of PHA, bioenergy, etc.). The sludge generated in STPs can be used for the production of acids and alcohols due to the high content in organic matter although the soluble COD concentration is from ten to hundred times lower than its total COD content (Lee *et al.* 2014). For this reason hydrolysis of insoluble compounds is required and is frequently the rate limiting step of the process.

Acids production under anaerobic conditions has been studied since a long time ago. Sans *et al.* (1995) selected the organic fraction of the municipal solid waste separated mechanically in a full-scale treatment plant as feedstock. They obtained 23.1 g VFA/L at an organic loading rate of 38.5 g VS/(L d) (Sans *et al.* 1995). Other study compared fixed bed and suspended biomass pilot-scale reactors for acid production from sewage. Better results were obtained for the fixed bed and a yield of almost 0.08 g HAc/g COD was achieved for 1.91 g COD/(L d) and a hydraulic retention time of 3.4 days (Colmenarejo *et al.* 2004). Recently, Longo *et al.* (2015) also reported the production of acids from sewage using an alkaline fermentation system using wallastonite. Different scenarios were tried -due to the different concentrations of wollastonite- with acids production yields in the range of 187–259 mg VFA_{COD}/g TVS (Longo *et al.* 2015). Bench scale fermentations were carried out obtaining yields of 0.10 g VFA_{COD}/g TCOD and between 0.15–0.46 g VFA_{COD}/g TCOD of waste activated sludge and thermal pre-treated sludge, respectively (Morgan-Sagastume *et al.* 2011).

Table 18.1 Summary of processes where value-added compounds are produced, by microbial mixed cultures, using sludge as feedstock.

Feedstock	Performance	Process/Scale	References
Final Product Acids and Alcohols			
Any biodegradable material (e.g. municipal solid waste and sewage sludge)	–	MixAlco® process Lab scale	Holtzapple *et al.* (1999)
Office paper waste and chicken manure	6.67 kg alcohols/100 kg waste	MixAlco® process Pilot scale	Vasquez *et al.* (2014)
Organic fraction of municipal solid wastes	23.1 g VFA/L por 38.5 g/(L d) of waste	Pilot scale	Sans *et al.* (1995)
Sewage	0.08 g HAc/g COD	Pilot scale	Colmenarejo *et al.* (2004)
Sewage	–	Pilot scale (alkaline fermentation)	Longo *et al.* (2015)
Waste activated sludge	0.10 g VFA_{COD}/g TCOD	Lab scale	Morgan-Sagastume *et al.* (2011)
Thermal pre-treated sludge	0.15–0.46 g VFA_{COD}/g TCOD	Lab scale	Morgan-Sagastume *et al.* (2011)
Final Product-PHA			
Excess sludge having treated MWW	56.5% PHBV	Lab scale	Mengmeng *et al.* (2009)
Sewage	25–32%	–	Chua *et al.* (2003)
Fermented excess sludge	59.47%	Pilot Plant (China)	Jia *et al.* (2014)
Fermented WAS	34%	Pilot Plant (Belgium)	Morgan-Sagastume *et al.* (2014)

Alcohol fuels have been proved to be produced by the MixAlco® technology. This patented process is able to transform any biodegradable material (municipal solid waste and/or sewage sludge among others) (Holtzapple *et al.* 1999) into biofuels. The feasibility of this process was tested at pilot scale (Vasquez *et al.* 2014) using a mixture of chicken manure and shredder waste office paper for the production of several types of alcohols. Vasquez *et al.* (2014) obtained 6.67 kg of alcohols (C3–C13) for each 100 kg of a mixture of chicken manure and paper waste in a ratio 2:98 w/w.

18.1.2.2 Biopolymers

Recent investigations have been focused in researching about the potential of using microbial mixed cultures and wastes as feedstock (Morgan-Sagastume *et al.* 2015). However, no industrial development has been done yet although some preliminary studies asses that PHAs are more economically and environmentally beneficial in comparison to biogas production (Gurieff & Lant, 2007).

Initial research works were performed using synthetic substrates as feeding but recently a wide range of different wastes have started to be used as feedstock for PHA production (Serafim *et al.* 2008). Even sewage and the solids produced in STPs have been found to be suitable for PHA production. At bench-scale, several authors have tested the feasibility of obtaining biopolymers using different streams available in STPs: PHA production using fermented excess sludge having treated municipal wastewater obtaining a PHBV copolymer content of 56.5% (Mengmeng *et al.* 2009); a PHA content of 25–32% was achieved using aerobic sludge with municipal wastewater as carbon source (Chua *et al.* 2003).

Just few pilot plants have been reported up to now using waste streams from STPs. One of them was located in Belgium at Brussels North STP (Aquiris, Belgium) where fermented sludge was used for VFA production and the maximum PHA amount reached a value of 34% (Morgan-Sagastume *et al.* 2014). A pilot plant was operated in China using the excess sludge of the WWTP of Wuxi to produce VFAs for PHA production, reaching values as high as 59.47% PHA in biomass. Although the obtained results are promising no data regarding the operation of larger plants are available up to date.

18.1.2.3 Methane

The conventional use of the anaerobic digestion is the production of biogas, mainly composed of methane (50–70 vol%). This biological treatment leads to the reduction of solids for disposal, destroys most of the pathogens and produces a valuable energy source. The annual potential of biogas production in Europe is estimated in excess in 200 billion m^3 (Appels *et al.* 2008). Despite all these advantages, the anaerobic digestion has some limitations like the slow reaction rate, sensibility to potential inhibitors, presence of other compounds (i.e., hydrogen sulphide) in the gas phase with the need of purification in case of further uses of the biogas, etc. (Appels *et al.* 2008).

18.2 PROCESSES AND TECHNOLOGIES

18.2.1 Acids and alcohols

One example of carboxylic acids, including acetic acid production, is described by Lopez-Garzon and Straathof (2014). These authors suggested that after the fermentation stage the most probable configuration of downstream processing includes clarification, primary recovery, counterion removal, concentration/purification, upgrading and formulation (Figure 18.2). The contribution of downstream processing in the production of carboxylic acids (citric, lactic, succinic and itaconic acids) is estimated for 30–40% of the total production costs (Lopez-Garzon & Straathof, 2014). The challenge is to design cost-effective and at the same time efficient recovery process.

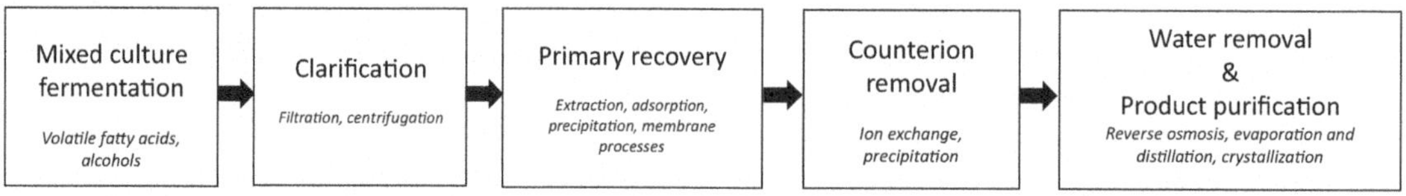

Figure 18.2 General sequence processing for the volatile fatty acids production and recovery (adapted from Lopez-Garzon and Straathof (2014)).

The purity of the final product is typically the key issue and the requirements will depend on the application. Most of the chemical applications of carboxylic acids require purity above 99.5%, although some exemption may occur. Another important requirement is the high extent of recovery as well as low chemicals and energy consumption during downstream processing (Lopez-Garzon & Straathof, 2014).

In the case of VFA (acids like acetic, propionic, butyric, valeric, ...) production an alternative is to operate a reactor in alkaline conditions to acidify the sewage sludge followed by a membrane filtration system for product recovery (Longo *et al.* (2015). Other alternatives are more complex and comprise more units including a sequence of steps such as: fermentation, descumming, dewatering and crystallization, ketonization, distillation, hydrogenation and finally oligomerization (Holtzapple *et al.* 1999). In this case the fermentation takes place in batch mode, the methanogenesis is inhibited by addition of iodoform and the liquid effluent is separated from the biomass using a screw press. Last steps of the process are devoted to products purification. The usual concentration of VFAs at the STP is around 2 g/L, which makes it very challenging for efficient recovery. Different techniques for separation of VFAs such as liquid-liquid extraction, membrane-based solvent extraction, pertraction, membrane separation were briefly described in Singhania *et al.* (2013).

Different approach was presented by Steinbusch *et al.* (2011) and Agler *et al.* (2012). According to Steinbusch *et al.* (2011) the VFAs themselves are not suitable for direct fuel use because of their quite high oxygen-to-carbon ratio and low energy density. However, they could be used as substrates for medium chain fatty acids production (MCFA), which might be more suitable for further processing compared to VFAs. The advantages of MCFA are their longer hydrophobic carbon chain and lower oxygen-to-carbon ratio. Consequently, MCFA have higher energy density and are easier separable from the fermentation broth, compared to VFAs. MCFA can be produced from VFAs and ethanol through β-revers oxidation. Steinbusch *et al.* (2011) achieved the highest concentration of caproic acid at the pH of 7 and temperature of 30°C, however an expensive inhibitor of methanogenesis must be applied. Agler *et al.* (2012), on the other hand, achieved promising results at the pH 5.5 (without the need of inhibitor addition) due to in-line extraction of caproic acid. Ge *et al.* (2015) proved that long-term caproic acid production is possible and that the microbial community during the process is stable. The obtaining of a percentage of 70% of n-caproate yield is feasible, which resulted in the concentration of ca. 3.4 g/(L · d). Grootscholten *et al.* (2013) was able to reach 57.4 g MCFA/(L · d) in an upflow anaerobic filter reactor using a two phase system: hydrolysis and acidogenesis in the first phase reactor and chain elongation in the second one.

Alcohols production via mixed culture fermentation has not been investigated in pilot or full-scale, even though, the metabolic pathways in anaerobic mixed culture include ethanol and butanol as products (Gonzalez-Cabaleiro *et al.* 2015). The downstream processing of alcohols recovery might follow the process configuration applied in lignocellulosic ethanol or butanol biorefineries (Kumar & Murthy, 2011).

18.2.2 PHA

PHAs are polyesters with thermoplastic properties and for this reason they are a promising alternative to conventional plastics, with the advantage of being fully biodegradable and biocompatible.

PHAs are produced by a large number of bacteria as internal reserves of carbon and energy. This ability is increasingly exploited by the biotechnological industry, and a recent report has estimated a market increase from 10,000 MT in 2013 to 34,000 MT in 2018 for PHA, (Markets & Markets Analysis, 2013). However, the costs of production and extraction are still too high to compete directly with synthetic plastics. The industrial production of PHAs has so far been based on fermentation with natural or recombinant bacterial strains using expensive chemically-defined feedstocks. Alternatively, the production of PHA using mixed microbial cultures (MMC) imply a substantial reduction in operating costs since there is no need to maintain aseptic conditions. Non-aseptic conditions also enable the use of wastewater or industrial sub-products with low commercial value as feedstock, which allows waste valorisation while further reducing costs (Du *et al.* 2012). The production of PHA from wastewater has been demonstrated for different municipal (e.g. Coats *et al.* 2011) and industrial effluents, including agro-industrial, brewery and paper mill (Khardenavis *et al.* 2007), achieving over 50% of the cell dry weight in PHA.

The use of mixed cultures requires a preliminary step of enrichment, where the microbial community is subjected to a selection pressure to maximise the number of PHA-storing organisms. Once the microbial culture is enriched, this biomass can then be forced to maximise their PHA storage by limiting their growth (most often through nutrient limitation). Since PHA producing organisms use VFA as substrate, a previous step of acidogenic fermentation is normally required. The three-stage process is thus comprised of an anaerobic fermentation step, a culture selection step and an accumulation step (Albuquerque *et al.* 2011).

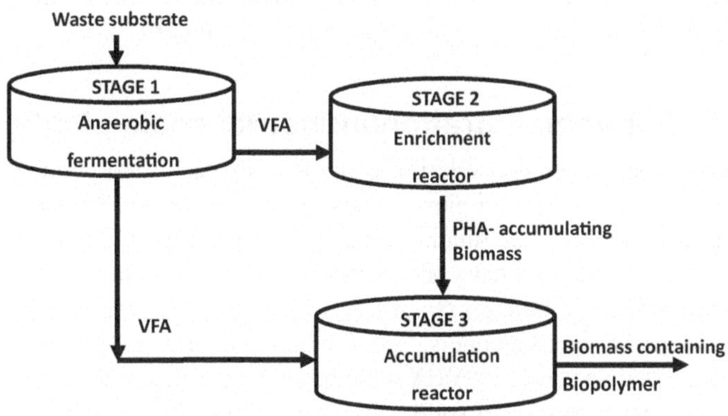

Figure 18.3 PHA production process scheme comprising three steps.

Microbial PHA synthesis occurs in growth limiting conditions. Growth limitation in pure cultures is imposed by limiting one of its essential nutrients. However, MMC systems have to guarantee the enrichment of the culture in PHA storing organisms. PHA production using MMC relies on imposing an internal growth limitation that shifts the carbon uptake for storage instead of growth. This limitation is often obtained by alternating substrate availability in feast and famine (FF) cycles (Majone *et al.* 1996) by applying the so called aerobic dynamic feeding (ADF), which compel the cells into a physiological adaptation that will result in slower growth rates when substrate is again available, in the following cycle. Under such conditions, external carbon is directed for storage by those microorganisms that have that capacity, giving them a competitive advantage in the subsequent famine phase.

A FF operation strategy is the most commonly employed fully aerated process to select an effective PHA storing mixed community. However, PHA storage also occurs when the limiting substrate is oxygen, i.e., in alternating anaerobic/aerobic conditions. This mechanism is carried out by e.g. polyphosphate accumulating organisms (PAO) and glycogen accumulating organisms (GAO), two microbial groups commonly found in STP. The production of PHA by GAOs was investigated by e.g. Bengtsson (2009), and a patent was issued (Bengtsson *et al.* 2010). Alternating anaerobic-aerobic conditions allow GAOs utilising their glycogen pool to produce PHA. A co-polymer is thereby obtained even when acetate is used as the sole carbon source, in contrast to the PHB mono-polymer obtained in fully aerated processes, which is known to be more brittle.

Another alternative to aerated FF processes consists on the use of photofermentation, where a photosynthetic mixed culture is used for PHA production in non-aerated FF conditions (Fradinho *et al.* 2013a). The process can potentially use natural sunlight as the energy source instead of aeration, increasing sustainability of PHA production. The PHA to substrate yield of this process can be as high as 0.9 C-mol PHA/C-mol acetate when operated with dark/light cycles (Fradinho *et al.* 2013b). This value is higher than most aerated FF processes reported due to the fact that carbon losses by respiration are minimised. When fed with VFA mixtures, the photosynthetic mixed culture was also capable of producing a hydroxybutyrate (HB): hydroxyvalerate (HV) co-polymer at similar PHA production yields (Fradinho *et al.* 2014). This approach is a promising alternative due to the potential savings in operational costs associated to aeration, which represents the highest fraction of the energetic costs of a WWTP.

Recently, Frison *et al.* (in press) demonstrated that PHA production can be linked to nitrogen removal in the nitrite pathway by alternating an aerobic feast phase with an anoxic famine phase. Selection of PHA storing and nitrifying organisms could be accomplished both, in the same reactor or separate reactors, although higher N-removal is achieved in the latter case (up to 90%). This process requires only approximately 36% of the oxygen costs of a conventional fully aerated FF process.

18.2.3 Reported pilot/demonstration/industrial scale plants

Although many organic acids are made by living cells, few are produced commercially. Citric, lactic, gluconic, itaconic, acetic, succinic and propionic acids are manufactured via large-scale bioprocesses. In all cases pure cultures of filamentous fungi, yeasts or bacteria are used. Lately, mixed fermentation processes via mixed bacterial consortia have been proposed for the simultaneous production of some of the above. During mixed fermentation short chain fatty acids (acetic, propionic, butyric and valeric) are cogenerated at different ratios whereas alcohols such as ethanol and butanol, are also produced.

However, many processes do not require VFA separation and purification, since they can make use of a complex product. In fact, fermentation of solid wastes (including sludge) and wastewater by undefined mixed cultures has long been carried out to produce mixtures of VFA that can then be applied for energy or PHA production, or for dosing in STP for improved biological nutrient removal (Thomas *et al.* 2003).

Longo *et al.* (2015) operated a pilot plant comprising a 500 L stirred reactor and the process was run at 35°C in a semi-continuous mode with HRT between 4.6 and 5.9 days (Table 18.2). The process of short chain fatty acids from sewage was carried out in alkaline conditions at pH of 9.5–10.5 by adding caustic soda or NaOH. The concentrations of VFAs reached a value of 7.4 g COD/L. The separation was carried out by employing two tubular cross-flow ultrafiltration membrane modules, operating in the inside–outside filtration mode. The membrane modules were made of polyvinylidene fluoride with internal diameter of 8 mm and molecular weight cut off of 15 kDa. The length of the membrane was 1 m and each module had a filtration area of 0.32 m^2. The maximum pressure of the module was 600 kPa and the maximum operating temperature was 40°C (Longo *et al.* 2015). The filtration process was operated in 10 h/d batch mode. The purpose of VFA production was to use separated fermentation liquid as a carbon source for biological nitrogen removal, therefore no further intensification of the product was necessary.

Another process configuration tested at pilot scale is the MixAlco™ process. Although data provided have been obtained in a pilot scale facility run on paper wastes and chicken manure as substrate (Vasquez *et al.* 2014), this is a patented technology applicable for alcohol fuels obtaining from any biodegradable material (Holtzapple *et al.* 1999). The fermentation pilot process run at the fed-batch mode with the C:N ration of 30 at the temperature of 40°C and the pH between 5.5 and 7.0. The fermenter was operated in non-sterile conditions and inoculated with marine soil. The methanogenesis process was inhibited by addition of iodoform on a daily basis. The retention time ranged between 7 and 10 days, depending on the carboxylic acids concentrations (10–15 g/L was the desired level). The effluent (rich in carboxylic acids) was separated from the fermentation broth by screw press. Authors reported that if the fermentation is prolonged to 30 days, almost 28 g/L of carboxylic acids would be possible to achieve, however near the end of the process the reaction rate was very low. The optimal conditions were achieved after 7–10 days of the batch experiment, reaching on average 12.5 g/L of carboxylic acids. The dominant acids were acetic (3.5 g/L) and butyric (3.5 g/L) as well as caproic (above 3 g/L) (Vasquez *et al.* 2014).

Table 18.2 Reported pilot plants for carboxylic acids and alcohols production from mixed microbial consortia.

Origin of Culture	Substrate	Reactor Volume (L)	Operational Mode	Products of Interest	Recovery Method	References
None	Sewage sludge	500	Semi-continuous, HRT 4.6–5.9 d	VFA	Membrane separation	Longo *et al.* (2015)
Marine soil	paper wastes and chicken manure	–	Batch, HRT 7–10 d	HAc, HPr, HCa	Dewatering, Crystalization, Ketonization, Distillation, Hydrogenation Oligomerization	Vasquez *et al.* (2014)

PHB and PHB-HV were firstly introduced to market by the British company Imperial Chemical Industries PLC (ICI) in the mid 80's under the trademark Biopol®. However, the product failed to be antagonistic in the market due to its high production cost. Since then PHB and its co-polymers have been commercialized successfully by other companies, including Biomer® (UK), Nodax® (UK), Bio-on (Italy), Metabolix (USA), Jiangsu Nantian Group (China), Mitsubishi Gas (Japan) and PHB Industrial (Brazil) (Chen, 2010; Serafim *et al.* 2008). In general pure cultures of genetically modified microorganisms are used as PHAs producers in industrial scale applications whereas considerable amounts of additives are blended with the recovered PHAs to achieve desired properties of the final product. Although production of PHAs from pure cultures is already fully industrialized since the 90's, it seems that the application of mixed cultures production at such scales has a long way to go. So far there are few studies/reports on pilot scale PHAs production from mixed cultures using municipal wastewaters as feedstocks. Process economics have revealed that inexpensive and renewable carbon substrates can be employed as PHA carbon feedstock, contributing as much as 40–50% reduction in the overall production cost, while the use of mixed bacterial cultures, with no need for aseptic conditions, could contribute to another 30–40% reduction (Serafim *et al.* 2008). These experiences are summarized in Table 18.3.

Table 18.3 Reported pilot plants for PHAs production from mixed microbial consortia.

Origin of Culture	Substrate	Operating Volume (L)	Operational Mode	Max PHAs Yield	PHAs Composition	References
STP activated sludge	Sludge fermentation liquid	70	Batch	0.59 g PHA/g VSS; 0.172 g PHA/g of COD	nd	Jia *et al.* (2014)
Municipal wastewater	Fermented WAS centrate	400	Continuous	0.38 g PHA/g VSS	copolymer HB-HV = 70:30	Morgan-Sagastume *et al.* (2015)

nd: not determined

18.3 QUANTITY, QUALITY AND APPLICATIONS

18.3.1 PHA

18.3.1.1 Feedstock requirements for sustainable productivity

To achieve a viable PHA production, several important requirements must be fulfilled such as: treating a stream with high COD concentration, controlling the cellular and hydraulic retention times and influent flow rate among others (Gurieff & Lant, 2007). Furthermore the correct selection process of the MMC is very important, which depends directly on the operational conditions of the enrichment reactor which is usually operated in ADF conditions (Valentino *et al.* 2015).

The conventional proposed system of PHA-production using MMC, as described in Figure 18.3 following the process proposed by Serafim *et al.* (2008), is commonly used when treating wastewater with high contents in organic matter (i.e., waste activated sludge, industrial wastewaters). Waste activated sludge can be treated in this way for PHA-production but, when just treating urban wastewater, the organic carbon content is much lower and this conventional method is not adequate. In the last years, new research has been done in this field. Recent investigation suggests the use of municipal wastewater for the MMC enrichment and the use of VFAs from the sludge fermentation just for the accumulation stage (Morgan-Sagastume *et al.* 2015; Valentino *et al.* 2015).

PHA production from sludge and municipal wastewater has been evaluated at lab-scale (Morgan-Sagastume *et al.* 2014) and pilot scale (Morgan-Sagastume *et al.* 2015). In these studies, wastewater with 290–570 mg COD/L was used to select the accumulating microorganisms from activated sludge under feast-famine regime. Then, the maximum accumulation of those selected microorganisms was evaluated by using fermented WAS with 200–700 mg COD/L as feeding, achieving up to 0.39 g PHA/g VSS.

PHA production from high organic matter content streams is preferable. The cost to kg of produced PHA ratio decreases as the concentration of COD increases and also depending on the flow rate (Gurieff & Lant, 2007). These authors stated that values of 3$/$kg_{polymer}$ can be achieved for an effluent of 20 g COD/L and a flow rate of 1000 m^3/d.

Substrate composition has been observed also to influence the storage yields as longer-chain fatty acids result in a lower amount of PHA formation (Serafim *et al.* 2008). Although, there is not full consensus and other authors established that this is not completely true as butyrate has a higher theoretical yield than acetate (Marang *et al.* 2013). In the opinion of these authors, to select the best substrate for PHA production is difficult, as the enrichment process plays the main role at this point. Mixed microbial communities are usually selected by using a defined substrate and are optimised for that composition. Besides this, the properties of the biopolymer can be roughly inferred by the substrate composition as VFAs with even number of carbon atoms lead to the production of PHB and VFAs with odd number of carbon atoms lead to the production of PHBV copolymers (Lemos *et al.* 2006). Depending on other factors, like the microbial composition of the MMC, the HB:HV ratio can change even when the substrate has the same composition (Carvalho *et al.* 2014; Lemos *et al.* 2006).

18.3.1.2 Effects of operation parameters on polymer quality

To drive the PHA production according to the final application is expected to contribute to the reduction of production costs. The physical and chemical properties of PHA shall be controlled and manipulated by: (i) altering the medium in which the bacteria grow (i.e. nutrient availability, carbon source concentration and type, temperature and pH) (Keshavarzand, 2010), (ii) using of dynamic conditions for substrate feeding (alternating carbon availability and unavailability i.e. feast-famine regimes, length of the cycle, etc.) (Villano *et al.* 2010) or (iii) manipulating the electron acceptor availability (aerobic or anaerobic conditions) (Morgan-Sagastume *et al.* 2014), among others.

When starting-up a reactor for enrichment of PHA-accumulating bacteria, one of the first parameters to decide is the C/N ratio or the SRT. It was found that the operation of SBR units with limitation of carbon source (molar C/N ratios between 6–13) provides high substrate uptake rates while nutrient limitation (molar C/N ratios between 15–24) leads to high ammonia uptake rates (Johnson *et al.* 2010). These authors established that the manipulation of the SRT and C/N molar ratio during the enrichment stage greatly influenced the amount of PHA stored in the maximization batch assays. The accumulation value was over 70 wt% when operating at SRT values of 4 and 1 days and applied C/N ratios of 11 and 8 mol/mol, respectively.

Also the cycle length applied in the SBR for the selection of the microbial mixed community influences parameters as: (i) the length of the feast phase over the famine phase, (ii) the calculated yields, (iii) the PHA content in both enrichment and accumulation stages, (iv) the biomass composition (Dionisi *et al.* 2007); and also the feed to microorganisms ratio which affects the biopolymer composition (Moralejo-Gárate *et al.* 2013). These authors found that successful selection of mixed microbial communities with accumulating capacity occurs when the feast phase is no longer than 20% of the total time of the cycle.

It is clear that the microbial population developed during enrichment process determines the polymer characteristics as different copolymer compositions can be achieved even when using the same substrate (Lemos *et al.* 2006). For this reason those parameters exerting and effect on this microbial selection will indirectly affect the obtained biopolymer. Among these parameters, the substrate composition and organic loading rate or even the cycle length have been identified to produce an effect (Albuquerque *et al.* 2013; Carvalho *et al.* 2014; Dionisi *et al.* 2007). Furthermore, a previous clarification of the media containing the VFAs leads to a higher PHA production yield referred to dry weight, although, the biopolymer composition does not change (Ntaikou *et al.* 2014)

Microbial population composition can be also affected by operational parameters. According to Albuquerque *et al.* (2013), *Azoarcus* is well-known for consuming acetate and propionate, *Thauera* does not consume acetate as it prefers propionate and butyrate, and *Paracoccus* has no preference among all of the VFAs (Albuquerque *et al.* 2013). The distribution and abundance of the different genera leads to different biopolymer compositions and it also influences the maximum amount of biopolymer inside the cells. Higher amount of *Azoarcus* leads to greater PHA production yields and lower biomass growth in comparison with reactors in which the presence of *Thauera* or *Paracoccus* are in bigger proportion (Carvalho *et al.* 2014). Janarthanan *et al.* (2014) performed the fractionation of a PHA accumulating mixed culture by Percoll assisted buoyant density separation, and they found that in spite of having common bacteria in the fractions, the relative abundances of species were found to be different between the two fractions. A higher average

PHA content (24 wt%) was achieved in one of the fractions than in the other (16 wt%). Furthermore, the PHA capacity of the enriched culture has been found to depend on the substrate used for the population selection. Moralejo-Gárate *et al.* (2014) produced an MMC grown on glycerol able to use other substrates as glucose, lactate and acetate and obtaining similar PHA accumulation percentages as with glycerol.

The effect of the pH value on the PHA-production with mixed microbial cultures has been studied. It influences both the enrichment of the microbial community and the accumulation of the biopolymer inside the cells. Villano *et al.* (2010) investigated the performance of three SBRs operated for the enrichment of a PHA-accumulating microbial community. They were operated under the same conditions and also with the same inoculum, but with different pH values (7.5, 8.5 and 9.5). On the one hand, the obtained results for the enrichment stage in each SBRs were similar (between 11 to 13 wt% of accumulated PHA) although obtained substrate removal rates decreased in a 20% when operating at pH of 9.5. On the other hand, accumulation assays, using the enriched biomass at pH 8.5, were performed at different pH values. Results indicated that substrate removal rates and PHA storage were higher at pH 8.5 and 9.5 than at pH 7.5. The pH values also influenced the biopolymer composition as the HV content increased with pH in such a way that 20% mol HV/mol of VSS was achieved at pH of 7.5 and 36% mol HV/mol of VSS at pH of 9.5 (Villano *et al.* 2010). Serafim *et al.* (2004) stated that lower PHA accumulation was achieved when controlling the pH at 7.0 and 8.3 during the batch accumulation assays in comparison with a batch assay without pH control (pH between 8.0–10.0). The values for PHA accumulation were of 27.5 wt% and 39.8 wt% for the assays at pH of 7.0 and 8.3, respectively; while the accumulation reached with no pH control was of 47.5 wt% (Serafim *et al.* 2004).

Feeding mode is another influencing parameter which affects the performance of PHA-accumulating mixed microbial cultures (MMCs). In this regard, several considerations have to be taken into account: (i) feeding regime or strategy (pulses, continuous, ADF, etc.), (ii) substrate composition (pure, mixed or waste substrates), (iii) concentration and organic loading rate (OLR) applied. Albuquerque *et al.* (2011) studied the effect of using continuous or pulse feeding and they found that continuous feeding increases the maximum storage capacity (77 wt% against 65 wt% in pulse feeding). On the contrary, Serafim *et al.* (2004) obtained higher PHA accumulation with a pulse feeding (78.5 wt%) than with a continuous feeding (56.2 wt%). However, they used a concentration of 180 Cmmol/L of acetate in the continuous feeding experiment that were proven to be inhibitory for the PHA-accumulation.

It is important to remark that all operational parameters are linked, being very difficult to individually assess the influence of each one. Other parameters cannot be discarded as affecting the composition of the obtained biopolymer however those mentioned above have been discussed as they have been already identified and studied.

18.3.1.3 Applications depending on polymer quality

There are over 100 types of PHA, each with specific physical and mechanical properties. Although post-production modification of the polymer (and its properties) can be performed with the incorporation of additives, the type of PHA obtained is often controlled by manipulation of the production conditions. The carbon composition of the substrate is one of the most evaluated parameters to obtain a polymer of a specific composition, since the carbon chain length of the substrate is known to be conserved during polymerisation.

Regarding the highly diverse structure and properties of PHA, associated with their biodegradability and biocompatibility, they have successfully replaced conventional plastics in various applications. These uses mainly include packaging (mostly for food services) and fiber production, as well as agricultural, biomedical and pharmaceutical applications. The thermochemical and mechanical properties of various PHAs are mainly responsible for categorizing different PHAs as suitable for different applications. Properties of a PHA as final product are highly dependent on its monomeric composition, the ratio of different monomers, the structural composition among monomers (i.e. forming co-polymers of blends of different homo-polymers) and the molecular weight of the polymer. To a less extent, properties can be affected by the degree of polymerization and crystallization whereas the presence of additives can be of extreme importance. The PHA monomeric composition is directly associated to the nature of the precursors comprising the substrate (Albuquerque *et al.* 2011). Moreover, the specificity of the PHA synthase seems to play a role in the composition of PHA produced by pure cultures (Fuller *et al.* 1999), and also in mixed cultures, the microbial composition of the enriched consortium has been correlated with the polymer composition (Carvalho *et al.* 2014).

In general, biomedical applications require the use of medium chain length PHAs of high purity (Brigham & Sinskey, 2012), which are an ideal scaffold material in tissue engineering due to their high elongation at break (about 300%) and hydrophilicity (Du *et al.* 2012).

Other specific application niches (Table 18.4) for PHA include waste bags (EcoWorks, Purac, USA, using PHA produced from Mirel, Metabolix, USA). The use of biodegradable bags facilitates downstream anaerobic digestion of urban solid waste, and PHA has improved resistance as compared to e.g. starch-based bags. PHA coated paper or cardboard increases the durability of e.g. paper beakers (Metabolix, USA) or card packages (Tetrapak) while still maintaining high biodegradability. PHA-based wrapping packaging material is used for e.g. in cosmetics (O Boticario, Biocycle).

Table 18.4 Possible applications for PHA and required properties.

Application	Process	Required Properties	References
Credit or gift cards Indoor/outdoor signage Plant pots and plant tags	Cast sheet extrusion	High melt strength material	http://www.matweb.com (Metabolix) http://www.biocycle.com.br
Packaging films Agricultural mulch films Laminates Diaper backsheet	Film	High melt strength	Du *et al.* 2012 http://www.biocycle.com.br
Bottles Cosmetic containers Pens and other stationery	Moulding	High toughness	Du *et al.* 2012 http://www.biocycle.com.br http://www.metabolix.com
Cold and hot cups, cup lids, yogurt containers, tubs and trays for meats and vegetables, other single-serve and disposable food packaging	Thermoforming	–	http://www.matweb.com
Caps and closures, House wares Tubs, trays, jars, and consumer product applications	Injection moulding	–	http://www.matweb.com (Metabolix)
Nonwoven fabrics Adhesives	Fibres	–	Du *et al.* 2012
Paper coating	Latex technology	–	Du *et al.* 2012 http://www.metabolix.com

18.3.2 Acids and alcohols

18.3.2.1 Feedstock requirements for sustainable productivity

Acids

A great variety of solid and liquid wastes have been studied as potential feedstock for organic acid production (Lee *et al.* 2014). Regarding municipal wastewater treatment plants, activated sludge process generates a great amount of sewage sludge that can be used for that purpose due to its high organic matter content (Table 18.1). In this way, both waste management and the obtaining of added value products will be achieved. Primary sludge and waste activated sludge are the two main solid wastes generated in municipal STPs. As it was previously mentioned they can be used to produce VFAs as sole sources (Morgan-Sagastume *et al.* 2014) or as a mixture of both, acids and alcohols (Ucisik & Henze, 2008).

Sometimes the low biodegradability of the sludge may require a pre-treatment prior the acidogenic fermentation to increase the fermentation yield. Different processes have been proposed to improve the solubilisation of the sludge, such as thermo-alkaline pre-treatment (Vlyssides & Karlis, 2004) or high pressure thermal hydrolysis (Morgan-Sagastume *et al.* 2011).

The composition of the effluent of the acidogenic fermentation in terms of organic acids will determine the suitability of the effluent for different purposes. Depending on the feed composition and the operational conditions the distribution of acids will change. Propionate and acetate were both dominant VFAs when fermentation of different primary sludge was tested, while acetate was the main organic acid when WAS was studied and propionate with a mixture of both (Ucisik & Henze, 2008). These authors also found higher specific VFA production (270 mg COD_{VFA}/g VSS) with primary sludge compared to the mixed and waste activate sludge experiments (114 and 62 mg COD_{VFA}/g VSS, respectively).

Product inhibition should also be taken into account. A threshold of 17 ± 1 g COD_{VFA}/L was found when a combination of VFA was obtained using pre-treated WAS (Pratt *et al.* 2012). Effects of several different inhibitors have been recently investigated on the performance of the anaerobic digestion of the sewage sludge (e.g. Lins *et al.* 2015). This research work was mostly focused on methane yield, methanogenesis inhibition and VFAs degradation and not on VFAs productivity.

Alcohols

Cheung *et al.* (1997) tested the use of municipal wastes (primary wastewater solids) as bioethanol feedstock due to its cellulose and lignin content and achieved overall conversion efficiencies of cellulose to ethanol in the range of 17–60%. Steinbusch *et al.* (2008) carried out the alcohol production by reduction of VFA with hydrogen as electron donor in the presence of granular sludge. They obtained concentrations of 3.7 mM of ethanol, 8.1 mM of propanol and 3.7 mM of n-butanol in batch experiments.

However, no results are available using VFA mixtures produced after acidification of activated sludge. In this line the production of alcohols (2-propanol principally) is been reported feasible by the MixAlco process (Holtzapple *et al.* 1999) using activated sludge, but no results in this sense are provided.

18.3.2.2 Effects of operation parameters

In natural conditions, the mixed culture fermentation will lead to methane production through anaerobic digestion process. To intensify VFA production, methanogenesis must be inhibited by different strategies such as addition of specific inhibitor (e.g. 2-bromoethanesulfonate), pH and temperature control, hydrogen concentration control, electric stimulation or by appropriate choice of micro- and macronutrients for microorganisms (Steinbusch *et al.* 2011).

The main operational conditions that affect the concentration, yield and composition of VFA produced from waste are: pH, temperature, HRT, SRT, OLR and additives (Lee *et al.* 2014). This control is important because downstream applications will depend on the composition of the rich-VFA stream generated.

One of the most crucial parameter for intensification of VFA concentration is the pH. The acidogenic microorganisms can survive in the range of pH between 3 and 12. According to Chen *et al.* (2007), VFA production from WAS under alkaline conditions is higher due to the hydrolysis of the sewage sludge by ionization of extracellular polymeric compounds. These compounds (mainly proteins and carbohydrates) are released through strong repulsion and become substrates for acidogenic microorganisms (Lee *et al.* 2014). Additionally, at pH above 8.5 the methanogenesis is fully inhibited.

Not only the total VFAs concentration but also the composition in particular acids depends on the pH value. The pH has impact especially on acetic, propionic and butyric acids ratios. Typically, at low pH ranges butyrate and acetate are the dominant products, whereas at high pH ranges, butyrate production should decrease. The shift from butyrate to acetate/ethanol production is typically noticed in the pH range between 6.5 and 8.0 (Jankowska *et al.* 2015). The exact operational conditions, when the shift from butyrate to acetate/ethanol occurs, are not well understood and they will depend on the specific mixed culture applied as inoculum.

Temperature is another factor which influences the VFAs productivity. Typically, higher temperature increases the fermentation rate. Thermophilic conditions increase the performance of the fermentation but decrease the stability of the process (Labatut *et al.* 2014). Due to the fast hydrolysis, organic overloading due to the intermediate products might occur which would lead to the unbalancing of the whole process.

Prolonging hydraulic (HRT) and solids retention time (SRT) will have positive impact on VFA concentration. However, the longer retention time leads to bigger fermentation tanks. The benefit of higher VFAs concentration has to overcome the higher cost of the bioreactors. The exact breaking point of costs vs. concentration has to be calculated for each specific case. Additionally the requirements of products concentration needed for efficient downstream processing has to be taken into account. Another factor is the risk of development of methanogens which typically need to operate at longer retention time compared to acidogenic organisms. Feng *et al.* (2009) increased the retention time from 4 to 12 days which resulted in 44% improvement of VFA yield (produced from primary sludge). However, further prolongation led to the decrease of the concentration, most likely caused by growth of methanogens.

In the case of the OLR, an overloading due to VFA accumulation may occur relatively fast when the substrates are easily hydrolysed while it takes longer in the case of more resistant substrates (e.g. rich in lignocellulose). However, in the latter case, sudden accumulation of inhibitors released during hydrolysis of lignocellulose might occur (Sawatdeenarunat *et al.* 2015). Slight overloading is favourable for acidogenesis because it inhibits the methanogenic step, but there is a risk of a complete inhibition of the process.

Organic carbon and inorganic nitrogen are the most essential nutrients for cell synthesis, growth, and metabolism. An optimal balance of carbon to nitrogen (C/N ratio) is an important parameter which has been used in optimizing anaerobic digestion. High C/N ratio means deficiency of nitrogen and low fermentation rate due to insufficient cells to maintain high microbial activity. On the other hand, low C/N ratio means high nitrogen content which leads to ammonia emission that is toxic to microorganisms and can inhibit and stop the process. Most studies are focused on investigating C/N ratio effects on methane productivity, with optimal values between 25 and 30 (Silvestre *et al.* 2015). Rughoonundun *et al.* (2012) studied different C/N ratios to determine the optimal conditions for VFAs production using as substrate a mixture of sewage sludge and bagasse (C-rich substrate to increase the C/N ratio). The optimal values were found in a range between 13 and 24 and an increase to a value of 32 caused a drop in production by 16%. The lowest production was obtained at the maximum investigated value (C/N = 65). According to Silvestre *et al.* (2015), increasing the C/N ratio to 87 by the addition of crude glycerine caused inhibition for both methane and VFAs production. High concentration of VFAs was achieved at C/N of 11, the shift between VFAs and methane production depended on organic loading rate and the pH value. Not only is the yield affected by the nutrients balance but also the composition of VFAs. The acetic acid is a major component regardless the C/N ratio, being dominant at high C/N ratio (e.g. 93% for C/N of 64.6).

The microbial community can also have impact on the qualitative and quantitative VFAs production. Forrest *et al.* (2012) researched the performance of the process using six different mixed microbial cultures as inoculum. Even though the microbial community was different, the acids concentrations varied around 12% and the conversion only 6%. Authors concluded that in the acidogenic fermentation carried out by mixed cultures, operational parameters have the main influence on the product spectrum and different microbial structures can perform similar functions at the same operational conditions (Forrest *et al.* 2012).

Moreover, several pre-treatment methods may be needed to enhance the production of VFAs prior to the acidogenic fermentation. To date, the main studies on sludge pre-treatment have been focused on enhancing the biogas production process and there is not too much research on the pre-treatment methods influencing directly the VFAs productivity. Pre-treatment methods of sewage sludge were recently reviewed by Cano *et al.* (2015). Highly alkaline environment may serve as pre-treatment for sludge and in consequence increasing the rate of hydrolysis (Pang *et al.* 2015).

In the commonly accepted network of metabolic mixed culture fermentation reaction there are also pathways for ethanol and butanol production. The shift from butyrate to acetate/ethanol production depends on the pH and was described above as well as proved in a recent research work of Gonzalez-Cabaleiro *et al.* (2015), stating that it was possible to achieve high yields of ethanol when high pH values are applied. So far, the process parameters favouring high butanol yield has not been determined yet in the mixed culture environments.

18.3.2.3 Applications depending on acid and alcohols quality

VFA-rich streams can be utilized as substrate to produce bioplastics, generate bioenergy or carry out biological nutrient removal, among other applications (Lee *et al.* 2014).

Polyhydroxyalkanoates

As mentioned in section 18.3.1.1, VFA-rich streams are suitable to be used as feedstock for PHA production. The biopolymer properties will vary depending on the composition in VFA. Mengmeng *et al.* (2009) obtained a copolymer of HB:HV in a 88:12 ratio using WAS as carbon source after an alkaline fermentation.

In this case it is important to generate a stream rich in VFA but regulate the ammonium and phosphorous content. High amounts of nutrients would favour the biomass growth and the yield of PHA production over VFA will be reduced (Morgan-Sagastume *et al.* 2010).

Biogas

The conventional way to utilize VFAs generated during acidogenesis is the biogas production through methanogenesis. The most common process configuration is a one-step anaerobic digestion where hydrolysis, acidogenesis and methanogenesis take place in one bioreactor. However, there are efforts in developing efficient two-phase systems to intensify hydrolysis and acidogenesis in one reactor and optimize the methanogenesis in the second one (Ge *et al.* 2011). In this case, the dominance of acetate is preferable; however, it should not exceed the inhibitory level for methanogens.

Lipids for biodiesel

Oleaginous microorganisms have the capacity of converting organic carbon into internal lipid storage. This provides a new source for biodiesel production, since those lipids can be extracted and converted to biodiesel through a transesterification process. To date, most studies have been focused on the lipid production with glucose as sole carbon source and pure cultures. In order to reduce the operational costs, the use of new alternative carbon sources such as VFAs is starting to get an increasing attention (Fei *et al.* 2011).

Biological nutrient removal

In some urban wastewater treatment plants the complete nutrient removal cannot be achieved due to the fact that the influent does not contain enough carbon source. Then, it is necessary to add an external carbon substrate. Since the municipal WWTPs generate a great amount of sludge with high organic carbon content, its valorisation becomes a good option. In this way, sludge fermentation can provide both the required carbon source and the treatment of large amounts of WAS generated (Li *et al.* 2011).

Alcohols

The main application of alcohols is their use as alternative fuels. Bioethanol has potential as a valuable replacement of gasoline in the transport fuel market (Holtzapple *et al.* 1999). Butanol shows high energy content, low volatility and is less corrosive than other fuels. It can be produced using mainly fungal species and renewable feedstock, such as lignocellulosic materials (Cheng *et al.* 2012).

18.4 FUTURE PERSPECTIVES

As explained in previous sections PHA (extracted from the biomass) and VFA (solubilized) appear as the most studied compounds to be produced in STP. However, at the moment, research interest is being focused also in another kind of polymers like lipids (Fei *et al.* 2011), alginates (Lin *et al.* 2013), enzymes, proteins, and so on (Table 18.5) to valorise either wastewater or waste sewage sludge. As a consequence they are a way of valorisation and management costs reduction of the large amounts of sewage sludge produced. However, in most cases research works are being performed at laboratory scale and only a few at pilot scale in the case of PHA production (Morgan-Sagastume *et al.* 2015).

Another example is the production of alginate-like polymers that has been studied in a Nereda pilot plant. These polymers are suggested to play an important role in providing hydrophobicity, compactness and elasticity to aerobic granular sludge and the obtained yield reaches a value of 160 ± 4 mg/g (VSS ratio) (Lin *et al.* 2010). Alginate produced is under study to be used as surface coating material, as it easily forms a film on hydrophilic surfaces, functioning as a water resistant barrier (Lin *et al.* 2015).

On the other hand, lipids are produced from waste activated sludge (Mondala *et al.* 2009) or from microalgae applied for sewage treatment. These obtained fatty acids present excellent characteristics for the production of biodiesel via transesterification. These specialized organisms are grown on the lipids from the wastewater which represent around 40% of the total organic content, being most of them triacylglycerols (TAGs). It is estimated that around 7% of the oil in the wastewater can be contained in the dry secondary sludge.

Table 18.5 Summary of additional organic compounds potentially produced in STP from the waste produced sewage sludge.

Organic Compound	Type	Source	Origin of the Source	Production mg/g	Product Use	References
Alginate-like polymers	EPS	Urban + slaughter house (25%)	Nereda® pilot plant	160 ± 4[a]	Surface coating material	Lin *et al.* (2010, 2015)
Fatty Acids Methyl Esters	Cell constituent	Primary sludge Secondary sludge	STP	14.5[b], 2.5[b]	Biodiesel	Mondala *et al.* (2009)
Triacylglycerols	Intracellular	Urban wastewater	STP	–	Biodiesel	Muller *et al.* (2014)
Proteins	Cell constituent	Sewage sludge	STP	10–13[b]	Fertilizer	Liu *et al.* (2009)

[a]VSS ratio.
[b]as w/w%.

Mondala *et al.* (2009) estimated a production price for biodiesel obtained from in-situ transesterified dried sludge similar to the petroleum diesel and soybean biodiesel. Furthermore several kinds of organisms are able to accumulate either triacylglycerols (TAGs) or wax inside the cells (Muller *et al.* 2014). This is possible due to the fact that the TAG accumulating organisms excrete extracellular lipases which catalyse the hydrolysis of lipids from wastewater previously to their accumulation (Muller *et al.* 2014). As an example, *M. parvicella* population has been found to be specialized in lipid consumption in anaerobic conditions which provides it with competitive advantages against bacteria able to take them up only under aerobic conditions (Nielsen *et al.* 2002). In both cases the challenge relies on the extraction of these compounds for further use. Again, for these lipids to be competitive the price should be similar or lower than the petro diesel cost. The use of sludge or microalgae as source for biodiesel production would reduce the existing pressure on lipid edible feedstocks. This biofuel will contribute to the development of the third generation biofuels.

Proteins constitute the 61% of the activated sludge composition and can be recovered to produce amino acid chelated trace elements (AACTE) useful for fertilizer production (Liu *et al.* 2009).

The main drawback found in the case of those organic products present as intracellular or cell constituents compounds is related to the fact that they have to be extracted. For this reason in many cases the use of the sludge enriched in a certain compound without treatment is under study.

18.5 CONCLUSIONS

At the moment no data are available regarding the environmental, economic and technical impacts of biopolymers production systems from wastewater. Technologies based on these processes are recent and no industrial scale applications are available to produce these data. The only thing that is clear is the fact that these processes can be compared to present technologies applied for the same purpose considered in this case as anaerobic digestion (AD) and activated sludge (AS) systems (Box 18.1, Table 18.6).

BOX 18.1 COMPARISON OF THE BIOPOLYMERS PRODUCTION SYSTEM WITH ACTIVATED SLUDGE AND ANAEROBIC DIGESTION SYSTEMS

Table 18.6 Advantages and disadvantages of organic value added products recovery from wastewater compared to AD and AS.

		Anaerobic Digestion	Activated Sludge
Biopolymers (PHA)	Advantages compared to	Less complex process Product easier to handle No methane production	Less CO_2 production Less biomass production Value added product
	Disadvantages compared to	No energy production More biomass production More difficult to separate the product Discontinuous process	More complex process Only for concentrated wastewater
VFA and alcohols	Advantages compared to	Less complex process Product easier to handle No methane production	Less CO_2 production Less biomass production Value added product
	Disadvantages compared to	No energy production More biomass production More difficult to separate the product	More complex process

Most of the technologies referred to in this document present technology readiness levels (TRL) of 4–5. Only the technologies Mixalco for alcohols production (Holtzapple *et al.* 1999) and the PHA production from sludge developed by Morgan-Sagastume *et al.* (2015) present a TRL value of 6–7. Therefore more research work is needed to implement these processes at industrial scale.

18.6 REFERENCES

Agler, M.T., Spirito, C.M., Usack, J.G., Werner, J.J. and Angenent, L.T. (2012). Chain elongation with reactor microbiomes: upgrading dilute ethanol to medium-chain carboxylates. *Energy & Environmental Science,* **5**, 8189–8192.

Albuquerque, M.G.E., Martino, V., Pollet, E., Avérous, L. and Reis, M.A.M. (2011). Mixed culture polyhydroxyalkanoate (PHA) production from volatile fatty acid (VFA)-rich streams: effect of substrate composition and feeding regime on PHA productivity, composition and properties. *Journal of Biotechnology,* **151**(1), 66–76.

Albuquerque, M.G., Carvalho, G., Kragelund, C., Silva, A.F., Barreto Crespo, M.T., Reis, M.A. and Nielsen, P.H. (2013). Link between microbial composition and carbon substrate-uptake preferences in a PHA-storing community. *The ISME Journal,* **7**(1), 1–12.

Bengtsson, S. (2009). The utilization of glycogen accumulating organisms for mixed culture production of polyhydroxyalkanoates. *Biotechnology and Bioengineering,* **104**(4), 698–708.

Bengtsson, S., Pisco, A.R., Johansson, P., Lemos, P.C. and Reis, M.A.M. (2010). Molecular weight and thermal properties of polyhydroxyalkanoates produced from fermented sugar molasses by open mixed cultures. *Journal of Biotechnology,* **147**(3–4), 172–179.

Brigham, C.J. and Sinskey, A.J. (2012). Applications of polyhydroxyalkanoates in the medical industry. *International Journal of Biotechnology for Wellness Industries*, **1**, 53–60.

Carvalho, G., Oehmen, A., Albuquerque, M.G.E. and Reis, M.A.M. (2014). The relationship between mixed microbial culture composition and PHA production performance from fermented molasses. *New Biotechnology,* **31**(4), 257–263.

Cano, R., Perez-Elvira, S.I. and Fdz-Polanco, F. (2015). Energy feasibility study of sludge pretreatment: a review. *Applied Engineering,* **149**, 176–185.

Chen, G.G.Q. (ed) (2010). *Plastics from Bacteria: Natural Functions and Applications.* Springer-Verlag, Berlin, Heidelberg, Germany.

Chen, Y., Jiang, S., Yuan, H., Zhou, Q. and Gu, G. (2007). Hydrolysis and acidification of waste activated sludge at different pHs. *Water Research,* **41**(3), 683–689.

Cheng, C.L., Che, P.Y., Chen, B.Y., Lee, W.J., Lin, C.Y. and Chang, J.S. (2012). Biobutanol production from agricultural waste by an acclimated mixed bacterial microflora. *Applied Engineering,* **100**, 3–9.

Cheung, S.W. and Anderson, B.C. (1997). Laboratory investigation of ethanol production from municipal primary wastewater solids. *Bioresource Technology,* **59**(1), 81–96.

Chua, A.S.M., Takabatakeb, H., Satoha, H. and Minoa, T. (2003). Production of polyhydroxyalkanoates (PHA) by activated sludge treating municipal wastewater: effect of pH, sludge retention time (SRT), and acetate concentration in influent. *Water Research,* **37**(15), 3602–3611.

Coats, E.R., VandeVoort, K.E., Darby, J.L. and Loge, K.J. (2011). Toward polyhydroxlkanoate production concurrent with municipal wastewater treatment in a sequencing batch reactor system. *Journal of Environmental Engineering,* **137**(1), 46–54.

Colmenarejo, M.F., Sánchez, E., Bustos, A., García, G. and Borja, R. (2004). A pilot-scale study of total volatile fatty acids production by anaerobic fermentation of sewage in fixed-bed and suspended biomass reactors. *Process Biochemistry,* **39**(10), 1257–1267.

Dionisi, D., Majone, M., Vallini, G., Gregorio, S.D. and Beccari, M. (2007). Effect of the length of the cycle on biodegradable polymer production and microbial community selection in a sequencing batch reactor. *Biotechnology Progress,* **23**(5), 1064–1073.

Du, C., Sabirova, J., Soetaert, W. and Lin, S.K.C. (2012). Polyhydroxyalkanoates Production From Low-cost Sustainable Raw Materials, *Current Chemical Biology,* **6**(1), 14–25.

Fei, Q., Chang, H.N., Shang, L. and Choi, J.-d.-r. (2011). Exploring low-cost carbon sources for microbial lipids production by fed-batch cultivation of *Cryptococcus albidus. Biotechnology and Bioprocess Engineering,* **16**, 482–487.

Feng, L., Wang, H., Chen, Y. and Wang, Q. (2009). Effect of solids retention time and temperature on waste activated sludge hydrolysis and short-chain fatty acids accumulation under alkaline conditions in continuous-flow reactors. *Bioresource Technology,* **100**, 44–49.

Forrest, A.K., Hollister, E.B., Gentry, T.J., Wilkinson, H.H. and Holtzapple, M.T. (2012). Comparison of mixed-acid fermentations inoculated with six different mixed cultures. *Bioresource Technology,* **118**, 343–349.

Fradinho, J.C., Domingos, J.M.B., Carvalho, G., Oehmen, A. and Reis, M.A.M. (2013a). Polyhydroxyalkanoates production by a mixed photosynthetic consortium of bacteria and algae. *Bioresource Technology,* **132**, 146–153.

Fradinho, J.C., Oehmen, A., Reis, M.A.M. (2013b). Effect of dark/light periods on the polyhydroxyalkanoate production of a photosynthetic mixed culture. *Bioresource Technology,* **148**, 474–479.

Fradinho, J.C., Oehmen, A. and Reis, M.A.M. (2014). Photosynthetic mixed culture polyhydroxyalkanoate (PHA) production from individual and mixed volatile fatty acids (VFAs): substrate preferences and co-substrate uptake. *Journal of Biotechnology,* **185**, 19–27.

Frison, N., Katsou, E., Malamis, S., Oehmen, A. and Fatone, F. (2015). Development of a Novel Process Integrating the Treatment of Sludge Reject Water and the Production of Polyhydroxyalkanoates (PHAs). *Environmental Science & Technology,* **49**, 10877–10885.

Fuller, R.C. (1999). Microbial inclusions with special reference to PHA inclusions and intracellular boundary envelopes. *International Journal of Biological Macromolecules,* **25**, 21–29.

Ge, H., Jensen, P.D. and Batstone, D.J. (2011). Relative kinetics of anaerobic digestion under thermophilic and mesophilic conditions. *Water Science and Technology,* **64**(4), 848–853.

Ge, S., Usack, J.G., Spirito, C.M. and Angenent, L.T. (2015). Long-term n-caproic acid production from yeast-fermentation beer in an anaerobic bioreactor with continuous product extraction. *Environmental Science & Technology,* **49**, 8012–8021.

Gonzalez-Cabaleiro, R., Lema, J.M. and Rodríguez, J. (2015). Metabolic energy-based modelling explains product yielding in anaerobic mixed culture fermentations. *PLoS ONE,* **10**(5): e0126739. doi:10.1371/journal.pone.0126739.

Grootscholten, T.I.M., Steinbusch, K.J.J., Hamelers, H.V.M. and Buisman, C.J.N. (2013). Improving medium chain fatty acid productivity using chain elongation by reducing the hydraulic retention time inanupflowanaerobic filter. *Bioresource Technology,* **136**, 735–738.

Gurieff, N. and Lant, P. (2007). Comparative life cycle assessment and financial analysis of mixed culture polyhydroxyalkanoate production. *Bioresource Technology,* **98**(17), 3393–3403.

Holtzapple, M., Davison, R., Ross, M.K., Aldrett-Lee, S., Nagwani, M., Lee, C.-M., Lee, C., Adelson, S., Kaar, W., Gaskin, D., Shirage, H., Chang, N.-S., Chang, V. and Loescher, M. (1999). Biomass conversion to mixed alcohol fuels using the MixAlco process. *Applied Biochemistry and Biotechnology,* **79**(1–3), 609–631.

Janarthanan, O.M., Yu, Y., Laycock, B., Werker, A. and Pratt, S. (2014). Fractionation of microbial populations in a PHA accumulating mixed culture and associated PHA content and composition. *International Journal of Biological Macromolecules,* **71**, 53–58.

Jankowska, E., Chwialkowska, J., Stodolny, M. and Oleskowicz-Popiel, P. (2015). Effect of pH and retention time on volatile fatty acids production during mixed culture fermentation. *Bioresource Technology,* **190**, 274–280.

Jia, Q., Xiong, H., Wang, H., Shi, H., Sheng, X., Sun, R. and Chen, G. (2014). Production of polyhydroxyalkanoates (PHA) by bacterial consortium from excess sludge fermentation liquid at laboratory and pilot scales. *Bioresource Technology,* **171**, 159–167.

Johnson, K., Kleerebezem, R. and van Loosdrecht, M.C. (2010). Influence of the C/N ratio on the performance of polyhydroxybutyrate (PHB) producing sequencing batch reactors at short SRTs. *Water Research,* **44**(7), 2141–2152.

Keshavarz, T. and Roy, I. (2010). Polyhydroxyalkanoates: bioplastics with a green agenda. *Current Opinion in Microbiology,* **13**(3), 321–326.

Khardenavis, A.A., Kumar, M.S., Mudliar, S.N. and Chakrabarti, T. (2007). Biotechnological conversion of agro-industrial wastewaters into biodegradable plastic, poly β-hydroxybutyrate. *Bioresource Technology,* **98**, 3579–3584.

Kumar, D. and Murthy, G.S. (2011). Impact of pretreatment and downstream processing technologies on economics and energy in cellulosic ethanol production. *Biotechnology Biofuels,* **4**(27), 1–19.

Labatut, R.A., Angenent, L.T. and Scott, R.N. (2014). Conventional mesophilic vs. thermophilic anaerobic digestion: a trade-off between performance and stability? *Water Research,* **53**, 249–258.

Lee, W.S., Chua, A.S.M., Yeoh, H.K. and Ngoh, G.C. (2014). A review of the production and applications of waste-derived volatile fatty acids. *Chemical Engineering Journal,* **235**, 83–99.

Lemos, P.C., Serafim, L.S. and Reis, M.A.M. (2006). Synthesis of polyhydroxyalkanoates from different short-chain fatty acids by mixed cultures submitted to aerobic dynamic feeding. *Journal of Biotechnology,* **122**(2), 226–238.

Li, X., Chen, H., Hu, L., Yu, L., Chen, Y. and Gu, G. (2011). Pilot-scale waste activated sludge alkaline fermentation, fermentation liquid separation, and application of fermentation liquid to improve biological nutrient removal. *Environmental Science & Technology,* **45**, 1834–1839.

Lin, Y., de Kreuk, M., van Loosdrecht, M.C.M. and Adin, A. (2010). Characterization of alginate-like exopolysaccharides isolated from aerobic granular sludge in pilot-plant. *Water Research,* **44**(11), 3355–3364.

Lin, Y.M., Nierop, K.G.J., Girbal-Neuhauser, E., Adriaanse, M. and van Loosdrecht, M.C.M. (2015). Sustainable polysaccharide-based biomaterial recovered from waste aerobic granular sludge as a surface coating material. *Sustainable Materials and Technologies,* **4**, 24–29.

Lins, P., Reitschuler, C. and Illmer, P. (2015). Impact of several antibiotics and 2-bromoethanesulfonate on the volatile fatty acid degradation, methanogenesis and community structure during themophilic anaerobic digestion. *Bioresource Technology,* **190**, 148–158.

Liu, Y., Kong, S., Li, Y. and Zeng, H. (2009). Novel technology for sewage sludge utilization: preparation of amino acids chelated trace elements (AACTE) fertilizer. *Journal of Hazardous Materials,* **171**(1–3), 1159–67.

Longo, S., Katsou, E., Malamis, S., Frison, N., Renzi, D. and Fatone, F. (2015). Recovery of volatile fatty acids from fermentation of sewage sludge in municipal wastewater treatment plants. *Bioresource Technology,* **175**, 436–444.

Lopez-Garzon, C.S. and Straathof, A.A. (2014). Recovery of carboxylic acids produced by fermentation. *Biotechnology Advances,* **32**, 873–904.

Marang, L., Jiang, Y., van Loosdrecht, M.C.M. and Kleerebezem, R. (2013). Butyrate as preferred substrate for polyhydroxybutyrate production. *Bioresource Technology,* **142**, 232–239.

Markets & Markets Analysis (2013). Polyhydroxyalkanoate (PHA) Market, By Application (Packaging, Food Services, Bio-medical, Agriculture) & Raw Material — Global Trends & Forecasts to 2018, http://www.marketsandmarkets.com/PressReleases/pha.asp

Majone, M., Massanisso, P., Carucci, A., Lindea, K. and Tandoi, V. (1996). Influence of storage on kinetic selection to control aerobic filamentous bulking. *Water Science and Technology,* **34**(5–6), 223–232.

Mengmeng, C., Hong, C., Qingliang, Z., Shirley, S.N. and Jie, R. (2009). Optimal production of polyhydroxyalkanoates (PHA) in activated sludge fed by volatile fatty acids (VFAs) generated by alkaline excess sludge fermentation. *Bioresource Technology,* **100**(3), 1399–1405.

Mondala, A., Liang, K., Toghiani, H., Hernandez, R. and French, T. (2009). Biodiesel production by in situ transesterification of municipal primary and secondary sludges. *Bioresource Technology,* **100**, 1203–1210.

Moralejo-Gárate, H., Palmeiro-Sánchez, T., Kleerebezem, R., Mosquera-Corral, A., Campos, J.L. and van Loosdrecht, M.C.M. (2013). Influence of the cycle length on the production of PHA and polyglucose from glycerol by bacterial enrichments in sequencing batch reactors. *Biotechnology and Bioengineering,* **110**(12), 3148–3155.

Moralejo-Gárate, H., Kleerebezem, R., Mosquera-Corral, A., Campos, J.L., Palmeiro-Sánchez, T. and van Loosdrecht, M.C.M. (2014). Substrate versatility of poyhydroxyalkanoate producing glycerol grown bacteria enrichment culture. *Water Research,* **66**, 190–198.

Morgan-Sagastume, F., Karlsson, A., Johansson, P., Pratt, S., Boon, N., Lant, P. and Werker, A. (2010). Production of polyhydroxyalkanoates in open, mixed cultures from a waste sludge stream containing high levels of soluble organics, nitrogen and phosphorus. *Water Research,* **44**(18), 5196–5211.

Morgan-Sagastume, F., Pratt, S., Karlsson, A., Cirne, D., Lant, P. and Werker, A. (2011). Production of volatile fatty acids by fermentation of waste activated sludge pre-treated in full-scale thermal hydrolysis plants. *Bioresource Technology,* **102**(3), 3089–3097.

Morgan-Sagastume, F., Valentino, F., Hjort, M., Cirne, D., Karabegovic, L., Gerardin, F., Johansson, P., Karlsson, A., Magnusson, P., Alexandersson, T., Bengtsson, S., Majone, M. and Werker, A. (2014). Polyhydroxyalkanoate (PHA) production from sludge and municipal wastewater treatment. *Water Science and Technology,* **69**, 177–84.

Morgan-Sagastume, F., Hjort, M., Cirne, D., Gérardin, F., Lacroix, S., Gaval, G., Karabegovic, L., Alexandersson, T., Johansson, P., Karlsson, A., Bengtsson, S., Arcos-Hernández, M.V., Magnusson, P. and Werker, A. (2015). Integrated production of polyhydroxyalkanoates (PHAs) with municipal wastewater and sludge treatment at pilot scale. *Bioresource Technology,* **181**, 78–89.

Muller, E.E.L., Sheik, A.R. and Wilmes, P. (2014). Lipid-based biofuel production from wastewater. *Current Opinion in Biotechnology,* **30**, 9–16.

Nielsen, P.H., Roslev, P., Dueholm, T.E. and Nielsen, J.L. (2002). Microthrix parvicella, a specialized lipid consumer in anaerobic-aerobic activated sludge plants. *Water Science and Technology,* **46**(1–2), 73–80.

Ntaikou, I., Peroni, C.V., Kourmentza, C., Illeva, V.I., Morelli, A., Chiellini, E. and Lyberatos, G. (2014). Microbial bio-based plastics from olive-mill wastewater: generation and properties of polyhydroxyalkanoates from mixed cultures in a two-stage pilot scale system. *Journal of Biotechnology,* **188**, 138–147.

Pang, L., Ni, J., Tang, X. and Chen, Q. (2015). Short-cut waste activated sludge fermentation and application of fermentation liquid to improve heterotrophic aerobic nitrogen removal by Agrobacterium sp. *LAD9. The Chemical Engineering Journal,* **259**, 911–917.

Pratt, S., Liew, D., Batstone, D.J., Werker, A.G., Morgan-Sagastume, F. and Lant, P.A. (2012). Inhibition of fatty acids during fermentation of pre-treated waste activated sludge. *Journal of Biotechnology,* **159**, 38–43.

Reis, M.A.M., Serafim, L.S., Lemos, P.C., Ramos, A.M., Aguiar, F.R. and van Loosdrecht, M.C.M. (2003). Production of polyhydroxyalkanoates by mixed microbial cultures. *Bioprocess and Biosystems Engineering,* **25**(6), 377–385.

Rughoonundun, H., Moher, R. and Holtzapple, M.T. (2012). Influence of carbon-to-nitrogen ratio on the mixed-acid fermentation of wastewater sludge and pretreated bagasse. *Bioresource Technology,* **112**, 91–97.

Sans, C., Mata-Alvarez, J., Cecchi, F., Pavan, P. and Bassetti, A. (1995). Volatile fatty acids production by mesophilic fermentation of mechanically-sorted urban organic wastes in a plug-flow reactor. *Bioresource Technology,* **51**(1), 89–96.

Serafim, L.S., Lemos, P.C., Oliveira, R. and Reis, M.A.M. (2004). Optimization of polyhydroxybutyrate production by mixed cultures submitted to aerobic dynamic feeding conditions. *Biotechnology and Bioengineering,* **87**(2), 145–160.

Serafim, L.S., Lemos, P.C., Albuquerque, M.G. and Reis, M.A. (2008). Strategies for PHA production by mixed cultures and renewable waste materials. *Applied Microbiology and Biotechnology,* **81**, 615–628.

Sawatdeenarunat, Ch., Surendra, K.C., Takara, D., Oechsner, H. and Khanal, S.K. (2015). Anaerobic digestion of lignocellulosic biomass: challenges and opportunities. *Bioresource Technology,* **178**, 178–186.

Silvestre, G., Fernandez, B. and Bonmati, A. (2015). Addition of crude glycerine as strategy to balance the C/N ratio on sewage sludge thermophilic and mesophilic anaerobic co-digestion. *Bioresource Technology,* **193**, 377–385.

Singhania, R.R., Patel, A.K., Christophe, G., Fontanille, P. and Larroche, C. (2013). Biological upgrading of volatile fatty acids, key intermediates for the valorization of biowaste through dark anaerobic fermentation. *Bioresource Technology,* **145**, 166–174.

Steinbusch, K.J.J., Hamelers, H.V.M. and Buisman, C.J.N. (2008). Alcohol production through volatile fatty acids reduction with hydrogen as electron donor by mixed cultures. *Water Research,* **42**, 4059–4066.

Steinbusch, K.J.J., Hamelers, H.V.M., Plugge, C.M. and Buisman, C.J.N. (2011). Biological formation of caproate and caprylate from acetate: fuel and chemical production from low grade biomass. *Energy & Environmental Science,* **4**, 216–224.

Thomas, M., Wright, P., Blackall, L., Urbain, V. and Keller, J. (2003). Optimisation of Noosa BNR plant to improve performance and reduce operating costs. *Water Science and Technology,* **47**(12), 141–148.

Ucisik, A.S. and Henze, M. (2008). Biological hydrolysis and acidification of sludge under anaerobic conditions: the effect of sludge type and origin on the production and composition of volatile fatty acids. *Water Research,* **42**(14), 3729–3738.

Valentino, F., Morgan-Sagastume, F., Fraraccio, S., Corsi, G., Zanaroli, G., Werker, A. and Majone, M. (2015). Sludge minimization in municipal wastewater treatment by polyhydroxyalkanoate (PHA) production. *Environmental Science and Pollution Research,* **22**(10), 7281–94.

Vasquez, S.T., Dunkelman, J., Chaudhuri, S.K., Bond, A.B. and Holtzapple, M.T. (2014). Biomass conversion to hydrocarbon fuels using the MixAlco process at a pilot-plant scale. *Biomass Bioenergy,* **62**, 138–148.

Villano, M., Beccari, M., Dionisi, D., Lampis, S., Miccheli, A., Vallini, G. and Majone, M. (2010). Effect of pH on the production of bacterial polyhydroxyalkanoates by mixed cultures enriched under periodic feeding. *Process Biochemistry* **45**(5), 714–723.

Vlyssides, A.G. and Karlis, P.K. (2004). Thermal-alkaline solubilization of waste activated sludge as a pre-treatment stage for anaerobic digestion. *Bioresource Technology,* **91**(2), 201–206.

Part 4

Economic, Environmental, Legal and Social Impacts

Chapter 19

The impact of innovation on wastewater treatment economics

María Molinos-Senante, Andrea Guerrini and Francesc Hernández-Sancho

19.1 INTRODUCTION

Since the establishment of the first wastewater treatment plants (WWTPs) in the UK a century ago, many technologies have been developed to produce treated water that meets legislative requirements (Sala-Garrido *et al.* 2012). Increasing demands for cleaner waters by citizens and environmental organizations have led to adopt several regulations concerning urban wastewater. Thus, in accordance with the EU Water Framework Directive 2000/60/EC, the European States should implement measures aimed to achieve good ecological status for all water bodies. Consequently, the Directive 2008/105/EC was adopted which lays down environmental quality standards for certain pollutants (Molinos-Senante *et al.* 2013a).

Economic implications of the resulting increasing number and complexity of WWTPs are very important from managers and water authorities' perspective. The implementation of innovative technologies to achieve an effluent with higher quality involves costs and benefits that should be evaluated. Sound decision-making involves adopting criteria based on policy, projects, and/or interventions, which are judged appropriate for the region, and subsequently performing an appraisal of how alternative options compare (Pearce *et al.* 2003). Under the current global economic crisis, an economic investment criterion has become essential to support improvements in wastewater management. Although several tools are available to evaluate the water programs, the Cost-Benefit Analysis (CBA) is the most widely applied to evaluate the feasibility of water programs.

Wastewater treatment in general and innovative wastewater treatment technologies involve noticeable environmental benefits which are defined in economic terms as positive externalities. However the value of these benefits is often not calculated because they are not determined by the market. Valuation of these benefits is nevertheless necessary to justify suitable investment policies and financing mechanisms.

The assessment of the economic feasibility of wastewater treatment is a key aspect to justify new investment projects with four main objectives: (i) implement basic wastewater treatment systems in developing countries; (ii) increase the quality of the effluent in order to reuse the effluent or to improve the quality of the receiving environment, (iii) construct new WWTPs in small towns where wastewater treatment is not a legal requirement and, (iv) implement new processes in WWTPs for the recovery of by-products.

19.2 COSTS OF IMPROVING/INNOVATION IN WWTPs

19.2.1 Internal costs

Internal costs are those directly linked with the project and can be calculated directly in monetary units because they have market prices. Hence, in wastewater treatment projects, internal costs involve investment costs (IC) and operation and maintenance costs (O&M) of the facility. On the one hand, IC involve investments in civil construction and equipment costs. On the other hand, O&M are integrated by the following cost items: staff costs, energy costs, waste management costs, reagents costs, maintenance costs and, other costs. To evaluate and compare the overall internal costs of several wastewater treatment technologies, the methodology Life Cycle Costing (LCC) provides a very useful tool of analysis. It is defined as the estimation of the total costs associated with an asset over time, including investment, operation and maintenance, and disposal. Hence, LCC enables comparison of different technologies and helps decision making (Termes-Rifé *et al.* 2013).

Information about the costs associated with wastewater treatment may be mainly obtained by two different ways namely engineering and parametric approaches:

19.2.1.1 Engineering approach

The WWTP is viewed as a system consisting of components or subsystems, each of which is simulated in detail (Panagiotakopoulos, 2004). In other words, the WWTP is divided in many different parts and a detailed cost analysis of each of these parts should be done. The sum of the obtained gives this process final cost. The detailed study of cost breakdown will allow to: (i) find out what are the factors of cost with major weight in global, through stages and for different processes, and make a detailed sensitivity analysis. Hotspots in costs can then be highlighted; (ii) the engineering method is more focused on cost drivers, so it may become a guide for facilities to help them better monitor their costs in different processes and; (iii) obtain an order of magnitude for each part in the cost breakdown.

To evaluate WWTPs internal costs, Termés-Rifé *et al.* (2013) identified the following cost categories:

- *First cost*: is the initial capital investment needed to build the WWTPs plus the cost of borrowing capital (i.e., interest and related expenses). Elements associated with capital investment are considered as assets with a useful life time of over one year. Hence, there is a need to define for how long these assets will be available for their use, and to annualize these costs throughout the lifespan.
- *Operating costs*: are annual costs of the activities related directly to wastewater processes. It is the cost obtained from periodic activities needed for the proper operation of the business. Hence, values will be estimated for every year.
- *Use and maintenance*: Costs related to the activities of maintenance of the different assets. These activities are basic to ensure a good performance of these assets and lengthen their lifespan. Sometimes it is difficult to estimate separately operating and maintenance costs and as a result, they are added together.
- *Major repairs, modernization and rehabilitation*: It takes into account investments to be done through a facility's lifespan.
- *Salvage value*: It is the value of investments at the end of their life time in the facility. This value may be negative (you may sell it and obtain some compensation for it) or positive (if the asset cannot be sold you must take into account all the costs related to its disposal).
- *Indirect and global costs*: Costs categories described above are directly related to processes which take place in WWTPs, which need complementary activities such as general management in human resources, administration and clerical services, information systems or accounting and finance controlling and activities related to clients such as billing, supervision or meter reading expenses.

19.2.1.2 Parametric approach

Parametric methods are represented mainly by cost functions, i.e, IC and O&M costs are predicted by using statistical and mathematical methods. Costs functions show the relationship between the dependent variable (costs) and a set of independent variables (a set of representative variables of the process). According to Molinos-Senante *et al.* (2013b), steps from the collection of the raw data to the generation of the cost functions can be summarized as follows:

- Collect data available about IC and O&M costs for a set of WWTPs already built.
- Sort data according to process. Sorting means distinguish between the various wastewater treatment technologies options.
- Choose a reference year for economic valuation. Due to the difficulty in getting economic data relating to the IC and O&M costs for WWTPs, sometimes the reference year of all available information is not homogeneous. In this case, it is necessary to choose a reference year which generally is the year of analysis. Subsequently, it is needed to update both IC and O&M costs to the current year by using an appropriate price index.
- Decide on the cost components that will be included in the cost functions. Usually, the treatment capacity of the facility is considered the most important factor to determine the cost. However, the operation of these facilities may be affected by other factors such as contaminants removed or age of facility.
- Choose the functional form of the cost function. The formulation of IC and O&M costs functions is based on the assessment of the relationship between the dependent variable (C) and the independent variables (X). For this purpose, different models can be used such as: inverse, power, logarithmic and quadratic.
    - Inverse: $C = a + b/Y$
    - Power: $C = a\,X^b$
    - Logarithmic: $C = a + b \ln X$
    - Quadratic: $C = a + bX + cX^2$
- Adjust all available data to comply with the choices regarding costs components.
- Use statistical methods to generate "best-fit" cost functions. A common method to get model parameters is ordinary least squares regression analysis. Subsequently, the significance of the independent variables should be tested.
- Assess the quality of the adjustment by estimating the coefficient of correlation. It measures the proportion of total variability of the dependent variables relative to its average according to the regression model. Its value is ranged within [0,1]. A R value of 1 means that the adjustment between actual and estimated data is perfect while a value of 0 indicates that there is no relationship between the variables.

19.2.1.3 Case studies

Engineering and parametric approaches are complementary since the parametric approach usually provides information about total IC and O&M costs while the engineering approach details the costs of each item. Hence, the sum them should be pretty similar to the total costs obtained through the parametric approach.

Empirical studies have developed to estimate IC and O&M costs of WWTPs using both approaches as follows:

Tsagarakis *et al.* (2003) estimate life cycle cost functions for wastewater treatment in Greece by means of the functional form: $y = ax^b$. The variable y represents costs of land use, construction and O&M

costs and x represents the capacity of the WWTPs in terms of population equivalents. Costs of sludge management are also taken into account. Estimates are provided for three different types of primary and secondary treatment. Based on the indicator of total annual estimated economic costs (sum of costs for land use, construction and O&M costs), the authors found that extended aeration with natural air drying is the most economical system, followed by extended aeration with mechanical drying and conventional secondary treatment. The poor economic performance of conventional treatment in relation to extended aeration treatment is attributed to high energy costs.

Several other analyses have used the parametric approach to develop cost functions of wastewater treatment. The cost drivers (explanatory variables) considered vary among studies. For example, Hernandez-Sancho *et al.* (2011) provide a comprehensive approach for estimating cost functions of wastewater treatment. Based on Spanish data, they estimate O&M costs (variable C in €/year) as a function of the volume of wastewater treated (variable V in m^3/year), the age of the plant (variable A in years) and the removal efficiency of the plants' pollutants (variables *SS, COD, BOD, N* and *P*) for removing suspended solids, organic components, nitrogen and phosphorus, respectively. The estimated cost functions for seven different treatment levels using 2008 costs as reference can be seen in Table 19.1.

Table 19.1 Examples of cost functions of different wastewater treatment systems.

Technology	Cost Functions
Extended aeration without nutrient removal	$C = 169.4844V^{0.4540}e^{(0.0009A+0.6086SS)}$
Activated sludge without nutrient removal	$C = 2.1165V^{0.7128}e^{(0.0174A+0.15122SS+0.0372BOD)}$
Activated sludge with nutrient removal	$C = 2.518V^{0.7153}e^{(0.007A+1.455COD+0.158N+0.243P)}$
Bacterial beds	$C = 17.3617V^{0.5771}e^{(0.1006A+0.6932COD)}$
Peat beds	$C = 1{,}510.84V^{0.2596}e^{(0.0171SS)}$
Biodisk	$C = 28.9522V^{0.4493}e^{(2.3771SS)}$
Tertiary treatment	$C = 3.7732V^{0.7223}e^{(0.6721COD+0.01958N+0.7603P)}$

Source: Adapted from Hernández-Sancho *et al.* (2011).

The cost functions presented in Table 19.1 show the relationship for each technology between the cost of annual operation and volume treated together with the percentage of pollutants extracted and age of the plant (in some cases). The parameter that accompanies each explanatory variable illustrates the level of influence of this variable on the cost of operation. In all cases, the relevance of the treated volume is shown; the percentage of pollutants removed and the age of the plant have a very heterogeneous influence, depending on technology. Using these functions helps determine the most adequate technologies according to the volume of wastewater to be treated and the objectives set for removal of contaminants. These results are consistent with some previous studies (Renzetti, 1999; Wen & Lee, 1999; Friedler *et al.* 2006; Nogueira *et al.* 2009).

Other studies have emphasized the assessment of the costs of suitable wastewater treatment for water reuse. Iglesias *et al.* (2010) analysed these costs in Spanish WWTPs. According to the requirements of the Spanish legislation (RD, 1620/2007), six wastewater treatments were defined as the following:

- *Type 1*: physical-chemical treatment with a lamella settling system, depth filtration , ultrafiltration and disinfection
- *Type 2*: physical-chemical treatment with a lamella settling system, depth filtration and disinfection
- *Type 3*: filtration and disinfection

- *Type 4*: depth filtration
- *Type 5.a*: physical-chemical treatment with a lamella settling system, depth filtration, ultrafiltration, reverse osmosis and residual chlorine removal
- *Type 5.b*: Physical-chemical treatment with a lamella settling system, double depth filtration, electrodialysis and disinfection.

Table 19.2 shows IC and O&M costs for each type of treatment train, calculated based on information from departments of water resources and operators of Spanish water reclamation plants. Ranges are due to the different sizes of water reclamation plants, climatic and geographical conditions and influent features.

Table 19.2 Investment costs and operation and maintenance costs for several wastewater treatment trains for 2010.

Treatment Train	Investment Costs (€/m³ day)	Operation Costs (€/m³ day)
Type 1	185–398	0.14–0.20
Type 2	28–48	0.06–0.09
Type 3	9–22	0.04–0.07
Type 4	5–11	0.04–0.07
Type 5.a	416–736	0.35–0.45
Type 5.b	310–506	0.35–0.45

Source: Adapted from Iglesias *et al.* (2010).

19.2.2 External costs

The benefits of wastewater treatment are obvious, however treatment process also involve environmental impacts mainly to the use of energy (Lassaux *et al.* 2007). Currently, there is growing interest in minimising energy consumption in WWTPs due to two reasons. First, energy consumption is an internal costs, i.e, a cost with market value. Second, energy consumption is a negative externality, i.e., a negative environmental impact with has no market price. WWTPs consume a significant amount of electricity which involves the indirect emission of greenhouse gases (GHG). Hence, energy consumption of WWTPs involves a negative externality which should not be overlooked (Molinos-Senante *et al.* 2015a).

The literature illustrates that there are two main approaches to estimate the economic value of the GHG emissions associated to WWTPs: (i) using CO_2 market value and (ii) calculating the shadow price of the CO_2.

The first approach involves that indirect GHG emissions are estimated based on WWTP energy demands. The first step is estimate the emission of each GHG emission based on national production mix. Subsequently, GHG emissions are converted to equivalent CO_2 emissions using 100-year global warming potential coefficients. The next step is to express GHG emissions in monetary units. For this purpose, the emissions trading system should be used. For example, in Europe, the European Uniońs Trading System (EU ETS) involves facilities accounting for the 40% of the total GHG emissions in the EU. The average price paid through the EU ETS (or other CO_2 market) during a time period may be used as a proxy to the price of GHG emissions. Thus, Molinos-Senante *et al.* (2013b) reported that from 2009 to 2012 the average market price of CO_2 was 11.9 €/t.

The second approach is based on the estimation of the CO_2 shadow price and therefore, it reflects the trade-off between the desirable output (treated water) and the undesirable output (GHG emissions) (further details about methodological issues are shown in Section 19.3.2). The CO_2 shadow price can be interpreted as the marginal abatement cost arising from regulations that prevent the free disposal of pollutants. By using a sample of 25 Spanish WWTPs, Molinos-Senante *et al.* (2015b) estimated that in average terms each kg of CO_2 equivalent that is emitted into the atmosphere as a result of wastewater treatment involves an environmental cost of 0.088 € which was around 17% of the value of the treated water).

19.3 BENEFITS OF IMPROVING/INNOVATION IN WWTPs

19.3.1 Internal benefit

Internal benefit involves revenues from the sale of the by-products that can be recovered during wastewater treatment process and have market price. For example, in water stress areas, the sale of the regenerated water may play a vital role to ensure the economic feasibility of some water reuse projects. Thus, the implementation of a novel wastewater treatment technology (maybe with higher IC and O&M costs than a traditional technology) may be economically justified by obtaining a higher quality effluent that can be reused. Moreover, if the reclaimed water is used for agriculture purposes, the nitrogen and phosphorus content in the water involves a saving in the fertiliser costs (Nogueira *et al.* 2013). Other incomes may be obtained from the sale of phosphorus recovered during wastewater treatment and the sale of stabilised sewage sludge to be used after composting.

Recently, some innovative technologies have been developed aimed to improve energy efficiency. Thus, some technologies involve a reduction in the consumption of energy and therefore, there is an economic saving that should be taken into account. Other technologies allow recovering energy from wastewater or from sewage sludge that can be used in the WWTP itself or sold. It involves an additional income that cannot be overlooked in the economic assessment of innovative wastewater treatment technologies.

Nowadays, there is a wide variation in the price of regenerated water depending on the type of use, flow rates, and local conditions – ranging from \$0/m^3 to \$0.52/m^3 (Morris *et al.* 2005). For example, irrigation rates vary from €0.02/ m^3 to €0.07/ m^3 in France, Spain, Jordan, Morocco, Tunisia to €0.16/m^3 to €0.23/m^3 in Israel and California (Lazarova & Bahri, 2008). In Australia, several tariffs are reported for regenerated water depending on use. For industrial purposes, the water rates is between \$0.4/m^3 to \$1.2/m^3. If the regenerated water is for residential supply for toilets and home gardens the tariff is between \$0.28/m^3 to \$0.83/m^3. Finally, the price of regenerated water for golf courses and park irrigation is around \$0.5/m^3 (Marsden, 2008).

Regarding phosphorus recovery as struvite, different studies have estimated its price for different countries as it is shown in Table 19.3.

Table 19.3 Price of the struvite from WWTPs.

Country	Price (€/ton)	Reference
Japan	250	Ueno and Fujii (2011)
Australia	188–314	Münch and Barr (2001)
–	464	Shu *et al.* (2006)
–	763	Dockhorn (2009)

19.3.2 External benefit

External benefits refer to the positive consequences of wastewater treatment that have no market prices. Hence, the quantification of external benefits from implementing innovative wastewater treatment technologies requires the use of economic valuation methods. The methodologies for the quantification and internalisation of environmental externalities arising from wastewater treatment can be grouped into two general categories: (i) conventional valuation methods and (ii) shadow price of pollutants approach (Molinos-Senante *et al.* 2015a).

19.3.2.1 Conventional valuation methods

They are based on the demand approach and therefore, they are used to determine the total economic value of goods and services that have no market (Hanley & Barbier, 2009) such as wastewater treatment. They are classified as:

- Indirect methods which rely on the use of data from actual transactions by individuals and therefore, the value of the good is deducted from the complementary relationship between it and other goods with market price. Some examples of this approach are the travel cost method and hedonic price method (Pearce & Turner, 1990).
- Direct methods also known as stated preference methods. This approach assumes that the economic value of a good arises from the interaction between an individual and an environmental asset as an expression of individual preference. The main categories of direct methods are the contingent valuation method and choice modelling techniques. The first method is based on the creation of a hypothetical market through a surveying process where individuals declare their willingness to pay for an improvement of the quality of the environmental good being analysed. The choice modelling techniques are based on ranking or rating a series of "product profits" that characterise products with specific attribute levels (Pearce & Özdemiroglu, 2002).

19.3.2.2 Shadow price of pollutants

Despite the popularity of the stated preference methods on water quality valuation, there are other methodologies that can shed light on this issue. In this context, using the concept of distance function (Färe *et al.* 1993) is possible to calculate the shadow prices for the undesirable outputs – which are the pollutants removed from wastewater – that have a clear negative impact on the environment. In this sense, wastewater treatment is considered as a productive process in which a desirable output (treated water) is obtained together with a series of undesirable outputs (suspended solids, heavy metals, nitrogen, phosphorus, COD, and BOD, etc.). A shadow price for these undesirable outputs can be considered as the equivalent of the environmental damage avoided since if they are dumped without control they would cause a negative impact on the environment (Hernández-Sancho *et al.* 2010). Thus, the avoided costs estimated with the methodology of the shadow prices, represent an estimation of the economic value of the environmental benefits obtained from the treatment process aimed to improve water quality.

The methodology for estimating the shadow prices of undesirable outputs is based on cost production. While initially this approach was used to estimate the economic value of atmospheric pollutants (Coggin & Swinton, 1996; Swinton, 1998) more recently, several empirical applications have been developed to estimate shadow prices of pollutants removed from wastewater as a result of treatment (Hernández-Sancho *et al.* 2010; Molinos-Senante *et al.* 2011). Contaminants extracted from wastewater are undesirable outputs, i.e. if pollutants are released into the environment they generate a negative impact. Therefore, shadow prices for pollutants are considered a proxy for the environmental benefits associated with wastewater treatment.

The suitability of each approach (conventional methods or shadow prices estimation) will depend upon several factors. The contingent valuation is a very flexible technique that can be applied by policy makers both before and after the construction of the WWTP in order to evaluate the environmental benefits from treating the wastewater. It is, however, very expensive to carry out this type of assessment. Funding can be a limiting factor, especially if representative samples of the entire population are needed (Randall, 1997). Shadow pricing, despite its more limited scope, may be useful to quantify environmental impacts derived from production processes. It does present an advantage since obtaining the necessary information is cheaper because expensive surveys are not direct but direct information about WWTPs is used (Färe *et al.* 2001).

19.3.2.3 Case studies

The literature on environmental benefits from wastewater treatment using the "shadow prices" methodology is quite recent, since as it was reported previously, the adaptation of the methodology developed by Färe *et al.* (1993) to wastewater framework was carried out by Hernández-Sancho *et al.* (2010). Nevertheless, some empirical applications have been developed since then as follows:

Hernández-Sancho *et al.* (2010) using a sample of 43 WWTPs in Spain, estimated the shadow prices of five pollutants for 2008: nitrogen (N); phosphorus (P), suspended solids (SS), biological oxygen demand (BOD) and chemical oxygen demand (COD) as shown in Table 19.4. The economic value of these pollutants differs depending on the type of the receiving water body and the different reference water prices assumed. The shadow prices are negative since they are associated with undesirable outputs that represent negative value in contrast to desirable outputs. The main environmental benefits for all four analysed destinations are the elimination of phosphorus followed by nitrogen. The shadow price of the nitrogen if discharged to wetlands is €-65.21/kg. This figure means that for every kilogram of this nutrient not dumped into a wetland, the damage prevented, or the environmental benefit generated, equals €65.21.

Table 19.4 Reference price of water treated (€/m^3) and shadow prices for pollutants removed from wastewater (€/kg) for 2008.

Effluent Destination	Reference Price of Water (€/m^3)	Estimated Shadow Prices for Undesirable Outputs (€/kg)				
		Nitrogen	Phosphorus	Suspended Solids	Biological Oxygen Demand	Chemical Oxygen Demand
River	0.7	–16.3	–30.9	–0.005	–0.03	–0.10
Sea	0.1	–4.6	–7.5	–0.001	–0.005	–0.01
Wetlands	0.9	–65.2	–103.4	–0.01	–0.12	–0.12
Reuse	1.5	–26.2	–79.3	–0.01	–0.06	–0.14

Source: Hernández-Sancho *et al.* (2010)

Based on this work, Molinos-Senante *et al.* (2013a) estimated the economic value of removing five pharmaceutical and personal care products (PPCPs) from wastewater that require intensive treatments. Developed countries with increasingly strict legislation governing water quality are increasingly interested in knowing the environmental benefits associated with removal of PPCPs from wastewater. The average values of shadow prices for the five PPCPs evaluated are shown in Table 19.5.

Table 19.5 Estimated shadow prices for PPCPs removed from wastewater (€/kg).

Scenarios	Diclofenac	Tonalide	Galaxolide	Sulfamethoxazole	Ethinyl Estradiol
Non-sensitive water bodies	−42.2	−11.0	−8.7	−35.0	−73.7
Sensitive water bodies	−53.5	−14.0	−11.1	−44.5	−93.8

Source: Molinos-Senante *et al.* (2013a).

19.4 NET PRESENT VALUE

Once the costs and benefits (internal and external) of implementing new processes to improve WWTPs have been quantified, they should be compared to evaluate if the project is economically feasible or not. In other words, the net present value (NPV) of each alternative should be estimated as follows:

$$NPV = \sum_{t=1}^{T} \frac{B_t - C_t}{(1+r)^t} - K \tag{19.1}$$

where: B_t is the total benefit of the project (internal benefits + external benefits) expressed in €/year; C_t is the total costs of the project (internal costs + external costs) expressed in €/year; t is each year; T is the project lifespan; r is the discount rate and K are the investment costs expressed in €.

NPV results determine the economic feasibility of the project. A project is economically feasible if, and only if, $NPV > 0$. If the result of the calculation is $NPV < 0$, then the project is not feasible from an economic point of view. The best option offers the highest NPV (Chen & Wang, 2009).

As it is known in Eq. (19.1) the NPV is estimated through the application of a discount rate. In this sense, the discount rate reflects the fact that people generally prefer having money in the present rather than in the future so, costs that will take place in coming years may have a lower value than those in present time. One difficulty in calculating the present value of an item is to obtain an appropriate value for the discount rate. Pearce *et al.* (2003) stressed the need for modification of the traditional assumption in discounting rates. In this context, an alternative approach is the use of a declining discount rate which replaces the exponential discount factor with a hyperbolic function. Hence, the viability of projects in which the costs occur early in the time horizon is improved. More details about this approach have been provided by Weitzman (2001) and Guo *et al.* (2006). Another approach suggested by Almansa and Martínez-Paz (2011) is the use of a dual-rate discount rate which involves the use of different discount factors for tangible and intangible goods. An appropriate discount rate can also be determined by the Delphi method, i.e., by consulting individually and anonymously to a panel of experts (Almansa & Martínez-Paz, 2011).

The estimation of the NPV involves some uncertainty associated to costs and benefits estimation which could influence on the economic valuation of wastewater treatment technologies. Hence, it is essential to carry out a sensitivity analysis which might be based on several methodologies such as Monte Carlo simulations, fuzzy logic models, Bayesian network models or statistical tolerances (Molinos-Senante *et al.* 2015a).

19.5 FUNDING OPPORTUNITIES

The realization of a WWTP, as other water infrastructures, could be funded with contracts alternative to traditional borrowing. Firstly, State or EU grants could provide interesting source of capital, even if in the light of the public debt crisis, this source has been dramatically reduced. Then, a further source of funds is represented by shareholders, who can keep more money in a water utilities for the realization of new investments. An alternative form of financing for public utilities is based on lease financing, which is described in this section.

Lease is a contract that allows to firms to purchase instrumental assets, useful for their operations. The financial lease contract include a borrowing and the purchase of an assets (Burrows, 1988; Chamberlain, 1975). As a matter of fact the "lessors" gives to the "lessee" the availability of an assets and the possibility to purchase it at the end of the contract period, paying a given price. During the contract period the lessee periodically paid a fee, which could be flat, or which could varies according to specific conditions (e.g. the free cash flow yearly generated by a firm).

In Europe there is the possibility to arrange a specific form of lease contract to realize public infrastructures. Usually, there are three actors involved in the lease contract:

- A public administration, as a municipalities or a public utilities, who needs the infrastructure;
- A lease firm, which takes part to the public auction for the realization of the infrastructure;
- A construction firm which take part to the public auction with the lease firm.

The lease firm and the construction firm are usually jointed in a temporary association in order to take part to the public auction. In case this association becomes responsible of the contract, the construction firm materially realize the infrastructure, while the lease firm purchases it from the constructor and gives it in use to the public administration who pays a periodic fee.

The main advantage of the lease contract is the absence of wide negative impact on the bank exposure of the public administration. The effect of a lease contract on financial ratios and performance of a firms are widely studied by literature (Abdel-Khalik *et al.* 1981; Altman, 1976; Bowman, 1980; Wilkins & Zimmer, 1983) and depends on the accounting standard rule adopted. Further the public administration receives a complete infrastructure ready to operate, without any risks linked to price adjustments or project variations.

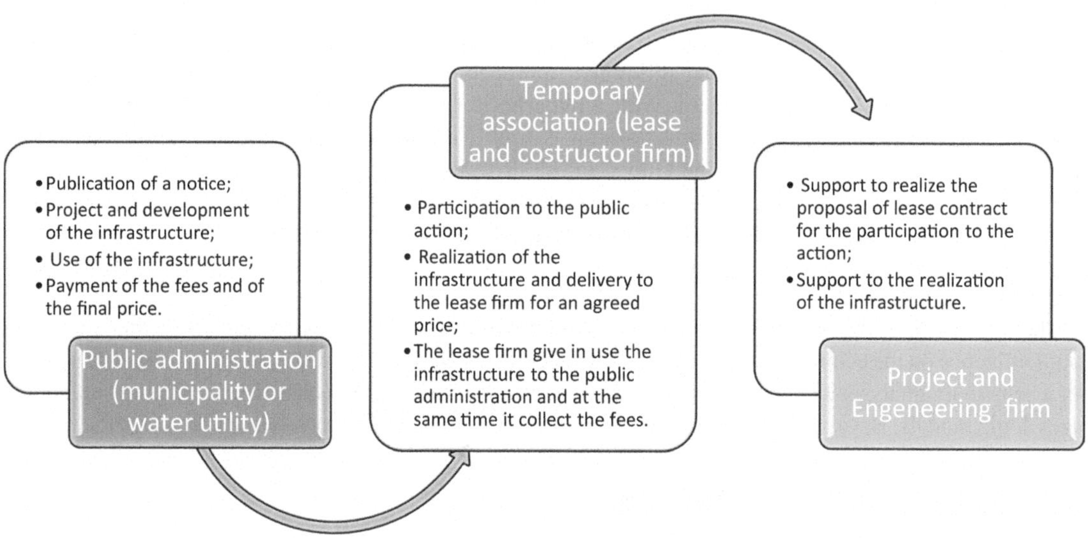

Figure 19.1 The relationship between the actors involved in a lease contract to realize a WWTP.

The public lease contract is quite similar to the financial lease, which shows the following differences if compared with operating lease (Clark, 1978):

- The contract assigns the ownership of an asset to the lessee only after the end of the contract, when the final price is paid;
- The lessee has the opportunity to purchase the asset after the end of the contract period, paying a price which is lower than the fair value;
- The duration of the contract is greater than half of the useful life of the asset;
- At the beginning of the contract the net present value of the fees is at least equivalent to the fair value of the asset;
- The asset has specific peculiarities that only the lessee can use it, without any technical variation.

The main differences with the project financing is that with the lease contract the asset could be immediately used by the public water utilities after the technical test and in the same time the private counterpart has concluded its tasks. Differently, with the project financing the private counterpart is involved also in the operations, as the delivery of water and wastewater services.

The procedure that a public water utilities can follows to arrange a lease contract requires to publish a notice with the main items that should be bargained during the auction among different competitors. These items are referred to the cost of the project and to the costs of the financial operation. The former includes:

- Cost for realization
- Costs for the project
- Interest paid for debt
- Maintenance expenditures
- Final price that the lessee has to pay to get the ownership of the asset
- Taxes
- Other technical and administrative costs.

The latter includes:

- The choice of the rate of interest of the leasing contract (fixed or variable)
- The spread that should be applied to IRS for a fixed rate and Euribor for a variable rate
- The duration of the contract
- The frequency of the fees (monthly, quarterly, yearly)
- The possibility to switch from variable to fixed rate
- Others.

It is quite surprising that the water and wastewater sectors was keep aside from these initiatives, notwithstanding it needs a huge amount of investments for infrastructure realization and maintenance (Guerrini & Romano, 2014). However, something is moving now and probably the alternative form of financing could revamp this mature business.

19.6 CONCLUSIONS

As the number and complexity of wastewater treatment plants (WWTPs) has increased, the interest of WWTPs managers and water authorities on economic issues has also growing. The assessment of the economic feasibility of wastewater treatment is a key issue to justify new investment projects including the implementation of innovative technologies. The implementation of alternative technologies to achieve high quality effluent involves costs and benefits – some of them without market value- that should be evaluated.

Life cycle costing is a useful tool of analysis to evaluate the total costs associated with WWTPs over time including investment and operation and maintenance costs. The two main approaches to estimate costs associated to wastewater treatment are engineering and parametric methods. The outcome of these methods are the costs functions which show the relationship between a dependent variable (costs) and a set of representative variables of the process. Hence, costs functions are a useful tool to predict wastewater treatment costs. Nevertheless, it should be noted that WWTPs also involved external costs, i.e., costs without market value, mainly associated to greenhouse gas emissions. Thus, the assessment of the total costs associated to wastewater treatment should also include a proxy of them such as CO_2 market value or CO_2 shadow price.

Wastewater treatment involves two types of benefits namely: (i) internal benefits associated to the sale of the by-products that can be recovered during wastewater treatment process and (ii) external benefits mainly linked to the environmental benefits of preventing the discharge of wastewater into waterbodies. While the latter ones have no market price, different methodologies for their quantification have been developed from economic theory. The internalisation of the environmental benefits of wastewater treatment is essential to justify the economic feasibility of implementing novel technologies which allow obtaining effluent with higher quality than traditional wastewater treatment technologies.

Lease financing is evidenced as an alternative form of financing for WWTPs. It is based on the fact that the "lessors" gives to the "lessee" the availability of an assets and the possibility to purchase it at the end of the contract period, paying a given price.

As a general conclusion, we emphasize that the economic analysis of implementing innovative technologies for wastewater treatment should not be based exclusively on costs assessment but also in the benefits – including the environmental ones- associated to obtain a high quality effluent. Otherwise, there is an under-estimation of the benefits and therefore, innovative technologies are not competitive because they are not fully evaluated.

19.7 REFERENCES

Abdel-Khalik A. R., Thompson R. B. and Taylor R. E. (1981). The Economic Effects on Leases of FASB Statement No. 13. Accounting for Leases. Financial Accounting Standards Board, Stamford, CT.

Almansa C. and Martínez-Paz J. M. (2011). What weight should be assigned to future environmental impacts? A probabilistic cost benefit analysis using recent advances on discounting. *Science of the Total Environment*, **409**, 1305–1314.

Altman E. I. (1976). Capitalization of leases and the predictability of financial ratios: a comment. *The Accounting Review*, **4**, 408–412.

Assilea (2011). Vademecum del Leasing Pubblico. Assilea, Roma, Italy.

Bowman R. G. (1980). The debt equivalence of leases: an empirical investigation. *The Accounting Review*, **4**, 237–253.

Burrows G. H. (1988). Evolution of a lease solution. *Abacus*, **2**, 107–119.

Chamberlain D. C. (1975). Capitalization of lease obligations. *Management Accounting*, **12**, 37–38.

Chen R. and Wang C. (2009). Cost-benefit evaluation of a decentralized water system for wastewater reuse and environmental protection. *Water Science & Technology*, **59**(8), 1515–1522.

Clark T.M (ed.) (1978). Leasing. McGraw Hill, London.

Coggin J. S. and Swinton J. R. (1996). The price of pollution: a dual approach to valuing SO_2 allowances. *Journal of Environmental Economic Management*, **30**(5), 58–72.

Dockhorn T. (2009). About the economy of phosphorus recovery. In: International Conference on Nutrient Recovery from Wastewater Streams, K. Ashley, D. Mavinic and F. Koch (eds), IWA Publishing, London, UK.

Färe R., Grosskopf S., Lovell C. A. and Yaisawarng S. (1993). Derivation of shadow prices for undesirable outputs: a distance function approach. *Review of Economics and Statistics*, **75**(2), 374–380.

Färe R. Grosskopf S. and Webwe W. L. (2001). Shadow prices of Missouri public conservation land. *Public Finance Review*, **29**(6), 444–460.

Friedler E. and Pisantly E. (2006). Effects of design flow and treatment level on construction and operating costs of municipal wastewater treatment plants and their implications on policy making. *Water Research*, **40**, 3751–3758.

Guerrini A. and Romano G. (2014). Water Management in Italy. Governance, Performance and Austainability. Springer, Cham, Heidelberg, New York, Dordrecht, London.

Guo J., Hepburn C. J., Tol R. S. J. and Anthoff D. (2006). Discounting and the social cost of carbon: a closer look at uncertainty. *Environmental Science and Policy*, **9**(3), 205–216.

Hanley N. and Barbier E. B. (2009). Pricing Nature: Cost-Benefit Analysis and Environmental Policy. Edward Elgar, Cheltenham.

Hernández-Sancho F., Molinos-Senante M. and Sala-Garrido R. (2010). Economic valuation of environmental benefits from wastewater treatment processes: an empirical approach for Spain. *Science of the Total Environment*, **408**(4), 953–957.

Hernández-Sancho F., Molinos-Senante M. and Sala-Garrido R. (2011). Cost modelling for wastewater treatment processes. *Desalination*, **268**, 1–5.

Iglesias R., Ortega E., Batanero G. and Quintas L. (2010). Water reuse in Spain: data overview and costs estimation of suitable treatment trains. *Desalination*, **263**(1–3), 1–10.

Lazarova V. and Bahri A. (2008). Water reuse practices for agriculture. In: Water Reuse: an International Survey of Current Practice, Issues and Needs, B. Jimenez. and T. Asano (eds.), IWA Publishing, London, pp. 199–227.

Marsden Jacob (2008). National snapshot of current and planned water recycling and reuse rates. *Marsden Jacob Associates*, 1–72.

Molinos-Senante M., Hernández-Sancho F. and Sala-Garrido R. (2011). Cost-benefit analysis if water-reuse projects for environmental purposes: a case study for Spanish wastewater treatment plants. *Journal of Environmental Management*, **92**, 3091–3097.

Molinos-Senante M., Reif R., Garrido-Baserba M., Hernández-Sancho F., Omil F., Poch M. and Sala-Garrido R. (2013a). Economic valuation of environmental benefits of removing pharmaceutical and personal care products from WWTP effluents by ozonation. *Science of the Total Environment*, **461–462**, 409–415.

Molinos-Senante M., Hernández-Sancho F. and Sala-Garrido R. (2013b). Cost modeling for sludge and waste management from wastewater treatment plants: an empirical approach for Spain. *Desalination and Water Treatment*, **51**(28–30), 5414–5420.

Molinos-Senante M., Hanley N., Hernández-Sancho F. and Sala-Garrido R. (2015a). The principles of economic valuation and cost-benefit analysis implemented in sewage treatment plants. In: Sewage Treatment Plants: Economic Evaluation of Innovative Technologies for Energy Efficiency, K. Stamatelatou and K. Tsagarakis (eds.), IWA Publishing, London, UK.

Molinos-Senante M., Hanley N. and Sala-Garrido R. (2015b). Measuring the CO_2 shadow price for wastewater treatment: a directional distance function approach. *Applied Energy*, **144**, 241–249.

Münch E. V. and Barr K. (2001). Controlled struvite crystallization for removing phosphorus from anaerobic digester sidestreams. *Water Research*, **35**, 151–159.

Nogueira R., Brito A. G., Machado A. P., Janknecht P., Salas J. J., Vera L. and Martel G. (2009). Economic and environmental assessment of small and decentralized wastewater treatment systems. *Desalination and Water Treatment*, **4**, 16–21.

Panagiotakopoulos D. (2004). Sustainable Municipal Solid Waste Management. 2nd edn. Zygos. Thessaloniki, Greece.

Pearce D. and Özdemiroglu E. (2002). Economic Valuation with Stated Preference Techniques. Summary Guide. Department for Transport, Local Governments and the Regions, London. http://www.communities.gov.uk/documents/corporate/pdf/146871.pdf.

Pearce D., Groom B., Hepburn C. and Koundouri P. (2003). Valuing the future: recent advances in aocial discounting. *World Economics*, **4**, 121–141.

Randall A. (1997). The NOAA panel report: a new beginning or the end of an era? *American Journal of Agricultural Economics*, **79**(5), 1489–1494.

Renzetti S. (1999). Municipal water supply and sewage treatment: costs, prices and distortions. *Canadian Journal of Economics*, **32**(3), 688–704.

Sala-Garrido R., Hernández-Sancho F. and Molinos-Senante M. (2012). Assessing the efficiency of wastewater treatment plants in an uncertain context: a DEA with tolerances approach. *Environmental Science and Policy*, **18**, 34–44.

Shu L., Schneider P., Jegatheesan V. and Johnson J. (2006). An economic evaluation of phosphorus recovery as struvite from digester supernatant. *Bioresource Technology*, **97**(17), 2211–2216.

Swinton J. R. (1998). At what cost do we reduce pollution? Shadow price of SO_2 emissions. *Energy Journal*, **19**(4), 63–83.

Termes-Rife M., Molinos-Senante M., Hernández-Sancho F. and Sala-Garrido R. (2013). Life cycle costing: a tool to manage the urban water cycle. *Journal of Water Supply: Research and Technology- AQUA*, **62**(7), 468–476.

Tsagarakis K. P., Mara D. D. and Angelakis A. N. (2003). Application of cost criteria for selection of municipal wastewater treatment systems. *Water Air Soil Pollution*, **142**, 187–210.

Ueno Y. and Fujii M. (2001). Three years experience of operating and selling recovered struvite from full-scale plant. *Environmental Technology*, **22**, 1373–1381.

Weitzman M.L. (2001). Gamma discounting. *American Economic Review*, **91**(1), 260–271.

Wen C.-G. and Lee C.-S. (1999). Development of a cost function for wastewater treatment systems with fuzzy regression. *Fuzzy Sets and Systems*, **106**, 143–153.

Wilkins, T. and Zimmer, I. (ed.) (1983). The effect of leasing and different methods of accounting for leases on credit evaluations. *The Accounting Review*, **10**, 749–764.

Chapter 20

Assessing environmental impacts and benefits of wastewater treatment plants

Christian Remy, Lluis Corominas, Almudena Hospido, Henrik Fred Larsen and Carmen Teodosiu

20.1 INTRODUCTION

The purpose of a wastewater treatment plant (WWTP) is to reduce the pollution of wastewater caused by human activities so as to minimize the negative influences on environmental quality and human health. Its effectiveness is normally measured in terms of contaminant removal per unit cost, assuming that legal requirements for effluent discharge or recycling/reuse are always fulfilled. Nowadays, however, society demands the application of wider sustainability criteria and therefore different tools and approaches have been put into practice to closely examine the total environmental performance of municipal WWTPs.

Ten years ago, the US National Decentralized Water Resources Capacity Development project (NDWRCDP) published a report entitled "Methods for comparison of wastewater treatment options" (Kirk *et al.* 2005). The motivation of this report was the fact that decisions in the US regarding waste and water systems were based primarily on economics and consider a very limited set of treatment options; by doing so, the true environmental and social cost of wastewater treatment were often not included in decision making (Kirk, 2005).

The report examined 18 methods, which were classified into the broad methodologies (Kirk, 2005):

- *Environmental Impact Assessment (EIA):* EIA is a framework for identifying, predicting, evaluating, and mitigating the biophysical, social, and other effects of proposed projects or plans and physical activities.
- *Life Cycle Assessment (LCA):* LCA is a method of accounting the environmental impacts of a product, service, or process over the course of its life cycle from extraction of materials to disposal or reuse of the final product.

Environmental Impact Assessment (EIA) is a widely used method both in the US and in Europe to assess the environmental impacts of a specific WWTP (Kirk, 2005; Marr *et al.* 1996; Trajkovska-Trpevska *et al.* 2009). At the European level, the EIA Directive (85/337/EEC) is in force since 1985, with further amendments (2011/92/EU & 2014/52/EU) and applies to a wide range of public and private projects, which are defined in its Annexes I and II. The Directive is mandatory for all projects listed in Annex I, which are considered to have significant effects on the environment and require a mandatory EIA, including wastewater treatment plants for more than 150,000 population equivalents (PE). For projects listed in Annex II, the national authorities have to decide whether an EIA is needed based on a "screening procedure". EIA may be applied to different types of activities (projects and/or existing production processes and services) and has as main deliverable a form of Environmental Impact Statement (EIS) (UNEP, 2002). The environmental impact assessment may be achieved by qualitative and quantitative impact evaluation methods that used to assign (qualitatively) or compute (quantitatively) numerical values for the environmental impacts (Singh *et al.* 2012).

In order to consider the local conditions related to the discharges of WWTP effluent so as to enable better decisions regarding wastewater management, the environmental impacts quantification (EIQ) may be realised in a single stage or by integrating also the environmental risks (environmental impact and risk assessment or EIRA) (Teodosiu *et al.* 2015; Teodosiu *et al.* 2016). The environmental impact is computed as a function of the impact magnitude (by considering the flows of discharged wastewaters), impact severity (considering the measured concentrations of water quality indicators) and the natural quality state of the receiving water. The environmental risk (ER) is calculated as a function of the magnitude of environmental impact (EI) and its probability of occurrence.

The origins of *Life Cycle Assessment (LCA)* can be traced back to the late 1960's (Bauman *et al.* 2004) but it was not until late 1990's that the International Standards organization (ISO) developed a standard procedure for LCA within the framework of its 14000 series. The current 14040 and 14044 ISO standards (ISO, 2006a,b) describe the general methodology consisting of four stages (see next section for details). Emmerson *et al.* (1995) has been cited as the first LCA study applied to WWTPs (Corominas *et al.* 2013), however 20 years before Antonucci and Schaumburg (1975) had already recognized the fact that "the production and utilization of all forms of consumable energy and treatment chemicals will degrade the environment somewhere in some way". The authors were the first in accounting for the direct and indirect impacts of a WWTP but only in a qualitative way, as they stated, "given the current technological knowledge and understanding, there is no feasible way to compare the environmental significance of the various types of contaminants resulting from wastewater treatment" (Antonucci *et al.* 1975). LCA is nowadays considering the framework that Antonucci and Schaumburg were asking for and its application to wastewater has been abundant (see Corominas *et al.* (2013) for a review of 45 international peer-reviewed papers published between 1995–2012 and Loubet *et al.* (2014) for an updated list of references and a major focus on the results of WWTP-related LCA studies). Due to the life-cycle perspective taken, LCA can identify inherent environmental trade-offs of WWTPs between different environmental areas of concern, e.g. avoiding water pollution at the cost of increased air emissions of power plants (Figure 20.1).

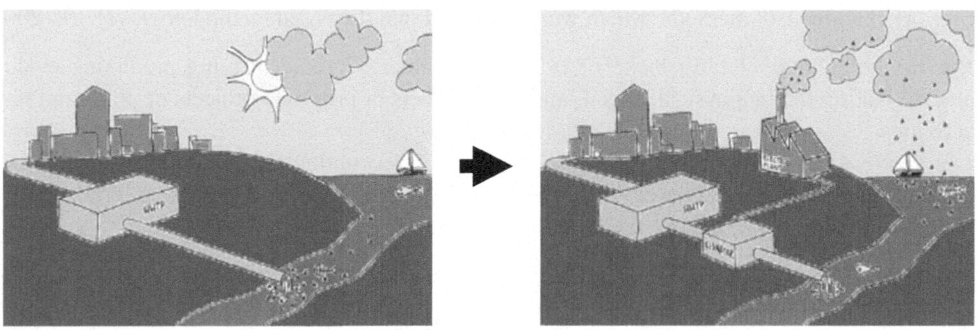

Figure 20.1 Avoiding an obvious problem in one place may induce a bigger problem somewhere else (sub-optimization).

In parallel to this extensive use, LCA has also been able to evolve and adapt itself to new requirements and demands. Of special relevance are the developments in toxicity-related impact categories for the inclusion of emerging contaminants and the calculation of spatially differentiated factors for local impact categories such as eutrophication (Corominas *et al.* 2013; Gallego *et al.* 2010; Alfonsin *et al.* 2014).

Another prominent instrument to evaluate the (un)sustainability of humans, nations, processes, products or activities are the "*footprint* family" analysis tools, such as environmental footprints (carbon, water, energy, nitrogen, etc.), social footprints, economic footprints, and composite or combined footprints (ecological footprint, sustainable process index, etc.) (Cucek *et al.* 2012). Many of those instruments share the life cycle approach of LCA but focus on a particular indicator.

In the field of water management, the *Water footprint* (WF) instrument was introduced by Hoekstra in 2002 as a spatially and temporally specific indicator, showing when, where and how the water is used (directly or indirectly) and polluted, helping managers to identify the main water users and to associate different kind of agricultural, municipal and industrial water users in the system (Hoekstra, 2003). In this concept, the blue water footprint is related to the surface and groundwater water resources use along the product supply chain; the green water footprint refers to consumption of rainwater (not as run-off), while the grey water footprint (GWF) refers to pollution and is defined as the volume of freshwater required to assimilate the load of pollutants based on existing ambient water quality standards (Hoekstra *et al.* 2011). The determination of these components (blue, green and grey) is realized in the water footprint accounting stage. The Water Footprint Assessment (WFA) is the next step of WF accounting and it supports better water management, including its use and allocation, by focusing on a complex analysis of the environmental sustainability, economic efficiency and social equity of freshwater use and allocation (Boulay *et al.* 2013).

Although the majority of studies concerning the assessment of wastewater treatment plants or technological alternatives are based on LCA, there are few studies that used the water footprint assessment in municipal wastewater treatment plants (Morera *et al.* 2016; Teodosiu *et al.* 2016; Garrido-Baserba *et al.* 2011) to identify complements between LCA and WF and compare the different assessment methods (Table 20.1).

While EIA is mandatory for large-scale WWTPs and evaluates the environmental impacts of a specific WWTP directly at a specific site, LCA takes a more global view and includes both direct and indirect impacts of WWTP and evaluates it with a dedicated set of environmental indicators that reflect *potential* environmental impacts. With this quantitative assessment, existing trade-offs can be identified while also discovering new environmental impact issues of concern that would not probably have been identified before the analysis at first sight. Unlike EIA, the assessments undertaken within the framework of LCA are not currently required by legislation. The WWTP impacts to surface waters (and subsequently their performance) are accounted by both water authorities and wastewater managers by using the legal requirements as a reference system which defines maximum allowed concentrations in the effluent, which water quality parameters to be monitored and with which frequency. LCA indicators have no direct correlation between the impact scores and the legal references, while EIA links the environmental impact directly to the legal requirements, thus providing a subsequent linkage to the objectives of the Water Framework Directive or other national or local water quality legislative (Teodosiu *et al.* 2016). The main difficulty with LCA is in defining what is to be included in the analysis and obtaining relevant data.

In this adaptation process, its role as an useful tool for the *decision making process* has been crucial in order to guarantee that the effort and time dedicated to carry out an LCA study is not in vain and results are effectively communicated. The combination of LCA with other tools such as Decision Support System (DSS) or Data Envelopment Analysis (DEA) illustrates this development.

DSSs, i.e. tools that enable a user to choose a consistent solution for a particular problem under reduced time frames, have gained interest in the wastewater management sector. An example of them is the Novedar-DSS that allows the selection between alternative WWTP process layouts, which includes analysis

Table 20.1 Environmental impact assessment procedures and outcomes (adapted from Teodosiu *et al.* 2016).

Assessment Instrument/ Procedure	Applications (Name of Procedure)	Types of Evaluated Impacts	Methods and Principles				Assessment Outcomes (Environmental Impact Statement)
			Impact Identification	Impact Analysis (Description)	Impact Evaluation		
					Qualitative Methods	Quantitative Methods	
Environmental impact assessment (mandatory procedure according to Directive 2014/52/ EU)	Plans, programs, policies (Strategic environmental assessment)	Potential impacts	Yes	Yes description of potential impact	Yes (e.g. interaction matrices, checklists,	No	(Strategic) Environmental Agreement
	New Projects (EIA)	Potential impacts	Yes	Yes, description of potential impact	Yes (e.g. interaction matrices, checklists	Yes	Environmental Agreement
	Existing, operational activities (EIA)	Actual impacts	Yes	Yes	Yes	Yes (impact indicators, indexes)	Environmental Permit
Life cycle assessment (voluntary procedure according to ISO 14040)	Traditional applied to Product systems Production processes and chains Water use cycle evaluations, Wastewater treatment	Potential (virtual)	Yes, through impact classifi-cation of LCIA	Yes, through impact characterization (normalization and weighting) of LCIA	No	Yes, through impact characterization (normalization and weighting) of LCIA	Interpretation of results phase: environmental profiles, recommendation and scenarios
Environmental impact quantification indexes (including Grey Water Footprint) (voluntary procedures)	Product systems, Production processes Potential (virtual)	Yes, (intrinsic)	Yes (intrinsic)	No	Yes, depending on impact definition (e.g. virtual volume of dilution water to reach objectives)	Environmental impact quantification indexes	

Source: Teodosiu *et al.* 2016.

of technical, economic, and social issues and operations (Garrido-Baserba *et al.* 2011). The synergistic combination of the two methodologies (DSS + LCA) to address the design and assessment of treatment facilities has served to identify the most sustainable options, embracing simultaneously a wide variety of analysis criteria, and enhancing the calculation of environmental savings (Garrido-Baserba *et al.* 2014).

Going beyond the standalone use of LCA for the definition of improvement actions that can be undertaken in individual WWTPs, the combination of LCA and DEA allows the estimation of specific operational benchmarks to monitor the performance (i.e., *eco-efficiency*) of a wide range of comparable units. The combined method has been applied to 113 WWTPs located in regions across Spain in order to determine the operational efficiency of each unit and obtain environmental benchmarks for inefficient plants (Lorenzo-Toja *et al.* 2015).

20.2 APPLICATION OF LIFE CYCLE ASSESSMENT TO WASTEWATER TREATMENT PLANTS AND PROCESSES

Method requirements and guidelines for conducting LCA studies are available through the ISO standards (Iso, 2006a) and (ISO, 2006b), and specific technical guidance is provided in (EC, 2010). Even though there are no specific guidelines for LCA in the field of wastewater treatment, three literature reviews (Corominas *et al.* 2013; Loubet *et al.* 2014; Yoshida *et al.* 2013) analyse the current situation and identify future challenges when applying LCA to this sector. Overall, an LCA includes the following phases: definition of goal and scope, setup of the life cycle inventory (LCI), a life cycle impact assessment (LCIA) and finally the interpretation of results. These phases are applied iteratively with different levels of data quality until the required level of accuracy in the model and precision of results is achieved (Figure 20.2).

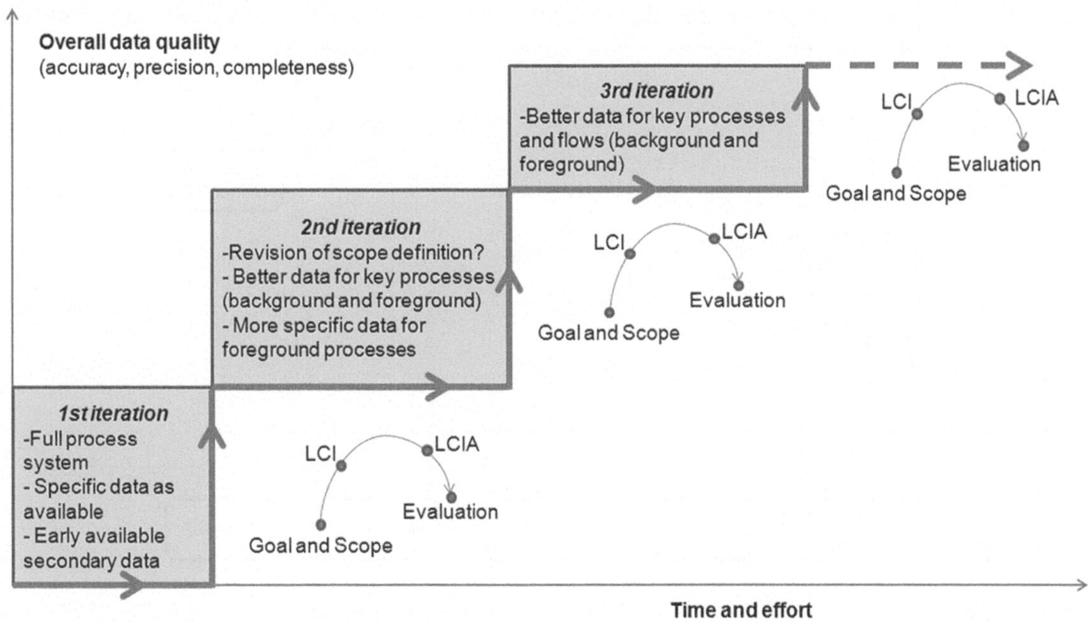

Figure 20.2 Iterative nature of LCA, involving at each iteration the 4 phases of LCA defined by ISO standards (LCI: life cycle inventory, LCIA: life cycle impact assessment). Adapted from EC (2010).

In the following sections, some recommendations are provided on how to adapt the LCA framework when dealing with prospective studies of "WWTP 2020" concepts and processes. It follows the indications provided in EC (2010):

a) Goal and scope definition

The starting point is to identify the process(es) or system(s) to be analysed in detail. An LCA is always anchored in a precise, quantitative description of the function(s) provided by the analysed system. The *functional unit* (FU) is the basis to which the impacts from different evaluated scenarios are referred to. The functional unit names and quantifies the qualitative and quantitative aspects of the function(s) along the questions "what", "how much", "how well", and "for how long". The most widely used functional unit for WWTPs has been the volume (m^3) of treated water for a certain period of time as the FU. However, the load associated with a one person equivalent (PE) tends to be used for comparative purposes, since it minimizes the differences associated with the influent composition and flow. When comparing several WWTPs with differences in the removal of nutrients, a functional unit that comprises the removal of both nutrients and organic matter, expressed in terms of kg PO_4 3 eq. removed is recommended (Rodriguez-Garcia *et al.* 2011).

The *system boundaries* define which parts of the life cycle and which processes belong to the analysed system, i.e. are required for providing its function as defined by its functional unit. Setting the system boundaries means deciding which life cycle stages (construction of the process, operation and dismantling) and specific systems/subsystems (e.g. primary, secondary, tertiary treatment, sludge treatment) to include and which to omit from the life cycle model. Examples of system boundaries are provided in Figure 20.3. Related to the construction stage, in recent years there is an increasing sense that the infrastructure of wastewater management services is more important that it seemed until now, and that it should be considered in LCA studies (Remy *et al.* 2008; Doka 2007). The most common practice is, as in (Foley *et al.* 2010) and for conventional WWTPs, estimating the volume of concrete used for the construction of the WWTP and relating this volume with the factors facilitated in Doka (2007) (to date, the most comprehensive construction inventory).

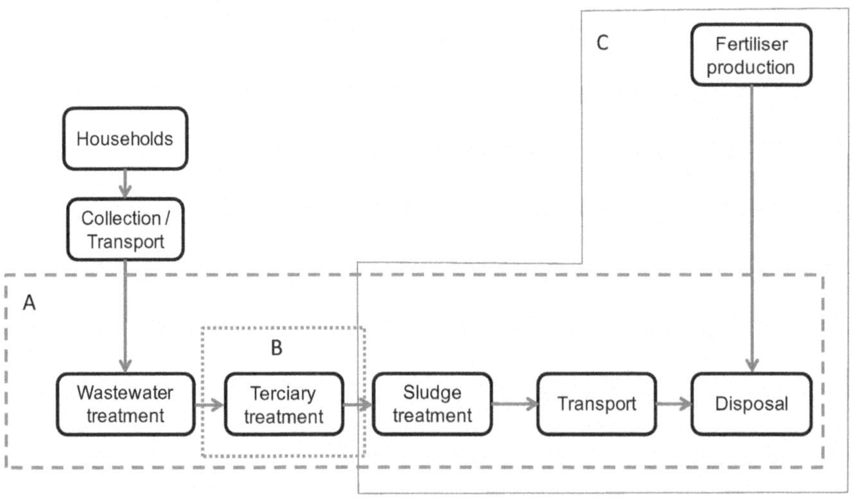

Figure 20.3 Boundaries of the urban wastewater system for the case-studies described in this chapter; NEPTUNE conventional WWTP (A), NEPTUNE organic micropollutants (B); Carismo (B); P-Rex (C).

The boundaries of the WWTPs have been expanded in some studies to include the whole urban water/ wastewater system, i.e. withdrawal of freshwater, drinking water production, distribution & use of drinking water, generation of wastewater and transport to the wastewater treatment plant. The boundaries of the WWTPs have also been expanded to consider the production of reclaimed water and the evaluation of potable water and desalinated water replacement.

b) Life cycle inventory

Figure 20.4 shows the input products, resources, emissions, (co-)products and waste which are to be quantified for the life cycle stages and systems considered as defined in the boundaries selection. The input product is the raw wastewater to be treated in the WWTP. Resources include energy and chemicals required to run the process and materials necessary to construct the process. Transportation of the resources to the process should be included as well as the transport of waste to the final disposal. Direct emissions to water, air and soil have to be quantified during the inventory phase. When evaluating conventional technologies (e.g. activated sludge process), proper quantification of nutrients discharge to the receiving water bodies is essential. Direct greenhouse gas emissions should ideally be considered and whenever possible be quantified by on-site measurement campaigns (Lorenzo-Toya *et al.* 2016; Yoshida *et al.* 2014a). The variability behind the published factors is large and proper measurements help in reducing the uncertainty in the obtained results.

Figure 20.4 Black box unit process. Examples for wastewater treatment are shown in brackets (adapted from ILCD handbook (EC, 2010)).

Within this phase, the studies often face problems associated with *data availability* and *data quality*. The foreground life cycle inventory (LCI) data is usually compiled directly from measurements, a calibrated mass balance model from the system, detailed design documents and vendor-supplied information. In the first iterations of the LCI exercise, emissions can also be assessed with the help of scientific models, literature values or specific emission factors. It is worth stating that all mass balances for calculating emissions should be closed to represent a reasonable mode of operation. Background datasets (e.g. electricity generation systems, concrete and chemicals production processes) are normally provided by publically available LCI databases, e.g. the EcoInvent database (www.ecoinvent.ch). It is important that published studies provide sufficient detail on the data sources and data quality to be able to reproduce the work. More information about the influence of different data collection schemes for WWTPs can be found in Yoshida *et al.* (2014b).

When evaluating new technologies tested in lab or pilot scale, the *correct scaling-up* has great importance for the LCI setup, especially with regards to energy consumption. Energy consumption at lab-scale is often not optimized and hence, over- or underestimations can mask the real trade-offs identified. The same applies for construction where proper scaling factors should be applied. It is highly recommended to cross-check any scaling factors with process engineers or technology suppliers to validate the input data for the LCA of new technologies or processes.

For each study, *co-products* of the process under study should be included by crediting avoided impacts of the primary product. A typical example is the agricultural application of WWTP sludge, which allows to take into account the positive effects of the nutrient value of the sludge and expanded the system to include the avoided production of synthetic fertilizers as well as the negative consequences associated with the heavy metals also present in the sludge. In the case study section, the technical recovery of phosphorus (P-REX) exemplifies this approach by taking into account the avoided impacts of mineral P fertilizer production.

Two main types of LCA modelling exist, i.e. *attributional or average and consequential or marginal.* Traditionally attributional modelling have been used, where the examined water system is analysed based on its existing structure typically by assigning unit processes based on average values disregarding consequences of the possible change of the system. However, the use of the more complex consequential modelling is increasing, where the consequences of changing the system (e.g. implementing new technology) is taken into account by e.g. using future energy production marginals (e.g. renewable energy). Whether attributional (average) or consequential (marginal) modelling should be used depends on the goal and scope of the study. For hot-spot analysis or comparison of environmental sustainability among existing technologies, attributional LCA may be sufficient. However, when introducing new technologies is part of the goal, a consequential approach including system expansion and identification of energy marginals is needed. In many existing studies a combination has been used, such as using an average approach as basis but doing system expansion regarding the application of sludge on agricultural land (consequence: fertilizer is substituted and its production avoided). In all cases transparency regarding the approach used is mandatory in order to obtain reliability of the results.

c) Life Cycle Impact Assessment

Within Life Cycle Impact Assessment (LCIA), midpoint indicators (e.g., global warming potentials) are *measures of potential impact*, defined at some intermediate point along the cause-effect chain from intervention (e.g. emissions and resource flows collected during the inventory phase) to the final damage at the area of concern (Hauschild *et al.* 2015) as shown in Figure 20.5.

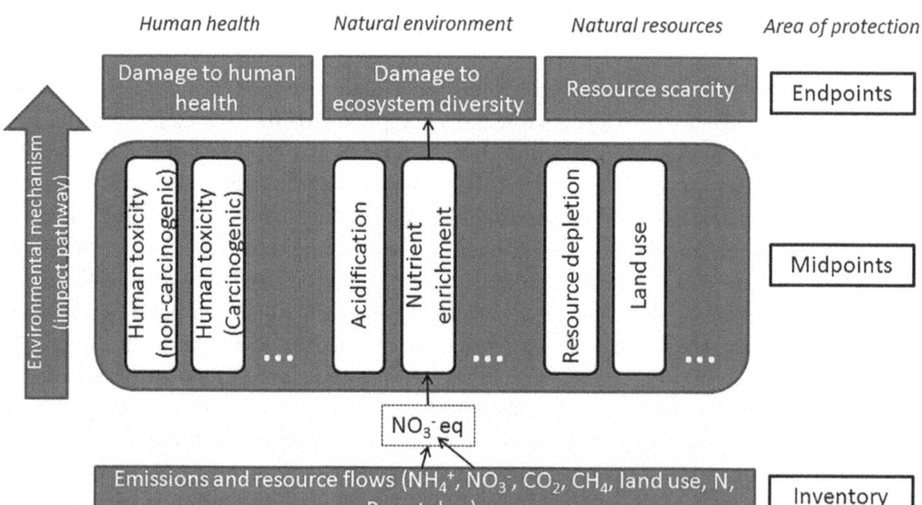

Figure 20.5 Life Cycle Impact assessment, from the inventory to mid-point and end-point indicators.

With regards to the selection of impact categories assessed, the ISO guidance indicates that the categories selected shall be consistent with the goal and scope, comprehensive, and have an identified effect (end-point). Traditionally, global warming potential, acidification, and eutrophication are the indicators that receive most attention in LCA of wastewater treatment. On a secondary level, photochemical oxidation and toxicity-related aspects are the issues of concern. Terrestrial ecotoxicity plays an important role when sludge disposal options are evaluated and heavy metals or micropollutants are considered (Corominas *et al.* 2013). Finally, ozone layer depletion is a matter of concern (especially when including N_2O emissions (Lane *et al.* 2012)) together with abiotic depletion (includes fossil energy and material depletion).

Concerning the set of impact models used for each impact category, there are several options within current LCA software. *Best-practice recommendations* for EU are available from a review of the best available models in 2009 (Hauschild *et al.* 2013) and further upgrades available in (EU, 2013). Furthermore, in order to achieve a robust LCA result, the ILCD approach can be combined with more complete models such as the ReCiPe suite (Goedkoop *et al.* 2009) which can normally be adopted by default in wastewater treatment studies, as it provides one of the most comprehensive and extensively used set of LCA impact categories. The ReCiPe method includes both midpoint, and endpoint (damage) categories addressing the safeguard objects ecosystems, human health and resource availability. With regards to toxicity, the best practice methodology for the moment is the consensus model USEtox (Rosenbaum *et al.* 2008) and USES-LCA (Huijbregts *et al.* 2000). New characterization factors for some organic micropollutants applicable to USEtox and USES-LCA can be found in (Alfonsin *et al.* 2014).

Normalization is a step that puts category indicator results into perspective by relating them to specific reference information. Although there are a number of normalization methods in use today, the method which has the greatest acceptance in the LCA community is external normalization to the total or per capita regional emissions/extractions. For instance in NEPTUNE results are normalized using the person-equivalents targeted (PET).

d) Interpretation and communication of LCA results

According to ISO 14040:2006 (ISO, 2006a), the interpretation should include a) identification of significant issues based on the results of the LCI and LCIA phases of an LCA b) evaluation of the study considering completeness, sensitivity and consistency checks and c) conclusions, limitations and recommendations.

The interpretation is usually based on methods such as *"hot spot" identification* in the life cycle of a specific technology/system or *comparison of the impact profile* of different technologies performing the same service, e.g. advanced phosphorus removal. Identification of significant issues can also follow the *"induced vs. avoided impacts"* approach, which is represented in the NEPTUNE case studies below. This approach is illustrated in Figure 20.1, and reflects a typical challenge in wastewater treatment, i.e. the achievement of higher effluent water quality on the expense of higher energy consumption or higher consumption of ancillaries, like precipitation chemicals. This approach has now been used in several studies and was introduced by Wenzel *et al.* (2008) and in a more comprehensive way by Larsen *et al.* (2007) and Larsen *et al.* (2010). The approach puts special demands on the toxicity related impact categories and the eutrophication potential.

Within the evaluation, completeness, sensitivity and consistency checks should be conducted. *Completeness* check attempts to ensure that the full required information and data from all phases have been used and are available for interpretation. *Sensitivity* analysis is executed to determine the influence of variations in assumptions, methods and data on the results. *Uncertainty* analysis can be applied to analyze the variability of results due to the variation of input parameters over their whole domain of uncertainty, normally using Monte-Carlo simulations. Consistency aims at ensuring that the methodology is applied uniformly to all components (e.g. consistent in data sources, data accuracy, time-related coverage, etc.).

The *communication* of the results is a challenging issue since multiple criteria are normally combined with multiple scenarios evaluated. This creates a space of large number of dimensions difficult to explain to the audience. One of the widely used ways of presenting the results is taking a reference scenario for which the impacts are calculated and relate the impacts of the other scenarios to that reference situation. In such a way induced and avoided impacts can be calculated for each scenario (e.g. NEPTUNE studies shown below and P-REX), being positive values induced impacts and negative values credits. For better understanding the results it is important to provide the contribution analysis of all processes to the overall impacts (see CARISMO case-study) through staked columns in which each processes show its own impacts. In addition, several graphical techniques are available such as sankey diagrams or treemaps. Finally, when interpreting the results *limitations* of the followed approach should be discussed (e.g. in terms of scope, data quality, excluded elements) and some recommendations should be provided.

20.3 CASE STUDIES

The case studies presented below in a fact-sheet layout should give an idea of the different nature, related goal and scope definitions, and type of results for several LCA studies in the field of wastewater treatment. Starting with the general LCA of a state-of-the-art WWTP in the first case study, more innovative processes such as ozonation for removal of organic micropollutants, improved energy recovery by advanced primary treatment or technical recovery of phosphorus from wastewater sludge are analysed with LCA.

20.3.1 Fact sheet: LCA of conventional WWTP

Source: EU-FP6 project NEPTUNE (www.eu-neptune.org).

Background: In the EU research project on "New sustainable concepts and processes for optimization and upgrading municipal wastewater and sludge treatment" (NEPTUNE, www.eu-neptune.org) the relative environmental sustainability of several waste water and sludge treatment technologies was assessed. Regarding nutrient removal conventional treatment, autotrophic anaerobic ammonium oxidation (anammox) and "instrumentation, control and automation strategies (ICA)"was included. As regards sludge disposal the inertization technologies on-site incineration, wet oxidation, middle temperature pyrolysis (gasification) and high temperature pyrolysis (pyrolysis) were investigated. Furthermore, different sludge triage systems (separate treatment of primary and secondary sludge) including mesophilic anaerobic digestion, incineration, aerobic thermophilic treatment, thermal disintegration, ultrasonic pre-treatment and application on agricultural land was assessed. Here the case of the reference scenario on conventional treatment will be described with the aim of illustrating that treating sewage water conventionally actually makes sense from a holistic environmentally point of view. Further details including all the other technologies may be found in (Larsen *et al.* 2010).

20.3.1.1 Goal and scope definition

Goal of LCA: The goal of this case study is to aim at answering the question: Does it make sense from a holistic environmental point of view to treat sewage water by conventional treatment before emission to the environment? In other words: do we avoid more potential environmental impact than we induce by treating the sewage?

Functional unit: Treatment of one m^3 raw sewage. The sewage water is defined by its composition regarding 45 substances including pharmaceuticals and other organics, metals, and nutrients (P and N). As described in details in Larsen *et al.* (2010) the composition and removal rates are based on measurements performed within NEPTUNE combined with literature values. The conventional treatment includes primary mechanical separation, biological nitrification/denitrification with an activated sludge loop, and finally chemical phosphorus removal.

System boundaries: System boundaries include the entire process of wastewater treatment and the sludge disposal Figure 20.6.

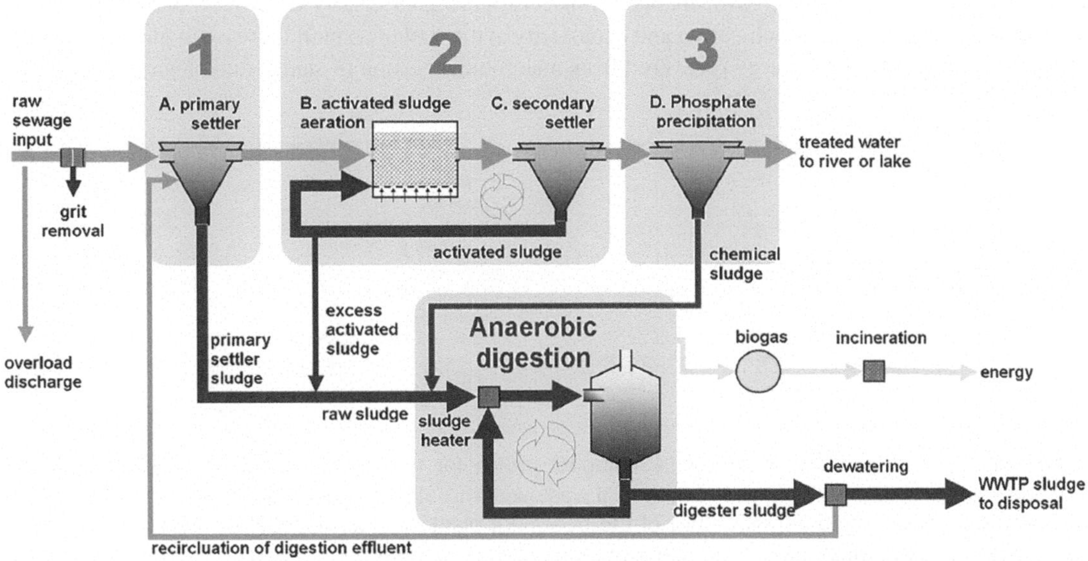

Figure 20.6 System boundaries of NEPTUNE case study 1 (All three steps of wastewater treatment and anaerobic digestion leading to incineration of digester sludge and biogas substituting fossil energy are included) (Larsen *et al.* 2010).

The impact assessment method used is EDIP97 (Wenzel *et al.* 1997) which includes 10 different categories of environmental impact: Acidification potential (AP), Global warming potential (GWP), Nutrient enrichment potential (NEP), Ozone depletion potential (ODP), Photochemical ozone formation potential – in low NOx areas (POP), Human toxicity in soil (HTS), Human toxicity in water (HTW), Human toxicity in air (HTA), Ecotoxicity in water, chronic (EWC), Ecotoxicity in soil, chronic (ESC).

Results are normalized according to the normalization references described in Strandorf *et al.* (2005) and a weighting factor of 1 is used for all impact categories. The weighted potential impacts are expressed in (micro) person equivalents targeted per cubic meter waste water ($\mu PET/m^3$) according to Wenzel *et al.* (1997). A consequential LCA approach has been used and all foreground processes are assumed to occur in Europe. Induced impacts are compared to avoided impacts.

Scenarios: Conventional treatment of sewage water as compared to the reference scenario, i.e. direct emission of sewage water without treatment.

Life Cycle Inventory: The modelling is consequential to the degree possible, e.g. the marginal technology for electricity production is assumed to be based on natural gas, which is also partly substituted by biogas production from anaerobic sludge treatment. The modelling tool used is GaBi 4 (www.gabi-software.com) with EcoInvent v2.0 database (www.ecoinvent.org).

Impact assessment: That it actually makes sense, in an overall environmental sustainability context, not to emit wastewater directly but to treat it beforehand by use of conventional technology, is illustrated in

Figure 20.7. The total environmental impact score is reduced by more than 50%, mostly because impacts from water discharge are greatly reduced in terms of eutrophication potential (nutrient enrichment) and ecotoxicity. On the other hand, global warming is slightly increased mainly due to the energy consumption of the treatment process. In order to get the full picture of energy demand and global warming, sludge treatment and disposal has to be included as well. By taking into account sludge disposal, avoided impacts for all categories can be off-set with induced impacts Figure 20.8 clearly showing that WWTP avoids environmental impacts in eutrophication and ecotoxicity at the cost of a small increase in global warming, which is partially off-set by energy recovered from anaerobic digestion of sludge and biogas valorisation.

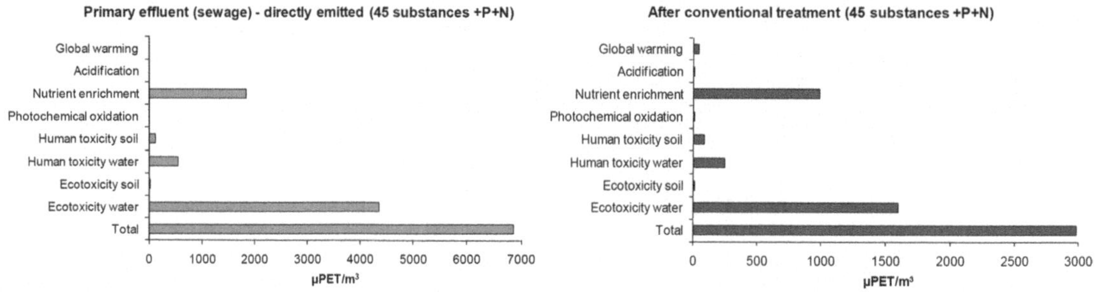

Figure 20.7 Normalized and weighed LCA impact profile for the direct emission of primary effluent (sewage water) to recipient (left) and a standard wastewater treatment plant (right), including the induced impact from operation and infrastructure of the treatment plant (sludge disposal (including incineration) not included). Calculation based on the raw wastewater concentration of 45 substances including pharmaceuticals, metals and other inorganics, and nutrients as described in Larsen *et al.* (2010). The EDIP97 LCIA methodology was used and weighting factor for all impact categories equals 1.

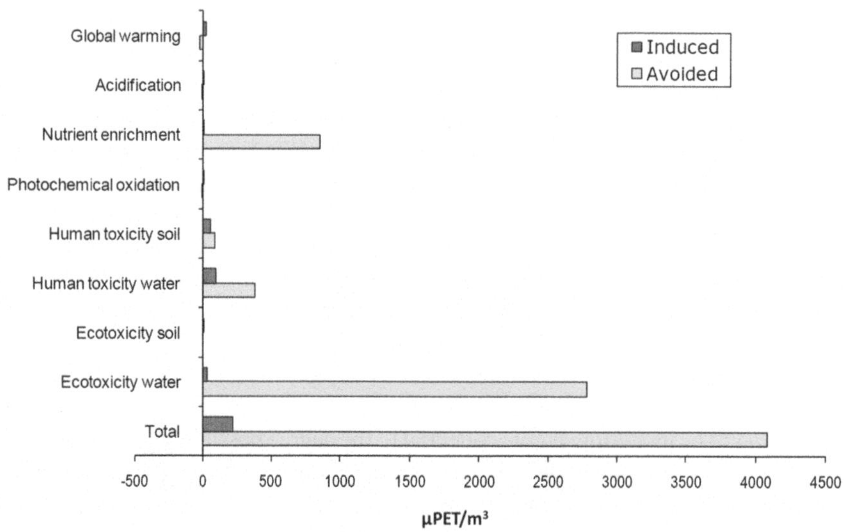

Figure 20.8 LCA impact profile based on Figure 20.7 showing induced and avoided impacts for conventional treatment of wastewater including sludge disposal (incineration).

20.3.2 Fact sheet: LCA study on WWTP upgrade for elimination of organic micropollutants

Source: EU-FP6 project NEPTUNE (www.eu-neptune.org).

Background: Removal of micropollutants was also part of the NEPTUNE project. In order to address the increasing focus on (eco)toxic substances like pharmaceuticals in waste water, different polishing technologies for removing micropollutants from secondary waste water were tested, and investigated for their relative sustainability. Included were pulverized activated carbon (PAC) addition, ozonation, sand filtration and combinations including final sand filtration. Secondary waste water is here defined as the treated waste water leaving a conventional treatment, i.e. primary mechanical separation, biological (an) aerobic (de)nitrification with an activated sludge loop, and finally chemical phosphorus removal. Here the case of ozonation will be described but further details including the other technologies may be found in Larsen *et al.* (2010) and Larsen *et al.* (2009).

20.3.2.1 Goal and scope definition

Goal of LCA: The goal of this case study is to aim at answering the question: Is ozonation an improvement in terms of environmental sustainability compared to conventional WWT, i.e. do we avoid more potential environmental impact than we induce by introducing the technology?

Functional unit: Treatment of one m^3 secondary wastewater (=effluent from clarifier). The secondary wastewater is defined by its composition regarding 22 pharmaceuticals (including one metabolite), 9 metals, total P and total N (Larsen *et al.* 2010). The metals and nutrients are only relevant if final sand filtration is included.

System boundaries: The impact assessment method used is EDIP97 (Wenzel *et al.* 1997) with comparable impact categories as in the first case study, also regarding normalization and weighting. A consequential LCA approach have been used and all foreground processes are mainly based on data produced by the NEPTUNE project (measurements on test sides etc.) and assumed to occur in Europe, whereas background data are taken from the EcoInvent v2.0 database (www.ecoinvent.org). Induced impacts are compared to avoided impacts as shown in Figure 20.9.

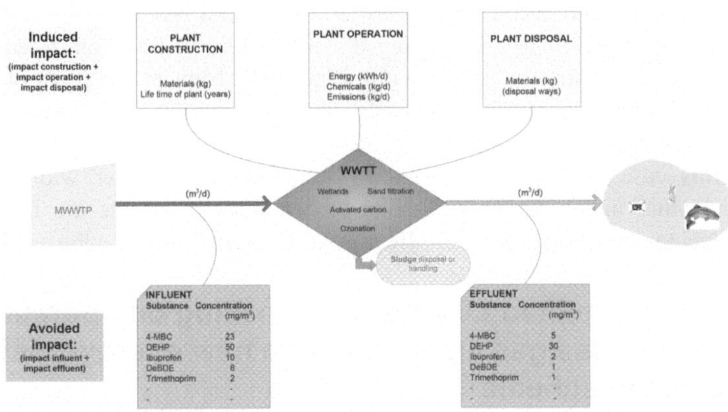

Figure 20.9 The scoping of the "avoided against induced impacts" approach. The municipal wastewater treatment plant (MWWTP) is outside the scope and the wastewater treatment technology (WWTT) is in this case ozonation and therefore no sludge is produced.

Scenarios: The reference scenario is direct emission of secondary wastewater with the composition described in Larsen *et al.* (2010). This scenario is compared to tertiary treatment with ozonation.

Life Cycle Inventory: The modelling is consequential to the degree possible, e.g. the marginal technology for electricity production is assumed to be based on natural gas. The modelling tool used is GaBi 4 (www.gabi-software.com) with EcoInvent v2.0.

Impact assessment: The results of the impact assessment regarding ozonation (3.2 g O_3/m^3) are shown in Figure 20.10. The weighted normalized potential impacts avoided by introducing ozonation at 3.2 g O_3/m^3 are seven times higher than the corresponding induced part.

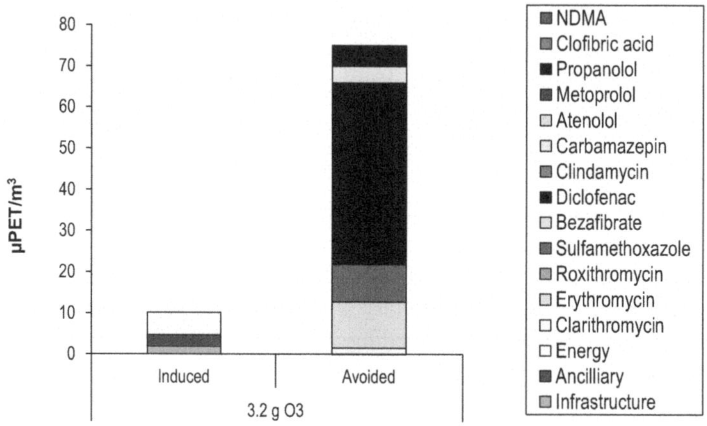

Figure 20.10 Induced and avoided impacts (expresses as µPET/m³) for the process of ozonation of secondary waste water at 3.2 g O_3/m^3. Ancilliary is here oxygen for ozone production. Insignificant pharmaceuticals are not shown.

It should be noted that the potential effect of metabolites (e.g. aldehydes) created during the ozonation and reflected in whole effluent toxicity of the ozonation effluent (not shown here) is not included in this LCA and therefore not taken into account in Figure 20.10. From a risk assessment point of view, this problem was addressed and solved by introducing downstream sand filtration. Results of an LCA of ozonation combined with sand filtration are shown in Figure 20.11 where the effect of sand filtration on metals and phosphorus in the effluent is also included. This combination results in about eighteen times more avoided than induced potential impact.

The results of this case study indicate that for municipal wastewater with typical content of pharmaceuticals and other micropollutants avoided environmental impacts are higher than induced impacts when introducing ozonation and especially ozonation in combination with sand filtration.

20.3.3 Fact sheet: Simplified LCA study focussing on operational energy demand and greenhouse gas emissions of a new energy-positive wastewater treatment scheme

Source: Research project CARISMO (http://www.kompetenzwasser.de/CARISMO.519.0.html).

Background: Raw wastewater contains much more energy in form of organic matter (COD) than is required for its treatment. To maximise the exploitation of this energy potential and construct an energy-positive

WWTP, the CARISMO project tested a new wastewater treatment scheme based on coagulation/flocculation of raw wastewater after mechanical screening and subsequent filtration in a drumfilter (100 μm mesh). Pilot trials with real wastewater (4–6 m^3/h) show that it is feasible to extract 70–80% of COD with this enhanced primary treatment, while the remaining effluent can be further treated in a simple biofilter for removal of residual COD and nitrogen. The new concept was compared to a conventional scheme with a simplified energy balance and carbon footprint study (Remy *et al.* 2014) to prove the energy-positive operation, including energy production from sludge via digestion and biogas valorisation.

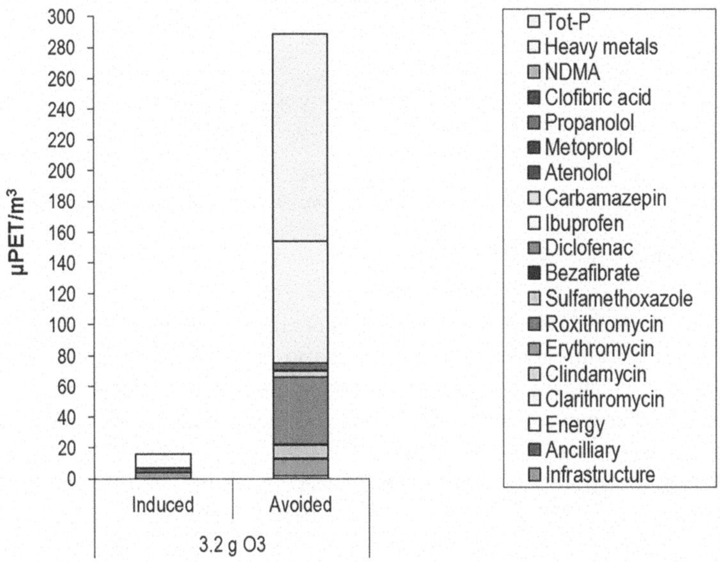

Figure 20.11 Induced and avoided impacts for the process of ozonation combined with post sand filtration.

20.3.3.1 Goal and scope definition

Goal of LCA: Simplified assessment of energy demand and greenhouse gas emissions of new wastewater treatment scheme and comparison to conventional process of activated sludge.

Functional unit: "m^3 of wastewater treated"

System boundaries: System boundaries include operation of wastewater treatment in terms of electricity and chemicals demand, as well as anaerobic digestion of sludge and biogas valorisation in CHP plant (Figure 20.12). Further sludge treatment (dewatering and disposal, return load) and infrastructure are excluded from the simplified assessment.

Scenarios: The reference scenario is defined as conventional activated sludge process with primary clarifier, and activated sludge with nitrification/denitrification and chemical P elimination (Fe dosing). The CARISMO scenario includes coagulation/flocculation of raw wastewater (after pre-screening), microsieve, and post-treatment in a biofiltration process. Both processes are designed to reach a comparable effluent quality (100 mg/L COD, 18 mg/L TN, 2 mg/L TP).

Life Cycle Inventory: Data for the conventional process was taken from reference models of WWTP. Data for CARISMO scheme was based on pilot trials (chemical dosing, removal of COD/N/P in primary treatment, biogas yield of sludge, demand of external carbon for denitrification) and design data for biofiltration.

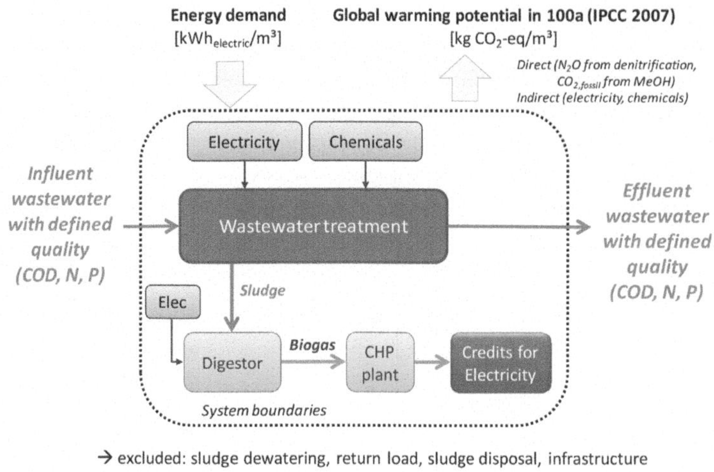

Figure 20.12 System boundaries of LCA in CARISMO.

Impact assessment: For impact assessment, two indicators have been used describing the energy demand and associated greenhouse gas emissions from the two scenarios. For energy demand, primary energy demand for chemicals production of coagulant, polymer and external carbon source (methanol) has been recalculated to kWh (electric) using the German energy mix to enable a comparison of electricity balances while including chemicals for both schemes. Electricity balances show that the CARISMO scheme is superior to the conventional scheme both in terms of net electricity balance, but also in terms of net energy balance including the production of chemicals (Figure 20.13). In fact, the CARISMO scheme produces +81% more biogas and thus has a net electriciy balance of –0.47 kWh/m^3, meaning that it can export electricity to the grid and clean the wastewater at the same time. Accounting also for chemicals production, the CARISMO scheme still has a positive energy balance equaling –0.17 kWh/m^3. Greenhouse gas emissions of the new concept indicate that it can be carbon-neutral even if expected N_2O emissions from biological N removal are included (Figure 20.14).

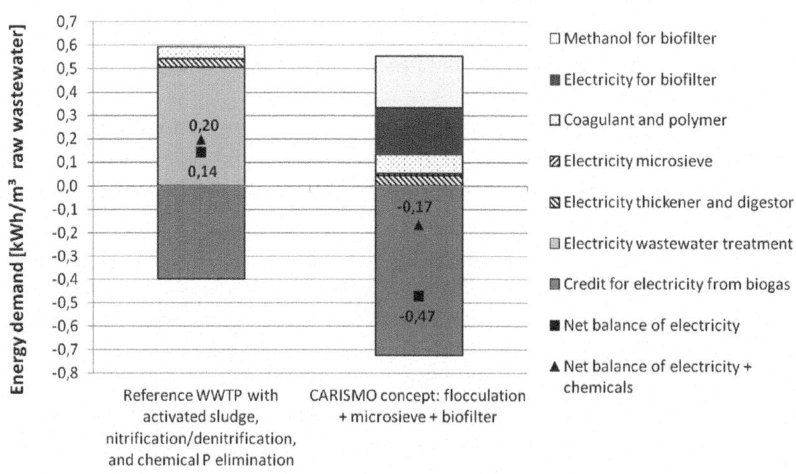

Figure 20.13 Primary energy demand of reference and CARISMO scheme (Remy *et al.* 2014).

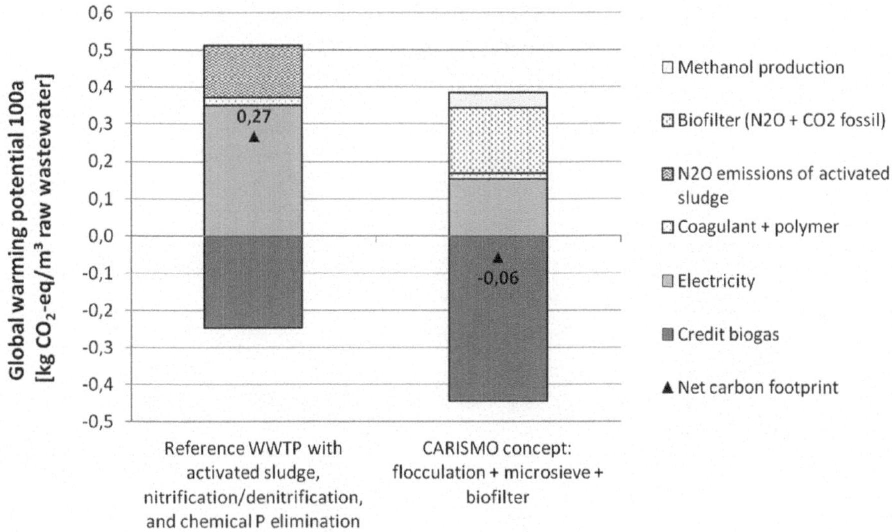

Figure 20.14 Global warming potential of reference and CARISMO scheme (Remy *et al.* 2014).

This study shows that it is feasible to design an energy-positive wastewater treatment scheme, using available technology and integrating it into a new concept. Up-scaling of the CARISMO process and full-scale trials for combining carbon extraction with downstream nitrogen removal will be the next step towards demonstrating the application of the CARISMO concept in real WWTPs.

20.3.4 Fact sheet: LCA study on phosphorus recovery from sewage sludge, sludge liquor, or incineration ash

Source: EU-FP7 project P-REX (www.p-rex.eu).

Background: The recovery of phosphorus (P) from sewage sludge, sludge liquor, or ash from mono-incineration can be realized with different processes which have been developed, tested or already realized in full-scale in recent years. However, these pathways and processes differ in their amount of P that can be recovered in relation to the total P content in sludge, in the quality of the recovered P product, and in their efforts in energy, chemicals, fuels, and infrastructure required for P recovery. This study analyses selected processes for P recovery from sludge, liquor, or ash in their potential environmental impacts (Remy *et al.* 2015).

20.3.4.1 Goal and scope definition

Goal of LCA: Assess selected processes for P recovery from municipal sewage sludge, liquor, or incineration ash in their environmental impacts, taking into account relevant side-effects on the sludge treatment or the WWTP.

Functional unit: "P recovery for a 1 Mio pe WWTP per year"

System boundaries: Include all processes that are related to sludge treatment and disposal, i.e. sludge digestion and biogas valorisation in CHP plant, sludge dewatering and transport, treatment of return liquor (simplified model for WWTP), mono-incineration and disposal of ash, infrastructure for P recovery, and substitution of mineral fertilizer production by recycling products (Figure 20.15).

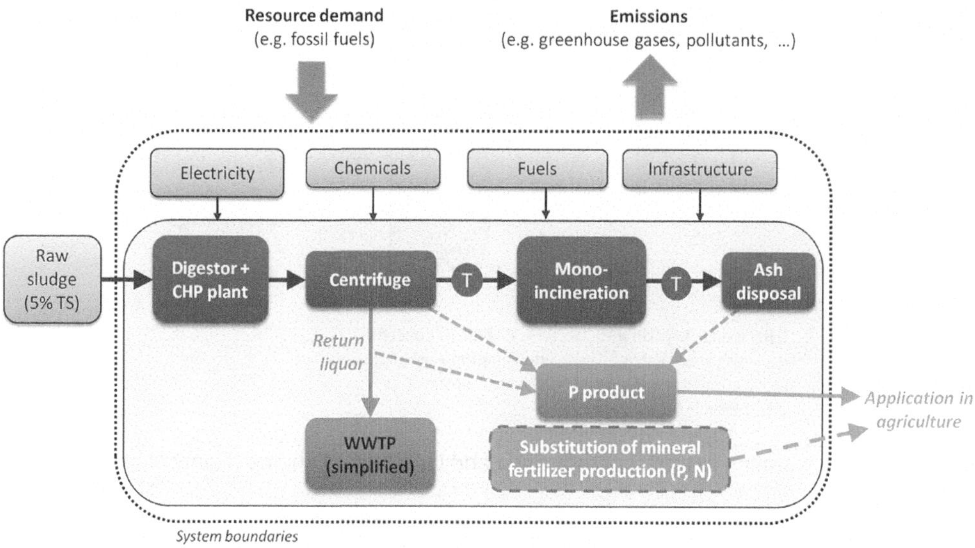

Figure 20.15 System boundaries for LCA of technical phosphorus recovery (P-REX).

Scenarios: 10 different processes for P recovery, including direct precipitation of struvite in sludge, struvite precipitation in liquor (2 processes), sludge leaching with acid (2 processes), metallurgic treatment of dried sludge or ash, ash leaching (2 processes), or thermo-chemical treatment of ash.

Life Cycle Inventory: Data for the reference WWTP system treating sludge from chemical P elimination (ChemP) or enhanced biological P elimination (EBPR) is defined according to mean values in Germany for a sludge line of a 1 Mio WWTP. Data for the P recovery processes is collected and validated from technology providers and case studies, relying on collected process data (status in 2014) of process efficiency, product quality, and energy and material demand from full-scale plants, pilot trials, or prospective modeling depending on the maturity of the investigated process. Background processes for electricity, chemicals, fuels, infrastructure and substitution of mineral fertilizer are taken from ecoinvent v3.1 database. For P substitution, plant availability in the different P products is not taken into account in this study, assuming a 100% substitution potential of recycled P.

Impact assessment: Environmental impacts are assessed with a set of 9 different indicators. Results show that pathways and processes for P recovery differ heavily in their amount of recovered P, but also in energy and related environmental impacts such as greenhouse gas emissions (Figure 20.16). As direct struvite precipitation in sludge or liquor relies on the dissolved amount of P in digested sludge, these processes are only applicable in wastewater treatment plants with biological P removal. Here, they can recover 4–18% of total P in sludge with a relatively low effort in energy and chemicals, reducing return load to the mainstream process and eventually improving sludge dewaterability in case of direct precipitation in sludge. Acidic leaching of P from digested sludge can yield up to 48% of P for recovery, but requires a significant amount of chemicals for control of pH (leaching and precipitation) and for minimizing heavy metal transfer into the product.

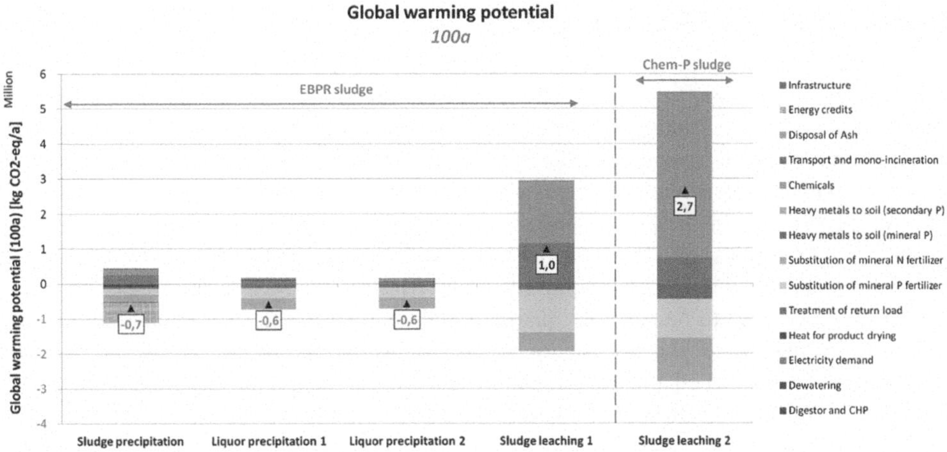

Figure 20.16 Global warming potential of P recovery from sludge or sludge liquor (additional impacts and benefits of P recovery shown in comparison to reference scenario).

Leaching of mono-incineration ash with sulphuric acid yields 70% P with moderate chemical demand, but the leached ash and co-precipitated materials have to be disposed. Complete digestion of ash in phosphoric acid and multi-stage cleaning with ion exchangers yields high recovery of 97% P in a high-quality product (H_3PO_4) and several co-products, having an overall low environmental impact (Figure 20.17). Thermo-chemical treatment of ash can recover up to 98% P with moderate energy input in case of integration into an existing mono-incineration facility. Metallurgic treatment of dried sludge or ash can also recover up to 81% of P, but the process has still to be tested in continuous pilot trials to validate product quality, energy demand, and energy recovery options.

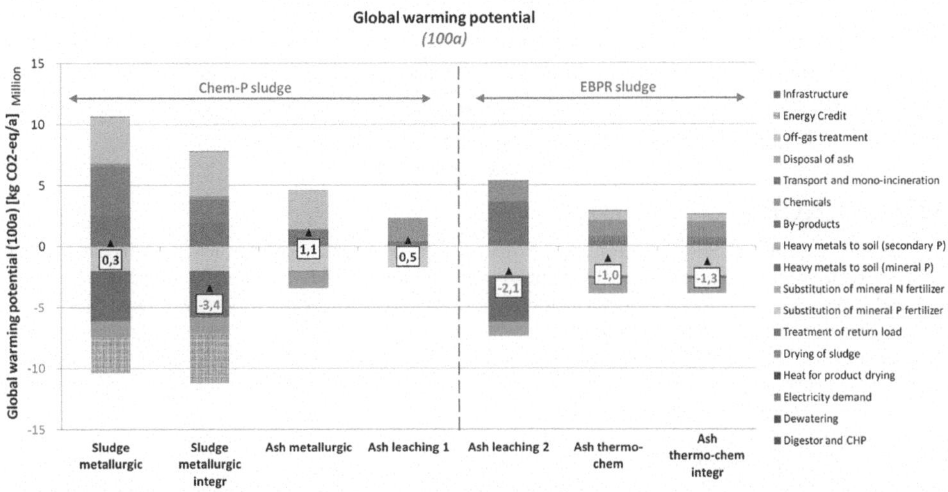

Figure 20.17 Global warming potential of P recovery from dried sludge or incineration ash (additional impacts and benefits of P recovery shown in comparison to reference scenario).

20.4 CONCLUSIONS AND OUTLOOK

For assessing environmental impacts of WWTPs, two types of methods are generally available and abundantly used: Environmental Impact Assessment and Life Cycle Assessment. While the former is mandatory for large-scale WWTPs and evaluates the additional environmental impacts of a specific WWTP directly at a specific site, LCA takes a more global view and includes both direct and indirect impacts of WWTP and evaluates it with a dedicated set of environmental indicators. With this quantitative assessment, existing trade-offs can be identified while also discovering new issues and aspects which have not been in focus before.

Following the definitions of ISO 14040/44, LCA is a standardized framework in which all environmental impacts of WWTP can be quantified and illustrated. However, the different steps within the general framework have to be carefully justified and adapted to the specific goal and scope of the LCA study to fit the purpose of the assessment. During interpretation of the results, a typical approach can be the comparison of induced and avoided environmental impacts (=net environmental benefit), thus showing the inherent trade-off of WWTPs which have to use additional resources (energy, chemicals) and create related emissions while improving the quality of discharged wastewater and thus protect the receiving waters.

Within the current paradigm shift of WWTPs from pollutant removal to resource recovery, LCA can play an important role by evaluating new technologies and processes in terms of overall environmental sustainability. This can help in identifying optimisation potentials of new or emerging processes and prevents (or at least illustrates) problem-shifting from one area of environmental concern to another.

Some future challenges for adapting the ISO-based LCA to the specific features of WWTPs include:

- The adaptation of impact assessment methods to include emerging pollutants (e.g. organic micropollutants, nanoparticles, microplastic) and hygienic parameters (e.g. bacteria, viruses, parasites).
- The development of more regional impact assessment factors (e.g. for water footprinting).
- The improvement of data quality for primary data (e.g. N_2O emissions from biological nitrogen removal).
- The reduction of uncertainty in LCI data and impact assessment.

Besides these particular challenges, stricter adherence to ISO methodological standards is needed to ensure quality and transparency of the studies and increase trust in the outcomes of LCA studies in the field of wastewater treatment. Initiatives working on this subject include the IWA working group on LCA and the initiative on Product Category Rules (PCR) for wastewater treatment.

20.5 REFERENCES

Alfonsín C., Hospido A., Omil F., Moreira M. T. and Feijoo G. (2014). PPCPs in wastewater – Update and calculation of characterization factors for their inclusion in LCA studies. *Journal of Cleaner Production*, **83**, 245–255.

Antonucci D. and Schaumburg F. (1975). Environmental effects of advanced wastewater treatment at South Lake Tahoe. *Water Pollution Control Federation*, **47**(11), 2694–2701.

Bauman H. and Tillman A. (2004). The Hitch Hiker's Guide to LCA. Studentlitteratur AB, Lund (Sweden).

Boulay A., Hoekstra A. and Vionnet S. (2013). Complementarities of water-focused life cycle assessment and water footprint assessment. *Environmental Science and Technology*, **47**, 11926–11927.

Corominas L., Foley J., Guest J., Hospido A., Larsen H., Morera S. and Shaw A. (2013). Life cycle assessment applied to wastewater treatment: state of the art. *Water Research*, **47**, 5480–5492.

Čuček L., Klemeš J. and Kravanja Z. (2012). A Review of Footprint analysis tools for monitoring impacts on sustainability. *Journal of Cleaner Production*, **34**, 9–20.

Doka G. (2007). Part IV: Life cycle inventory of wastewater treatment. Life cycle inventories of waste treatment services – EcoInvent report No. 13. Swiss Center for Life Cycle Inventories, Dübendorf, Switzerland, December 2007.

EC (2010). European Commission - Joint Research Centre - Institute for Environment and Sustainability: International Reference Life Cycle Data System (ILCD) Handbook - General guide for Life Cycle Assessment – Detailed guidance. 1st edn. March 2010. EUR 24708 EN. Publications Office of the European Union, Luxembourg.

EU (2013). COMMISSION RECOMMENDATION of 9 April 2013 on the use of common methods to measure and communicate the life cycle environmental performance of products and organisations. *Official Journal of the European Union*. **56**, 1–210.

Emmerson R., Morse G., Lester J. and Edge D. (1995). The Life-Cycle Analysis of small-scale sewage-treatment processes. *Water and Environment Journal*, **9**(3), 317–325.

Foley J., de Haas D., Yuan Z. and Lant P. (2010). Nitrous oxide generation in full-scale biological nutrient removal wastewater treatment plants. *Water Research*, **44**(3), 831–844.

Gallego A., Rodriguez L., Hospido A., Moreira M. T. and Feijoo G. (2010). Development of regional characterization factors for aquatic eutrophication. *The International Journal of Life Cycle Assessment*, **15**, 32–43.

Garrido-Baserba M., Reif R., Rodríguez-Roda I., Poch M. (2011). A knowledge management methodology for the integrated assessment of WWTP configurations during conceptual design. *Water Science and Technology*, **66**(1), 165–172.

Garrido-Baserba M., Hospido A., Reif R., Molinos-Senante M. and Comas J. (2014). Including the environmental criteria when selecting a wastewater treatment plant. *Environmental Modelling and Software*, **56**, 74–82.

Goedkoop M., Heijungs R., Huijbregts M., De Schryver A., Struijs J. and Van Zelm R. (2009). ReCiPe 2008, A life cycle impact assessment method which comprises harmonised category indicators at the midpoint and the endpoint level. First edition Report I: Characterization. http://www.lcia-recipe.net.

Hauschild M. and Huijbregts M. (2015). Life Cycle Impact Assessment. Springer, Dordrecht, Netherlands.

Hauschild M., Goedkoop M., Guinée J., Heijungs R., Huijbregts M., Jolliet O., Margni M., De Schryver A., Humbert S., Laurent A., Sala S. and Pant R. (2013). Identifying best existing practice for characterization modeling in life cycle impact assessment. *International Journal of Life Cycle Assessment*, **18**(3), 683–697.

Hoekstra A. (2003). Virtual Water Trade: Proceeding of the International Expert Meeting on Virtual Water Trade. Value of Water Research Report Series No. 12. UNESCO-IHE, Delft, Netherlands.

Hoekstra A., Chapagain A., Aldaya M. and Mekonnen M. (2011). The Water Footprint Assessment Manual: Setting the Global Standard. Earthscan, London, UK.

Huijbregts M., Thissen U., Guinee J., Jager T., Kalf D., van de Meent D., Ragas A., Sleeswijk A. and Reijnders L. (2000). Priority assessment of toxic substances in life cycle assessment. Part I: calculation of toxicity potentials for 181 substances with the nested multi-media fate, exposure and effects model USES-LCA. *Chemosphere*, **41**(4), 541–573.

https://www.linkedin.com/grp/home?gid=4921250

http://www.environdec.com/en/PCR/Detail/pcr2013-14#.VhE81Pntmko

ISO (2006a). ISO 14040: 2006 Environmental Management – Lice Cycle Assessment – Principles and Framework, International Standardisation Organisation, Geneva, Switzerland.

ISO (2006b). ISO 14044: 2006 Environmental Management – Lice Cycle Assessment – Requirements and Guidelines, International Standardisation Organisation, Geneva, Switzerland.

Kirk B. (2005). Peering beyond the end of the pipe: non-monetary methods for comparing wastewater treatment options. EcoEng Newsletter No. 11 (http://www.iees.ch/EcoEng051/EcoEng051_Kirk.html)

Kirk B., Etnier C., Kärrman E. and Johnstone S. (2005). Methods for Comparison of Wastewater Treatment Options. Project No. WU-HT-03-33. Prepared for the National Decentralized Water Resources Capacity Development Project. Washington University, St. Louis, MO by Ocean Arks International, Burlington, VT (US).

Lane J. and Lant P. (2012). Including N2O in ozone depletion models for LCA. *The International Journal of Life Cycle Assessment*, **17**(2), 252–257.

Larsen H., Hauschild M., Wenzel H. and Almemark M. (2007). Homogeneous LCA methodology agreed by NEPTUNE and INNOWATECH. Deliverable 4.1. EC Project "NEPTUNE", contract No.: 036845. www.eu-neptune.org

Larsen H., Olsen S., Hauschild M. and Laurant A. (2009). Methodology for including specific biological effects and pathogen aspects into LCA. Deliverable 4.2. EC Project "NEPTUNE", contract No.: 036845. www.eu-neptune.org

Larsen H., Hansen P. and Boyer-Souchet F. (2010). Decision support guideline based on LCA and cost/efficiency assessment. Deliverable 4.3. EC Project "NEPTUNE", contract No.: 036845. www.eu-neptune.org

Lorenzo-Toja Y., Vazquez-Rowe I., Chenel S., Marín-Navarro D., Moreira M. and Feijoo G. (2015). Eco-efficiency analysis of Spanish WWTPs using the LCA + DEA method. *Water Research*, **68**, 651–666.

Lorenzo-Toja Y., Alfonsin C., Amores M., Aldea X., Marin D., Moreira M. and Feijoo G. (2016). Beyond the conventional life cycle inventory in wastewater treatment plants. *Science of the Total Environment*, **553**, 71–82.

Loubet P., Roux P., Loiseau E. and Bellon-Maurel V. (2014). Life cycle assessments of urban water systems: a comparative analysis of selected peer-reviewed literature. *Water Research*, **67**, 187–202.

Marr K. and Wood C. (1996). A comparative analysis of EIA practice for wastewater treatment plants in Great Britain and Germany. *International Planning Studies*, **1**(2), 217–238.

Morera S., Corominas L., Poch M., Aldaya M. and Comas J. (2016). Water footprint assessment in wastewater treatment plants. *Journal of Cleaner Production*, **112**, 4741–4748.

Remy C. and Jekel M. (2008). Sustainable wastewater management: life cycle assessment of conventional and source-separating urban sanitation systems. *Water Science and Technology*, **58**(8), 1555–1562.

Remy C. and Jossa P. (2015). Life Cycle Assessment of selected processes for P recovery from sewage sludge, sludge liquor, or ash. Deliverable 9.2, EC project "P-REX" . Kompetenzzentrum Wasser Berlin, Berlin, Germany, (www.p-rex.eu).

Remy C., Boulestreau M. and Lesjean B. (2014). Proof of concept for a new energy-positive wastewater treatment scheme. *Water Science and Technology*, **70**(10), 1709–1716.

Rodriguez-Garcia G., Molinos-Senante M., Hospido A., Hernandez F., Moreira M. T. and Feijoo G. (2011). Environmental and economic profile of six typologies of wastewater treatment plants. *Water Research*, **45**, 5997–6010.

Rosenbaum R., BachmannT., Gold L., Huijbregts M., Jolliet O., Juraske R., Koehler A., Larsen H., MacLeod M., Margni M., McKone T., Payet J., Schuhmacher M., van de Meent D. and Hauschild M. (2008). USEtox-the UNEP-SETAC toxicity model: recommended characterisation factors for human toxicity and freshwater ecotoxicity in life cycle impact assessment. *International Journal of Life Cycle Assessment*, **13**(7), 532–546.

Singh R., Murty H., Gupta S. and Dikshit A. (2012). An overview of sustainability assessment methodologies. *Ecological Indicators*, **15**, 281–299.

Stranddorf H., Hoffmann L. and Schmidt A. (2005). Impact categories, normalisation and weighting in LCA. Environmental News no. 78 from the Danish Environmental Protection Agency, Copenhagen.

Teodosiu C., Robu B., Cojocariu C. and Barjoveanu G. (2015). Environmental impact and risk quantification based on selected water quality indicators. *Natural Hazards*, **75**, 89–105.

Teodosiu C., Barjoveanu G., Robu Sluser B., Ene Popa S. and Trofin O. (2016). Environmental assessment of municipal wastewater discharges: a comparative study of evaluation methods. *The International Journal of Life Cycle Assessment*, **21**(3), 395–411.

The Decentralized Water Resources Collaborative (DWRC), more formally known as the National Decentralized Water Resources Capacity Development Project (NDWRCDP), is a cooperative effort funded by the U.S. EPA to support research and development on decentralized wastewater and stormwater systems. The DWRC is committed to advancing knowledge, science and training in decentralized systems in order to build the capacity of organizations and individuals to appropriately implement them. For more information: http://www.ndwrcdp.org/

Trajkovska-Trpevska M. and Dodeva S. (2009). Environmental Impact Assessment of the Waste Water Treatment Plant for City of Skopje. International Symposium on Water Management and Hydraulic Engineering. Ohrid/Macedonia, 1–5 September 2009.

UNEP (2002). Environmental Impact Assessment Training Resource Manual. UNEP Division of Technology, Industry and Economics: Economics and Trade Branch, Geneva, Switzerland.

Wenzel H., Hauschild M. and Alting L. (1997). Environmental Assessment of Products. 1st edn. Chapman & Hall, London, Great Britain, Vol. **1**.

Wenzel H., Larsen H., Clauson-Kaas J., Høibye L. and Jacobsen B. (2008). Weighing environmental advantages and disadvantages of advanced wastewater treatment of micro-pollutants using environmental life cycle assessment. *Water Science and Technology*, **57**(1), 27–32.

Yoshida H., Christensen T. H. and Scheutz C. (2013). Life cycle assessment of sewage sludge management: a review. *Waste Management & Research*, **31**(11), 1083–1101.

Yoshida H., Mønster J. and Scheutz C. (2014a). Plant-integrated measurement of greenhouse gas emissions from a municipal wastewater treatment plant. *Water Research*, **61**, 108–118.

Yoshida H., Clavreul J., Scheutz C., Christensen T. (2014b). Influence of data collection schemes on the Life Cycle Assessment of a municipal wastewater treatment plant. *Water Research*, **56**, 292–303.

Chapter 21

Determining benchmarks in wastewater treatment plants using life cycle assessment

Gumersindo Feijoo, Yago Lorenzo- Toja, Desirée Marín-Navarro, Maria Teresa Moreira and Ian Vázquez-Rowe

21.1 INTRODUCTION

Wastewater treatment has become an important industry since the early 20th century, when Ardent and Lockett first described the activated sludge process in the United Kingdom. Ever since the early 20th century, wastewater treatment technologies have been implemented in urban, rural and industrial environments all over the world, as a way to prevent public health hazards, as well as avoiding the gradual degradation of natural ecosystems (Metcalf & Eddy, 2003; Corominas *et al.* 2013). In fact, wastewater treatment plants (WWTPs) have slowly become complex systems, in which flexibility within their operation has become an important factor to be taken into consideration (Hopkins *et al.* 2001). For instance, it is a common issue that WWTPs may suffer important changes in flow rate entering the plant, as well as gradual or drastic changes in the composition of the wastewater.

In this context, analysing the efficiency of WWTPs has become a challenging and desired task for stakeholders, wastewater researchers and policy makers. These circumstances have led to a situation in which numerous studies have dealt with the efficiency analysis of WWTPs. However, traditionally this analysis has focused on economic and productivity aspects of the venues (Sala-Garrido *et al.* 2012). On the one hand, Hernández-Sancho and Sala-Garrido (2009) worked on the assessment of the economic efficiency of several WWTPs along the Spanish Mediterranean coast. The findings in this study emphasized the usefulness of this type of analysis, since it can offer a detailed description of possible reductions in the use of operational inputs that are appropriate with a certain output (i.e., the effluent with fixed quality standards). On the other hand, some other studies analysed the changes in the productivity of WWTPs in urban environments (Marques & Monteiro, 2003; Byrnes *et al.* 2010), or the impact of privatisation and regulation processes in the water industry (Saal & Parker, 2000, 2001).

Given their complexity and the challenges that their correct operation poses, a wide range of factors may influence the performance of WWTPs. For instance, one of the most important driving forces affecting the efficiency of these plants is linked to be the characteristics of the influent. Influents that present high load usually tend to have adequate levels of pollutant removal. However, if the influent is diluted, these characteristics are likely to cause operational issues.

Despite a series of obvious environmental benefits linked to the main purposes of WWTPs (Macleod & Haygarth, 2010; Gracia-Lor *et al.* 2012), including the removal of coarse solids and organic pollutants (e.g., dissolved organic matter, solids and nutrients), there are some important threats linked to their operation in terms of environmental and economic sustainability (Molinos-Senante *et al.* 2011). These threats or limitations include aspects that affect environmental protection, as well as social and economic development (Balkema *et al.* 2002). Hence, some studies have identified that certain operational inputs in WWTPs, such as the use (and type) of energy or the nature of the chemicals used, as well as the treatment of the wastes produced result in a rise of the total environmental impact linked to the treatment process (Hospido *et al.* 2004).

Based on this discussion, WWTPs have been a recurrent topic in environmental sustainability studies in recent years. Although numerous environmental management tools and methods have been applied to give answer to different problems in this sector, and internationally standardized methodology, named Life Cycle Assessment (LCA) has become the most used tool for the environmental assessment of these systems (Emmerson *et al.* 1995; Hospido *et al.* 2004; Pasqualino *et al.* 2009; Rodriguez-Garcia *et al.* 2011). Ever since the first research applying LCA to wastewater treatment occurred in the late 1990s, studies have suffered a gradual transformation, becoming increasingly specialized, focusing on the main hotspots of wastewater treatment: type of treatment technology (Høybe *et al.* 2008), nutrient removal (Rodriguez-Garcia *et al.* 2014) sludge management (Hospido *et al.* 2010) or direct greenhouse gas (GHG) emissions (Lorenzo-Toja *et al.* 2015b). In this context, as mentioned previously, even though the main function of a WWTP, which is to remove solids and dissolve organic matter and nutrients from water, leads to an obvious environmental benefit, the energy and chemicals consumed in the overall process, as well as the direct GHG emissions can compromise to a great extent the sustainability of the entire process (Lorenzo-Toja *et al.* 2015).

Another important factor that influences the sustainability of wastewater treatment is the operational efficiency of the WWTPs. A commonly used methodology to study production frontiers in operations research is Data Envelopment Analysis (DEA), a nonparametric linear programming technique (Cooper *et al.* 2007). A series of previous DEA studies that delved into the efficiency analysis of this kind of facilities have focused on economic (Sala-Garrido *et al.* 2012) and productivity aspects (Marques & Monteiro, 2003). In contrast, environmental impacts and gains due to wastewater treatment have not been considered within the efficiency analysis of WWTPs.

21.2 JOINT APPLICATION OF LIFE CYCLE ASSESSMENT AND DATA ENVELOPMENT ANALYSIS TO WASTEWATER TREATMENT PROCESSES

The current research study proposes the combination of LCA and DEA in order to analyse the eco-efficiency of WWTPs in a similar way to how it has been done with other production systems (Vázquez-Rowe *et al.* 2011; Iribarren *et al.* 2013; Vázquez-Rowe *et al.* 2012). In fact, the LCA + DEA method has shown to be a useful methodology when it comes to analysing the eco-efficiency in scenarios with multiple inventory datasets available (Vázquez-Rowe & Iribarren, 2014).

For this study, a total of 113 plants located across the Spanish territory were analysed with the aim of:

- providing target values for the inputs involved in the operation of the different plants;
- calculating the current environmental impacts, as well as the environmental benefits of applying the target operational values;
- identifying a series of specific factors that control the eco-efficiency of WWTPs.

21.3 MATERIALS AND METHODS

21.3.1 The five-step LCA + DEA method

The five-step LCA + DEA methodology was presented for the first time in Vázquez-Rowe *et al.* (2010) and further methodological developments were advanced in a series of studies, such as Iribarren *et al.* (2011) or Vázquez-Rowe and Iribarren (2015). This combination of methods allows the minimization of operational inputs to achieve reductions in operational inefficiencies and, consequently, the environmental impacts linked to the operation of the system. The five-step method, as shown in Figure 21.1, is structured as follows:

- individual data collection for each of the units (i.e, decision making units – DMUs) in the study and construction of he life cycle inventory (LCI) and DEA matrices;
- calculation of the environmental burdens of each WWTP individually by means of a selected life cycle impact assessment (LCIA) method (e.g., ReCiPe, IPCC 2013; Impact 2002+, etc);
- computation of DEA matrices to obtain the efficiency scores and the virtual operational values of each unit;
- estimation of the new environmental burdens for the inefficient plants based on the virtual operational values obtained in step 3;
- eco-efficiency verification and benchmark set up. In addition to the conventional five-steps, it was necessary to do a slight modification of the method in order to compute one of the selected inputs (use of chemicals) in a homogeneous form. To do so, step 2 was divided into two blocks. The first one keeps the original mission of step 2, while the second one allows the calculation of the environmental load related to chemical use through the weighted ReCiPe endpoint LCIA method (Goedkoop *et al.* 2009), as described in Avadí *et al.* (2014).

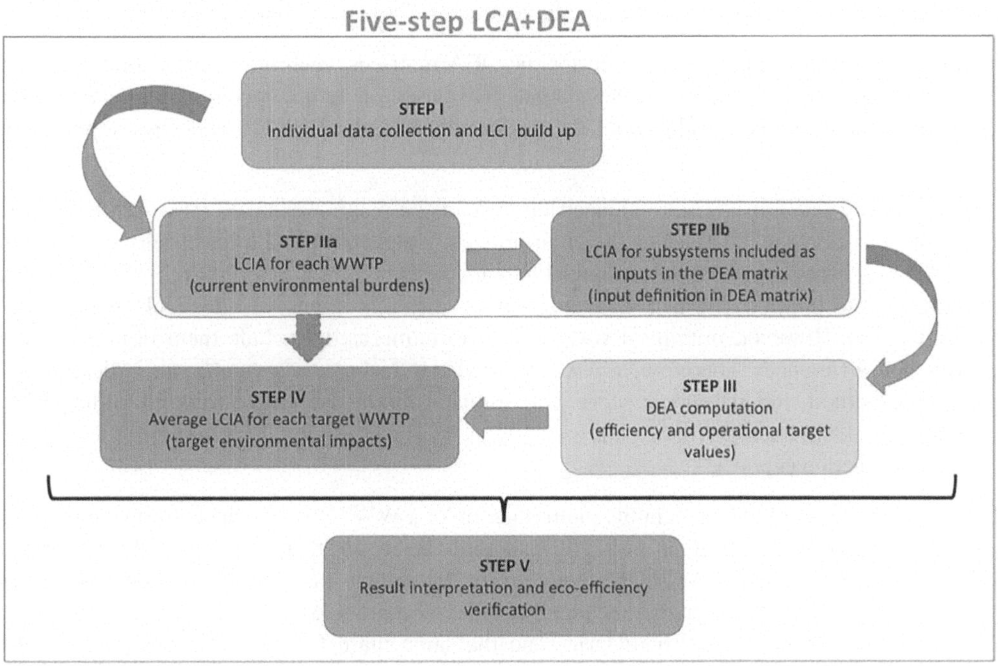

Figure 21.1 Schematic representation of the LCA + DEA method (Five-step approach).

21.3.2 DEA model selection and matrices build up

The model chosen for this case study was the slacks-based measure of efficiency (SBM), as described by Cooper *et al.* (2007). The rationale behind this selection was linked to the elasticity concerning the calculation of inefficiencies, the non-radial characteristics of the inputs and outputs approach and, finally, the computation of any type of inefficiency rather than only those linked to technical issues (Cooper *et al.* 2007). Moreover, the SBM model provides operational reductions of inputs and outputs for the inefficient units based on the calculated slacks, providing a convenient framework for the setting up of the benchmarking. Regarding other details that should be mentioned of the model, it is important to bear in mind that is was run under the input-oriented approach, which allows minimizing the use of inputs without compromising the final output, in this case the quality of the effluent per cubic metre of treated water. Finally, the constant returns-to-scale approach was selected as the segregation of the DEA matrices proposed below reduces the effects of scale factor among the WWTPs (Banker *et al.* 1984). In other words, the division of the WWTPs depending on their size guarantees that they can be treated as comparable units of production.

Consequently, three different matrices were created to provide a more accurate assessment. Based on a previous study that was performed within the AQUAENVEC (2012) project, analysing 80 Spanish WWTPs, it was determined that these can be classified in three groups when taking into consideration their operational size. In fact, the main function and technological solutions adopted remain similar for all the units inside each group. Therefore, three matrices were created based on the real person equivalent (p.e.) treated in year 2011:

- WWTPs between 0 and 20,000 population equivalent (Small WWTPs);
- WWTPs between 20,001 and 50,000 population equivalent (Medium WWTPs);
- WWTPs treating wastewater above 50,000 population equivalent (Large WWTPs).

Regarding the input and output selection for the DEA matrices, each matrix was composed of three inputs and one output. The selection of these inputs was based on actual operational aspects that can be objectively minimized without influencing the performance of the WWTP. The three inputs included were:

- electricity use (kWh/m^3). The total electricity use of the plants was taken into consideration in the year of operation (i.e., 2011). For the environmental impacts linked to electricity production, the Spanish electricity mix for 2011 was taken into account;
- chemical consumption (Pt/m^3). Consumption of different types of chemicals was taken into consideration. However, initially it was difficult to numerically include them in one single input given their differences. Therefore, as a way to be able to include them in one sole indicator, an LCA endpoint method, ReCiPe single score, was used to obtain one single value to homogenize this group;
- sludge production (kg/m^3).

The output was selected to represent the main function of a WWTP: removing eutrophication potential from the aquatic environment (Larsen *et al.* 2007). It could be argued that the removal of microorganisms is also an important function of WWTPs. However, in the sample that was chosen for this case study, WWTPs with tertiary treament were not included, in order to guarantee that the DMUs (i.e., the individual WWTPs) have comparable system boundaries and the same main function. For this purpose, the net environmental benefit (NEB) methodology for the eutrophication category proposed by Godin *et al.* (2012) was adopted.

21.4 RESULTS AND DISCUSSION

21.4.1 Inventory data and DEA computation

Data from a total of 470 facilities were gathered thanks to the collaboration with one of the main water technology centres in Spain (CETAQUA). A wide range of operational aspects of the plants, such as water quality parameters, chemical and energy consumption, waste generation and other operational parameters were obtained from the CETAQUA databases and the record books of each plant, as reflected in the system boundaries (see Figure 21.2). Nevertheless, due to data gaps, unreliable data quality and the presence of tertiary treatments, a great number of WWTPs were discarded and finally only 113 (see Figure 21.3) were considered in the study, representing approximately 7% of the total amount of p.e. treated in Spain in 2011 according to the Ministry of the Environment (2014). Concerning data for background processes, these were obtained from the ecoinvent® database, except for the Spanish electricity mix that was modelled following the report from Red Electrica Española (2011), as described in Vázquez-Rowe *et al.* (2015). The functional unit (FU) selected, in line with the function of the system, was 1 m^3 of wastewater treated by the WWTP in the year 2011.

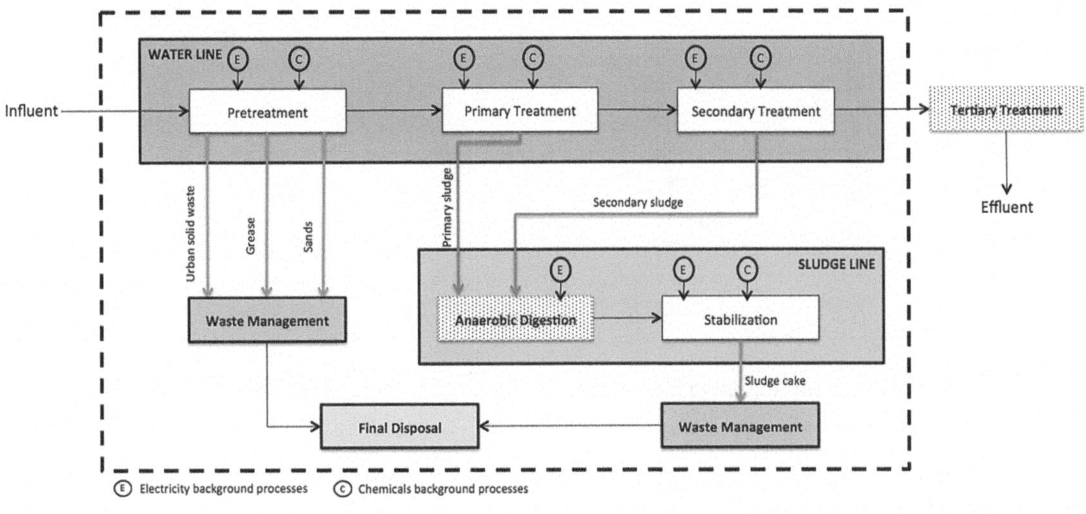

Figure 21.2 Schematic representation of the system boundaries of the production system for each wastewater treatment plant. (Adapted from Lorenzo-Toja *et al.* 2015.)

The three identified matrices were computed in the DEA-Solver Professional Release 10.0 software (Saitech, 2014) using the abovementioned SBM model. The results obtained in this first iteration showed that only 11 facilities (approximately 10%) operated at fully efficiency (i.e., efficiency score of 100%). For the rest of the sample, most DMUs presented efficiencies between 25% and 75%, which suggests that the operational differences between plants are considerably high. This characteristic allowed a substantial reduction in the input target values proposed by the SBM model without hindering the output quality (see Table 21.1).

21.4.2 Environmental and operational performance

The environmental impacts found for the units under assessment were in line with previous results extracted from LCA studies that delved into the environmental assessment of WWTPs. For instance, in terms of the freshwater eutrophication potential (EP) category, the average value was 2.16 · 10–3 kg P eq/m^3, in line with the results presented in Rodríguez-Garcia *et al.* (2011). In the case of another commonly discussed category, climate change (CC), the average outcome 4.5 · 10–1 kg CO_2 eq/m^3 was also in line with previous research in the field (Pasqualino *et al.* 2009; Rodríguez-Garcia *et al.* 2011).

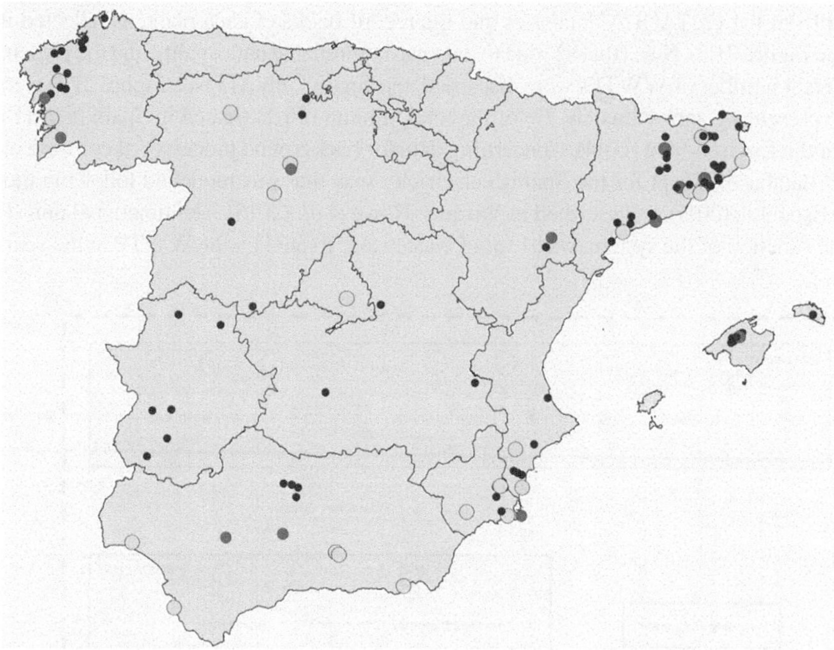

Figure 21.3 Map of Spain representing the wastewater treatment plants considered in the study.

Table 21.1 Average target reduction percentages for operational inputs and efficiency scores (%) per size range.

Range of WWTP	Inputs			Efficiency (%)
	Electricity (%)	Chemical Consumption (%)	Sludge Production (%)	
Small	58.9	75.0	71.4	31.6
Medium	50.3	59.6	42.3	49.3
Large	31.9	62.3	48.0	52.6

Finally, in terms of the calculated NEB, which was estimated for all the WWTPs to use it as the output of the DEA matrix, no negatives results were found. Thus, concerning eutrophication, the existence of a WWTP always appeared as a more adequate option than the direct withdrawal of the influents to the corresponding water basins.

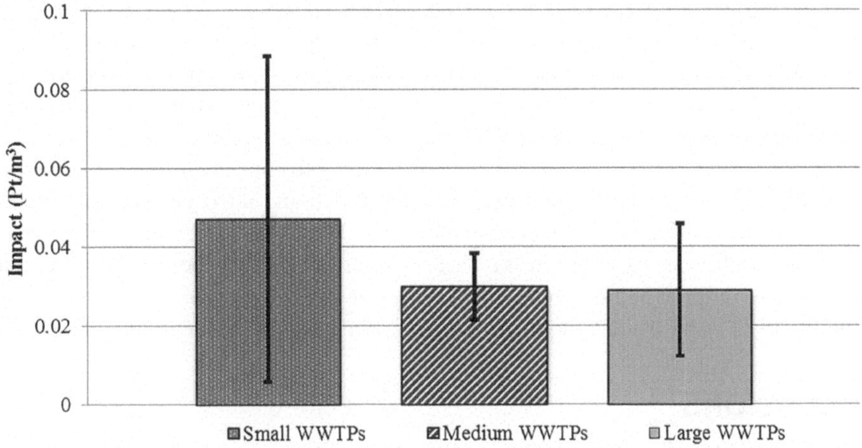

Figure 21.4 Average single score endpoint value per each of the three WWTPs ranges.

Efficiency differences were considerable between DMUs in all three groups, as shown in Table 21.2. The large WWTPs group, composed by 22 plants, averaged the highest efficiency values (52.6%). The medium WWTPs matrix contained 14 facilities and behaved similar to the large WWTPs group with an average efficiency of 49.3%. Finally, small WWTPs, with 77 plants computed, presented the lowest results in terms of efficiency, with only 31.6%. The main reasons behind the lower efficiency of this range are the lack of a rigorous operational protocol and the lower plant flexibility.

Table 21.2 WWTPs benchmark. Average inputs and output values for efficient WWTPs.

Range of WWTP	Inputs			Output
	Electricity (kWh/m^3)	Chemical Consumption (Pt/m^3)	Sludge Production (kg sludge/m^3)	NEB (kg N eq/m^3)
Small	$8.97 \cdot 10^{-1}$	$8.79 \cdot 10^{-4}$	$5.76 \cdot 10^{-1}$	$5.48 \cdot 10^{-2}$
Medium	$3.59 \cdot 10^{-1}$	$5.96 \cdot 10^{-4}$	$9.28 \cdot 10^{-1}$	$4.69 \cdot 10^{-2}$
Large	$2.81 \cdot 10^{-1}$	$2.62 \cdot 10^{-3}$	$4.43 \cdot 10^{-1}$	$2.80 \cdot 10^{-2}$

21.4.3 Factors affecting WWTPs efficiency

Several factors were found to influence the efficiency of WWTPs besides plant size: influent characteristics, over and undersized installations and type of treatment technology.

Regarding influent characteristics, organic load tend to be an important operational parameter within WWTPs (Garnier *et al.* 2013). For instance, low organic loads may cause operational drawbacks related to sludge retention inside biological reactors, whereas high loads tend to lead to higher sludge production rates and increased energy consumption. However, no clear tendency was observed relating WWTP efficiency and influent load. Nevertheless, results suggest that higher efficiency levels are attained whenever the influent EP ranges from 4 · 10–2 to 9 · 10–2 kg Neq/m^3. Beyond this range, efficiency scores tend to be rather poor.

When the over or undersized nature of the installations is considered, 71 out of the 113 plants under assessment were found to be operating below 50% of their design capacity. Even though the design and construction of WWTPs consider seasonal variation, population growth and a wide range of climatic possibilities, the number of oversized plants resulted substantially higher than expected. With regard to undersized facilities, 8 plants operated above 120% of their design capacity.

Available data permitted a classification of the plants taking into consideration their removal objectives. Hence, they were segregated in: (i) carbon removal; (ii) carbon and nitrogen removal; and (iii) carbon, nitrogen and phosphorous removal. Interestingly, simple technologies (i.e., only carbon removal) were found to attain higher efficiencies levels in the case of small WWTPs, whereas more complex systems resulted more efficient in large and medium WWTPs, a line of thought that is in line with the common believe that small WWTPs have a deficit of operational supervision in most cases.

21.5 CONCLUSIONS

The study provides a deep analysis regarding eco-efficiency evaluation of a relevant fraction of the Spanish wastewater treatment sector. Dividing the WWTPs in three different sizes allowed an accurate assessment based on a scale factor. However, large inefficiencies were identified in all WWTPs ranges. The main tendency observed is that large WWTPs achieved higher efficiency levels, followed by medium WWTPs and finally small WWTPs. The main reasons behind the higher performance of large plants are scale factor, plant flexibility and continuous monitoring.

Despite the fact that they average the lowest efficiency levels, small WWTPs are the range that presents higher opportunities to attain environmental improvements in all impact categories assessed. Several factors influencing plant efficiency, such as climate, influent load or plant over- and under-use have been found to be time-dependent and, thus, future research should include this variable in the assessment. In fact, current research is focusing on interannual behaviour of WWTPs, to determine if yearly performance follows a regular pattern.

21.6 REFERENCES

Avadí Á., Vázquez-Rowe I. and Fréon P. (2014). Eco-efficiency assessment of the Peruvian anchoveta steel and wooden fleets using the LCA + DEA framework. *Journal of Cleaner Production*, **70**, 118–131.

Banker R., Charnes A. and Copper W. (1984). Some models for estimating technical and scale inefficiencies in data envelopment analysis. *Management Science*, **30**, 1078–1092.

Cooper W. W., Seiford L. M. and Tone K. (2007). Data Envelopment Analysis: A Comprehensive Text with Models, Applications, References and DEA-solver Software. Springer, New York.

Emmerson R. H. C., Morse G. K., Lester J. N. and Edge D. R. (1995). The life-cycle analysis of small-scale sewage-treatment processes. *Journal of Water Environment*, **9**, 317–325.

Garnier J., Brion N., Callens J., Passy P., Deligne C., Billen G., Servais P. and Billen C. (2013). Modeling historical changes in nutrient delivery and water quality of the Zenne River (1790s–2010). *The role of land use, waterscape and urban wastewater management. Journal of Marine Systems*, **128**, 62–76.

Godin D., Bouchard C. and Vanrolleghem P. A. (2012). Net environmental benefit: introducing a new LCA approach on wastewater treatment systems. *Water Science and Technology*, **65**, 1624–1631.

Goedkoop M., Heijungs R., Huijbregts M., de Schryver A., Struijs J. and van Zelm R. (2009). ReCiPe 2008. A Life Cycle Impact Assessment Method Which Comprises Harmonised Category Indicators at the Midpoint and the Endpoint Level. Report I: Characterisation. Ministry of Housing, Spatial Planning and Environment (VROM), Ministry of Housing, Spatial Planning and Environment (VROM). www.lcia-recipe.info

Høibye L., Clauson-Kaas J., Wenzel H., Larsen H. F., Jacobsen B. N. and Dalgaard O. (2008). Sustainability assessment of advanced wastewater treatment technologies. *Water Science and Technology*, **58**, 963–968.

Hospido A., Moreira M. T., Fernández-Couto M. and Feijoo G. (2004). Environmental performance of a municipal wastewater treatment plant. *International Journal of Life Cycle Assessment*, **9**, 261–271.

Hospido A., Carballa M., Moreira M., Omil F., Lema J. M. and Feijoo G. (2010). Environmental assessment of anaerobically digested sludge reuse in agriculture: potential impacts of emerging micropollutants. *Water Research*, **44**, 3225–3233.

Iribarren D., Martín-Gamboa M. and Dufour J. (2013). Environmental benchmarking of wind farms according to their operational performance. *Energy* **61**, 589–597.

Marques R. C. and Monteiro A. (2003). Measuring productivity and efficiency of water and sewerage utilities. A Factor Productivity Approach, II International Conference: Efficiency Use and Management of Urban Water Supply. Tenerife, Spain.

Ministry of the Environment – Spain (2014). Banco público de indicadores ambientales. Depuración de aguas residuales urbanas. Grado de conformidad de la carga contaminante de aguas residuales tratadas en estaciones depuradoras según Directiva 91/271/CE.

Lorenzo-Toja Y., Vázquez-Rowe I., Chenel S., Marín-Navarro D., Moreira M. T. and Feijoo G. (2015). Eco-efficiency analysis of Spanish WWTPs using the LCA+ DEA method. *Water Research*, **68**, 651–666.

Lorenzo-Toja Y., Alfonsín C., Amores M. J., Aldea X., Marin D., Moreira M. T. and Feijoo G. (2016). Beyond the conventional life cycle inventory in wastewater treatment plants. *Science of The Total Environment*, **553**, 71–82.

Pasqualino J. C., Meneses M., Abella M. and Castells F. (2009). LCA as a Decision Support Tool for the Environmental Improvement of the Operation of a Municipal Wastewater Treatment Plant. *Environmental Science and Technology*, **43**, 3300–3307.

Red Eléctrica Española – REE (2011). El sistema eléctrico español 2011 (Spanish Electric System 2011). Red Eléctrica de España, Alcobendas, Madrid (Spain).

Rodríguez-Garcia G., Molinos-Senante M., Hospido A., Hernández-Sancho F., Moreira M. T. and Feijoo G. (2011). Environmental and economic profile of six typologies of wastewater treatment plants. *Water Research*, **45**, 5997–6010.

Rodriguez-Garcia G., Frison N., Vázquez-Padín J. R., Hospido A., Garrido J. M., Fatone F. and Feijoo G. (2014). Life cycle assessment of nutrient removal technologies for the treatment of anaerobic digestion supernatant and its integration in a wastewater treatment plant. *Science of the Total Environment*, **490**, 871–879.

Sala-Garrido R., Hernández-Sancho F. and Molinos-Senante M. (2012). Assessing the efficiency of wastewater treatment plants in an uncertain context: a DEA with tolerances approach. *Environmental Science and Pollution Research*, **18**, 34–44.

Vázquez-Rowe I. and Iribarren D. (2014). Life-cycle benchmarking approaches oriented towards energy policy making: launching the CFP+ DEA method. *The Scientific World Journal*. http://dx.doi.org/10.1155/2015/813921

Vázquez-Rowe I. and Tyedmers P. (2013). Identifying the importance of the “skipper effect” within sources of measured inefficiency in fisheries through data envelopment analysis (DEA). *Marine Policy*, **38**, 387–396.

Vázquez-Rowe I., Iribarren D., Moreira M. T. and Feijoo G. (2010). Combined application of life cycle assessment and data envelopment analysis as a methodological approach for the assessment of fisheries. *International Journal of Life Cycle Assessment*, **15**, 272–283.

Vázquez-Rowe I., Iribarren D., Hospido A., Moreira M. T. and Feijoo G. (2011). Computation of operational and environmental benchmarks within selected Galician fishing fleets (NW Spain). *Journal of Industrial Ecology*, **15**, 776–795.

Vázquez-Rowe I., Villanueva-Rey P., Iribarren D., Moreira M. T. and Feijoo G. (2012). Joint life cycle assessment and data envelopment analysis of grape production for vinification in the Rías Baixas appellation (NW Spain). *Journal of Cleaner Production*, **27**, 92–102.

Vázquez-Rowe I., Reyna J. L., García-Torres S. and Kahhat R. (2015). Is climate change-centrism an optimal policy making strategy to set national electricity mixes? *Applied Energy*, **159**, 108–116.

Chapter 22

Public perceptions of recycled water

Angela J. Dean and Kelly S. Fielding

22.1 INTRODUCTION

22.1.1 Public perceptions – a road block on the journey to recycled water schemes?

Implementation of recycled water schemes are increasingly viewed as a vital strategy to ensure future water security. Alternative water sources are especially important in the context of population growth and environmental change, driving variations in rainfall patterns and increased water demand. Emerging technologies continue to improve our capacity to deliver high-quality recycled water in a range of settings. Yet, despite these technological advances, a challenge for successfully implementing recycled water schemes relates to community support for these schemes. Many individuals have negative views about recycled water, and community perceptions can have a significant impact on the political will of governments to adopt new water policies or infrastructure. A number of potable water recycling schemes across the world have faced public opposition that has prevented their successful introduction (CH2MHILL, 2004; Hurlimann & Dolnicar, 2010; Po *et al.* 2003). Because of this, governments and water authorities may view potential community opposition as a significant risk associated with recycled water.

22.1.2 How perceptions are formed – the importance of emotions

Those working in the areas of science and technology may assume that individuals are influenced by data, research evidence and rational argument – and feel frustrated when they perceive that people are resisting 'the facts' about recycled water. But research indicates that pathways to perceptions are more complicated than this.

Traditionally, research focused on the role of personal values and the likelihood of expected outcomes as important factors influencing risk perceptions. Within this context, perceptions of recycled water would be based on both beliefs about recycled water and the likelihood of any expected outcomes. So, if a person's beliefs about recycled water are predominantly negative (e.g. recycled water is a risk to human health) and negative health outcomes are considered to be highly likely, then this person will have negative attitudes to recycled water. In contrast, if a person's beliefs are predominantly positive (e.g. recycled water will provide a secure water supply or recycled water will prevent the need for future water restrictions) and this person expects these outcomes to be highly likely, this will result in positive attitudes to recycled water.

Other research shows that attitudes are not always influenced by reasoned thinking processes, but may result from fast, intuitive responses based on emotions (Slovic, 1999). For example, judgements about risk are commonly made quickly; these judgements rely on emotion and past experience to assess an issue as ‘good’ or ‘bad’, and are not initially influenced by logic or fact. Perceptions of risk that are driven by emotions – sometimes referred to ‘risk as feelings’ – play an important role in decision making. Within this framework, individuals may express opposition to recycled water – not as a result of weighing up advantages and disadvantages, but because they ‘feel’ disgust towards recycled water.

An understanding of public perceptions of recycled water can also draw on research examining public perceptions of the meanings that people associate with *water*. It has been argued that in our everyday lives, water has come to represent cleanliness (Shove, 2003). Availability of water for regular washing, showering and bathing means that water now symbolises what it means to be clean, and household water use is an important way for people to ‘feel clean’. This has implications for domestic water demand management programs, where reduced domestic water use may symbolise reduced cleanliness for some individuals. The role of water as a symbol of cleanliness may also influence perceptions of recycled water, where sourcing water from wastewater may challenge society’s need to perceive water as ‘pure’ and ‘clean’.

22.1.3 Importance of considering public perceptions

Understanding public perceptions is an essential foundation for cultivating public support for recycled water schemes. Research shows that perceptions about recycled water is the strongest predictor of willingness to use recycled water and acceptance of recycled water schemes (Dolnicar *et al.* 2011). This chapter aims to review the research exploring public perceptions of recycled water: what do people think about it, who is willing to use it, and why, or why not? Understanding public perceptions about recycled water can improve our capacity to engage effectively with the public about these issues. Importantly, we also review intervention studies that provide guidance about how to cultivate this support through *enhancing* public perceptions of recycled water. The literature suggests that when building public support for recycled water schemes, we should aim to achieve not just passive acceptance, but to foster active community support.

22.2 WHAT DO THE PUBLIC THINK ABOUT RECYCLED WATER?

Water can be reclaimed from diverse sources – household wastewater, stormwater or oceans – and treated using diverse technologies. For the purposes of this review, we will focus on reclaimed domestic wastewater that is recycled for domestic use. The early part of our review considers many uses of recycled water in domestic settings; the latter part of the review focuses on potable use of recycled water.

22.2.1 Are people willing to use recycled water?

Much of the research examining perceptions of recycled water focuses on whether people would be ‘willing to use’ recycled water, or ‘how comfortable’ the person is with recycled water. Overall, these studies indicate that a significant proportion of the public express some form of concern about recycled water. The type of concern expressed and the proportion of people expressing this concern vary across studies. One of the earliest studies assessing public perceptions of recycled water was conducted in ten towns in California (United States) (Bruvold & Ward, 1972). More than half of the respondents (56%) reported being opposed to use of ‘reclaimed water’ for drinking. Since this time, many studies from diverse regions have reported similar findings. An early review of US studies reports that rates of opposition to use of potable recycled water vary from 44% to 63% (Dishman *et al.* 1989). More recently, large surveys suggest that approximately one third of community members are opposed to recycled water schemes (Dolnicar & Hurlimann, 2010a).

Much of the research surveys individuals who are not directly affected by recycled water schemes but community perceptions may differ when a recycled scheme is being planned. A proposed recycled water scheme for Toowoomba (Australia) that was put to a referendum was voted down. Although almost all residents who were surveyed agreed that water conservation measures were important because of water scarcity (Hurlimann & Dolnicar, 2010) one third thought that recycled water was 'disgusting' and that it was wrong to supply it to people's homes (Hurlimann & Dolnicar, 2010).

Other surveys examine acceptance of recycled water in communities with existing recycled water schemes, usually for non-potable use. A study conducted in California (United States) comparing five towns with recycled water schemes to five towns without recycled water schemes showed that rates of acceptance for various uses were similar between towns with and without recycled water (Bruvold & Ongerth, 1974). Surveys of Australian communities who already have non-potable recycled water programs report high levels of acceptance, with one study reporting that 78% of respondents had no problems with the recycled water scheme (Hurlimann, 2008). In contrast, some studies conducted in the United States report that communities with non-potable recycled water schemes have lower acceptance of recycled water than communities without such schemes (Marks *et al.* 2006).

22.2.1.1 The role of context – different levels of support for different types of water uses

Negative perceptions about recycled water are most associated with potable recycled water programs. Willingness to use recycled water varies considerably, depending on the purpose for which water is used (Figure 22.1). In general, people are more willing to use recycled water for non-contact activities such as toilet flushing, cleaning outdoor areas, firefighting, or irrigation (Flack & Greenberg, 1987; Marks *et al.* 2006). Concern about recycled water is greatest for activities that involve body contact, such as drinking, brushing teeth and bathing.

A survey of 1055 visitors to a recycled water demonstration project in London's Millennium Dome reported high acceptance of recycled water for toilet flushing: 95% agreed that dual water supply for bathrooms public areas was acceptable (Hills *et al.* 2002). Support for recycled water was lower for activities such as watering vegetables, although the majority remained positive. Early research on recycled water demonstrated that the difference between high contact and low contact uses became less important if people were told about the beneficial effects of recycled water, such as benefits to the environment, health or water security (Po *et al.* 2003).

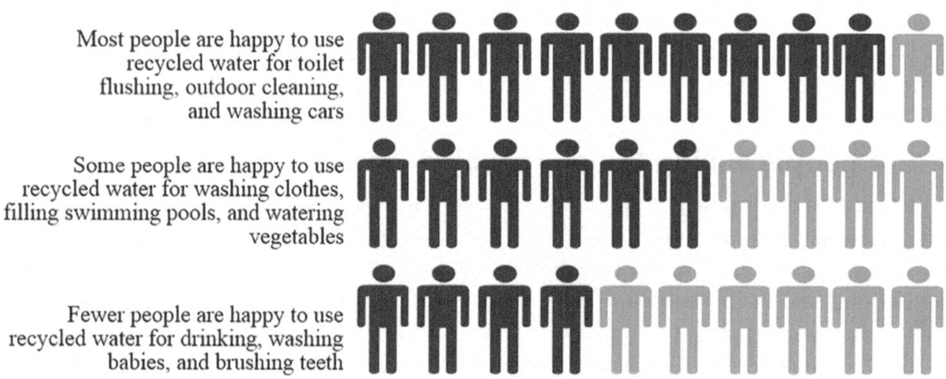

Figure 22.1 Acceptance and willingness to use recycled water varies depending on the intended use.

22.2.1.2 The role of language – different levels of support for different types of descriptions

The way we describe recycled water may influence people's perceptions of the water, and willingness to use recycled water for purposes such as drinking (Dolnicar *et al.* 2014). Words that emphasise the source of recycled water – such as 'sewage' or 'wastewater' – may increase the likelihood of a negative emotional response and unwillingness to drink (Po *et al.* 2005). In contrast, words that emphasise the treatment processes and quality of the final water product – such as 'purified water' are more likely to generate positive responses.

An Australian survey asked what types of water the respondents would be willing to drink. More than three-quarters of people were willing to drink 'bottled water', 'tap water' and 'purified water'. In contrast, less than a third of respondents were willing to drink 'recycled water', 'reclaimed wastewater' or 'recycled wastewater', even when information was provided. Other terms, such as 'potable water' or 'purified recycled water' resulted in approximately half of respondents reporting willingness to drink (Simpson & Stratton, 2011).

In the same study, respondents were asked to rate different terms for recycled water, based on how 'reassuring' the term was. The terms most likely to be rated as 'reassuring' include:

- 'very pure water'
- 'water that is a standard higher than drinking water'
- 'purified water'
- 'water that is purer than drinking water'

Terms that were rated as 'less reassuring' include: 'reclaimed water', 'recycled water' and 'reverse osmosis water' (Simpson & Stratton, 2011).

22.2.2 Why are some people unwilling to use recycled water?

22.2.2.1 Association with sewage and human waste

Many of the negative emotional responses to recycled water result from it being sourced from human waste. This has been referred to as the 'yuck factor'. Surveys suggest that at least one third of the public think that recycled water would be 'disgusting' or would contain human waste (Miller, 2012). For example, in a survey of more than 6000 Australians, 52% of respondents agreed with the statement that 'because the water cycle is closed, recycled water contains human waste' (Dolnicar *et al.* 2011).

Understanding how people think about 'contagions' can help us understand the power of this association between recycled water and human waste. It has been argued that people have 'mental short-cuts' that help them make decisions about everyday behaviours. One of these has been called the law of contagion: if a 'pure' object touches something disgusting, then this 'pure' object also becomes disgusting. For example, if a cockroach is dipped into a beverage, then that beverage will be considered 'contaminated' and disgusting. In the context of recycled water, the association between the final product and human wastewater generates a strong feeling of disgust (Miller, 2012).

In some cases, highlighting the physical distance between recycled water and its source can reduce this 'contagion effect'. For example, indirect potable water reuse returns recycled water to reservoirs, where it mixes with the existing water supply, prior to entering water reticulation systems. Surveys suggest that public support for indirect reuse schemes is higher than support for direct potable reuse schemes (Nancarrow *et al.* 2007). It is likely that in these cases, water reservoirs act as a psychological buffer, creating a sense of distance between wastewater and recycled water (Miller, 2012).

22.2.2.2 General safety and health risks

Another commonly-expressed concern about recycled water relates to whether recycled water is safe to drink. Surveys suggest between one and two thirds of the public think that drinking recycled water may pose a risk to human health (Dolnicar *et al.* 2014; Marks *et al.* 2006; Po *et al.* 2003). In one survey (Dolnicar *et al.* 2014), 60% agreed with the statement that recycled water 'could be a health concern, for instance if people would drink it'. Recycled water was viewed as less safe than other forms of water. However, when interpreting survey responses, it is important to put this in the context of how other water sources are perceived. The same survey reports that rates of agreement that water could pose a health concern were 45% for rainwater from a tank, 36% for desalinated water and 21% for tap water.

Nonetheless, expressing concerns about the impact of recycled water on human health was also one of the strongest predictors of willingness to drink (Dolnicar *et al.* 2014), highlighting the importance of addressing community concerns about health.

22.2.2.3 Microbial and chemical contamination

The potential for microbial or chemical contaminants to be present in recycled water is another area of concern. Almost three quarters of Australians (70%) agreed with the statement that recycled water 'might contain pathogens, such as bacteria or viruses'. Similarly, 68% agreed that 'it might contain industrial chemicals' and 53% agreed that it 'may contain substances such as hormones, which can affect human fertility' (Dolnicar *et al.* 2014; Marks *et al.* 2006; Po *et al.* 2003).

The perceived risk of contaminants present in water is not restricted to recycled water. For example, in the same survey, the potential for water to contain pathogens was also reported for rainwater (73% of respondents) and tap water (54%). Nonetheless, ratings related to potential contamination were highest for recycled water. Concerns about microbial and chemical contamination of recycled water are most salient when considering using recycled water for drinking purposes and other close-contact activities (e.g. showering).

Some community members have expressed concern about the potential for salt contamination of recycled water. As expected, concerns about high salt concentration were highest for desalinated water, expressed by 52% of survey respondents, compared to 38% of respondents about recycled water. Concern about saltiness is usually expressed in the context of watering gardens and irrigation (Hurlimann, 2008).

22.2.2.4 Aesthetic features – colour, taste and odour

How recycled water appears, or is perceived to appear, also is important for public perceptions. When considering non-potable uses, the colour of recycled water is important to the public in the context of washing clothes, whereas smell is considered important for toilet flushing (Hurlimann, 2008). Surveys indicate that more than half of the public think than recycled water would be clear (63%) and odourless (54%). However, the same survey reports that 52% agreed with the statement that recycled water 'does not taste good', and 31% agreed that recycled water would 'stain the washing' (Dolnicar *et al.* 2014).

A survey of Australians using non-potable recycled water schemes reported that 60% perceived no difference between recycled water and their normal water source. Those who did perceive a difference were most likely to report that recycled water was a different colour–typically a brown colour (Hurlimann, 2008). Surveys conducted in the United States indicate that residents of areas with recycled water used for irrigation purposes also report that the water sometimes exhibits a change in colour or smell; this is attributed to damaged pipes or maintenance issues at the water treatment site (Marks *et al.* 2006). It should be noted that recycled water for non-potable use is likely to differ in quality to recycled water for potable use.

22.2.2.5 Environmental benefits and impacts

While much research focuses on the negative perceptions of recycled water, it is important to remember that some groups have positive perceptions about recycled water. Recycled water is seen as environmentally friendly, especially compared to bottled water, and to a lesser extent, tap water (Harris-Lovett *et al.* 2015; Marks *et al.* 2003; Nancarrow *et al.* 2007). For example, a large survey invited respondents to indicate whether they agreed or disagreed with a range of statements about recycled water (Dolnicar *et al.* 2014). The majority (84%) agreed that recycled water was environmentally responsible, and more than half (54%) agreed that recycled water was the most environmentally responsible form of water to use. Potential benefits with high rates of endorsement were reductions in the amount of wastewater discharged to the environment (endorsed by 84%), and reduced contamination of beaches (63%).

Some respondents indicated the potential for recycled water to have negative environmental impacts. For example, 79% thought that recycled water required chemicals to be produced, and almost half thought that production of recycled water utilised a lot of energy (although energy needs of recycled water were significantly lower than ratings for desalinated water) (Dolnicar *et al.* 2014).

Importantly, much research indicates that the majority of the public recognise that recycled water can improve water security, reduce the need for water restrictions, and generate benefits for the city or region (Dolnicar *et al.* 2014; Po *et al.* 2005).

22.2.2.6 Price

Perceptions of price impacts of recycled water vary from study to study. Some surveys suggest that community members perceive that recycled water will have lower costs than other forms of water (Dolnicar & Hurlimann, 2010a; Marks *et al.* 2003). Although one survey reported that 69% of respondents thought that recycled water would be expensive to produce, this was lower than rates for desalinated water (rated as expensive by 82%) and bottled water (80%). According to another Australian survey, 73% of respondents thought that the price of recycled water would not affect their decision about whether or not to drink recycled water (Po *et al.* 2005).

A survey of residents using non-potable recycled water schemes reported that 25% of respondents were not satisfied with recycled water, and this was related to it being more expensive than anticipated (Hurlimann, 2008). It has been suggested that costs of recycled water do not rate highly as a concern in surveys asking individuals about a hypothetical recycled water scheme. When a particular recycled water scheme is more imminent, it is possible that cost implications may become more important.

22.3 WHAT INFLUENCES PERCEPTIONS ABOUT RECYCLED WATER?

Perceptions of recycled water are not observed equally across the population – some social groups may exhibit greater support for recycled water, whereas other groups may express greater concerns. Identifying how individual characteristics or features of the social context influence perceptions of recycled water improves our capacity to better understand these concerns, and address them more effectively (Figure 22.2).

Figure 22.2 Summary of factors that can influence support for recycled water.

22.3.1 Socio-demographics

22.3.1.1 Gender

Many studies reported that levels of acceptance of recycled water and other alternative water sources are higher in men than in women (Dolnicar *et al.* 2011; Fielding *et al.* 2015; Miller & Buys, 2008; Nancarrow *et al.* 2008). One study examined gender differences in perceptions of risk related to recycled water, and found that women perceive greater risk for particular activities – washing clothes, growing vegetables, and showering (Hurlimann & Mckay, 2007). Women tend to be responsible for many water-related activities in the home (Dean *et al.* 2016b), highlighting the importance of considering the effect of gender on acceptance of household recycled water schemes.

This gender difference is interesting because it contrasts with other research showing that women exhibit higher rates of pro-environmental behaviours and broad-based engagement in water-related issues (Dean *et al.* 2016b). This suggests that from a gender perspective, recycled water is viewed as a 'risky behaviour', rather than a 'pro-environmental behaviour'. This is consistent with much research in which females report higher perceptions of risk (Slovic, 1999). It is thought that this is related to social factors such as the experience of power and control over one's life, where males may perceive lower risks 'because they create, manage, control, and benefit from many of the major technologies and activities.' (Slovic, 1999). Research also suggests that women report lower rates of trust of recycled water technology (Miller & Buys, 2008).

22.3.1.2 Age

Age is also a factor that may influence acceptance of recycled water. Interestingly, much research shows that older adults report greater acceptance of recycled water (Dolnicar *et al.* 2011; Fielding *et al.* 2015; Nancarrow *et al.* 2008). Specifically, one study reports that people aged over 50 years were much more comfortable using recycled water for personal uses such as drinking and washing hands (Hurlimann & Mckay, 2007).

If we view acceptance of recycled water as a 'risky behaviour' – where acceptance of recycled water relates to perceptions of risk – the finding that older people are more accepting of recycled water goes against popular stereotypes that young people are 'risk takers' and become more risk averse as they age. In fact, the relationship between risk-taking and age is complex, and the specific pattern about age and risk-taking may depend on the type of risk being discussed. Some research suggests that older individuals may have greater trust in scientists and technology (Fielding *et al.* 2015).

On the other hand if we view recycled water as a 'pro-environmental behaviour', then the relationship between older age and acceptance is consistent with other findings. Research examining how Australians' engage in water-related issues – their knowledge about water, their support for policies and the extent to which they adopt behaviours which conserve water or reduce waterway pollution – found that even when controlling for factors that may correlate with age, such as home ownership, older Australians report greater engagement in water-related issues (Dean *et al.* 2016b). Research suggests that as they age, people accumulate greater information and experience about a range of issues, which then generates uptake of pro-environmental behaviours.

22.3.1.3 Education

Higher levels of education is associated with greater acceptance of recycled water and lower perceptions of risk (Dolnicar & Saunders, 2006). This difference in risk perceptions applies for different water uses, where individuals with lower rates of education report that recycled water is riskier for drinking, showering, clothes washing, and watering vegetable gardens.

The influence of education on perceptions of recycled water is likely to occur via a range of pathways. Greater educational experience can enable individuals to acquire and retain new knowledge across a range of topic areas. People with university degrees are also more likely to report greater trust in government and scientists than those without degrees; conversely, those without university degrees report being less likely to be influenced by others about recycled water (Dolnicar *et al.* 2011).

22.3.2 Experience of water shortages

Direct life experiences shape perceptions of recycled water. Research demonstrates that experience of water restrictions is associated with greater support for recycled water, and many individuals acknowledge that such experience would influence their support for recycled water (Dolnicar *et al.* 2011). This is consistent with other research showing that experience of water shortages, or of water restrictions, increases adoption of household water-saving behaviours. Experience of water shortages, even for short periods, can promote awareness of water scarcity and the need for change to ensure availability of freshwater into the future.

Similarly, a review of existing recycled water schemes internationally reported that support is high when there are limited other water sources available (Australia Water Recycling Centre of Excellence, 2014). A survey of Australians living in an area where a recycled water scheme was being planned indicated that support was linked to the perceived need for recycled water: although most respondents would rather not drink recycled water, willingness to use recycled water was higher if it was seen as a necessity (Nancarrow *et al.* 2007).

22.3.3 Knowledge

Research shows that knowledge is an important predictor of support for recycled water. Specific knowledge about recycled water (Dolnicar *et al.* 2011) and broad knowledge about water management (Dean *et al.* 2015) have been associated with support for recycled water or alternative water sources more generally.

Topic knowledge is an important predictor of policy support across a range of different issues, such as waterway protection or water conservation initiative. Although knowledge about recycled water and related issues is associated with greater support for recycled water, it is important to acknowledge that other factors may influence whether this knowledge is translated into 'active' support within a real-life setting. Knowledge has been conceptualised as a 'necessary, but not sufficient' ingredient to promote pro-environmental behaviours, which may also be influenced by many other factors (Dean *et al.* 2015). These may include psychological factors such as values or social norms, or contextual factors such as perceived price impacts of new water schemes.

There are many potential reasons why greater knowledge is associated with greater support for recycled water schemes. Greater knowledge about water treatment processes is likely to generate fewer negative perceptions of recycled water, and thereby greater support. But knowledge may also influence support because it allows individuals to more easily seek out and interpret information about recycled water. Or they may be more confident having conversations about recycled water, which may consolidate both knowledge and social norms about recycled water.

22.3.4 Exposure to information and expertise

Exposure to information has a strong influence on support for recycled water. It is well-established that having access to information has an important influence on water-related attitudes and behaviours (Dean *et al.* 2016b). Often, professional groups in the water sector will hope to enhance community support for recycled water by providing information as part of a community engagement campaign. The effectiveness of providing information will be discussed in Section 22.4.

It is important to recognise that community members may also source information from popular media or informal social networks. Individuals who are influenced by a greater number of sources, are more likely to support recycled water (Dolnicar *et al.* 2011). Although survey respondents state that they are influenced by research findings, other popular sources of information include family (influencing 77%), partners (69%), friends (54%), and neighbours (41%) (Dolnicar & Hurlimann, 2010b). In this study, more than two thirds reported being influenced by scientists (77%) and water authorities (68%), but fewer were influenced by the government (38%) or politicians (15%). Seventeen percent reported being influenced by no one; this group also reported the lowest likelihood of using recycled water (Dolnicar & Hurlimann, 2010b).

Communities are likely to look towards information sources that are perceived to be unbiased and scientific (Hurlimann & Dolnicar, 2010). However, information that is widely available – i.e. the mainstream media – may not provide accurate or unbiased information. A review of newspaper reporting of the planned recycled water scheme in Toowoomba, Australia, reported that only 10% of articles in newspapers were unbiased. Research suggests that newspaper coverage of recycled water in Singapore, which has greater support for recycled water, contains more positive words, such as 'pure water', compared to Australian newspapers which reported more negative words, such as 'wastewater' (Ching, 2010).

22.3.5 Trust in institutions and technology

Much research shows that trust is a strong predictor of willingness to use recycled water. There are two key groups relevant to assessing trust. These are (i) trust in organisations, usually government and water authorities responsible for management processes and regulations; and (ii) trust in the technology itself – involving the scientific processes used to treat water.

22.3.5.1 Organisational trust – governments and water authorities

Some individuals have expressed concerns or suspicions about the motives of politicians and organisations involved in recycled water schemes (Dolnicar & Saunders, 2006). Trust in the government is a strong predictor of being comfortable with recycled water (Fielding *et al.* 2015; Nancarrow *et al.* 2007). The issue of trust appears to be at least partly related to whether an individual feels that they have been treated fairly as part of a broader process, where perceptions of fair procedures were found to relate to greater trust, and greater acceptance of recycled water. Individuals with lower ratings of trust in water authorities also report higher ratings of perceived health risks associated with recycled water, and greater risk of technological failure (Nancarrow *et al.* 2009).

Surveys of communities with existing recycled water schemes also links support for recycled water with trust: higher rates of satisfaction with recycled water were associated with a sense of being well informed, trusting the water authority, and feeling that they had been treated fairly (Hurlimann, 2008).

22.3.5.2 Scientific trust – water-treatment technology and scientists

In contrast to lack of trust in organisations, which typically reflects concerns about institutional motivations, lack of trust in recycled water technology, or the scientists responsible for developing this technology, reflects a concern about the effectiveness and reliability of water treatment processes. For example, an early survey conducted in Colorado (United States) (Flack & Greenberg, 1987), reported that of those who supported use of recycled water, 83% believed that it was technically possible to treat waste water to a high quality. In contrast, two thirds of those who did not support recycled water for drinking, did not believe it was possible to create high quality water from treated wastewater.

Comfort with recycled water is associated with both trust in scientists and comfort with technology (Fielding *et al.* 2015). Some surveys indicate that only half of respondents think that water suppliers can be trusted to ensure adequate water quality, and 73% agreed with the statement that recycled water was prone to technology failure; in contrast, 82% thought that desalinated water was prone to technology failures (Dolnicar *et al.* 2011).

22.3.6 Values and social norms

Psychological research explores how issues such as personal values and social norms influence support for recycled water.

22.3.6.1 Environmental values

The values most frequently examined in the context of recycled water are pro-environmental values. Individuals who report high pro-environmental values typically place high importance on protection of nature and natural ecosystems, preventing pollution, minimising use of certain resources, and respecting the earth. Research suggests that individuals who report greater pro-environmental values or concern for environmental problems report greater support for recycled water schemes and greater willingness to use recycled water (Dolnicar *et al.* 2011; Harris-Lovett *et al.* 2015). This aligns with findings presented earlier in this chapter that some people consider recycled water to be an 'environmentally-friendly' source of water. It is not clear whether this relates specifically to lower energy requirements (compared to sources such as desalinated water) or reduced wastewater flows entering the environment.

22.3.6.2 Social norms

Social norms reflect common and accepted behaviours (or attitudes) within a group – what other people do, what they think about certain issues, and what they think others should do or think. Norms are known to influence and guide behaviour and attitudes, where people adopt attitudes and behaviours that conform to group norms, even though they may not recognise that others are having this influence. Social norms have been shown to influence water-related attitudes and behaviours (Dean *et al.* 2016b). Research has also shown that those who think others in their community support recycled water are more likely to report similar support (Nancarrow *et al.* 2009).

22.4 INTERVENING TO IMPROVE PUBLIC PERCEPTIONS OF RECYCLED WATER

Understanding public perceptions of recycled water and factors that influence these perceptions provides a range of 'footholds' for intervening with communities' acceptance of recycled water. For example, factors such as knowledge and trust may be enhanced via a range of interventions and engagement techniques. Understanding the role of other factors, such as age and gender, although not modifiable, can provide guidance on how messages or interventions should be targeted to particular audiences. Although much of the research focuses on negative perceptions about recycled water, it is important to recognise that research has identified a range of techniques that improve public perceptions of recycled water. These approaches include (Green *et al.* 2010):

- providing clear and accessible information targeted to diverse audiences;
- targeting beliefs and attitudes of the target audience, and framing messages carefully;
- creating genuine dialogue with communities about their needs and concerns, ensuring processes are viewed as fair and equitable; and
- providing opportunities for communities to experience recycled water, and view water treatment processes.

22.4.1 Providing information

A key element of initiatives that aim to improve acceptance of recycled water is providing information (Dolnicar & Saunders, 2006) given that community knowledge about recycled water may be poor (Po *et al.* 2005). Information interventions typically target community concerns about the safety of recycled water for drinking, focusing on the effectiveness of water treatment processes in ensuring that water is safe for drinking and other close-contact activities. A number of experimental studies have demonstrated the potential effectiveness of providing information about recycled water (Fielding & Roiko, 2014; Price *et al.* 2015).

Some studies have provided comprehensive information about the recycled water process and shown positive effects of this information. For example, participants in a study by Simpson and Stratton (2011) viewed a 47-page online information booklet and this increased their knowledge and acceptance of recycled water compared to a control group who did not receive the information. Similarly, another study provided comprehensive information about water recycling (e.g. why recycle water, the process, safety and use of recycled water) and participants exposed to this information had more positive attitudes and greater acceptance of the information (Roseth, 2008).

Other studies have provided briefer information and shown increases in support for this water source (Dolnicar *et al.* 2010; Fielding & Roiko, 2014). The finding that brief information can also be effective is important when considering the extent to which media and information campaigns have to compete in an information-rich time-poor world. One set of studies provided brief information that addressed health risk concerns, a key issue for many people. Compared to a control group who received no information, those receiving information reported more positive emotions and less negative emotions about recycled water, lower risk perceptions, and greater support for recycled water schemes, including increased likelihood of voting in favour of such a scheme (Fielding & Roiko, 2014). In fact, the provision of information doubled the number of people willing to vote in favour of the introduction of a potable recycled water scheme: 45–56% of participants in the information conditions indicated they would vote in favour of a recycled water scheme compared to 26% of those not receiving any information (Fielding & Roiko, 2014). There was also some indication that the provision of information influenced behaviour in that participants who received the information drank approximately twice as much recycled water as those who did not receive the information.

Moving beyond simply considering the effects of information, other research has shown that in situations where participants were assumed to be motivated to pay attention to the issue of water recycling, providing justifications that support the positive aspects of water recycling was more effective at increasing support than simply stating the positive aspects of the water source (Price *et al.* 2015). The research also showed that risk information was more effective at increasing support for potable recycled water schemes than benefit information and that initial attitudes to recycled water influenced how participants responded to the information – consistent with the arguments that information needs to be targeted to diverse audiences (Dean *et al.* 2016a; Green *et al.* 2010). Using language and images that focus people on what the water becomes (i.e., suitable for drinking) rather than where it has come from results in more positive responses to potable recycled water (Greenaway, 2013)

To maximise understanding of information provided, communicators should use visual aids and avoid use of jargon. Information initiatives are likely to be more effective if information is provided via diverse channels, and made accessible and relevant for diverse community members. When providing information about recycled water, it is important to recognise that information detection, uptake, and recall is not just a product of information exposure. Information is more likely to be noticed and retained if it is relevant (Price *et al.* 2015).

22.4.2 Psychological approaches to communication

Many communication approaches extend beyond provision of facts. Psychological approaches to communication involve trying to shift people's opinion not just with information, but also using emotion and logical arguments, considering the existing values and beliefs of the target group. These approaches are sometimes called 'persuasive communication' or 'community-based social marketing'.

As discussed in *Section 22.3.6.2*, social norms guide behaviour and attitudes in the community. Importantly, social norms can be activated using a range of interventions, which involve highlighting the number of people who have adopted the behaviour (or attitude). For example, communicating that the majority of people are willing to support a policy initiative can increase support for that policy. Research shows that activating social norms may be effective for many pro-environmental behaviours, although people are often unaware or even deny the effects of this type of normative information (Steg *et al.* 2014). Few studies have examined the effectiveness of social norm interventions to enhance public support for recycled water. Endorsement of recycled water by public figures and community leaders may also help to communicate norms and thereby contribute to support for recycled water, when used in the context of broader campaigns (Guan & Toh, 2012; Water Corporation, 2013).

A key aspect to consider when sending normative messages is whether the messengers are considered to be members of one's ingroup. People who are considered 'one of us' are more trusted and persuasive, and their messages are more likely to be accepted (Fielding & Hornsey, 2016). For example, one study provided information about recycled water, and varied whether or not a scientist who endorsed the safety of recycled water lived and worked in the same region as participants (Schultz & Fielding, 2014). For those residents who were highly identified with their region, having a scientist who was a 'local' come out in support of recycled water increased their support for the water source relative to receiving this same endorsement from a scientist whose identity (i.e. regional location) was not mentioned. (Schultz & Fielding, 2014).

Another element of persuasive communication is message framing. Message framing may emphasise different parts of a message, such as the benefits of policy support, or the negative consequences of not supporting a policy. As noted above, research has shown that framing that emphasises the low risks of recycled water was found to be more effective than framing that emphasised the benefits (Price *et al.* 2015). Effective framing aims to create a favourable sense of the issue (e.g. water scarcity) and the solution (e.g. recycled water). One example of message framing relates to the terms used to describe recycled water. In Singapore, the Public Utilities Board used careful message framing that shifted attention away from the source of the water, to the state-of-the-art technology. Terms that had a negative connotation, such as wastewater or sewage, were avoided (Guan & Toh, 2012; Leong *et al.* 2011).

One of the challenges in communicating about recycled water is that opposition groups will also attempt to frame the message, linking recycled water with sewage and wastewater, using phrases such as 'toilet to tap' (CH2MHILL, 2004). Media coverage of recycled water may also frame messages in negative or counter-productive ways. Once recycled water has been framed in terms of sewage, it is very difficult to reverse this association in the short term. For this reason, plans for recycled water schemes should initiate communication and engagement early in the planning process – proactively, rather than reactively (CH2MHILL, 2004).

22.4.3 Community dialogue

Because risk perceptions relating to recycled water are often strong and some individuals do not have strong trust in the organisation providing information, more intensive communication approaches may be necessary. These intensive approaches usually involve two-way, face-to-face communication.

22.4.3.1 Dialogue targeting risk perceptions

An example of more intensive communication about risk is the Hawkesbury Water Recycling Scheme (Australia) (Attwater & Derry, 2005). This scheme uses treated wastewater and stormwater for irrigation of crops and sporting fields. Rather than targeting the broader community, the team focused on individuals who were likely to have direct contact with recycled water via their work roles. Two social groups were identified: horticulture workers and childcare workers who took children to visit horticultural facilities. Each professional group participated in a series of focus groups to identify concerns about recycled water, and potential strategies for addressing these concerns. Both groups emphasised the importance of simple and accessible information about recycled water that identified suitable risk management practices. Horticultural workers requested support to develop risk assessment protocols and occupational health and safety (OH&S) materials and hygiene information. The focus was on improving risk awareness, and demonstrating adherence to OH&S responsibilities. In contrast, the concern of childcare workers were related to child health and potential risk of gastroenteritis. These concerns led to the development of tools to interpret and communicate monitoring outcomes, using a simple 'traffic light' system, where red light indicated significant bacterial health risk (Attwater & Derry, 2005).

22.4.3.2 Dialogue targeting community needs

Ensuring that recycled water schemes specifically address needs of community members or groups can contribute to support for recycled water. One initiative describes how community consultation assisted implementation of a recycled water program (Friend & Coutts, 2006). A series of information workshops involved community members representing eight different organisations. After considering a series of proposals for recycled water initiatives, the selected project – recycled water for irrigation of sporting or recreational ovals – was chosen based on its social orientation which was a key priority for the community. The community was supportive of this recycled water scheme (Friend & Coutts, 2006). Although not specifically examined as part of this project, it is likely that both the organisational process targeting community needs and preferences, and the subsequent experience of a recycled water program would foster trust in the organisation and support for broader recycled water initiatives.

22.4.4 Ensure fair and transparent processes for planning and decision making

Fairness is an important element of acceptance of recycled water, where recycled water schemes need to be seen as fair to all users (Po, 2003). There are two elements of fairness that can be considered: procedural justice and environmental justice (CH2MHILL, 2004; Po *et al.* 2003).

Procedural justice refers to having processes that are fair, transparent and accountable. This usually involves many aspects of comprehensive community engagement: provision of good information at all stages of project planning, providing opportunities for community members to contribute to decision making processes, and providing feedback about how community preferences have been incorporated into the project planning. The most suitable approach for involving community members in project planning may vary from project to project. Nonetheless, it is important to provide opportunities for meaningful involvement at all stages of the project; tokenistic involvement once decisions have been made can actually worsen community acceptance of water projects (Dean *et al.* 2016a). A case study from the United States highlights how implementing engagement processes can reduce conflict and improve outcomes (Ingram *et al.* 2006). To improve future water security, Redwood City (California, U.S.) decided to extend a small recycled water project into a larger area. A community opposition group emerged, which opposed the use of recycled water in children's play areas. Despite reassurances from technical experts about the safety of the proposal, many in the community remained opposed to the scheme. To resolve this conflict, the Council decided to establish a community-based task force, whose purpose was to identify alternatives to use of recycled water in children's parks, while continuing to meet water supply goals. This task force comprised ten community members in favour of recycled water, ten community members opposed to recycled water, and an independent facilitator. After six months of ongoing dialogue, the task force delivered a series of recommendations to council which sanctioned some use of recycled water, and identified alternatives to recycled water in playgrounds that also enabled reduced water demand. This process allowed the community to learn, develop trust in each other and resolve conflict (Ingram *et al.* 2006).

Environmental justice seeks to ensure that the benefits or burdens of recycled water projects are shared evenly across geographic or socio-economic gradients (CH2MHILL, 2004; Po *et al.* 2003). After public opposition stopped plans for a recycled water scheme in Toowoomba, Australia, feedback suggested that residents of this area did not want to be the first or only location in Australia drinking recycled water (Hurlimann & Dolnicar, 2010). If residents knew that other cities were also considering recycled water schemes, this may have alleviated some residents' concerns. One way to potentially reduce problems related to environmental justice of recycled water schemes is to introduce it in higher status communities first, such as those with greater education or income (Dolnicar & Saunders, 2006).

22.4.5 Provide opportunities to experience recycled water

Some research focuses on communities where recycled water schemes are being introduced. This research shows that the experience of recycled water schemes can lead to greater satisfaction with recycled water, greater trust in water authorities, and lower ratings of concern about recycled water (Hurlimann, 2008). Familiarity may lead to a decrease in perceived risk. Once an issue becomes more 'routine' it may also be less likely to trigger negative media coverage (Harris-Lovett *et al.* 2015).

One survey (Hurlimann, 2008) reported that the proportion of residents 'in favour' of recycled water was less than 10% prior to introduction of the recycled water scheme; this increased to 58% in the final survey after introduction. The proportion of residents 'willing to use' recycled water increased from less than 50% to 74% in the final evaluation (Hurlimann, 2008). Three quarters of respondents said that recycled water had met or exceeded their expectations. Those who were not satisfied with recycled water indicated that it was more expensive than originally anticipated.

Experience may provide a stronger shift in attitudes than information alone. It has been suggested that associating recycled water with pleasant activities – such as providing opportunities for people to swim in recycled water – would counter the negative associations with recycled water (Dolnicar & Saunders, 2006). Gradual introduction of recycled water provides time for people to become comfortable with different uses, and allows norms to shift towards accepting recycled water (Dishman *et al.* 1989). An early program in Santee, California (United States) first allowed people to see recycled water in a reservoir. After a period, boating was allowed, then fishing, followed by its use in a public swimming pool. It is thought this incremental exposure generated willingness to use potable recycled water (Dishman *et al.* 1989).

Allowing the public to visit water treatment facilities can promote understanding and trust in water treatment processes. Examples of successful recycled water schemes have utilised public site visits to water treatment facilities (Guan & Toh, 2012; Water Corporation, 2013).

22.4.6 Building public support – features of successful programs

Examining recycled water programs that have been successfully implemented can identify the elements of what is needed to promote support for recycled water initiatives (See Box 2.1 for an overview of principles for building public support for recycled water).

22.4.6.1 Groundwater Replenishment System – Orange County Water District, United States

Orange County Water District began its first potable reuse program in 1976. A range of strategies were used to build community support and trust. Significant resources were put into explaining the rationale for the scheme and how it addressed community interest (Harris-Lovett *et al.* 2015). For example, more than 1200 presentations were given to community and business leaders. Presentations were also provided in Spanish, Vietnamese and Chinese, to ensure immigrant communities in the region had access to information and could raise their concerns with the project team (Miller, 2012). Presentations were targeted to the needs of specific audiences, integrating information addressing their needs with core information about the importance of safe and reliable water supplies. For example, mothers were identified as potentially having greater concerns about recycled water, so mothers groups were targeted for engagement activities (Miller, 2012). Public meetings were run by staff of the water treatment facility, rather than consultants, which contributed to the reputation of the organisation.

Communities had the opportunity to raise concerns via citizen's advisory committees, focus groups, and ongoing discussions with community leaders, and project planning and implementation were adapted to address issues raised by communities. Support was gained from pro-environmental groups, who viewed a water recycling program as aligned with other pro-environmental recycling initiatives.

Over time, as the program expanded, Orange County Water District extended protocols for monitoring of contaminants, and developed a range of standard operating procedures that were shared with the public. The program was called a 'Groundwater Replenishment System'. This was chosen to build on public familiarity with other long-standing programs that replenished groundwater systems to prevent salt water intrusion.

In 2000, a problem arose that could have triggered major opposition to the project: a contaminant was detected in the water supply that was generated via the water treatment process. Rather than concealing the problem until after a solution was identified, the management decided to inform the public. Regular public updates demonstrated to both the public and regulators that the water authority was competent in identifying and resolving problems. It has been suggested that this honesty contributed towards greater trust between the water authority and the community (Harris-Lovett *et al.* 2015). Since this time, Orange County Water District has won the US Environmental Protection Authority's Water Efficiency Leader Award for integrating advanced treatment technology with a successful public education campaign.

22.4.6.2 Aquifer recharge trial – Perth, Australia

Western Australia is a state experiencing ongoing periods of reduced rainfall that are related to a drying climate rather than intermittent droughts. The principle water supplier for the region, Water Corporation, developed a plan to address future water security in the context of a drying climate. This plan included using highly-treated wastewater to replenish, or 'recharge', groundwater aquifers.

A series of engagement strategies were selected to build trust in the community (Water Corporation, 2013). This strategy utilised face-to-face approaches rather than traditional mass media campaigns, in a two-step process. The first step focused on engaging with experts and opinion leaders; the second step focused on engaging with communities. This was based on theories stating that most individuals form their opinions based on the views of opinion leaders in the media. Step 1 involved briefings with more than 160 health, environment and local government stakeholder groups.

Face-to-face approaches used to engage communities focused on tours of the Visitor Centre for the replenishment trial, and presenting to numerous community events. Almost 400 Visitors centre tours were conducted during the trial, involving more than 7,400 visitors. Support for groundwater replenishment increased from 74% before a Visitors Centre tour to 93% at the end of a tour. These strategies were supplemented by information packs, website, a social media campaign, advertising, and media releases. Similar to Orange County, monitoring data was also made publically available using a traffic light system (e.g. green light means no problems). School-based engagement was promoted by extending an existing Water Education Program to include content on groundwater replenishment. Water quality reports were regularly provided to the public.

This campaign regularly sought feedback from the community, using annual community surveys, surveys of visitor centre and events, email surveys, focus groups and online forums. Importantly, the information gathered was then provided to engagement experts and fed back into the engagement campaign. Regular community surveys indicate that support for groundwater replenishment remains at 70–76%. Groundwater replenishment is not a key concern for community, with unprompted awareness being only 5%. In contrast, the rate of prompted awareness in the community surveys was 44%. Indicators of trust in Water Corporation remained high.

22.4.6.3 Introduction of NEWater – Singapore

Another example of successful introduction of potable recycled water is the Singapore NEWater initiative (Guan & Toh, 2012; Leong *et al.* 2011). During the 1970s, recycled water was promoted for toilet flushing, but this was unsuccessful due to negative feedback about the smell and appearance of the water. As time progressed, increasing water demand persuaded the government to revisit the plan for recycled water. In addition to optimising and testing water treatment technologies, the Public Utilities Board developed a deliberate communications campaign that aimed to overcome the public's psychological barrier towards drinking recycled water. This involved careful message framing that shifted attention away from the source of the water, to the state-of-the-art technology. As noted earlier in this chapter, words with a negative connotation, e.g. wastewater or sewage, were avoided. The term 'NEWater' was chosen for the recycled water product, emphasising its 'new and improved' characteristics. Sewerage treatment plants were renamed 'water reclamation plants'. In addition to message framing, there were many other elements of the successful campaign, including:

- significant media coverage
- associating the issue with water security, where the campaign was framed as ensuring a secure water supply for Singapore and reducing dependence on Malaysia
- endorsement by leaders, with government officials drinking NEWater publicly
- public displays and exhibitions
- public sampling, with bottled NEWater distributed to the public at various events.
- ongoing public education, via the NEWater Visitors Centre which has had more than 700,000 visitors.

Public acceptance of NEWater is very high: 82% of respondents indicated that they would drink NEWater directly; 16% indicated that they would drink it when mixed with reservoir water. NEWater now contributes to 30% of Singapore's water supply (Guan & Toh, 2012).

BOX 22.1 PRINCIPLES FOR BUILDING PUBLIC SUPPORT FOR RECYCLED WATER SCHEMES (ADAPTED FROM DEAN *ET AL.* 2016A)

- ***Know your community:*** gauge existing public support and identify awareness, preferences and concerns about the issue. The most suitable approach to build public support will depend on the degree of public concern with the proposed scheme. A scheme with low public support will usually need more intensive strategies that extend beyond media campaigns.

- ***Provide good information at all stages of project planning:*** ensure benefits, costs and impacts are clearly explained. Avoid using jargon, and ensure the information is easily accessible and relevant to your target audiences
- ***Use diverse mechanisms to reach diverse communities:*** using a small number of engagement techniques will limit the number of people exposed to your message. Consider diverse outreach pathways to maximise your reach.
- ***Gain the support of community leaders and use them as spokespersons:*** public figures can raise the profile of your issue and enhance the persuasiveness of your message. Consider community leaders such as business leaders, local teachers, religious leaders, local opinion leaders, or individuals with high community credibility. Recognise community members who contribute to the issue or the campaign.
- ***Frame the issue effectively:*** define the issue and the solutions in a way that will generate the greatest number of people supporting it. This includes: defining it clearly, frame it as mainstream (rather than extreme or radical), don't make unsupported claims, and where possible, emphasise universal or near-universal values (e.g. we all want water security for our community). Use language that emphasises the quality of the water (e.g. pure water) rather than the source.
- ***Build supportive partnerships:*** working with other organisations and departments can ensure your messaging is consistent and reinforced by multiple partners; it can also extend the networks of active supporters able to indirectly build public support. Individuals within these networks may be asked to contribute to activities such as hosting meetings. Recognise organisations or staff members who are pivotal in supporting your engagement program.
- ***Demonstrate existing support:*** high rates of demonstrated support can mobilise additional support via activating social norms. In general, the public like to 'back a winner'. Regularly gauge changes in support; as your support grows, share this as part of the campaign.
- ***Consider more intensive engagement approaches to build trust:*** public participation programs or use of community task forces can build relationships and trust with the community. These are not 'quick fix' solutions; it is important that these processes are given adequate time and budgeting. Allow the community to contribute in a meaningful way to these programs, and avoid 'tokenistic' gestures.

22.5 CONCLUSIONS

Public perceptions of recycled water are not always positive. Many individuals have a negative emotional response to recycled water, and others may express concerns about the health implications. But understanding why people have these concerns, and identifying who is most likely to have these concerns provides a foundation for building public support. Public education and information campaigns have a role in building public support: providing information about the importance of water security, and water purification processes can enhance support for recycled water schemes. The successful introduction of recycled water schemes around the world also highlight the need for ongoing engagement with communities to develop transparent relationships with communities. Dialogue is needed at all stages of project planning and is essential for fostering the trust necessary for widespread community support.

22.6 REFERENCES

Attwater R. and Derry C. (2005). Engaging communities of practice for risk communication in the Hawkesbury Water Recycling Scheme. *Action Research*, **3**(2), 193–209.

Australia Water Recycling Centre of Excellence (2014). Module 1: Community perceptions and barriers to water reuse, Victoria University.

Bruvold W. H. and Ward P. C. (1972). Using reclaimed wastewater: public opinion. *Journal of the Water Pollution Control Federation*, **44**(9), 1690–1696.

Bruvold W. H. and Ongerth H. J. (1974). Public Use and Evaluation of Reclaimed Water. *Journal of American Water Works Association,* **66**(5), 294–297.

CH2MHILL (2004). Southern California Water Recycling Projects Initiative: Successful Public Information and Education Strategies Technical Memorandum, U.S. Department of the Interior, Bureau of Reclamation, California, U.S.

Ching L. (2010). Eliminating 'Yuck': a simple exposition of media and social change in water reuse policies. *International Journal of Water Resources Development,* **26**(1), 111–124.

Dean A.J., Fielding K.S., Newton F. and Ross H. (2015). Community Knowledge about Water: Who has Better Water-Related Knowledge and is this Important? Cooperative Research Centre for Water Sensitive Cities., Melbourne, Australia.

Dean A. J., Fielding K. S., Ross H. and Newton F. (2016a). Community Engagement in the Water Sector: An outcome-focused Review of Different Engagement Approaches. Cooperative Research Centre for Water Sensitive Cities, Melbourne, Australia.

Dean A. J., Lindsay J., Fielding K. S. and Smith L. D. G. (2016b). Fostering water sensitive citizenship – Community profiles of engagement in water-related issues. *Environmental Science & Policy,* **55**, Part 1, 238–247.

Dishman C. M., Sherrard J. H. and Rebhun M. (1989). Gaining support for direct potable water reuse. *Journal of Professional Issues in Engineering,* **115**(2), 154–161.

Dolnicar S. and Hurlimann A. (2010a). Desalinated versus recycled water: what does the public think? *Sustainability Science and Engineering,* **2**, 375–388.

Dolnicar S. and Hurlimann A. (2010b). Water alternatives – who and what influences public acceptance? *Journal of Public Affairs,* **11**, 49–59.

Dolnicar S. and Saunders C. (2006). Recycled water for consumer markets – a marketing research review and agenda. *Desalination,* **187**(1–3), 203–214.

Dolnicar S., Hurlimann A. and Nghiem L.D. (2010). The effect of information on public acceptance – The case of water from alternative sources. *Journal of Environmental Management,* **91**(6), 1288–1293.

Dolnicar S., Hurlimann A. and Grün B. (2011). What affects public acceptance of recycled and desalinated water? *Water Research,* **45**(2), 933–943.

Dolnicar S., Hurlimann A. and Gruen B. (2014). Branding water. *Water Research,* **57**, 325–338.

Fielding K., Gardner J., Leviston Z. and Price J. (2015). Comparing public perceptions of alternative water sources for potable use: the case of rainwater, stormwater, desalinated water, and recycled water. *Water Resources Management,* **29**(12), 4501–4518.

Fielding K. S. and Hornsey M. J. (2016). A social identity analysis of climate change and environmental attitudes and behaviors: insights and opportunities. *Frontiers in Psychology,* **7**.

Fielding K. S. and Roiko A. H. (2014). Providing information promotes greater public support for potable recycled water. *Water Research,* **61**, 86–96.

Flack J. E. and Greenberg J. (1987). Public attitudes towards water conservation. *Journal American Water Works Association,* **79**(3), 46–51.

Friend H. D. and Coutts S. S. (2006). Achieving sustainable recycled water initiatives through public participation. *Desalination,* **187**(1–3), 159–166.

Green M., Fielding K., Leviston Z. and Price J. (2010). Communicating scientific and technical information about alternative water supply options: A review of the literature.

Greenaway T. (2013). Public responses to drinking recycled water: the effect of affect and social identity, Thesis, The University of Queensland, Australia.

Guan Y. K. and Toh S. (2012). Water Sensitive Cities. Howe, C. and Mitchell, C. (eds.), IWA Publishing, London, UK, pp. 139–146.

Harris-Lovett S. R., Binz C., Sedlak D. L., Kiparsky M. and Truffer B. (2015). Beyond user acceptance: a legitimacy framework for potable water reuse in California. *Environmental Science & Technology,* **49**(13), 7552–7561.

Hills S., Birks R. and McKenzie B. (2002). The Millenium Dome "Watercycle" experiment: to evaluate water efficiency and customer perception at a recycling scheme for 6 million visitors. *Water Science and Technology,* **46**(6–7), 233–240.

Hurlimann A. (2008). Community attitudes to recycled water use: An urban Australian case study – part 2.

Hurlimann A. and Dolnicar S. (2010). When public opposition defeats alternative water projects – The case of Toowoomba Australia. *Water Research,* **44**(1), 287–297.

Hurlimann A. and Mckay J. (2007). Urban Australians using recycled water for domestic non-potable use – An evaluation of the attributes price, saltiness, colour and odour using conjoint analysis. *Journal of Environmental Management,* **83**(1), 93–104.

Ingram P. C., Young V. J., Millan M., Chang C. and Tabucchi T. (2006). From controversy to consensus: the Redwood City recycled water experience. *Desalination,* **187**(1–3), 179–190.

Leong C. C., Jarvis D., Howlett M. and Migone A. (2011). Controversial science-based technology public attitude formation and regulation in comparative perspective: the state construction of policy alternatives in Asia. *Technology in Society,* **33**(1–2), 128–136.

Marks J., Cromar N., Fallowfield H. and Oemcke D. (2003). Community experience and perceptions of water reuse. *Water Science and Technology,* **3**(3), 9–16.

Marks J. S., Martin B. and Zadoroznyj M. (2006). Acceptance of water recycling in Australia: national baseline data. *Water,* **33**(2), 151–157.

Miller E. and Buys L. (2008). Water-recycling in south-east Queensland, Australia: what do men and women think? *Rural Society,* **18**(3), 220–229.

Miller G. (2012). Getting Minds out of the Sewer. *Science*, **337**(6095), 679–680.

Nancarrow B. E., Leviston Z., Tucker D., Po Greenhill M., Price J. and Dzidic P. (2007). Community acceptability of the indirect potable use of purified recycled water in South East Queensland and preferences for alternative water sources: A baseline measure.

Nancarrow B. E., Leviston Z., Po M., Porter N. B. and Tucker D. I. (2008). What drives communities' decisions and behaviours in the reuse of wastewater. *Water Science and Technology*, **57**(4), 485–491.

Nancarrow B. E., Leviston Z. and Tucker D. I. (2009). Measuring the predictors of communities' behavioural decisions for potable reuse of wastewater. *Water Science and Technology*, **60**(12), 3199–3209.

Po M., Kaercher J. D. and Nancarrow B. E. (2003). Literature review of factors influencing public perceptions of water reuse. CSIRO Land and Water.

Po M., Nancarrow B. E., Leviston Z., Porter N. B., Syme G., J and Kaercher J. D. (2005). Predicting community behaviour in relation to wastewater reuse: What drives decisions to accept or reject? Country, N.R.F.W.f.a.H. (ed).

Price J., Fielding K. S., Gardner J., Leviston Z. and Green M. (2015). Developing effective messages about potable recycled water: the importance of message structure and content. *Water Resources Research*, **51**(4), 2174–2187.

Roseth N. (2008). Community views on recycled water – The impact of information.

Schultz T. and Fielding K. (2014). The common in-group identity model enhances communication about recycled water. *Journal of Environmental Psychology,* **40**, 296–305.

Shove E. (2003). Converging conventions of comfort, cleanliness and convenience. *Journal of Consumer Policy,* **26**, 395–418.

Simpson J. and Stratton H. (2011). Talking about water: words and images that enhance understanding. Waterlines Report Series No 49, National Water Commission, Canberra.

Slovic P. (1999). Trust, emotion, sex, politics, and science: surveying the risk-assessment battlefield. *Risk Analysis*, **19**(4), 689–701.

Steg L., Bolderdijk J. W., Keizer K. and Perlaviciute G. (2014). An Integrated Framework for Encouraging Pro-environmental Behaviour: the role of values, situational factors and goals. *Journal of Environmental Psychology,* **38**, 104–115.

Water Corporation (2013). Groundwater Replenishment Trial: Final Report. Water Corporation, Perth, Australia.

Chapter 23

Greenhouse and odour emissions

Raquel Lebrero, Raul Muñoz, Adrian Oehmen, Jose Porro, Eveline Volcke and Maite Pijuan

23.1 GREENHOUSE GAS EMISSIONS DURING WASTEWATER TREATMENT

23.1.1 Introduction

Carbon dioxide (CO_2), methane (CH_4) and nitrous oxide (N_2O) are the most important greenhouse gases (GHG) related to wastewater treatment processes. CO_2 is emitted directly, originating from the consumption of organic matter in the bioreactor and indirectly, as a result of the electricity consumption of the plant. According to the Intergovernmental Panel on Climate Change (IPCC, 2013), the direct emissions of CO_2 derived from wastewater treatment are considered of biogenic origin and therefore are excluded from the greenhouse gas inventory. However, a recent study found that between 4 to 14% of the organic carbon present in the wastewater could be from fossil origin (Law *et al.* 2013), and therefore, if it is converted into CO_2 during its treatment it should be taken into account in the overall GHG inventory.

On the other hand, CH_4 and N_2O have to be accounted, having a global warming potential which is about 25 and 265 times larger than the one attributed to CO_2 respectively, in a 100-years scope (IPCC, 2013). Nowadays, the IPCC accounting reports are used worldwide to estimate GHG emissions from wastewater treatment facilities and to assess their carbon permit liabilities. However, this method is based on fixed emission factors (i.e. 0.035 g N_2O N k^{-1} TKN influent for the case of N_2O) calculated from limited data sets (Czepiel *et al.* (1995)) which may not be representative for the broad diversity in wastewater treatment plants (WWTPs) configurations. To overcome this limitation, the online quantification of GHG emissions from full-scale municipal WWTPs of different configurations has been the purpose of studies conducted in recent years (Ahn *et al.* 2010; Aboobakar *et al.* 2013; Daelman *et al.* 2012, 2013; Rodriguez-Caballero *et al.* 2014, 2015). Ahn *et al.* (2010) presented N_2O emissions data from 12 different WWTPs located in the United States with results ranging from 0.01 to 1.8% of the influent total Kjeldahl nitrogen (TKN). More recently, Aboobakar *et al.* (2013) reported 0.036% of the total nitrogen (TN) load being released as N_2O in a full-scale nitrifying WWTP. On the other hand, long-term research performed by Daelman *et al.* (2013) delivered values as high as 2.3% of the incoming N being released as N_2O, while a 2 months monitoring campaign conducted in a full-scale sequencing batch reactor reported a 3.4% of the N treated emitted as N_2O (Rodriguez-Caballero *et al.* 2015). All these reports have contributed to put into perspective the high relevance of N_2O emissions from wastewater treatment facilities. The majority of these campaigns are actually research related, and their objectives vary from understanding potential emissions under different WWTP conditions to mechanistic modelling of the N_2O production and emissions from full-scale WWTPs. The latter objective requires a minimum level of full-scale data and several datasets for gaining consensus on N_2O model validity. This becomes difficult when there are gaps in the data as well as inconsistencies with how the data is collected and expressed.

On the other hand, reports on CH_4 emissions from full-scale domestic facilities are more scarce. A very comprehensive monitoring study comprising all the parts of a WWTP was conducted by Daelman *et al.* (2012). This study reported that 1.13% of the influent chemical oxygen demand (COD) was emitted as CH_4. About three quarters of these emissions were originated during primary and secondary sludge digestion. In another study where emissions were only monitored in the plug-flow bioreactor of a WWTP, CH_4 emissions accounted for 0.016% of the influent COD, occurred in the first aerated side and were strongly related with the influent and reject wastewater flows entering the bioreactor (Rodriguez-Caballero *et al.* 2014).

To put into context these emissions, recent studies are starting to assess the relevance that direct GHG emissions (N_2O and CH_4) might have on the overall carbon footprint of the plant, calculated as the indirect CO_2 generated associated with the energy consumption. In that sense, the CH_4-related carbon footprint found by the previous cited study from Daelman and co-workers (2012), was larger than the CO_2 emissions that were avoided by using biogas for energy generation. On the other hand, the study from Rodriguez-Caballero (2015) found that 83% of the total C-footprint of the WWTP was related with the N_2O emitted from the SBRs, and only 17% was linked to the C-footprint generated as a result of the plant's electricity consumption.

Overall, the high variability of GHG emissions reported in full-scale studies has spread the general idea of those emissions being strongly bounded to the differences in configuration, operation and performance (Law *et al.* 2012). The difficulty in comparing emissions from different systems suggests that the implementation of mitigation strategies may have to be approached on individual basis.

23.1.2 Operational factors affecting direct GHG emissions during wastewater treatment

In WWTPs, numerous relevant operational and environmental factors can impact the production of N_2O under either aerobic or anoxic conditions. N_2O is emitted largely from aerated zones in WWTPs, since aeration acts as a stripping gas that causes the transfer of N_2O from the liquid to gas phases, whereby in anoxic zones the lack of gas addition to the mixed liquor leads to most of the N_2O remaining dissolved, not contributing significantly to the total N_2O emitted by the plant (Ahn *et al.* 2010; Law *et al.* 2012). Nevertheless, heterotrophic denitrification under anoxic conditions, and subsequent stripping and emission under aerobic conditions, can in some situations be a significant contributor to the total N_2O emissions of the WWTP. The factors impacting N_2O production under aerobic conditions by nitrifiers, and under anoxic conditions by denitrifiers, will now be discussed.

23.1.2.1 Factors affecting N_2O production during aerobic conditions by nitrifiers

Limiting levels of dissolved oxygen (DO) in aerobic zones has been found to significantly increase N_2O production by ammonium oxidising bacteria (AOBs), particularly through the nitrifier denitrification pathway, where nitrite is denitrified to N_2O by AOB. It is significant that the lack of N_2O reductase in the genomes of AOB such as Nitrosomonas (Chain *et al.* 2003; Stein *et al.* 2007) and Nitrosospira (Norton *et al.* 2008) shows that these organisms are not capable of performing N_2O reduction, which is consistent with experimental findings showing that N_2O is indeed the final product of nitrifer denitrification (Law *et al.* 2012). Since low aeration levels are increasingly applied at WWTPs to promote energy-savings (and correspondingly, reduced indirect GHG emissions), a balance must be achieved regarding the aeration level in order to avoid situations of excessive N_2O production. This is of particular importance due to the fact that N_2O is approximately 300 times more potent a GHG as compared to CO_2, and can far outweigh the indirect GHG emissions in cases where it accumulates (Daelman *et al.* 2013; Rodriguez *et al.* 2015).

It should also be noted that, in certain cases, high aeration levels have been correlated with increased N_2O emissions, both in a nitritation system (Kampschreur *et al.* 2009) and full nitrification at a full-scale activated sludge system in Brazil with varying aeration, where the maximum N_2O emission correlated with a DO level of ~2mg/L (Brotto *et al.* 2015). At high DO concentrations, the relative contribution of the hydroxylamine pathway has been shown to be of increased importance as compared to low DO levels (Law *et al.* 2013; Ni & Yuan, 2015). In nitritation, ammonia is converted first to hydroxylamine and then to nitrite by AOBs. In the hydroxylamine pathway, NO and N_2O can be formed as by-products (biologically and/or chemically) during hydroxylamine oxidation to NOH followed by nitrite (Law *et al.* 2012). Metabolic models exist describing the production of N_2O via both the nitrifier denitrification and hydroxylamine pathways (Ni *et al.* 2014).

Furthermore, transient conditions is also an important factor leading to increased N_2O production, both from the point of view of rapid changes to process conditions employed during wastewater treatment as well as transitions between anoxic and aerobic phases (Kampschreur *et al.* 2009; Ahn *et al.* 2010). The change from anoxic to aerobic conditions, also observed in full-scale WWTPs (Ahn *et al.* 2010), has been linked with the shift of low specific metabolic activity in AOBs to high specific metabolic activity, necessitating a recovery period of the organism to the shift in operational condition (Yu *et al.* 2010). Other factors causing metabolic imbalances in AOBs, such as the imposition of shock loads of ammonium or toxic compounds (Burgess *et al.* 2002; Kampschreur *et al.* 2009), as well as dynamics in DO or nitrite levels (Tallec *et al.* 2006; Kampschreur *et al.* 2009), have also been shown to lead to N_2O production.

While transient conditions can be difficult to avoid in WWTPs, some proposed control strategies to minimise N_2O have found that short periods of aerobic and anoxic conditions can lower total N_2O production (Rodriguez-Caballero *et al.* 2015). The so-called intermittent aeration strategy was successfully shown to reduce N_2O emissions at a full-scale sequencing batch reactor (SBR), due to the prevention of high nitrite accumulation by frequent heterotrophic denitrification under anoxic conditions of the nitrite formed aerobically during nitrification. While the SBR plant was shown to typically lead to higher N_2O emissions as compared to continuous-flow WWTPs, likely due to the frequent transition periods inherent to the SBR configuration, Rodriguez-Caballero *et al.* (2015) showed that the N_2O emission factor could be reduced by an order of magnitude after applying intermittent aeration (from 50–70 g N_2O-N/kg NH_4-N removed to 5 g N_2O-N/kg NH_4-N removed).

Perhaps the most important factor impacting N_2O production by AOBs is the accumulation of nitrite, inherent to partial nitrification processes as an end product as well as in full nitrification systems under certain situations. Nitrite accumulation (in the ~1–50 mgN/L range) has been frequently shown to increase N_2O production in lab-scale and full-scale systems (Kampschreur *et al.* 2009; Law *et al.* 2012). In nitritation systems, very high nitrite levels (>50 mgN/L) has been shown to actually decrease the N_2O production rate, due to the inhibition of the nitrifier denitrification pathway (Law *et al.* 2013). The hydroxylamine pathway became an increasingly relevant N_2O production mechanism as the nitrite and/or DO concentration increased during this study.

Combinations of relevant factors can have simultaneous and contrasting impacts on N_2O production at WWTPs. Also, some factors can have an indirect impact on N_2O emissions since they can result in nitrite accumulation, often by leading to higher AOB growth versus nitrite oxidising bacteria (NOB). Such factors include high temperature, short SRT, high salinity and the presence of sulphides, heavy metals and other toxic compounds (Kampschreur *et al.* 2009). The DO level and aeration control strategy also impacts the relevant abundance of AOB vs NOB, thus the impact of each factor in combination can often be synergistic.

While NOBs have generally been found to contribute far less to N_2O production as compared to AOBs, it should be noted that NO production by the NOB Nitrobacter has been shown to be of relevance in granular systems in certain situations (Winkler *et al.* 2015). NO is generally not a stable intermediate of the denitrification pathway and can be rapidly converted to N_2O in WWTPs. Winkler *et al.* (2015) found that during mixotrophic (i.e. autotrophic and heterotrophic) growth of NOB, NO can be formed particularly at high temperatures and/or low DO concentrations.

23.1.2.2 Factors affecting N_2O production during anoxic conditions by denitrifiers

N_2O production has been more frequently attributed to metabolic activity by nitrifiers as compared to denitrifiers, but numerous cases have been reported where the heterotrophic denitrification pathway played a significant or even dominant role with respect to the total N_2O emissions.

Some factors that also impact nitrifiers, such as DO and nitrite accumulation, also impact the N_2O production of denitrifiers. The presence of oxygen in the anoxic zone inhibits denitrification activity, where N_2O reductase is more sensitive to oxygen than the other denitrification enzymes, requiring a longer lag phase prior to being synthesised when transitioning from aerobic to anoxic conditions, and is more quickly inhibited by oxygen when transitioning from anoxic to aerobic conditions (Otte *et al.* 1996; Kampschreur *et al.* 2009; Law *et al.* 2012). Nitrite accumulation causes inhibition of the denitrification rate, whereby NO and N_2O reduction are more severely impacted than nitrite reduction (von Schulthess *et al.* 1995).

Limitations of organic carbon (i.e. low COD/N ratios) has also been shown to lead to increased N_2O production (von Schulthess & Gujer, 1996; Itokawa *et al.* 2001; Kishida *et al.* 2004), mainly since a competition for electron donors takes place, where NO and N_2O reductases generally have a lower affinity for electrons as compared to nitrate and nitrite reductases (Law *et al.* 2012). Electron competition has also been found to take place in denitrifying cultures fed with different carbon sources, even at non-limiting COD levels, where the N_2O reduction rate again appeared to be more negatively affected as compared to the other denitrification steps (Pan *et al.* 2013; Ribera *et al.* 2015).

Competition for electron donor can also be relevant in systems where polyhydroxyalkanoate (PHA)-driven denitrification takes place (Kampschreur *et al.* 2009), such as in polyphosphate accumulating organisms (PAOs) and glycogen accumulating organisms (GAOs). Previous studies have found substantial N_2O production by PAO or GAO in granular sludge systems (Zeng *et al.* 2003; Lemaire *et al.* 2006; Zhou *et al.* 2008). These situations of N_2O production were also linked with nitrite accumulation, where Zhou *et al.* (2008) identified that free nitrous acid (FNA - the acidified form of nitrite) was the true inhibitor of N_2O production, thus making it difficult to assess the impact of PHA-driven denitrification in isolation. In an enriched denitrifying PAO culture, Zhou *et al.* (2012) found that at low (<1.25) COD/N ratios fed with similar nitrite levels that external carbon-driven denitrification in fact led to higher N_2O production than PHA-driven denitrification, while at higher COD/N ratios (1.875 and 2.5) the N_2O production by both carbon sources was not significantly different.

Heterotrophic denitrification can also take place in partial nitrification reactors treating sludge digestion liquors, and contribute to the N_2O production therein. Under conditions of low DO concentrations heterotrophic denitrification was observed to be a significant or even dominant pathway in these nitritation reactors, which could be minimised after increasing the DO concentration of the reactor (Wang *et al.* 2014; Mampaey *et al.* 2015).

23.1.2.3 Factors affecting CH_4 production

Methane, having a global warming potential of 34 CO_2-equivalents over a 100 year time horizon (IPCC, 2013), is expected to be formed in the sewer system (Guisasola *et al.* 2008) and in those parts of the WWTP where anaerobic conditions prevail.

The dissolved methane which is formed in the sewer system, can be stripped from the incoming wastewater, in the headworks, the primary settler and/or the biological reactor. Daelman *et al.* (2012) estimated the influent methane load as 1% of the influent COD load. They quantified the contribution from methane in the influent to the total methane emission from a WWTP from less than 30 up to 85%, corresponding to 3.5–6.7 gCH_4.(kg $COD_{influent}$). The methane emissions associated with the influent are affected by the way in which the raw wastewater enters the WWTP: screw conveyors imply a more intense contact between the wastewater and the air than centrifugal pumps, leading to a higher stripping rate of methane. This should be taken into consideration when designing new wastewater treatment plants.

Methane present in wastewater can be aerobically oxidized in the activated sludge tanks. Daelman *et al.* (2012) found that about 80% of the methane entering the activated sludge reactor was biologically oxidized. The methane oxidizing capacity of activated sludge could be exploited as a means to further decrease methane emissions from wastewater treatment. This potential mitigation route was investigated by Daelman *et al.* (2014) through simulation. Accurate aeration control should ensure a sufficient supply of oxygen. If the oxygen level is too low to sustain methanotrophic growth, methane will only be stripped without being converted, while aerating too much will benefit stripping at the expense of conversion. Dissolved methane could be effectively removed in an aeration tank at an aeration rate that is in agreement with optimal effluent quality. Aiming at the same effluent quality in terms of COD and nitrogen, more dissolved methane is converted as the aerator equipment is installed deeper below the liquid surface, so subsurface bubble aeration is better than surface aeration. A continuous stirred tank reactor performs better than a plug flow reactor with regard to methane conversion. In case of a plug flow reactor, excessive aeration in the first part of the aeration tank should be avoided to limit methane stripping. The conversion efficiency of methane in the activated sludge tank further benefits from higher methane concentrations in the WWTP's influent. Finally, if an activated sludge tank is aerated with methane containing off-gas, only a limited amount of methane (about 12% maximum) is absorbed and converted in the mixed liquor.

Sludge storage contributes significantly to the methane emissions. In their measurement campaign on a full-scale wastewater treatment plant, Daelman *et al.* (2012) found that three quarters of the total methane emission originated from the anaerobic digestion of primary and secondary sludge. The digested sludge has a considerable residual methane potential because the conversion of the influent sludge to biogas in a completely mixed anaerobic digester is never complete. Consequently, during digested sludge storage a significant amount of methane can still be produced. In the study of Daelman *et al.* (2012); the methane emissions associated with anaerobic digestion exceeded the carbon dioxide emission that was avoided by utilizing the resulting biogas for energy generation. Nonetheless it should be emphasized that the emission of methane could be significantly reduced by a better design and good housekeeping. For instance, the ventilation air of sludge handling facilities could be used as combustion air in the gas engines of the cogeneration plant.

23.1.3 GHG monitoring methodologies

In the past few years there have been important efforts to quantify and investigate GHG emissions from WWTP. These monitoring campaigns initially started with grab-sampling methods but due to the large fluctuations in emissions occurring in many cases, the continuous online monitoring methodology is being now extensively used. This section describes the most commonly used GHG monitoring methodology, based on gas hood collection systems connected to online gas analysers. Also, it is presented two novel monitoring approaches which might be a good alternative to the conventional methodology.

23.1.3.1 The Floating hood + gas analyser approach

Numerous GHG measurement campaigns have been conducted by various groups using floating hoods to collect the greenhouse gases coming off of the water surface in the wastewater treatment plant biological reactors. Traditionally, this method has been employed to measure VOCs from wastewater treatment plants (Tata *et al.* 2003) using a surface emission isolation flux chamber (SEIFC). These SEIFCs are one of the few devices approved by the USEPA and are used in USEPA method EPA/600/8–86/008 (1986) for measuring gaseous emission rates from land surfaces (1986). Chandran (2009, 2011) adapted this method for measuring N_2O emissions from biological nutrient removal (BNR) wastewater treatment plants, which has resulted in a comprehensive field measurement protocol certified by the USEPA (Chandran, 2009). Although specifics for gas collection with the SEIFC and a tracer method for flux determination are prescribed in the protocol, which is recommended whenever it can be followed closely, others have successfully used the protocol as a guide in alternate methods to achieve the same fundamentally in WWTP N_2O measurement campaigns (Rodriguez-Caballero *et al.* 2014; Aboobakar *et al.* 2013; Guo *et al.* 2013). These methods include the following fundamentals: (1) collect greenhouse gas with floating hood; (2) measure greenhouse gas concentrations with gas analyser; and (3) determine flux. Each of these are discussed in this section.

The site-specific measurement plan should address floating hood locations/placement depending on the biological reactor configuration and the objectives of the measurement campaign. But as far as hood types, Porro *et al.* (2014) detailed the effect of different hood types, comparing side-by-side measurements of off-gas N_2O concentrations with a custom-made, large floating hood and the SEIFC hood. Figure 23.1a is a photograph of the custom-made, large hood and the SEIFC hood. Results comparing the hood types indicated that hood size/design should not significantly impact measurements assuming they are properly vented to prevent pressure build up. Of course, the larger surface area covered by the hood will provide better averaging of emissions in a given zone if there is significant spatial variability within the zone, since it provides greater surface area to capture emissions. On the other hand, the smaller hood is more portable and easier to move around for measuring in different places. Figure 23.1b shows another side-by-side measurements comparison, this time between a smaller, custom-made hood (using a plastic storage bin) and the same custom-made, large hood used by Porro *et al.* (2014). Results are not yet published because this comparison was only made recently in a measurement campaign in The Netherlands; however, it further confirmed that hood size and design do not significantly impact measurements if they are vented properly to prevent pressure inside of the hood from exceeding atmospheric pressure and causing accumulation/concentrating of the gas. However, in non-aerated zones, since there is no airflow coming into the hood from the water surface, it is important to have the vent port allow ambient air in to prevent pulling a vacuum in the hood, which can cause it to be pulled below the water surface. If this 'ambient method' is used in lieu of the Chandran WWTP N_2O field measurements protocol, which uses a sweep gas that is forced into the hood rather than pulling in ambient air, ambient GHG concentrations should be checked frequently to account for any potential background N_2O and/or CH_4 coming into the hood. These comparisons of different hood types present numerous possibilities for hood design/construction, which can be tailored depending on measurement objectives and what materials and funding are available.

(a) (b)

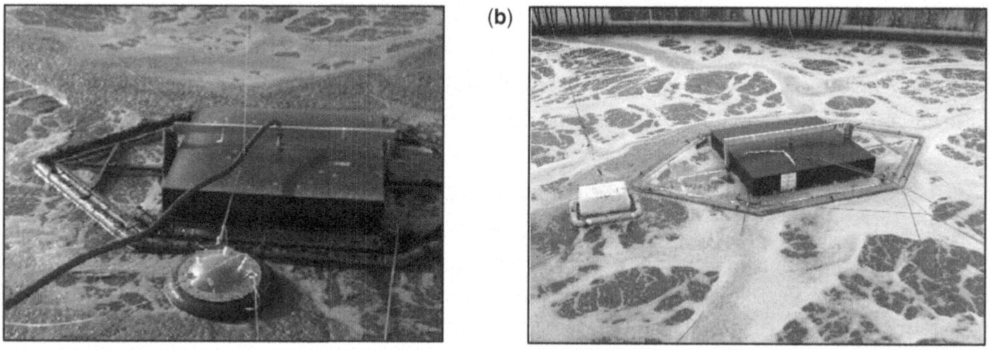

Figure 23.1 Different gas collection hoods used for GHG monitoring in WWTP. (a) Large custom made gas collection hood and the SEIFC hood. (b) Large custom made gas collection hood and small custom made gas collection hood.

In some configurations, strong spatial variation in emissions, mainly N_2O, has been reported in numerous studies (Ahn *et al.* 2010a,b; Aboobakar *et al.* 2013; Rodriguez-Caballero *et al.* 2014). This spatial variation becomes relevant in particularly in plug-flow reactors which present spatial gradients in concentrations of dissolved oxygen and nitrogen species along the reactor path. In these cases there is the need to characterise spatial variability in emissions to achieve a more accurate quantification. To accomplish a good quantification in these type of systems, Pan and co-workers proposed the use of a multiple hood system connected to an online gas analyser to simultaneously measure GHG from multiple locations within the bioreactor (Pan *et al.* 2016).

Gas analysers that have been typically used for measuring N_2O emissions from WWTPs utilize the infrared red (IR) principle and have included N-Tox® (by Water Innovate, UK), VA-3000 (by Horiba, Japan), API Model 320E (Teledyne, USA), and X-STREAM X2GP (Emerson, USA). Typically, CH_4 emissions from activated sludge biological reactors have been measured concurrently with N_2O emissions using either the VA-3000 (by Horiba, Japan), or the X-STREAM X2GP (Emerson, USA), which can also measure CH_4 and CO_2 via IR, and O_2 via electrochemical paramagnetic sensors. To carry the gas from the floating hood to the analyser, air tight PTFE tubing and fittings are used, while a moisture trap is used to deal with condensation and remove moisture content from the gas before it is introduced into the analyser. The moisture traps, or sampling prep units can be built-in as part of the analyser cabinet and can include a moisture sensor to shut the analyser off in the event that moisture passes the moisture trap. This is recommended for longer un-manned campaigns, because turbulence or other extreme conditions can lead to wastewater/sludge being sucked into the PTFE tubing from the hood being submerged.

Ultimately, determination of the flux, or advective gas flow rate, through the floating hood or SEIFC is needed to quantify the GHG emissions. Chandran (2009) details a helium tracer method, adapted from American Society for Testing and Materials (ASTM) method D1946. The method consists of introducing a helium tracer gas into the hood at a known flow rate and concentration and measuring the concentration coming out the hood with a field gas-chromatograph equipped with a thermal conductivity detector (GC–TCD). The difference in concentration due to the dilution can then be used to calculate the flux. Regardless of the method to calculate the dilution of the GHG off-gas, the net flux of GHG is based on the gas flow rate out of the hood, the hood headspace gas concentration, which is directly measured in aerobic zones, or calculated in anoxic or non-aerated zones based on dilution from sweep gas or ambient air, and the cross-sectional area of the hood. This flux (mass/area) can then be applied to the part of the biological treatment train that has been characterized by the measurements to calculate the total emissions.

Alternative methods to the helium tracer method for calculating the advective gas flow rate include using the measured airflow in the blower air piping if it is measured at the air header leading to the diffuser grid where the measurements are performed. The total air flow into the grid can then be divided by the total diffuser grid area to determine the air flow rate per area, which can then be applied to the surface area of the hood to estimate the gas flow rate out of the hood. The accuracy of this method has not been tested, but it can serve to provide a good estimate of the emissions. Alternatively, the air flow rate can be measured coming out of the hood by installing an air mass flow rate sensor on the vent port. For non-aerated zones, the sweep gas method and tracer method can be used as specified by Chandran (2009, 2011) can be followed, but alternatively, if the VA-3000 (by Horiba, Japan), or the X-STREAM X2GP (Emerson, USA) are used or some other means of measuring off-gas O_2, and the hood vent port is used to pull ambient air into the hood to carry the gas to the analyser, a mass balance on O_2 can be performed to determine the flux because the airflow rate is known (from analyser rotameter or pump specs), and zero O_2 concentration can be assumed in hood because of anoxic conditions. In the case of WWTP with surface aerators, the method prescribed by Ye *et al.* (2014) is recommended.

23.1.3.2 Estimating N_2O emissions through N_2O dissolved data

Concentrations of dissolved gases such as N_2O can be measured based on gas phase measurements, according to the method proposed by Mampaey *et al.* (2015). The proposed method relies on a gas stripping device, consisting of a stripping flask and a scum trap flask, as displayed in Figure 23.2.

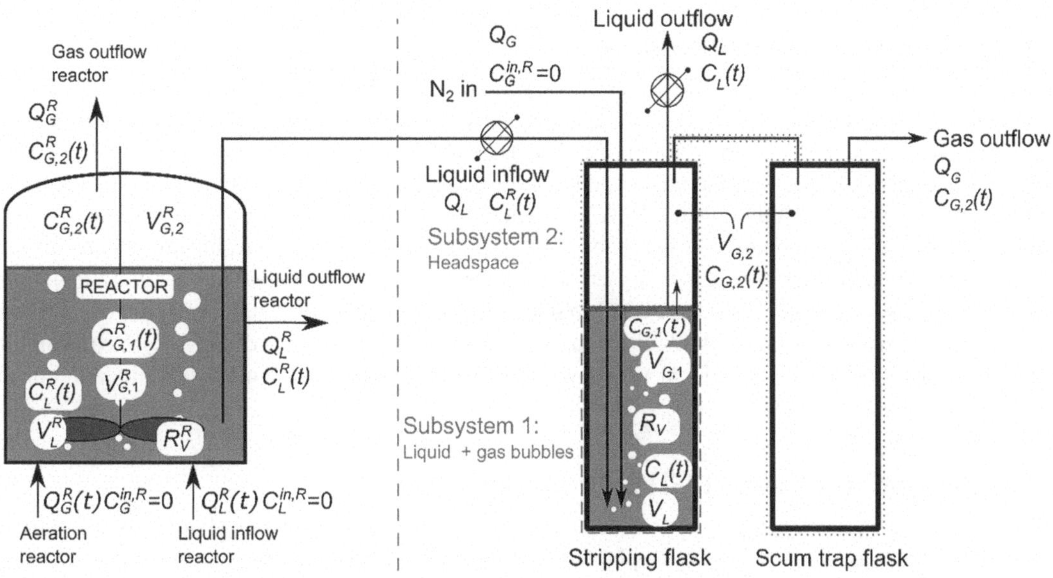

Figure 23.2 Lay-out of the reactor (left) and the gas stripping device (right) for monitoring dissolved gases (Mampaey *et al.* 2015). Note that the superscript R refers to the reactor.

A liquid sample stream from the reactor is continuously supplied to the stripping flask at a constant flow rate QL, while maintaining a constant liquid volume VL in the stripping flask. The scum trap flask is an empty bottle to collect entrained scum from the stripping flask. Nitrogen is used as stripping gas in the stripping flask through fine bubble aeration at a constant flow rate Q_G. The gas outflow of the gas stripping device is analysed by an online gas phase analyser. The dissolved concentration in the reactor, $C_L^R(t)$, is calculated from the measured gas concentration $C_{G,2}(t)$ according to Equation 23.1.

$$C_L^R(t) = \frac{Q_G}{Q_L} \cdot \left(1 + \frac{Q_L}{a_3 \cdot V_L}\right) \cdot C_{G,2}(t) - \frac{Q_G}{Q_L} \cdot a_1 \tag{23.1}$$

in which Q_G represents the stripping gas flow rate.

The parameters a_1 and a_3 are determined from a batch stripping test. During this test, the stripping flask is filled batch-wisely with a liquid sample from the reactor under study from which the dissolved N_2O is subsequently stripped with N_2. The monitored gas phase profile $C_{G,2}$ from the stripping device is then described by Equation 23.2:

$$C_{G,2}(t) = a_1 + a_2 \cdot \exp(-a_3 \cdot t) - a_4 \cdot \exp(-a_5 \cdot t) \tag{23.2}$$

The gas stripping device provides an adequate method to indirectly measure dissolved gases (N_2O or other) in the liquid phase, for aerated as well as non-aerated conditions/reactors, following variations both in time and in space. Its application to an intermittently aerated (on/off) partial nitritation (SHARON) reactor was demonstrated by Mampaey *et al.* (2015). Castro-Barros *et al.* (2015) applied the method to a one-stage partial nitritation – anammox reactor, subject to alternating high and low aeration. In both cases, the mass balance approach on which the liquid N_2O concentration measurement method is based also allowed the determination of the N_2O formation rate.

23.1.3.3 Plant integrated measurements

Monitoring of fugitive GHGs such as N_2O and CH_4 for the entire plant, as opposed to only the activated sludge tank can be advantageous in order to account for essentially all plant emissions simultaneously, since other unit processes not always monitored, such as the secondary clarifiers for example (Mikola *et al.* 2014), can also contribute to the GHG budget of the plant. Plant-wide measurements of N_2O and CH_4 can be readily employed at fully-covered WWTPs, where the off-gas from most or all components of the plant are centralised to one outlet that is treated prior to its release into the environment, which can then be withdrawn and fed to an online gas analyser, as performed in the work of Daelman *et al.* (2013). From open WWTPs, Yoshida *et al.* (2014) have employed a tracer dispersion method to record gas concentration measurements downwind from a WWTP through a cavity ring-down spectroscopy (CRDS) sampling device, to evaluate both N_2O and CH_4 at very low levels (low ppb range) and enable quantification of the total plant emissions over time.

23.1.4 Mitigation of direct GHG emissions

Full-scale GHG monitoring campaigns conducted at different locations worldwide show emission patterns which seem to be characteristic from each plant. However, studies carried out in control lab-scale reactors together with the results from full-scale installations have allowed identifying several factors which increase the risk of having CH_4 and N_2O emissions if not taken into account. These factors are summarized in Table 23.1.

Table 23.1 Main causes and minimization opportunities in WWTPs facilities.

GHG	Main Factors Contributing to its Formation	Minimization Practices
CH_4	Anaerobic conditions (lack of O_2 or NO_x), biodegradable COD and presence of methanogenic archaea	– Preventive measures to sewer systems to avoid formation of methane on those location and arrival as dissolved CH_4 in the WWTP. Chemical addition (nitrate or iron salts), air or oxygen will avoid its formation in the sewer network Gutierrez *et al.* (2008); Auguet *et al.* (2015). – Promoting biological conversion over stripping in WWTP Daelman *et al.* (2014).
N_2O	– Presence of nitrite Kampscheur (2008) – Low dissolved oxygen levels Aboobackar (2013) – Sudden transitions from anoxic to aerobic phases Yu (2010) – High Nitrogen loading rates Ahn (2010a) – Long aeration periods in SBR systems	– Promoting full nitrification Ahn *et al.* (2011); Rodriguez-Caballero *et al.* (2013). – Avoiding aeration drops in aerobic zones Rodriguez-Caballero *et al.* (2014). – Avoiding addition of reject wastewater to the inlet of the plant during those periods when influent wastewater contains more nitrogen. – Alternating short aerobic with short anoxic phases in SBR systems. Rodriguez-Caballero *et al.* (2015).

Although it is still a challenge at this stage to identify a common operating protocol for WWTP that ensures minimization of their direct GHG emissions, there are some practices that have been proven to reduce emissions on those full-scale facilities that have been implemented. These minimization strategies are also presented in Table 23.1. It is recommended, however, that a first diagnosis of the type and magnitude of emissions of the plant is conducted before implementing any mitigation measure, to ensure the effectiveness of the implemented measure. Also important is to take into account the possible indirect CO_2 emissions linked to the strategy implemented. If minimization of direct GHG emissions can only be achieved by increasing energy consumption (i.e. by increasing aeration) an overall assessment taking into account both type of emissions needs to be done to ensure that a decrease on direct GHG does not incur a substantial increase on indirect CO_2 emissions.

23.2 ODOUR EMISSIONS DURING WASTEWATER TREATMENT

23.2.1 Introduction

Odourous emissions rank among the major causes of public environmental complaints received by local authorities worldwide. Particularly, malodours from WWTPs are included among the most unpleasant ones (Henshaw *et al.* 2006). This increase in public discomfort together with the harmful consequences derived from polluted gas emissions have recently resulted in the enforcement of stricter environmental legislation (Aatamila *et al.* 2011). In this context, the control of these malodourous emissions constitutes nowadays one of the main challenges during WWTPs design and operation.

Sources of odourous compounds in WWTPs are very diverse and include wastewater collection, transfer, treatment units and the resulting sludge and its disposal. Among them, the plant headworks receiving the raw wastewater and the sludge treatment units (thickeners, centrifuges and disposal units) have been ranked as the main odour sources within a WWTP (Zarra *et al.* 2008). The emitted malodours can result from the stripping of either components present in the receiving municipal wastewater or in the industrial wastewater discharges, or from by-products generated in the degradation of organic matter by microorganisms generally under anaerobic conditions (Lebrero *et al.* 2011).

Odour management in WWTPs considerably differ from conventional air pollution control due to the unique characteristics of malodourous emissions: high air flowrates with numerous compounds from a wide range of chemical families (organic and inorganic sulphides, amines, organic acids, aldehydes, ketones, etc.), odourants interactions, low pollutant concentrations (ranging from ng m^{-3} to mg m^{-3}), and spatial and temporal variability (Zarra *et al.* 2008; Lebrero *et al.* 2011). Recent studies on the composition of WWTPs emissions have confirmed their complex nature: more than 30 different substances have been detected and identified in each study with different analytical techniques, including organic sulphides, organic N-based compounds, oxygenated organic compounds, monoaromatics, halogenated compounds, organic acids, etc. (Dincer & Muezzinoglu, 2008; Zarra *et al.* 2008; Zarra *et al.* 2014) (Table 23.2).

Table 23.2 Main components and maximum concentrations detected in WWTP emissions.

Chemical Group	Examples	Maximum Concentration ($\mu g\ m^{-3}$)
Sulfur compounds	H_2S Mercaptans	$5–100 \times 10^3$ 0.04–0.7
Aromatics	Toluene, benzene	0.01–0.5
Aldehydes and ketones	Acetone, benzaldehyde	0.02–0.07 (aldehydes) 0.4–4.5 (ketones)
Nitrogen-derived compounds	Indole, ammonia	–
Volatile fatty acids	Acetic, butyric	0.01–0.1
Terpenes	Limonene, α-pinene	0.01–0.1

Source: Zarra *et al.* (2008).

New regulations and guidelines have been introduced in the past decade in the EU in an attempt to achieve convergence in environmental protection, satisfy the increasing public demands of a uniform level of safety and life quality standards, and meet the industrial need for clear performance criteria regarding environmental management. This legislation increasingly depends on quantification of impacts and acceptable exposure to odours, and most of them refer to the dynamic olfactometry regulation EN 13725:2003 (Van Harreveld, 2003; Lebrero *et al.* 2011). Interesting examples are found in certain European Commission directives (i.e. BREFs for Intensive Rearing of Poultry and Pigs, Waste Treatment Industries, Waste Management in the Chemical Sector), the Reference Report on monitoring emissions from IED-installations (IED, 2010/75/EU), or a number of current national, regional and municipal odour regulations in Europe (Germany, Ireland, Spain or United Kingdom). Standards based on the EN 13725 are also found in Australian, Chilean or Colombian legislation, among others. In contrast, odour regulation policies in the United States are based on H_2S concentration standards or dilution to threshold ratios, whereas in Japan legislation relies on detection thresholds obtained using the Japanese Triangle Method (Van Harreveld, 2003; Lebrero *et al.* 2011). A detailed revision of recent odour regulation and policies can be found elsewhere (Sironi *et al.* 2013).

That said, it is important to remark that the availability of an objective and reproducible odour measurement technique with a suitable low uncertainty to be used in the legal framework constitutes an essential prerequisite for this quantitative approach in odour regulation (Van Harreveld, 2003; Nicell, 2009).

23.2.2 Odour characterization: sensorial and chemical analysis

Although the first sensorial and gas-chromatography odour measurements date back from 1886 and the late 1970s, respectively, it was not until the last decade of the 20th century when an important step forward on odour characterization occurred with the development of capillary chromatography and solid phase microextraction techniques, gold film H_2S sensors, and the standardization of olfactometric techniques (Van Harreveld, 2004). Nowadays, odour characterization can be addressed by different techniques classified into three groups: analytical, sensorial and mixed. However, none of these techniques still provides a complete characterization of the malodourous emission, and up to date there is no universally accepted method for the characterization of odours and the assessment of their impacts (Zarra *et al.* 2014). In this sense, while sensorial characterization provides important information regarding the impact of a specific odour on an individual or a community (information required to assess compliance to regulations), a detailed chemical characterization is necessary to evaluate and optimize the performance of odour abatement technologies (Lebrero *et al.* 2011) (Table 23.3).

Table 23.3 Advantages and limitations of analytical and sensorial odour monitoring techniques.

Description	Advantages	Limitations	Examples
Analytical techniques			
Identification and quantification of the chemical compounds	• Objectivity • Repeatability • Reproducibility • Accuracy • Useful for the identification of odour sources	• No indication of the actual odour nuisance • The instrumental detection limit might be higher than odourant concentration	Gas chromatography (GC), colorimetric reaction-based assay, sensors (catalytic, infrared and electrochemical), absorption spectroscopy, fluorescence spectrometry
Sensorial techniques			
Quantitative and qualitative characterization of the sensorial component of the odour	• Determination of odour concentration (OU m^{-3}) • Description of the odour effect on receptors	• Highly costly • Time consuming • Subjective • Sensitive to receptor characteristics and external factors	Dynamic olfactometry
Mixed techniques			
Use both a sensorial and an analytical approach	• Simulate human sense • eNose allows for continuous, in-site monitoring	• Insufficient measurement accuracy and repeatability	Electronic Nose (eNose), gas chromatography and olfactometry (GC-O)

Sources: Gostelow *et al.* (2001); Littarru, (2007); Muñoz *et al.* (2010); Capelli *et al.* (2013).

23.2.2.1 Analytical techniques

Among the different instrumental techniques for the identification and quantification of the chemical compounds present in malodourous emissions in WWTPs, gas chromatography (GC) and electrochemical sensors are the most widely used techniques (*Muñoz et al.* 2010).

Gas chromatography is the best technique to separate complex mixtures of odourous compounds. Several detectors can be used: mass spectrometry, flame ionization detector, thermal conductivity detector, etc. Usually, the concentration of the compounds present in the odourous emission is lower than the detection limit of the instrumental technique, which requires a pre-concentration system (adsorption tubes, cryogenic traps, etc.) before their analysis by GC. The desorption of the concentrated odourants is performed by temperature increase or using solvents before injection. While providing a complete picture of all the components present in the sample, the main disadvantage of this technique relies on the difficulty of correlating the chemical composition to the olfactory properties of the emission.

The use of electrochemical sensors provides a rapid and economical measurement of the concentration of some target odourants (such as H_2S or NH_3), which are used as odour indicators. However, their application is limited when there is no certainty about the correlation between H_2S and odour concentration, which is influenced by the presence of other compounds (Gostelow *et al.* 2001; Muñoz *et al.* 2010). For instance, Dincer and Muezzinoglu (2008) observed that not only H_2S, but also propanal and toluene, were the main estimators explaining 99% of the variability in odour concentration.

23.2.2.2 Sensorial techniques

In spite of the advantages of analytical measurements, most of the existing odour regulations impose limits of odour concentration (based on dynamic olfactometry) and/or maximum time of exposure or compliance frequency (Littarru, 2007; Nicell, 2009). Dynamic olfactometry, the most commonly employed sensorial technique, quantifies the concentration of an odour (odour units (OU) m^{-3}) as the number of dilutions with odourless air needed to reduce the odourant concentration to its detection threshold. The detection threshold is determined by presenting the sample to panelists at increasing concentrations by using a dilution device called olfactometer (Figure 23.3a). The odour thresholds are estimated by two different methods: the yes/no method, where the individual has to decide if the sample is clean or odourous air, and the forced choice method, where the panelist is forced to choose which of the different streams contains the odour. The selection of the panel members, as well as the construction and operation of the olfactometer, has to fulfill specific criteria to ensure the repeatability and accuracy of the measurements, as specified in the standard EN 13725 "*Air quality-Determination of odour concentration by dynamic olfactometry*" (Nicell, 2009; Muñoz *et al.* 2010; Sironi *et al.* 2010).

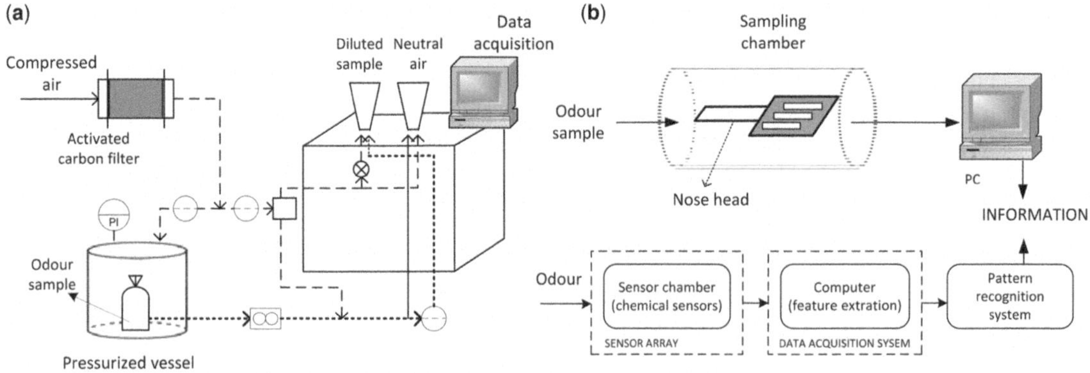

Figure 23.3 Schematic representation of (a) a dynamic olfactometer and (b) an electronic nose. *Source:* Adapted from Muñoz *et al.* (2010), Di Francesco *et al.* (2001) and Davide *et al.* (2001).

Little development on qualitative sensorial analysis has been performed to date, except for some innovative methodologies such as the odour profile method, consisting on linking the odour to some descriptors and intensity scales (Burlingame, 1999) or odour field measurements.

23.2.2.3 Mixed sensorial and analytical techniques

In the last decades, an important effort towards the development of electronic instruments able to mimic the human olfactory apparatus has been made, with the subsequent advances on electronic nose (eNose) technology. An eNose is a device composed of (1) an array of sensors for the classification of odours (conducting polymers, metal oxide semiconductors or quartz microbalances sensors immerse into a chamber containing the target odourant) with partially overlapping sensitivities, (2) a data acquisition system that stimulates electrically each sensor and collects the response, and (3) a pattern recognition system which provides a hypothesis about the odourant being analyzed (Figure 23.3b). An eNose has the potential to combine odour perception and continuous field monitoring. While largely applied in food or cosmetic industries for quality control, tailoring the properties of chemical sensors, the signal processing and the effects of external operating conditions is still necessary for the application of eNoses in the environmental field (Capelli *et al.* 2008; Zarra *et al.* 2013). Recent studies have demonstrated the similarity of results obtained with dynamic olfactometry and eNose analysis, revealing the potential of eNose for continuous monitoring of emissions due to their lower costs, rapid response and on-line analysis (Littarru, 2007; Zarra *et al.* 2014).

Gas chromatography-olfactometry (GC-O) is the most recently developed technique combining the sensorial analysis of the odour sample by olfactometry with the chemical analysis of its constituents by GC-MS, giving both the quantification and identification of the odour character. This technique still presents several limitations such as the strong dependence on the quality and storage conditions of the sample, its limited practical use for continuous process monitoring, or its inability to predict interactions among the odourants, together with the common limitations associated with olfactory analysis (nose saturation, fatigue, etc.) (Muñoz *et al.* 2010).

23.2.2.4 Field and laboratory applications of analytical and sensorial techniques

The application of analytical and sensorial techniques significantly differs in the lab and field scale. This was clearly shown in a study conducted by Muñoz *et al.* (2010) including 68 pilot-scale facilities treating real odourous emissions from wastewater and solid waste facilities, livestock farms and food industry, and 47 laboratory-scale setups treating synthetic odourous mixtures. The results showed that GC and olfactometry techniques are the most commonly applied techniques at full scale (81% and 69%, respectively) since they provide valuable data in terms of odour policy compliance and specific odourant removal. Sensors (40%) and detection tubes (24%) are also commonly used at field scale due to their low cost, simplicity and short analysis time, although they only provide a partial characterization of the emission. On the other hand, olfactometric analyses were rarely used for lab-scale monitoring (9%) due to their high costs, while GC and specific sensors appeared as the most frequently employed techniques (54%). Electronic noses and GC-O are scarcely used in both full and laboratory-scale (Table 23.4).

In terms of the main parameters measured, volatile organic and volatile sulfur compounds (VOCs and VSCs, respectively) and odour concentration accounted for 66 and 69% of the total analyzed full and pilot-scale plants, respectively, followed by H_2S and NH_3. On the contrary, H_2S was the major model odourant selected in laboratory-scale experiments (65%), together with VOCs/VSCs (48%) and, to a lesser extent, NH_3 (33%). The technical difficulties and the high costs of olfactometric measurements reduce the use of odour concentration as a model parameter in laboratories (Table 23.4).

Table 23.4 Main odour monitoring techniques and model odourants measured in full and pilot-scale and laboratory-scale processes.

		Full and Pilot-scale	Laboratory-scale
Monitoring techniques	GC	81%	54%
	Olfactometry	69%	9%
	Specific sensors	40%	54%
	Detection tubes	24%	24%
	eNose and GC-O	Minor/complementary	Scarce
Model odourant	VOCs/VSCs	66%	48%
	Odour concentration	69%	9%
	H_2S	46%	65%
	NH_3	39%	33%

Source: Adapted from Muñoz *et al.* 2010.

23.2.3 Impact assessment

The specific characteristics of odourous emissions, together with the subjective nature of human perception, make necessary a different and unique approach towards the assessment and management of odourous pollution as compared to other air pollutants. In this sense, source characterization alone does not account for the effective impact of odours on the exposed community, thus other methods to evaluate actual exposure are necessary. Two main approaches for odour impact assessment have been developed: measurement of odour exposure by monitoring at the receptor location and evaluation of odour impact from sources by dispersion modelling (Naddeo *et al.* 2013).

23.2.3.1 Measuring odour impact at the receptor location

Different assessment tools can be applied: sensorial direct field inspection (field sniff testing VDI 3940), complaints and odour diaries, odour annoyance surveys, or continuous in-site monitoring (i.e. by eNoses). The most appropriate evaluation technique to be used depends on the application, the activity and its location. Some limitations that must be taken into account in field measurements are the dilution of the emissions at the receptor site below the instrument detection limit, the variability of the emission or the dispersion conditions and the short term of some sampling method. Social participation may allow identifying the origin, determining the degree of annoyance and the effects, or the exact time/location of an odour episode (Capelli *et al.* 2013; Naddeo *et al.* 2013).

23.2.3.2 Evaluation of odour impact from source by dispersion modelling

In addition to calculate the ground odour concentration in a given space-time domain, odour dispersion models are also predictive, which constitutes an advantage over field measurements. Among the existent models, analytical stationary plume models are the simplest, with Gaussian models as the most traditional and less costly (e.g. AER-MOD). Turbulent dispersion is parameterized with empirical coefficients in basic versions, while hybrid models include dispersion parameterized from meteorological data. Puff models include a representation of the spatial and temporal variation of meteorological conditions (e.g. CALPUFF). 3-D models such as Lagrangian particle (e.g. NAME, AUSTAL) or Eulerian grid models (ex. CMAQ, CALGRID) incorporate more advanced tools for atmospheric dispersion simulation. Finally, Fluid Dynamic Models (ex. CFD), which solve three dimensional equations for wind, temperature, humidity and concentration, have been also applied for odour dispersion modelling (Lebrero *et al.* 2011; Capelli *et al.*

2013). Despite the wide availability of models for odour mapping, only few studies have been developed to assess their validation. In this sense, uncertainty when using dispersion modelling results not only from model selection but also from the necessary input data (meteorological, topographical, odour emission rates, etc.). Besides, the result obtained from the model still needs to be correlated with the level of annoyance.

23.2.3.3 Odour impact assessment

These limitations suggest that techniques for odour measuring should be used in combination with odour dispersion models for odour impact assessment. Similarly, odour surveys and questionnaires are usually employed as a tool to link dispersion and perception. For example, Sironi *et al.* (2010) assessed the odour impact of several sources based on the emission data resulting from olfactometric surveys conducted in different periods of the year. These data, together with meteorological and topographical data, allowed for the evaluation of the odour dispersion by modelling, which was further validated through a questioning survey. The correspondence between the simulated and the perceived odour impacts was of 86.5%. Littarru (2007) proposed a combination of dynamic olfactometry and eNose measurements for an objective estimation of odour annoyance. This author obtained a correlation between the data given by both methods, which allowed for a quantitative estimation of the odour intensity in the environment samples and its subsequent comparison with the standards of the air quality. The correlation also provided a way to quantitatively estimate the odourous emissions by simply using the eNose, lowering the time and costs required. Henshaw *et al.* (2006) developed a methodology that consisted on an odour impact model (OIM, a set of dilution-response relationships for a particular odour source given by a group of panelists) combined with the dispersion modelling results. The results could satisfactorily establish which of the odour sources was the main contributor to the community impact as well as predict the effectiveness of odour reduction strategies prior to implementation. Another interesting example of comparison between odour dispersion modelling by Calpuff and a grid field inspection is given in a study by Ranzato *et al.* (2012), in which the two techniques were applied to the assessment of the olfactory nuisance caused by an anaerobic treatment plant for municipal solid waste. Both techniques assessed similar spatial extents of odour nuisance in terms of frequency of odour episodes.

In general, there is a need of rigorous information regarding odours characterization, measurement standardization and impact assessment, resulting still in the cornerstone for odour management and techniques performance control.

23.2.4 Minimization, mitigation and treatment of odourous emissions

Odour management can be undertaken by acting at different levels: (1) Sewage system and wastewater plant design, (2) Good operation practices and proper maintenance of sewer systems and wastewater treatment plant units, (3) Impact mitigation, (4) Odourants abatement. The technical and economic viability of these approaches decreases as the actuation level increases. In this sense, impact prevention and minimization (levels 1 and 2) represents the most cost-effective solution for odour control.

23.2.4.1 Minimization of odour formation

Offensive localized malodours often result from an incorrect plant design or poor operational practices. Avoiding malodourous formation essentially consists on preventing the development of anaerobic conditions, which are the main responsible of odourants generation such as H_2S and volatile fatty acids. This often involves good operational practices (minimize the sludge retention time, ensure sufficient aeration and mixing), the addition of chemicals (nitrate, O_2, Fe salts, hydrogen peroxide, O_3) or good maintenance (frequent cleaning to prevent the accumulation of organic matter) (Estrada *et al.* 2015). Table 23.5 summarizes the main causes and minimization opportunities in WWTPs facilities and sewer networks.

Table 23.5 Odour causes and minimization alternatives in WWTPs facilities.

Unit	Odour Cause	Minimization Practices
Sewage system	Lack of O_2 or NO_3^- and presence of sulfur-reducing or fermentative bacteria	– Use of gravity networks instead of hydraulic pumps – Adequate design of gravity networks – Avoid waterfalls and pronounced curves when designing the network – Airtight sealing of existing access parts (manholes) – Chemical addition (nitrate or iron salts), air or oxygen
Pumping stations	Odourants release from wastewater Organic matter accumulation in screens and filters	– Periodical cleaning – Adequate gradient design in receiving lines – Reduce the height of hydraulic falls – Avoid high turbulence in collection and discharge channels
Storm tanks	Decomposition of wastewater and organic residues due to a prolonged storage	– Minimize wastewater storage time – Periodical cleaning to avoid organic matter accumulation – Reduce the height of hydraulic falls in collection and discharge points to avoid emissions
Plant headworks	Reception of septic waters Decomposition of accumulated wastewater Open channels	– Avoid accumulation of organic matter – Periodical cleaning of screens and filters – Reduce the height of discharge locations
Primary settling	Long wastewater residence times (2–4 hours) Emission from surface, overflows or purge piping	– Avoid settling during low-flow periods – Reduce the height/use submerged overflow weirs – Installation of an automatic sludge purge system
Biotrickling filters	Overload operation (DBO_5 >400–500 mg L^{-1}) Packing material or biofilm deterioration Wastewater accumulation	– Increase wastewater recycling when packing material is clogged – Minimize height between bed and wastewater distribution – Ensure adequate ventilation – Cover units, forced ventilation and odour removal systems
Aeration tanks	Reception of septic waters Operation at low dissolved oxygen (<1 mg L^{-1})	– Ensure proper aeration and mixing of the mix liquor – Use air diffusers instead of mechanical agitation – Reception of septic waters from the lower part
Sludge thickeners and dehydrates	Emission from sludge and return wastewaters	– Avoid sludge accumulation: proper design of the units capacity – Reduce height in sludge collection and discharge points – Cover units, forced ventilation and odour removal systems
Anaerobic digesters	Fugitive emissions (H_2S, VFA) Emissions in overflows and discharge points Start-up and operating problems	– Reduce height in overflows and sludge discharge points – Addition of iron salts or oxygen to the digester to avoid H_2S release
Sludge storage tanks	Emission during filling, emptying and mixing	– Reduce storage time and discharge points height – Cover existing units – Isolate tanks from odour pollution sensitive areas

Two economic alternatives for odour prevention are activated sludge recycling (based on recycling the settled activated sludge or the activated sludge from the aeration tank to the inlet of the WWTP headworks) and oxidized ammonium recycling (which relies on recycling streams with high nitrate or nitrite concentration to the inlet works or upstream in the sewer system). Both technologies present a high application potential using readily available plant by-products with a minimum plant upgrading, and low investment and operating costs (Estrada *et al.* 2015b).

The installation of covers in specific odour units not only prevents the emission but also assists odour abatement by generating lower flowrates and more concentrated emissions. Odour containment is increasingly applied in WWTPs: headworks, clarifiers or sludge management units are commonly covered to prevent malodourous emissions. Depending on the source superficial area, the emitted flowrate or the required unit maintenance, low or high-level covers can be installed. While low-level covers reduce the space between wastewater / sludge and the cover itself increasing the emission concentration and reducing the flowrate emitted, visual inspection and sampling is highly hindered. On the contrary, high-level covers allow personal and machinery access, and are designed with a slight vacuum to avoid malodours diffuse leaks, which is an important factor in highly-concentrated atmospheres. However, unwanted side effects from covering process units, such as corrosion and reduced operator access, usually require proper ventilation strategies (Tchobanoglous *et al.* 2003; Estrada *et al.* 2015).

23.2.4.2 Impact minimization

The nuisance caused on the nearby population can be mitigated by dispersion or dilution of the emission. Buffer zones separate the odour source from the affected receptors, and rely on the fact that odour perception decreases exponentially with the distance due to its dilution in the open atmosphere. Buffer distances are stablished by regulatory agencies, although parameters such as rainfall, moisture conditions, wind direction and wind speed must be also considered (Estrada *et al.* 2013). Turbulent-inducing structures such as trees or high barrier fences and chimneys promote dispersion and dilution by redirecting the wind trajectory and decreasing its speed. The reduction in odour concentration and thus annoyance is highly subjected to a proper selection of the structure type and location according to meteorological conditions and space availability (Estrada *et al.* 2013).

Masking and neutralizing agents offer a short-term solution for impact minimization, usually pulverized or atomized over diffused sources. Their interaction with the odourant is different: while masking agents are aromatic oils that cover the malodour, inhibitory or neutralizing compounds (terpenes or aliphatic or aromatic aldehydes, respectively) chemically react with the odourant. The hedonic tone of the emission can be substituted by a more pleasant one, decreased or eliminated. Nevertheless, the efficiency of these chemicals has been widely questioned, and some studies have associated their implementation with an increase in odour concentration even above regulatory limits (Deccottignies *et al.* 2007; Lebrero *et al.* 2011). In addition, the addition of masking and neutralizing chemicals entail high operation costs and must be only considered as a temporary approach when no other mitigation strategies can be implemented.

23.2.4.3 Odour abatement

End of the pipe technologies are implemented when neither prevention nor mitigation are viable or sufficiently efficient. These treatment technologies are based either on physical-chemical or biological principles (Chapter 11).The capital and operating costs associated with odour treatment using traditional technologies (biofilters, biotrickling filters, adsorption and chemical scrubbing) might represent from 5% up to 15% of the total costs of WWTPs, thus selection of the most appropriate technology requires special attention. In this sense, Estrada *et al.* (2015) concluded that physical-chemical technologies offer an inferior performance than their biological counterparts in terms of energy consumption, economic and environmental efficiency. Among biological technologies, biotrickling filtration exhibited the lowest energy requirements and better overall efficiency when process economic, environmental impacts and robustness are taken into account. A comprehensive economic, sustainability, sensitivity, and robustness analysis of the most frequently implemented technologies can be found elsewhere (Estrada *et al.* 2011, 2012, 2015; Alfonsín *et al.* 2015).

23.3 CONCLUSIONS

While traditionally regulations applied to wastewater treatment systems focused on effluent water quality discharges, growing attention and new regulations are starting to emerge for the odour and greenhouse gas emissions originating in the plants. The first ones create a problem to the local community exposed to odourous compounds and their control is a priority by many local water authorities worldwide, receiving many complaints from the local communities.

On the other hand, greenhouse gas emissions are often neglected by the water authorities. They contribute to global warming which will ultimately affect the whole society but since their contribution is still uncertain and their effects are not immediate, their reduction is not a priority for wastewater practitioners. This chapter has summarised the current techniques used for an accurate quantification of these emissions as well as providing mitigation strategies that can be easily applied to reduce them. Ongoing research will deliver new tools and operational strategies to reduce gas emissions but their wide implementation will ultimately depend on the policies implemented to regulate them.

23.4 REFERENCES

Aatamila M., Verkasalo P. K., Korhonen M. J., Suominen A. L., Hirvonen M. R., Viluksela N. K. and Nevalainen A. (2011). Odour annoyance and physical symptoms among residents living near waste treatment centres. *Environmental Research*, **111**, 164–170.

Aboobakar A., Cartmell E., Stephenson T., Jones M., Vale P. and Dotro G. (2013). Nitrous oxide emissions and dissolved oxygen profiling in a full-scale nitrifying activated sludge treatment plant. *Water Research*, **47**(2), 524–534.

Ahn J. H., Kim S. P., Park H. K., Rahm B., Pagilla K. and Chandran K. (2010a). N_2O emissions from activated sludge processes, 2008–2009: results of a national monitoring survey in the United States. *Environmental Science & Technology*, **44**, 4505–4511.

Ahn J. H. Kim S. P., Park H. K., Rahm B., Pagilla K. and Chandran K. (2010b). Spatial and temporalvariability in atmospheric nitros oxide generation and emission from full-scale biological nitrogen removal and non- BNR processes. *Water Environment Research*, **82**, 2362–2372.

Ahn J. H., Kwan T. and Chandran K. (2011). Comparison of partial and full nitrification processes applied for treating high-strength nitrogen wastewaters: microbial ecology through nitrous oxide production. *Environmental Science & Technology*, **45**(7), 2734–2740.

Alfonsín C., Lebrero R., Estrada J. M., Muñoz R., Kraakman N. J. R., Feijoo G. and Moreira M. T. (2015). Selection of odour removal technologies in wastewater treatment plants: a guideline based on Life Cycle Assessment. *Journal of Environmental Management*, **149**, 77–84.

Auguet O., Pijuan M., Guasch-Balcells H., Borrego C. and Gutierrez O. (2015). Implications of Downstream Nitrate Dosage in anaerobic sewers to control sulfide and methane emissions. *Water Research*, **68**, 522–532.

Brotto A. C., Kligerman D. C., Andrade S. A., Ribeiro R. P., Oliveira J. L. M., Chandran K. and de Mello W. Z. (2015). Factors controlling nitrous oxide emissions from a full-scale activated sludge system in the tropics. *Environmental Science and Pollution Research*, **22**, 11840–11849.

Burgess J. E., Stuetz R. M., Morton S. and Stephenson T. (2002). Dinitrogen oxide detection for process failure early warning systems. *Water Science and Technology*, **45**, 247–254.

Burlingame G. A., Suffet I. H., Khiari D. and Bruchet A. L. (2004). Development of an odour wheel classification scheme for wastewater. *Water Science and Technology*, **49**, 201–209.

Capelli L., Sironi S., Céntola P., Del Rosso R. and Il Grande M. (2008). Electronic noses for the continuous monitoring of odours from a wastewater treatment plant at specific receptors: focus on training methods. *Sensors and Actuators B Chemical*, **131**, 53–62.

Capelli L., Sironi S., Del Roso R. and Guillot J. M. (2013). Measuring odours in the environment vs. *dispersion modelling: a review. Atmospheric Environment*, **79**, 731–743.

Castro-Barros C. M., Daelman M. R. J., Mampaey K. E., van Loosdrecht M. C. M. and Volcke E. I. P. (2015). Effect of aeration regime on N_2O emission from partial nitritation-anammox in a full-scale granular sludge reactor. *Water Research*, **68**, 793–803.

Czepiel P., Crill P. and Harriss R. (1995). Nitrous oxide emissions from municipal wastewater treatment. *Environmental Science & Technology*, **29**(9), 2352–2356.

Chain P., Lamerdin J., Larimer F., Regala W., Lao V., Land M., Hauser L., Hooper A., Klotz M., Norton J., Sayavedra-Soto L., Arciero D., Hommes N., Whittaker M. and Arp D. (2003). Complete genome sequence of the ammonia-oxidizing bacterium and obligate chemolithoautotroph Nitrosomonas europaea. *Journal of Bacteriology*, **185**, 2759–2773.

Chandran K. (2009). Characterization of nitrogen greenhouse gas emissions from wastewater treatment BNR operations – Field protocol with quality assurance plan (U4R07), WERF.

Chandran K. (2011). Protocol for the measurement of N_2O fluxes from biological wastewater treatment plants: Chapter 16. *Methods in Enzymology*, **486**, ISSN 0076-6879, doi: 10.1016/S0076-6879(11)86016.

Daelman M. R. J., van Voorthuizen E. M., van Dongen U. G. J. M., Volcke E. I. P. and van Loosdrecht M. C. M. (2012). Methane emission during municipal wastewater treatment. *Water Research*, **46**(11), 3657–3670.

Daelman M. R. J., van Voorthuizen E. M., van Dongen L. G. J. M., Volcke E. I. P. and van Loosdrecht M. C. M. (2013). Methane and nitrous oxide emissions from municipal wastewater treatment: results from a long-term study. *Water Science and Technology*, **67**(10), 2350–2355.

Daelman M., van Eynde T., van Loosdrecth M. and Volcke E. (2014). Effect of Process design and operating parameters on aerobic methane oxidation in municipal WWTP. *Water Research*, **66**, 308–319.

Davide F., Holmberg M. and Lundström I. (2001). Virtual olfactory interfaces: electronic noses and olfactory displays. In: Communications Through Virtual Technology: Identity Community and Technology in the Internet Age, G. Riva. and F. Davide (eds), IOS Press, Amsterdam, pp. 193–220.

Decottignies V., Filippi G. and Bruchet A. (2007). Characterisation of odour masking agents often used in the solid waste industry for odour abatement. *Water Science and Technology*, **55**, 359–364.

Di Francesco F., Lazzerini B., Marcelloni F. and Pioggia G. (2001). *An electronic nose for odour annoyance assessment. Atmospheric Environment*, **35**, 1225–1234.

Dincer F. and Muezzinoglu A. (2008). Odour-causing volatile organic compounds in wastewater treatment plant units and sludge management areas. *Journal of Environmental Science and Health, Part A*, **43**, 1569–1574.

Estrada J. M., Kraakman N. J. R. B., Muñoz R. and Lebrero R. (2011). A comparative analysis of odour treatment technologies in wastewater treatment plants. *Environmental Science & Technology*, **45**, 1100–1106.

Estrada J. M., Kraakman N. J. R. B., Lebrero R. and Muñoz R. (2012). A sensitivity analysis of process design parameters, commodity prices and robustness on the economics of odour abatement technologies. *Biotechnology Advances*, **30**, 1354–1363.

Estrada J. M., Lebrero R., Quijano G., Kraakman N. J. R. and Muñoz R. (2013). Strategies for odour control. In: Odour Impact Assessment Handbook, V. Belgiorno, V. Naddeo. and T. Zarra (eds), John Wiley & Sons, Ltd, New Delhi, India, pp. 85–124.

Estrada J. M., Lebrero R., Quijano G., Kraakman N. J. R. B. and Muñoz R. (2015a). Odour abatement technologies in WWTPs: energy and economic efficienty. In: Sewage Treatment Plants: Economic Evaluation of Innovative Technologies for Energy Efficiency, K. Stamatelatou. and K.P. Tsagarakis (eds), IWA Publishing, London, pp. 163–185.

Estrada J. M., Kraakman N. J. R., Lebrero R. and Muñoz R. (2015b). Integral approaches to wastewater treatment plant upgrading for odour prevention: activated Sludge and Oxidized Ammonium Recycling. *Bioresource Technology*, **196**, 685–693.

Gostelow P., Parsons S. A. and Stuetz R. M. (2001). Odour measurements for sewage treatment works. *Water Research*, **35**, 579–597.

Guisasola A., de Haas D., Keller J. and Yuan Z. (2008). Methane formation in sewer systems. *Water Research*, **42**(6–7), 1421–1430.

Gutierrez O., Mohanakrishnan J., Sharma K. R., Meyer R. L., Keller J. and Yuan Z. (2008). Evaluation of oxygen injection as a means of controlling sulfide production in a sewer system. *Water Research*, **42**(17), 4549–4561.

Henshaw P., Nicell J. and Sikdar A. (2006). Parameters for the assessment of odour impacts on communities. *Atmospheric Environment*, **40**, 1016–1029.

IPCC (2013). Climate Change 2013: The Physical Science Basis. Working Group I Contribution to the IPCC 5th Assessment Report – Changes to the underlying Scientific/Technical Assessment, IPCC, Cambridge, United Kingdom and New York, NY, USA.

Itokawa H., Hanaki K. and Matsuo T. (2001). Nitrous oxide production in high-loading biological nitrogen removal process under low COD/N ratio condition. *Water Research*, **35**, 657–664.

Kampschreur M. J., van der Star W., Wielders H. A., Mulder J. W., Jetten M. and van Loosdrecht M. C. M. (2009). Dynamics of nitric oxide and nitrous oxide emission during full-scale reject water treatment. *Water Research*, **42**, 812–826.

Kishida N., Kim J. H., Kimochi Y., Nishimura O., Sasaki H. and Sudo R. (2004). Effect of C/N ratio on nitrous oxide emission from swine wastewater treatment process. *Water Science and Technology*, **49**, 359–365.

Klenbusch M. (1986). Measurement of gaseous emission rates from land surfaces using an emission isolation flux chamber, Users Guide. U.S. Environmental Protection Agency, Washington, D.C., EPA/600/8-86/008.

Law Y., Ye L., Pan Y. and Yuan Z. (2012). Nitrous oxide emissions from wastewater treatment processes. *Philosophical Transactions of the Royal Society B*, **367**, 1265–1277.

Law Y., Jacobsen G. E., Smith A. M., Yuan Z. and Lant P. (2013a). Fossil organic carbon in wastewater and its fate in treatment plants. *Water Research*, **47**(14), 5270–5281.

Law Y., Lant P. and Yuan Z. (2013b). The Confounding Effect of Nitrite on N_2O Production by an Enriched Ammonia-Oxidizing Culture. *Environmental Science & Technology*, **47**, 7186–7194.

Lebrero R., Bouchy L., Stuetz R. and Muñoz R. (2011). Odour assessment and management in wastewater treatment plants: a review. *Critical Reviews In Environmental Science And Technology,* **41**, 915–950.

Lemaire R., Meyer R., Taske A., Crocetti G. R., Keller J. and Yuan Z. G. (2006). Identifying causes for N_2O accumulation in a lab-scale sequencing batch reactor performing simultaneous nitrification, denitrification and phosphorus removal. *Journal of Biotechnology*, **122**, 62–72.

Littarru P. (2007). Environmental odours assessment from waste treatment plants: dynamic olfactometry in combination with sensorial analysers "electronic noses". *Waste Management*, **27**, 302–309.

Mampaey K. E., van Dongen U. G. J. M., van Loosdrecht M. C. M. and Volcke E. I. P. (2015). Novel method for online monitoring of dissolved N_2O concentrations based on gas phase measurements. *Environmental Technology*, **36**(13), 1680–1690.

Mampaey K. E., De Kreuk M. K., van Dongen U. G., van Loosdrecht M. C. and Volcke E. I. (2016). Identifying N_2O formation and emissions from a full-scale partial nitritation reactor. *Water Research*, **88**, 575–585.

Mikola A., Heinonen M., Kosonen H., Leppanen M., Rantanen P. and Vahala R. (2014). N_2O emissions from secondary clarifiers and their contribution to the total emissions of the WWTP. *Water Science and Technology*, **70**, 720–728.

Muñoz R., Sivret E. C., Parcsi G., Lebrero R., Wang X., Suffet I. H. and Stuetz R. M. (2010). Monitoring techniques for odour abatement assessment. *Water Research*, **44**, 5129–5149.

Naddeo V., Belgiorno V. and Zarra T. (2013). Procedures for odour impact assessment. In: Odour Impact Assessment Handbook, V. Belgiorno, V. Naddeo. and T. Zarra (eds), John Wiley & Sons, Ltd, pp. 187–203.

Ni B.-J. and Yuan Z. (2015). Recent advances in mathematical modeling of nitrous oxides emissions from wastewater treatment processes. *Water Research*, **87**, 336–346.

Ni B.-J., Peng L., Law Y., Guo J. and Yuan Z. (2014). Modeling of nitrous oxide production by autotrophic ammonia-oxidizing bacteria with multiple production pathways. *Environmental Science & Technology*, **48**, 3916–3924.

Nicell J. A. (2009). Assessment and regulation of odour impacts. *Atmospheric Environment*, **43**, 196–206.

Norton J. M., Klotz M. G., Stein L. Y., Arp D. J., Bottomley P. J., Chain P. S. G., Hauser L. J., Land M. L., Larimer F. W., Shin M. W. and Starkenburg S. R. (2008). Complete genome sequence of Nitrosospira multiformis, an ammonia-oxidizing bacterium from the soil environment. *Applied and Environmental Microbiology*, **74**, 3559–3572.

Otte S., Grobben N. G., Robertson L. A., Jetten M. S. M. and Kuenen J. G. (1996). Nitrous oxide production by Alcaligenes faecalis under transient and dynamic aerobic and anaerobic conditions. *Applied and Environmental Microbiology*, **62**, 2421–2426.

Pan Y., Ni B.-J., Bond P. L., Ye L. and Yuan Z. (2013). Electron competition among nitrogen oxides reduction during methanol-utilizing denitrification in wastewater treatment. *Water Research*, **47**, 3273–3281.

Pan Y., van den Akker B., Ye L., Ni BJ., Watts S., Reid K. and Yuan Z. (2016). Unravelling the spatial variation of nitrous oxide emissions from a step-feed plug-flow full scale wastewater treatment plant. *Scientific Reports*, **6**, article number 20792.

Ranzato L., Barausse A., Mantovani A., Pittarello A., Benzo M. and Palmeri L. (2012). A comparison of methods for the assessment of odour impacts on air quality field inspection (VDI 3940) and the air dispersion model CALPUFF. *Atmospheric Environment*, **61**, 570–579.

Ribera-Guardia A., Kassotaki E., Gutierrez O. and Pijuan M. (2014). Effect of carbon source and competition for electrons on nitrous oxide reduction in a mixed denitrifying microbial community. *Process Biochemistry*, **49**, 2228–2234.

Rodriguez-Caballero A., Aymerich I., Poch M. and Pijuan M. (2014). Evaluation of process conditions triggering emissions of green-house gases from a biological wastewater treatment system. *Science of the Total Environment,* **493**, 384–391.

Rodriguez-Caballero A., Aymerich I., Marques R., Poch M. and Pijuan M. (2015). Minimizing N_2O emissions and carbon footprint in a full-scale activated sludge sequencing batch reactor. *Water Research*, **71**, 1–10.

Sironi S., Capelli L., Céntola P., Del Rosso R. and Pierucci S. (2010). Odour impact assessment by means of dynamic olfactometry, dispersion modelling and social participation. *Atmospheric Environment*, **44**, 354–360.

Sironi S., Capelli L., Dentoni L. and Del Roso R. (2013). Odour regulation and policies. In Odour Impact Assessment Handbook, Belgiorno, V., Naddeo, V. and Zarra, T. (eds), John Wiley & Sons, Ltd, pp. 175–186.

Stein L. Y., Arp D. J., Berube P. M., Chain P. S. G., Hauser L., Jetten M. S. M., Klotz M. G., Larimer F. W., Norton J. M., den Camp H. J. M. O., Shin M. and Wei X. (2007). Whole-genome analysis of the ammonia-oxidizing bacterium, Nitrosomonas eutropha C91: implications for niche adaptation. *Environmental Microbiology*, **9**, 2993–3007.

Tallec G., Garnier J., Billen G. and Gousailles M. (2006). Nitrous oxide emissions from secondary activated sludge in nitrifying conditions of urban wastewater treatment plants: effect of oxygenation level. *Water Research*, **40**, 2972–2980.

Tata P., Witherspoon J. and Lue-Hing C. (eds), (2003). VOC Emissions from Wastewater Treatment Plants. Lewis Publishers, Boca Raton, FL.

Tchobanoglous G., Burton F. L. and Stensel H. D. (2003). Wastewater Engineering: Treatment and Reuse. McGraw Hill, New York, USA.

Van Harreveld A. P. (2003). Odour Regulation and the History of Odour measurement in Europe. Odour Measurement Review. *Japan Ministry of the Environment*. 54–61. http://www.env.go.jp/en/air/odour/measure/02_1_3.pd.

Van Harreveld A. P. (2004). Odour management tools – filling the gaps. *Water Science and Technology*, **50**, 1–8.

Von Schulthess R. and Gujer W. (1996). Release of nitrous oxide (N_2O) from denitrifying activated sludge: verification and application of a mathematical model. *Water Research*, **30**, 521–530.

Von Schulthess R., Kuhni M. and Gujer W. (1995). Release of nitric and nitrous oxides from denitrifying activated-sludge. *Water Research*, **29**, 215–226.

Wang Q., Jiang G., Ye L., Pijuan M. and Yuan Z. (2014). Heterotrophic denitrification plays an important role in N_2O production from nitritation reactors treating anaerobic sludge digestion liquor. *Water Research*, **62**, 202–210.

Winkler M.-K. H., Le Q. H. and Volcke E. I. P. (2015). Influence of Partial Denitrification and Mixotrophic Growth of NOB on Microbial Distribution in Aerobic Granular Sludge. *Environmental Science & Technology*, **49**, 11003–11010.

Ye L., Ni B. J., Law Y., Byers C. and Yuan Z. (2014). A novel methodology to quantify nitrous oxide emissions from full-scale wastewater treatment systems with surface aerators. *Water Research*, **48**, 257–268.

Yoshida H., Monster J. and Scheutz C. (2014). Plant-integrated measurement of greenhouse gas emissions from a municipal wastewater treatment plant. *Water Research*, **61**, 108–118.

Yu R., Kampschreur M. J., van Loosdrecht M. C. M. & Chandran K. (2010). Mechanisms and specific directionality of autotrophic nitrous oxide and nitric oxide generation during transient anoxia. *Environmental Science and Technology*, **44**(4), 1313–1319.

Zarra T., Naddeo V., Belgiorno V., Reiser M. and Kranert M. (2008). Odour monitoring of small wastewater treatment plant located in sensitive environment. *Water Science and Technology*, **58**, 89–94.

Zarra T., Naddeo V., Belgiorno V., Higuchi T., Dutta N. and Bhuyan M. (2013). Instruments and methods for odour sampling and measurement. In: Odour Impact Assessment Handbook, Belgiorno, V., Naddeo, V. and Zarra, T. (eds), John Wiley & Sons, Ltd, New Delhi, India, pp. 31–83.

Zarra T., Reiser M., Naddeo V., Belgiorno V. and Kranert M. (2014). Odour emissions characterization from wastewater treatment plants by different measurement methods. *Chemical Engineering Transactions,* **40**, 37–42.

Zeng R. J., Lemaire R., Yuan Z. and Keller J. (2003). Simultaneous nitrification, denitrification, and phosphorus removal in a lab-scale sequencing batch reactor. *Biotechnology and Bioengineering*, **84**, 170–178.

Zhou Y., Pijuan M., Zeng R. J. and Yuan Z. (2008). Free Nitrous Acid Inhibition on Nitrous Oxide Reduction by a Denitrifying-Enhanced Biological Phosphorus Removal Sludge. *Environmental Science & Technology*, **42**, 8260–8265.

Zhou Y., Lim M., Harjono S. and Ng W.J. (2012). Nitrous oxide emission by denitrifying phosphorus removal culture using polyhydroxyalkanoates as carbon source. *Journal Of Environmental Sciences-China,* **24**, 1616–1623.

Chapter 24

The impact and risks of micropollutants in the environment

Paola Verlicchi, Damia Barcelò, Dragana Mutavdžić Pavlović, Matteo Papa, Mira Petrovic, Nick Voulvolis and Elena Zambello

24.1 INTRODUCTION

In recent years, many investigations have highlighted the presence of an increasing number of micropollutants (MPs) – substances occurring in the range ng/L-µg/L – in water and soil and the need to better analyze the behavior of "critical" compounds – those exhibiting the highest concentrations in water and the potential negative effects once released into the environment.

Increasing attention has been paid to pharmaceutical compounds (PhCs) in water compartments due to their widespread (increasing) consumption worldwide and their high excretion rate, quite often as parent compounds (Verlicchi *et al.* 2012b). Hospital wastewaters (HWWs) have been the subject of specific investigations providing in-depth chemical and physical characterizations (Verlicchi *et al.* 2010), environmental risk assessments of untreated and treated effluents (Verlicchi *et al.* 2012a; Santos *et al.* 2013) and suggestions for proper management and treatment (Al Aukidy *et al.* 2014). Also surface water and groundwater have been deeply investigated (Fatta-Kassinos *et al.* 2011; Meffe & de Bustamante, 2014; Jurado *et al.* 2012).

Not only the water matrix, but also sludge from municipal wastewater treatment plants (WWTPs) and sludge-amended soil (Verlicchi & Zambello, 2015) have been analyzed.

This chapter aims to provide a snapshot of the occurrence of a selection of MPs in different water compartments and in sludge, and their fate and behavior in the environment. The main issues related to legal, analytical and ecotoxicological aspects are also discussed along with the application of bioassays in the water environment. Finally, the most critical compounds are presented, based on a common approach used worldwide (the risk quotient RQ).

The selection of compounds includes substances suggested by European legislation (Directive, 2013/39/EU, Decision 2015/495/EU), compounds largely administered worldwide, and contaminants frequently detected in water environments at high concentrations (>10 µg/L). Table 24.1 shows those that are mostly studied, together with their CAS number and chemical formula, as well as their main chemical and physical properties (molecular weight M_w, solubility in water S_w, dissociation constant pK_a, coefficients of water-octanol partition K_{ow}, solid-water distribution K_d, soil organic carbon-water partitioning K_{oc}, biological degradation rate k_{biol}, and bioconcentration factor BCF). These will be cited throughout the

Table 24.1 Main physico-chemical properties of the selected PhCs[a].

Class	PhACs	CaS No.	Formula	M_w, g/mol	S_w^b, mg/L	pK_a	log K_{ow}	K_{oc}, L/kg	K_d, L/kg	BCF	k_{biol} L/(g_{SS} d)[a]
Analgesics and antinflammatories (A)	Acetaminophen	103-90-2	$C_8H_9NO_2$	151.17	4150	9.38[b]	0.46	20.94	42[e]	3.16	58–80
	Diclofenac	15307-86-5	$C_{14}H_{11}Cl_2NO_2$	296.15	4.47	4.15[b]	4.51	404.58[a], 158.5–2630[d]	18.2–151.4[d]	3.16	0.04-1.2
	Ibuprofen	15687-27-1	$C_{13}H_{18}O_2$	206.29	68.4	4.91[b]	3.97	224.91[a], 66.1–1318[c]	10.6–60.3[d]	3.16	1.5–35
	Mefenamic acid	61-68-7	$C_{15}H_{15}NO_2$	241.29	13.7	4.2[b]	5.12	881.05[a], 461[d]	18916.98[d]	10	–
	Naproxen	22204-53-1	$C_{14}H_{14}O_3$	230.27	51.1	4.15[b]	3.18	93.54[a], 100–1000[d]	10.72–51.3[d]	3.16	0.2–2.9
	Salicylic acid	69-72-7	$C_7H_6O_3$	138.12	11300	2.97[b]	2.26	37.41[a], 404[d]	0.3–67.9[d]	3.16	–
Antibiotics (B)	Amoxicillin	26787-78-0	$C_{16}H_{19}N_3O_5S$	365.41	958	3.20 ± 0.26, 7.34 ± 0.05, 10.47 ± 0.11[c]	0.87	5.17		3.16	–
	Azithromycin	83905-01-5	$C_{38}H_{72}N_2O_{12}$	749.00	514	8.96 ± 0.04[c]	4.02	47.42	376[e]	208.45	<0.1
	Clarithromycin	81103-11-9	$C_{38}H_{69}NO_{13}$	747.97	217	8.79 ± 0.06[c]	3.16	23.49[a], 150[e]	262[e]	56.49	<0.4
	Erythromycin	114-07-8	$C_{37}N_{67}NO_{13}$	733.95	459	8.83 ± 0.05[c]	3.06	25.47[a], 570[e]		48.53	0.15–6
	Metronidazole	443-48-1	$C_6H_9N_3O_3$	171.16	5920		−0.02	13.18		3.16	–
	Ofloxacin	82419-36-1	$C_{18}H_{20}FN_3O_4$	361.38	1440	5.97, 8.28[e]	−0.39	0.9908	100[d]	3.16	–
	Roxithromycin	80214-83-1	$C_{41}H_{76}N_2O_{15}$	837.07	187	8.82 ± 0.04[c]	2.75	7.21		30.39	0.2-9
	Sulfamethoxazole	723-46-6	$C_{10}H_{11}N_3O_3S$	253.28	459	1.86 ± 1.01, 6.04 ± 0.04[c]	0.89	34.36[a], 72[e]		3.16	0.3
Beta-blockers (C)	Propranolol	525-66-6	$C_{16}H_{21}NO_2$	259.35	79.4	9.42[b]	3.48	282.49[a], 4405 ± 378; 2803 ± 135[d]	16.3 ± 1.4/ 199 ± 9.6[d]	51.05	0.36–0.46
Lipid regulators (D)	Clofibric acid	882-09-7	$C_{10}H_{11}ClO_3$	214.65		3.2[d]	2.57	42.95[a], 14.125[d]	200.73[d]		0.3-0.8
	Fenofibrate	49562-28-9	$C_{20}H_{21}ClO_4$	360.84	0.707	5.19[d]	5.19	9594.01		322.11	–
	Fenofibric acid	42017-89-0	$C_{17}H_{15}ClO_4$	318.76	–	–	2.9				7.2–10.8
	Gemfibrozil	25812-30-0	$C_{15}H_{22}O_3$	250.34	27.8	4.7[d]	4.77	707.95		322.11	6.4–9.6
Psychiatric drugs (E)	Carbamazepine	298-46-4	$C_{15}H_{12}N_2O$	236.28	152	2.45[d]	2.45	168.66[a], 3870[d]	25.52[d]	19.23	0.1
	Diazepam	439-14-5	$C_{16}H_{13}ClN_2O$	284.75	12.2	3.4[b]	2.82	274.16[a], 192-630[e]	1.9-24.8[e]	33.73	<0.16
	Fluoxetine	54910-89-3	$C_{17}H_{18}F_3NO$	309.33	1.7		4.05	1122.02		153.82	5–9

(Continued)

Table 24.1 Main physico-chemical properties of the selected PhCs[a] (*Continued*).

Class	PhACs	CaS No.	Formula	M_w, g/mol	S_w^b, mg/L	pK_a	log K_{ow}	K_{oc}, L/kg	K_d, L/kg	BCF	k_{biol} L/(g_{SS} d)[a]
Hormones (F)	17-β estradiol (E2)	362-07-2	$C_{19}H_{26}O_3$	302.41	9.68	10.71[d]	4.01	792.50[a], 794[d]	35.2[d], 476[e]	205.59	175–460
	Estriol (E3)									–	
	Estrone (E1)	53-16-7	$C_{18}H_{22}O_2$	270.37	3.94	10.5; 10.77[d]	3.13	1044.72[a], 1047[d]	26[d]	53.95	10–162
	17-α-ethinylestradiol (EE2)	57-63-6	$C_{20}H_{24}O_2$	296.41	6.77	10.5[d]	3.67	512.86[a], 501[d]	77[d], 584[e]	122.46	0.4–20
Contrast agent (G)	Iopromide	73334-07-3	$C_{18}H_{24}I_3N_3O_8$	791.12	23.75	−2.05	0.0213	3.16	1.6-2.5		
Antiseptics (H)	Triclocarban	101-20-2	$C_{13}H_9Cl_3N_2O$	315.59			4.90	5321.08[a], 12000–71687[e]	438[e]	796.16	
	Triclosan	3380-34-5	$C_{12}H_7Cl_3O_2$	289.55	6.05	7.9[b]	4.76	8413.95[a], 11397–15892[e]	127[e]	642.69	

[a]EPI Suite™ (see website reported among references), Verlicchi *et al.* (2012b).
[b]DRUGBANK database (see Wishart *et al.* 2006).
[c]Zrnčić *et al.* (2015).
[d]Ying *et al.* (2013).
U.S. National Library of Medicine, TOXNET (see website reported among references).

chapter to predict behavior in the environment. Other compounds are also included regarding to some aquatic compartment (mainly surface water and groundwater).

24.2 LEGAL AND ANALYTICAL ASPECTS

The main source of MPs within the water cycle is the *wastewater of urban, industrial and agricultural origins*. In spite of legislative actions seeking to ensure the proper treatment of wastewater (i.e. EU Urban Wastewater Directive, EU Council Directive 91/271/EEC) no regulation and consensus exist regarding the organic MPs in the effluents from WWTPs and their discharge to the aquatic environment. Similarly, there is no consensus and no legal action is taken to limit or regulate the levels of MPs in solids disposed to the agricultural land in spite of attempts to address this issue and ensure that the agronomic benefits ensuing from the use of biosolids in agriculture are not offset by the chemical contamination that could follow. In some countries there is particular attention to improving the removal of specific pollutants. This is the case in Sweden, where a new regulation is under discussion in order to impose a sanitation step including chemical and thermal treatments for all those sludges allocated to agriculture purposes. The aim is to avoid the risk of spreading microorganisms into the environment through sludge.

Currently, in the EU, water pollution is regulated under the Water Framework Directive (Directive, 2000/60/EC, 2000), which established a framework for Community action in the field of water policy. The most recent European regulation set Environmental Quality Standards (EQS) for 45 priority substances (Directive, 2013/39/EU, 2013) and establishes a watch list with additional 10 groups of compounds (17 individual substances) of possible concern that require targeted EU-wide monitoring in order to support the prioritization process in future reviews of the priority substances list (Table 24.2).

Table 24.2 A watch list of substances established in the Commission implementing Decision (EU) 2015/495 for Union-wide monitoring in the field of water policy as set out in Directive 2008/105/EC.

Name of Substance/group of Substances
17-alpha-ethinylestradiol (EE2)
17-beta-estradiol (E2) and Estrone (E1)
Diclofenac
2,6-di-tert-butyl-4-methylphenol
2-Ethylhexyl 4-methoxycinnamate
Macrolide antibiotics (Erythromycin, Clarithromycin, Azithromycin)
Methiocarb
Neonicotinoids (Imidacloprid, Thiacloprid, Thiamethoxam, Clothianidin, Acetamiprid)
Oxadiazon
Tri-allate

In addition to those regulated pollutants, in water compartments there is a wide range of *emerging pollutants* defined as pollutants that are currently not included in routine monitoring programs at the European level and which may be candidates for future regulation, depending on research on their (eco)toxicity, potential health effects and public perception and on monitoring data regarding their occurrence in the various environmental compartments (Norman network available at the website: http://www.norman-network.net).

In view of future research and monitoring dates, the issue of emerging micropollutants is closely related to the analytical capabilities. In recent years, we have experienced a dramatic progress in analytical techniques translated to an increased sensitivity and selectivity of modern analytical methods that have enabled detection of many new or previously ignored and/or unrecognized contaminants.

In the last two decades, numerous analytical methods have been developed for the determination of different classes of micropollutants in wastewater and receiving surface waters. Generally, the identification and quantification of micropollutants in complex environmental matrices requires analytical methods of high sensitivity and selectivity and typically relies on liquid or gas chromatography (LC or GC) coupled to mass spectrometry (MS).

The preferred analytical approach is based on target analysis of preselected compounds of interest, using LC or GC tandem MS instruments (Krauss *et al.* 2010). Over the years, a gradual shift from class specific methods to multi-residue methods for simultaneous analysis of a large number of target compounds, belonging to different classes, is observed (Petrovic *et al.* 2013a). A number of multi-residue methods have been developed for the analysis of multi-class pharmaceuticals (Gros *et al.* 2012), illicit drugs (Baker & Kasprzyk–Hordern, 2011), pesticides (Masiá *et al.* 2013).

However, the major drawback of this approach is that the majority of current analytical methods focus only on the parent target compounds and rarely include metabolites and transformation products, which sometimes can be more toxic and persistent than the original compounds. One of the reasons is that the majority of transformation products are not known and many of those that are known are not commercially available or they are too expensive. But the main reason is that all relevant contaminants, metabolites and transformation products that may be encountered in the aquatic environment are impossible to be included in any targeted multi-residue method, making therefore a strong case for the application of non-target screening protocols using high resolution mass spectrometry (HR-MS) (Eichhorn *et al.* 2012). However, general screening for unknown substances is time-consuming and expensive, and is often shattered by problems, such as lack of mass spectral libraries. Therefore, the main challenge still remains how to prioritize micropollutants and decide on the significance of the chemical data.

24.3 OCCURRENCE OF MICROPOLLUTANTS IN TREATED EFFLUENTS, SLUDGE, SURFACE AND GROUND WATER

The occurrence of a selection of MPs is here addressed and shown for WWTP effluents and surface water (SW, Fig. 24.1), sludge and sludge-amended soil (Fig. 24.2), and for groundwater (Table 24.3). Some compounds or groups of compounds have been more thoroughly investigated than others, whereas some substances are reported for a few compartments.

Regarding Fig. 24.1, effluent data refer to secondary biological effluents (from conventional activated sludge processes) in different countries all over the world, while surface water refers to rivers and canals generally affected by the presence of catchment areas and the release of treated urban and industrial effluents. The cited literature, reported directly under the figures may provide all the necessary information.

The most investigated compounds are analgesics and anti-inflammatories, antibiotics and the psychiatric drug carbamezepine in WWTP effluents and in surface water. A comparison between the ranges observed for WWTP effluents and SW highlights that for some compounds, concentrations in SW are unfortunately higher. This is the case of the antibiotics clarithromycin, sulfamethoxazole, the hormones estradiol and estriol and the disinfectant triclosan. This can be due to different reasons, including extremely high consumption of these compounds, scarce or no removal during wastewater treatment, the release of untreated wastewater, and a modest dilution effect in surface water.

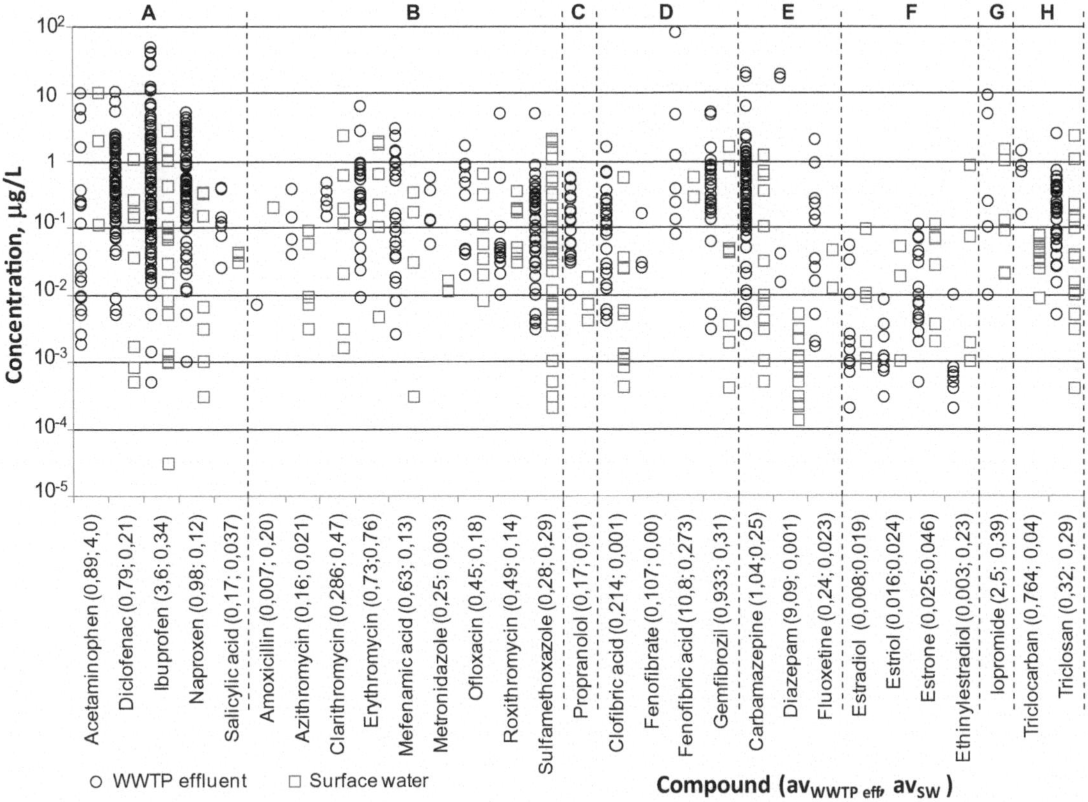

Figure 24.1 Occurrence of the selected compounds in urban WWTP effluent and surface water. Numbers in brackets correspond to average concentration in WWTP effluent and in surface water. Data from: Verlicchi *et al.* 2012b; 2014; Calamari *et al.* 2003; Fatta-Kassinos *et al.* 2011; Ferrari *et al.* 2011; Hernando *et al.* 2007; Kolpin *et al.* 2002; Kumar *et al.* 2010; Loos *et al.* 2007; Luo *et al.* 2014; Pal *et al.* 2014; Schwab *et al.* 2005; Zuccato *et al.* 2000.

The highest *maximum* values in the treated effluent were reported for the analgesics ibuprofen, diclofenac and acetaminophen, the psychiatric drugs carbamazepine and diazepam together with the lipid regulator fenofibric acid - these were between 10 and 100 µg/L. With regard to surface water, the highest values were found for acetaminophen, followed by clarithromycin, erythromycin, sulfamethoxazole and triclosan. However, while the level of acetaminophen was around 10 µg/L, the others occurred at levels of between 1 and 10 µg/L.

The maximum *average* concentrations were found for fenofibric acid (11 µg/L) and diazepam (9.1 µg/L) for WWTP effluents, and for acetaminophen (4.0 µg/L) for surface water.

An idea of the concentrations observed in digested sludge is provided in Fig. 24.2, together with concentrations measured in sludge-amended soil. Unfortunately investigations are not as numerous as those referring to water environments, and available data are limited and not for all the selected compounds. The most analyzed compounds are analgesics/anti-inflammatories, antibiotics and carbamazepine.

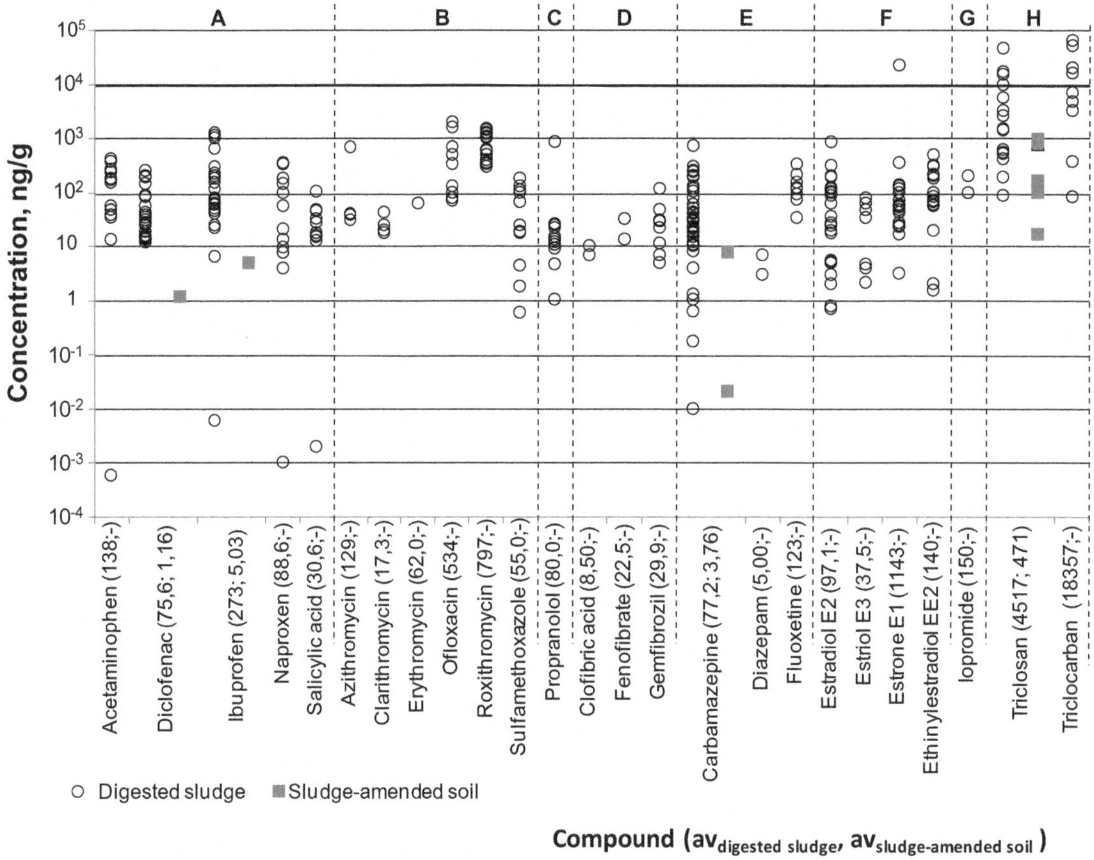

Figure 24.2 Occurrence of selected compounds in digested sludge and sludge-amended soil. Numbers in brackets correspond to average concentration in digested sludge and in sludge-amended soil. Data from: Verlicchi and Zambello, 2015.

Triclosan and triclocarban (antiseptics) and estrone (hormone) exhibited the *maximum* concentrations, higher than 10,000 ng/g, followed by ibuprofen (analgesic), and ofloxacin and roxithromycin (antibiotics), whose maximum values were in the range of 1000–10,000 ng/g. All the other compounds were mostly found in the range of 10–1000 ng/g.

Data referring to soil-amended sludge is scarce, and refers to only four compounds: diclofenac, ibuprofen, carbamazepine and triclosan. The highest concentration was found for the disinfectant triclosan.

With regard to observed concentrations in groundwater, recent studies showed that they are generally lower than in surface water as well as the detection frequency of the analyzed compounds (Loos *et al.* 2010).

This could likely reflect the more direct pathways for transport of MPs into surface waters (e.g. direct discharge of wastewater effluent), as well as other factors such as differences in environmental fate and transport processes (e.g. sorption, volatilization, degradation, etc.) as these contaminants are transported along surface- versus subsurface pathways. The main entry routes for groundwater include diffuse sources such as agricultural land, leakage from landfills or sewer and also ground water/ surface water interaction.

Table 24.3 Maximum concentrations observed in groundwater worldwide.

Group	Compound	Groundwater, [μg/L]	Reference
Antiseptic	Triclosan	2.11	a
Industrial compounds	Chloroform	4.8	b
	Bisphenol A	2.3;9.3	c; a
Personal care	Propylparaben	5.5	a
Pesticides	Atrazine	2.7	b
	Bentazone	16	b
	Dichloran	650	b
	Diuron	100	b
	Metolachlor	12.5	b
	Procymidone	820	b
	Propoxur	90	b
	Propyzamide	450	b
	Simazine	221	b
	Terbuthylazine	29.05	d
Pharmaceuticals	Azithromycin	1.62	e
	Carbamazepine	0.1;0.136;3.6	d; b; a
	Ciprofloxacin	0.443	e
	Crotamiton	0.002;2	d
	Danofloxacin	0.252	e
	Diclofenac	0.38	e
	Ibuprofen	0.998	e
	Ketoprofen	2.9	c
	Norfloxacin	0.462	e
	Ofloxacin	0.367	e
	Oxybenzone	70.4	a
	Propyphenazone	0.6	d
	Spyramycin	2.98	e
	Tamoxifen	0.088	e
Stimulant	Caffeine	0.002;0.02;4.5	d; a

a = Stuart *et al.* 2012; b = Meffe and de Bustamante, 2014; c = Jurado *et al.* 2012; d = Kuroda *et al.* 2012; e = Lopez-Serna *et al.* 2013.

24.4 FATE OF SELECTED COMPOUNDS

The way in which MPs enter the environment depends on their usage pattern and their application.

Depending on the receiver, different transformations can take place, which create products that are so different with regard to their behavior in the environmental and ecotoxicological profile. This means that degradation products of some pollutants can often be even more resistant than their original compounds and show greater toxicity (Farré *et al.* 2008). Reason for that lies in the fact that the complexity of a PhC in the environment with all its transformations, interactions with other pollutants and with environmental compounds can never be described completely, especially as the environment varies very much (Stamm *et al.* 2008).

Among physico-chemical properties, the octanol–water partition coefficient (K_{ow}), water solubility (S_w) and sorption coefficient (K_d) are very valuable parameters. K_{ow} value defines whether a substance is hydrophilic ($\log K_{ow} < 4$) or hydrophobic ($\log K_{ow} > 4$). K_d is defined as the ratio between the concentration of the micropollutant in the soil/sludge and its concentration in the aqueous phase. Normalizing the coefficient K_d with respect to organic carbon content leads to coefficient K_{oc} which is a useful indicator of the micropollutants binding capacity on organic matter of soil and sewage sludge and allows comparisons to be made between different micropollutants. Hydrophobic contaminants are expected to occur at higher concentration in primary sludge than secondary ones, because they have a greater affinity to solids and hence concentrate in the organic-rich sewage sludge (Jurado *et al.* 2012; Le-Minh *et al.* 2010). Conversely, contaminants with low $\log K_{ow}$ values tend to have high S_w and both lower bioaccumulation potential and soil/sludge sorption coefficients. However, care must be taken because $\log K_{ow}$ does not always correlate with sorption capacity onto mineral sediments (Jurado *et al.* 2012).

In addition to the mentioned physico-chemical properties, there is the ionization constant (pK_a) which describes the degree of ionization. Since the degree of micropollutant ionization depends on pH-value, and in some cases physico-chemical properties of ionic and nonionic micropollutant forms are different, it follows that their behavior in the environment depends on the degree of ionization. For this reason it is necessary to know pK_a values. On the basis of their knowledge, it is possible to predict ionic or neutral form of investigated micropollutants because the number of molecules with charged cationic groups increases at pH values below the molecules pK_a value, and decreases at pH above their pK_a. For micropollutant anionic forms situation is opposite.

With regards to PhCs presented in Table 24.1, few compounds are expected to present hydrophobic behaviour as diclofenac, triclocarban and mefenamic acid. The most of remaining PhCs, according to their physico-chemical properties (Table 24.1), should present more hydrophilic or less hydrophobic behaviour being more frequently detected in the aquatic environment. In fact, some PhCs like carbamazepine, are qualified as suitable markers for anthropogenic impacts in the aquatic environment since they are highly resistant towards elimination in water/sediment. It was found that estrogens can sorb to aquifer sediments. For this reason, it is not surprising that they have not been detected in the Spain groundwater (Jurado *et al.* 2012). Finally, acetaminophen, amoxicillin, metronidazole showed hydrophilic behaviour with considerably high water solubility (Table 24.1).

24.4.1 Biodegradation

Biodegradation depends on the characteristics of PhCs, specific biomass, microbial activity, and microbes present in the soil or sewage sludge. In addition to PhC inherent chemical structure, environmental conditions such as temperature, pH, redox potential are crucial factors influencing biodegradation of a compound in the environment (Ying *et al.* 2013).

In order to evaluate the biodegradation potential of PhCs and therefore predict their behaviour in the environment, common parameters are used. They are: bioconcentration and biological degradation constant (k_{biol}).

Bioconcentration is the accumulation of a micropollutant in or on the fish or specified tissues when the source of contaminant is solely water. This information could express by bioconcentration factor (BCF) which is correlated to K_{ow} (can be predicted from $\log K_{ow}$). This factor represents the division of chemical between an organism and the surrounding environment (Hemond, 2000) so hydrophobic component could bioaccumulate into the lipids and lipid membranes of organisms. BCF > 1 is indicative for hydrophobic chemicals (Landis *et al.* 2011), but BCF < 1000 means that the substance is not bioaccumulative, if BCF is in the range 1000–5000 the compound is considered bioaccumulative and if BCF > 5000, the substance is very bioaccumulative.

According to Stamm *et al.* (2008) for polar PhCs with log $K_{ow} < 4.5$, bioaccumulation is irrelevant in general. For example, using sewage inoculum in case of EE2 Carlsson *et al.* (2006) suggest biodegradation is not an important environmental fate process in water which is a proof to the fact that synthetic steroids such as EE2 can be more persistent in the aquatic environment, especially under anaerobic condition. Most natural steroids such as El and E2 can be degraded by microorganisms within several hours or days. Some microalgae in addition to bacteria and fungi, can also transform steroids (e.g. EE2) in water. From other side, acidic drugs are biodegradable in aquatic environments. For example, ibuprofen was biologically transformed in microcosms prepared with sediment and fortified lake water. Ibuprofen transformation products were detected in a cultivated river water reactor, confirming the possibility of the partial biodegradation of ibuprofen (Maeng *et al.* 2011). Ibuprofen is degradable in water–sediment systems with its half-lives of 6–10 days, while carbamazepine is quite resistant to biodegradation with its half-lives up to 328 days. A bench-scale biodegradation study showed effective biodegradation for four acidic pharmaceuticals diclofenac, ibuprofen, naproxen, and gemfibrozil with the half-lives of 2.5–18.6 days with moving sediment (aerobic conditions), but no removal for clofibric acid. Microbial degradation in estuarine and coastal waters was determined for sulfamethoxazole with its half-lives of 85–100 days (Ying *et al.* 2013).

24.4.2 Sorption

Removal of PhCs by sorption is a relevant mechanism for some compounds. Tolls (2001) reviewed sorption behavior of PhCs in soil. Some of this information might be helpful in judging PhC sorption onto sewage sludge and sediments. Important differences between soil and sewage sludge are: lower mineral content, higher lipid concentration, thus higher concentrations of nonpolar, less polar and cationic material in sewage sludge (Kümmerer, 2009). The organic carbon content of soils greatly contributes to the sorption of non-polar PhCs. However, K_{oc} may not properly describe the distribution behaviour between soil and water for some PhCs that contain charged sites and that exist as ionic compounds in the aquatic environment (Maeng *et al.* 2011). According to physico-chemical parameters presented in Table 24.1, estimated K_{oc} values determined from log K_{ow} indicates that diclofenac, ibuprofen, mefenamic acid, E1, E2, EE2 and gemfibrozil are expected to have moderately to low mobility in solid matrix in contrast to triclocarban, triclosan, propranolol and fenofibrate which are practically immobile. Components such as acetaminophen, salicylic acid, amoxicillin and sulfametoxazole, represent a potential risk for groundwater, due to their great mobility through solid matrices. Diclofenac, ibuprofen and naproxen have moderately high log K_{ow}, and sorption would likely be the main mechanism of their removal. They can also be removed from aqueous solution onto solid particulates by ion exchange, complex formation with metal ions, and polar hydrophilic interactions. PhCs, adsorbed on flocks, suspended solids and/or activated (microbial) sludge, will be removed from the aqueous phase by sedimentation and subsequent disposal of excess sludge (Le-Minh *et al.* 2010).

In the case of sulfamethoxazole it has been found that removal by sorption to soil particles is a significant process (Tolls, 2001; Kümmerer, 2009). From Table 24.1 it is evident that almost all compounds exist in ionic form during the soil passage because their pK_a values are higher or lower than pH of the environment where the compounds end up.

The degree of ionization becomes a key factor in sorption mechanisms for acidic PhCs such as non-steroidal anti-inflammatory drugs and lipid regulators. Therefore, it is important to know PhC pK_a as it controls the degree of ionization and the pH must also be considered when estimating parameters that influence the fate of acidic PhCs (Maeng *et al.* 2011). For example, ofloxacin is an amphoteric compound and based on its pK_a values, it will have an ionic charge at any environmental pH. Because of that, it appears to occur through ionic binding, and the high sorption to soil prevents leaching into ground or surface waters (Nowara *et al.* 1997). pK_a value of diclofenac indicates that this compound will exist almost entirely in the dissociated form in the environment and anions generally do not adsorb more strongly to organic carbon and clay than their neutral counterparts (Doucette, 2000).

An important difference between erythromycin and other macrolides, such as clarithromycin and roxithromycin, is the sensitivity of erythromycin to pH. Under acidic conditions, erythromycin is unstable and it is transformed into an inactive anhydro-form by the loss of one H_2O molecule (Gobel *et al.* 2004). At ambient operational pH ranges (6.5–8), erythromycin can exist in both its active original form and as the inactive erythromycin-H_2O. But, macrolides may also adsorb to biomass via cation exchange processes due to the fact that under typical wastewater conditions, many of them are positively charged ($pK_a > 8.8$) and the surface of activated sludge is predominantly negatively charged (Le-Minh *et al.* 2010).

All these facts suggest that ionic interactions may have more influence on the removal of such micropollutants than sorption does.

24.4.3 Photodegradation: direct and indirect

Photodegradation reactions can be classified as direct or indirect. With direct photodegradation reaction, compound is directly excited by absorbing the energy from solar radiation (290–800 nm), while the indirect photodegradation takes place through the formation of reactive intermediates of the matrix components which immediately interact with the substrate. Both processes depend on the chemical structure of the micropollutant and may occur simultaneously (Tong *et al.* 2011).

Some PhCs are sensitive to sunlight, so photochemical decomposition can play an important role in surface water as an additional elimination pathway (Kümmerer, 2009). Photolysis of a chemical is influenced by chemical structure, water pH, water depth, dissolved organic matter (DOM) and inorganic ions and climate (Ying *et al.* 2013). However, not all compounds are photo-degradable. The significance and extent of direct and indirect photolysis of PhCs in the aquatic environment are different for each compound (Kümmerer, 2009).

PhCs such as diclofenac, mefenamic acid, and naproxen are photodegradable under sunlight. Propranolol and carbamazepine were found to be photodegradable with half-lives of up to 730, 8.3, and 2100 h (Ying *et al.* 2013). However, one research in Germany (Kunkel & Radke, 2012) shows that elimination by photolysis is of minor importance for most drugs in rivers. Only under optimal river conditions photolysis contributes up to 50% to the total elimination for a highly photolabile drug diclofenac.

Antiepileptic drug carbamazepine is capable of photolyzing and undergoes photochemical transformation in distilled water and river waters with a half-life of up to 907 sunlight hours. Nitrate and humic acid have opposite effects on its degradation, the latter inhibiting and the former promoting. Steroids such as E1, E2 and EE2 undergo a rapid photodegradation in natural waters with their half-lives of a few minutes to several hours (Ying *et al.* 2013). Azithromycin photodegradation followed first-order reaction kinetics in environmental waters under simulated solar radiation. A half-life of 20 hours was measured in deionized water (pH 5.3), whereas in artificial freshwater (pH 6.8), the half-life was 7 hours. Addition of nitrate, humic acid, and nitrate with humic acid resulted in half-lives of 3.7, 1.2, and 1.1 hours (Tong *et al.* 2011).

Except direct photochemical degradation, micropollutants may be susceptible to a wide array of indirect photochemical pathways, including reaction with singlet oxygen ($^1O_2$), hydroxyl radical (•OH), peroxy radicals (•OOR), photo-excited organic matter, and other reactive species. Photolytic reactions are often complex and lead to multiple reaction products (Farré *et al.* 2008).

24.4.4 Hydrolysis

Hydrolysis is another important non-biotic process for some PhCs in the aquatic environment. But not all PhCs can be hydrolyzed in water, because all PhCs do not have functional groups which are hydrolyzable. Reason for that probably lies in the fact that PhCs were designed for oral use, so they are not sensitive to chemical reactions like hydrolysis (Andreozzi *et al.* 2003). For example, fluoroquinolones, steroids and acidic drugs cannot undergo hydrolysis. For sulfonamides (sulfametoxazole), an acidic pH solution is most favorable to hydrolysis, followed by neutral and alkaline solutions. Hydrolysis rates for roxithromycin and clarithromycin in the presence of iron (III) were low with their half-lives calculated to be 1.99 and 2.67 days (Ying *et al.* 2013).

24.5 ECOTOXICOLOGICAL ASPECTS

Drugs receive extensive pharmacological and clinical testing during development. However, information on the ecotoxicity of these substances is generally much more limited. Any environmentally deleterious effects observed for most pharmaceuticals have, so far, only been under laboratory conditions.

In principle, as PhCs are designed to be biologically active compounds, they have the potential to be particularly troublesome (Christensen, 1998). Levels are generally low (usually below 1 µg /L) making acute toxic effects unlikely. However, because of their continual input, there is the possibility of chronic effects that may not become apparent for many years after they are first looked for. Recently, risk assessments using these toxicity tests have been used to show that measured, or predicted, environmental concentrations of some of these compounds are sufficient to pose a threat to organisms such as Daphnia magna (EMEA, 2006; Bound & Voulvoulis, 2005; Webb, 2001). Some PhCs can cause effects on bacteria and animals at concentrations well below those usually used in safety and efficacy tests. In addition, breakdown products or the combination of different biologically active compounds may have unanticipated effects on the environment. This issue is further complicated by the fact that exposure to only one drug/toxicant at a time is probably a rare event (Tyler *et al.* 1998). In the aquatic environment, most organisms are continually exposed to a range of toxic substances, with possibly only slight temporal and spatial variations in concentration levels (Schowanek & Webb, 2002). Research is beginning to demonstrate the significance of exposure to mixtures of chemical (and non-chemical) stressors at low concentrations, and this raises the question of whether additive effects might occur or whether synergy could magnify the effects of certain pharmaceuticals under study (Stuer-Lauridsen, 2000). The levels to date have been at least one (often several) order(s) of magnitude below the standard doses of environmental PhCs applied in human (or veterinary) medicine. Nevertheless, while the individual concentrations of PhCs in natural waters might be low, the combined concentrations from the variety and numbers of substances in active use could prove to be significant with regard to effects on aquatic life. This would be especially true when multiple PhCs share the same mode of action raising the possibility of synergistic effects.

Many drugs have side effects in humans and some may also have unexpected effects on non-target organisms. For instance, acetaminophen (paracetamol) has been shown to be useful in controlling the brown tree snake (*Boiga irregularis*) in Guam, at doses of 40 mg/L of the active compound (Johnston *et al.* 2002). It has also been shown to inhibit oestrogen-induced vitellogenin production in isolated trout liver cells. Approximately 50% inhibition was achieved with 0.05 mM acetaminophen, while using 0.3 mM inhibited secreted vitellogenin levels to undetectable levels. The drug may, therefore, alter other oestrogen regulated processes (Miller *et al.* 1998). At low mg/L concentrations, the betablocker propranolol can affect the growth and reproduction of Japanese medaka (*Oryias latipes*), a small freshwater fish (Huggett, 2002). It may also cause germinal vesicle breakdown of full-grown folliculated oocytes of the catfish

(*Clarias batrachus*), cultured in vitro, in a dose-dependent manner (Haider & Baqri, 2000). Certain antidepressants (including fluoxetine) have been show to effect spawning in shellfish concentrations in the μg/L range (Daughton & Ternes, 1999; Brooks *et al.* 2003a).

Brooks and co-workers (2003b) also studied the waterborne and sediment toxicity of fluoxetine to several species. Average LC50 values for the Daphnia species Ceriodaphnia dubia and Daphnia magna, and the fathead minnow (Pimephales promelas), were 234, 820 and 705 μg/L, respectively. Growth of the alga *Pseudokirchneriella subcapitata* and *C. dubia* fecundity were decreased by fluoxetine treatments of 14 and 223 μg/L, respectively. *Oryias latipes* survival was not affected by fluoxetine exposure up to a concentration of 8.9 μg/L. An LC50 of 15.2 mg/kg was estimated for the non-biting midge *Chironomus tentans*. Survival of the amphipod *Hyalella azteca* was not affected up to 43 mg/kg fluoxetine sediment exposure. The lowest observed effect concentrations for *C. tentans* and *H. azteca* were 1.3 and 5.6 mg/kg, respectively. This indicates fluoxetine, and possibly related compounds, can have effects on organisms living on/in the sediment and the water column at relatively low concentrations.

Other, innocuous compounds may also affect organisms. Dramatic inhibition of sperm activity has been observed in certain aquatic animals exposed to calcium-channel blockers. The COX-inhibitor indomethacine has been shown to promote egg-shell thinning in birds at doses of 50–100 mg/L. This is comparable to the effects reported as consequences of environmental contamination with DDT (Lundholm, 1997). The ability of dissected polyps of the cnidarian *Hydra vulgaris* to regenerate their hypostome, tentacles and foot was inhibited by diazepam, digoxin and amlodipine at a concentration of only 10 μg/L (Pascoe, 2003).

Plants have also been shown to be adversely affected by drugs in soil or in sewage sludge used to amend soil or through irrigation with contaminated wastewater (Migliore *et al.* 1997). Growth rate, nitrogen fixation, heterocyst frequency and bioaccumulation have been investigated and shown to be adversely affected if an appropriate concentration is reached (Forni *et al.* 2002; Migliore *et al.* 2003). This is unlikely to ever occur in the wild though, since the concentrations required for an effect are several orders of magnitude higher than those so far observed in the environment. It is worth bearing in mind that, should livestock feed on plants that have absorbed pharmaceuticals, there is a small possibility that lipophilic substances may be retained in their body tissue and/or milk, opening up a potential route to the human food chain.

There is another facet to the problem posed by exposure to mixtures of chemicals. Some compounds, which on their own present no inherent risk, may contribute to risk by increasing the toxicity of others. An example of this is the effect of efflux pump-inhibiting drugs in compromising aquatic health (Kurelec, 1992). Now recognised for enabling a significant portion of the increasing incidence of antimicrobial resistance among bacteria, efflux pumps also play a critical role in protecting many different types of cells from xenobiotics and are a common defensive strategy for aquatic biota (Bard, 2000). By minimising the intracellular concentrations of harmful compounds, they prevent the accumulation of pollutants and so allow many aquatic organisms to survive in contaminated waters that might otherwise prove toxic (Bamdad, 1999). Any of a diverse array of chemicals (some of the more potent being verapamil, reserpine and cyclosporine) can inhibit these pumping systems, thereby allowing toxins to cause adverse effects at lower concentrations than normal. For instance, the incidence and severity of developmental abnormalities observed in embryos and larvae of the mussel *Mytilus edulis* exposed to vinblastine, methyl methanesulfonate, chloroquine, mitomycin-C, cadmium chloride and colchicine have been shown to be significantly increased when each toxin is added in the presence of 20 μM of verapamil compared to clean seawater (McFadzen, 2000). Organisms in less-polluted aquatic environments may be at higher risk to newly introduced toxicants because of their lower induced levels of efflux pumps (Daughton & Ternes, 1999).

There is also concern that broad-spectrum antiseptics, such as triclosan, may promote widespread antibiotic resistance simply by inducing bacteria to produce more efflux pumps. Backhaus and co-workers (2011) demonstrated, for example, mixture effects to be significantly higher than those of the individual medicinal products, even when those were only at low, individually insignificantly toxic concentrations. Significant mixture effects from low-effect individual concentrations (EC50) were also observed in a study by Fent and co-workers (2006) for a mixture of cimetidine, fenofibrate, furosemide and phenazone. A mixture of fluoxetine and clofibric acid killed more than 50% of a water-flea (*Daphnia*) population after an exposure of 6 days, although the components were present at concentrations that did not provoke significant effects individually (Flaherty & Dodson, 2005). In the same study, a significant shift in sex ratio was observed after an exposure to a three-component mixture of erythromycin, triclosan and trimethoprim, again at a mixture concentration at which all components were present at concentrations that did not provoke significant individual effects.

Ecotoxicity data are used to develop a Predicted No Effect Concentration (PNEC) whereas the PNEC value combines the ecotoxicity data with an assessment factor. This factor reflects the confidence in the data. Risk characterisation involves assessing risk by comparing the Measured or Predicted Environmental Concentration (MEC or PEC) with the PNEC. If the MEC (or PEC) is greater than PNEC, indicates that the substance may cause harm. Therefore the ratio of MEC (or PEC) to PNEC is taken as a measure of the probability that harm will occur (UNEP/IPCS, 1999).

Occurrence data have to be used not only to confirm the presence of a compound in the environment, but also in combination with relevant ecotoxicoly test data to allow the refinement of risk assessments. For these reasons, further research is necessary in order to:

- analyse the occurrence of scarcely investigated PhCs in the influent and effluent of urban WWTPs,
- define PNECs for a wider spectrum of compounds,
- evaluate the environmental impact of mixtures of different PhCs,
- evaluate the chronic effect of authentic PhCs mixtures on the aquatic life,
- evaluate the best end-of-pipe measures for the existing WWTPs to guarantee better removal of the most persistent compounds,
- suggest source control options to reduce the quantity and variety of PhCs in the water cycle.

Furthermore, most PhCs assessed so far by the Swedish Association of the Pharmaceutical Industry show PEC/PNEC ratios less than one, suggesting that they pose no risk (FASS, 2008). As PhCs and personal care products are considered less harmful to the environment than other pollutants including industrial chemicals and pesticides they are not considered as priority pollutants in the context of the Water Framework Directive (European Council, 2000) and are not included among those compounds to be monitored, notwithstanding their occurrence having been documented for more than 20 years in many European countries.

In general, the potential ecological and human-health effects associated with the presence of pharmaceuticals in the environment have been largely ignored, despite the numerous concerns regarding the hazards of pharmaceutical compounds in the environment. A more diverse range of animal species with defined genders and physiological status should be tested. However, to avoid animal testing, in vitro tests with detailed correlation to in vivo tests could be undertaken. This is of benefit as it should help gain an understanding of the mechanisms involved in the responses observed. Moreover, properties such as exposure routes, timing, frequency and duration, and the presence and absence of other possible contaminants, are necessary for detailed dose-response assessments.

24.5.1 Whole effects approach

The assessment of the impacts and risks of micropollutants in the environment can also be switched from a compounds-oriented to an effects-oriented approach. Indeed, the chemical analysis of micropollutants allows to detect single priority pollutants of concern, but it is not clear which fraction of the overall pollutant burden can be covered by chemical analysis (e.g. metabolites and parent compounds are often missing): therefore, this analysis can be complemented with bio-analytical tools, in order to appreciate the exerted effects as well as not to miss unknown compounds and transformation products (Escher & Leusch, 2012). In this way, the direct measure of the biological activity can help to overcome the constraints and limitation of the application of the single-substance approach, taking into account the so-called "mixture/cocktail" effect.

Among the variety of bioassays that are currently studied in the field of wastewater research (an example of multi-tiered monitoring is provided in Table 24.4), in this section plenty of attention is dedicated to two specific modes of toxic action, the estrogenicity and the mutagenicity; indeed, they represent the indicators of the main diseases linkable to wastewater discharge (endocrine disorders and malignant neoplasms, respectively), next to the diarrhoeal diseases measured by means of the conventional microbiological assays, e.g. *E. coli* determination (Papa *et al.* 2013).

Table 24.4 Example of bio-analytical test battery for wastewater monitoring.

Target Mode of Toxic Action	Example of Assay	Involved Micropollutants
Non-specific toxicity		
Baseline toxicity	Bioluminescence inhibition in *Vibrio fischeri*	All micropollutants (e.g. diclofenac, erythromycin, carbamazepine, …)
Specific toxicity		
Estrogenicity	Luciferase activity quantification in human cells (E-SCREEN)	Natural and synthetic estrogens (estradiol, estrone, estriol; ethinylestradiol) estrogenic industrial chemicals (triclosan, octylphenol, nonylphenol, bisphenol A)
Mutagenicity	Point mutations in bacteria *Salmonella typhimurium* (Ames tests)	Chlorinated byproducts (e.g. trihalomethanes) polycyclic aromatic hydrocarbons (e.g. anthracene) some pharmaceuticals (e.g. ofloxacin, …)
Antibiotic-resistence	PCR amplification systems for the detection of antibiotic resistance genes	Antibiotics (e.g. sulphamethoxazole, roxithromycin, …)

Source: Adapted from Macova *et al.* 2010.

24.5.1.1 Estrogenic activity

Estrogenic activity assays are based on the interaction between Endocrine Disrupting Compounds (EDCs: molecules that can mimic, block or interfere with hormonal activities in living organisms) and estrogenic receptors and can measure the total estrogenic activity of a sample, regardless of the individual compounds identification (Bicchi *et al.* 2009) and also accounting for compounds which exert a biological effect at concentration less than analytical detection limits: indeed, though at very low concentrations, every xenoestrogen, even a weak one, may increase the total estrogenic effects. In particular, the main responsible of estrogenic activity are several micropollutants such as: the natural (estrone, estradiol, estriol) and synthetic (ethynil-estradiol) estrogens, alkylphenols (octyl- and nonyl-phenol) and the bisphenol A. In general, alkylphenols occur at 1000 or more times greater concentration than hormones, but exert an estrogenic potency 1000 lower.

Synthetically, Table 24.5 provides a matrix summarizing the most common estrogenic bioassays (E-SCREEN, based on proliferation of estrogen-dependent breast cancer cells; ER-CALUX, MELN and KBluc, which are mammalian-based reporter gene assays; yeast estrogen screen (YES), a yeast-based reporter gene assay), together with performances, strengths and weaknesses (adapted from Escher & Leusch, 2012).

Table 24.5 Summary of the most common bioassays applied for estrogenic activity determination, together with performances, strengths and weaknesses.

	E-SCREEN	ER-CALUX	MELN	KBluc	YES
Performances					
Likeness to other assays	↑	↑	↓	↑	↑
Method quantification limit	↑	↑	↑	↑	↓
Strengths and weaknesses					
Analysis of environmental samples	↑	↑	↔	↑	↓
Ease of use	↔	↔	↔	↔	↑
Simple training	↓	↓	↓	↓	↑
Sensitivity	↑	↑	↑	↑	↓
Robustness	↑	↑	↑	↑	↓
Reproducibility	↑	↑	↔	↑	↑
Maturity (widespread use)	↑	↑	↔	↔	↑
Quick results	↓	↑	↑	↑	↑

As regards the environmental risk for the receiving water bodies, the main adverse impact related to this kind of biological activity is represented by impair reproductive performance in wildlife: for example, Routledge *et al.* (1998) documented that steroidal hormones can induce feminization in fish and other aquatic organisms at concentrations of 1 ng/L or less of estradiol equivalent; Witters *et al.* (2001) reported that significant reproduction effects in male and female fish might appear at levels above 10 ng/L of estradiol equivalent. Moreover, due to the physiological role of estrogens, any interference in their homeostasis has been shown to be related to a number of human diseases (breast, ovary and prostate tumors, cardiovascular diseases, infertility, etc.).

We successfully applied an E-SCREEN-like assay (based on the human breast cancer cell line MCF-7) to assess the effect of different treatments on the estrogenic activity of treated effluents, as well as on surplus sludge. As example, we found that the MBR treatment was more efficient in estrogenic activity reduction than the activated sludge, despite similar removal were recorded for target EDCs (detailed results in Bertanza *et al.* 2011).

24.5.1.2 Mutagenic activity

Among all the toxic effects that a substance (or a mixture of substances) can exert on living organism, mutagenicity is one of most critical and complex. Independently to its final phenotypic effect, a mutation consists in a permanent change in the amount or structure of genetic material of a cell or organism, involving either a single gene or whole chromosome. Decades of research have shown that mutagenesis is a critical component and a high correlation between carcinogenicity and mutagenicity exists (Claxton *et al.* 2010).

Several assays have been developed, standardized and improved to be applied on complex matrixes like wastewater. They can be classified by the biological system employed and the genetic endpoint detected. Among the most common, the following can be listed:

- the Ames test. It is applied to detect point mutations in *Salmonella typhimurium* bacterial strains and it is specifically designed to detect chemically induced mutagenesis;
- the *Allium cepa* micronuclei test. It is adopted to detect the chromosomal mutation in root cells of common onion bulbs and it is considered as a biomarker of chromosomic damage and genome instability;
- the Comet assay (single cell gel electrophoresis). It allows to measure the primary DNA damage on eukaryotic cells (typically human leukocytes) in form of single/double strand breakages. It is used as an indicator of interaction of tested matrix and DNA, thus mutations can be induced.

We applied this integrated set of assay to rate the mutagenic strength of different WWTPs effluents, both liquid and solid. As an example, we pointed out that a tertiary ozonation system was not able to further reduce the mutagenic activity of wastewater, compared to the activated sludge system (detailed results in Bertanza *et al.* 2013).

24.6 RISK ASSESSMENT OF MICROPOLLUTANTS: THE MOST CRITICAL COMPOUNDS

In recent years, attention has been paid to the environmental risk posed by the occurrence of residues of MPs in water environments as well as in soil. Quite often, environmental risk assessments have been carried out by means of the risk quotient approach (RQ), which consists of evaluating the ratio between the concentrations (MEC or PEC) of the compound of interest measured in the specific water compartment, in sludge or in sludge-amended soil, and its corresponding predicted-no-effect concentration PNEC (for water, sludge and soil) (EMEA, 2001) as remarked in section 24.5. A commonly used ranking criterion was applied, according to Hernando *et al.* (2006): $RQ < 0.1$ low risk to aquatic/terrestrial organisms, $0.1 \leq RQ \leq 1$ medium risk; $RQ \geq 1$ high risk. In this risk assessment average measured concentrations are assumed while in other studies the maximum concentrations have been preferred to take the worst situation into consideration.

PNEC values generally refer to acute toxicity and consider neither the effect of a mixture of micropollutants occurring in the same environment, nor the chronic effect. Values adopted in the present risk assessment are those reported in Table 24.6.

The main results of the environmental risk assessments are reported in Figures 24.3 and 24.4. Referring to the selected compounds, those posing a high risk in all three of the investigated water compartments are the antiseptic triclosan, the hormones ethinylestraiol, estrone and estradiol, and the antibiotics ofloxacin, amoxicillin, sulphamethoxazole and clarithromycin. In secondary effluents there is a wider group of compounds with $RQ > 1$ compared to HWW and SW. Once released into the receiving water, the dilution effect may reduce the concentration of PhCs, resulting in a risk reduction. Other sources may introduce micropollutants into surface water (e.g. land runoff, unauthorized releases, untreated wastewater, etc.) causing an increment in the concentrations of specific contaminants. These could explain the presence of contaminants exhibiting a high risk in SW but not in WWTP effluents (e.g. erythromycin and acetaminophen).

Table 24.6 PNEC values for the different matrices. Values are taken from Verlicchi *et al.* (2012b) (PNEC water) and Verlicchi and Zambello, 2015 (PNEC sludge and soil). Further references are reported at the bottom of the table.

Compound	Predicted no Effect Concentration PNEC		
	Water, [μg/L]	Digested Sludge, [ng/g]	Soil, [ng/g]
Acetaminophen	1	19	
Diclofenac	9.7	820	0.013
Ibuprofen	1.65	48.5	0.730
Naproxen	2.62	51.7	
Salicylic acid	1.28	29.4	
Amoxicillin	0.0037		
Azithromycin	0.15	55.05	
Clarithromycin	0.07	50.05	
Erythromycin	0.02	3.8	0.0041
Mefenamic acid	0.43[a]		
Metronidazole	2.5		
Ofloxacin	0.016	282	
Roxithromycin	4	194	
Sulfamethoxazole	0.027	0.44	0.025
Propranolol	0.244	80.7	
Clofibric acid	40.2	0.92* ()	
Fenofibrate	0.1	366*	
Fenofibric acid	7.7	1.36*	
Gemfibrozil	0.9	0.02835*	
Carbamazepine	13.8	201	
Diazepam	2		
Fluoxetine	0.05		
Estradiol	0.002[b]		0.061
Estriol	0.06[b]	2421	0.050
Estrone	0.006[b]		
Ethinylestradiol	0.0001[b]	690	44.0
Triclosan	0.01[c]	2200*	0.096–2.1
Triclocarban	0.39[c]		
Iopromide	3.7×10^5	2553000*	

[a]Luo *et al.* (2014).
[b]Al Aukidy *et al.* (2014).
[c]Zhu and Cheng (2013).

With regard to digested sludge (Figure 24.4), a high environmental risk is posed by antibiotics (sulfamethoxazole, erythromycin, roxithromycin, azithromycin, and ofloxacin), hormones (E1, E2, and EE2), analgesics and anti-inflammatories (acetaminophen, ibuprofen, naproxen and salicylic acid) and the beta-blocker propranolol.

According to the few studies available, the highest risk posed by the application of digested sludge on land seems to be caused by the hormone ethinylestradiol, the antibiotics ofloxacin and tetracycline, and the antiseptics triclosan and triclocarban (Table 24.7).

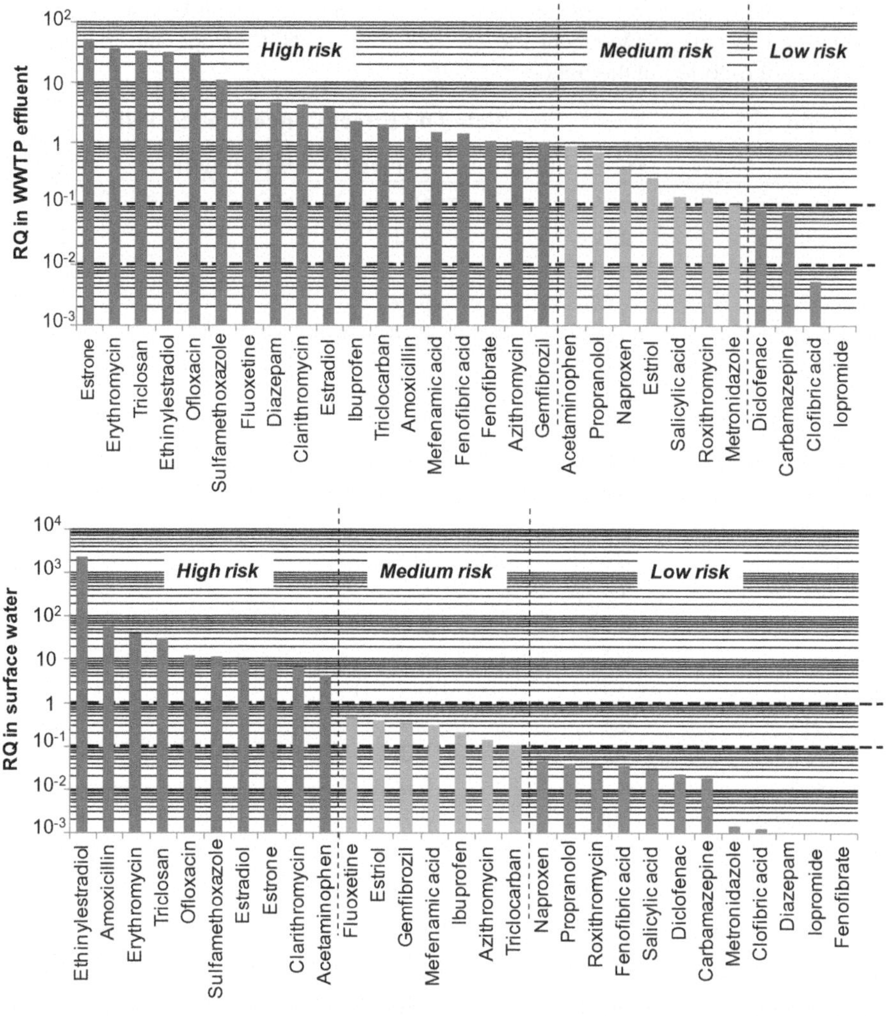

Figure 24.3 Risk quotient WWTP effluent and surface water with regard to the selected compounds.

With regard to groundwater, a MP contamination may pose a serious risk for decades to come due to expected long residence times and persistence due to relatively scarce microbial degradation and reducing chemistry.

Finally MP contamination of groundwater may pose a serious environmental risk for decades to come to the aquifer due to expected long residence times and persistence due to relatively scarce microbial degradation and reducing chemistry.

24.7 FINAL REMARKS AND CONCLUSIONS

A wide spectrum of PhCs occurs in the different environmental compartments. Monitoring campaigns carried out in the last few decades have documented the presence of many compounds in different water environments and soil, and analyzed the fate and behavior of some common substances during wastewater treatments and after the release of the treated effluent into the environment, mainly surface water. Increasing attention has been paid in recent years to the ecotoxicological effects of the residues of single compounds (their metabolites or transformation products) in water and soil and attempts are being made to evaluate the acute and chronic effects of a mixture of micropollutants occurring in water and soils, on the environment.

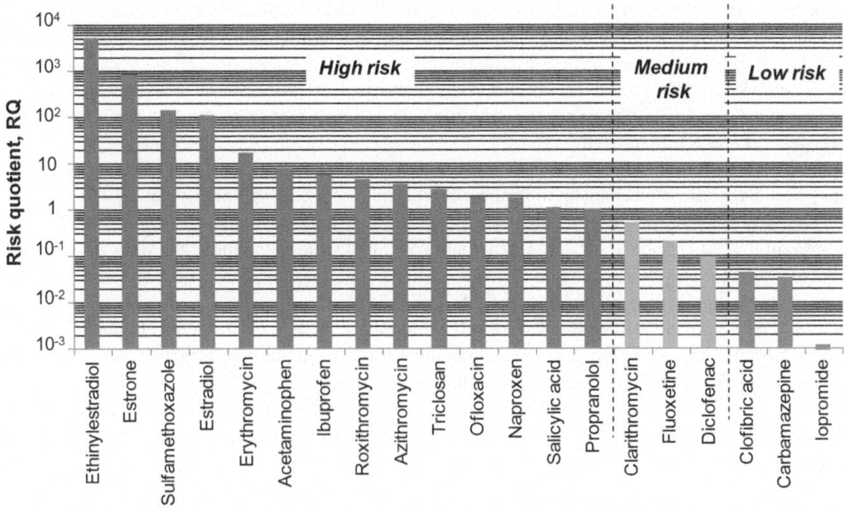

Figure 24.4 Risk quotient for the selected compounds in digested sludge.

Table 24.7 Risk quotient for sludge-amended soil.

References	RQ ≥ 1	0.1 < RQ < 1	RQ ≤ 0.1
Martin *et al.* 2012	Estradiol	Ethynilestradiol	Ibuprofen
McClellan and Halden, 2010	Ofloxacin, tetracycline, triclosan, triclocarban		

To date, these compounds are still unregulated, but at EU-level a watch list has been created, including compounds in the running for becoming priority substances and thus the subjects of regular monitoring in the environment. For them, monitoring campaigns and ecotoxicological studies are necessary to provide all the information required in order to make an informed decision regarding their inclusion in, or exclusion from the list of priority compounds.

Note that this watch list is a "dynamic" list, as new compounds may be included if there is a suspicion that they could cause serious negative effects on the environment.

In this context, decision-supporting systems require the development and validation of new analytical methods able to detect specific *new* substances, and ecotoxicological investigations able to evaluate the potential impact of the occurrence of such compounds and their mixture on the different environmental compartments (water and soil). Biological assays are fundamental for the correct evaluation of the "mixture/cocktail" effect and a multi-tiered approach is recommended for a holistic evaluation of WWTP performances and assessment of the effects of discharged effluents on ecosystem quality and human health. Wastewater can be considered a puzzle, in which many pieces should be combined in order to obtain meaningful information: a combination of chemical (to target the compounds responsible for) and biological assays (to measure the effects) represents a winning tool to properly rate treatment technologies.

24.8 REFERENCES

Al Aukidy M., Verlicchi P. and Voulvoulis N. (2014). A framework for the assessment of the environmental risk posed by pharmaceuticals originating from hospital effluents. *Science of the Total Environment*, **493**, 54–64.

Andreozzi R., Raffaele M. and Nicklas P. (2003). Pharmaceuticals in STP effluents and their photodegradation in aquatic environment. *Chemosphere,* **50**, 1319–1330.

Backhaus T., Porsbring T., Arrhenius A., Brosche S., Johansson P. and Blanck H. (2011). Single substance and mixture toxicity of five pharmaceuticals and personal care products to marine periphyton communities. *Environmental Toxicology and Chemistry,* **30**, 2030–2040.

Baker D. R. and Kasprzyk–Hordern B. (2011). Multi-residue analysis of drugs of abuse in wastewater and surface water by solid-phase extraction and liquid chromatography-positive electrospray ionisation tandem mass spectrometry. *Journal of Chromatography A*, **1218**(21), 1620–1631.

Bamdad M., Brousseau P. and Denizeau F. (1999). Identification of a multidrug resistance-like system in Tetrahymena pyriformis: evidence for a new detoxication mechanism in freshwater ciliates. *FEBS Letters*, **456**(3), 389–393.

Bard S. M. (2000). Multixenobiotic resistance as a cellular defense mechanism in aquatic organisms, *Aquatic Toxicology*, **48**(4), 357–389.

Bertanza G., Pedrazzani R., Dal Grande M., Papa M., Zambarda V., Montani C., Steimberg N., Mazzoleni G. and Di Lorenzo D. (2011). Effect of biological and chemical oxidation on the removal of estrogenic compounds (NP and BPA) from wastewater: an integrated assessment procedure. *Water Research*, **45**, 2473–2484.

Bertanza G., Papa M., Pedrazzani R., Repice C., Mazzoleni G., Steimberg N., Feretti D., Ceretti E. and Zerbini I. (2013). EDCs, estrogenicity and genotoxicity reduction in a mixed (domestic + textile) secondary effluent by means of ozonation: a full-scale experience. *Science of the Total Environment*, **458–460**, 160–168.

Bicchi C., Schiliro T., Pignata C., Fea E., Cordero C., Canale F. and Gilli G. (2009). Analysis of environmental endocrine disrupting chemicals using the E-screen method and stir bar sorptive extraction in wastewater treatment plant effluents. *Science of the Total Environment*, **407**(6), 1842–1851.

Bound J. P. and Voulvoulis N. (2005). Household disposal of pharmaceuticals as a pathway for aquatic contamination in the United Kingdom, *Environmental Health Perspectives,* **113**(12), 1705–1711.

Brooks B. W., Foran C. M., Richards S. M., Weston J., Turner P. K., Stanley J. K., Solomon M. Slattery K. R. and La Point T. W. (2003a). Aquatic ecotoxicology of fluoxetine, *Toxicology Letters,* **142**, 169–183.

Brooks B. W., Turner P. K., Stanley J. K., Weston J. J., Glidewell E. A., Foran C. M., Slattery M., La Point T. W. and Huggett D. B. (2003b). Waterborne and sediment toxicity of fluoxetine to select organisms. *Chemosphere*, **52**(1), 135–142.

Calamari D., Zuccato E., Castiglioni S., Bagnati R. and Fanelli R. (2003). Strategic survey of therapeutic drugs in the rivers po and Lambro in northern Italy. *Environmental Science & Technology*, **37**(7), 1241–8.

Carlsson C., Johansson A. K., Alvan G., Bergman K. and Kuhler T. (2006). Are pharmaceuticals potent environmental pollutants? Part I: environmental risk assessments of selected active pharmaceutical ingredients. *Science of the Total Environment*, **364**(1/3), 67–87.

Christensen F. M. (1998). Pharmaceuticals in the Environment—A Human Risk? *Regulatory Toxicology and Pharmacology*, **28**(3), 212–221.

Claxton L. D., Umbuzeiro G. D. and DeMarini D. M. (2010). The Salmonella Mutagenicity Assay: the Stethoscope of Genetic Toxicology for the 21st Century. *Environmental Health Perspectives*, **118**(11), 1515–1522.

Council Directive 91/271/EEC of 21 May 1991 concerning urban waste-water treatment.

Daughton C. G. and Ternes T. A. (1999). Pharmaceuticals and personal care products in the environment: agents of subtle change? *Environmental Health Perspectives,* **107**(6), 907–938.

Commission Implementing Decision (EU) 2015/495 of 20 March 2015 establishing a watch list of substances for Union-wide monitoring in the field of water policy pursuant to Directive 2008/105/EC of the European Parliament and of the Council, OJ L 78, 24.3.2015, pp. 40–42.

Directive 2000/60/EC of the European Parliament and of the Council of 23 October 2000 establishing a framework for Community action in the field of water policy, OJ L 327, 22.12.2000, pp. 1–73.

Directive 2008/105/EC of the European Parliament and of the Council on environmental quality standards in the field of water policy, amending and subsequently repealing Council Directives 82/176/EEC, 83/513/EEC, 84/156/EEC, 84/491/EEC, 86/280/EEC and amending Directive 2000/60/EC of the European Parliament and of the Council.

Directive 2013/39/EU of the European Parliament and of the Council of 12 August 2013 amending Directives 2000/60/EC and 2008/105/EC as regards priority substances in the field of water policy, OJ L 226, 24.8.2013, pp. 1–17.

Doucette W. J. (2000). Soil and sediment sorption coefficient. In: Handbook of Property Estimation Methods for Chemicals: Environmental and Health Sciences, Robert S. Boethling and D. Mackay (eds), CRC Press, Florida, pp. 141–188.

Wishart D. S., Knox C., Guo A. C., Shrivastava S., Hassanali M., Stothard P., Chang Z. and Woolsey J. (2000). DrugBank: a comprehensive resource for in silico drug discovery and exploration. Nucleic Acids Res. 2006 Jan. 1;34 (Database issue):D668-72. 16381955. See http://www.drugbank.ca (access February 2017).

Eichhorn P., Pérez S. and Barceló D. (2012). Time-of-flight mass spectrometry versus orbitrap-based mass spectrometry for the screening and identification of drugs and metabolites: is there a winner? In: Rodriguez Fernandez-Alba, A. (ed.), *TOF-MS within Food and Environmental Analysis*, Elsevier, Comprehensive Analytical Chemistry, **58**, 217–272.

EMEA, (2001). European Agency for the Evaluation of Medicinal Products. Draft Discussion Paper on Environmental Risk Assessment of Non-Genetically Modified Organism (non-GMO) Containing Medicinal Products for Human Use; London.

EMEA (2006). "Guideline on the environmental risk assessment of medicinal products for human use", Committee for Medicinal Products for Human Use.

EPI Suite™, Estimation Program Interface (EPI Suite™) Program Modifications & New Features in v4.11 (November 2012). See https://www.epa.gov/tsca-screening-tools/estimation-program-interface-epi-suite-tm-program-modifications-new-features (accessed February 2017).

Escher B. and Leusch F. (2012). Bioanalytical Tools in Water Quality Assessment. IWA Publishing, London, UK.

Farré M., Sandra Pérez S., Kantiani L. and Barcelo D. (2008). Fate and toxicity of emerging pollutants, their metabolites and transformation products in the aquatic environment. *Trends in Analytical Chemistry*, **27**(11), 991–1007.

FASS (2008). Swedish environmental classification of pharmaceuticals: The Swedish Association of the Pharmaceutical Industry, http://www.fass.se (accessed 25 September 2015).

Fatta-Kassinos D., Meric S. and Nikolaou A. (2011). Pharmaceutical residues in environmental waters and wastewater: current state of knowledge and future research. *Analytical and Bioanalytical Chemistry*, **399**(1), 251–275.c

Fent K., Escher C. and Caminada D. (2006). Estrogenic activity of pharmaceuticals and pharmaceutical mixtures in a yeast reporter gene system, *Reproductive Toxicology*, **22**(2), 175–185.

Ferrari F., Gallipoli A., Balderacchi M., Ulaszewska M. M., Capri E. and Trevisan M. (2011). Exposure of the main Italian river basin to pharmaceuticals. *Journal of Toxicology*, 989270.

Flaherty C. M. and Dodson C. I. (2005). Effects of pharmaceuticals on Daphnia survival, growth and reproduction. *Chemosphere*, **61**, 200–207.

Forni C., Cascone A., Fiori M. and Migliore L. (2002). Sulphadimethoxine and Azolla filiculoides Lam.: a model for drug remediation. *Water Research*, **36**(13), 3398–3403.

Gobel A., McArdell C. S., Suter M. J. F., Giger W. (2004). Trace determination of macrolide and sulfonamide antimicrobials, a human sulfonamide metabolite, and trimethoprim in wastewater using liquid chromatography coupled to electrospray tandem mass spectrometry. *Analytical Chemistry*, **76**(16), 4756–4764.

Gros M., Rodríguez-Mozaz S. and Barceló D. (2012). Fast and comprehensive multi-residue analysis of a broad range of human and veterinary pharmaceuticals and some of their metabolites in surface and treated waters by ultra-high-performance liquid chromatography coupled to quadrupole-linear ion trap tandem mass spectrometry. *Journal of Chromatography A*, **1248**, 104–121.

Haider S. and Baqri S. S. R. (2000). beta-Adrenoceptor antagonists reinitiate meiotic maturation in Clarias batrachus oocytes. *Comp. Biochem. Phys. A: Molecular & Integrative Physiology*, **126**, 517–525.

Hemond H. (2000). Chemical Fate and Transport in the Environment. Elsevier, San Diego, 156–157.

Hernando M. D., Mezcua M., Fernandez-Alba A. R. and Barceló D. (2006). Environmental risk assessment of pharmaceutical residues in wastewater effluents, surface waters and sediments. *Talanta*, **69**, 334–42.

Hernando M. D., Agüera A. and Fernández-Alba A. R. (2007). LC-MS analysis and environmental risk of lipid regulators. *Analytical and Bioanalytical Chemistry*, **387**(4), 1269–1285.

Huggett D. B., Brooks B. W., Peterson B., Foran C. M. and Schlenk D. (2002). Toxicity of select beta adrenergic receptor-blocking pharmaceuticals (B-blockers) on aquatic organisms. *Archives of Environmental Contamination and Toxicology*, **43**(2), 229–235.

Johnston J. J., Savarie P. J., Primus T. M., Eisemann J. D., Hurley J. C. and Kohler D. J. (2002). Risk assessment of an acetaminophen baiting program for chemical control of brown tree snakes on Guam: evaluation of baits, snake residues, and potential primary and secondary hazards, *Environmental Science Technologies*, **36**(17), 3827–3833.

Jurado A., Vàzquez-Suñé E., Carrera J., López de Alda M., Pujades E. and Barceló D. (2012). Emerging organic contaminants in groundwater in Spain: a review of sources, recent occurrence and fate in a European context. *Science of the Total Environment*, **440**, 82–94.

Kolpin D. W., Furlong E. T., Meyer M. T., Thurman E. M., Zaugg S. D., Barber L. B. and Buxton H. T. (2002). Pharmaceuticals, hormones, and other organic wastewater contaminants in U.S. *streams, 1999–2000: a national reconnaissance. Environmental Science & Technology*, **36**(6), 1202–1211.

Krauss M., Singer H. and Hollender J. (2010). LC-high resolution MS in environmental analysis: from target screening to the identification of unknowns. *Analytical and Bioanalytical Chemistry*, **397**(3), 943–951.

Kumar K. S., Priya S. M., Peck A. M. and Sajwan K. S. (2010). Mass loadings of triclosan and triclocarbon from four wastewater treatment plants to three rivers and landfill in Savannah, Georgia, USA. *Archives of Environmental Contamination and Toxicology*, **58**(2), 275–285.

Kümmerer K. (2009). Antibiotics in the aquatic environment – A review – Part I. *Chemosphere*, **75**, 417–434.

Kunkel U. and Radke M. (2012). Fate of pharmaceuticals in rivers: deriving a benchmark dataset at favourable attenuation conditions. *Water Research*, **46**, 5551–5565.

Kurelec B. (1992). The multixenobiotic resistance mechanism in aquatic organisms. *Critical Reviews in Toxicology*, **22**(1), 23–43.

Kuroda K., Murakami M., Oguma K., Muramatsu Y., Takada H. and Takizawa S. (2012). Assessment of groundwater pollution in tokyo using PPCPs as sewage markers. *Environmental Science & Technology*, **46**(3), 1455–1464.

Landis W. G., Sofield R. M. and Yu M. H. (2011). Introduction to Environmental Toxicology: Molecular Structures to Ecological Landscapes. 4th edn. CRC Press, Florida, 117–162.

Le-Minh N., Khan S. J., Drewes J. E. and Stuetz R. M. (2010). Fate of antibiotics during municipal water recycling treatment processes. *Water Research*, **44**, 4295–4323.

Loos R., Wollgast J., Huber T. and Hanke G. (2007). Polar herbicides, pharmaceutical products, perfluorooctanesulfonate (PFOS), perfluorooctanoate (PFOA), and nonylphenol and its carboxylates and ethoxylates in surface and tap waters around lake maggiore in northern italy. *Analytical and Bioanalytical Chemistry*, **387**(4), 1469–78.

Loos R., Locoro G., Comero S., Contini S., Schwesig D., Werres F., Balsaa P., Gans O., Weiss S., Blaha L., Bolchi M. and Gawlik B. M. (2010). Pan-European survey on the occurrence of selected polar organic persistent pollutants in ground water. *Water Research*, **44**(14), 4115–26.

López-Serna R., Jurado A., Vázquez-Suñé E., Carrera J., Petrović M. and Barceló D. (2013). Occurrence of 95 pharmaceuticals and transformation products in urban groundwaters underlying the metropolis of Barcelona, Spain. *Environmental Pollution*, **174**, 305–15.

Lundholm C. D. (1997). DDE-induced eggshell thinning in birds. *Comparative Biochemistry and Physiology*, **118**, 113–128.

Luo Y., Guo W., Ngo H. H., Nghiem L. D., Hai F. I., Zhang J., Liang S. and Wang X. C. (2014). A review on the occurrence of micropollutants in the aquatic environment and their fate and removal during wastewater treatment. *Science of the Total Environment*, **473–474**, 619–641.

Macova M., Escher B., Reungoat J., Carswell S., Chue K. L., Keller J. and Mueller J. F. (2010). Monitoring the biological activity of micropollutants during advanced wastewater treatment with ozonation and activated carbon filtration. *Water Research*, **44**(2), 477–492.

Maeng S-K., Sharma S. K., Lekkerkerker-Teunissen K. and Amy G. L. (2011). Occurrence and fate of bulk organic matter and pharmaceutically active compounds in managed aquifer recharge: a review. *Water Research*, **45**, 3015–3033.

Martin J., Camacho-Muñoz D., Santos J.L., Aparicio I. and Alonso E. (2012). Occurrence of pharmaceutical compounds in wastewater and sludge from wastewater treatment plants: removal and ecotoxicological impact of wastewater discharges and sludge disposal. *Journal of Hazardous Materials*, **239–240**, 40–47.

Masiá A., Ibáñez M., Blasco C., Sancho J. V., Picó Y. and Hernández F. (2013). Combined use of liquid chromatography triple quadrupole mass spectrometry and liquid chromatography quadrupole time–of–flight mass spectrometry in systematic screening of pesticides and other contaminants in water samples. *Analytica Chimica Acta*, **761**, 117–127.

McClellan K. and Halden R. U. (2010). Pharmaceuticals and personal care products in archived U.S. *biosolids from the 2001 EPA national sewage sludge survey. Water Research*, **44**, 658–668.

McFadzen I., Eufemia N., Heath C., Epel D., Moore M. N. and Lowe D. M. (2000). Multidrug resistance in the embryos and larvae of the mussel Mytilus edulis. *Marine Environmental Research*, **50**(1–5), 319–323.

Meffe R. and de Bustamante I. (2014). Emerging organic contaminants in surface water and groundwater: a first overview of the situation in Italy. *Science of the Total Environment*, **481**(1), 280–295.

Migliore L., Civitareale C., Brambilla G. and di Delupis G. D. (1997). Toxicity of several important agricultural antibiotics to Artemia. *Water Research*, **31**(7), 1801–1808.

Migliore L., Cozzolino S. and Fiori M. (2003). Phytotoxicity to and uptake of enrofloxacin in crop plants. *Chemosphere*, **52**(7), 1233–1244.

Miller M. R., Wentz E. and Ong S. (1999). Acetaminophen alters estrogenic responses in vitro: inhibition of estrogen-dependent vitellogenin production in trout liver cells. *Toxicological Sciences*, **48**, 30–37.

Nowara A., Burhenne J. and Spiteller M. (1997). Binding of fluoroquinolone carboxylic acid derivatives to clay minerals. *Journal of Agricultural and Food Chemistry*, **45**(4), 1459–1463.

Pal A., He Y., Jekel M., Reinhard M. and Gin K. Y. H. (2014). Emerging contaminants of public health significance as water quality indicator compounds in the urban water cycle. *Environment International*, **71**, 46–62.

Papa M., Pedrazzani R. and Bertanza G. (2013). How green are environmental technologies? A new approach for a global evaluation: the case of WWTP effluents ozonation. *Water Research*, **47**(11), 3679–3687.

Pascoe D., Karntanut W. and Muller C. T. (2003). Do pharmaceuticals affect freshwater invertebrates? A study with the cnidarian Hydra vulgaris. *Chemoshere*, **51**(6), 521–528.

Petrovic M., Farré M., Eljarrat E., Díaz-Cruz M. S. and Barceló D. (2013). Environmental analysis: emerging pollutants. In: *Liquid Chromatography: Applications*, Fanali, S. Haddad, P. R. Poole, C. Schoenmakers, P. and Lloyd, D. (eds.), Elsevier, Massachusetts, USA, pp. 389–410.

Routledge E. J., Sheahan D., Desbrow C., Brighty G.C., Waldock M. and Sumpter J. P. (1998). Identification of estrogenic chemicals in STW effluent. 2. In vivo responses in trout and roach. *Environmental Science & Technology*, **32**(11), 1559–1565.

Santos L. H. M. L. M., Gros M., Rodriguez-Mozaz S., Delerue-Matos C., Pena A., Barceló D. and Montenegro M. C. B. S. M. (2013). Contribution of hospital effluents to the load of pharmaceuticals in urban wastewaters: identification of ecologically relevant pharmaceuticals. *Science of the Total Environment*, **461–462**, 302–316.

Schowanek D. and Webb S. (2002). Exposure simulation for pharmaceuticals in European surface waters with GREAT-ER. *Toxicology Letters*, **131**(1–2), 39–50.

Schwab B. W., Hayes E. P., Fiori J. M., Mastrocco F. J., Roden N. M., Cragin D., Meyerhoff R. D., D'Aco V. J. and Anderson P. D. (2005). Human pharmaceuticals in US surface waters: a human health risk assessment. *Regulatory Toxicology and Pharmacology*, **42**(3), 296–312.

Stamm C., Alder A. C., Fenner K., Hollender J., Krauss M., McArdell C. S., Ort C. and Schneider M. K. (2008). Spatial and temporal patterns of pharmaceuticals in the aquatic environment: a review. *Geography Compass*, **2**(3), 920–955.

Stuart M., Lapworth D., Crane E. and Hart A. (2012). Review of risk from potential emerging contaminants in UK groundwater. *Science of the Total Environment*, **416**, 1–21.

Stuer-Lauridsen F., Birkved M., Hansen L. P., Lützhøft H. C. and Halling-Sørensen B. (2000). Environmental risk assessment of human pharmaceuticals in Denmark after normal therapeutic use. *Chemosphere*, **40**(7), 783–793.

Tolls J. (2001). Sorption of veterinary pharmaceuticals in soils: a review. *Environmental Science & Technology*, **35**, 3397–3406.

Tong L., Eichhorn P., Pérez S., Wang Y. and Barceló D. (2011). Photodegradation of azithromycin in various aqueous systems under simulated and natural solar radiation: kinetics and identification of photoproducts. *Chemosphere*, **83**, 340–348.

U.S. National Library of Medicine, TOXNET Toxicology Data Network. See https://toxnet.nlm.nih.gov/ (accessed February 2017).

Tyler C. R., Jobling S. and Sumpter J. P. (1998). Endocrine disruption in wildlife: a critical review of the evidence. *Critical Reviews in Toxicology*, **28**(4), 319–361.

UNEP/IPCS (1999). Chemical Risk Assessment: Human Risk Assessment, Environmental Risk Assessment And Ecological Risk Assessment, Training Module No. 3, United Nations Environment Programme/International Programme On Chemical Safety.

Verlicchi P. and Zambello E. (2015). Pharmaceuticals and personal care products in untreated and treated sewage sludge: occurrence and environmental risk in the case of application on soil - A critical review. *Science of the Total Environment*, **538**, 750–67.

Verlicchi P., Galletti A., Petrovic M. and Barcelò D. (2010). Hospital effluents as a source of emerging pollutants: an overview of micropollutants and sustainable treatment options. *Journal of Hydrology*, **389**(3–4), 416–428.

Verlicchi P., Al Aukidy M., Galletti A., Petrovic M. and Barceló D. (2012a). Hospital effluent: investigation of the concentrations and distribution of pharmaceuticals and environmental risk assessment. *Science of the Total Environment*, **430**, 109–118.

Verlicchi P., Al Aukidy M. and Zambello E. (2012b). Occurrence of pharmaceutical compounds in urban wastewater: removal, mass load and environmental risk after a secondary treatment-A review. *Science of the Total Environment*, **429**, 123–155.

Verlicchi P., Al Aukidy M., Jelic A., Petrović M. and Barceló D. (2014). Comparison of Measured and Predicted Concentrations of Selected Pharmaceuticals in Wastewater and Surface Water: a Case Study of a Catchment Area in the Po Valley (Italy). *Science of the Total Environment*, **470–471**, 844–854.

Verlicchi P., Al Aukidy M. and Zambello E. (2015). What have we Learned from Worldwide Experiences on the Management and Treatment of Hospital Effluent? – an Overview and a Discussion on Perspectives. *Science of the Total Environment*, **514**, 467–491.

Webb S. F. (2001). A data based perspective on the environmental risk assessment of human pharmaceuticals II: aquatic risk characterisation. In: Pharmaceuticals in the Environment. Sources, Fate, Effects and Risks K. Kümmerer (ed.), Springer-Verlag, Berlin, pp. 319–343.

Witters H. E., Vangenechten C. and Berckmans P. (2001). Detection of estrogenic activity in Flemish surface waters using an in vitro recombinant assay with yeast cells. *Water Science and Technology*, **43**(2), 117–123.

Ying G. G., Zhao J. L., Zhou L. J. and Liu S. (2013). Fate and Occurrence of Pharmaceuticals in the Aquatic Environment (Surface Water and Sediment). In: *Analysis, Removal, Effects and Risk of Pharmaceuticals in the Water Cycle Occurrence and Transformation in the Environment Comprehensive Analytical Chemistry*, Elsevier, Poland, 453–557.

Zhu S. and Cheng H. (2013). The fate and risk of selected pharmaceutical and personal care products in wastewater treatment plants and a pilot-scale multistage constructed wetland system. *Environmental Science and Pollution Research*, **21**(2), 1466–1479.

Zrnčić M., Babić S. and Mutavdžić Pavlović D. (2015). Determination of thermodynamic pKa values of pharmaceuticals from five different groups using capillary electrophoresis. *Journal of Separation Science*, **38**, 1232–1239.

Zuccato E., Calamari D., Natangelo M. and Fanelli R. (2000). Presence of therapeutic drugs in the environment. *Lancet*, **355**(9217), 1789–1790.

Chapter 25

Legal and policy frameworks for the management of wastewater

Sarah Hendry and Jamie Benidickson

25.1 INTRODUCTION

Legal frameworks for the management of wastewater involve, on the one hand, the development and enforcement of detailed and often technical rules relating to environmental protection, public health and hazardous substances. On the other hand, wastewater management can be seen as a way of closing loops (in the water cycle, for nutrient management or as a source of energy) and contributing to a wide range of global and local policy goals around resource management and sustainable development. This chapter will analyse the former, in the context of the latter, and beginning with that policy context.

At the time of writing, the United Nations has just met to agree the Sustainable Development Goals (SDGs, UN, 2015). Goal 6 is to 'Ensure availability and sustainable management of water and sanitation for all'. It is unsurprising that the 'water goal' prioritises water and sanitation services; the management of the water resource is built into the detailed targets, but, it is the availability of safe drinking water, and acceptable ways of disposing of human and industrial waste, that has kept water at the top of the global policy agenda. From perspectives of both sustainable human development and environmental protection, we need to manage wastewater better. It is estimated that in developing countries, as much as 90% of sewage and 70% of industrial effluent wastes are discharged without treatment (UNEP, 2010). Despite the Millennium Development Goals that preceded the new SDGs, UNICEF/WHO (2015) estimate some 2.4 billion people still lack an acceptable sanitation service, Further, sanitation, unlike drinking water, is not seen as a vote-winner but is still a taboo subject, which politicians prefer not to discuss. Yet lack of adequate sanitation affects life, health and human dignity; an estimated 10% of the total global burden of disease could be prevented by access to improved water, sanitation and hygiene (Prüss-Üstün *et al.* 2008). Meantime, failure to close loops on the water cycle is resource mismanagement that we can no longer afford.

Currently, the global population stands at an estimated 7.2 billion, predicted to rise to 9.6 billion by 2050 (UNDESA, 2012); by 2030, we will need 30% more water, 50% more food and 50% more energy – and water is a major input to both food and energy. Wastewater is a vital resource; developed countries may treat wastewater to a very high level, but then discharge out to the marine environment rather than returning directly to the freshwater cycle. Essential nutrients in sludge, or in untreated wastewater, are treated as a solid waste problem; alternatively, untreated wastewater may be used for irrigation or aquaculture without proper safeguards for human health (WHO, 2006). Further, climate change will exacerbate other environmental and social pressures; water and wastewater management will be affected by both mitigation and adaptation measures (UN-Water, 2010).

Finally, in the policy context, is the question of the human right to sanitation. That is less widely accepted by states than the right to water; but the UN General Assembly, the Human Rights Commission and the Special Rapporteur have all taken the view that these rights are indivisible (see e.g., de Albuquerque, 2013). Although it will not be possible to explore the human rights dimension, it is important to note it as a possible emerging customary right.

25.1.1 Structures for ownership and regulation

Water services can be delivered in a wide variety of structures for both ownership and regulation (Hendry, 2015). In many countries, water services (including drinking water, wastewater and sanitation services) are a municipal function, and this may be specified under the constitution. In some countries there may be some private sector participation, which may be through a variety of contractual arrangements including concessions or leases of the whole system, with various allocations of risk; by Build-Own-Operate, Build-Own-Operate-Transfer, or some other variant, especially for treatment plant; or by short-term contracts, including for management advice (World Bank/PPIAF, 2006; Delmon, 2001). It is usual for the assets to be and remain in public ownership, although they were divested in England and Wales. Water and wastewater services may be managed together, or separately; if separately, then it is more likely that wastewater and sanitation will remain with a public authority. In rural areas, it is very likely that there will be private or community-owned systems which bring particular management problems (section 3 below). Finally, as wastewater is both a potential source of water and a potential source of nutrients and energy (thermal or biogas), there may be competition for it as a resource, and indeed wastewater management may be a more commercially attractive activity than water supply, but is not necessarily supported by adequate health or environmental safeguards.

25.1.2 Regulation and liability

Regulation of water services generally is attracting increasing interest (OECD, 2015). Networked utilities will be natural monopolies; economic theory would see regulation as a substitute for a competitive market. Both public and private sectors can provide a better service where there is a functioning system for economic regulation (OECD, 2011). Prices of services must allow for maintenance and investment in the system, and also provide for meeting relevant technical standards and other environmental and social requirements. The former would include abstraction of raw water and discharges of wastewater, whether treated or not; the latter would include protection for those who cannot pay the full cost, usually by some cross-subsidy or subsidy from general taxation. The need for investment in water services, and especially wastewater and sanitation, is recognised in both the developed and the developing world. The World Bank estimates that developing countries alone need to spend $103 billion per year on water services (Rodriguez *et al.* 2012).

The regulation of all dimensions of wastewater services is a public law matter, and may be carried out by a Government department, municipality or an independent utilities regulator. Often overall responsibility for water is split, for example between departments managing public health, public works, and the environment; though in some countries all aspects of water are managed together (for example in South Africa, with the recent integration of the Department of Water and Sanitation, covering water resources management, drinking water supply and sanitation). However there may also be private law dimensions, around liability for nuisance, negligence or environmental harm, such as flooding from sewerage systems or failure to meet treatment standards before discharge. In the UK, in *Marcic v Thames Water,* the final court of appeal held that decisions on investments were a matter for public policy within the state's discretion, and rejected the view of the courts below that Mr Marcic's human rights (to property and family life) had been breached by failure to upgrade the sewerage system, causing regular sewage flooding. More importantly in some ways, they held that there was no action in nuisance or negligence, as his remedy was to approach the regulator and seek intervention. In an era of unprecedented global change, including climate change, and the widespread acceptance of the theory of the precautionary principle, liability standards for negligence actions between private parties, as well as the liability of the state, may shift.

Whilst this chapter will not be able to explore these issues in depth, it will seek to recognise both the challenges of the failure to manage wastewater and sanitation effectively (not just in the developing world) as well as the opportunities brought by a more innovative and holistic approach to the resource. In outline, the chapter will review the regulation of wastewater treatment facilities at the municipal and industrial level (section 2); and in relation to on-site or contained treatment arrangements, including septic tanks and non-waterborne options (section 3). Handling, use and disposal of residual matter in the form of sludge or biosolids is then discussed (section 4) followed by wastewater re-use and resource recovery (section 5), and energy and climate change issues (section 6). We then turn to the challenges presented by emerging pollutants such as pharmaceuticals (section 7) before offering a brief conclusion (section 8).

25.2 REGULATION OF WASTEWATER TREATMENT FACILITIES

In the 19th century, urbanization, accompanied by the introduction of municipal water supply arrangements and rapid advances in the new science of epidemiology, stimulated regulatory oversight of wastewater treatment and discharge from the perspectives of environmental protection and public health (Benidickson, 2007).

Wastewater has been described by the United Nations Environment Programme (UNEP) and UN-Habitat as '*a combination of one or more of: domestic effluent consisting of blackwater (excreta, urine and faecal sludge) and greywater (kitchen and bathing wastewater); water from commercial establishments and institutions, including hospitals; industrial effluent, stormwater and other urban run-off; agricultural, horticultural and aquaculture effluent, either dissolved or as suspended matter*' (Corcorran *et al.* 2010).

Historically, and still today in many countries, if there is a networked system, wastewater of all types may be collected together and either discharged directly, or taken to a treatment facility. That facility may be regulated by the general environmental law, or by specific rules, or both. Alternatively, different streams of wastewater may be collected and managed separately, through dual (separate) systems, which should improve the opportunities for appropriate reuse. If there is no network then a variety of on-site technologies may be used for the management of human waste, which may or may not be waterborne. It is likely that industrial wastewater going through a public system will be subject to a system of trade effluent consents (possibly involving pre-treatment requirements) and charges (section 2.2.4). Stormwater in many countries is the responsibility of roads and transport authorities and therefore not properly integrated into the water cycle, although this is changing, for example by the use of sustainable urban drainage systems (see e.g., Government of Queensland, 2013), and/or by collecting and treating stormwater along with other wastewater streams (see e.g., Government of Australia, 2009; and section 5 below).

25.2.1 General environmental law

Adverse impacts associated with the discharge or release of environmental contaminants have encouraged regulatory intervention to control wastewater flows in a variety of forms, through general environmental law.

The regulatory framework may encompass approvals and inspections; prohibitions accompanied by penalties; regulated standards applicable to treatment technology or environmental conditions; and in some cases the use of economic instruments to facilitate pollution credit exchanges or to promote improved performance. Ancillary requirements may include reporting requirements, public disclosure, or financial mechanisms such as insurance. The level of development will be highly relevant to the specification of these rules.

The WHO recommends a 'step-wise' approach to managing water pollution and water quality especially in developing countries (Helmer & Hespanol, 1997). This would involve identifying the most likely contaminants in a country or local context and developing a set of realistic and enforceable parameters for these. Although not focused on human waste, faecal indicators and nutrients (from human settlements or agriculture) are always likely to be core parameters. In South Africa, for example, there is a set of basic emission values which would be applied to any discharge containing these.

25.2.2 Specific regulation of wastewater treatment

25.2.2.1 The European Union urban waste water treatment directive

The European Union (EU) has legislated extensively on water from the 1970's onwards, including emission standards for industry and quality standards for receiving waters, and drinking water quality (Kallis & Nijkamp, 2000). In the 1990s, it introduced a specific Directive on Urban Waste Water Treatment (UWWTD). More recently the Water Framework Directive has rationalised and brought together much of the EU's water laws, but the UWWTD continues in force as part of the suite of relevant rules.

The UWWTD has entailed significant capital investment by Member States. Essentially, it requires collection systems and treatment at various levels, for biodegradeable wastewater (human waste and also, for example, food waste). Collection systems are not required where this would be excessively costly or would produce no environmental benefit (Art. 3) but the same level of environmental protection must be provided (for example by individual systems).

There was a staged system of implementation beginning with the largest 'agglomerations' by population equivalent. Small communities (or industries) with population equivalent of less than 2000 do not need to implement the UWWTD but instead must have 'appropriate treatment' sufficient to meet all other relevant EU quality standards (Art. 7). Larger communities discharging to coastal or estuarine waters may be designated as 'less sensitive' and be subject to only 'primary treatment' (Art. 6).

The usual level of treatment is 'secondary treatment' as defined. Annex I sets standards for 3 parameters, biochemical oxygen demand, chemical oxygen demand and total suspended solids. Where the discharge is made into 'sensitive' waters (subject to eutrophication) additional 'more stringent' (tertiary) treatment may be required, such as ultraviolet treatment, and standards are also set for total phosphorous and total nitrogen. The Directive specifies sampling frequencies and related permitted failures. The UWWTD also prohibited the dumping of sewage sludge at sea by ships or pipes (Art. 14). Both sludge and treated wastewater should be reused 'wherever appropriate' (Arts. 14, 12).

The UWWTD has been a very controversial instrument. Its implementation took place at a time of polarised political debate around the liberalisation of water services, still ongoing today. In the UK for example, in England and Wales water services had been divested in 1989; but in Scotland, where the service remained in public hands, a new generation of wastewater treatment plant were built under the 'private finance initiative', using Build-Own-Operate schemes. It is now accepted that over the lifetime of these plant they have been significantly more expensive than if Government had met the capital costs, albeit not public debt, whilst the argument that the UWWTD 'requires' large end-of-pipe solutions in all cases is not really borne out by careful examination of its contents. Arguably, the implementation and the use of private capital gave preference to established and large-scale technology, rather than fostering innovation.

Nonetheless, implementation of the UWWTD remains challenging. Many enforcement actions have been brought by the Commission against the original 15 Member States (the EU-15). New members in Eastern Europe (the EU-12) are given extended deadlines relative to their date of accession, but find the requirements complex and burdensome. In the latest implementation report (European Commission, 2013), whilst almost all of the EU-15 states complied regarding collection systems (97% of agglomerations), not all had met the requirements for either secondary (88%) or enhanced treatment (90%). Some 20 'horizontal' enforcement actions were still underway (that is, grouped actions against several related breaches). For the EU-12, 72% of agglomerations have collection systems, but just 39% have complied regarding secondary treatment and 14% with more stringent treatment requirements. Thus even in the EU, securing compliance with a specific instrument has been economically and politically challenging.

25.2.2.2 The USA

In 1972 the Clean Water Act established the over-arching framework within the United States 'to restore and maintain the chemical, physical, and biological integrity of the Nation's waters.' To that end the 1972 legislation articulated several specific national goals or policies:

> *'(1) … that the discharge of pollutants into the navigable waters be eliminated by 1985; (2) … ; (3) … that the discharge of toxic pollutants in toxic amounts be prohibited; (4) … that Federal financial assistance be provided to construct publicly owned waste treatment works; (5) … that area-wide treatment management planning processes be developed and implemented to assure adequate control of sources of pollutants in each State; (6) … that a major research and demonstration effort be made to develop technology necessary to eliminate the discharge of pollutants into the navigable waters, waters of the contiguous zone and the oceans; and (7) … that programs for the control of nonpoint sources of pollution be developed and implemented …'* (s.101(a)).

Notwithstanding continuing challenges, especially in relation to nonpoint pollution, the Clean Water Act has contributed to marked improvement in municipal and industrial wastewater treatment and water quality (Andreen, 2004).

A permit-based system emphasizing technology-based effluent standards alongside monitoring and reporting requirements was initially implemented by officials of the Federal Environmental Protection Agency (EPA) with administrative responsibility for permits – still subject to federal requirements – now largely in the hands of state officials. The National Pollutant Discharge Elimination System permit programme is applicable to a range of municipal discharge sources including publicly owned treatment works which are expected to provide secondary treatment, as well as to sanitary sewer systems which are intended to transport sewage flows to treatment centres and combined sewer systems collecting runoff, domestic sewage and industrial wastewater. The latter are operated in nearly 800 municipalities across the country. Enforcement mechanisms, including EPA, state level, and citizen-initiated prosecutions, as well as administrative compliance measures and civil suits, are an important feature of the Clean Water Act framework for maintaining wastewater standards (Andreen, 2004; at 549–50).

Onsite or decentralized arrangements – typically septic treatment facilities – service nearly one quarter of US households and are, accordingly, a further focus of EPA attention from the perspective of wastewater management.

25.2.2.3 Canada

Municipal wastewater treatment systems in Canada, while operated as local government services, have historically fallen under the constitutional authority of provincial governments, which have exercised permitting and standard-setting authority over wastewater treatment facilities.

In the context of persistent criticism – especially of certain coastal municipalities – the Canadian Council of Ministers of the Environment (CCME) has more recently pursued initiatives to formulate a Canada-wide strategy for municipal wastewater effluents and biosolids. In connection with wastewater effluents, most jurisdictions agreed to specific commitments including risk level determinations, control of overflows from development, regulatory harmonization, bilateral agreements involving the federal government, and the creation of a national database on regulatory reporting. The strategy is subject to regular review at five year intervals (CCME, 2014).

New wastewater systems effluent regulations under Canada's federal Fisheries Act introduce mandatory minimum effluent quality standards that can be achieved through secondary wastewater treatment (s.6). In addition, monitoring, record-keeping, reporting and toxicity testing requirements are now applicable for 3,500 treatment facilities across the country. Approximately one in four of these facilities will require substantial upgrades – typically from primary to secondary treatment – as a consequence.

While definitions vary and sub-classifications are common, production, handling and use of biosolids, municipal sludge, and treated septage are subject to further regulation at all jurisdictional levels. Again, the CCME has developed guidance on sustainable management practices including composting and land application (CCME, 2012).

25.2.2.4 Trade effluents

In many countries, there will be a system for the management of trade effluents. It is very likely that discharges from industry to the public sewer will be licensed and charges levied based on the volume and strength of the effluent (see, for example, the UK 'Mogden formula'). It may also be possible for an industry located close to a waterway to treat their own wastewater and seek a discharge consent from the environmental authority to discharge directly and not to sewer. If both these options are available, there is a risk that if charges for trade effluents are set too high, industry will choose the latter option and therefore revenues to support the public system will fall.

Arguably, discharge to the public sewer (if appropriate standards and enforcement are in place) will ensure better treatment and a higher and more consistent level of protection. However, given the complexity of modern industrial processes and the wide range of chemical compounds that may be discharged, some authorities are now taking the view that it is more appropriate for industry (or some industries) to be required to treat their own wastewater, including as pre-treatment for discharge to sewer. In Scotland, recent legislative change enables the public service provider to enforce this, with particular reference to the diminution or elimination of 'priority substances' as well as other pollutants. Priority substances are designated under an EU Directive setting quality standards, and this Directive also makes provision for a 'watch list' of substances that are not yet controlled but which may require regulation as further evidence of their impacts emerges (see also section 7). The management of pollutants prior to discharge is also relevant to the reuse of treated wastewater (section 5).

In a different context, in the US in New York State, analysis has been made of wastewater treatment options for water used in hydraulic fracturing to extract gas from the Marcellus Shale. It has been argued that the capacity of public treatment plant is inadequate, both in terms of volume and chemical constituents, and that it would be more effective for dedicated treatment plant to be established by the industry operators (Rahm & Riha, 2012).

25.3 REGULATION OF ONSITE SANITATION

Onsite facilities – not attached to a network, or only to a small local network – can be provided in various ways. These include septic tanks, and other small waterborne systems; and non-waterborne sanitation.

These are likely to be found in rural areas in both developed and developing countries; but there may also be opportunities here to manage human wastes differently in urban environments.

The UN defines 'improved' sanitation as flush or pour-flush toilets to sewer; septic tanks; ventilated pit latrines; pit latrines with slab; and composting toilets (UN-Water, 2006). Unimproved sanitation is public or shared latrines; pit latrines without slabs; hanging toilets or latrines; buckets; or no facility at all. In terms of regulation, much depends on whether there is a waterborne system of any type, or not.

Small rural communities in developed countries may use individual or community septic tanks at different scales. These are likely to have flush toilets and fully piped systems. Regulatory and practical problems tend to concern their siting and construction (especially in relation to drinking water sources), discharge to soakaways, and maintenance (including routine emptying and eventual closure). Thus in Scotland, septic tanks serving ≤15 population equivalent must be registered with the environmental authority, and larger tanks must be licenced (SEPA, 2015); all septic tanks should be designed and constructed in accordance with building control and after an assessment of the underlying geology, to protect groundwater. Older tanks especially (but not exclusively) are likely to be causing pollution problems, either because they are badly sited; or poorly maintained; or simply because of cumulative effects of multiple structures. Again in Scotland, recent legislation has enabled co-owners to proceed with essential maintenance and recover the costs even where not all owners have agreed to do the works. It is also possible that discharges from such community systems will be filtered through reedbeds or other constructed ponds or wetlands, subject to agreement from the relevant environmental agencies. Similar systems can be used for agricultural wastewater and in both cases, it will be important to avoid contaminants that cannot be treated in these ways (heavy metals, some pathogens, pesticides, and emerging pollutants). Septic tanks are particularly sensitive to bleach and other household chemicals, needing a high degree of consumer responsibility.

Septic tanks also need emptying periodically, giving rise to a sludge management problem (section 4); other non-waterborne forms of sanitation also create sludge or solid waste. In South Africa, ventilated pit latrines are specifically identified as a 'basic' service which municipalities are encouraged to provide (DWAF, 2003). However, there has been considerable concern as to the management of these facilities, especially in an urban context (Tissington, 2008, 2011; Mjoli, 2009), with some municipalities taking the view that emptying the facility should be a household responsibility. That seems unrealistic, and a public authority really needs to deal with the sludge management; and was one of the reasons for moving sanitation into the Department for Water Affairs.

It is also possible, and may become increasingly desirable, for composting toilets to be designed which are acceptable and useable in any environment. These are widely used, for example in rural Scandinavia or Australia, but should perhaps be given increased attention. We can no longer afford to continue to mix small volumes of waste with large volumes of (usually potable) water, simply in order to wash the waste to a treatment plant and expend energy cleaning the water back up again. Whilst the reuse of greywater to flush toilets is one approach (section 5.1), taking the water out entirely is surely another.

25.3.1 Impacts on groundwater

Discharges from any waterborne system will have potential impacts on receiving waters, regardless of scale or treatment type. The management of small systems such as septic tanks may have a less onerous licencing system for discharges compared to large-scale municipal plant, for example under the EU regime discussed above, or the simple registration required in Scotland. As a minimum there should be *ex ante* control to ensure proper siting and construction, but ongoing maintenance will always be an issue.

However, onsite systems of all types have particular consequences for groundwater. Septic tanks discharging to soakaways, or reedbed systems or other filtration methods, need assessment of the soil type and structure to ensure groundwater will not be contaminated. Non-waterborne systems such as pit latrines produce sludges, which may be stored in facilities which are not properly sealed and therefore contaminants may leach into groundwater. Especially if these systems are controlled by a municipality, or a department responsible for public works or housing, there may be inadequate links between environmental authorities and water services authorities. As with the management of any solid waste, leachate requires a joined-up approach across authorities with special reference to groundwater.

Ideally, some part of the policy framework will note these linkages and provide a system for cross-regulatory referral, addressing links to both wastewater and solid waste as well as to planning and landuse generally (see e.g., SEPA, 2009). In some states there may be explicit systems for cross-referral of licence applications; for example in Queensland, Australia, one authority may be the 'assessment manager' and another the 'referral' or 'concurrence' agency under their Sustainable Planning Act, which controls many small-scale environmental impacts.

25.4 SLUDGE DISPOSAL AND REUSE

Wastewater treatment procedures generate substantial volumes of sludge or biosolids. Ideally these will be utilised in some fashion, to extract the value (nutrients, or energy); otherwise, they will be a solid waste problem needing disposal.

The two terms, sludge and biosolids, represent generally broad and somewhat flexible categories of treatment residuals. The Canadian Council of Ministers of the Environment, for example, defines municipal wastewater biosolids as: '*organic products produced from the treatment of wastewater sewage sludge and septage to reduce pathogens and vector attraction (odours). Municipal wastewater biosolids may be solid, semi-solid or liquid and come primarily from the treatment of domestic wastewater and municipal sludge, although municipal wastewater treatment plants may also treat some commercial and industrial sewer effluents*'. (CCME, 2010). The EU UWWTD defines 'urban waste water' as 'domestic waste water or the mixture of domestic waste water with industrial waste water and/or run-off rain water'; and 'domestic waste water' as 'waste water from residential settlements and services which originated predominantly from the human metabolism and from household activities' (Art. 2).

Certain historic approaches, notably discharge to surface waters by pipeline or ocean dumping, have been dramatically reduced or eliminated through regulatory intervention. As noted above, for example, the EU UWWTD, prohibited ocean dumping as of 31 December 1998 but controversial pipeline discharges continue in some other jurisdictions.

25.4.1 Solid waste disposal

Treatment options vary according to the final use or disposal procedure that has been selected, and might include aerobic or anaerobic digestion, dewatering, stabilization (with lime), and thermal drying or incineration. Final disposal by means of landfill or incineration is widespread, subject to regulatory oversight, usually as part of a solid waste management regime. This may be the responsibility of a different agency or department from the management or regulation of the water service, again requiring good institutional links. In the EU, large waste facilities are regulated under an integrated industrial permitting system addressing discharges to any environmental medium. Landfills have specific rules including the progressive removal of biodegradable waste from landfill.

25.4.2 Agricultural use

The regulation of sludge or biosolids in an agricultural context addresses composition and utilization. Composition standards may be established, for example, in relation to the presence of trace metals, pathogens and organic chemicals. Considerations respecting the use or application of biosolids include restrictions on the frequency or seasonality of application, separation requirements (setbacks or minimum distances from residential or other occupied structures, watercourses, or other sensitive areas such as parks and playgrounds) and stability and application rate. Similar provisions are also relevant for agricultural wastes; for example EU member states are required to control the application of manures, to protect freshwater.

Sewage sludge with its nitrogen and phosphorus content may also be used for its nutrient or fertilizer value, again, subject to regulation. Indeed, there are some important examples – for example Milwaukee, in the US – of commercial fertilizer production based on sludge recovery and treatment. The Milwaukee experience resulted in 'Milorganite' (Milwaukee organic nitrogen) which was widely used and marketed. Increasing levels of detergent phosphates in wastewater enriched phosphorus concentrations in fertilizer, but the introduction of toxics from industrial and household sources (solvents and paints, for example) raised health issues. Concerns around cadmium in particular led to warnings about the use of Milorganite on vegetable gardens, resulting in a significant drop in sales. Regulatory controls and pre-treatment requirements sharply reduced cadmium levels in the fertilizer (Schneider, 2011, 2012). Recovery of phosphorous especially is becoming critical, and many water services providers are developing relevant technology; many of the technical chapters in this book are relevant and it is important that legislation enables this whilst managing the potential contra-indicators.

The EU legislated specifically for sludge management some years ago, and these rules were implemented in Member States. Essentially this requires analysis of the composition of the sludge, focusing on metals, and records to be kept by the producer and the occupier of the land. In the UK the regulations are supplemented by a 'sludge matrix', which gives information regarding the types of crops on which sludges with different treatment levels can be applied.

The Sludge Directive dates back several decades, and for a number of years the EU has been looking at revisions within the context of their Soil Thematic Strategy (European Commission, 2006, 2012). There was a proposal for a Soil Directive (European Commission, 2006a), but it proved difficult to secure political agreement and this did not progress. Sewage sludge is only one small part of a soil strategy and again this indicates the links to broader frameworks, in this case for landuse, when seeking to manage wastewater or its residues. Taking a broader perspective, the difficulties of managing sludge may again indicate that the waterborne route is the root of the problem; alternatively new technical approaches may enable the recovery of nutrients in a form which is safe for users and critically, acceptable to consumers.

25.4.3 Marine wastewater discharge from vessels

The vulnerability of marine and coastal environments, including shoreline settlements, to various forms of pollution (oil, ballast water, sewage, hazardous cargo and so on) from ocean-going vessels is addressed at the international level within a complex legal framework involving the law of the sea (UNCLOS) and the International Convention for the Prevention of Pollution from Ships (MARPOL), among other conventions.

MARPOL regulates discharges of harmful substances on the basis of a definition of discharge that includes 'any escape, disposal, spilling, leaking, pumping, emitting or emptying' and an understanding of harmful that involves human health hazards, harm to marine live, or damage to amenities and legitimate uses of the sea. Regulatory and enforcement authority respecting marine matters, including pollution, is dispersed amongst 'flag' states, port states, and coastal states.

A dedicated Annex to MARPOL addresses sewage pollution on the basis of proposed surveys of designated ships (new vs existing vessels; tonnage and passenger capacity) to provide for certification. In addition, certain discharges are prohibited depending on the type of discharge, its location and operation of an approved sewage treatment plant, among other considerations.

Cruise ships – as floating human communities – present particular challenges, whether traveling the high seas or in coastal waters. In some ports, the number of passengers on board may represent a notable increment to the local population. In response to a petition from the Bluewater Network, the US EPA surveyed waste streams from cruise ships, including sewage and greywater (while Friends of the Earth now produces an annual cruise ship report card on the sewage treatment systems of sixteen major cruise lines. Inventories and assessments of this nature support provide indirect support to regulatory and enforcement initiatives while simultaneously providing prospective passengers with a level of transparency concerning vessel performance.

25.5 REUSE OF WASTEWATER

The need to reuse our wastewater is obvious and a priority; we no longer have access to sufficient raw resource. The uses to which effluent can be put depend on two linked factors: its source and composition, and the treatment it has had. The US EPA (undated) explains that treated effluent can be used for *'beneficial purposes such as agricultural and landscape irrigation, industrial processes, toilet flushing, and replenishing ground water'*. In South Africa, General Authorisations permit use of wastewater within certain limits (both volumetric and in terms of composition) for irrigation especially.

In section 2 we noted the frequently used classification for (domestic wastewater), as 'greywater' *(kitchen and bathing wastewater)* or 'blackwater', *(excreta, urine and faecal sludge)* and such a distinction – oriented generally around the level of risk – is highly relevant to legal frameworks for reuse. There may also be specific provision for some industrial or commercial users of water, and for stormwater. However, once wastewater discharges of any kind enter combined municipal systems, it will be necessary to treat them all as 'blackwater'.

The World Health Organization (WHO) has comprehensively examined wastewater use, including black and greywater. Regulatory intervention will vary between jurisdictions and according to policy frameworks, but in general WHO lists the following technical elements to be addressed: hazard identification; determination of health risks and possible health protection measures; establishment of health-based targets for risk management; implementation of protection measures; assessment and monitoring. (WHO, 2006; Vol. 1).

Much of the focus of their guidance relates to agricultural use of wastewater in the developing world. As with the EU's sludge use regime, they propose health protection measures including restrictions against usage for particular crops and techniques of wastewater application. Food preparation issues are relevant for consumers, but there are distinctive risks of exposure for fieldworkers engaged in agricultural activity. There are also preferred management strategies to reduce environmental impacts on soils, groundwater and surface water supplies (WHO, 2006; Vol. 4).

The EU has also been addressing wastewater reuse. In 2012 the European Commission undertook a review of EU water law including the implementation to date of the Water Framework Directive (WFD) and more generally, and produced a 'Blueprint' for the future (EC, 2012a). They considered that the requirement in the WFD to recover the full costs of water services was not being well-implemented. This would affect charges for public wastewater supply, but also, charges for raw water abstractions. The availability of recycled water for specific uses would contribute to both of these elements, and therefore there is a proposal to bring forward standards for resuse of treated wastewater in the near future (TYPSA, 2012, 2013; EC/JCR, 2014; Bio-Deloitte, 2015). These are likely to distinguish between different types of effluent and also, different permitted uses.

Most commercial and industrial premises will discharge some wastewater that is effectively domestic in content. In addition, there may be specialist 'process' effluents. In the developed world, reuse of process water is an important way for industries to reduce their water-and-wastewater costs and ensure greater sustainability in the coming decades. Ideally, regulation should encourage this by appropriate pricing structures for industry for both water supply and wastewater treatment, otherwise the service provider risks losing revenue from both services. As with other policy dimensions, the practice of the sector leaders is likely to be far ahead of that of many smaller businesses but there is no doubt that larger corporations with high water usage are becoming increasingly aware of the need to address this (see e.g., WBCSD/IUCN, 2012).

25.5.1 Regulation of greywater reuse

The specific reuse of greywater is rather different, as it assumes that greywater is separated from blackwater and can therefore be more easily reused. Over the past twenty-five years or so the re-use of greywater has become more common and has been regulated to varying degrees in different jurisdictions. In the interests of controlling risk and exposure, legislative intervention addresses such matters as the definition of greywater, authorized and prohibited uses including the manner in which greywater may be utilized, considerations of scale, as well as requirements for prior approval of greywater systems or exemption from an approval process, and, in some cases, incentives to promote greywater use. Likely uses with limited potential health impacts include flushing toilets, watering gardens and public parks, and car washes. This may be in a domestic context or for commercial premises, but for the latter is distinct from reuse of process effluents. These are likely to require specialist pre-treatment and as noted above this is may be a condition of discharging to sewer (Gross *et al.* 2015; Vapnek *et al.* 2009).

Many countries seek to enable some use of greywater safely by use of plumbing and building controls. For example in Queensland, the Building Code provides standards where such systems are being used. The Plumbing Code authorises small-scale greywater systems as well as onsite sewerage. There is national and Commonwealth legislation for the registration of water efficient products. Just a decade ago, Queensland prohibited the onsite use of any wastewater where a public sewer was available; the legislation was reviewed and greywater use enabled, to reduce water consumption.

In the UK by comparison, the government has been reluctant to impose the best available technologies or practices for water saving, even on newbuild, preferring a deregulatory approach. The Sustainable Housing Code (in England) establishes relevant standards, but is not mandatory. Building standards are mandatory and provide *inter alia* for greywater use, rainwater systems, composting toilets and use of non-potable water in some circumstances, as well as effluent from septic tanks and similar systems. However there is no real regulatory or policy 'push' to implement these even in the south east, which is relatively short of water. Local authority planning departments are often reluctant to be early adopters in terms of approving innovations, especially if there is any potential risk to public health, and this is likely to be true of municipalities in many countries. Low-flush toilets and other plumbing standards are also required, though low flush toilets are not mandatory. Shower heads, like washing machines and dishwashers, are not regulated in the UK, though there are labelling schemes.

For businesses, charges for water supply are relevant to changing behaviour. Specialist advice is increasingly available from water service providers and the private sector may be well-placed here; for example in Scotland, limited licensed competition for retail services for business users has meant better water efficiency advice for customers. Queensland requires businesses to have water efficiency plans.

In California, an elaborate Nonpotable Water Reuse Systems regulation within the context of the state Plumbing Code illustrates elements of the legal framework by distinguishing amongst system types (clothes washer system, simple system and complex system) and providing, *inter alia*, guidance on authorized discharges and technical specifications. Further elements of the California regulation address requirements for irrigation and disposal fields as well as the construction of mulch basins. Arrangements for testing and maintenance are also specified. Greywater systems should be clearly labelled, and prohibited contents diverted to sewer. Generally, labelling, and proper training for plumbers and other installers, is important. In developing countries especially, symbols and colour-coded pipes will assist in ensuring that connections are correctly made and users protected. California insists, for example, that a specific warning be placed at regular intervals stating: 'Caution: Nonpotable Gray Water, Do Not Drink'.

At the level of individual structures and facilities, encouragement of sustainable water and wastewater arrangements may take a variety of forms. Some of these are comprehensive in nature, such as metering programmes that promote conservation by means of a pricing mechanism, targets for reduction of leakage, or building code amendments that can introduce performance requirements such as low-flush toilets. Innovation and advances may also be encouraged more indirectly through certification programmes which are in varying degrees integrated with legal requirements. Of these, the Leadership in Energy and Environmental Design certification programme (LEED) is most prominent. Supported through national Green Building Councils in a growing number of jurisdictions, the LEED creates recognition-based incentives for sustainability enhancement, including wastewater measures, alongside sustainable energy, materials, and transportation-related building features. Thus, for example, design measures that reduce wastewater production or that facilitate on-site re-use of wastewater, such as the re-circulation of greywater for toilet flushing, may support an application for certification subject to independent third party certification. Graded designations apply on a silver – gold – platinum scale to new or renovated buildings in residential, commercial and industrial categories.

Although LEED certification is a fundamentally voluntary decision, legal re-enforcement might arise within the context of landlord-tenant relations where certain standards are contractually-required, or by means of supply-chain arrangements whereby purchasers of goods and services are seeking assurances of sustainability. In addition, on the basis of their by-law making authority, municipalities may require LEED certification or some equivalent demonstration of sustainability as a condition of construction permitting within their jurisdictions.

25.5.2 Reuse as drinking water

Perhaps most controversial is the possibility that treated effluent can be discharged directly back into freshwater resources that will subsequently be used for, *inter alia*, drinking water supply. The potential for social opposition is very high. Singapore (NEW Water) is a well-known example of best practice here (Lin Heng, 2008); Windhoeck in Namibia, maybe less so.

A recent report for the National Water Research Institute at Fountain Valley, California considered Direct Potable Reuse (DPR): '*DPR is a technically feasible method of stabilizing water supplies for municipalities and agriculture; preventing, minimizing, or correcting environmental damage resulting from inter-basin water transfers; and conserving energy*'. The study cautions, however, that '*the application of DPR on a large scale, such as Southern California, will raise significant political issues related to the ownership of water that will need to be resolved*' (Schroeder, 2012). Ownership of the effluent stream is one political issue; the general acceptability of reusing the treated effluent is another.

In 2008, Orange County, California, began operating the world's largest, most modern reclamation plant – a facility that can turn 70 million gallons of treated sewage into drinking water every day. The 'groundwater replenishment system' (GWRS) takes sewage water through treatment procedures, including microfiltration, reverse osmosis and ultraviolet light exposure. A portion of the treated water is delivered to injection wells where it contributes to a barrier against saline intrusion while the remainder is discharged to percolation basins *en route* through clay, sand, and rock into aquifers in the groundwater basin where it re-enters the water supply. From a water quality perspective and in relation to distribution, GWRS operations are subject to the approval of the Santa Ana Regional Water Quality Control Board and California's Department of Public Health. The latter has been engaged in development of water recharge regulations in response to this growing phenomenon (Dadakis *et al.* 2004; Asano & Cotruvo, 2004).

Water recycling targets have been established in several Australian jurisdictions and the Commonwealth government has set a 2015 national wastewater recycling target at 30%, whilst regulatory frameworks for direct potable use have been developed in some states (ATSE, 2013). In Queensland, there is detailed provision for the reuse of treated wastewater, including its return to a drinking water supply, within a Drinking Water Quality Management Plan and a Recycled Water Management Plan, with public reporting. It is important that all such schemes are developed by engaging with water users and the public at large, to ensure acceptability. The EU is also considering regulating for water reuse by setting standards for different types of subsequent uses; see EC-JCR (2014); TYPSA 2012, TYPSA 2013.

25.6 CLIMATE CHANGE AND ENERGY IN THE WASTEWATER SECTOR

At a macro-level, energy is a major user of water, competing with agricultural, domestic and other industrial uses. Globally, macro-level policy and law is framed by the UN Convention on Climate Change and related initiatives. Climate change presents challenges from both a mitigation and adaptation perspective.

25.6.1 Mitigation considerations

Insofar as mitigation-oriented initiatives seek to reduce GHG emissions, it is relevant to consider the extent of those emissions attributable to wastewater management (and water supply systems) which require significant energy inputs and therefore contribute substantially on a global scale to greenhouse gas emissions (UNESCO/UN-Water, 2012).

With respect to wastewater treatment, the International Energy Agency anticipates a 44% increase in energy use between 2006 and 2030 (UNESCO/UN-Water, 2012), with much of that demand coming from non-OECD countries where wastewater is largely untreated today. For these reasons, there is great interest in increasing energy use efficiency in the wastewater sector and further, in recovering energy from wastewater.

Efforts to recover energy from human waste and sewage flows exist at various scales. At the domestic level, for example, more than 15 million rural households in China reportedly make use of biogas digestion technology (George, 2008; Chapter 5) Various technologies operate at larger scales to extract energy or heat from sewage. (District heating via sewage is available without gas production through traditional heat pump technology as employed at the False Creek Energy Centre in Vancouver (Moralles & Oberg, 2012).

Biogas plants, including sewage-based facilities, are common throughout Europe and have been systematically pursued in several jurisdictions. In Sweden, production arrangements have evolved over several decades. Of 230 biogas plants operating as of 2009, 136 were associated with sewage treatment operations and were responsible for over 60% of national biogas production (Swedish Gas Association, 2011). Possibilities for utilization include immediate consumption in wastewater treatment, vehicle fuel for automobiles or public transit systems, and district heating (International Gas Union, 2015). The use of biogas off-site, as vehicle fuel, for example, or for injection into the natural gas grid involves regulated quality standards.

San Antonio, Texas, operates in partnership with a major renewable energy company to process or upgrade biogas into natural gas for delivery via pipeline to the commercial market. Proceeds in the form of royalties to the city are directed to the expenses of municipal water operations; this type of financial provision is an essential incentive. Several utility operators in the UK (Thames Water, British Gas, and Scotia Gas) launched a programme to connect bio-methane produced from municipal sewage to the national power grid. This 'flush to flame' pilot provides the heating requirements for roughly 200 homes. It is anticipated that similar ventures would have potential to meet the renewable gas requirements of 200,000 households. At minimum, legislation should ensure that service providers own the effluents in the sewers and can make appropriate use of them. A number of U.S. cities including Indianapolis, Nashville, Hartford, Buffalo and Jacksonville have documented substantial fossil fuel savings, of up to 50%, through sewage sludge incineration. Plant for this would need to meet standards for industrial pollution emissions to air, under general environmental law; and might need to be approved by any price-setting mechanism for delivery of the sanitation service.

25.6.2 Adaptation considerations

We must equally contemplate the many and complex adaptation dimensions of wastewater regulation and management that arise from climate change impacts, as historic assumptions about natural water flows, temperature and availability come into question, with implications for waste collection, treatment and removal. Heavy rainfall can overwhelm the existing capacity of storm and wastewater infrastructure, with the result that sewage flows will by-pass treatment facilities or accumulated contaminants such as animal waste or chemicals applied to land will be transported and discharged to nearby waters. As the IWA explains: '*Treatment works will be challenged with more variable sewage flows, either diluted (rainstorms) or concentrated (droughts), sewers may not be able to cope with rainstorms, leading to sewer overflows and backflushing of sewage to street level, etc. This may lead to serious public health issues.*' (IWA Specialist Group on Climate Change, undated). Thus again policy and law needs to make connections, this time with flood management and disaster planning.

Sewage overflows due to flooding increase the prevalence of harmful bacteria in the water and may be associated with more widespread illness amongst those who are exposed. Extreme weather events in Canada – flooding in Calgary and in Toronto – highlighted widespread vulnerabilities which are beginning to influence waste planning and management decisions. (Gandia, 2013; Vitello, 2013; Kerr *et al.* 2008).

In 2012, Hurricane Sandy's impact on the U.S. east coast illustrated the extreme vulnerability of urbanized coastal regions to infrastructure failings. Through storm surges and power outages, Sandy was responsible for sewage overflows lasting in some cases for weeks and amounting to 11 billion gallons. One third of this volume was entirely untreated (Kenward *et al.* 2013). Water management experts have identified a broad range of approaches to such adaptation challenges in the water and wastewater sector: more flexible operational responses; structural changes; sociological transformation such as demand management; technological innovations to facilitate water conservation; political decisions affecting allocation; and institutional changes to promote collaborative initiatives, for example (IWA Specialist Group on Climate Change, undated). Virtually all of these, it may be observed, call for the participation of informed legal advisors whether in relation to consultation and approvals procedure, liability and insurance against emerging risks, general questions of governance and institutional design, or land use planning and environmental assessment (see e.g., US EPA, 2012).

Overall, the management of climate change is one of the greatest global challenges that we face. Along with population growth and urbanisation, it is driving our water (and wastewater) crisis. As with other sections in this chapter, policy and law must recognise both the need to innovate and the need to manage basic human needs.

25.7 REGULATION OF CONTAMINANTS OF EMERGING CONCERN

Contaminants of emerging concern, sometimes (but less accurately) described as emerging contaminants or emerging pollutants), include, but are not limited to, pharmaceutical and personal care products. These have been defined by the US Geological Survey as *'any synthetic or naturally occurring chemical or any microorganism that is not commonly monitored in the environment but has the potential to enter the environment and cause known or suspected adverse ecological and (or) human health effects.'* They add: *'In some cases, release of emerging chemical or microbial contaminants to the environment has likely occurred for a long time, but may not have been recognized until new detection methods were developed. In other cases, synthesis of new chemicals or changes in use and disposal of existing chemicals can create new sources of emerging contaminants.'* Given the thousands of chemicals in everyday use, and the multiplicity of possible reactions, it is unsurprising that the field of research is very wide.

Much of that research has concerned the presence of contaminants in the water environment, and their effects on aquatic life; and attention is increasingly being paid to potential effects on human health (Bergman *et al.* 2013). In the EU, there are new controls on chemicals generally under the 'REACH' initiative; but regulation is always subject to various financial imperatives. For example, the WHO has stated that *'Close to 800 chemicals are known or suspected to be capable of interfering with hormone receptors, hormone synthesis or hormone conversion. However, only a small fraction of these chemicals have been investigated in tests capable of identifying overt endocrine effects in intact organisms.'* Yet when there was a proposal to include contraceptives in the Priority Substances Directive (which would then require tertiary treatment at wastewater treatment plant) this was rejected as being too expensive (EEA/JRC, 2012; Chapter 6). Instead they were placed on the Directive's 'watch list' for possible future action, along with certain common painkillers.

In 2011; Deblonde *et al.* reviewed the literature for studies analysing the inputs and outputs of emerging pollutants from wastewater treatment plant. They looked at a range of common pharmaceuticals, as well as disinfectants, plasticisers and other agents also found in cosmetics and personal care products (but the pharmaceuticals did not include contraceptives). They found that there were seasonal factors, in that degradation was more likely in warm weather; more importantly, and unsurprisingly, concentrations of some substances increased significantly when there was a hospital discharging into the plant. They found significantly more studies had been done on pharmaceuticals than on other contaminants. Where contaminants are biodegraded, they will end up in the effluent stream; and where they are not, they will be present in the sludge. They also noted that a wide range of pollutants including pharmaceuticals are emitted through veterinary medicines and especially (though not included in this review) via agricultural activities, which may find pathways directly to either surface or groundwater but will not be treated at municipal works.

An analysis of the regulatory frameworks in the US specific to pharmaceuticals (Eckstein, 2015) looked at the Resource Conservation Act, Clean Water Act and Safe Drinking Water Act, as well as the National Environmental Policy Act and legislation specific to food and drugs (all at federal level), as well as some state rules. He found that there were a wide variety of approaches and standards, but no systematic management of most pharmaceuticals.

Indeed there is no evidence in the legal literature of any state or jurisdiction (not just in the US) making comprehensive provision for the management of pharmaceuticals or other emerging substances. They may (but will probably not) be 'caught' by rules on wastewater treatment, or indeed general controls on discharges to fresh or marine waters or disposal of solid wastes (not just sewage sludge) to landfill. It is thought arguable that most of these substances have been widely used for decades; it is the understanding of the *effects* which is now emerging. As Epstein notes, many of these substances have not had a hazard assessment, let alone an assessment of the potential cumulative and reactive effects. Yet although citizens may be concerned about substances and products, in the environment or impacting on health, tertiary treatment is still very costly whilst source management is particularly difficult for pharmaceuticals or personal care products. Any suggestion of limiting either is likely to be very challenging for policymakers, although initiatives to promote return of unused drugs to pharmacies and other suppliers can reduce the volume of discharge.

25.8 CONCLUSIONS

In November 2000, the Water Supply and Sanitation Collaborative Council endorsed the 'Bellagio Principles for Sustainable Sanitation' with a view to promoting awareness of waste as a resource, and encouraging more holistic, integrated and decentralized approaches to management (Bellagio Principles, EAWAG, 2000). Such approaches can reduce water demand and also water pollution. Efforts are now accelerating to identify mutually reinforcing developments including integration of multiple objectives such as water sustainability and human rights.

Whilst such approaches tend to focus on the unmet needs of the world's poor, and especially the emerging urban environments in which most people will live, it has lessons for us all at whatever stage of development. This chapter has attempted to outline the many ways in which policy and legislative and regulatory frameworks, seek to address the technical, environmental and social aspects of wastewater management. It has noted throughout the gulf between the most advanced (countries, businesses, technologies) and the least; and the need to (still) provide for the most basic needs of many, whilst recognising the increasing science evidence of the complexity of the problem in terms of industrial use and new contaminants. On the one hand, the basic problems are well-understood but often poorly managed. On the other, new innovations, to close loops and recover value from the resource, can provide solutions but also need to be regulated. The focus of this handbook is on new innovations in treatment options that allow resources to be reused and value recovered; innovation is high on the policy agenda. This chapter explored legal and policy frameworks applying currently in the developed and the developing world, as well as legal and policy initiatives to enable and encourage innovative solutions. Inevitably, as the law is often a reactive force, the policy context may be more developed than the legislation; but there are numerous instances where the policy, as well as the legislation, is insufficiently joined up.

It is possible that smaller-scale, decentralised and less water-intensive approaches will facilitate innovation and in time be recognised as the best way forward for both developed and developing world, to meet the Sustainable Development Goals and the rights of all (including the environment) to clean water and to safe and acceptable systems for managing human wastes. Such solutions should be one priority, along with continuing technological research into closing loops at different scales, and, the science research into the impacts of different waterborne wastes on human health and ecosystems. Ultimately, political will is what is needed.

25.9 REFERENCES

Andreen W. (2004). Water quality today: has the clean water act been a success? *Alabama Law Review*, **55**, 537–593.

Asano T. and Cotruvo J. A. (2004). Groundwater recharge with reclaimed municipal wastewater: health and regulatory considerations. *Water Research*, **38**, 1941–1951.

ATSE (Australian Academy of Technological Sciences and Engineering) (2013). Drinking Water Through Recycling: The benefits and costs of supplying direct to the distribution system. http://www.atse.org.au/Documents/Publications/Reports/Water/drinking-water-through-recycling-full-report.pdf

Baker J. (ed.) (2012). Climate Change, Disaster Risk and the Urban Poor: Cities Building Resilience for a Changing World. World Bank, Washington DC.

Benidickson J. (2007). The Culture of Flushing: A Social and Legal History of Sewage. University of British Columbia Press.

Bergman Å., Heindel J. J., Jobling S., Kidd K. A. and Thomas Zoeller R. (2013). State of the Science of Endocrine Disrupting Chemicals 2012 (WHO / UNEP). http://www.unep.org/pdf/WHO_HSE_PHE_IHE_2013.1_eng.pdf

Bio-Deloitte (2015). Optimising water reuse in the EU Public consultation analysis report (DG Environment). http://ec.europa.eu/environment/water/blueprint/pdf/BIO_Water%20Reuse%20Public%20Consultation%20Report_Final.pdf

California Building Standards Code, Part 5 California Plumbing Code, Ch.16 Non-potable Water Reuse Systems.

CCME (2010). A Review of the Current Canadian Legislative Framework for Wastewater Biosolids.

CCME (2012). Canada-Wide Approach for the Management of Wastewater Biosolids.

CCME (2014). Canada-Wide Strategy for the Management of Municipal Wastewater Effluent: 2014 Progress Report.

Clean Water Act (USA) 1972, 33 U.S.C. §1251 et seq.

Corcoran E., Nellemann C., Baker E., Bos R., Osborn D. and Savelli H. (eds) (2010). Sick Water? The Central Role of Wastewater Management in Sustainable Development. A Rapid Response Assessment. UNEP/UN HABITAT. See http://www.unep.org/pdf/SickWater_screen.pdf

Dadakis J., Patel M. and Fitzsimmons S. (2011). Orange County's Groundwater Replenishment System: Water Quality Monitoring and Facility Expansion in the Face of Changing Regulations'. Proceedings of the Water Environment Federation, WEFTEC 2011: Session 21 through Session 30, pp. 1355–1366.

de Albuquerque C. (2013). Water and sanitation are human rights: why does it matter? In: International Law and Freshwater: The Multiple Challenges, Boisson de Chazournes, L. *et al.* (eds), Edward Elgar, London, UK.

Deblonde T., Cossu-Leguille C. and Hartemann P. (2011). Emerging pollutants in wastewater: a review of the literature. *International Journal of Hygiene and Environmental Health*, **214**, 442–448.

Delmon J. (2001). Water Projects A Commercial and Contractual Guide. Kluwer Law International, London.

Department of Communities and Local Government (2010).

Department of Housing and Public Works (2013).

Department of Housing and Public Works (2013a) Queensland Development Code Part MP4.2.

Directive 1991/271/EC Concerning Urban Waste Water Treatment.

Directive 1991/676/EEC concerning the Protection of Waters against Pollution by Nitrates.

Directive 1999/31/EC on the Landfill of Waste.

Directive 2000/60/EC Establishing a Framework for Community Action in the Field of Water Policy.

Directive 2010/75/EU on Industrial Emissions.

Directive 2013/39/EU as regards Priority Substances in the Field of Water Policy.

Directive 86/278/EEC on the Use of Sewage Sludge in Agriculture.

DWAF (2003). Strategic Framework for Water Services. http://www.sswm.info/sites/default/files/reference_attachments/KASRILS%202003%20Strategic%20Framework%20for%20Water%20Services.pdf

EAWAG (2000). Summary Report of Bellagio Expert Consultation on Environmental Sanitation in the 21st Century. http://www.sswm.info/sites/default/files/reference_attachments/EAWAG%20SANDEC%20WSSCC%202000%20Bellagio%20Principles%20Report.pdf (accessed 1–4 February 2000).

EC Regulation 1907/2006 on the Registration, Evaluation, Authorisation and Restriction of Chemicals.

EC-JCR (2014). Water Reuse in Europe: Relevant guidelines, needs for and barriers to innovation JRC92582. http://bookshop.europa.eu/is-bin/INTERSHOP.enfinity/WFS/EU-Bookshop-Site/en_GB/-/EUR/ViewPublication-Start?PublicationKey=LBNA26947

Eckstein G. (2015). Drugs on tap: managing pharmaceuticals in our nation's waters. *NYU Environmental Law Journal*, **23**, 37–90.

EEA/JRC (2012). Environment and Human Health Report No.5/2013 EUR 25933 EN chapter 6.

European Commission (2006). Thematic Strategy for Soil Protection COM (2006) 231 final.

European Commission (2006a). Proposal for a Framework Directive on Soil Protection COM (2006) 232.

European Commission (2012). Policy Report on the implementation of the Thematic Strategy for Soil COM (2012) 46.

European Commission (2012c). A Blueprint to Safeguard Europe's Water Resources COM (2012) 673.

European Commission (2013). Seventh Report on the Implementation of the Urban Waste Water Treatment Directive (91/271/EEC) COM (2013) 574 final.

Fisheries Act (Canada) R.S.C., 1985, c. F-14.

For an example of a legislative provision, see Water Environment and Water Services (Scotland) Act 2003 s.33.

For example, in the UK, the Sludge (Use in Agriculture) Regulations 1989 SI 1263.

Gandia R. "Evacuation of Bonnybrook waste water plant means untreated sewage allowed to flow in Bow River," Calgary Sun 27 June 2013.

George R. (2008). The Big Necessity: Adventures in the World of Human Waste. Portobello Books, London, Ch.5.

Government of Australia DoE (2009). National Water Quality Management Strategy Australian Guidelines for Water Recycling: Managing Health and Environmental Risks (Phase 2) Stormwater Harvesting and Reuse. http://www.environment.gov.au/water/quality/publications/nwqms-australian-guidelines-water-recycling-managing-health-phase2-stormwater

Government of Queensland Department of Energy and Water Supply (2013). Urban Drainage Manual. https://www.dews.qld.gov.au/water-supply-regulations/urban-drainage

Gross A. *et al.* (2015). Greywater Reuse. Taylor and Francis, CRC Press, UK.

Helmer and Hespanol (eds) (1997). Water Pollution Control. WHO Publication. http://www.who.int/water_sanitation_health/resources/watpolcontrol/en/

Hendry S. (2015). Frameworks for Water Law Reform CUP/IHP Series chapter 5, Cambridge, England.

HM Government Building Regulations (2010) Parts G and H.

http://www.foe.org/cruise-report-card.

International Convention for the Prevention of Pollution from Ships, 1973, 1978, in force, 1983.

International Gas Union (2015). Biogas – from refuse to energy. http://www.igu.org/sites/default/files/node-page-field_file/IGU%20Biogas%20Report%202015.pdf

IWA Specialist Group on Climate Change (undated). Climate Change and the Water Industry: practical responses and actions. http://www.worldwatercouncil.org/fileadmin/wwc/Library/Publications_and_reports/Climate_Change/PersPap_10._Water_Industry.pdf

Kallis G. and Nijkamp P. (2000). Evolution of EU water policy: a critical assessment and a hopeful perspective. *Journal of Environmental Law and Policy*, **3**, 301–355.

Kenward A., Yawitz D. and Raja U. (2013). Sewage Outflows from Hurricane Sandy. Climate Central. http://www.climatecentral.org/pdfs/Sewage.pdf

Kerr, Wood and Leidel Consulting Engineers (2008). Vulnerability of Vancouver Area Sewerage Infrastructure to Climate Change (Final Report).

Local Government etc. (Scotland) Act c.39 Part II.

Lye L. H. (2008). A Fine City in a Garden – Environmental Law and Governance in Singapore. *Singapore Journal of Legal Studies*, **8**(1), 68–117.

MARPOL 73/78, Annex IV, Regulations for the Prevention of Pollution by Sewage from Ships.

Marcic v Thames Water PLC [2003] UKHL 66.

Mjoli N. (2009). Review of Sanitation Policy and Practice in South Africa 2001–2008 WRC 1741/1/09.

Moralles G. and Oberg G. (2012). The Idea of Sewage as a Resource: An Integrated Study of Knowledge and Decision Making in Liquid Waste Management in Metro Vancouver, B.C. Canada. Institute of Resources, Environment and Sustainability. http://watergovernance.ca/wp-content/uploads/2012/03/Gunilla-Oberg-Sewage-as-a-resource-10April2012.pdf

OECD (2011). Water Governance in OECD Countries: A Multi-Level Approach, OECD Studies on Water. OECD Publishing, Paris, http://dx.doi.org/10.1787/9789264119284-en

OECD (2015). The Governance of Water Regulators, OECD Studies on Water. OECD Publishing, Paris.

Prüss-Üstün A., Bos R., Gore F. and Bartram J. (2008). Safer Water Better Health WHO. http://whqlibdoc.who.int/publications/2008/9789241596435_eng.pdf

Rahm B. and Riha S. (2012). Toward strategic management of shale gas development: Regional, collective impacts on water resources. EnvSciPol **17**, 12–23.

Schneider D. (2011). Hybrid Nature: Sewage Treatment and the Contradictions of the Industrial Ecosystem. MIT Press, Cambridge, Massachusets.

Schneider D. (2012). Purification for profit: milwaukee and the contradictions of sludge. In: Histories of the Dustheap, S. Foote, and E. Mazzolini (eds), MIT Press, Cambridge, Massachusets.

Schroeder E., Tchobanoglous G., Leverenz H. L. and Asano T. (2012). Direct Potable Reuse: Benefits for Public Water Supplies, Agriculture, the Environment and Energy Conservation NWRI White Paper. http://www.nwri-usa.org/documents/NWRIWhitePaperDPRBenefitsJan2012.pdf

See ADAS 2001, Safe Sludge Matrix, http://adlib.everysite.co.uk/resources/000/094/727/SSMatrix.pdf.

See http://presscore.ca/2012/us-cities-reducing-their-fossil-fuel-usage-by-50-by-using-sewage-sludge-as-a-fuel.html

See http://toxics.usgs.gov/regional/emc/

See http://water.epa.gov/polwaste/npdes/<imoco[a;ootoes-amd-Wastewater-Treatment-Plants.cfm.

See http://www.gizmag.com/human-waste-to-gas-project-goes-live/16572/

See http://www.saws.org/Your_Water/Recycling/Biogas/.

See http://www.usgbc.org/leed

See http://www.usgbc.org/leed

See http://www.waternexussolutions.org/284/the-perfect-storm-scenario-and-nexus-thinking-to-meet-increasing-demand-for-energy-food-water.html#.VhOGE-xViko

See https://www.ofwat.gov.uk/nonhousehold/yourwaterbill/hownonhousehold/trade/mogden

SEPA (2009). Groundwater Protection Policy for Scotland. http://www.sepa.org.uk/media/34371/groundwater-protection-policy-for-scotland-v3-november-2009.pdf

SEPA (2015). The Water Environment (Controlled Activities) (Scotland) Regulations 2011 (as amended) A Practical Guide Version 7.2. http://www.sepa.org.uk/media/34761/car_a_practical_guide.pdf

South Africa General Authorisation No.665 of 2013, National Water Act 1998 s.39; Schedule 1.

South Africa General Authorisation No.665 of 2013, National Water Act 1998 s.39; Table 2.1.

Sustainable Planning Act (Qld) 2009 No.36, Chapter 6.

Swedish Gas Association (2011). Biogas in Sweden. file:///C:/Users/jbenidi/Downloads/BiogasInSweden.pdf

Tissongton K., Dettmann M., Langford M. and Conteh S. (2008). Water Services Fault Lines: An Assessment of South Africa's Water and Sanitation Provision across 15 Municipalities. CALS/COHRE/NHCR Johannesburg, South Africa.

Tissington K. (2011). Basic Sanitation in South Africa: A Guide to Legislation, Policy and Practice SERI. http://www.nwu.ac.za/sites/www.nwu.ac.za/files/files/p-fl/documents/SERI_Guide_to_Basic_Sanitation_Jul11.pdf

TYPSA (2012). Wastewater Reuse in the European Union (DG Environment). http://ec.europa.eu/environment/water/blueprint/follow_up_en.htm

TYPSA (2013). Wastewater Reuse in the European Union (Update) (DG Environment). http://ec.europa.eu/environment/water/blueprint/follow_up_en.htm

UN (2015). Transforming Our World: the 2030 Agenda for Sustainable Development. https://sustainabledevelopment.un.org/content/documents/7891Transforming%20Our%20World.pdf

UNDESA (2012). World Population Prospects 2012 Revision. http://esa.un.org/unpd/wpp/Documentation/pdf/WPP2012_Volume-I_Comprehensive-Tables.pdf

UNEP (2010). The Greening of Water Law: Managing Freshwater Resources for People and the Environment. UNEP, Nairobi.

UNESCO/UN-Water (2012). World Water Development Report 4, Managing Water under Uncertainty and Risk (Volume 1). http://www.unesco.org/new/en/natural-sciences/environment/water/wwap/wwdr/wwdr4-2012/

UNICEF/WHO (2015). Progress on Sanitation and Drinking Water 2015 Update and MDG Assessment. http://www.unicef.org/publications/files/Progress_on_Sanitation_and_Drinking_Water_2015_Update_.pdf

United Nations Convention on the Law of the Sea, 1982, in force, 1994.

United Nations Framework Convention on Climate Change, 1992, in force, 1994.

UN-Water (2006). World Water Development Report 2: Water, a Shared Responsibility. http://www.unesco.org/water/wwap/wwdr/

UN-Water (2010). Climate Change Adaptation: The Pivotal Role of Water Policy Brief. http://www.unwater.org/downloads/unw_ccpol_web.pdf

US EPA (2008). Cruise Ship Discharge Assessment Report EPA 842-R-07-005. http://water.epa.gov/type/oceb/upload/0812cruiseshipdischargeassess.pdf

US EPA (2012). National Water Programme Strategy: Response to Climate Change. http://www2.epa.gov/sites/production/files/2015-03/documents/epa_2012_climate_water_strategy_full_report_final.pdf

US EPA (undated). Water Division Region IX, 'Water Recycling and Reuse: The Environmental Benefits' EPA 909-F-98-001.

Van Den Berg C., Goksu A. J. and Rodriguez D. J. (2012). Investing in Water Infrastructure: Capital, Operations and Maintenance. World Bank, Washington DC.

Vapnek J., Aylward B., Popp C. and Bartram J. (2009). Law for Water Management: A Guide to Concepts and Effective Approaches. FAO, Rome.

Vitello J. "Toronto beaches experience 'record levels of E.coli' after storm pumps up to one billion litres of sewage into lake," National Post 13 July 2013. news.nationalpost.com/2013/07/12/toronto-beaches-experience-record-levels-of-e-coli-after-storm-pumps-up-to-one-billion-litres-of-sewage-into-lake/

Water Act (UK) 1989 c.15.

Water Efficiency Labelling and Standards (Qld) Act 2005 No.69; Water Efficiency Labelling and Standards (Cwlth) Act 2005 No.4.

Water Resources (Scotland) Act 2013 s.35.

Water Resources (Scotland) Act 2013 s.37.

Water Supply (Safety and Reliability) (Qld.) Act 2008 No.34, Chapter 3.

Water Supply (Water Fittings) Regulations SI 1999/1148 Sch.2 para.25.

WBCSD/IUCN (2012). Sharing Water: Engaging Business. http://www.wbcsd.org/Clusters/Water/Resources/Sharing-water-Engaging-business

WHO (2006). Guidelines for the Safe Use of Wastewater, Excreta and Greywater. http://www.who.int/water_sanitation_health/wastewater/gsuww/en/

World Bank/PPIAF (2006). Water and Sanitation Privatization Toolkit. http://siteresources.worldbank.org/FINANCIALSECTOR/Resources/282044-1307652042357/Water_Full.pdf

Part 5

Conceiving, Comparing and Selecting Efficient Processes

Chapter 26

Environmental decision support systems

Alba Castillo, Joaquim Comas, Manel Garrido-Baserba, Francesc Hernández-Sancho, Ulf Jeppsson, Ignasi Rodríguez-Roda and Manel Poch

The academic version of the NOVEDAR_EDSS can be downloaded for free from http://lequia.udg.cat/novedar/

26.1 INTRODUCTION

A decision (from the Latin *decidere*, to cut) is the end product of a cognitive process that involves choosing just one action or option from a set of possibilities or alternatives that are mutually exclusive and not necessarily known *a priori*. This choice may be conscious or not. Decisions may be individual or collective and may be made by an intelligent machine or a set of them and by groups where individuals and machines interact.

The amount of time required to make a decision affects the type and quality of the information used to make this decision. This means that individuals who make decisions quickly use aspects readily available in the environment to make them or base them on previous experience. On the other hand, if someone is considering a situation where important features may be hidden among thousands or millions of pieces of data, then making a speedy judgement is not a good idea. Such a situation requires an information search and processing that results in a more coherent piece of better quality information so we can make a choice that is more based on measurements and/or empirical knowledge.

Decision-making theory is based on the development of methods and criteria for making decisions in these environments.

- Risk indicates that we do not know what the outcome of certain decisions will be, but we know what could happen and – to some extent – the probability of this.
- Uncertainty indicates that we do not know what the outcome of certain decisions will be, but we know what could happen among various possibilities.

To summarise, we can say that a decision, which is a mental output, may be made to be executed as a task and may modify the environment in which the task is performed. As the complexity of the process to be controlled increases, so does the complexity of decisions and how they are made. Thus, the decision-making process is a reasoned or emotional process that can be rational or irrational – based on explicit or tacit assumptions – allowing individuals and/or groups to achieve goals that let them interact with the environment or with other individuals and groups successfully and, ultimately, to survive.

The inherent complexity of making a decision related to the '3R' (reduce, reuse and recover) concept in wastewater treatment, as in many other environmental systems, make the appearance of automatic decision-making systems necessary. Even when using such systems, in most cases all the possible alternatives cannot be analysed even with the very latest analytical techniques and computers available. Therefore, the goal is to make an approach towards the most acceptable solutions possible. There is an inverse correlation between the amount (and the certainty) of the information available and the relevance of the decision, as shown in Figure 26.1.

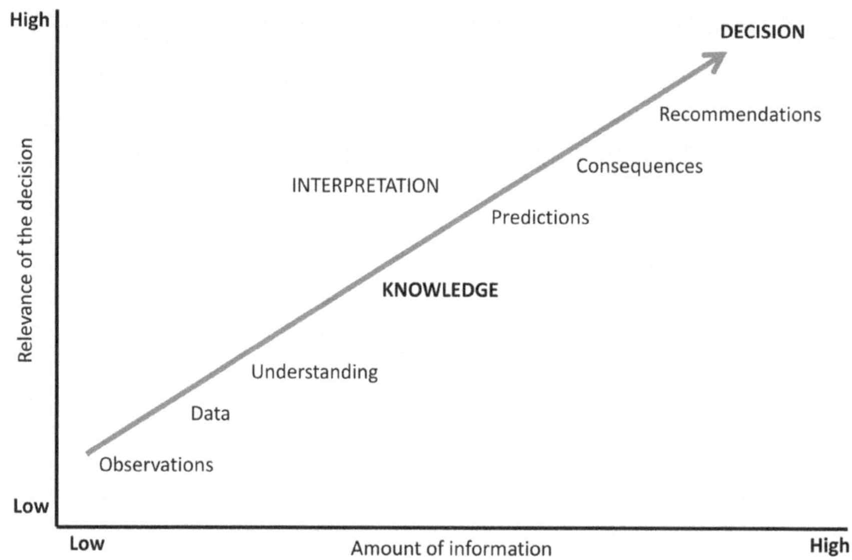

Figure 26.1 Relation between amount of information and relevance of decision.

26.2 LEVELS OF DECISION

There are different decision-making levels where to include the '3R' concept in the design of urban water systems. The first level corresponds to strategic decisions, which, mostly, are not directly related to the urban water cycle, but condition its design and operation. Usually the administration responsible for water management has little influence at this level. Among other strategic decisions, we can highlight:

- Urban planning, which may be more sustainable. Cities can be more or less densely populated, leading to *concentrated* or *extensive* city models and conditioning the amount of wastewater generated, its concentration, the distribution of collection systems, the possibility of using more extensive or intensive treatment systems, etc.
- The funding policy for the costs associated with the water system. The European Directive sets that the costs should be recovered directly from the water system policy, which should be self-sustained. This model is still far from being applied in the countries affected by the Directive. Actually, the countries have different funding models with different weights given to public-private investments.

- Industrial development and its link to the urban water system, with the existence and size of industry in town centres or the definition of the integrated treatment policy (domestic/industrial) for wastewater. This affects the quantity and quality of the water entering the sewer system.
- General environmental policies, which make reference to the existence of collection and/or re-use regulations relative to rainwater, to source separation within buildings, or to the recycling of grey water in buildings, with the existence of adequate piping.

At the second level, decisions are made concerning the selection of the configuration of collection and treatment systems. At this level, and from the information provided at the previous level, the configuration and technologies most suitable for achieving the objectives and restrictions defined in the previous level are selected. This is a level that includes new elements that influence decisions, particularly related to the technologies available to achieve the objectives defined in the previous level. In terms of treatment, the following information must be included for each technology:

- *Treatment capacities for different types of pollutants.* We have to take into account the fact that different technologies have been developed over a period of time during which priorities and objectives have changed. Therefore, there are different characteristics for different pollutants; this must be taken into account.
- *Investment, operation and energy costs.* Using merely the cost of construction as a proxy of the plant cost does not account for the total costs incurred during the plant life. Instead, an actualized investment cost should be added up to the expected cost of operation. Simultaneously, and with the gradual increases in the cost of energy, the energy cost must be considered at this stage.
- *Compatibility with other treatment operations.* Not all treatment operations are compatible with each other or have the ability to adapt to the changes that the system may experience over time, such as the need for possible future extensions. Therefore, this aspect should be taken into account as part of the decision-making process.
- *Secondary effects, impacts and generation of by-products.* Urban water systems are designed to reduce the environmental impact of wastewater, but their operation has other environmental impacts due to construction, energy and chemicals consumption, etc. Today, we do not just have economic costs but also by-products and impacts associated with their activity to consider, such as the emission of greenhouse gases.

The third level corresponds to the design and optimisation of the equipment and is a fundamentally technical level. Previous levels have identified the sequence of operations to be included in the urban water system, which achieves the first level objectives. The third level is necessary to identify the dimensions associated with each of the units involved. At this stage, we need to identify the volumes, surfaces and power etc. of the pumps, pipes and reactors involved. We should also consider aspects of plant operation to ensure optimal performance later, so we must include the control elements and define operating conditions for maximum process efficiency.

26.3 COMPLEXITY OF THE DECISIONS

An interesting aspect of the establishment of three levels in the design of urban water systems is that we can better visualise some of the characteristics that have evolved throughout the design process. Somehow we are identifying the basic elements that identify the complexity of the decisions made at each stage, allowing the simultaneous determination of the suitability of the agents involved in decision-making at each stage. Three levels of complexity are considered associated with the three levels of decision in the design of urban water systems.

The first level corresponds to really complex systems where there is high epistemological or ethical uncertainty, and where what is at stake may involve conflicts of interest between the parties involved in the process, as well as a significant risk. In this case, it is important to recognise the need to consider a plurality of perceptions and perspectives. In urban water systems, this corresponds to the management of a unit mass of water, where different factors – economic, technical and ecological – come into play and each factor is associated with different goals. Therefore, there is a need for collaboration between the different actors, which also implies different objectives and different experiences that must be integrated.

The second level relates to the selection of the configuration and would correspond to systems with a lower level of uncertainty, but that are difficult to represent in a satisfactory manner, by applying a standard model that can be reproduced anywhere and by any competent practitioner. In this instance, the personal element and acquired experience is important, so the presence and participation of an expert is significant. This selection will vary by location and the assessment that the person responsible made of the importance of the various phenomena involved. Certain quality standards have to be maintained at the output and there are different options available to achieve them. The selection of one or another configuration depends on the person responsible for the design and his/her own experience.

The third level of design is for less complex systems where uncertainty is reduced and what is at stake has less importance, since the degrees of freedom and investments have been reduced. These are systems which can be represented using a single perspective and where we can find a model that provides a satisfactory description. The input is perfectly defined, the number of alternatives is limited and the available information is sufficient to discriminate between them.

There are two characteristics that evolve significantly throughout the design of the process:

- The impact of the decisions on the project costs as well as the number of considered options, will decrease as the project nears completion. This finding, which seems obvious, is, curiously enough, not always related to the effort spent in the different stages. Often the time and resources required to make decisions do not match the impact they may have on the final cost of the project. It is one of the obvious truths that is sometimes forgotten.
- The amount of information available increases as the definition of the project evolves. Not only are there fewer options to be dealt with, but there is usually more information with less uncertainty. The uncertainty aspect is important as it affects the confidence in decisions made; therefore, the minimisation of uncertainty should be one of the key elements in decision support systems.

As can be seen in Figure 26.2, the integration of the variation of the impact of decisions with the information available at each level allows us to identify both the existence of the different skills involved in making decisions and the different agents. In addition, each of these levels establishes different relationships between the agents and with the decision-making process. The first decision level comprises important political components (in a broad sense) which, as noted, may have some clear intentions, which are difficult to quantitatively translate, so decision support systems will be needed to manage this characteristic. At the second level, the key characteristic is experience; the decisions will mostly be made by experts, who may come from various fields. Each can be an expert in their own field, but the decision support system must be able to manage the "paradox of expertise", which reminds us that the more you know about a subject, the harder it is to explain the reasons behind the proposed decision. At the third level, instrumental behaviour becomes more important. There is less uncertainty and more tools are available, which are able to quantify the processes that take place in the system. It is the environment in which the engineers and operators can evolve with more comfort and efficiency.

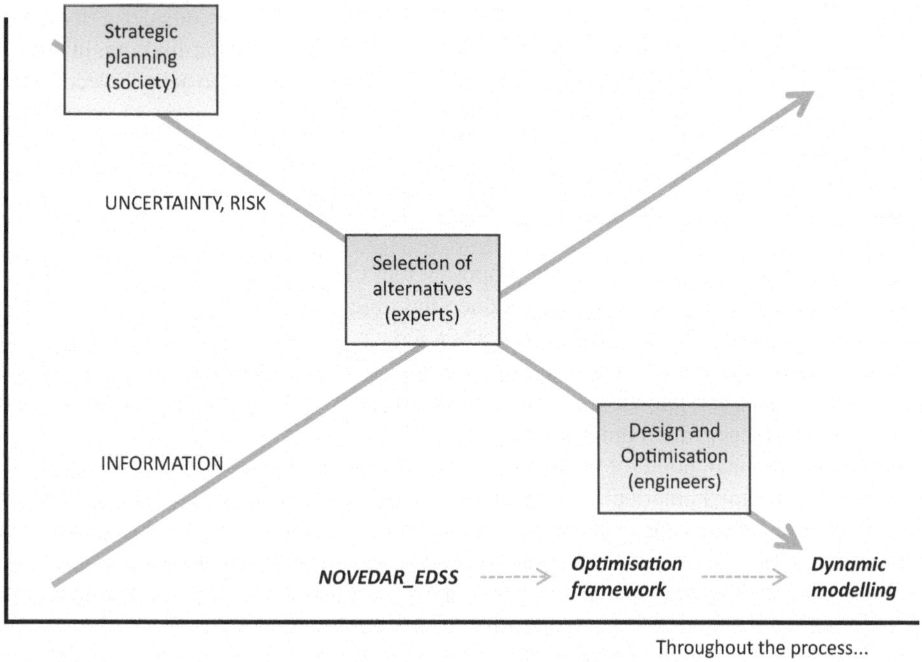

Figure 26.2 Integration of the uncertainty and risk in decision-making with the amount of information required.

The three tools presented within Chapters 26 to 28 support decision making at the second level, i.e. for the selection of alternatives (NOVEDAR_EDSS, Chapter 26), while the optimisation framework (see Chapter 27) and plant wide dynamic modelling (see Chapter 28) assist at the plant and operation design and optimisation level (Figure 26.2).

26.4 WHAT IS AN EDSS?

Over the years, artificial intelligence has been developing tools able to mimic human behaviour relative to perception, learning and reasoning abilities. Tools that have been applied to the management of complex problems, and that have demonstrated their ability to cope with them, especially when integrated with numerical tools, as they complement the limitations present when applied to complex problems with unstructured domains where expert knowledge is significant. In terms of urban water systems, as in other complex systems, the use of these techniques has evolved over time, from their initial applications with single tools (for example expert systems) to the use of more deliberative tools applied in a more integrated manner (like decision support systems, DSS).

DSS applied to environmental systems (i.e., environmental decision support systems, EDSS) were born in the 80s with the aim of providing decision-making support, help that was simultaneously beyond that offered by the mathematical models that had restrictions relative to incorporating qualitative knowledge and beyond what would be a simple accumulation of experiences difficult to manage. Since their inception, they are systems that bring different tools from different fields together. This versatility, and their recent development, also means that there is no single definition for them.

Thus, Fox and Das, in their book "Safe and Sound" (Fox & Das, 2000) consider a decision support system to be a computational system that helps anyone responsible for decision-making, in the process of deciding between alternatives or actions, applying knowledge about the field to achieve recommendations relative to the different options. The system includes an explicit decision process based on a set of theoretical principles justifying the "rationality" of the process. In this case, we focus our attention on the need to include the justification of the proposal as a significant element, but without referring to elements such as DSS response time, which may be important in the case of application to linear process management.

However, this second aspect focuses on the proposal from Cortés *et al.* (2000), who defines a DSS as an intelligent information system, which helps reduce the time needed to make decisions and improves their consistency and quality. Decisions are made when a deviation from the expected or desired state of the system is observed (or predicted). This implies awareness of the problem which, in turn, must be based on information, experience and knowledge of the process. Besides, EDSS also have the ability to integrate different types of knowledge, which must be present in a DSS.

From a more operational point of view, we can define an EDSS as an interactive, flexible and adaptable system able to link numerical and algorithmical methods with artificial intelligence techniques, geographic information systems (GIS) and environmental ontologies. This definition reinforces the idea of an EDSS as an integrative tool that incorporates methodologies from different fields as elements capable of describing the complexity of the systems studied – in our case, urban water systems – because the EDSS simultaneously manages numerical data, qualitative knowledge and ontologies, as well as incorporating spatial (with GIS) and temporal dimensions (with mathematical models).

EDSS developers must choose which tools should be integrated for each case or objective. In our opinion, the following elements cover the minimum requirements, namely:

- They must enable data management, but also, and importantly, knowledge gained from experience.
- They must incorporate results and knowledge from different areas and different experts and different levels of description.
- They must allow recovery of data and knowledge in a manner that is easy and useful to the user.
- They must be able to justify the proposals, indicating what and who supports them, that is to say the reliability of the results provided by the EDSS.

26.5 WHY USING AN EDSS?

Since the early DSS proposals in the 80s, they have been applied to different environmental problems, especially those related to water management. It is far from the objectives of this book to carry out a thorough review of the EDSSs applied to environmental management, but different classification criteria can be set up that may help us to identify their potential.

On the one hand, EDSSs have been applied to planning, where they have demonstrated their ability to incorporate qualitative knowledge from different agents that may intervene at this stage. In this sense, the progressive increase in the incorporation of participatory processes in decision-making leads to the existence of an ever growing set of information and details. The consideration of different types of expertise and interest in these processes is not often spelled out specifically; there is dispersed knowledge that makesdifficult to use traditional numerical techniques. Therefore, the use of ontologies and artificial intelligence techniques specialising in the emulation of human behaviour have allowed EDSSs to be used as systems capable of integrating all this knowledge and providing – in a manner that is easy to understand by users – reasoned proposals that are used as elements of discussion to reach consensual solutions.

Simultaneously, there has been an evolution in the incorporation of EDSSs into dynamic management systems. This has involved the incorporation of monitoring technologies, including data acquisition, their validation and use in real time, in order to be able to provide answers and actions according to operational needs. The incorporation of expert knowledge to complement classic control systems can be considered as one of the major challenges in the use of an EDSS. Through their application, the following features have been shown:

- Their ability to acquire, represent and structure knowledge, being able to process uncertainty relative to both data and knowledge.
- The ability to separate data models, and therefore, the possibility of working in more general and broader spectra.
- The ability to work with temporal and spatial dimensions.
- The ability to provide expert knowledge integrating specific knowledge bases.
- The ability to provide objective off-line and on-line responses.
- The ability to be used for diagnosis, planning, management and optimisation.
- The ability to help the user when formulating the problem and the selection of methods and models to solve it, enabling different alternatives to be assessed.

26.6 HOW TO BUILD AN EDSS?

Given that there are different proposals for what can be considered as a DSS, there are also different proposals for how they can be built. Although there may be different nuances from case to case, depending on the characteristics of each problem, we propose the realisation of five sequential steps. These begin with the approach to the problem, which will be to identify what is expected of the EDSS and the different elements that come into play for its building, including material aspects, but, especially, the definition of the relations between the agents involved (who coordinates, who asks for information, what its work flow will be or who can access which parts of the system). In the second stage, we believe it is important to stress the complementarity between data collection and knowledge acquisition. From our experience, we understand it is important to collect knowledge from experts and well as the use of data mining tools, which allow knowledge to be extracted from databases. It will be specific knowledge from the system under study, but important to complement the theoretical knowledge that can be obtained from literature or the experience of the experts interviewed. The third stage of analysing the results of the cognitive analysis is important because it is often a turning point in the building of the system. This is a point at which some of the initial objectives can be revisited, based on the reality observed in this stage. The fourth stage, model selection, is conditioned by the type of knowledge acquired and the previous experience of the developers, but in any case, one of the EDSS strengths is its ability to integrate quantitative and qualitative aspects. The final integration and implementation stage is more technical. If in previous stages, leadership can be by more experienced people in the area of sanitation or water in general, leadership in this part corresponds to engineers or scientists with expertise in computer systems.

There are two remaining tasks that should be executed in parallel although most of the time these tasks are (unfortunately) carried out at the end. On the one hand, there is the validation of each of the tasks to be performed. It is not necessary to wait until the end to rethink some objectives or evaluate the reliability of some results. On the other hand, there is the required transfer of the product built to the end user. An EDSS is a product that is usually built with input from many people, but that will surely be used by a smaller number of people, and those should be involved in the project from the start to feel at home.

26.7 NOVEDAR_EDSS: AN EDSS FOR SELECTION OF WWTP CONFIGURATIONS

NOVEDAR_EDSS tackles the difficult task of the wastewater treatment technologies selection and integration, i.e. the aim of this EDSS is to select the optimal integration of technologies for the treatment of a specific wastewater in a specific location. There is not a single solution but a combination of different technologies for the water (pre-treatment, primary treatment, secondary treatment and advanced treatment) and the sludge lines (thickening, enhanced digestion and dewatering). 250 different technologies are included in the NOVEDAR_EDSS, with a very high number of potential alternatives to compare. At the same time, the criteria to be used to evaluate the suitability of the design are increasingly elaborated. It is not just about achieving adequate water quality levels at the effluent, but also additional aspects such as operational safety, costs (with special attention to energy requirement) and the environmental impact of the plant relative to greenhouse gases and other emissions are considered.

The knowledge used to build the knowledge base of the NOVEDAR_EDSS was extracted from technical and scientific literature as well as expert interviews from academia and engineers from relevant water companies. Therefore, both empirical and theoretical knowledge was considered.

All this knowledge was collected and organized in two main knowledge bases (KBs): The first KB (Skb, for Specificities knowledge base) includes the main features of the different treatment technologies, based on technical, economical and environmental information. For every technological unit or process the following information can be found:

- *Influent Information:* Parameters that define the water quality expected at the input of the unit process in order to perform properly its function within the overall process (maximum COD, TN, etc.).
- *Effluent Information:* Information about the expected water quality at the output of the unit, indicating its performance (process efficiencies for a series of pollutants, nutrient removal, etc.)
- *Impacts/Subproducts:* Information about the whole range of possible impacts that a WWTP can generate either from social issues (odors, noises, visual impact, etc.) to environmental parameters resulting from Life Cycle Analysis, Environmental Avoided Impact, etc.
- *Operation:* Data defining design issues and more technical characteristics of the units such as maintenance, process stability, problem frequency, etc.
- *Costs:* Mathematical equations that allow an objective quantification of the main costs in the treatment process (investment, operation costs, energy consumption, etc.).

The second KB (Ckb, for Compatibilities knowledge base) contains information about the degree of compatibility among the different technologies. Both KBs are linked to another database with additional information about legislation to determine the effluent requirements. The knowledge-based methodology provides a platform able to generate an extensive WWTP alternatives response surface according to the treatment requirements and the decision maker desires. Such information is represented as a structure in the form of a network. All possible WWTP alternatives can be encapsulated in a single network figure, composed by nodes (technologies) and edges (connectivity properties between technologies).

A scenario is first defined by means of the influent data, effluent requirements and the priority setting criteria selection. Then the generation of WWTP alternatives is carried out by the interaction between the two knowledge bases. When the EDSS is executed, secondary treatments feasible for the specific scenario are selected and, based on the compatibility knowledge base, the complete process flow diagram (PFD) is generated (Figure 26.3). When the compatible PFDs have been created, feasible solutions that meet the overall degree of satisfaction of the user have to be selected. After this step, the multiple technological combinations can be evaluated based on technical, economical and environmental criteria. All PFDs that do not reach all the specified user requirements will be directly removed.

The EDSS shows firstly the most appropriate secondary technologies from highest to lowest degree of satisfaction. Once the list of most suitable options has been obtained, the user can click upon each of the options. This will allow seeing which combinations of primary, tertiary and sludge treatment are recommended for that particular secondary option. Once a whole line of technologies has been selected, composing a complete WWTP alternative, the EDSS program displays the results that would support the decision-making. All technical, environmental and operational results are presented for the selected treatment alternative in every operational unit. Finally, a summary is presented showing global information for the total flow diagram selected. Moreover, if further information about any technology is required in order to assist in the decision-making, the user can access the library just by clicking on the desired technology.

Therefore the EDSS allows: (i) the synthesis of multiple flow process diagrams including different treatment schemes, and (ii) the analysis of these diagrams from an environmental, economical, social and technical point of view. Thus, the sequence of unit processes that maximizes the degree of satisfaction of the different objectives by using multi-criteria and sensitivity analysis tools can be selected (Garrido-Baserba *et al.* 2012).

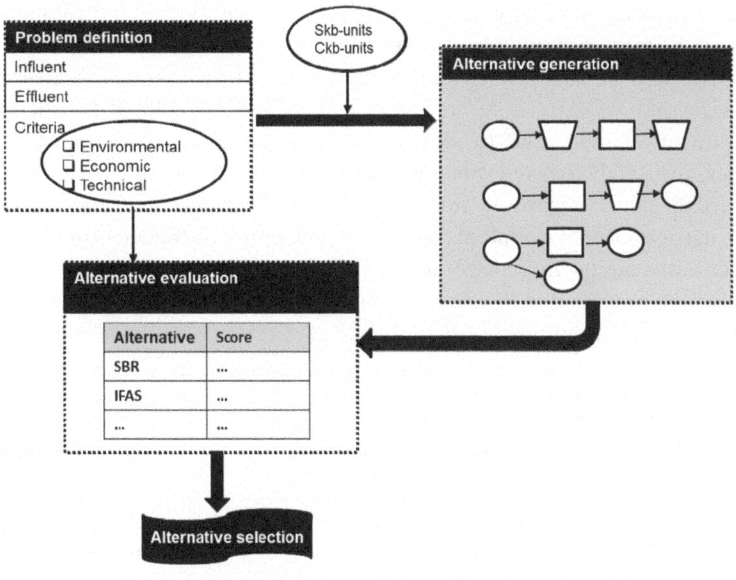

Figure 26.3 Scheme showing the NOVEDAR_EDSS alternative process selection (Castillo *et al.* 2015). Skb stands for specific knowledge base while Ckb for compatibility knowledge base.

The NOVEDAR_EDSS can be used for the selection of the most adequate process flow diagram for both greenfield (new) facilities and for existing WWTP retrofitting, ensuring its widespread application. The NOVEDAR_EDSS interfaces have been designed to become an easy to use instrument, and are intended to be very intuitive for any user while operating.

NOVEDAR_EDSS was developed in the framework of the NOVEDAR Consolider research project and validated and upgraded within the Water_2020 COST Action. The NOVEDAR consortium involved 11 universities while Water_2020 integrated a consortium of 85 institutions (73% universities, 9% technological centres, 9% companies and 9% water authorities). Currently, a version of NOVEDAR_EDSS is in the stage of being commercialised after a licensing agreement was signed with the company Aqualogy (Suez, France).

26.8 NOVEDARPLUS_EDSS: AN EDSS FOR THE '3R' PARADIGM

Nowadays, population growth, urbanisation and a stronger social awareness of environmental aspects (including climate change) are driving a new approach towards wastewater treatment. This new paradigm imposes new challenges that can be condensed in the '3R' (reduce, reuse and recover) objectives for wastewater treatment. More specifically, there is an increasing demand for systems more efficient in reducing energy and space requirements, costs and environmental impacts (greenhouse gases and emerging pollutants such as micropollutants and odor emissions), while there is also a need/interest for reusing water and sludge as well as for recovering energy, nutrients and organic added value products from wastewater. This new paradigm partly replaces the wastewater treatment conception (mostly focused on pollutant removal) by recovery of resources from wastewater, so that WWTPs can be considered water resource recovery facilities (WRRFs).

To satisfy these new challenges, NOVEDARplus_EDSS is being developed in the framework of the Water_2020 COST Action. This new version of the EDSS includes an upgraded knowledge base with the latest technologies as well as new functionalities and criteria to deal with the '3R' objectives. On the other hand, some of the technologies already included in NOVEDAR_EDSS were updated with the new information obtained.

The knowledge acquisition step involved a detailed survey, interviews and bibliographic review for innovative and more efficient technologies. A specific questionnaire to collect data and information about each technology and its stage of development was sent to the Water_2020 experts from WG1 (efficient technologies) and WG2 (resource recovery technologies) (Table 26.1). The questionnaire was complemented with a set of interviews to identify design and operational parameters, removal and/or recovery efficiencies, costs and other relevant parameters for each technology.

The new knowledge gained regarding innovative processes covers: (i) more efficient technologies for the removal of nutrients, micropollutants and odors, in terms of higher removal efficiencies, lower sludge production or lower capital and/or operation costs, (ii) technologies for resource recovery, (iii) new configurations both in the water and in the sludge treatment line, and (iv) information on the compatibility of these new processes with conventional ones (Table 26.1).

Table 26.1 Fields of the new knowledge acquired.

Efficient Technologies	Resource Recovery Technologies	New Configurations
Energy efficient nutrient removal	Nitrogen compounds	Anaerobic treatment as mainstream in the water line
Anaerobic treatments	Phosphorus compounds	Reject water
Biofiltration and biotrickling filters (odors)	Biopolymers and bioplastics	Sludge line
Granular systems	Struvite precipitation	Odor removal
Biolectrochemical systems	Energy	
Micropollutants	Water (tertiary treatment)	
Membrane bioreactors	Pre-treatment technologies for enhanced biogas production	

As an example of the data acquired in the interviews and questionnaires, Table 26.2 presents the most relevant design and operational parameters concerning one of the proposed configurations for biopolymers production, as an organic matter recovery strategy.

Table 26.2 Data for the recovery strategy based on biopolymers.

	Parameter	Data
Influent	Adequate COD conc. (mg/l)	more than 1500
	organic loading (COD kg/d)	1.5 (ER)/5.0 (AR)
	Adequate VFA concentration (mg/L)	<60 Cmmol/L (as VFA) (AR)
	Adequate TS concentration (mg/L)	As low as possible
	Adequate Total Nitrogen, TKN (mg/L)	Enough for biomass growth (ER)/as low as possible (AR)
	Adequate Ammonium (NH4; mg/L)	Enough for biomass growth (ER)/as low as possible (AR)
	C:N ratio	Similar as for sewage treatment (ER)/as high as possible (AR)
	Adequate phosphorus concentration (mg/l)	Enough for biomass growth (ER)/not required (AR)
Resource recovery	COD recovery	10% of the COD contained in the acidified sludge/1 kg of COD gives 0.62 kg COD as VFA that produces 0.11 kg of PHA
Operational parameters	HRT(h)	1–2 d (ER)/Discontinuous 8–24 h (AR)
	SRT (d)	1–6 d (ER)
	MLSS (g/l)	1,000–3,000 (ER)/1,000–7,000 (AR)
	Substances and/or conditions that could cause disturbances during performance	Bad performance of the acidifying reactor causing very variable composition of the effluent in VFA/Too much nitrogen in the accumulation reactor/High solids concentrations in influent due to poor anaerobic sludge dewatering/Not sufficiently high DO concentration to perform the process/Too low temperature conditions. Fluctuations of the pH value.
Economic	CAPEX	*Acidifying reactor*: Similar to anaerobic digester. ER and AR similar to activated sludge (AS)
	OPEX	Similar to AS
Impacts	Energy consumption	Similar to AS
	Greenhouse release (CO_2, kg/m^3)	37% (ER)/45% (AR)
	Other impacts, by-products generation	Residual solids after PHA extraction and treated wastewater.
Process characteristics	Suitable for retrofitting	Yes. Placed in the side-stream line it is needed a previous settler or a membrane to remove anaerobic sludge is required between the acidifying reactor and the SBR and discontinuous reactor.
	Configuration	Sludge acidification reactor + SBR for PHA accumulating mixed culture enrichment + discontinuous reactor for PHA accumulation maximization
	Stage of development	Under development at pilot scale
Technical qualitative parameters	Need for specialized staff	Only during the design and start-up processes.
	Problem frequency	Low (once stable conditions were reached in the acidifying reactor)
	Control	Medium (Required for the control of the operational cycles)
	Safety and health risk	No specific risk associated
	Robustness	Low
	Flexibility	High
	Stability	Medium

ER: enrichment reactor; AR: accumulation reactor.
Source: Mosquera (2015).

Regarding the new functionalities, they refer to the possibility to include: (i) technological options for resource recovery in the mainstream line, (ii) the reject water treatment line (supernatant from anaerobic sludge digestion, thickening and dewatering), also including resource recovery options, (iii) resource recovery strategies in the sludge treatment line, as well as (iv) the odor control systems, which will be included in order to allow the selection of proper treatment for gaseous emissions in a WWTP (Figure 26.4). Furthermore, the prioritised list of alternative process flow diagrams is based on the assessment of the complete treatment line (currently NOVEDAR_EDSS only considers the secondary treatment technology for ranking the different alternatives) and new criteria are included to perform the multi-criteria analysis (for example, regarding ecotoxicology, micropollutants and odors removal or nutrient recovery efficiencies).

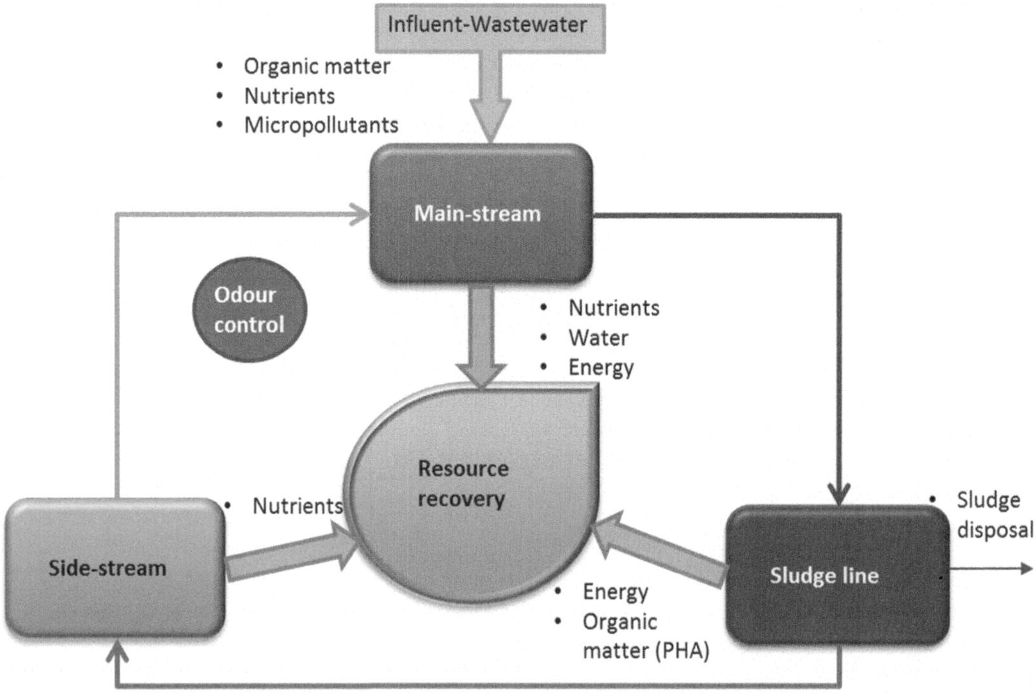

Figure 26.4 Flow diagram considered in the new NOVEDARplus_EDSS for the '3R' WWTP concept.

An EDSS development must be understood as a continuous process since new advances should be included to ensure the usefulness and longevity of the tool. In our case, NOVEDARplus_EDSS is being developed as a flexible tool able to tackle the new paradigm by including new specific functionalities and knowledge.

26.9 CASE STUDIES

In this subsection, three different case studies will demonstrate the NOVEDAR_EDSS capabilities to create suitable Process Flow Diagram (PFD) alternatives. Case studies 1 and 2 are related to a new (greenfield) and a retrofitted WWTP, where the scenario defined in the EDSS is based on real data from the corresponding facility. The EDSS generates a list with the most feasible alternatives to be applied in each case and, based on the selected criteria, secondary treatment options are scored through a multi-criteria analysis (Table 26.3). Finally, results obtained are compared to the treatment alternatives applied in the two real WWTPs selected.

On the other hand, case study 3, based on the Benchmark Simulation Model No. 2 (Gernaey *et al.* 2014), has been selected to illustrate the integration of three decision support tools for conceiving, comparing and selecting efficient wastewater treatment processes: NOVEDAR_EDSS, a superstructure-based optimization tool (Chapter 27) and simulation models (Chapter 28). In order to perform it, in the first stage the EDSS, based on expert knowledge and artificial intelligence techniques, is applied to select the most adequate alternatives, taking into account environmental, technical and economic criteria (Table 26.3). Secondly, the optimization tool, based on steady state simplified models and the definition of an objective function, allows selecting an optimal solution from the shortlisted candidates obtained with the EDSS. Finally, the identified optimal alternatives are evaluated using rigorous and dynamic models in order to obtain the optimal process design and operational parameters. The BSM2 platform is a general simulation protocol for benchmarking of operational and control strategies at WWTPs. It consists of: (i) a general plant layout including both the water and sludge lines; (ii) a set-up of sub-models for the included processes; (iii) models for sensors, controllers and actuators to allow for implementations of various control strategies; (iv) a specified simulation procedure including an influent profile; and, (v) an evaluation procedure including three aggregated indices: Effluent Quality Index (EQI), Operational Cost Index (OCI) and Operational Risk Index. More information about the BSM platform can be found in Gernaey *et al.* (2014).

Table 26.3 Selected environmental, technical and economic criteria considered for every case study (and their corresponding weights).

	Criteria	BSM2	WWTP Retrofitting	New WWTP
Environmental	LCA	5.0		20.0
	Visual Impacts	2.5	5.5	
	Odor Potential	2.5	5.5	
	Noise Potential	2.5	5.5	
	Energy Consumption		8.3	
	Sludge Production		8.3	
	Space Requirements			40.0
Technical	Operation Simplicity	11.7	4.7	6.6
	Control Over the Process	11.7	4.7	
	Flexibility	11.7	4.7	6.6
	Reliability	11.7	4.7	6.6
	Problem Frequency	11.7	4.7	
	Need for Skilled Staff	11.7	4.7	
	Innovation Degree		4.7	
Economic	Investment	7.5	16.5	10.0
	Operation and Maintenance (O&M)	7.5	16.5	10.0

Numbers for each criterion are referring to the percentage for the total score.

26.9.1 Case study#1: design of a greenfield WWTP under different conditions

The case study#1 is based on a real WWTP situated in Cork (Ireland). The aim of this case study is to illustrate the usefulness of the EDSS when a greenfield WWTP for urban wastewater treatment is designed. The WWTP was designed for organic matter removal and with an effluent quality required to discharge in a sensitive area. One of the most important constraints in this case was space limitation, and therefore a compact solution was required.

Objective

In this case, the objective is to determine whether the EDSS recommendations for a specific case are similar to the selected treatment in the real project when a greenfield facility should be developed. Moreover, different conditions and criteria are applied in order to determine the EDSS response.

Scenario definition

Influent data as well as effluent requirements are presented in Table 26.4. It is a medium-size plant with a high contaminant load (based on Metcalf & Eddy, 2003).

Table 26.4 Influent wastewater characteristics and effluent requirements in Cork WWTP.

Parameter	New WWTP	Effluent Requirements	Units
Q	71230		m^3/d
Pop.-eq.	440000		
COD	841	125	g/m^3
BOD	418	25	g/m^3
TSS	393	35	g/m^3
TKN	58	10	$g\ N/m^3$
TP	11.5 (default)	1	$g\ P/m^3$

Source: Degrémont, Water Treatment Handbook, 2007.

Since space limitation was an important constraint, this criterion will be weighted as the most important one. In Table 26.3 the selected criteria (and weights) for this specific case study are presented.

Results

After running the EDSS, a list of wastewater treatment alternatives able to achieve the effluent requirements is generated, where the main difference is the secondary treatment. Table 26.5 presents the different feasible alternatives together with the score for each of the criteria as well as the total one.

It can be observed that the EDSS ranks the SBR as the best treatment alternative when the selected criteria are those presented in Table 26.3. SBR is obtaining the highest score when a combination of economic, environmental and operational criteria is selected (Figure 26.5). The main advantage of SBR systems is their lower investment costs, since they do not require a secondary sedimentation tank, which is also related to reduced space requirements. The EDSS recommendation is in agreement with the solution adopted in reality since SBR was also the treatment technology selected for the real project.

Table 26.5 List of feasible treatment alternatives with the score when the criteria selected in this case study are applied.

Alternative	Total Score	Economic	Environmental	Technical
SBR + N and P removal	**8.23**	2.0	5.23	1.0
Conventional plug flow + N and P removal	**7.22**	0.83	5.05	1.33
UCT + N removal	**5.88**	0.83	3.89	1.17
VIP + N removal	**5.88**	0.83	3.89	1.17
PhoStrip	**5.72**	0.83	3.89	1.0
MLE + P removal	**5.54**	0.83	3.38	1.33
Ludzack-Ettinger + P removal	**5.21**	0.83	3.38	1.0
MBR-Flat Sheet + P removal	**4.98**	0.32	3.66	1.0
MBR-Hollow Fiber + P removal	**4.52**	0.32	3.2	1.0
Bardenpho + P removal	**4.51**	0.83	2.68	1.0
Double Stage + P removal	**3.15**	0.83	1.32	1.0

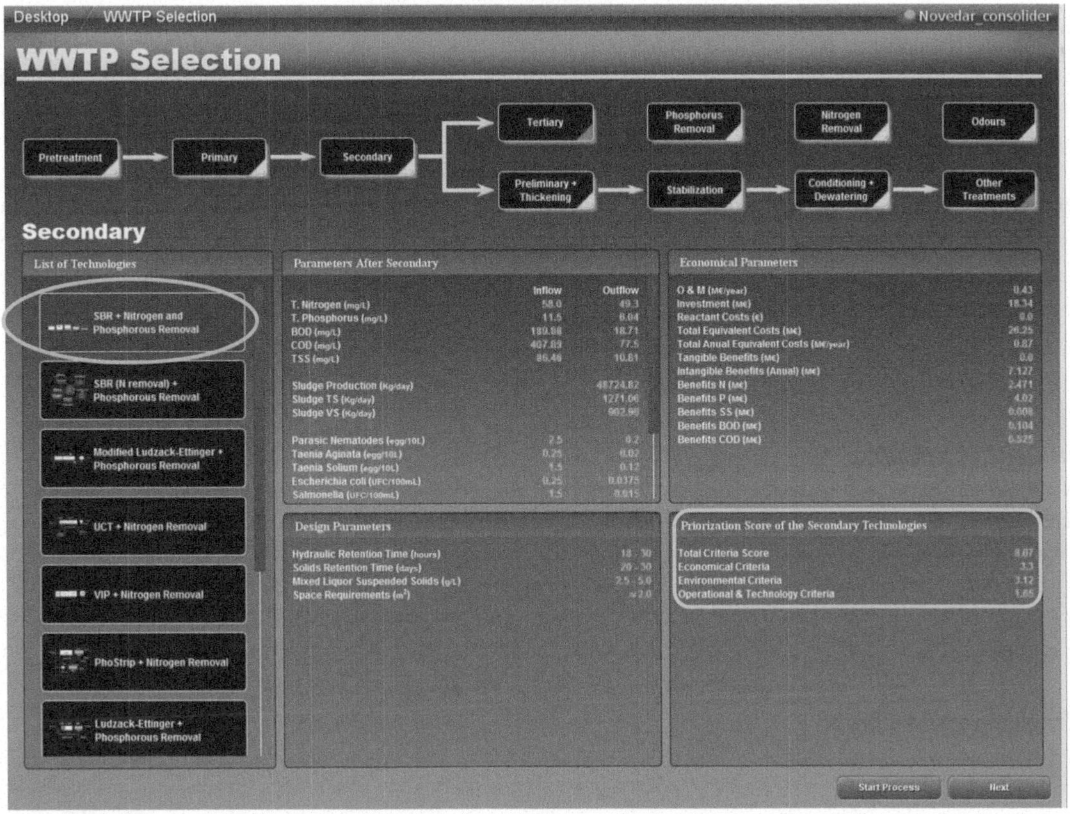

Figure 26.5 EDSS screen with the list of feasible secondary treatment alternatives for case study#1, where the score for the best ranked alternative is presented for the economic, environmental and operational criteria.

The use of the EDSS for the selection of process alternatives presents several advantages. First, it allows performing an integrated assessment based on economic, environmental and operational parameters within the same tool, representing significant time savings, and thus economic savings since normally different tools are needed for the different criteria. Moreover, it implements a multi-criteria analysis to score the different alternatives based on the user needs and it calculates some LCA-based criteria allowing the environmental impact for each alternative to be determined.

Analysis performed under different criteria

Environmental criteria: when only space requirements are considered, SBR is the best ranked option, followed by MBR. Moreover, when the aim is to select the alternative which represents the lower environmental impact, EDSS ranks a conventional plug flow reactor higher, followed by SBR and MLE.

Economic criteria: Since SBR systems do not require a secondary settler, even though they require a buffer tank, the investment costs are lower than those for the other alternatives considered, while presenting similar operational costs. MBR is the worst option when economic criteria are prioritized.

Operational criteria: MLE together with conventional plug flow are the best ranked treatment systems when operational criteria become more important. This could be related to the fact that MLE systems are the most widespread wastewater configuration and their reliability has been proven. In this case, the SBR obtains a similar mark as the MBR, since both are treatments requiring more process control.

In order to achieve the required removal efficiencies for nutrient removal, the EDSS recommends different alternatives for phosphorus and nitrogen removal: methanol addition, Anammox, Canon, Oland or deammonification for nitrogen removal and chemical phosphorus removal strategies when there is no anaerobic stage in the secondary alternative selected.

Operation and maintenance (O&M) costs as well as Cost Benefit Analysis (CBA) data, including or not environmental externalities (which take into account for the economic analysis that prevent environmental damage when pollutants are removed from wastewater), for each technology is also provided by the EDSS.

Results under different conditions

In this subsection, the EDSS solutions sensitivity is analysed under different variable conditions, such as the influent flow rate or effluent requirements.

Influent conditions

When an increase of 20% in the flow rate is applied, the treatment alternatives considered by the EDSS are different from those in the scenario with the original flow rate. In fact, SBR is no longer among the feasible treatments, since it is recommended for medium to small facilities. A plug flow is the best ranked alternative in this case, if we keep the criteria in Table 26.3 (Figure 26.6). However, when space constraints are prioritized as being the only selected criterion, then the best ranked alternative is MBR (Figure 26.7).

Effluent requirements

If reutilization of the effluent is needed, then the EDSS will recommend a tertiary treatment after the secondary. The list of feasible secondary treatments consists essentially of the same treatments as recommended in the original scenario but applying a tertiary treatment (such as ozone, ultraviolet radiation…). However, if MBR is selected it would not be necessary to apply any additional treatment since this technology already allows for a better effluent quality.

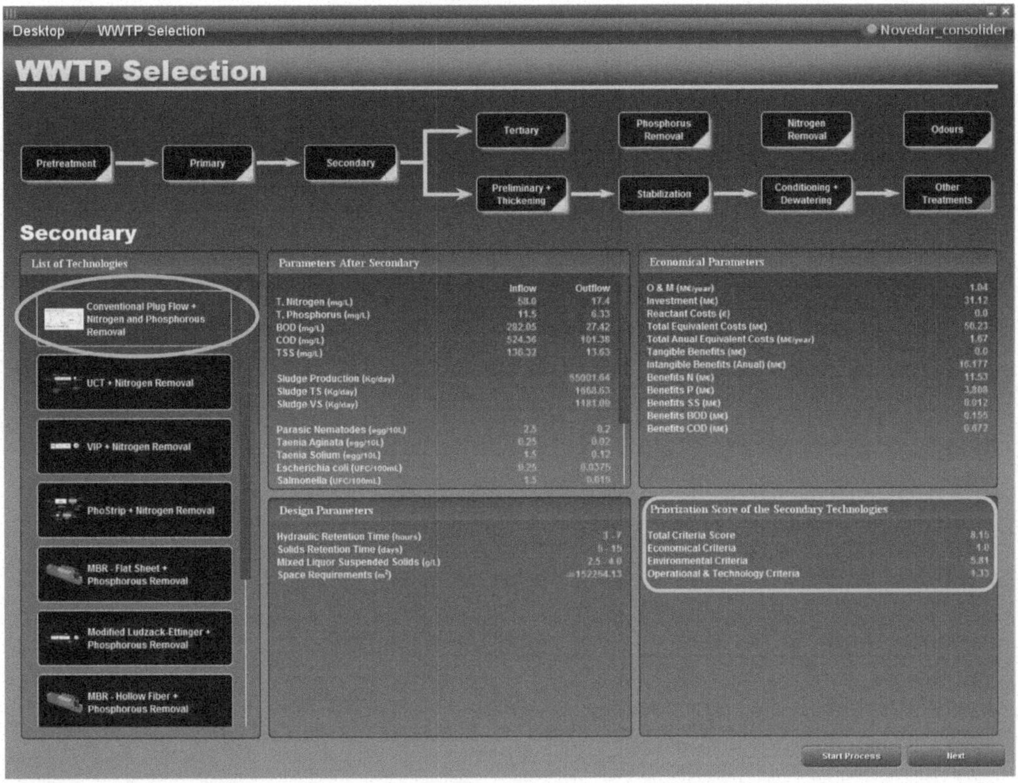

Figure 26.6 EDSS screen with the list of feasible secondary treatment alternatives for case study#1, when influent flow rate is 20% higher than the real value. The score for the best ranked alternative is presented for the economic, environmental and operational criteria.

On the other hand, when the effluent discharges in a river, instead of in a sensitive area, the EDSS recommends some of the secondary treatments in the list (Table 26.5) but no additional nutrient removal treatment is required.

Conclusion

EDSS recommends the same secondary treatment as the actual treatment adopted in the facility. The EDSS also allows performing a multi-criteria analysis where, depending on the selected criteria, different scores are obtained for the secondary treatment.

26.9.2 Case study#2: retrofitting of a real WWTP under different conditions

Case study#2 is based on a real WWTP situated in Girona (Catalonia, Spain). The aim of this case study is to illustrate the usefulness of the EDSS when an existing WWTP for municipal wastewater has to be upgraded. The original facility was built in 1986 and it was retrofitted in 2008 since new effluent requirements needed to be achieved.

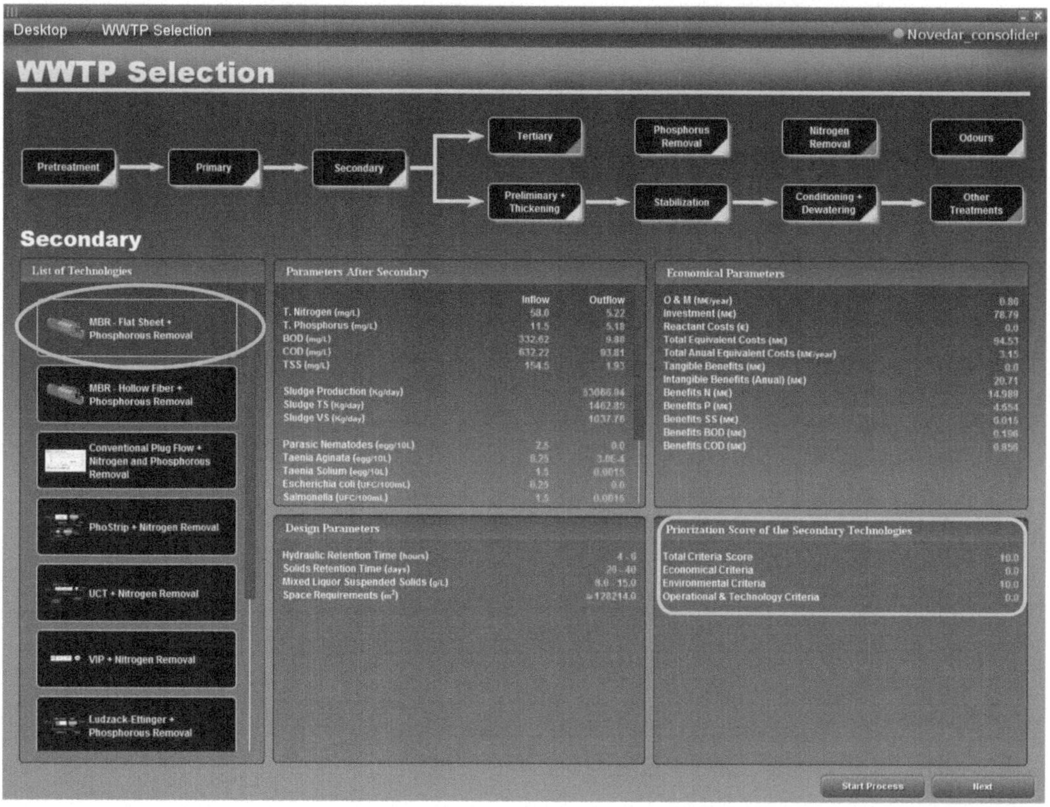

Figure 26.7 EDSS screen with the ranked alternatives when only the criterion regarding space requirements is prioritized.

Objective

The objective in this case is to assess the EDSS recommendations for a WWTP upgrade and also to compare it with the actual upgrading performed.

Scenario definition

Data related to the influent characteristics for both the original plant as well as for the upgraded one are presented in Table 26.6.

Three different situations have been identified when a WWTP has to be upgraded:

(1) The whole plant needs to be rebuilt; therefore it becomes like a new/greenfield plant.
(2) The treatment line is duplicated, so that the existing treatment technology is also applied to the new treatment line.
(3) The treatment technology needs to be modified in one or in all treatment lines.

Currently, the EDSS can already be used for those retrofitting cases where a whole secondary treatment was rebuilt, as in the case of Girona. The application of the NOVEDAR to the other situations requires some specific functionalities that are under development. Future work is focused on allowing the existing treatment system to be fixed so that, depending on the needs (space availability, etc.), the EDSS will recommend those treatment alternatives that are compatible with the existing configuration/technology.

Table 26.6 Influent data for the original and the upgraded facility and effluent requirements to be achieved by the upgraded plant.

Parameter	WWTP 1986	Upgraded WWTP 2008	Effluent Requirements	Units
Q	42000	55000		m^3/d
Pop.-eq	140000	206250		
COD	515	550	125	g/m^3
BOD	200	225	25	g/m^3
TSS	220	250	35	g/m^3
NKT	51	58	10	$g\ N/m^3$
TP	8	8	1	$g\ P/m^3$

Results

Two scenarios are generated in the EDSS for this case study.

(a) Data regarding influent concentrations and effluent requirements for the original plant are used to generate the first scenario in the EDSS.

The EDSS generates a list with feasible treatment alternatives able to achieve the effluent requirements, where the main difference is the selected secondary treatment (Figure 26.8). It should be noted that the original configuration of the plant coincides with one of the proposed technologies (in bold):

- Biofilter
- SBR
- SBR (nutrient removal)
- **Conventional plug flow**
- Virginia Inititative Plant (VIP) (anaerobic/anoxic/aerobic)
- Modified Ludzack-Ettinger (MLE) (anoxic/aerobic)
- University of Cape Town (UCT) (anaerobic/anoxic/aerobic)
- Bardenpho (anoxic/aerobic/anoxic/aerobic)
- PhoStrip (combines biological and chemical phosphorous removal)
- Ludzack-Ettinger (LE) (anoxic/aerobic)
- Double Stage (anoxic/aerobic)
- SBR (N removal)

(b) Data regarding influent concentrations and effluent requirements when the WWTP has to be retrofitted are used to generate the second scenario.

In this case, since the main difference from the old facility is the nutrient concentrations at the effluent, the feasible treatments are essentially the same as in scenario 1 but **adopting some nutrient removal technologies**. The recommended treatment depends on the criteria selected when defining the scenario (Table 26.3). In order to consider different aspects, economic, environmental and operational criteria are selected to score secondary treatments in this case study. Results are presented in Table 26.7.

At this point the EDSS scores only secondary treatments, i.e. it is not considering the whole process flow diagram to rank the different alternatives. Therefore sometimes the best scored secondary treatment alternative implies additional strategies for nutrient removal and it would obtain a lower score if the whole process flow diagram was considered (because of higher investment costs etc.), becoming better scored than some other secondary treatment alternatives, which would not need any additional strategy.

When only operational criteria are considered to rank the treatment alternatives, those that can implement organic matter and nutrient removal within the same technology (i.e. they do not need additional treatment) are better scored than others (Figure 26.9). For instance, MLE and UCT are better ranked, and UCT was the treatment alternative finally adopted for the upgrade of the Girona WWTP.

O&M costs as well as Cost Benefit Analysis (CBA) data, including or not environmental externalities, for each technology is also provided by the EDSS.

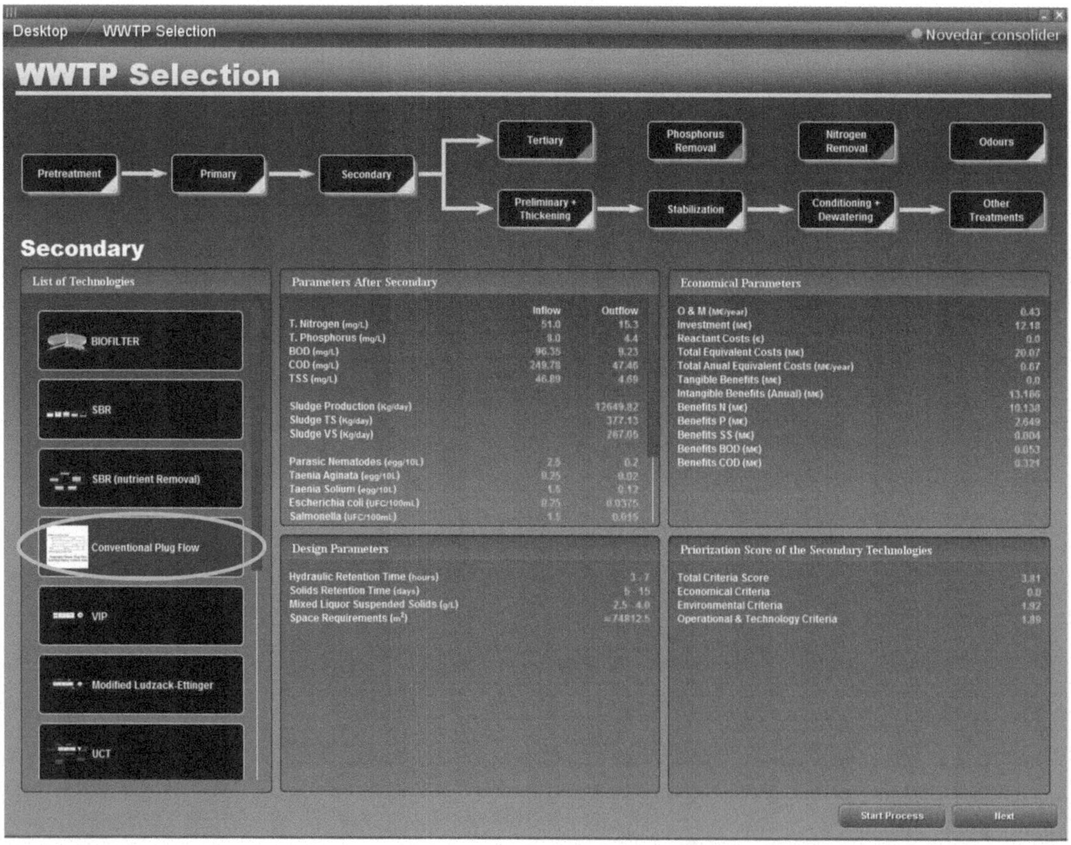

Figure 26.8 EDSS screen with the list of secondary treatment alternatives feasible for case study#2. They are scored based on economic, environmental and operational criteria.

Table 26.7 List of secondary treatment alternatives with the corresponding score when the criteria selected in Table 26.3 are applied.

Alternative	Total Score	Economic	Environmental	Technical
Biofilter + P removal	**6.71**	3.3	1.99	1.41
SBR + P removal + N removal	**6.02**	3.3	1.31	1.41
SBR (nutrient removal) + N removal	**5.26**	3.3	0.55	1.41
SBR (N removal) + P removal	**3.89**	1.65	0.82	1.41
Conventional plug flow + P removal + N removal	**3.81**	0.0	1.92	1.89
VIP + N removal	**3.05**	0.0	1.17	1.89
MLE (anoxic/aerobic) + P removal	**3.01**	0.0	1.0	2.0
UCT + N removal	**2.91**	0.0	0.91	2.0
Bardenpho + P removal	**2.85**	0.0	0.96	1.89
PhoStrip + N removal	**2.81**	0.0	1.04	1.77
Ludzack-Ettinger + P removal	**2.77**	0.0	1.0	1.77
Double Stage + P removal	**2.34**	0.0	0.69	1.65

Conclusion

EDSS selects as a feasible alternative the treatment that was implemented in the original WWTP when scenario 1 is performed. After that, when a retrofitting scenario is considered, technologies with better removal efficiencies are required and the technology adopted in the real facility is one of the best scored when selecting operational parameters.

26.9.3 Case study#3: BSM2 case study

The BSM2 plant configuration is described by Gernaey *et al.* (2014) and consists on a primary clarifier followed by a secondary treatment system based on a Modified Ludzack-Ettinger configuration and a secondary clarifier. The sludge line includes a thickener, an anaerobic digester and a dewatering unit. Finally, reject water from thickening and dewatering is returned to the inlet of the primary clarifier. The plant layout is shown in Figure 26.10.

The influent contributions considered are: households (62%), industry (13%) and rain (25%) sources. The average flow-weighted influent concentrations calculated for one year of influent data generated with the BSM2 influent model are presented in Table 26.8.

Objective

As stated before, in this case, the objective is to run the EDSS to identify the most adequate wastewater treatment alternatives for the defined BSM2 influent, which will be sequentially introduced in the superstructure optimization tool. It will also allow to determine if the actual BSM2 configuration (MLE, Modified Ludzack-Ettinger) is among the ones recommended by the EDSS, while the multi-criteria analysis will allow to identify which are the criteria priorities when MLE is the best scored treatment.

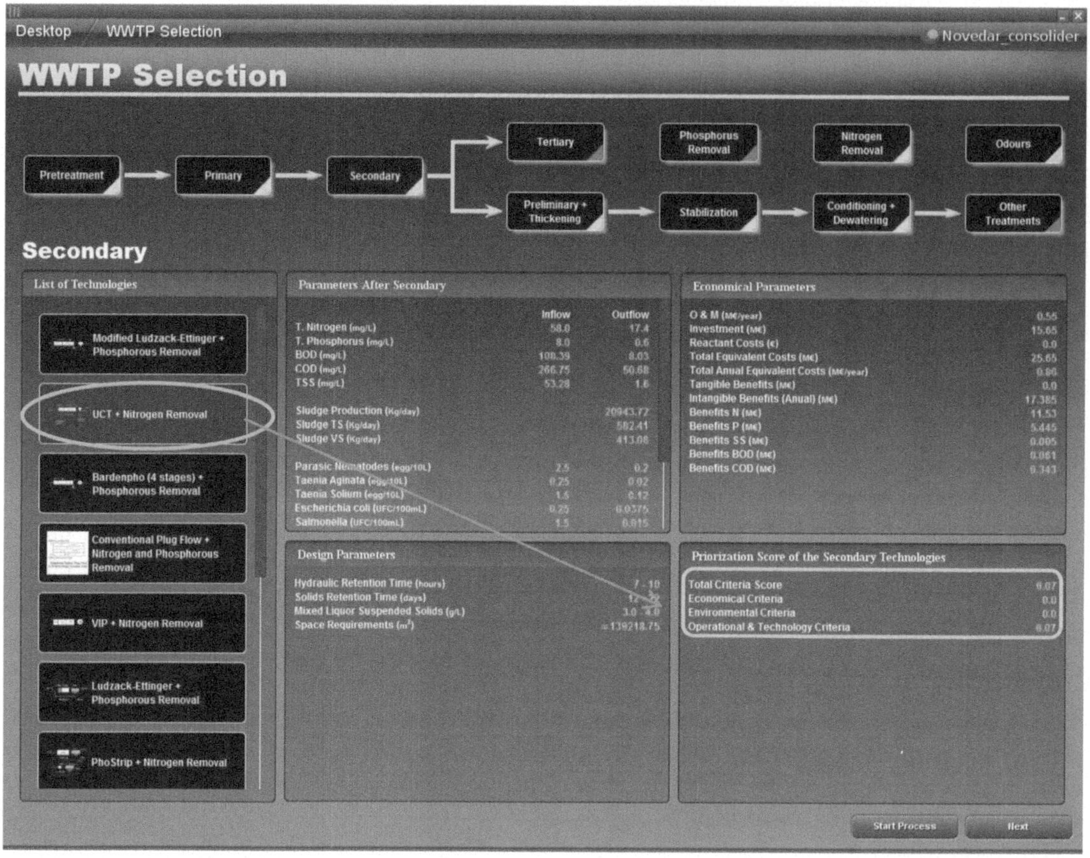

Figure 26.9 EDSS screen with the list of feasible secondary treatment alternatives for case study#2, with the score for the best ranked alternative when operational criteria are selected.

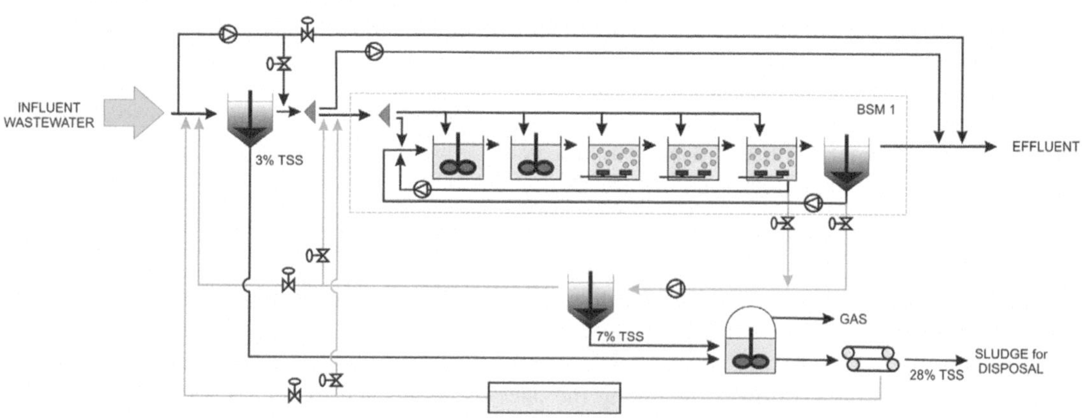

Figure 26.10 Schematic representation of the BSM2 plant design (Gernaey *et al.* 2014).

Table 26.8 Influent concentrations from the BSM2 influent model (nomenclature from ASM1, Henze *et al.* 2000).

Variable	Concentration	Unit
Q	20648.36	m^3/d
S_I	27.23	g COD/m^3
S_S	58.18	g COD/m^3
X_I	92.50	g COD/m^3
X_S	363.94	g COD/m^3
$X_{B,H}$	50.68	g COD/m^3
S_{NH}	23.86	g N/m^3
S_{ND}	5.65	g N/m^3
X_{ND}	16.13	g N/m^3
TSS	380.34	g/m^3

Scenario definition

Concentrations for the BSM2 influent (Table 26.9) can be calculated from Table 26.8, assuming a BOD/COD ratio equal to 0.6. Effluent requirements (Table 26.9) are those for a sensitive area, and therefore nutrient removal is needed. Note that phosphorus is not considered in the standard BSM2 framework but added here for the purpose of the EDSS.

Table 26.9 BSM2 influent data and effluent requirements to discharge in a sensitive area.

Variable	Influent	Effluent Requirements	Unit
Q	20668.44		m^3/d
COD	592.25	125	g/m^3
BOD	355.35	25	g/m^3
TSS	380.17	35	g/m^3
TKN	45.61	10	g N/m^3
TP	11.5 (default)	1	g P/m^3

Source for the effluent requirements: European Community, 2000.

Results

Table 26.10 compiles the list of feasible secondary treatment alternatives for the BSM2 case study after running the NOVEDAR_EDSS. Scores obtained in the multi-criteria analysis represent when integrated assesment is applied, i.e. taking into account environmental, economic and operational costs (Table 26.3).

The MLE configuration obtains the best score. The difference between this alternative and the second best ranked alternative (conventional plug flow) is due to the environmental and technical aspects, while economic parameters should be quite similar (the same economic score is obtained). MBR obtains the lowest score, due to its bigger investment and operational costs and because it is a more technologically demanding.

Table 26.10 List of feasible secondary treatment alternatives with the corresponding score when the criteria selected in Table 26.3 are applied.

Alternative	Total Score	Economic	Environmental	Technical
MLE + P removal	**6.44**	1.24	0.53	4.67
Conventional Plug Flow + P removal	**6.28**	1.24	0.66	4.38
Oxidation ditch + N and P removal	**6.15**	1.24	0.53	4.38
UCT	**6.12**	1.24	0.5	4.38
Ludzack-Ettinger + P removal	**5.86**	1.24	0.53	4.08
Bardenpho (4 stages) + P removal	**5.85**	1.24	0.52	4.08
VIP	**5.83**	1.24	0.5	4.08
Double Stage + Premoval	**5.56**	1.24	0.52	3.79
PhoStrip	**5.54**	1.24	0.5	3.79
IFAS + N and P removal	**5.1**	0.56	0.75	3.79
SBR + N and P removal	**4.92**	1.5	0.5	2.92
SBR (nutrient removal) + P removal	**4.89**	1.5	0.47	2.92
SBR (N removal) + P removal	**4.85**	1.4	0.53	2.92
MBR-Flat Sheet + P removal	**3.93**	0.67	0.34	2.92
MBR-Hollow Fiber + P removal	**3.83**	0.67	0.25	2.92

Shortlist selection of alternatives

Based on the rank of the feasible treatment alternatives recommended by the EDSS and the current superstructure database, a shortlist of alternatives is selected. This set of alternatives is presented in Table 26.11 with better detail regarding investment and O&M costs as well as Cost Benefit Analysis (CBA) data, including or not externalities (which gives an economic value to the environmental damage avoided when pollutants are removed from wastewater).

Table 26.11 Shortlist of alternatives to be considered in the superstructure optimization tool.

Treatment alternative	Investment Cost (M€)	O&M Cost (M€/year)	CBA (M€)	CBA with ext. (M€)
MLE	11.16	0.4	–51.29	–30.56
OxD	11.16	0.4	–51.96	–34.62
IFAS	11.71	1.57	–82.76	–65.82
SBR	8.58	0.4	–49.39	–28.66
MBR	21.01	0.36	–59.83	–39.76

With respect to the second objective for this case study, it can be stated that the actual secondary treatment configuration defined in the BSM2 is among the secondary treatment alternatives also recommended by the EDSS. Since MLE allows achieving the required nitrogen limits in the effluent, no additional nitrogen removal is needed. However, a strategy for phosphorus removal is needed and EDSS recommends a chemical phosphorus precipitation.

In order to determine which criteria would prioritize MLE among the other feasible treatment alternatives in this case study, a criteria selection process has been performed. Results indicate that MLE is the best ranked alternative (Figure 26.11) when a set of environmental, economic and technical criteria is considered. The specific criteria together with their corresponding weights are presented in Table 26.3. However, when different criteria are prioritized, different configurations may be better suited. For instance, when only economic aspects are selected, other alternatives such as SBR are better ranked by the EDSS.

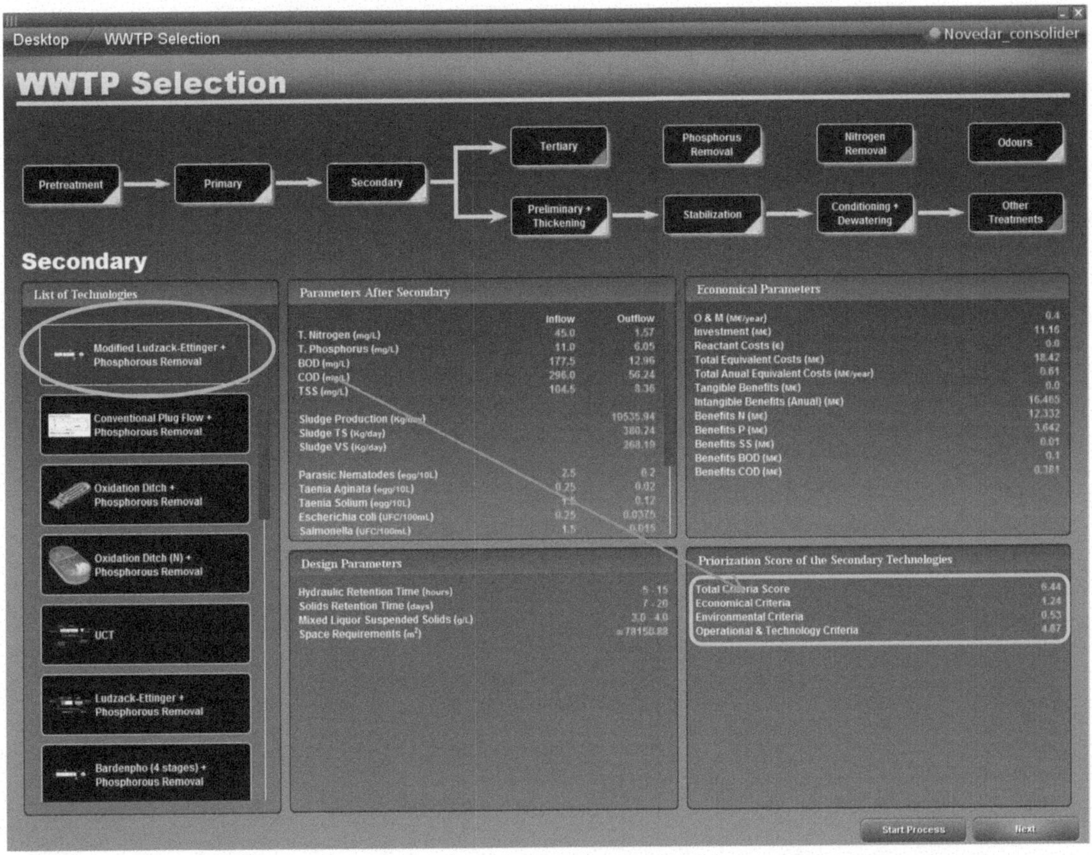

Figure 26.11 EDSS screen with the list of secondary treatment alternatives for case study#3, where the score for the best ranked alternative (MLE) is based on economic, environmental and operational criteria.

26.10 CONCLUSIONS

The EDSS recommends a MLE configuration for the wastewater treatment system when a scenario based on the BSM2 influent and effluent requirements is defined, which is indeed the actual BSM2 configuration. Results from the multi-criteria analysis show that the MLE is the best ranked alternative when an integrated assessment is applied, i.e. considering economic, environmental and operational criteria.

26.11 REFERENCES

Castillo A., Cheali P., Gómez V., Comas V., Poch M. and Sin G. (2016). An integrated methodology for the optimal selection of sustainable and economic wastewater treatment plants (WWTP): intelligent expert system and superstructure-based optimization tool. *Environmental Modelling and Software*, **84**, 177–192.

Cortés U., Sànchez-Marrè M., Cecaronni L., Rodríguez-Roda I. and Poch M. (2000). Artificial intelligence and environmental decision support systems. *Applied Intelligence*, **13**(1), 77–91.

Degrémont (2007). Water Treatment Handbook, 7th edn, Lavoisier, Paris, France.

European Community (2000). Water framework directive (2000/60/EC). http://ec.europa.eu/environment/water/water-framework/index_en.html (accessed 13 November 2015).

Fox J. and Das S. (2000). Safe and sound. Artificial Intelligence in Hazardous Applications, AAAI Press/The MIT Press, Cambridge, MA, USA.

Garrido-Baserba M., Reif R., Hernández F. and Poch M. (2012). Implementation of a knowledge-based methodology in a decision support system for the design of suitable wastewater treatment process flow diagrams. *Journal of Environmental Management*, **112**, 384–391.

Gernaey K. V., Jeppsson U., Vanrolleghem P. A. and Copp J. B. (2014). Benchmarking of Control Strategies for Wastewater Treatment Plants, Scientific and Technical Report No. 23, IWA Publishing.

Henze M., Gujer, W., Mino, T. and van Loosdrecht M. (2000). Activated Sludge Models ASM1, ASM2, ASM2d and ASM3, Scientific and Technical Report No. 9, IWA Publishing, London, UK.

Metcalf & Eddy (2003). Wastewater Engineering: Treatment and Reuse, 4th edn, McGraw-Hill, New York.

Mosquera A. (2015). Knowledge gained through a survey and personal interviews during the Water_2020 COST Action meetings.

Chapter 27

Superstructure-based optimization tool for plant design and retrofitting

Hande Bozkurt, Krist V. Gernaey and Gürkan Sin

27.1 INTRODUCTION

Many wastewater treatment processes and technologies have been developed since the beginning of the 20th century to meet increasingly stringent performance demands. Still the wastewater industry has been facing many new challenges such as increasing energy costs, presence of trace organics which has become more critically investigated, depletion of the resources, water conservation as well as more stringent regulations (Reardon *et al.* 2013).

As a result, domestic wastewater is now being considered more as a resource than as a waste with the recovery possibilities for clean water, energy and various materials satisfied by novel approaches and emerging technological developments (McCarty *et al.* 2011). Therefore, in parallel to these recent research trends into energy efficiency, GHG emissions, resource recovery, the WWTP process selection and network design problem has moved from being a simple technical design problem into a complex problem that requires an integrated approach for making effective decisions (Bozkurt *et al.* 2015; Hamouda *et al.* 2009).

Economics has always been a key driver when making decisions on which treatment methods and technologies to be applied. As a consequence of the fact that wastewater treatment plants are significant energy consumers, principally for aeration, mixing, pumping, sludge disposal etc., together with the rising energy costs and concerns and restrictions on emissions of greenhouse gases; wastewater specialists focus more on effective energy management and alternative energy strategies. Current initiatives include increasing the biogas production, managing oxygen demand by controlling oxygen concentration or by decreasing the oxygen demand of the microbial activities (e.g. Partial nitritation described in Hellinga *et al.* (1998), and anaerobic ammonium oxidation described in Mulder *et al.* (1995)). Moreover, the increasing construction cost and decreasing space availability stimulated the development and use of more compact systems which require less footprint (for instance, biological aerated filters and granular activated sludge.

Hence both the multi-criteria nature of the decision-making process and the large number of alternatives of wastewater treatment technologies bring us to the question 'How do we take strategic decisions on the wastewater treatment technologies and process networks to choose for a given wastewater treatment project?'

(a) *Problem formulation for WWTP process design:* One of the most challenging steps in wastewater treatment plant design is the selection of the treatment technologies, defined as a combination of unit operations and processes capable of meeting effluent permit requirements (Tchobanoglous, 2003). This particular task is referred to as Wastewater treatment process synthesis and defined as:

"*Wastewater treatment process synthesis* is the step in the design or retrofitting of a wastewater treatment plant (WWTP) where the design engineer selects unit processes (separation and/or reaction including physical, chemical and biological processes) from numerous alternatives and interconnects them to create the process flow diagram."

Hence, the objective of process synthesis is to find the optimal process flow diagram, among numerous alternatives, for treating a given influent wastewater with its flow rate and composition to meet predefined performance criteria including effluent permit requirements as well as cost and technical requirements.

(b) *Motivation for an optimization framework:* In order to realize the full potential of the process synthesis approach and its use in a wastewater treatment context, there are a number of barriers that need to be tackled and overcome, including the representation of the increasing number of unit operations and processes being used in wastewater treatment as well as tackling the resulting multi-disciplinary complexity of the optimization problem, which requires both competences and methods from optimization together with the wastewater engineering discipline. In particular, for formulating a realistic wastewater treatment design problem (i.e. a large number of alternatives representing the market together with their feasible combinations as well as necessary constraints that should be taken into account), the complexity of the mixed integer nonlinear programming problem (MINLP) can grow exponentially, which needs an effective formulation and analysis method, which is the focus of this contribution.

Therefore, in this section, a superstructure based optimization framework based on mathematical programming is presented as methodology to manage the complexity of the WWTP design and retrofitting problems. The purpose of the framework is to support and complement the expertise of the design engineers/specialists in the process of making early stage design decisions by allowing them to compare several different treatment technologies at their optimality with respect to many different criteria. The framework contains a superstructure method for representing the design space, and a systematic method for modelling and data collection to constitute the database covering design data for individual treatment technologies. The framework also includes a generic and effective formulation of a mixed integer nonlinear programming problem to address the wastewater treatment process synthesis problem. This section is complementary to the methodology presented in Chapter 26, in which the results from the environmental decision support system for the expert-based selection of alternatives can be used as input to define in a systematic and intelligent way the superstructure. The integration of NOVEDAR_EDSS and the superstructure based optimisation is presented elsewhere (Castillo *et al.* 2016).

27.2 SUPERSTRUCTURE-BASED OPTIMIZATION FRAMEWORK

An earlier developed process design framework for addressing the chemical process synthesis problem (Quaglia *et al.* (2013) is extended and adapted for use in the context of WWTP design and retrofit problems. The framework is illustrated in Figure 27.1. The framework consists of six steps and results in the optimal solution for the network design. It contains a superstructure method for representing the design space and a systematic method for data collection and modelling of the treatment alternatives, which further enables effective formulation and solution of a MI(N)LP problem under both deterministic conditions and uncertainty.

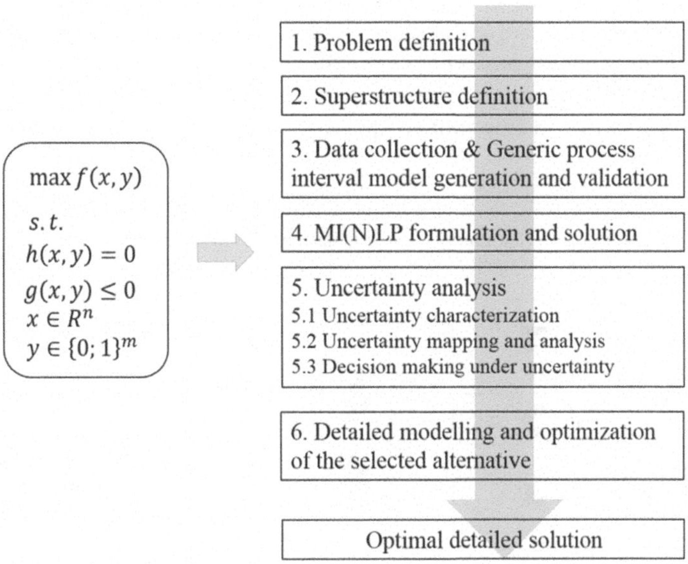

Figure 27.1 Framework for the superstructure based optimization methodology. Bozkurt *et al.* (2015).

(a) *Methodology*: Step by step: After defining the wastewater characterization, sink limitations and objective function in the first step, the superstructure is generated consisting of the base case of the existing WWTP (i.e. the existing treatment units, only in the retrofitting studies) and other alternative treatment technologies under the existing or new treatment tasks. This step can be carried out with the support of the NOVEDAR_EDSS. The formulation of the superstructure is then finalized by defining the feasible connection streams between the treatment tasks. Each treatment unit in the superstructure is defined using a generic model based on mass input-output. The parameters of the generic model are maintained in a database, e.g. process performances (yields, stoichiometry etc.), utility consumptions, volumes and sludge production. The optimal wastewater network problem is then formulated as a mixed integer (non)linear programming (MI(N)LP) problem and solved for deterministic conditions and under uncertainty by defining several scenarios.

(b) *Superstructure definition, data collection and design of individual treatment technologies:* A superstructure, as shown in Figure 27.2, is a compact representation of different process alternatives (i.e. treatment technologies in the wastewater treatment case). Process steps, which are represented in the columns of the superstructure, comprise wastewater sources and sinks for the effluent streams (effluent water, sludge, by-products etc.) as well as different tasks to be carried out throughout the network in order to establish the wastewater and sludge treatment network in between sources and sinks. In each process step, in the rows of the superstructure, alternative treatment process intervals (e.g. separation – primary clarifier, secondary clarifier, membrane reactor etc.; or reaction – activated sludge for C, N and P removal, anammox, anaerobic digestion etc.) responsible for a specific task are placed. The superstructure definition is finalized by defining the feasible connection streams between treatment tasks. As presented earlier, the information needed to define the superstructure concerning relevant technological alternatives to consider for a wastewater treatment problem of interest can be obtained in several ways: (a) by performing literature research and considering all alternatives available and linking them using all possible combinations, (b) by means of knowledge-based decision support systems utilizing process engineering knowledge as well as experiences (presented in Chapter 26), or (c) through a bottom up approach using a phenomena based approach (Quaglia, 2013).

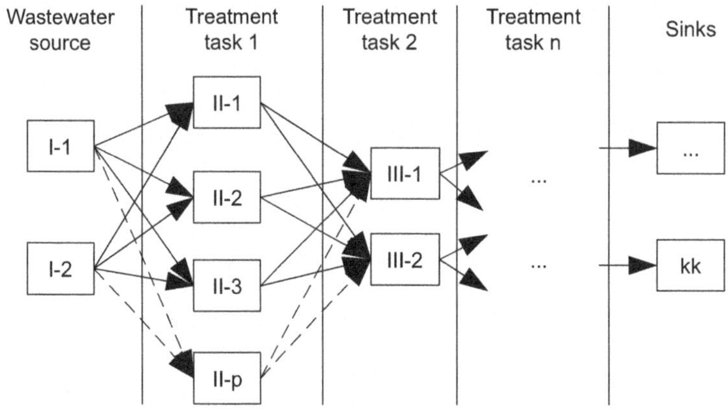

Figure 27.2 A representative superstructure for wastewater treatment networks.

For each technology alternative in the superstructure, system specific data and information about the specified treatment technology are collected such as; sludge retention time (SRT) and hydraulic retention time (HRT), temperature dependent biokinetic constants, settling data and recycle ratios if applicable, stoichiometry information for the reactions together with volumetric loading and/or removal rates and process performances in terms of removal efficiencies of the key contaminants. Here, the treatment technologies are designed at fixed temperature, at their optimality and at steady state by fixing the design parameters (SRT, HRT, efficiencies etc.) rather than optimizing them; later, in a second step, more rigorous models can be used for optimization once the number of alternative technologies is reduced. This two-tiered approach for optimization is chosen on purpose to manage the complexity of the optimization problem which becomes otherwise intractable. The output from the design includes volumes of the units, utility consumption (electricity, chemicals, aeration etc.) and sludge production data which are used to calculate the capital and operational costs. When the steady state design is optimized, the design parameters and performance values are compared with the steady state results obtained from a simulation carried out using a rigorous model (e.g. Activated Sludge Model No.1 (ASM1) (Henze *et al.* 2000)). Finally, the cost data for utilities, electricity, products and construction of the treatment units are collected from the open literature.

(c) *Generic process interval model generation and validation:* Each process interval in the superstructure is structured using a generic model which is illustrated in Figure 27.3.

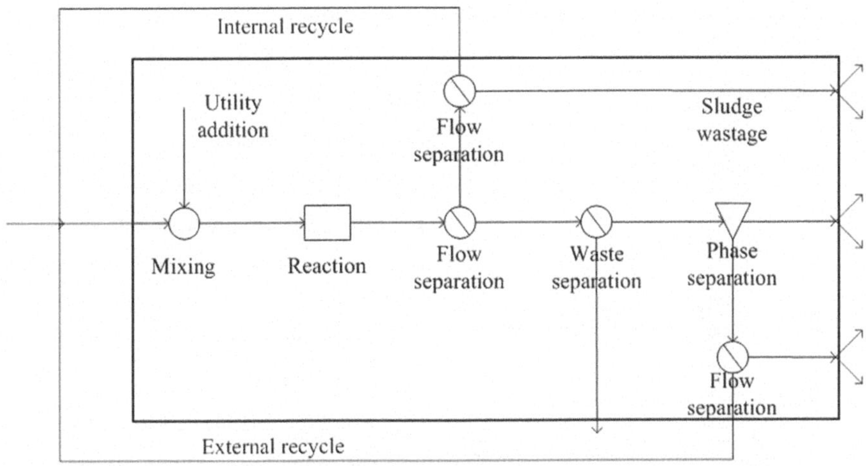

Figure 27.3 Generic process interval structure: definition of phenomena.

The generic model is composed of a number of phenomena namely: mixing of all the flows entering the interval and the utilities added, reaction, flow separation, waste separation and phase separation. Thus, the process interval can perform as a combination of different units (reaction and separation with external and internal recycle flows) as well as a simple reactor with one inlet and one outlet. Using the generic model, the treatment alternatives are described based on input-output mass balances. The flow of components is represented in mass flux units (M/t) whereas the total flow rate of the stream is given in volumetric flow rate units (L^3/t). Below, the definition of all the phenomena existing within the generic process interval and the associated mathematical model are explained in Table 27.1.

(d) *Mixed integer (non)-linear programming for optimal solution:* The formulation of the MI(N)LP problem and its deterministic solution is done in the fourth step of the framework. The solution of the optimization problem results in the optimal network at steady state, the fate of pollutants/components throughout the selected treatment network and the value of the objective function together with the cost breakdown into the components of the objective function (i.e. utility cost, product cost, capital cost etc.). With respect to the nature of the problem, the optimization problem can result in a linear (MILP) or non-linear (MINLP) formulation which depends on the definition of the objective function (linear with respect to continous variables or not) and the constraints (e.g. fractional splitting of outflows allowed).

In this step, the MI(N)LP problem is formulated and solved. The models represent the mass input-output model for each treatment technology (i.e. each process interval in the superstructure), process constraints, structural constraints, effluent limit constraints, and economic models together with the objective function. The adapted MI(N)LP formulation for the specific case of a WWTP design/retrofit study is described below with the equations 27.1–27.4. In these equations, the objective function is formulated as the minimization of the sum of operational and capital (annualized) costs which is subject to steady-state mass balances described in ***h***, the superstructure information contained in ***g*** and integer variable ***y*** stipulating only one process alternative per interval can be selected. All the other notations are described in Table 27.1.

Table 27.1 Mathematical equations representing the generic process model in each interval of the superstructure.

Phenomena	Equation	Explanations
Mixing	$F_{i,kk}^{in} = \sum_{f} F_{i,k,kk}$	*i,ii:* Component index
		k,kk: Process interval index
	$F_{i,kk}^{mix} = F_{i,kk}^{in} + \alpha_{i,kk} * R_{i,kk}$	$F_{i,ikk}^{in}$: Inflow to the process interval
		$F_{i,k,kk}$: Inflow of component *i* to process *kk* coming from *k*
Utility Addition	$R_{i,kk} = \sum_{ii} (\mu_{i,ii,kk} * F_{ii,kk})$	$F_{i,kk}^{mix}$: Flow after mixing
Reaction	$F_{i,kk}^{reac} = F_{i,kk}^{mix} + \sum_{rr,react} \left(\gamma_{i,kk,rr} * \theta_{react,kk,rr} * F_{react,kk}\right)$	$R_{i,k,kk}$: Utility flow
		$\alpha_{i,kk}$: Fraction of utility consumed
		$\mu_{i,ii,kk}$: Specific consumption of utility
		$F_{i,kk}^{reac}$: Flow after reaction
Waste separation	$F_{i,kk}^{w} = F_{i,kk}^{reac} * (1 - W_{i,kk})$	$\gamma_{i,kk,rr}$: matrix representing reaction stoichiometry
Flow Separation	$F_{i,kk}^{out1} = F_{i,kk}^{w} * Split_{i,kk}$	$\theta_{react,kk,rr}$: Conversion efficiency of the key reactant *react*
	$F_{i,kk}^{out2} = F_{i,kk}^{w} - F_{i,kk}^{out1} - F_{i,kk}^{rec}$	$F_{i,kk}^{w}$: Flow after waste separation
		$W_{i,kk}$: Waste split factor
	$F_{i,kk}^{out3} = F_{i,kk}^{reac} * SW_{kk}$	$F_{i,kk}^{out1}, F_{i,kk}^{out2}, F_{i,kk}^{out3}$: Outlet streams from interval
	$F_{i,kk}^{rec} = \left(F_{i,kk}^{w} - F_{i,kk}^{out1}\right) * rec_{kk}$	$Split_{i,kk}$: Flow split factor
		SW_{kk}: Sludge wastage flow rate ratio
	$FX_{i,k,kk} \le F_{i,kk}^{outX} * S_{k,kk}$	$F_{i,kk}^{rec}$: External recycle flow rate
Activation	$y_{kk} * x_{k}^{LO} \le x_{k} \le y_{kk} * x_{k}^{UP}$	rec_{kk}: External recycle ratio
		X: 1,2,3 (representing three different outlet flow streams)
Logical Cuts	$\sum_{kk} y_{kk} \le 1$	$S_{k,kk}$: Binary variables containing superstructure information
Effluent Limits	$Lim_{i,kk} \ge \sum_{f} F_{i,kk}^{in}$	y_{kk}: Binary variable describing the process interval
		x_k: Variable bounded by x_k^{LO} and x_k^{UP}
		$Lim_{i,kk}$: Effluent limit value of component *i*

$$\text{Min } OBJ = \sum_{kk} OPEX_{kk} + \frac{CAPEX_{kk}}{t} \tag{27.1}$$

subject to;

$$\boldsymbol{h}\left(\alpha_{i,kk}, \mu_{i,ii,kk}, \gamma_{i,kk,rr}, \theta_{react,kk,rr}, W_{i,kk}, Split_{i,kk}, rec_{kk}\right) = 0 \tag{27.2}$$

$$g\left(S_{k,kk}\right) \leq 0 \tag{27.3}$$

$$\sum_{kk} y_{kk} \leq 1 \quad \textbf{where } y \in \{0;1\}^n \tag{27.4}$$

27.3 CASE STUDY APPLICATION

To illustrate the potential combined use of the EDSS and the optimisation framework, BSM2 is selected as the case study to highlight the application of the framework. Thus, wastewater characteristics and effluent requirement as defined in BSM2 are used in this case study.

(a) *Problem and superstructure definition & generic process interval model generation:* The problem is defined as treatment of domestic wastewater comprising mainly COD, nitrogen and solids as pollutants. The wastewater composition is shown in Table 27.2. Accordingly, the influent total COD is 592.25 g COD/m^3 and the total nitrogen content is 61 g N/m^3. The objective is to design the WWTP network against the lowest operational cost (aeration cost, sludge disposal cost, pumping and mixing cost as well as biogas price) and capital cost possible; while satisfying the effluent limitations for organic material and nitrogen in the water stream to be discharged into the environment.

Table 27.2 Influent wastewater characterization, average composition.

Component	Value	Unit	Component	Value	Unit
S_I	27.21	g COD/m^3	S_O	0	g COD/m^3
S_S	58.15	g COD/m^3	S_{NO}	0	g N/m^3
X_I	92.46	g COD/m^3	S_{NH}	23.85	g N/m^3
X_S	363.77	g COD/m^3	S_{ND}	5.64	g N/m^3
$X_{B,H}$	50.66	g COD/m^3	X_{ND}	16.12	g N/m^3
$X_{B,A}$	0	g COD/m^3	S_{ALK}	7	mole/m^3
X_P	0	g COD/m^3	Q	20,648.36	m^3/d

Source: Gernaey *et al.* (2014).

The superstructure, as shown in Figure 27.4 consists of a domestic wastewater source interval and sinks for effluent water, sludge and biogas. Treatment technologies are located in between the source and the sinks, and include the following (for the sake of this tutorial, the number of process alternatives is kept small intentionally):

- For the *primary treatment*, the only technology put into the superstructure is the primary clarifier (PC).
- *For the secondary treatment*, based on the EDSS output, three different biological treatment units are present: (1) Modified Ludzack Ettinger (MLE) system with its low SRT, (2) Oxidation ditch, which has a relatively higher SRT and (3) Upflow anaerobic sludge blanket (UASB) technology, which works under anaerobic conditions and produces biogas as a result of degradation of organics. While the first two technologies are proven to be effective in terms of organic matter and nitrogen removal, the UASB type of reactors only perform organic matter destruction.

- *For this reason, a tertiary treatment* is added into the superstructure with the aim of performing nitrogen removal. The two different technologies put here are: Sharon, which is responsible for nitritation – denitritation and a Sharon reactor coupled with an Anammox reactor to perform partial nitritation and anaerobic ammonia oxidation, respectively.
- *Disinfection* technologies comprise different means of treatment with respect to the disinfection agent or method applied: UV, O_3 and chlorine.
- The *sludge treatment* line consists of sludge stabilization options namely anaerobic digestion and aerobic digestion, which are receiving the sludge from a thickener.
- For each task, there is an additional process interval defined as the By-pass interval – BP1-BP5 (except for the sludge thickener).

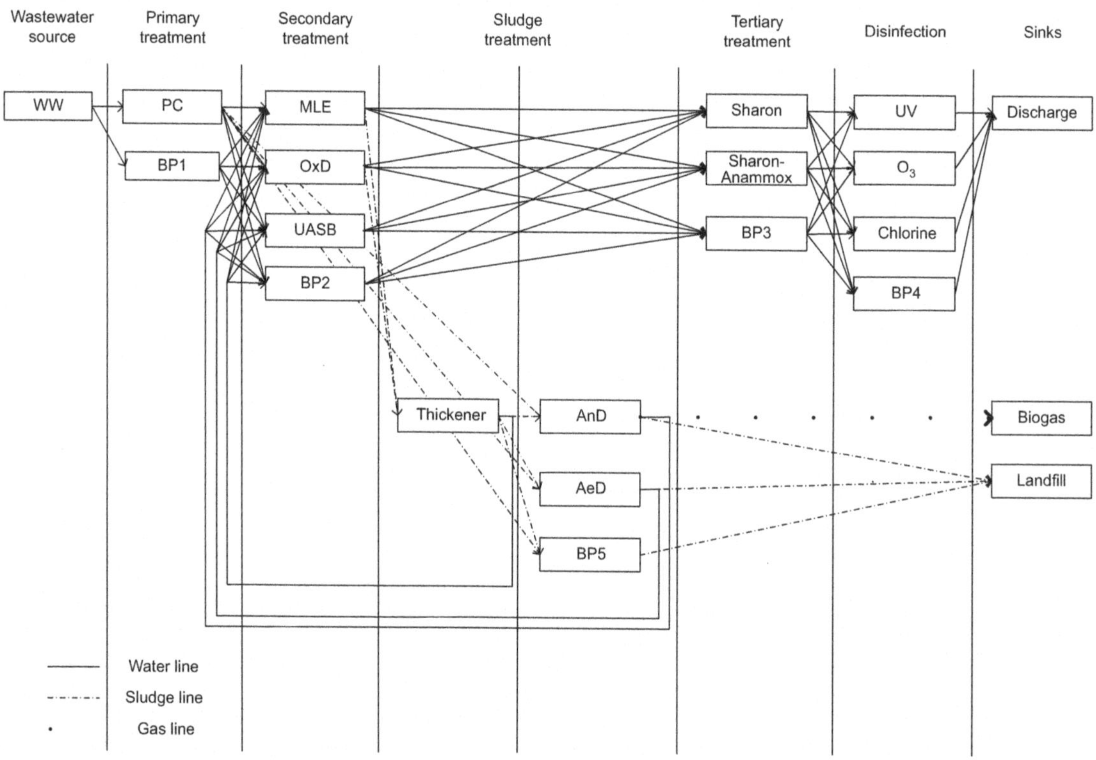

Figure 27.4 Superstructure developed for the case study.

Design of primary treatment alternatives

In the primary sedimentation, 99.3% of the soluble and 50% of the particulate components by mass leave the primary sedimentation tank with the water effluent stream while the rest is sent to underflow (Gernaey *et al.* 2014). The volume of the primary sedimentation tank is calculated to be 1,537 m^3 by setting the Surface Overflow Rate (SOR) and retention time (td) values from literature (Tchobanoglous, 2003).

Design of secondary treatment alternatives

The wastewater characterization of BSM2 represents a raw wastewater source; in other words, it constitutes a high amount of solids. For this purpose, in order to design the secondary treatment units, the pre-settled wastewater characterization of Benchmark Simulation Model No.1 (BSM1; Copp, 2002) is used. The BSM1 platform represents the predecessor of BSM2 and is based on the same principles and simulation protocol. However, the BSM1 only describes the acticated sludge and secondary clarification processes of the WWTP whereas BSM2 respresents a complete plant. The BSM1 average dry weather wastewater composition comprises approximately 50% less total suspended solids (TSS) as compared to the raw wastewater composition of BSM2 (Copp, 2002; Gernaey *et al.* 2014). The pollutants are represented by ASM1 components. The effluent limits given in the Urban Wastewater Treatment Directive (91/271/EEC) are taken as a reference for the sink intervals. The treatment plant has been designed for operating at 15°C. System specific design data has been collected from several sources (Tchobanoglous, 2003 & WEF, 2010) together with the temperature dependent biokinetic constants and stoichiometry information (Copp, 2002).

The secondary settler is incorporated into the secondary treatment process intervals (MLE and OxD) as the phase separation. Within the settler, 48% of the soluble components and 0.2% of the particulates by mass leave with the water stream (Copp, 2002), while the rest is assumed to settle in the sludge zone. Moreover, the volume of the sedimentation basin is calculated by assuming specific surface overflow rate (SOR) and solids loading rate (SLR) values as well as a certain depth of the tank from a range given for circular clarifiers (WEF, 2010). The main outcomes of the design procedure for secondary treatment alternatives are summarized in Table 27.3.

Table 27.3 Process information for process intervals under the secondary treatment task.

Properties		**Process Interval**	
Corresponding Technology	**Modified Ludzack-Ettinger (MLE)**	**Oxidation ditch (OxD)**	**Upflow Anaerobic Sludge Blanket (UASB)**
Temperature (°C)	15	15	15
SRT (days)	14	28	120
HRT (hours)	12	24	14
Reactor volume (m^3)	9,223	18,446	12,956
Settler volume (m^3)	3,774	3,411	–
Anoxic/Aerobic volume ratio	0.6	0.6	–
MLSS (g/m^3)	3,410	3,032	18,590
Sludge wastage flow (% influent flow)	3.5	3.5	–
Sludge recycle flow (% influent flow)	100	100	–
COD removal efficiency (%)	88.4	87.78	68.5
Total N removal efficiency (%)	77.2	78.48	–

Design of tertiary treatment alternatives

The tertiary treatment alternatives are placed in the superstructure for removing the nitrogen from the wastewater stream. For this purpose two technologies were selected: Sharon and a two-stage Sharon/Anammox reactor.

- For the Sharon system, the SRT is equal to its HRT and selected as 1 d (Hellinga *et al.* 1998). The volume of the reactor is then calculated as 18,446 m^3. For the two-stage Sharon/Anammox system, the sizing of Sharon reactor is maintained and an Anammox reactor is designed with the nitrogen removal rate of 10 kg N/$m^3 \cdot d$ (van der Star *et al.* 2007). While calculating the nitrogen load into the tertiary treatment process intervals, the influent nitrogen is assumed not to be changing in the secondary treatment task; so that tertiary treatment can handle the maximum amount of nitrogen that is potentially entering the system. With regard to that assumption, the volume of the anammox reactor is calculated as 90 m^3.
- The removal efficiencies are taken as 86% and 60% (Hellinga *et al.* 1998) for the nitritation and denitritation mechanisms occurring in the Sharon reactor, whereas partial nitritation and anammox are assumed to proceed with 50% and 80% efficiencies, respectively (van Dongen *et al.* 2001).

Design of sludge stabilization alternatives

- *For the anaerobic digestion, the following observations have been made*: (1) the change in the concentrations of S_I, S_S, S_O, S_{NO}, S_{ND}, X_I and X_{ND} are either very low or there is no change at all (some of these components are not present in the influent sludge stream); (2) the increase in the concentrations of S_{NH} and X_P are proportional to the decrease in the heterotrophic biomass concentration – X_{BH}; and, (3) the increase in the methane gas production is proportional partly to biomass degradation and partly to particulate biodegradable COD (i.e. X_S) reduction. Thus, the design has been made generic by defining three reactions based on three components – X_{BH}, X_{BA} and X_S where they are removed with 100% efficiency; and all the other components (S_{NH}, X_P and methane gas) are produced accordingly.
- For the design of aerobic digestion, the observations were similar. Only this time, instead of methane production, oxygen consumption is observed to be proportional to the destruction of biomass and particulate biodegradable COD.
- The thickener unit is added into the superstructure as a separate process interval. The dewatering is incorporated into the sludge digestion units as phase separation. The separation factors for the thickening unit are calculated as 0.02 and 0.94 for particulates and soluble components, respectively. In other words, 2% of the particulates and 94% of the soluble components by mass, leave with the water stream while the rest are concentrated into the underflow. These values are set to 1% and 95% for the particulate and soluble components, respectively in the dewatering unit.

Design of disinfection alternatives

All the disinfection units are assumed to remove the pathogens in the wastewater stream with 100% efficiency. At this point, the utility (chemical and/or electricity) addition requirements are defined based on information from the open literature.

Collection of operational and capital cost data

The objective function represents the total annualized cost (TAC) and it is formulated to be minimized. OPEX corresponds to the operational cost and is composed of aeration, electricity consumption, chemical addition, pumping, mixing and landfill cost as well as biogas price. CAPEX on the other hand, represents the capital cost. All the cost data are collected from information available in the open literature and summarized in Table 27.4.

Table 27.4 Cost information for operational and capital cost items.

Cost item	Unit	Value/ Range	Description/Assumption
Electricity consumption of oxygen transfer[a]	kg O_2/kwh	1.9–3.2	Coarse bubble diffusor
Sodium hypochlorite cost[b]	euro/kg	0.12	–
Energy requirement for ozone[c]	kwh/kg O_3	21–35.2	Sum of ozone generation (air feed), ozone contacting and all other uses (on the average)
Electricity cost[d]	euro/kwh	0.0978	In Denmark for industry
Landfill cost[e]	euro/t	107	In Denmark, on the average
Biogas price[f]	eurocent/m^3 methane	40.3	In Denmark (assumptions: 1 mole of methane is 24 L and 1 mole of methane accounts for 64 g of COD)
Capital cost–UASB[7]	US$/$m^3$	425	Based on the price level of 2006, for 100,000 PE
Capital cost–Aeration tank[g]	US$/$m^3$	175	Based on the price level of 2006, for 100,000 PE
Capital cost–Secondary settler[g]	US$/$m^3$	290	Based on the price level of 2006, for 100,000 PE
Capital cost–Primary settler[g]	US$/$m^3$	375	Based on the price level of 2006, for 100,000 PE
Capital cost–Sludge thickener[g]	US$/$m^3$	400	Based on the price level of 2006, for 100,000 PE
Capital cost–Anaerobic digester[g]	US$/$m^3$	350	Based on the price level of 2006, for 100,000 PE

[a]Siemens (2009)
[b]AWWA Michigan Section (2006)
[c]Tchobanoglous (2003)
[d]Europe's Energy Portal (2013)
[e]Confederation of European Waste-to-Energy Plants (2013)
[f]Hahn *et al.* (2010)
[g]van Haandel and van der Lubbe (2012)

(b) *Deterministic solution:* The MILP problem is formulated as previously shown in equations 27.1–27.4. The objective of the optimization problem is to select among the treatment alternatives so that the resulting treatment process flow diagram has the minimum TAC and at the same time satisfies the effluent limits given in the Urban Wastewater Treatment Directive (91/271/EEC). The formulated optimization problem is solved by using GAMS and specifically the solver CPLEX. The details of the deterministic solution are analyzed and discussed for three scenarios corresponding to different objective function formulations:

- *Scenario 1*: The objective function covers only OPEX and the effluent limit for nitrogen is set to 15 mg N/L.
- *Scenario 2*: The objective function covers TAC with the effluent nitrogen limit set to 15 mg N/L.
- *Scenario 3*: The objective function covers TAC; however, this time the effluent nitrogen limit is set to 10 mg N/L.

The optimal process selection and value of the objective function after solution of the optimization problem are given in Table 27.5 for three different scenarios. Under the conditions of the first scenario, tertiary treatment and disinfection tasks are by-passed (this is mainly because the secondary treatment is effective enough to meet the effluent nitrogen limit and there is no pathogen defined in the influent wastewater composition); the water stream is sent to the water sink interval after being treated by the primary clarifier and the low SRT MLE system whereas the sludge is stabilized in the anaerobic digester and sent to the sink interval. When the capital cost is also added into the formulation of the objective function (i.e. scenario 2), the network selection does not change; however an expected increase in the objective function is observed due to capital cost.

In the third scenario, the optimizer once again selects the MLE technology coupled with a primary clarifier to treat the wastewater, and an anaerobic digester for sludge stabilization purposes.

Table 27.5 Summary of results for different scenarios.

Scenario	Objective Function	Total N Limit (mg/L)	Selected Process Flow Diagram	Value of Objective Function (Unit Cost)
1	OPEX	15	WW-PC-MLE-BP3-BP4-Thickener-AnD-Discharge-Sludge-Biogas	219.051
2	TAC	15	WW-PC-MLE-BP3-BP4-Thickener-AnD-Discharge-Sludge-Biogas	912.080
3	TAC	10	WW-PC-MLE-BP3-BP4-Thickener-AnD-Discharge-Sludge-Biogas	912.080

Although the anaerobic treatment alternative of the UASB coupled with the nitrogen rich wastewater treatment alternatives of the tertiary treatment task can satisfy the effluent total nitrogen limit, the UASB alone cannot generate an effluent stream complying with the COD effluent limit criteria when operated at such low temperatures. Experimental work also showed the decrease in the efficiency of the UASB reactor with a decreasing temperature (Lew *et al.* 2004). Although anaerobic treatment of domestic wastewater has been successfully demonstrated in full scale all over the world, the conclusion here, however, is that in order to comply with the regulations, the system should either be operated at higher temperatures or it should be integrated with effluent polishing steps which are not included in the current superstructure (i.e. facultative ponds, sand filtration, constructed wetlands, trickling filters, physico-chemical treatment and activated sludge treatment) (Henze *et al.* 2008). Therefore, this treatment alternative has not been selected by the optimizer for any of the scenarios as expected from the above-mentioned process engineering expertise.

The cost summary and the performance evaluation for the scenarios are presented in Table 27.6. MLE is favored both for its low operational cost (low utility requirement and high sludge production resulting in high biogas production) and relatively low capital cost. These results are in agreement with the EDSS results and MLE was indeed the actual configuration selected for BSM2.

Table 27.6 Cost summary and performance evaluation for the different scenarios.

	Unit	Scenario 1	Scenario 2	Scenario 3
Objective function	–	OPEX & Total N limit of 15 mg/L	TAC & Total N limit of 15 mg/L	TAC & Total N limit of 10 mg/L
Aeration cost	unit cost	229.187	229.187	229.187
Landfill cost	unit cost	614.191	614.191	614.191
Biogas price	unit cost	695.058	695.058	695.058
Electricity cost	unit cost	–	–	–
Pumping cost	unit cost	43.687	43.687	43.687
Mixing cost	unit cost	27.044	27.044	27.044
Capital cost	unit cost	–	693.029	693.029
OBJ	unit cost	219.051	912.080	912.080
Effluent COD	g COD/m^3	39.16	39.16	39.16
Effluent Total N	g N/m^3	9.82	9.82	9.82

(c) *Uncertainty analysis and sensitivity analysis:* Uncertainty in this specific case study is characterized with respect to cost related parameters (oxygen transfer efficiency, electricity and landfill prices), effluent total nitrogen limits and influent wastewater characterization. The parameters that are considered uncertain and their probability distribution together with mean, minimum and maximum values are given in Table 27.7. The alpha (α), beta (β) and fouling factor (F) parameters are used in the equation 5 (Tchobanoglous, 2003), in which the correction factor is calculated to convert the standard oxygen transfer rate in tap water (SOTR) to the actual oxygen transfer rate (AOTR) by taking into account the effects of salinity-surface tension, temperature, elevation etc. The other variables in the equations are as follows C_L, concentration of oxygen in the liquid, $C_{s,20}$, the staturation concentration of oxygen at 20°C and $C_{s,T,H}$ is the staturation concentration at a given temperature (T) and humidity (H). This affects the electricity consumption needed to supply the oxygen demand to the WWTP. The standard aeration efficiency value is given as a range as shown in Table 27.7. The uncertain domain is defined accordingly. The electricity price is taken as the end-user energy price for industrial consumers in Denmark and a variation of 20% is assumed over the average price given. Landfill cost, given for Denmark by the Confederation of European waste-to-energy Plants as a range, is used while defining the uncertain domain. The effluent total nitrogen limitation is assumed to change between 15 and 10 mg N/L. Moreover, the possible change in the COD fractionations is taken into account together with the change in influent ammonium nitrogen concentration. Four different COD fractions (S_I, S_S, X_I and X_{BH}) were sampled and the resulting X_S concentration was calculated assuming that the total COD in the influent wastewater is constant.

Table 27.7 Uncertain parameters and their domain definition.

Data	Probability distribution	Mean	Min	Max	Unit
Alpha (α)	Uniform	0.75	0.30	1.20	dimensionless
Beta (β)	Uniform	0.965	0.95	0.98	dimensionless
Fouling factor (F)	Uniform	0.775	0.65	0.9	dimensionless
Standard aeration efficiency (SAE)	Uniform	2.55	1.9	3.2	kg O_2/kwh
Price of electricity	Uniform	0.0977	0.08	0.12	Euro/kwh
Landfill cost	Uniform	127	75	179	Euro/ton of sludge
Limit–Total N	Uniform	12.5	10	15	g N/m^3
Soluble inert organic matter (S_I)	Uniform	30	27	33	g COD/m^3
Readily biodegradable substrate (S_S)	Uniform	63.18	56.86	69.5	g COD/m^3
Particulate inert organic matter (X_I)	Uniform	51.2	46.08	56.32	g COD/m^3
Active heterotrophic biomass ($X_{B,H}$)	Uniform	28.17	25.35	30.99	g COD/m^3
$NH_4^+ + NH_3$ nitrogen (S_{NH})	Uniform	31.56	28.4	34.72	g N/m^3

$$AOTR = SOTR\left(\frac{\beta * C_S, T, H - C_L}{C_{S,20}}\right)(1.024^{T-20})(\alpha)(F) \tag{27.5}$$

Once the uncertain parameters are selected and their uncertain domain is defined, the domain is sampled through Monte Carlo sampling by using Latin Hypercube Sampling (LHS) as the sampling technique. Here, the uncertain domain is sampled to create 50 future scenarios with respect to realization of 12 uncertain parameters given in Table 27.7. It is assumed that among the uncertain parameters there is no correlation existing.

Later, the optimization problem was solved for 50 scenarios created in the preceding sampling step. The analysis of the optimization results indicated that two different WWTP networks were identified as optimal with different frequencies as shown in Table 27.8. The majority of the future solutions with respect to future realizations of uncertain parameters resulted in the selection of the same WWTP network as the deterministic solution (MLE). Although the probability is very low, in 16% of the future scenarios the UASB is selected to treat the organic content of the wastewater together with the 2-stage Sharon-Anammox reactor for further nitrogen removal. This network does not comply with the effluent COD limit; however, in the realization of those scenarios, paying the effluent penalty for COD limit violation is found to be more feasible than choosing any other technology – from the limited search space – by the optimizer.

Table 27.8 Uncertainty mapping results.

Network	Probability of Realization	Selected Intervals
1	84%	WW-BP1-MLE-BP3-BP4-Thickener-AnD-Discharge-Sludge-Biogas
2	16%	WW-BP1-UASB-Shar-An-BP4-BP5-Discharge-Sludge-Biogas

The cumulative distribution of the objective function is illustrated in Figure 27.5 where the x-axis shows the objective function value, which represents operational and capital cost, and the y-axis represents the probability that the value of the objective function will be lower than the stated value on the x-axis. This indicates that there is a significant uncertainty on the treatment cost due to the cost related parameters selected to be uncertain; and the objective function value ranges from 693 to 1,606 unit cost. Compared with the deterministic solution case, it can be seen that 78% of the scenarios result in a higher objective function value, and in 16% of the scenarios a different network configuration is selected. Although the output from uncertainty analysis very much depends on the defined domain of input uncertainties, this comparative analysis already indicates the significance of considering uncertainty analysis for better informed decision making, at least compared to single-point analysis (the case of the deterministic solution).

(d) *Optimal process configuration:* The optimizer gives the flow of components through the selected process flow diagram which is illustrated in Figure 27.6. The stream table of the components for the solution of Scenario 1 is shown in Table 27.9 where the flows are given as the flows after reaction – F_{reac} – in each process interval. By using this information, the selected process flow diagram is evaluated in terms of its performance presented also in the same table. It can be seen that both COD and total nitrogen concentrations are below the limits set by the regulations. This analysis shows that the selected process flow diagram is capable of decreasing the concentrations of the key pollutants below the effluent discharge limits while using a minimum amount of utilities.

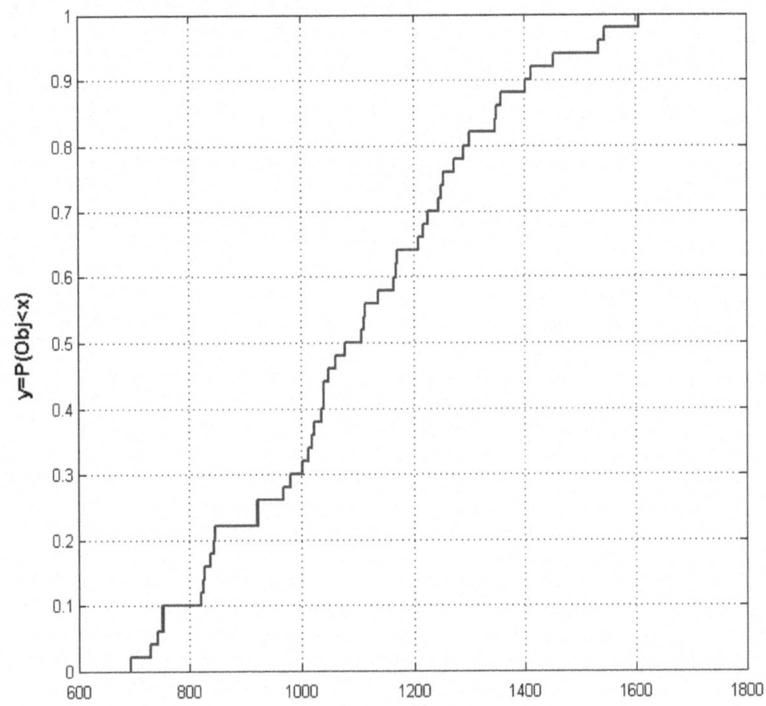

Figure 27.5 Cumulative distribution of the objective function.

Table 27.9 Scenario 1 stream table.

	Unit	WW	PC	MLE	BP3	BP4	Thickener	AnD	Discharge	Landfill
Q	m^3/d	20,648	20,648	42,633	20,553	20,553	763	190	20,553	95
S_I	kg/d	562	562	1,160.38	559.42	559.42	20.77	5.16	559.42	2.58
S_S	kg/d	1,201	1,201	–	–	–	–	8.41	–	4.20
S_O	kg/d	–	–	85,266	41,106	41,106	1,526	0	41,106	0
S_{NO}	kg/d	–	–	341.20	164.49	164.49	6.11	0.36	164.49	0.18
S_{ND}	kg/d	117	117	–	–	–	–	–	–	–
S_{NH}	kg/d	492	492	22.06	10.63	10.63	0.39	126.27	10.63	63.13
X_I	kg/d	1,909	1,909	52,521.86	105.04	105.04	1,811.66	2,743.81	105.04	2,724.60
X_S	kg/d	7,514	7,514	116.23	0.23	0.23	4.01	752.19	0.23	746.93
X_{BH}	kg/d	1,046	1,046	34,557.85	69.12	69.12	1,192.02	–	69.12	–
X_{BA}	kg/d	–	–	6,344.53	12.68	12.68	218.85	–	12.68	–
X_P	kg/d	–	–	29,234.09	58.47	58.47	1,008.39	1,031.71	58.47	1,024.48
X_{ND}	kg/d	333	333	5,167.21	10.33	10.33	178.24	342.54	10.33	340.14
CH_4	kg/d							4,599.23		

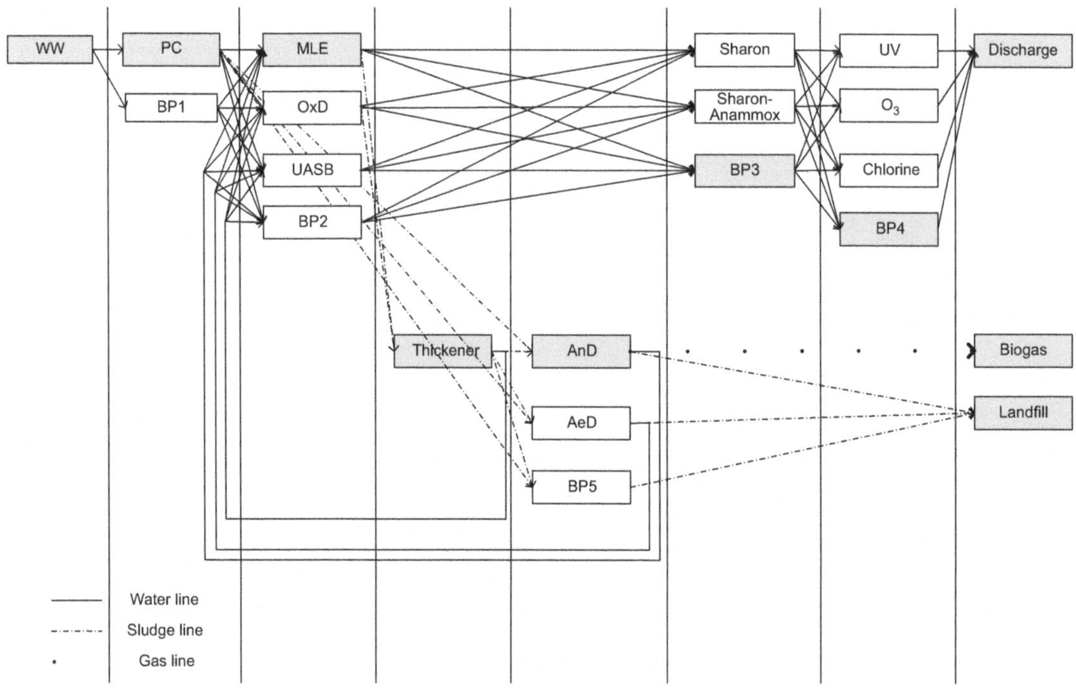

Figure 27.6 Process flow diagram and interconnections.

27.4 CONCLUSIONS AND FUTURE PERSPECTIVES

In this study, a superstructure based optimization methodology is presented with the aim of supporting optimal treatment process selection – a critical and challenging step in the early stage of wastewater treatment plant design and retrofitting studies. The underlying theory and mathematical concepts, the required methods for its solution and analysis and its practical implementation as a tool is presented using a simple case study as an example. The following are the main features of the methodology:

- The primary focus is to support the early stage decision making studies for optimal process concept selection for a wastewater treatment problem of interest. The framework is composed of several steps and brings together methods and tools from different engineering practices.
- The design space is represented as a superstructure. The superstructure approach can cover numerous treatment alternatives. The definition is flexible in the sense that any task, any treatment alternative and any connection can be included or excluded in the search space. Eventually, the search space allows for covering a high number of WWTP network alternatives including well-known as well as innovative combinations of treatment units in a compact representation.
- A generic process interval is defined and described using a mass input-output model. This formulation of a generic model to define any unit or process in a treatment plant is important in order to handle the complexity of the MINLP problem formulation and its solution.
- Uncertainty and sensitivity analysis methods are included in the framework to help robustify the decision making problem.

The superstructure based optimization approach presented here is novel and has the potential to facilitate early stage design and retrofitting decisions. The framework aims to fill the gap between optimization-based design methods and wastewater engineering community. The proposed framework and the resulting tools are successfully evaluated and applied for full-scale WWTP retrofitting problems in a separate study (Bozkurt *et al.* 2016) which confirms its promising potential for real-life applications.

Therefore the framework is well positioned to support the process design experts and engineers in their continuous efforts to identify and design novel, optimal and efficient WWTPs for a given wastewater treatment problem.

The content and therefore impact of the proposed framework can be increased. In the current framework, the step 2, where the superstructure is generated is done manually by incorporating expert knowledge and previous experiences. However, this step can also be done through more systematic methods by for instance combining the superstructure optimization based tool with the knowledge-based environmental decision support systems (EDSS) developed successfully by Comas *et al.* (2003). In fact, as the result of a recent joint collaboration, Castillo *et al.* (2016) has demonstrated the feasibility of integrating the EDSS with the superstructure based optimization methodology. This integration allows for an exhaustive and objective search of all possible combinations of technologies, avoiding omissions or biased preferences. Moreover the outcome from the superstructure based optimization approach is a shortlist of optimal WWTP process concepts which can be further validated, refined and optimized using more rigorous dynamic models as presented in the Chapter 28.

27.5 REFERENCES

AWWA Michigan Section (2006). Sodium Hypochlorite and Chlorine Gas: Design, Handling and Storage Issues. Water Treatment Practices Committee, AWWA Michigan Section, Spring Regionals.

Bozkurt H., Quaglia A., Gernaey K. V. and Sin G. (2015). A mathematical programming framework for early stage design of wastewater treatment plants. *Environmental Modelling & Software*, **64**, 164–176.

Bozkurt H., van Loosdrecht M. C. M., Gernaey K. V. and Sin G. (2016). Optimal WWTP process selection for treatment of domestic wastewater – a realistic full-scale retrofitting study. *Chemical Engineering Journal*, **286**, 447–458.

Castillo A., Cheali P., Gómez V., Comas V., Poch M. and Sin G. (2016). An integrated methodology for the optimal selection of sustainable and economic wastewater treatment plants (WWTP): intelligent expert system and superstructure-based optimization tool. *Environmental Modelling and Software*, **84**, 134–138.

Comas J., Alemany J., Poch M., Torrens A., Salgot M. and Bou J. (2003). Development of a knowledge-based decision support system for identifying adequate wastewater treatment for small communities. *Water Science and Technology*, **48**(11–12), 393–400.

Copp J. B. (2002). The COST Simulation Benchmark: description and Simulator Manual. Office for Official Publications for the European Communities, Luxembourg. In: Benchmarking of Control Strategies for Wastewater Treatment Plants, Gernaey, K. V. Jeppsson, U. Vanrolleghem, P. A. and Copp, J. B. (eds), (2014) IWA Scientific and Technical Report No. 23, IWA Publishing, London, UK.

Hahn H., Rutz D., Ferber E. and Kirchmayer F. (2010). Examples for financing of biogas projects in Germany, Austria, The Netherlands, Denmark and Italy. IEE Project BiogasIN. Report no. D.3.2.,WP3.

Hamouda M. A., Anderson W. B. and Huck P. M. (2009). Decision support system in water and wastewater treatment process selection and design: a review. *Water Science and Technology*, **60**(7), 1757–1770.

Hellinga C., Schellen A. A. J. C., Mulder J. W., van Loosdrecht M. C. M. and Heijnen J. J. (1998). The SHARON process: an innovative method for nitrogen removal from ammonium-rich wastewater. *Water Science and Technology*, **37**(9), 135–142.

Henze M., van Loosdrecht M. C. M., Ekama G. and Brdjanovic D. (2008). Biological Wastewater Treatment Principles, Modelling and Design. IWA Publishing, Glasgow.

Lew B., Tarre S., Belavski M. and Green M. (2004). UASB reactor for domestic wastewater treatment at low temperatures: a comparison between a classical UASB and hybrid UASB-filter reactor. *Water Science and Technology*, **49**(11–12), 295–301.

McCarty P. L., Bae J. and Kim J. (2011). Domestic wastewater treatment as a net energy producer – Can this be achieved? *Environmental Science & Technology*, **45**(17), 7100–7106.

Mulder A., van de Graaf A. A., Robertson L. A. and Kuenen J. G. (1995). Anaerobic ammonium oxidation discovered in a denitrifying fluidized bed reactor. *FEMS Microbiology Ecology*, **16**, 177–184.

Quaglia A. (2013). An Integrated Business and Engineering Framework for Synthesis and Design of Processing Networks. PhD thesis, Technical University of Denmark, p. 220.

Quaglia A., Sarup B., Sin G. and Gani R. (2013). A systematic framework for enterprise-wide optimization: synthesis and design of processing networks under uncertainty. *Computers & Chemical Engineering*, **59**(5), 47–62.

Reardon R., Davel J., Baune D., McDonald S., Appleton R. and Gillette R. (2013). Wastewater treatment plants of the future: current trends shape future plans. *Florida Water Resource Journal*, **10**, 8–14.

Siemens (2009). Providing Answers to the Wastewater Aeration Market: Air Compression Solutions. Siemens AG 91058 Erlangen, Germany.

Tchobanoglous G., Burton F. L. and Stensel H. D. (2003). Wastewater Engineering: Treatment and Reuse. McGraw-Hill Publishing, New York, USA.

Europe's Energy Portal (2013). www.energy.eu (accessed October 2013).

Confederation of European Waste-to-Energy Plants (2013). www.cewep.eu (accessed October 2013).

Van der Star W. R. L., Abma W. R., Blommers D., Mulder J. W., Tokutomi T., Strous M., Picioreanu C. and van Loosdrecht M. C. M. (2007). Startup of reactors for anoxic ammonium oxidation: experiences from the first full-scale anammox reactor in Rotterdam. *Water Research*, **41**, 4149–4163.

Van Dongen U., Jetten M. S. M. and van Loosdrecht M. C. M. (2001). The SHARON-Anammox process for treatment of ammonium rich wastewater. *Water Science and Technology*, **44**(1), 153–160.

Van Haandel A. C. and van der Lubbe J. G. M. (2012). Handbook of Biological Wastewater Treatment: Design and Optimisation of Activated Sludge Systems. IWA Publishing, London, UK.

WEF and ASCE/EWRI (2010). Design of Municipal Wastewater Treatment Plants. WEF Press, Virginia, USA.

Chapter 28

Model-based comparative assessment of innovative processes

Tamara Fernández-Arévalo, Xavier Flores-Alsina, Paloma Grau, Ulf Jeppsson, Miguel Mauricio-Iglesias, Darko Vrecko and Eduardo Ayesa

28.1 INTRODUCTION

As discussed in previous sections, the generalised concern regarding the sustainability and scarcity of natural resources is leading to a change in how we understand wastewater. From this changing perspective, a new paradigm is emerging in which urban wastewater - traditionally considered as a pollution problem – is starting to be understood as a continuous and sustainable source of resources. Thus, Wastewater Treatment Plants (WWTPs) are now being considered to be Wastewater Resource Recovery Facilities (WRRFs), from which valuable products such as chemicals, nutrients, bioenergy, and bioproducts can be obtained (Keller, 2008; Guest *et al.*, 2009). In order to address this shift, new and innovative combinations of emerging and conventional technologies and configurations in WWTPs can offer sustainable solutions for obtaining the required effluent quality and, simultaneously, optimising the recovery of valuable by-products and energy. Such is the case of the presence of energetically more efficient technologies (nitritation, anammox, deammonification, etc., for nitrogen (N) removal or an increase in biogas production by including sludge pretreatment technologies).

However, prior to exploring any full-scale implementation, an analysis of the interrelations between the "leading-edge" technologies existing, under development, or being researched is needed. In this context, mathematical modelling has for many years been proven to be a valuable tool for assessing integrated solutions for conventional or innovative WWTP layouts. The complexity of the configurations and processes with recirculations and interrelations between the water/sludge line makes it necessary to consider a plant-wide perspective in order to establish an optimum solution for the design or operation of the entire plant and therefore avoid local optimisation (Jeppsson *et al.* 2007; Grau *et al.* 2007).

Among the existing approaches in the literature, the Plant-Wide Modelling (PWM) methodology proposed by Ceit-IK4 (Grau *et al.* 2007; Fernández-Arévalo *et al.* 2014; Lizarralde *et al.* 2015) constitutes a very suitable tool for a rigorous and global assessment of incorporating new leading-edge technologies in conventional plant layouts (Fernández-Arévalo *et al.* 2017b) or for selecting the most appropriate operational strategies at existing full-scale facilities (Fernández-Arévalo *et al.* 2017b).

28.2 E-PWM METHODOLOGY

The Extended Plant-Wide Modelling (E-PWM) methodology (Grau *et al.* 2007; Fernández-Arévalo *et al.* 2014; Lizarralde *et al.* 2015; Fernández-Arévalo *et al.* 2017b) allows for rigorous and systematic construction of compatible unit-process models (UPM) for describing the dynamic behaviour of the water and sludge lines in an integrated manner. The E-PWM methodology is based on selecting the set of process transformations required to model all unit-processes incorporated in each specific WWTP. In this manner, with a unique standard model it is possible to simulate the plant that is needed in each case. The unification of these sets of transformations permits the definition of a unique component vector for the whole plant, without the need to develop specific transformers for interfacing unit-process models. An accurate definition of the stoichiometry ensures the elemental mass (in terms of C, N, O, H, P, and other elements) and charge continuity in all these transformations, while the definition of the enthalpies of the formation of each component enables an estimation of the reaction heat of each transformation (Fernández-Arévalo *et al.* 2014). Thus, this methodology allows for straightforward construction of compatible mathematical models that are especially suitable for a comparative assessment of any combination of existing or under development technologies.

Based on this E-PWM methodology, a new model library (Figure 28.1) was developed to bring together all the models in a structured manner (Fernández-Arévalo, 2016). The library compiles information arranged by aspect to be considered in the construction of the model, so that the modeller can select those that are the most interesting for the case at issue in accordance with the following three steps:

(i) Category selection: the library contains different categories that compile the model required for the representation of biochemical, chemical, and physico-chemical reactions;
(ii) Unit-process models selection: wherein the mass and heat transports are defined depending on the phases considered in the unit-processes under study (e.g. for CSTR reactors, primary or secondary settlers, solid separation systems, etc.);
(iii) Actuator models selection, which are required in the costs estimation.

Once the categories, unit-processes, and the actuator models have been selected, the model is constituted and ready for use.

28.2.1 Category selection

The selection of the category must ensure the correct description of the biochemical, chemical, and physico-chemical transformations that may take place throughout the plant. In the first step, the modeller must select the biochemical and physico-chemical transformations deemed necessary to describe the plant, considering the aim of and accuracy required in the study. For instance, the modeller must decide, among other aspects, whether the model needs to describe the nitrogen removal process in one step or two steps, incorporate chemical or biological phosphorus removal transformations, or analyse the recovery of compounds. In the second step, the modeller is able to select those chemical components, species, and transformations (acid-base and ion-pairing equilibrium reactions) that are relevant for the definition of the plant, depending on the biochemical and physico-chemical transformations selected in the previous step (Grau *et al.* 2007; Lizarralde *et al.* 2015).

In order to facilitate the work, the library contains different categories that include all the transformations present in characteristic plant. Consequently the modeller can select one category or another depending on the requirements and goals (Figure 28.1). The encoding used in the categories is as follows: "C" and "N" are used to describe biological organic matter and nitrogen biodegradation in aerobic and anoxic conditions at low and high temperatures (thermal hydrolysis reactions); "P" is used to define biological and chemical P removal; "2N" is used to detail two step N removal and Anammox reactions; "Pchem" is used to consider chemical P removal, but not biological P removal; "prec" is used to include precipitation reactions; and finally, "AnD" is used to describe anaerobic conditions at low and high temperatures.

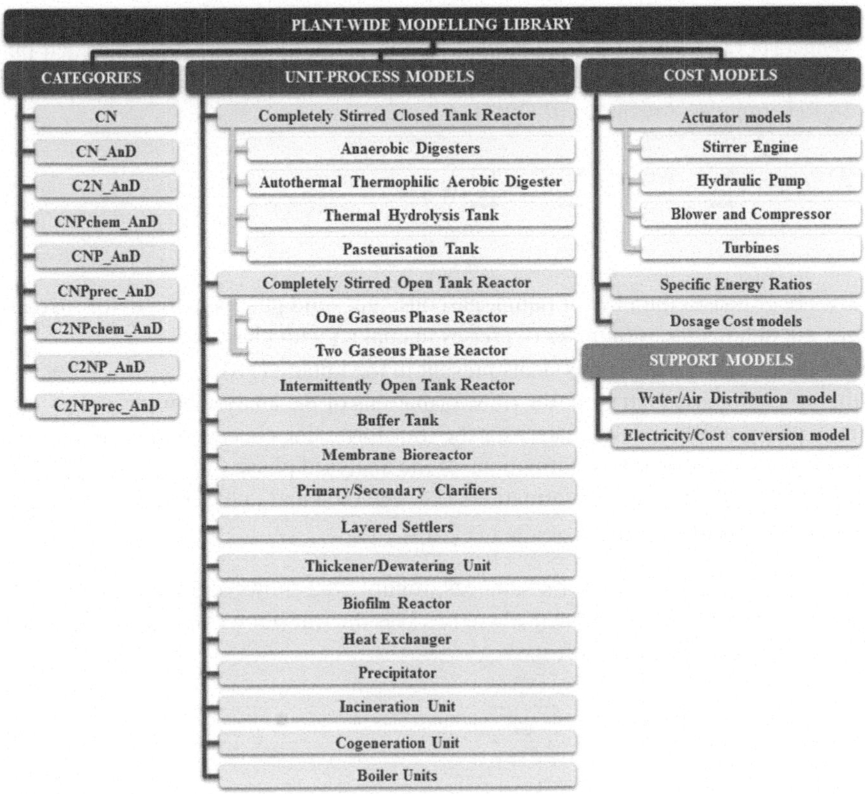

Figure 28.1 Schematic representation of the Ceit PWM library.

The organised structure that presents the methodology enables a straightforward development of categories, allowing for the continuous updating of the library.

28.2.2 Unit-process models selection

The library contains a comprehensive set of unit-process models, as shown in Figure 28.1. At this stage, all unit-process models (UPM) needed to complete the plant under study must be selected. The first step consists in deciding on the number of phases that each UPM contains. In some of them such a decision is not necessary since the processes are defined for a given number of phases (e.g. for separation processes, incineration units, cogeneration units). In the case of completely stirred open tank reactors (O-CSTR) and completely stirred closed tank reactors (C-CSTR), one must decide between one or two gaseous phases to describe the liquid-gas transfer and if the process needs the definition of a solid phase due to the uncontrolled precipitation of solids (e.g. struvite precipitation in anaerobic digesters).

Usually, although there may be exceptions, the unit-processes present in a WWTP are composed of a maximum of 4 phases: (1) an aqueous phase, or "w"; (2) a solid phase representing the precipitates formed during the process, or "s"; (3) a first gas phase, which describes the gaseous phase in contact with the free surface of the mixed liquor (an off-gas phase for closed units and the atmosphere for open reactors), or "g1"; and (4) a second gas phase or a gas hold-up phase symbolising the gas phase or bubbles contained in the liquid phase, or "g2".

Each UPM incorporates the mathematical description of the mass transport for each phase (liquid, gaseous, solid) and the transformations designated in the category by following the matrix structure shown in Equation 28.1 (Fernández-Arévalo *et al.* 2014).

$$\left(\frac{d\bar{M}}{dt}\right)_i = \tilde{E}_{i,i}^T \bar{\rho}_{i,i} + \sum_{j=1}^{\text{No. adj. phase}} \tilde{E}_{i,j}^T \bar{\rho}_{i,j} + \sum_{in} \bar{\dot{m}}_i - \sum_{out} \bar{\dot{m}}_i \tag{28.1}$$

where:

$i = 1, 2\ldots$ no. of phases (number of liquid, gaseous, and solid phases in the unit-process);
$j = 1, 2\ldots$ no. of adj. phases (number of phases in contact with phase i);
$\bar{M}$ = the mass vector for the components present in the i phase (gE);
$\tilde{E}_{i,i}$ = the stoichiometric matrix for the transformations of the analysed phase;
$\tilde{E}_{i,j}$ = the i phase stoichiometric matrix for the transformations between the i and j phases;
$\bar{\rho}_{i,i}$ = the kinetic rate for the transformations of the analysed phase ($gE_{removed\ component}\ d^{-1}$);
$\bar{\rho}_{i,j}$ = the kinetic rate for the transformations between the i and j phases ($gE_{removed\ component}\ d^{-1}$);
$\dot{m}_i$ = the inlet and outlet i phase mass flux ($gE\ d^{-1}$).

For a correct interpretation of the mass balances, Figure 28.2 shows a graphical representation of the mass balances of the four phases mentioned where $\bar{\dot{m}}_{g2,g1}$ is the transition from the first gaseous phase (gas hold-up) to the second gaseous phase (off-gas).

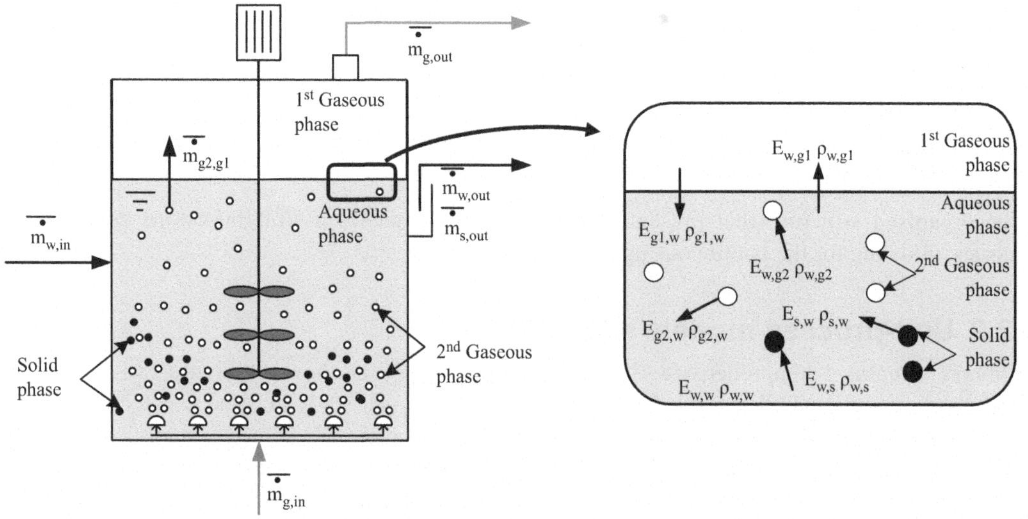

Figure 28.2 Schematic representation of the mass balance in a C-CSTR for four phases: aqueous (w), solid (s), off-gas (g1), and gas hold-up (g2) phases.

In the second step, the modeller decides whether it is necessary to ensure the overall heat balance of the plant or the balance of some units in particular. Thus, each UPM incorporates the mathematical description of the heat balance for each phase (liquid, gaseous, solid), in addition to the transport model and transformations designated in the category. Based on the energy conservation principle, the general one-dimensional dynamic heat transfer model includes biochemical, chemical, or physico-chemical heat transfers (H_{trans}), conduction and convection fluxes ($H_{Cond/Conv}$), heat energy fluxes transmitted by the actuators (H_{Act}), and short-wave (solar) and long-wave (atmospheric) radiation fluxes (H_{solrd}, H_{atmrd}), as shown in Equation 28.2.

$$\left(\frac{dH_T}{dt}\right)_i = \sum H_{trans} + \sum H_{Cond/Conv} + \sum H_{solrd} + \sum H_{atmrd} + \sum H_{Act} + \sum \bar{H}_{in} + \sum \bar{H}_{out} \qquad (28.2)$$

A detailed graphical representation of the heat balances of the four typical phases can be seen in Figure 28.3.

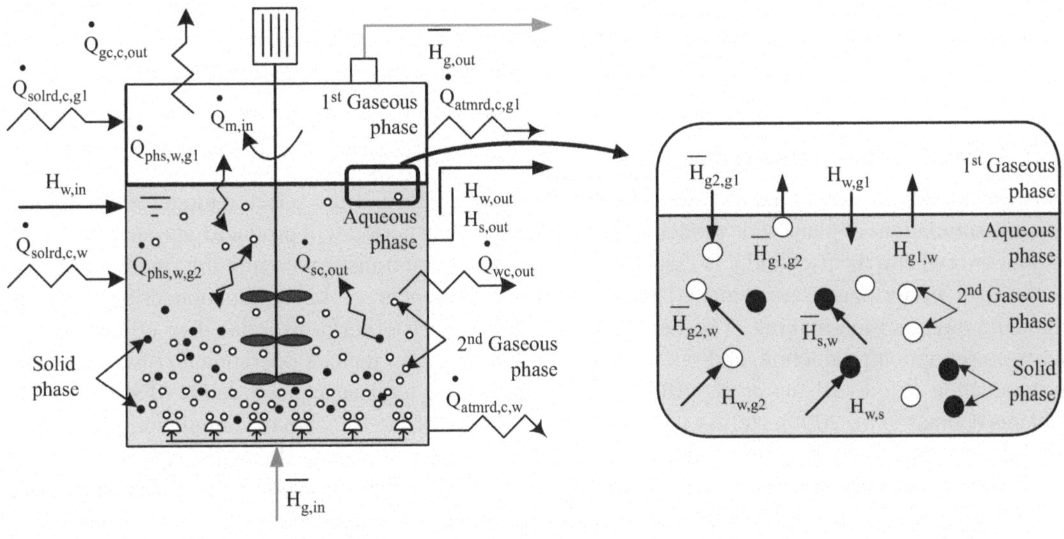

Figure 28.3 A schematic representation of the heat balance in a C-CSTR (the heat of the reactions have not been plotted) for four phases: aqueous (w), solid (s), off-gas (g1), and gas hold-up (g2) phases.

where,

$\bar{H}_{i,in}$ = the inlet i phase enthalpy flux (kJ d^{-1});
$\bar{H}_{i,out}$ = the outlet i phase enthalpy flux (kJ d^{-1});
$\dot{Q}_{gc,c,out}$ = the enthalpy lost/gained through walls and pipes by conduction in the off-gas phase (kJ d^{-1});
$\dot{Q}_{ic,out}$ = the enthalpy lost/gained through walls and pipes by conduction in the i phase (kJ d^{-1});
$\dot{Q}_{solrd,c,i}$ = the solar radiation flux for fluids covered by solid materials (kJ d^{-1});
$\dot{Q}_{atmrd,c,i}$ = the atmospheric radiation flux for fluids covered by solid materials (kJ d^{-1});
$\dot{Q}_{phs,i,j}$ = the enthalpy flow transfer between the i phase and the j phase (kJ d^{-1});
$\dot{Q}_{m,in}$ = the heat transmitted from the actuators (kJ d^{-1});
$\bar{H}_{i,j}$ = the enthalpy associated with mass transfer (kJ d^{-1}).

A detailed characterisation of the components (elemental mass characterisation) enables an estimation of the formation enthalpies for each model component and enables a systematic calculation of the heat released or absorbed by each transformation, guaranteeing heat energy continuity at any point in the plant. Thus, the specific enthalpy change in the reaction (Δh°_{r} in kJ $g_{stoich.\ unit}^{-1}$) due to biochemical, physico-chemical, or chemical transformations can be defined as the difference between the enthalpy of formation of the products and the enthalpy of formation of the reactants (applying Hess's law). Additional information on the heat balance with an exhaustive definition of each term can be found in Fernández-Arévalo *et al.* (2014).

28.2.3 Actuator models selection

Finally, the library includes a set of actuator models for detailed estimation of the costs of each actuator (Fernández-Arévalo *et al.* 2017b). Almost all actuator models are developed based on engineering expressions, instead of directly using cost curves or fixed values. The goal is to have more accurate models in which the oversimplifications and low standardisation of some expressions are avoided. As shown in Table 28.1, the models are standardised, so these can be used interchangeably in any category. The standardisation of the models has been pursued to prevent future model adaptations due to the incorporation of new components and to continue with the standardisation philosophy of the Plant-Wide Modelling methodology.

28.2.4 Evaluation criteria

A strict methodology is also required to evaluating the performance of the selected model configurations. Dynamic simulations of plant-wide models over an extensive time period will produce huge amounts of data that may overwhelm the user and it is essential for these data to be transformed into information providing an objective performance assessment based on a limited number of key evaluation criteria, thereby enhancing any decision-making. The most fundamental of such criteria are related to effluent quality, energy consumption/production, and operational cost, but may also include operational risks, greenhouse gas production (Flores-Alsina *et al.* 2014), microbiology-rated total suspended solids (TSS) separation problems (Comas *et al.* 2008), etc. As part of the development of the Benchmark Simulation Model No. 2 (BSM2) platform, an extensive list of such priority evaluation criteria has been defined (Gernaey *et al.* 2014). The final decision-making process will always be a multi-dimensional problem where different priorities must be weighted and set against each other. Keeping the number of dimensions as low as possible, without giving up any essential information, will enhance the probability of arriving at good decisions.

The most fundamental criteria used in the BSM2 protocol are the effluent quality index (EQI) and the operational cost index (OCI). These are based on dynamic output values recorded every 15 minutes and assuming constant values between samples over a period of 364 days. The reason for such an extended evaluation period is to enable the inclusion of all diurnal, weekly, holiday and seasonal effects (load, temperature, different energy prices, etc.) and is recommended for other simulation studies as well. EQI (in kg pollution units.d^{-1}) is calculated as a weighted averaged sum of all relevant effluent loads. Weights may naturally be modified in accordance with national/regional/local legislation and priorities. The EQI is accompanied by specific effluent limits for certain key compounds, for which 95th-percentiles, the number of violations, and the percentage of time the plant is in violation are provided. The constraints of meeting legislative effluent requirements must, of course, also be considered. OCI is calculated as a weighted sum of different costs related to aeration energy, pumping energy, sludge production for disposal, the addition of external carbon and other chemicals, mixing energy, methane production, and heating energy needed to increase the temperature of the sludge in the anaerobic digester, etc.

Table 28.1 Description of actuator models.

Actuator	Equations	Definition of the Parameters
Agitation engine model	$W_{agit} = \frac{N_P \varphi_S N_{js}^3 D_{stir}^5}{\eta_{agit}} F_{oversize}$ $N_{js} = S\left(\frac{g(\varphi_S - \varphi_W)}{\varphi_W}\right)^{0.45} \frac{X_{TSS}^{0.13} d_p^{0.2} \upsilon_W^{0.1}}{D_{stir}^{0.85}}$	$\mathbf{W_{agit}}$ (stirring electric consumption [kJ d^{-1}]); $\mathbf{F_{oversize}}$ (oversize factor [-]); $\mathbf{N_P}$ (Power number [-]); $\boldsymbol{\varphi_S}$ & $\boldsymbol{\varphi_W}$ (solid and aqueous phase densities [g m^{-3}]); $\mathbf{N_{js}}$ (impeller rotational speed required to just suspend the particles [Hz]); $\mathbf{D_{stir}}$ (impeller diameter [m]); η_{agit} (agitation engine efficiency [%]); **S** (impeller/tank geometry factor [-]), $\mathbf{X_{TSS}}$ (weight percentage of solids in the suspension [%]); $\mathbf{d_p}$ (particle size [m]); **g** (gravitational acceleration [m s^{-2}]); υ_W (kinematic viscosity of the aqueous phase [m^2 s^{-1}]).
Combined heat and power (CHP) unit model	$W_{therm,w} = (\dot{m}_{g,in})_{G_{CH4}} \Delta h^{\circ}_r \, \eta_{therm,w}$ $W_{therm,g} = (\dot{m}_{g,in})_{G_{CH4}} \Delta h^{\circ}_r \, \eta_{therm,g}$ $W_{elect} = (\dot{m}_{g,in})_{G_{CH4}} \Delta h^{\circ}_r \, \eta_{elect}$	$\mathbf{W_{therm,i}}$ (water and gas phase thermal energy production [kJ d^{-1}]); $\mathbf{W_{elect}}$ (electrical energy production [kJ d^{-1}]); $\mathbf{\dot{m}_{g,in}}$ (inlet gas phase mass flux [gE d^{-1}]); $\mathbf{\Delta h_r}$ (specific reaction enthalpy [kJ gE^{-1}]); $\boldsymbol{\eta_{therm,i}}$ (thermal degree of efficiency to produce water and gas phases [-]); $\boldsymbol{\eta_{elect}}$ (degree of electrical efficiency [-]).
Blower or compressor model	$W_{blow} = \sum_{i=1}^{m}\left[\frac{(\dot{m}_{g,in})_i R\, T_{g,in}}{(MW)_i \left(\frac{\gamma_{g,i-1}}{\gamma_{g,i}}\right)\eta_{blow}}\right]\left[\left(\frac{P_{g,out}}{P_{g,in}}\right)^{\frac{\gamma_{g,i}-1}{\gamma_{g,i}}} - 1\right]$	$\mathbf{W_{blow}}$ (electrical consumption of blowers or compressors [kJ d^{-1}]); **R** (ideal gas constant [kJ mol^{-1} K^{-1}]); $\mathbf{T_{g,in}}$ (gas phase inflow temperature [K]); $\mathbf{MW_i}$ (molecular weight of i gaseous phase components [g mol^{-1}]); $\boldsymbol{\gamma_{g,i}}$ (heat capacity ratio of the i gaseous phase components [-]); η_{blow} (efficiency of blowers/compressors [-]); $\mathbf{P_{g,in}}$ and $\mathbf{P_{g,out}}$ (absolute gas pressure at the blower inlet and outlet [bar]).
Pump model	$W_{pump} = \varphi_W \, g \, Q_W \, HL \, \eta_{pump}$ $HL = HL_S + HL_f + HL_l$ $HL_f = f_{coeff}\left(\frac{L_{pipe}}{D_{pipe}}\right)\left(\frac{u_W}{2g}\right)$	$\mathbf{W_{pump}}$ (electrical pump consumption [kJ d^{-1}]); $\mathbf{Q_w}$ (water flow rate [m^3 d^{-1}]); $\boldsymbol{\eta_{pump}}$ (pumping efficiency [-]); **HL** (total head loss [m]); $\mathbf{HL_s}$ (static head [m]); $\mathbf{HL_f}$ (friction head loss [m]); $\mathbf{HL_l}$ (minor losses [m]); $\mathbf{f_{coeff}}$ (friction coefficient [-]); $\mathbf{L_{pipe}}$ & $\mathbf{D_{pipe}}$ (pipe length and diameter [m]); $\mathbf{u_W}$ (liquid velocity [m s^{-1}]).

(Continued)

Table 28.1 Description of actuator models (*Continued*).

Actuator	Equations	Definition of the Parameters
Boiler model	$W_{steam} = (\dot{m}_{g,in})_{GCH4}\,\Delta h^{\circ}_{r}\,\eta_{steam}$	$\mathbf{W_{steam}}$ (steam thermal energy production [kJ d^{-1}]); $\boldsymbol{\eta_{steam}}$ (thermal degree of efficiency to produce steam [-]).
Dissolved Air Flotation (DAF) model	$W_{DAF} = W_{blow} + W_{pump} + W_{scraper}$ $(\dot{m}_{g,in})_i = \frac{A}{S} TSS\ Q_W\ \varphi_{air} \bar{X}_i$ $Cost_{DAF} = Cost_{coagulant} C_{coagulant} Q_W$	$\mathbf{W_{DAF}}$ (DAF electric consumption [kJ d^{-1}]); $\mathbf{W_{scraper}}$ (scraper electric consumption [kJ d^{-1}]); **A/S** (air to solids ratio [m^3_{air} g^{-1}_{solids}]); **TSS** (total suspended solids concentration [g m^{-3}]); $\boldsymbol{\varphi}$**air** (air density [g m^{-3}]); $\bar{X}_1$ (mass fraction of the gaseous phase components [$gE_i\ gE_{phase}^{-1}$]); $\mathbf{Cost_{DAF}}$ (DAF raw material cost [€ d^{-1}]); $\mathbf{Cost_{coagulant}}$ (coagulant cost [€ kg^{-1}]); $\mathbf{C_{coagulant}}$ (coagulant concentration [kg m^{-3}]).
Dewatering model	$W_{dewat} = E_{consump} TSS\ Q_W$ $Cost_{dewat} = Cost_{poly} Q_W \left[TSS_{PS}\ k_{Poly,PS} + TSS_{SS} (k_{Poli})_{SS}\right]$	$\mathbf{W_{dewat}}$ (dewatering electric consumption [kJ d^{-1}]); $\mathbf{E_{consump}}$ (energy consumption [kWh tDS^{-1}]); $\mathbf{Cost_{dewat}}$ (dewatering raw material cost [€ d^{-1}]); $\mathbf{Cost_{poly}}$ (polyelectrolyte cost [€ g^{-1}]); $\mathbf{TSS_{PS}}$ & $\mathbf{TSS_{SS}}$ (primary and secondary sludge TSS concentration [kg m^{-3}]); $\mathbf{k_{poli,PS}}$ & $\mathbf{k_{poli,SS}}$ (polymer dosage [$g_{poly}\ kg_{TSS}^{-1}$])

Sources: Adapted from Metcalf and Eddy 2014; Perry *et al.* 1999; Weisbach, 1845; Zweitering, 1985.

Risk assessment models provide another avenue of plant performance criteria related to the risk of various operational problems (examples in BSM2 include the risk of microbiology-related settling problems in activated sludge systems, operational problems related to the anaerobic digestion process, and increased greenhouse gas production). Although not as straightforward and easily validated as the effluent quality index and the operational cost index, the risk indices can be used for relative comparisons of different control strategies as an indicator of potential problems that an otherwise excellent control strategy may lead to.

If operational strategies are a main focus of the evaluation, then also criteria related to controller performance should be included. Such assessment quantifies the control strategy's effect on controller performance and can be divided into two categories:

- controlled variable performance (various tracking statistics regarding the controlled variables);
- manipulated variable performance (deviation statistics).

A number of controller performance assessment criteria can be found in Gernaey *et al.* (2014).

However, in order to assess the global environmental impact, not only the direct emissions of different pollutants from the WWTP into water, land, and air are relevant, but also the up and downstream processes need to be taken into account. These include external processes, such as the production of input goods like power and chemicals, but also the impacts of the remaining effluent load on the recipient and the different utilisation of bio-solids and bio-methane. These types of effects have successfully been assessed for WWTPs using life cycle assessment (LCA) (Baresel *et al.* 2016; Corominas *et al.* 2013a,b).

The ISO standard for LCA (ISO, 14044) provides a structured procedure for performing LCA where, after defining the study, a life cycle inventory (LCI) is carried out whereby the environmental loads from the whole system are calculated, followed by a life cycle impact assessment (LCIA) wherein the loads are characterised by the selected impact categories to obtain aggregated measures of the potential environmental impacts expressed in equivalent units (Baumann & Tillman, 2004). In a similar fashion, also the life cycle cost (LCC) can be included in the evaluation, including any necessary investments in new processes and equipment – although numerous studies have shown that the operational costs are the predominant costs over the full life expectancy of a WWTP.

In order to capture both the dynamic performance of the WWTP and the global environmental impact, a combination of dynamic mechanistic process models and LCA has been explored on generic benchmark-type WWTP layouts (Corominas *et al.* 2013b; Meneses *et al.* 2016; Bisinella de Faria *et al.* 2015). These studies use combinations of different WWTP models and LCA to assess control strategies, load variations, and local recipient conditions. However, all of these studies are limited in their coverage of WWTPs, either just the water line is modelled mechanistically (Corominas *et al.* 2013b; Meneses *et al.* 2016) or, as in Bisinella de Faria *et al.* (2015), highly dynamic GHG production is modelled using static emission factors.

Arnell (2016) explored the hypothesis that it is possible to combine a plant-wide WWTP process model – with detailed energy and GHG models – and LCA in order to evaluate the overall performance of operational strategies at WWTPs, capturing both the dynamic effects at the plant and the global environmental impact due to external resource use. A model framework is presented in the paper and also tested in a case study performed at a full-scale Swedish WWTP by comparing two operational strategies.

28.3 MODEL-BASED COMPARATIVE ASSESSMENT OF CONVENTIONAL AND INNOVATIVE PLANT LAYOUTS

In subsection 26.7, the ability of the NOVEDAR_EDSS tool to select suitable general plant configuration alternatives considering environmental, technical, and economic criteria was shown. Based on some proposed plants, the superstructure-based optimisation tool allowed the selection of optimal solutions using steady-state simplified models (Chapter 27). Therefore, these two tools can enable an initial selection of the most appropriate plant configurations for specific situations. However, if the goal is to conduct a more thorough analysis of closer to real situations, including dynamic influent variations and with consideration of changing environmental conditions, it is necessary to perform a dynamic simulations with more comprehensive and accurate tools. In this case, a higher degree of detail is required to demonstrate that previous selections are correct or are operationally viable, and, in turn, to provide an optimum process design and operating guidelines. As for the above-mentioned tools, in order to analyse plant upgrades, or any new WWTP configuration, a complete model library is required. Under this framework, the PWM library offers everything needed comforts to carry out these analyses in a fast and straightforward way.

Following the structure of the preceding Chapters, this section also discusses three different scenarios. As a reference plant for the study, a conventional plant based on the widely known BSM2 configuration (Jeppsson *et al.* 2007; Gernaey *et al.* 2014) was selected. The other two configurations correspond to an upgraded or retrofitted WWTP, where advanced technologies were incorporated in the reference plant; and a new wastewater treatment (WWT) concept, with the same requirements and restrictions as in the previous cases, based on a new treatment philosophy, more suited to the paradigm shift that is affecting the wastewater treatment.

28.3.1 Conventional WWTP

Until recently, the philosophy or the goal of conventional waste(water) treatment plants was to remove the residual material containing the water, waste(water), without considering the possibility of using or recovering resources and energy, or considering this as a secondary objective. It was for this reason that wastewater was considered an environmental issue rather than as a potential source of resource. Having a configuration based on removal, the energy self-sufficiency of the plant was not an issue to be considered. The degree of self-sufficiency mainly depends on the $COD/N_T/P_T$ influent ratio and the optimal plant operation. Nevertheless, this treatment philosophy can only take advantage of a small part of the influent energy potential.

In Figure 28.4, the layout of the conventional plant used in the study is shown, which comprises by an activated sludge process for carbon and nitrogen removal based on a Modified Ludzack-Ettinger configuration, chemical phosphorus removal by chemical agents (ferric chloride, ferric sulphate, etc.), and anaerobic digestion of sludge to recover energy as biogas. Besides the addition of the chemical agents for phosphorus removal, ferric chloride is also added to enhance the settling characteristics of the primary sludge, for cases in which the production of primary sludge needs to be maximised; two other chemical additions required in flotation and dewatering processes to obtain the solids removal efficiencies proposed by Gernaey *et al.* (2014).

In order to measure the full potential of the plant configuration, an overall heat and cost balance analysis of the plant was carried out. This heat balance primarily enabled an estimation of the heat required to maintain the anaerobic digester at 35 degrees. The heat produced in cogeneration is not always enough to supply the heat requirements of the digester. In these cases, it is necessary to divert a portion of the biogas produced in the anaerobic digestion to a boiler to carry out combustion without electricity production. Regarding the cost balance, the most representative power consumption/productions and chemical agent costs were estimated. These include the power consumption for the aeration, pumping, flotation, mixing, and dewatering processes, as well as the costs of the energy produced in the cogeneration unit and the chemical agent costs.

In a preliminary simulation performed for this plant configuration, it was found that the aeration power represents approximately 61% of the overall operational power requirements of the plant, resulting in a specified aeration electricity use of 0.19 kWh/m^3 (see Table 28.2). Regarding the use of the influent water energy content (energy associated with the composition of water), it is known that wastewater contains an energy content equivalent to 10 times that required for treatment (McCarty *et al.* 2011; Shoener *et al.* 2014; Kokabian *et al.* 2015). However, this plant configuration uses only 11% of the influent COD to produce electricity, representing a low utilisation of the energy contained in the COD. Considering the high aeration cost, the need for chemical agents to carry out the phosphorus precipitation, and inefficient resource utilisation in the energy production, the energy self-sufficiency and neutral cost balance of the plant is difficult to achieve (Table 28.2), confirming the suspicion that the plant was not designed to be self-sustaining (Fernández-Arévalo *et al.* 2017b).

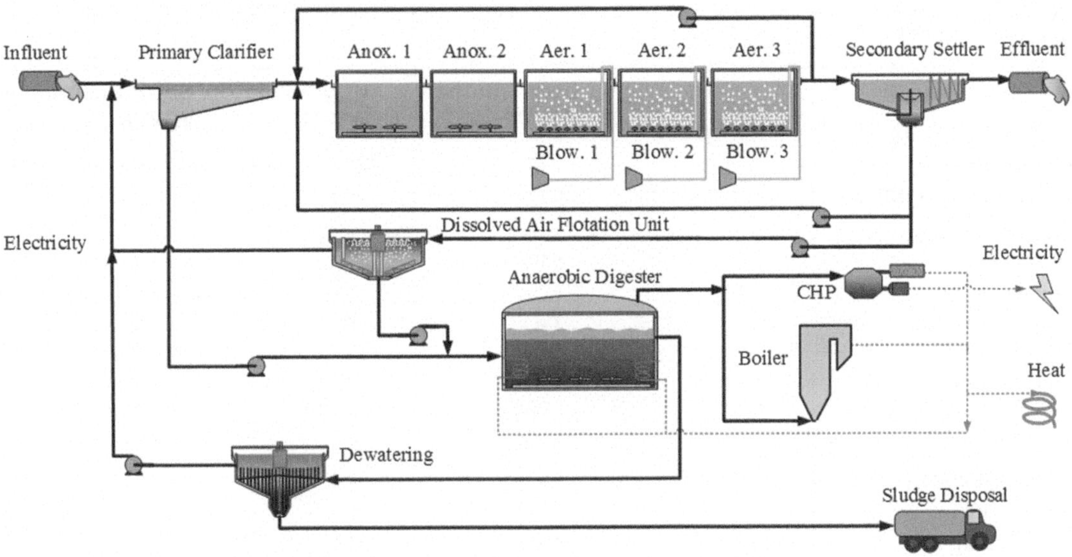

Figure 28.4 Conventional wastewater treatment plant (based on a BSM2 layout).

28.3.2 Upgraded WWTP

The second case study focused on a techno-economic analysis of an upgraded plant. The plant is based on the reference case (conventional WWTP) with the incorporation of two advanced technologies in the sludge line: a thermal hydrolysis (TH) reactor and a nitritation/Anammox process for treating the rejected supernatants (Figure 28.5).

The thermal hydrolysis process will increase the biogas production by improving the biodegradability of the sludge. Increased biogas production will be approximately 20%, but this production will depend on the proportions of primary and secondary sludge fed into the digester. For this, pressurized steam must be fed into the reactor to maintain the chamber at 170 degrees (Fernández-Polanco *et al.* 2008). As was done in the case study above, but with a different purpose, part of the biogas produced in the anaerobic digestion is diverted to a boiler to carry out combustion and produce the required steam in the TH process. The sludge biodegradability increase also involves the supplemental release of ammonium, which must be treated in the plant, representing in this study a 20% increase in the dewatering supernatant flow or an increase of 2% in the mainstream, entailing a 1–3% increase in the aeration costs. To remove this extra nitrogen and the ammonium released in the anaerobic digestion, a nitritation/Anammox process can be an interesting alternative. In the Anammox process, part of the ammonium is oxidised with the nitrites formed in the nitritation process, without oxygen and COD consumption, gaining theoretical aeration cost savings in the nitrogen removal process of up to 63% (Volcke *et al.* 2006). Analysing the aeration from a global perspective, the incorporation of a nitritation/Anammox process for treating the rejected supernatants represents an overall aeration cost savings of approximately 12% without sludge pre-treatment processes, and somewhat lower, around 9%, when a thermal hydrolysis process is incorporated into the layout (see Table 28.2) (Fernández-Arévalo *et al.* 2017b).

Table 28.2 Comparative analysis of the three configurations (Conventional, upgraded, and C/N/P decoupling WWTP).

	Conventional WWTP	**Upgraded WWTP**	**C/N/P Decoupling WWTP**
Total Power (kW)	274	259	202
Global balance (kW)	37	−22	−78
Power distribution (%):			
– Aeration	61%	59%	68%
– Mixing	12%	13%	12%
– Pumping	7%	7%	2%
– Dewatering	14%	13%	15%
– Flotation	6%	6%	3%
– Energy production	−86%	−108%	−139%
Total Cost (€ d^{-1})	1170	1127	669
Global balance (€ d^{-1})	510	344	−111
Costs distribution (%):			
– Aeration	40%	38%	57%
– Mixing	8%	9%	10%
– Pumping	4%	5%	2%
– Dewatering	9%	9%	13%
– Flotation	4%	4%	2%
– Ferric chloride ($FeCl_3$) dosage	35%	36%	16%
– Energy production	−56%	−69%	−117%
Aeration specific ratio	0.19 kW m^{-3}	0.18 kW m^{-3}	0.16 kW m^{-3}
Influent COD used in electricity production	11%	15%	14%

The philosophy of this second configuration is based primarily on increasing energy production in order to achieve a net overall energy balance closer to self-sufficiency. Analysing the operating costs, it can be seen that the decrease in aeration costs is not significant (0.18 kWh/m^3), similar energy is required for stirring and in the sludge separation processes, so that the differentiating factor, with respect to the reference configuration, is the increase in biogas production (25%) (see Table 28.2).

Regarding the influent energy content use, the configuration enables the transformation of 15% of the influent potential energy into electrical energy. This increase will allow the configuration to be closer to self-sufficiency, especially when the COD/N_T and COD/P_T influent ratios are high.

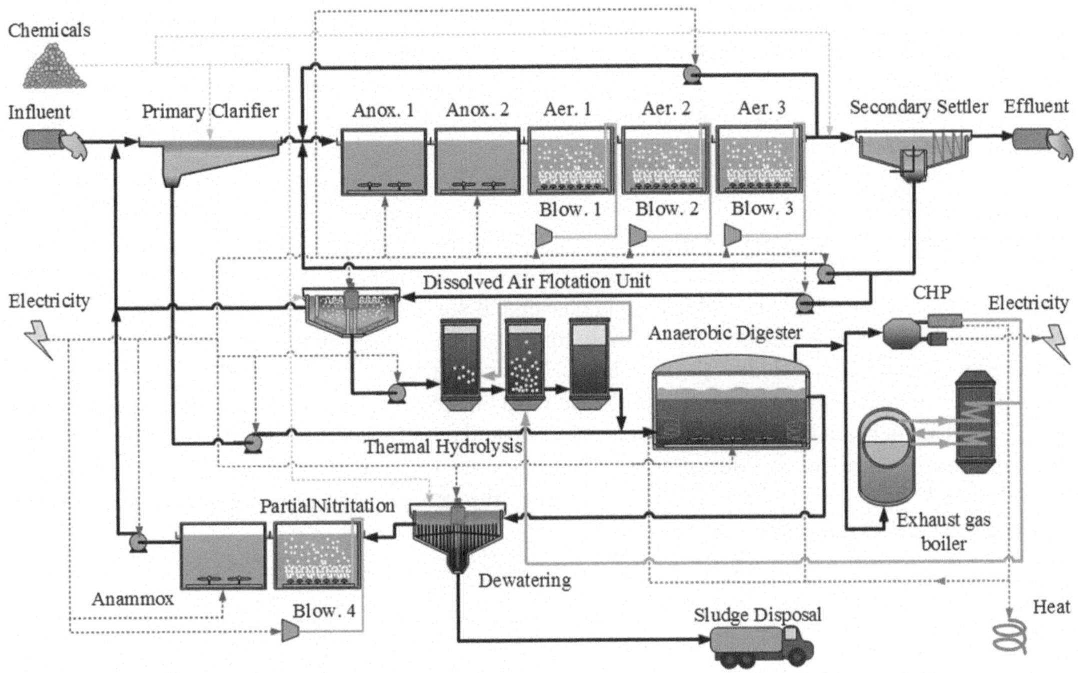

Figure 28.5 Upgraded wastewater treatment plant.

The only real drawback of this plant configuration is the effect of water temperature. If the influent water temperature is low, the steam requirements will be higher, and the biogas portion allocated to the boiler will also be higher. In this case, the differentiating factor of this plant configuration will be lost, and the plant should be analysed energetically in detail, in order to decide whether the configuration as a whole provides benefits with respect to the reference configuration or not.

28.3.3 A new WWT concept: C/N/P decoupling WWTP

Partition-release-recover (PRR) concept proposed by Batstone *et al.* (2015) was used as an example of the new WWT concept. The PRR configuration focuses on a first stage of organic matter and nutrient accumulation in the solids, a second stage of release through the digestion process, and ending with digestate treatment (Verstraete *et al.* 2009). The proposed configuration completely decouples carbon and nutrient treatments in order to seek greater process performance. In the work of Gao *et al.* (2014), the decoupling of carbon and nitrogen removal was already mentioned as a promising strategy to minimise energy requirements. As shown in Figure 28.6, the configuration consists of a Phoredox (A/O) process for biological phosphorus accumulation, thermal hydrolysis technology to increase the biodegradability and dewaterability of the sludge, an anaerobic digestion process for carbon removal and phosphorus and nitrogen release, a crystalliser for phosphorus precipitation as struvite, and a partial nitritation/Anammox process in the mainstream and sidestream to treat the nitrogen.

In the Phoredox process, soluble carbon and ortho-phosphates are accumulated in the solids (heterotrophic organisms, polyphosphate-accumulating organisms, polyhydroxyalkanoates, and polyphosphates) thanks to a combination of anaerobic and aerobic reactors. If these reactors are operated at a very short sludge retention time (2–4 days, depending on the temperature), it is possible to avoid excessive accumulation of inerts, nitrification reactions, and the subsequent nitrogen removal. Thus, only the strictly necessary nitrogen for the growth of the microorganisms will be removed. The secondary sludge produced, with low content of non-biodegradable material, is sent to the thermal hydrolysis process to transform part of this non-biodegradable material in biodegradable. As in the above configuration, the aim of the thermal hydrolysis process is to maximise the biogas production through an increase in the sludge biodegradability. In this case, it is also very important to analyse the effect of influent temperature and decide whether the process is profitable or not. The primary sludge and the pre-treated secondary sludge are sent to the anaerobic digester, where the biodegradation of the organic matter occurs. The carbonaceous matter is released as biogas (gas phase) and simultaneously the nitrogen and phosphorus become soluble matter in the form of ammonium and ortho-phosphates (liquid phase). After performing a dehydration of the digested sludge, the simulations show that the resulting rejected supernatant flow has a high nutrient content – 30% of the influent nitrogen and 60–65% of the phosphorus (Fernández-Arévalo *et al.* 2017b). The high nutrient content and concentration make it a suitable stream to perform the recovery of nutrients such as struvite. According to a global mass balance, the 40–45% of influent P and 5% of influent N could be recovered as struvite. The remaining nitrogen joins the mainstream and is treated with a partial nitritation/Anammox process.

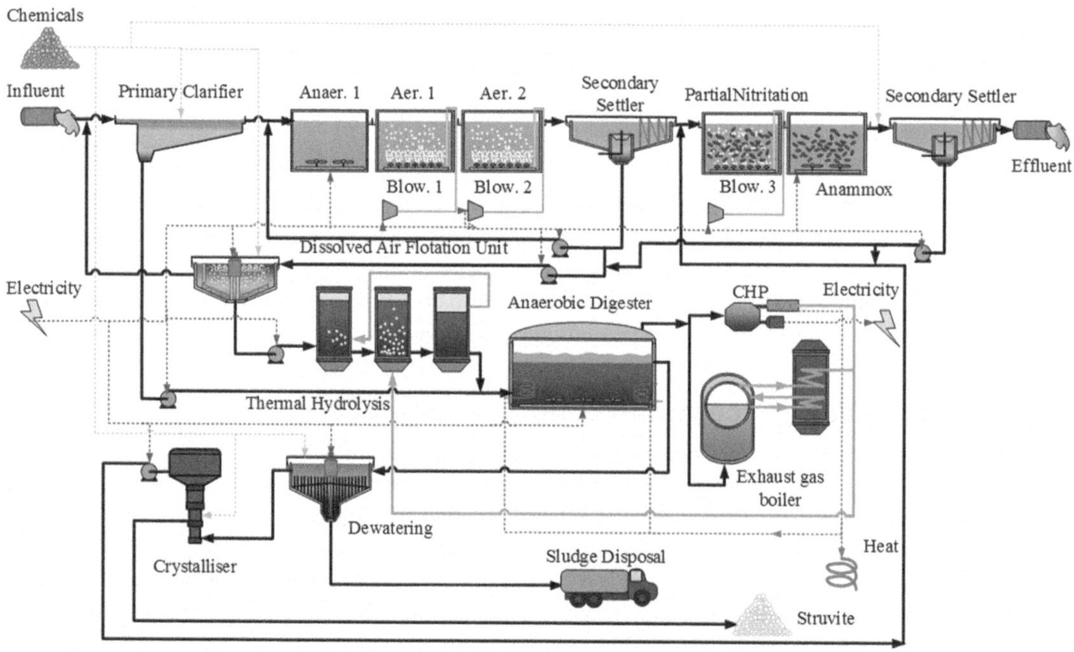

Figure 28.6 Carbon, nitrogen, phosphorus decoupling wastewater treatment plant.

Unlike the upgraded WWTP configuration, this configuration allows low energy treatment, and in this case the self-sufficiency of the plant is possible for almost any urban wastewater influent.

The decoupling of carbon, nitrogen, and phosphorus treatments allows the incorporation of advanced technologies based on recovery and low-energy use. Thus, the decoupling of carbon removal and nutrients removal allows for the maximising of biogas production or energy production, and in turn enables the incorporation of low-energy nitrogen treatment, such as the Nitritation-Anammox process. Moreover, the combination of a biological phosphorus removal process with an anaerobic digestion process (where all the accumulated phosphorus is released), allows the use of recovery technologies, thanks to the nutrient rich stream resulting from the anaerobic digestion.

The results related to energy and costs balances from the three configurations are compiled in Table 28.2. It can be seen that with an upgrade of the conventional plant, the energy consumption could be significantly reduced and that the C/N/P decoupling could clearly imply a self-sufficiency of the plant and energy production.

28.4 MODEL BASED ANALYSIS AND OPTIMISATION OF PLANT OPERATION

Every wastewater treatment plant (WWTP) deals with multiple requirements, in particular with effluent and discharge limits that constitute, obviously, the first priority of the plant. While maintaining compliance with the effluent limits, a WWTP operation is frequently targeted at an economic objective (e.g. minimising operating costs), leading to complex trade-off solutions. Designing the operation of an innovative process is an even more challenging task given that: (i) experience with the operation of the process is scarce or non-existent at plant level and, (ii) innovative processes are often conceptually more complex, feature more sub-units, or have diverse objectives. For instance, the novel decoupling WWT *(*Figure 28.6*)* previously presented aims not only at treating the wastewater but also at recovering phosphate as struvite and valorising COD as biogas. In the future, other challenging objectives can be expected in dealing with the emission of greenhouse gas (GHG) or the removal of pharmaceutical and personal care products (Corominas *et al.* 2013a; Kim *et al.* 2015).

How to define an optimal operating policy for multiple objectives under constraints while possibly taking uncertainty into account is a central topic in process and control engineering. The rigorous formulation of mathematical optimisation often gives rise to an intractable problem, which has led researchers to try to provide simplified problem formulations (Moon *et al.* 2011). A widely accepted methodology for defining the operating policy together with the control design of a whole plant based on a heuristic decomposition of the objectives and constraints was presented by Larsson and Skogestad (2000). The main feature is to link suitable variables to comply with the constraints and use the remaining free variables to pursue the plant objectives. In general, the constraints are to be satisfied at all times during operation, whereas the objectives (a proxy for cost or production) are to be optimised in the long term. This time scale separation helps to classify the possible decision variables into those that can be changed in real time or with relative flexibility (to regulate the constraints) and those with slow variations which can be scheduled to improve the plant objectives.

In this subsection, the authors present and demonstrate a methodology to design the operation of a WWTP, taking into account the operating constraints (effluent limits) and objectives. The goal of the methodology is to find the best operating policy and to decide how to keep the operation at the desired state despite disturbances, i.e. the design of the control system. In this context, the authors consider as decision variables (i.e. variables that can be changed) all those that can be adjusted during the operation at any time scale, whether by manual adjustment or by automatic control. The methodology is presented stepwise and partially based on Larsson and Skogestad's (2000) plant-wide control design method but adapted to WWTP specifications. The steps are summarised below:

Step 1. Definition of the operation objective and requirements. This step consist of defining the operation objective(s), which is normally expressed as the minimisation of a cost. The requirements or constraints are related to the effluent discharge limits and operational limits, such as the aeration capacity and the maximum concentration of solids.

Step 2. Determination of the degrees of freedom for operation and control. For a given plant there are a number of decision variables that can be changed independently, which constitute the degrees of freedom (DOFs). For designing and operation policy, the DOFs include variables that can be adjusted during the operation, manually or automatically.

Step 3. Study of the effect of the degrees of freedom on the objectives and the constraints. The previously identified DOFs can be used to drive the plant towards the desired operation provided there is an effect between the DOF and the objectives and/or constraints. This step can be carried out ideally using a model of the plant, although it is possible and often desirable to incorporate expert knowledge.

Step 4. Match the degrees of freedom with the constraints. Based on the previous study, it is possible to pair operational DOFs and constraints to ensure that the process is in compliance with the requirements defined in Step 1. As a process must be compliant despite disturbances, it is often necessary that the DOFs paired with constraints can be manipulated in real time and have a fast effect on the constrained variables.

Step 5. Use the remaining degrees of freedom to optimise the process. The DOFs that are not used to keep the process in compliance can be used to optimise the process following the objective(s) previously defined. Contrary to compliance with the constraints, the objectives are often to be followed for a long-term period; therefore, variables that have a slow effect on the process can be used at this stage.

28.5 CASE STUDY DEMONSTRATION: ANALYSIS AND OPTIMISATION OF A CONVENTIONAL WASTEWATER TREATMENT PLANT

The definition of an operating policy for a conventional wastewater treatment plant (Figure 28.4) is used here as a case study for demonstrating the methodology and followed step by step.

For the study, a conventional plant optimized at a critical temperature of 13 degrees with a mixed-liquor total suspended solids concentration of 3,500 mg/l was used. Influent concentrations were extracted from the BSM2 influent model (Table 26.8) and adapted to model the components of the PWM methodology. To avoid possible interference from environmental conditions, such as the effect of low influent water temperatures on digester heating, simulations were carried out for a constant temperature of 15°C.

Step 1: Definition of the operational objective and requirements

The objective of the plant is to minimise the operating costs expressed as equivalent kW. The operating expenses are the net combination of: (i) the energy consumption for mechanical mixing in non-aerated reactors, pumping, compression work for blowers, and dewatering electrical energy consumption; (ii) electricity production by the CHP unit, and (iii) the equivalent energy cost of added chemicals (in this case, $FeCl_3$ and polyelectrolyte). Equations for each of the terms can be seen in Table 28.1. Compliance with the effluent limits was evaluated here by steady state simulation, requiring the maximum concentration in the effluent to be:

Constraint 1: Total nitrate and nitrite species (NO_x) in the effluent <8 mgN/L
Constraint 2: Total ammonium species (NH^+_4 and NH_3) in the effluent <1 mgN/L
Constraint 3: Total phosphorus (P) in the effluent <1 mgP/L

Step 2: Determination of the degrees of freedom for operation and control

An analysis of the plant must indicate the degrees of freedom (DOFs) for the operation and control. There is a commonly accepted practical distinction between operation and control. In general, whether an actuator is to be kept constant or manipulated in real-time it is not essential in this work. When carrying out a model-based analysis of a DOF it is especially important to ascertain the possible range of the DOF for the process as it has been designed, e.g. an aerated tank must maintain a minimum aeration flow in order to keep sludge from settling.

For the plant in Figure. 28.4 the following DOFs were identified:

(i) The internal recirculation flow rate. At nominal conditions, its value is approximately 3 times the influent flow, but the range of operation was considered to extend between 1.5 and 5.5 times the influent flow. This is a dynamic DOF that can be manipulated in real time and therefore it can be used for operation and control.

(ii) The three dissolved oxygen (DO) set-points in the aerated tanks. At nominal conditions, the DO setpoint is 2 mg O_2/L. Although not mandatory, it is common practice to keep the three tanks at the same DO level; therefore, they represent one single DOF (Åmand *et al.* 2013). Note that, the air flow rate could be considered to be the DOF. However, setting a constant value for the air flow rate is much less robust to model mismatch. The range for DO was chosen to be between 0.8 and 6 mg O_2/L in order to ensure a minimum of mixing in the aerated tanks.

(iii) The addition of chemicals to enhance phosphorus removal. The possibility of adding coagulating agents for postprecipitation of phosphorus is a DOF as it can be manipulated independently.

(iv) The concentration of total suspended solids (TSS) in the last aerated tank. At nominal conditions, the concentration of TSS in the last tank is kept at 3,000 mg TSS/L. This value can be manipulated by changing the purge flow rate, therefore it acts as a proxy for the solid retention time (SRT). As previously noted, setting a given value on the concentration of TSS is more robust to model mismatch than setting a given purge flow rate, even though they are equivalent DOFs.

(v) The primary clarifier solids removal efficiency (η). It was considered here that the efficiency of the primary clarifier can also be manipulated and, therefore, be a DOF. In effect, if the settler is operated at a different retention time, it is possible to reduce its efficiency. Likewise, using coagulating agents, it is possible to increase to a certain extent the primary clarifier efficiency, albeit at the expense of a higher coagulant cost. At nominal conditions, the removal efficiency was set at 60%, while the addition of coagulant agent is required to obtain higher efficiencies (Tik and Vanrolleghem 2017).

Step 3: Study the effect of the degrees of freedom on the objectives and the constraints

The pairing between the degrees of freedom and the constraints/objectives is based on studying their relationships and interactions. In particular, Figures 28.7 and 28.8 show the derivative of the variable, which is to be kept controlled with respect to the most relevant DOF. This is equivalent to plotting the *gain* of the DOF with respect to the constrained variable or objective. In order to obtain comparable magnitudes, the resulting derivatives were multiplied by the potential range of the DOF. In effect, each DOF can only be varied within certain boundaries. Hence, the interpretation of the figures at each point is the following: for a linear system, what the variation of the constrained variable would be if each DOF varies the maximum allowable range. The advantage of this expression is that it gives a clear view of the change that can be reached with each of the DOFs, allowing their effects to be compared.

Effect on constraint 1. Total nitrate in the effluent. The internal recirculation has a great effect on NO_x-N removal (Figure 28.7). At low recirculation flow rates, the removal of NO_x-N is greatly increased as more nitrate is available at the anoxic region for heterotroph denitrification. However, as the recirculation increases, it reaches a point where the amount of NO_x-N actually increases as too much oxygen is recirculated, reducing the anoxic volume and hindering denitrification. In contrast, the level of oxygen in the aerobic zone has a marginal effect: the increase in oxygen increases the level of NO_x-N due to further nitrification of the ammonium. At constant dissolved oxygen, an increase in the concentration of solids also leads to an increase in NO_x-N due to further oxidation of ammonia. Finally, increasing the efficiency of the primary clarifier leads to a severe increase in the NO_x-N concentration as less COD is available for denitrification.

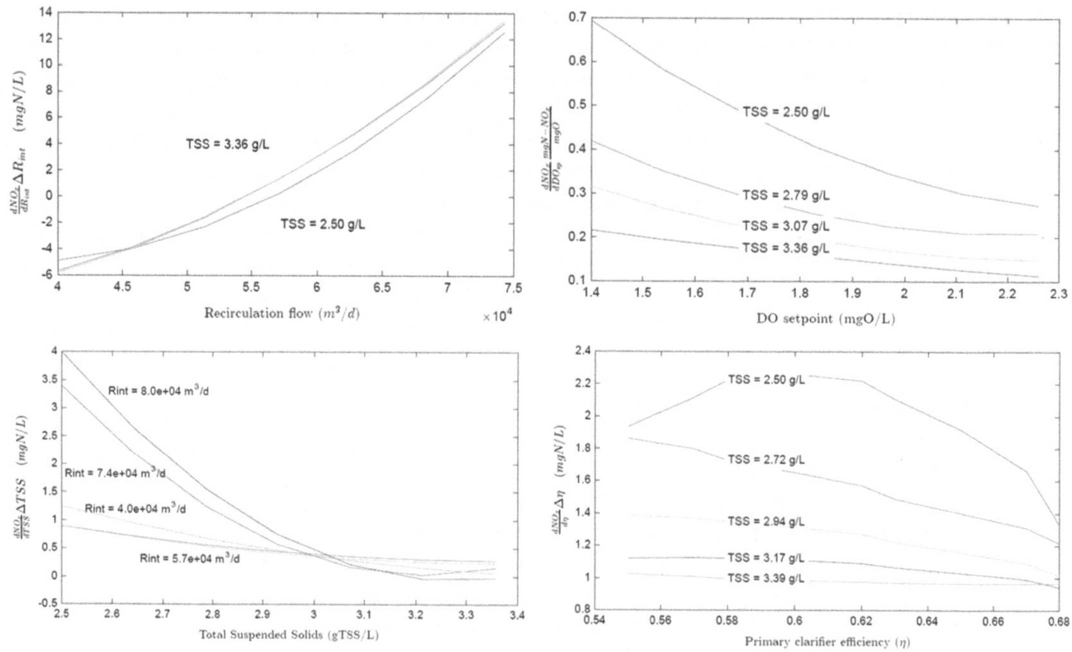

Figure 28.7 The influence of four DOFs on NO_x-N concentration in the effluent for different TSS values. From top left and clockwise, the influence of the recirculation flow rate, the DO set-point, primary clarifier efficiency, and TSS on NO_x-N concentration.

Effect on constraint 2. Total ammonium in the effluent. The largest effect on ammonium is caused by the variables that directly affect the activity of nitrifiers, i.e. the level of oxygen needed for nitrification and the TSS, which is proportional to the amount of biomass in the reactors (Figure 28.8). The trends describing the effect of increasing the DO or the TSS in the reactor are similar. At low levels, not all the ammonium is converted and therefore it is possible to greatly reduce the concentration of ammonium by increasing the DO and/or TSS. As there is less ammonium left in the reactor (in particular, at concentration lower than the required 1 mg N/L), it becomes more difficult to remove the residual ammonium and the effects of increasing DO or TSS is reduced.

Effect on objective. Total operating expenses. As an example, Figure 28.9 explores the impact of the TSS concentration in the activated sludge reactors and primary clarifier TSS removal efficiency on the total operating expenses. The TSS concentration in the activated sludge line has a major impact on the cost: low TSS concentration leads to a larger production of sludge and a higher rate of recovery of resources by anaerobic digestion, lower endogenous respiration, and lower air requirements; but a very low concentration can lead to a loss of bacterial activity (under the bottom line). The internal recirculation has a negligible direct effect on the cost and ammonium elimination; on the contrary, it is essential to fulfil the requirements regarding the total nitrogen in the effluent. Hence, it can be used to control the nitrate and total nitrogen removal. The TSS removal efficiency in the clarifier also has a high impact on operational costs: high solids removal efficiencies lead to higher production of primary sludge and biogas, but with a higher coagulant cost (at more than 60% solids removal). As shown in Figure 28.9, operational costs considerably changed from 60% efficiency due the addition of a new expense into the global balance (the cost of chemical agent), nevertheless, the trend continues. On the other hand, very high efficiencies can lead to a lack of COD for denitrifying and a violation of quality requirements (above the top line).

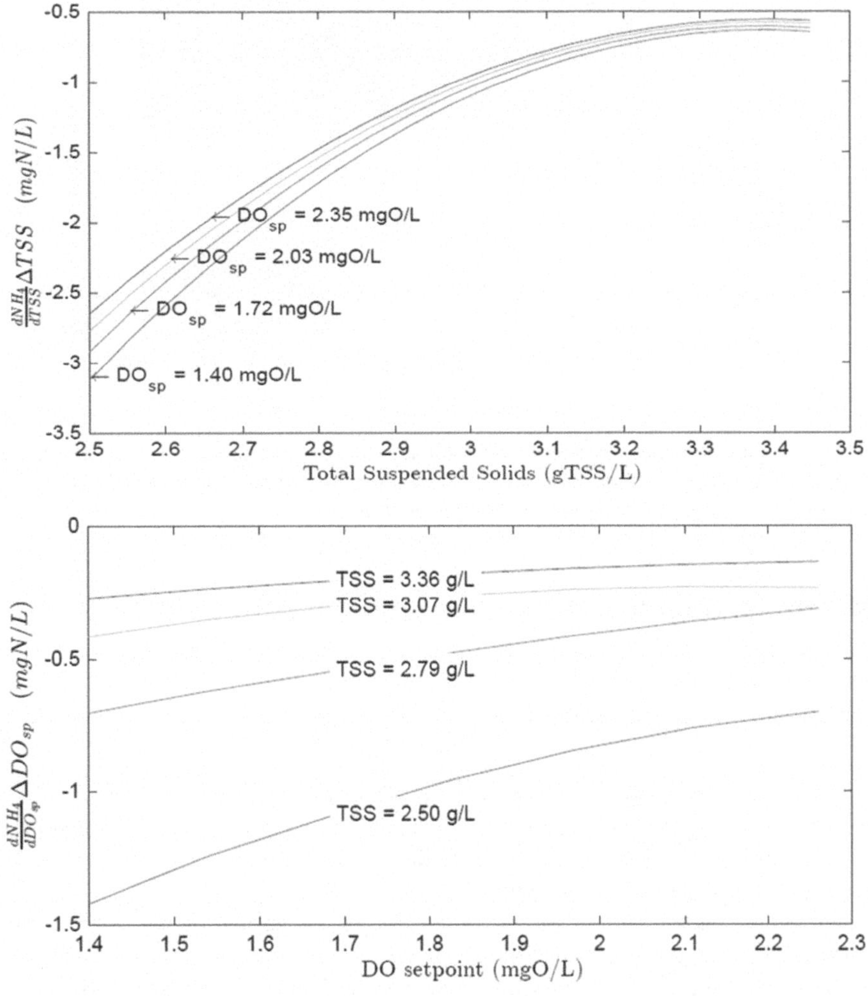

Figure 28.8 The influence of TSS (top) and DO setpoint (bottom) on NH_4-N concentration in the effluent.

Manipulating both the settler efficiency and the TSS concentration is the best way to manage the COD in the plant, so that the biogas production is maximised but there is still enough COD to carry out the removal of nitrogen at the desired level.

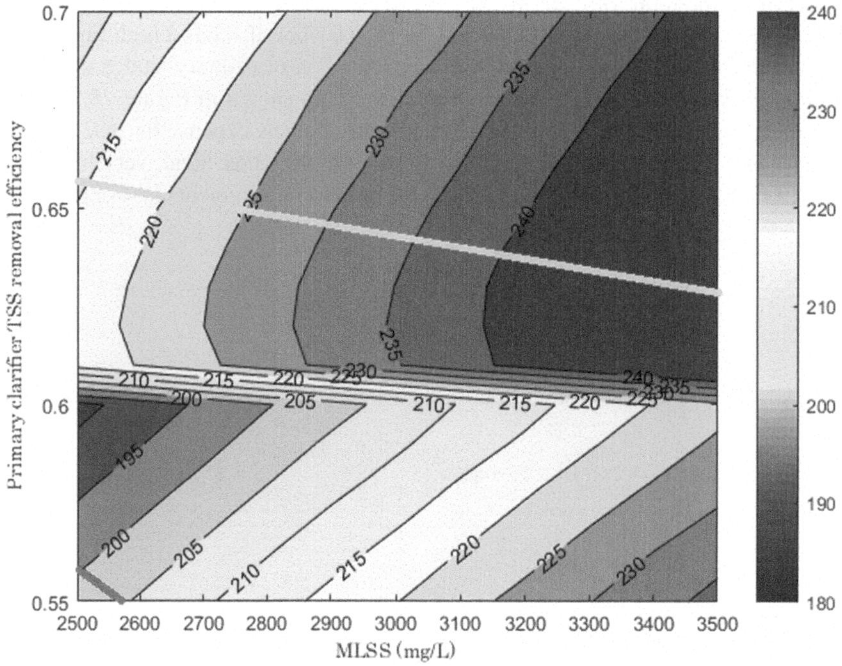

Figure 28.9 Total equivalent operating cost (in kW) with respect to the total suspended solids in the aerobic tank and the primary clarifier efficiency. Below the bottom line, constraint 2 (ammonium limit) is violated; over the top line, constraint 1 (NO_x-N limit) is violated.

Step 4: Match the degrees of freedom with the constraints

The operation must fulfil three constraints that relate to the effluent discharge limits of the plant during operation. Therefore, at least three DOFs must be linked with each of the constraints to ensure operational compliance at all times. From the information shown in the previous step it is possible to assign each DOF:

- *Constraint 1:* The concentration of NO_x-N in the effluent is regulated by the internal recirculation flow rate. It can be implemented in practice by a control loop where the controlled variable is the NO_x-N concentration (commonly measured at the end of the anoxic region) and the manipulated variable is the internal recirculation flow rate.
- *Constraint 2:* The concentration of NH_4-N in the effluent is regulated by the DO level in the aerobic tanks, which is by itself a variable controlled by manipulation of the air flow rate. The resulting control structure is a cascaded control where the master loop acts as a supervisory controller keeping the NH_4-N level below the effluent discharge limit and the slave loop keeps the DO at the level set by the master loop.
- *Constraint 3:* The concentration of P in the effluent is regulated by the addition of coagulant after the biological treatment (postprecipitation). This DOF is not dependent on other constraints but interacts with them (through the formation of inorganic components after precipitation, which modify the concentration of TSS). Therefore, its assignment to the regulation of the phosphorus level appears self-evident.

Step 5: Use the remaining degrees of freedom to optimise the process

The remaining degrees of freedom are the TSS in the last aerated tank and the primary clarifier TSS removal efficiency (η). Figure 28.10 shows the operating expenses with respect to these two variables while the three constraints are regulated by control loops according to the assignment from step 4. Analysing the results, it can be seen that control loops provide stability to the process, ensuring the quality of the effluent for all the values of TSS and η. Furthermore, the control loops enable the plant to operate without exceeding the limits, i.e. to avoid concentrations of NH_4-N above 1 mg N/L and to adjust the internal recycle to maximise the denitrification potential of the anoxic zone. Hence, this operation reduces the need for air and pumping costs, minimising the operating expenses of the plant (up to 10% for this particular example).

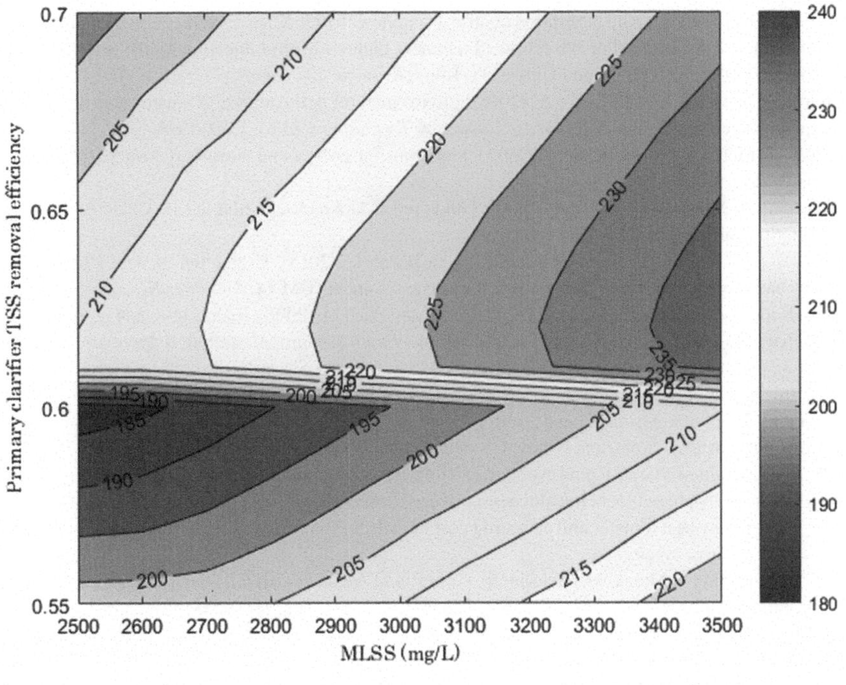

Figure 28.10 Total equivalent operating costs (in kW) with respect to the total suspended solids in the aerobic tank and the primary clarifier TSS removal efficiency with control loops to comply with the three effluent limits.

28.6 CONCLUSIONS

The tools presented in Chapters 26 and Chapter 27 allowed an initial selection of the most appropriate plant configurations for specific situations, considering environmental, technical, and economic criteria. The aim of the E-PWM methodology presented in this Chapter is to use these preselected configurations to carry out a more thorough, comprehensive, and accurate analysis, considering changing environmental conditions, dynamic influent variations, and general system disturbances. Hence, it is a suitable tool for the design and operation of plants. In the examples shown herein, the potential of the tool to compare traditional and advanced configurations, as well as to create regimen maps that aid in the interpretation of the process results has been demonstrated. Furthermore, following the steps of the methodology, it has been possible to explore the effects of the variables that can be changed to drive the operation towards the desired optimum, while satisfying the requirements of the process. Nevertheless, the methodology presented here is generic and can be used for any other plant, provided that a model is available. The use of plant-wide models is, in this context, essential to ensure that complex plants featuring different technologies can be analysed reliably and that the model faithfully reproduces the plant behaviour, also in terms of energy and chemical consumption. It is expected that the methods presented here can assist process and design engineers in the design and operation of novel plants implementing upcoming technologies.

28.7 REFERENCES

Åmand L., Olsson G. and Carlsson B. (2013). Aeration control – a review. *Water Science & Technology*, **67**(11), 2374–2398.

Arnell M. (2016). Performance Assessment of Wastewater Treatment Plants – Multi-Objective Analysis Using Plant-Wide Models. PhD thesis, ISBN 978-91-88934-72-7, Lund University, Lund, Sweden.

Baresel C., Dalgren L., Almemark M. and Lazic A. (2016). Environmental performance of wastewater reuse systems: impact of system boundaries and external conditions. *Water Science & Technology*, **73**(6), 1387–1394.

Batstone D. J., Hülsen T., Mehta C. M. and Keller J. (2015). Platforms for energy and nutrient recovery fron domestic wastewater: a review. *Chemosphere*, **140**, 2–11.

Baumann H. and Tillman A. M. (2004). The Hitch Hiker's Guide to LCA. An Orientation in Life Cycle Assessment Methodology and Application. Studentlitteratur, Lund, Sweden.

Bisinella de Faria A. B., Spérandio M., Ahmadi A. and Tiruta-Barna L. (2015). Evaluation of new alternatives in wastewater treatment plants based on dynamic modelling and life cycle assessment (DM-LCA). *Water Research*, **84**, 99–111.

Comas J., Rodríguez-Roda I., Gernaey K.V., Rosen C., Jeppsson U. and Poch M. (2008). Risk assessment modelling of microbiology-related solids separation problems in activated sludge systems. *Environmental Modelling & Software*. **23**(10–11), 1250–1261.

Corominas L., Foley J., Guest J. S., Hospido A., Larsen H. F., Morera S. and Shaw A. (2013a). Life cycle assessment applied to wastewater treatment: state of the art. *Water Research*, **47**(15), 5480–5492.

Corominas L., Larsen H. F., Flores-Alsina X. and Vanrolleghem P. A. (2013b). Including Life Cycle Assessment for decision-making in controlling wastewater nutrient removal systems. *Journal of Environmental Management*, **128**, 759–767.

Fernández-Arévalo T., Lizarralde I., Grau P. and Ayesa E. (2014). New systematic methodology for incorporating dynamic heat transfer modelling in multi-phase biochemical reactors. *Water Research*, **60**, 141–155.

Fernández-Arévalo T. 2016. New heat transfer and operating cost models for the Plant-Wide simulations of full-scale WWTPs. PhD thesis, University of Navarra, Spain.

Fernández-Arévalo T., Lizarralde I., Maiza, M., Beltrán, S., Grau P. and Ayesa E. (2017a). Diagnostic and optimization of WWTPs using the PWM library: full-scale experiences. *Water Science & Technology*, **75**(3), 518–529.

Fernández-Arévalo T., Lizarralde I., Fdez-Polanco F., Pérez-Elvira S.I., Garrido J.M., Puig S., Poch M., Grau P. and Ayesa E. (2017b). Quantitative assessment of energy and resource recovery in wastewater treatment plants based on plant-wide simulations. *Water Research* (in press).

Fernández-Polanco, F., Velázques, R., Pérez-Elvira, S.I., Casas, C., Del Barrio, D., Cantero, F.J., Fernández-Polanco, M., Rodríguez, P., Panizo, L., Serrat, J., Rouge, P., 2008. Continuous thermal hydrolysis and energy integration in sludge anaerobic digestion plants. *Water Science & Technology*, **57**(8), 1221–1226.

Flores-Alsina X., Arnell M., Amerlinck Y., Corominas L., Gernaey K. V., Guo L., Lindblom E., Nopens I., Porro J., Shaw A., Snip L., Vanrolleghem P. A. and Jeppsson U. (2014). Balancing effluent quality, economic cost and greenhouse gas emissions during the evaluation of (plant-wide) control/operational strategies in WWTPs. *Science of the Total Environment*, **466**, 616–624.

Gao H., Scherson Y. D. and Wells G. F. (2014). Towards energy neutral wastewater treatment: methodology and state of the art. *Environmental Science: Processes & Impact*, **16**(6), 1223–1246.

Gernaey K. V., Jeppsson U., Vanrolleghem P. A. and Copp J. B. (2014). Benchmarking of Control Strategies for Wastewater Treatment Plants, Scientific and Technical Report No. 23, IWA Publishing, London, UK.

Grau P., de Gracia M., Vanrolleghem P. and Ayesa E. (2007). A new Plant-Wide modelling methodology for WWTPs. *Water Research*, **41**(19), 4357–4372.

Guest, J.S., Skerlos, S.J., Barnard, J.L., Beck, M.B., Daigger, G.T., Hilger, H., Jackson, S.J., Karvazy, K., Kelly, L., MacPherson, L., Mihelcic, J.R., Pramanik, A., Raskin, L., van Loosdrecht, M.C., Yeh, D., Love, N.G., 2009. A new planning and design paradigm to achieve sustainable resource recovery from wastewater. *Environmental Science & Technology* **43**, 6126–6130.

ISO 14044 (2006). Environmental management – Life cycle assessment – Requirements and guidelines.

Jeppsson U., Pons M. N., Nopens I., Alex J., Copp J. B., Gernaey K. V., Rosen C., Steyer J. P. and Vanrolleghem P. A. (2007). Benchmark simulation model No 2 – general protocol and exploratory case studies. *Water Science & Technology*, **53**(8), 67–78.

Keller, J., 2008. Paradigm shift from wastewater treatment to resource recovery system. *Australian-German workshop on sustainable urban water management*, Melbourne, Australia.

Kim D., Bowen J. D. and Ozelkan E. C. (2015). Optimization of wastewater treatment plant operation for greenhouse gas mitigation. *Journal of Environmental Management*, **163**, 39–48.

Kokabian B. and Gude V. G. (2015). Role of membranes in bioelectrochemical systems. *Membrane Water Treatment*, **6**(1), 53–75.

Larsson T. and Skogestad S. (2000). Plantwide control – a review and a new design procedure. *Modeling, Identification and Control*, **21**(4), 209–240.

Lizarralde I., Fernández-Arévalo T., Brouckaert C. J., Vanrolleghem P. A., Ikumi D. S., Ekama G. A., Ayesa E. and Grau P. (2015). A new general methodology for incorporating physico-chemical transformations into multi-phase wastewater treatment process models. *Water Research*, **74**, 239–256.

McCarty P. L., Bae J. and Kim J. (2011). Domestic wastewater treatment as a net energy producer – can this be achieved? *Environmental Science & Technology*, **45**, 7100–7106.

Meneses M., Concepcion H. and Vilanova R. (2016). Joint environmental and economical analysis of wastewater treatment plants control strategies: a benchmark scenario analysis. *Sustainability*, **8**(4), doi: 10.3390/su8040360

Metcalf & Eddy (2014). Wastewater Engineering: Treatment and Resource Recovery, 5 edn, revised by Tchobanoglous G., David Stensel H., Tsuchihashi R. and Burton F. McGraw Hill Education, New York, USA.

Moon J., Kim S. and Linninger A. A. (2011). Integrated design and control under uncertainty: embedded control optimization for plantwide processes. *Computers & Chemical Engineering*, **35**(9), 1718–1724.

Perry R. H., Green D. W. and Maloney J. O. (1999). Perry's Chemical Engineers' Handbook. 7th edn. McGraw-Hill: New York, Chicago, San Francisco, Lisbon, London, Madrid, Mexico City, Milan, New Delhi, San Juan, Seoul, Singapore, Sydney, Toronto. McGraw-Hill Companies Inc, EE.UU.

Porro J., Comas J., Rodriguez-Roda I., Pijuan M., Corominas L., Guo L., Daelman M., Volcke E.I.P., van Loosdrecht M.C.M., Vanrolleghem P.A., Nopens I. (2014). Risk assessment modelling of N2O production in activated sludge systems: Quality not Quantity. In: *Proceedings of 4th IWA/WEF Wastewater Treatment Modelling Seminar* (WWTmod2014), Spa 2014.

Shoener B. D., Bradley I. M., Cusick R. D. and Guest J. S. (2014). Energy positive domestic wastewater treatment: the roles of anaerobic and phototropic technologies. *Environmental Science: Processes & Impacts*, **16**, 1204–1222.

Tik S. and Vanrolleghem P.A. (2016). Chemically enhancing primary clarifiers: model-based development of a dosing controller and full-scale implementation. Water Science & Technology, **75**(5), 1185–1193.

Verstraete W., Van de Caveye P. and Diamantis V. (2009). Maximum use of resources present in domestic "used water". *Bioresource Technology*, **100**, 5537–5545.

Volcke E. I. P., Gernaey K. V., Vrecko D., Jeppsson U., van Loosdrecht M. C. M. and Vanrolleghem P. A. (2006). Plant-wide (BSM2) evaluation of reject water treatment with SHARON-Anammox process. *Water Science & Technology*, **54**(8), 93–100.

Weisbach J. (1845). Lehrbuch der Ingenieur und Maschinen Mechanik, Vol. 1. Theoretische Mechanik. Vieweg und Sohn, Braunschweig, Germany, 535 pages (in German).

Zwietering Th. N. (1985). Suspending of solid particles in liquid by agitators. *Chemical Engineering Science*, **8**, 244–253.

Annex 1

E-course: Micropollutants in water

Alette A.M. Langenhoff and Süreyya Meriç

Since the last 10–20 years the quality of surface water systems and their interacting groundwater systems is increasingly under pressure as a result of new groups of chemicals entering these natural water systems. Especially the occurrence of organic micropollutants (OMPs) such as pharmaceuticals, synthetic hormones and personal care products have an increasing attention of drinking water companies and water resource institutions. Furthermore, the development of analytical techniques to measure these compounds at low concentrations have accelerated this awareness. The last 40 years, technologies for removal and recovery of bulk substances from wastewater effluents, such as organics and nutrients like phosphorus and nitrogen, have been developed and implemented into the water cycle infrastructure. As these technologies are designed to deal with bulk load emissions, many OMPs are not removed during passage through these systems (chapter 9).

Organic micropollutants, such as pharmaceuticals, enter the water cycle either as the original compound or in a metabolized form at low loads, resulting in relatively low concentrations, i.e. in the ng/L – µg/L range. The recent development of new analytical techniques has allowed us to detect such low concentrations in the environment. These concentrations have most likely been present in the water since many years, and their levels have only recently been quantified and acknowledged to be associated with a potential ecological risk. Various studies dealing with the quality and ecological functioning of water systems show that these chemicals form a potential new problem impacting quality of ecological and possibly human life through estrogenic and carcinogenic toxicity.

This e-course covers various aspects that are related to micropollutants, mainly organic micropollutants, in our water (Figure A1.1). The e-course is prepared by the members of the COST Action Water_2020. It is developed to teach scientists/trainees on organic micropollutants in our water, and to open this topic to a wider audience. The course is organised in 12 modules and includes 5 case studies in parallel to the activities of COST Action Water_2020, and is accessible at the following website http://water2020.eu/micropollutants-in-water-e-course.

The contributing authors and their affiliations is given in Table A1.1.

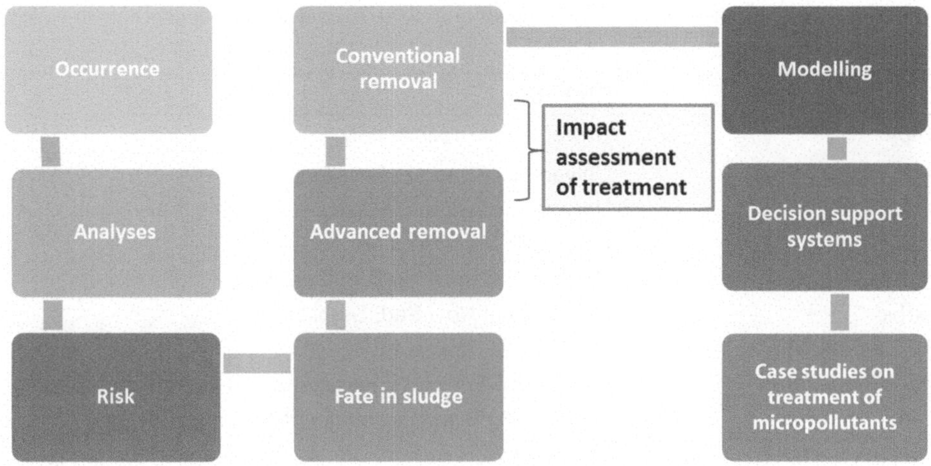

Figure A1.1 Overview of the different topics of the e-course 'Micropollutants in water'.

Table A1.1 Overview of the authors and their affiliations of the e-course 'Micropollutants in water'

Name	Affiliation
Alette Langenhoff	Wageningen University & Research, the Netherlands
Giorgio Bertanza	University of Brescia, Italy
Ivan Brnardic	Faculty of Metallurgy, Croatia
Marta Carballa	University of Santiago de Compostela, Spain
Alba Castillo	University of Gerona & Aqualogy, Spain
Feiran Chen	Helmholtz Centre Munich, Germany
Joaquim Comas	University of Gerona & ICRA, Spain
Jolanta Dvarioniene	Kaunas University of Technology, Lithuania
Sandra Contreras	Rovira i Virgili University, Spain
Nataša Đurišić-Mladenović	University of Novi Sad, Serbia
Zişan Emre	Saraykoy Nuclear Research and Training Centre, Turkey
Xavier Flores Alsina	Technical University of Denmark , Denmark
Antonio García Fernández	University of Murcia, Spain
Manel Garrido-Baserba	University of California, USA
Mark Healy	National University of Ireland, ireland
Brigitte Helmreich	Technical University Munich, Germany
Giusy Lofrano	University of Salerno, Italy
Emma Martinez-Lopez	University of Murcia, Spain
Giuseppe Mascolo	Water Research Institute (IRSA), Italy
Süreyya Meriç	Namık Kemal University, Turkey
Dragana Mutavdžić Pavlović	Faculty of Chemical Engineering and Technology, Croatia

(*Continued*)

Table A1.1 Overview of the authors and their affiliations of the e-course 'Micropollutants in water' (*Continued*).

Name	Affiliation
Matteo Papa	University of Brescia, Italy
Sanja Papic	University of Zagreb, Croatia
Manuel Poch	University of Girona, Spain
Marie-Noëlle Pons	University of Lorraine, France
Peter Schröder	Helmholtz Centre Munich, Germany
Biljana Škrbić	University of Novi Sad, Serbia
Mohamed Tawfic Ahmed	Suez Canal University, Egypt
Konstantinos Tsagarakis	Faculty of Chemical Engineering and Technology, Greece
Christian Vogelsang	Norwegian Institute for Water Research, Norway
Jelena Živan ev	University of Novi Sad, Serbia

Annex 2

Implementing an ecoefficiency tool for the holistic design and assessment of the water cycle

Gumersindo Feijoo, Maria Teresa Moreira, Joan Rieradevall, Francesc Hernández, Desirée Marín-Navarro

A2.1 INTRODUCTION

The use of water for human activities represents a growing concern, since its management, especially in urban environments, has grown in complexity in recent decades. The selection of key sustainable performance indicators is essential in the design and assessment of water facilities. The management of the urban water cycle (UWC), which includes wastewater treatment and potable water supply, requires its evaluation for a wide range of sustainability indicators in order to assess impacts and costs and, thereafter, include potential alternatives. However, it should be noted that current sustainability assessment methods tend to use a life-cycle perspective in order to identify the environmental, economic and/or social hotspots of implementing complex managerial systems. Of these, Life Cycle Assessment (LCA) for environmental sustainability and Life Cycle Costing (LCC) for economic assessment are chosen based on the fact that they are standardized assessment methods which evaluate the life-cycle impacts of the abovementioned processes in a rigorous and integrated manner. The objective of this work was to develop an eco-efficiency tool, especially devoted to policy makers and public and private water managers to provide them a set of environmental, economic and eco-efficiency indicators in a very comprehensive way for a more sustainable decision-making.

After a first revision of existing tools and benchmark values for environmental or economic assessment tools, the user requirements were set to develop the prototype of the eco-efficiency tool. Detailed information of water facilities and networks for water use and treatment were required for that purpose. Once the structure applied at the prototype was proved to meet the basic requirements, the different modules were implemented: data input, calculation and results. The tool was finally online and opened to the public in May 2015. Below it follows a description of the tool features, with special emphasis on the different actions to perform a project for a single facility or for the whole urban water cycle (Figure A2.1).

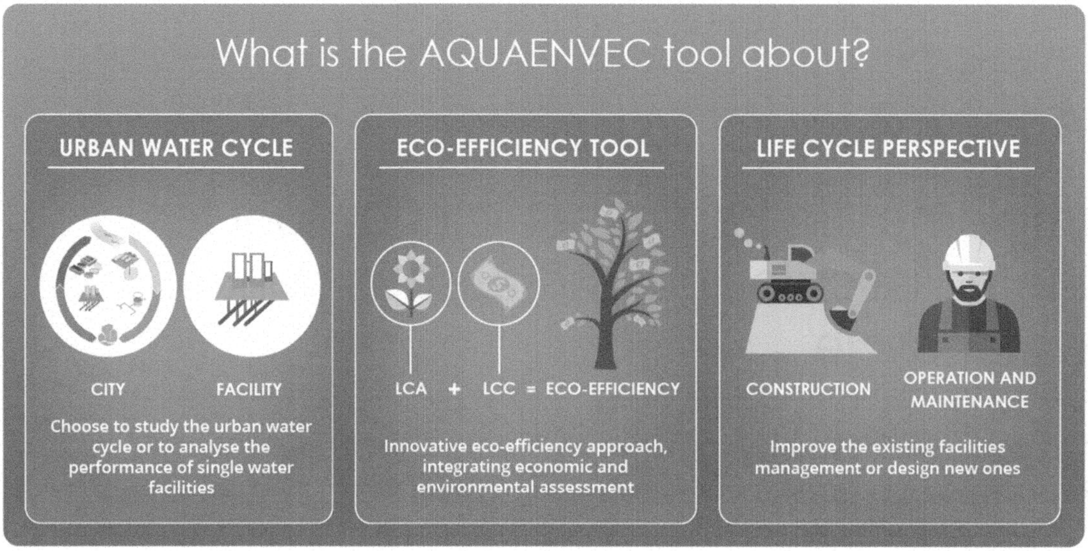

Figure A2.1 What is the Aquaenvec tool about?

A2.2 TOOL FEATURES

The AQUAENVEC tool is a user-friendly web tool developed to support decision-makers to assess and improve the eco-efficiency of water activities in the urban context. The tool is available online at: http://www.life-aquaenvec.eu/the-aquaenvec-tool/ or at the direct link http://tool.life-aquaenvec.eu/en. There is no installation package as it is a web tool. The tool has been developed both in English and Spanish.

To begin with the AQUAENVEC tool, the user must register with username, password, contact details, organization and field of expertise and he must accept legal terms of use. At the right side of the sign in option, the link to log in (email address and password) is found in the main screen.

There are different possibilities to create a new project; the user can choose between an urban water cycle project (including all the stages of the cycle) and a single water facility (Figure A2.2). When choosing single water facility, the stages of the water cycle will be specified: DWTP (drinking water treatment plant), supply network, sewer network and WWTP (wastewater treatment plant).

A screen with the different stages will appear. By clicking in each one of them, the user will be able to navigate across the pages to introduce or visualize the data. In the bottom right corner of every stage there is a little information icon. If it appears in **green**, the stage has all the required data and it is valid. If it appears in **red** instead, the stage is invalid because there is some required data that has not been introduced yet. If the stage is invalid, you can click the icon and a message will appear, indicating the reasons why it is not valid.

Now it follows a description on how to introduce the data associated to each one of the stages.

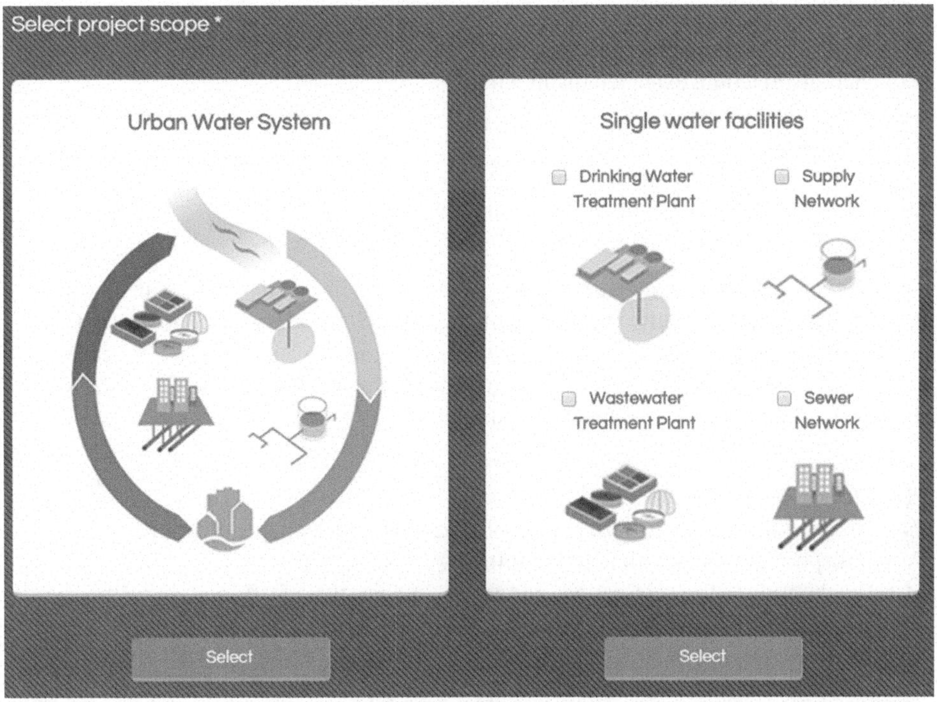

Figure A2.2 Selection of the project scope: single facilities (right) or the complete urban cycle (left).

A2.3 DRINKING WATER TREATMENT PLANT (DWTP)

The *General data* section comprises the required data of the installation: facility's identification name, average treated flow, number of served municipalities, served population, etc. In the *Facility definition* the user can select different treatments and modify the configuration, by adding a category or process (water treatment or sludge treatment) as well as one of their subcategories (treatment group). When doing so, a new icon is added to the scheme, representing the selected treatment. An element can be deleted by clicking on the trashcan (in the top right corner of its icon). To link two elements, the user must click on the first unit (highlighted in **orange**) and thereafter, click on the second element. This link can be deleted by double clicking the line.

A wide range of units can be added for water and sludge treatment:

- Water treatment
  - Water abstraction: open intake, well intake.
  - Pre-treatment: screening, grit removal.
  - Clarification: flotation, coagulation/flocculation + settling.
  - Filtration: bed filter, pressurised filter.
  - Disinfection: disinfection.
  - Other treatment: stripping, activated carbon.

- Chemical treatment: chemical dosing.
- Membrane treatment: membrane treatment.
- Extended disinfection: extended disinfection.
- Sludge treatment
  - Thickening: thickening.
  - Dewatering: filtration, centrifugation.
  - Drying: thermal drying, atomization, solar drying.
  - Sludge to final disposal: sludge to landfill, sludge to energy recovery and sludge to recycling.

Clicking on *Validate* the system will provide a set of recommendations about the processes and treatments, as a user guide. Finally, the user will be able to introduce data about the operation and maintenance. In the *Treatments* section, the designed scheme can be visualized and, for each element, you will be able to modify its specific characteristics clicking on the notebook at the top of the icon.

The user can also introduce more information about:

- Waste disposal
  - Plastics: to incineration, to landfill.
  - Paper & cardboard: disposal, packaging of hazardous waste, paper disposal.
  - Wood: disposal, to incineration, to landfill.
  - Metals: aluminium to incineration, copper to incineration, ferro metals to incineration, ferro metals to landfill, steel to incineration, steel to landfill.
  - Lamps: fluorescent, light bulbs
  - Glass: disposal, to incineration, to landfill, to sorting plant
  - Other waste: aerosols, batteries, biodegradable, building, electronic devices, laboratory liquid, municipal solid
- Water quality (chemical parameters, indicator parameters, microbiological parameters and radioactivity)
  - Inlet water quality
  - Drinking water quality
- Drinking water impulsion: electricity consumption, drinking water impulsion
- Renewable energy: photovoltaic cell, wind power

A2.4 SUPPLY NETWORK

The section *General data* requires information related to the network and its connections: lifespan of the network, peak population, water connection, etc. In the *Construction* section, the user will be able to add pipes and other materials of the network. You can add one by one these elements with the button *Add* or use the *Add default values* (Figure A2.3), which will only require entering the total length of the network. When considering the appurtenances by default, you will see that each type has different frequency assigned that will be used to calculate the required number of units. If the elements are added one by one, the user will be able to specify some data for pipes (material, size, length and trench) and for appurtenances (type, number of units and volume in the case of the water tanks).

In the *Summary information* a summary of the appurtenances added and a graph of the pipe materials are presented, including information about the type of material, size and length (Figure A2.4).

In the Operation and maintenance section, four subsections are presented: consumables (electricity and sodium hypochlorite disinfection), network replacement, appurtenances replacement and other operation and maintenance issues: leakages, personnel costs, laboratory, cleaning, etc.

Default data

Material	Pipe size (mm)	% of total network length (default)	% of total network length
Ductile iron	150	5	
Fibrocement	80	5	
GFRP	300	5	
HDPE	50	10	
LDPE	90	50	
PVC	110	20	
Reinforced concrete with metal sleeve	1100	5	

Figure A2.3 Default data for the selection of materials for the piping network.

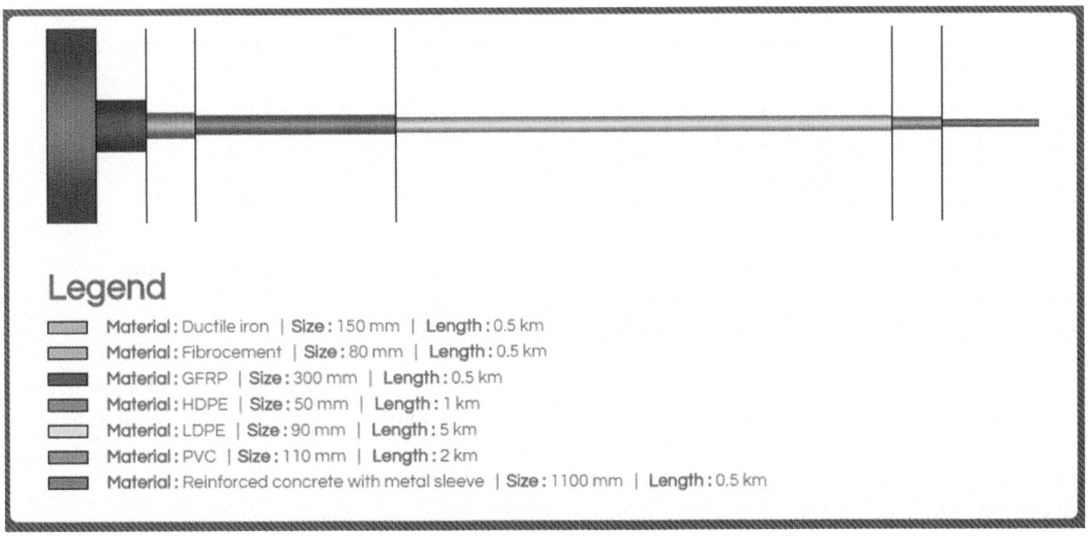

Figure A2.4 Summary of the pipe materials selected, including information about the type of material, size and length.

A2.5 SEWER NETWORK

The configuration of the sewer network is very similar to the one of supply network. In the *General data* section, the user will have to enter the required data for the network and the type of sewer network: served population, collected water, climatic conditions, etc. Depending on the selected type of sewer network, in the Construction and Operation and Maintenance sections there is a set of tabs with different networks to select: combined sewer, separate sewer (single or double trench) and mixed sewer. While for combined sewer, the information to fill up corresponds to the combined network, in the separate sewer with double trench, the information of the storm sewer network and the sanitary sewer network will be required. The most comprehensive approach is the mixed sewer, which comprises the storm sewer network, the sanitary sewer network and the combined sewer network.

In *Construction* the user will be able to add pipes and appurtenances as described in the supply network section and in *Operation and Maintenance* the following data will be required: energy consumption, network replacement, appurtenances replacement, sewer cleaning and inspection: electricity consumption, petrol, diesel, cost of cleaning, waste management, etc. and other operation and maintenance issues: personnel cost, maintenance, laboratory and analysis.

A2.6 WASTEWATER TREATMENT PLANT (WWTP)

The configuration of this stage is very similar to the one in the DWTP stage. In *General data* the user will be able to introduce the name of the facility, flow related data, the number of served municipalities and the population equivalent. The Design section is specific to the treatments present in a WWTP, which comprises the following sections:

- Water treatment
  - Pre-treatment: flotation, screening, sand/grit removal
  - Primary treatment: clarification
  - Secondary treatment: perfectly mixed reactor (CSTR), extensive treatment, plug-flow reactor (PFR), settling, biofilter, sequenced biological reactor (SBR), membrane biological reactor (MBR)
- Sludge treatment
  - Thickening: thickening
  - Stabilization: composting, anaerobic digestion
  - Dewatering: filtration, centrifugation
  - Drying: atomization, solar drying, thermal drying
  - Sludge final disposal: energy recovery, landfill, agriculture, gasification, composting, recycling
  - Energy recovery: combined heat and power
  - Other treatments: gasification, flaring

In the *Treatments* section the user will visualize the designed configuration and, for each element, you will be able to modify its specific characteristics clicking on the notebook at the top of the icon. Regarding the water quality (inlet or outlet) you will be able to enter data related to the directive urban wastewater discharge (91/2717ECC), metals, PPCPs and priority substances. In the *Other operation and maintenance issues section* it is possible to compute the diesel and natural gas consumption.

A2.7 VIEWING THE RESULTS OF A PROJECT

By the direct access to the Projects section in the home screen, the user can obtain a list with some basic data about each Project: title, scope (urban water system or single water facility), city, country, creation date and author. At the right side of every row there is a menu of actions for editing or deleting a project.

Once the user has completed all the required information in the different stages of a Project, the Results section will present the main outcomes of the assessment in terms of environmental, economic or eco-efficiency results (Figure A2.5).

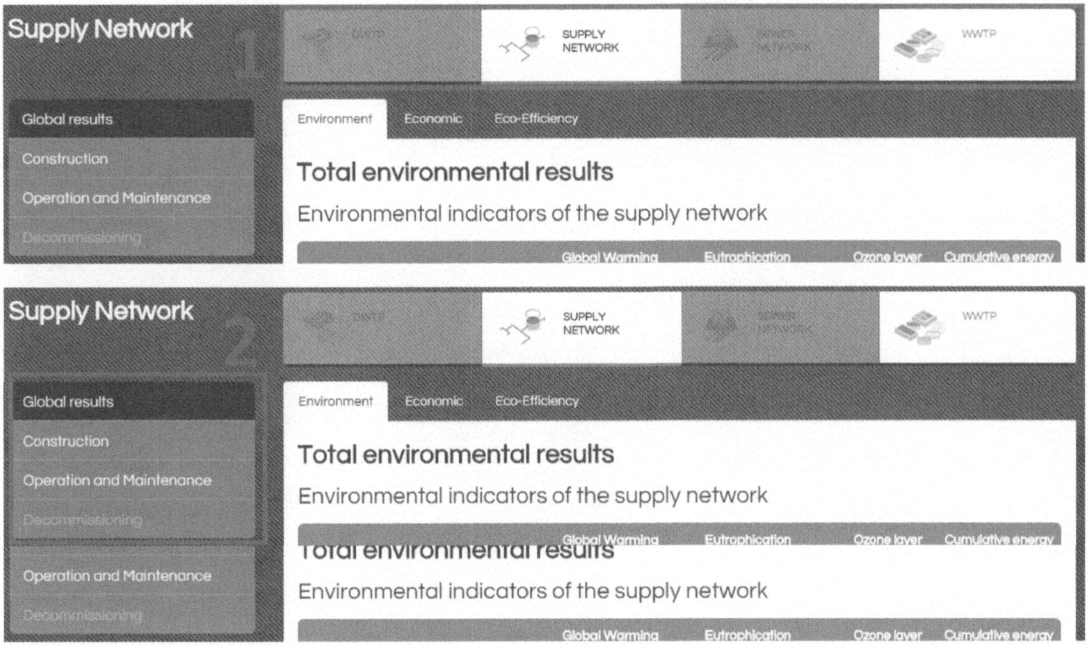

Figure A2.5 Visualization of the results of the assessment. 1: Selection of facility; 2: Global results for construction, operation and maintenance stages; 3: Environmental, economic and eco-efficiency results.

The environmental results will present data about the global warming potential, the eutrophication potential, the ozone layer depletion potential and the cumulative energy demand (Figure A2.6).

The economic results present data about the costs in the lifecycle, volumetric costs, annual costs in the operation and maintenance, annual costs per citizen and more information depending on the stage and section selected. The figures also change depending on the stage selected (Figure A2.7).

Concerning the eco-efficiency results, a cross view of the environmental and economic data is presented, in order to display graphics that combine this information (Figure A2.8)

Now follows a summary of the tables and graphics available although they may vary according to the selected project.

	Global Warming potential	Eutrophication potential	Ozone layer depletion potential	Cumulative energy demand
	[kg CO_2 eq.]	[kg PO_4^{3-} eq.]	[kg CFC-11-eq]	[MJ-eq.]
Annual impact of supply network [/year]	1.10 E+06	4.65 E+02	3.26 E-02	6.17 E+06
Life cycle impacts	1.10 E+08	4.65 E+04	3.26 E+00	6.17 E+08
Annual citizen impacts [/inhabitant * year]	8.92 E+01	3.78 E-02	2.65 E-06	5.01 E+02
Volumetric impacts [/m³ supplied]	4.78 E+04	2.02 E+01	1.42 E-03	2.68 E+05

Figure A2.6 Environmental indicators of the supply network.

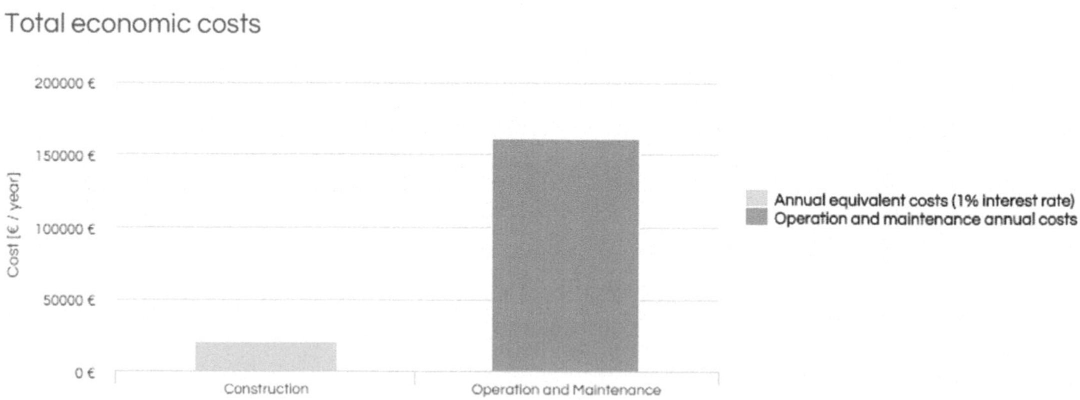

Figure A2.7 Economic indicators of the supply network.

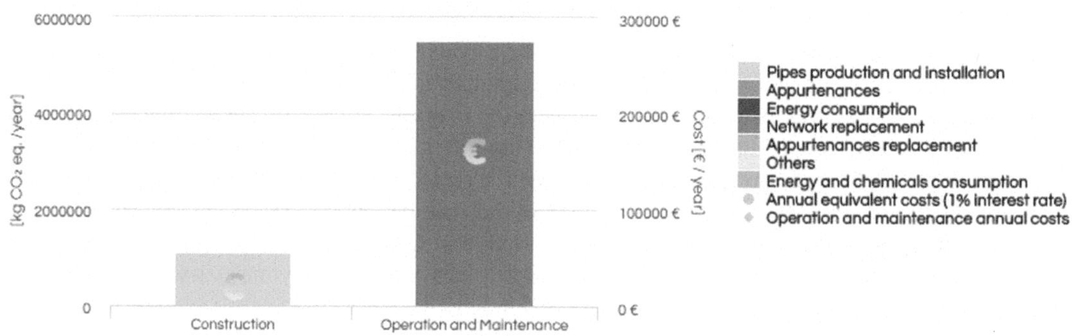

Figure A2.8 Ecoefficiency results in terms of Global Warming potential per life cycle stage.

- DWTP
  - Global results
    - Environment: Environmental indicators of the DWTP plant, Distribution of impacts per life cycle stage of the DWTP
    - Economic: Economic indicators of the DWTP, Distribution of the costs in the life cycle stages of the DWTP
    - Eco-efficiency: Eco-efficiency indicators of the DWTP, Distribution of impacts per life cycle stage of the DWTP
  - Operation and maintenance
    - Environment: Environmental indicators of DWTP Operation and Maintenance, Water quality
    - Economic: Economic indicators of the DWTP Operation and Maintenance
    - Eco-efficiency: Eco-efficiency indicators of Operation and Maintenance
- Supply network
  - Global results
    - Environment: Environmental indicators of the supply network, Distribution of impacts per life cycle stage of the supply network
    - Economic: Economic indicators of the supply network, Distribution of the costs in the life cycle stages of the supply network
    - Eco-efficiency: Eco-efficiency indicators of the supply network, Supply network eco-efficiency
  - Construction
    - Environment: Environmental indicators of supply network construction, Results per type of pipe: environmental impact vs. pipe length
    - Economic: Construction: Economic costs
    - Eco-efficiency: Eco-efficiency of supply network construction, pipes, appurtenances
  - Operation and maintenance
    - Environment: Environmental indicators of supply network operation and maintenance
    - Economic: Economic indicators of supply network operation and maintenance
    - Eco-efficiency: Eco-efficiency in supply network operation and maintenance
- Sewer network
  - Global results
    - Environment: Environmental indicators of the sewer network, Distribution of impacts per life cycle stage of the sewer network
    - Economic: Economic indicators of the sewer network, Distribution of the costs in the life cycle stages of the sewer network
    - Eco-efficiency: Eco-efficiency indicators of the sewer network, Sewer network eco-efficiency
  - Construction
    - Environment: Environmental indicators of sewer network construction, Results per type of pipe: environmental impact vs. pipe length
    - Economic: Construction: Economic costs
    - Eco-efficiency: Eco-efficiency of sewer network construction, pipes, appurtenances
  - Operation and maintenance
    - Environment: Environmental indicators of sewer network operation and maintenance
    - Economic: Economic indicators of sewer network operation and maintenance
    - Eco-efficiency: Eco-efficiency in sewer network operation and maintenance

- WWTP
  - Global results
    - Environment: Environmental indicators of the WWTP plant, Distribution of impacts per life cycle stage of the WWTP
    - Economic: Economic indicators of the WWTP, Distribution of the costs in the life cycle stages of the WWTP
    - Eco-efficiency: Eco-efficiency indicators of the WWTP, Distribution of impacts per life cycle stage of the WWTP
  - Operation and maintenance
    - Environment: Environmental indicators of WWTP Operation and Maintenance, Water quality
    - Economic: Economic indicators of the WWTP Operation and Maintenance
    - Eco-efficiency: Eco-efficiency indicators of Operation and Maintenance

The user can hover with the cursor on the different elements of any graphic in order to obtain more information. By clicking on the elements in the graphics legend, it is possible to hide or show different data. When clicking on the icon with three horizontal bars at the top right corner of any graphic, you will be able to access some options to print or download the graphic in different formats. Additionally, a file in CSV format with all the data of the report can be downloaded, clicking on the Results CSV button.

Annex 3

NOVEDAR_EDSS: Intelligent/expert screening of process technologies

Alba Castillo, Joaquim Comas, Manel Garrido-Baserba, Francesc Hernández-Sancho, Ulf Jeppsson, Ignasi Rodríguez-Roda and Manel Poch

The academic version of the NOVEDAR_EDSS can be downloaded for free from http://lequia.udg.cat/novedar/

A3.1 INTRODUCTION

NOVEDAR_EDSS integrates different knowledge-based techniques to perform the pre-selection of the process flow diagram in a WWTP. This tool was developed following the five steps proposed by Poch *et al.* 2004: analysis of the problem, data and knowledge acquisition, cognitive analysis, model selection and integration. The software was verified and validated by different experts from academia and from companies participating in the Consolider and Water_2020 COST Action projects. The results were considered satisfactory after an exhaustive qualitative model evaluation, when simulated data was compared with real WWTPs (Bennett *et al.* 2013).

The generation of WWTP network configurations is carried out by means of the interaction of two knowledge bases. The first one, specification knowledge base (Skb-units) (information represented in Table A3.1), summarizes the main features of the different treatments (primary, secondary, and tertiary) and sludge line technologies, while the second one contains information regarding the degree of compatibility amongst the different technologies; both are linked to another database with additional information about legislation. The combination of these databases results in the generation of a network of possible alternative process flow diagrams, including different treatment schemes. These process flow diagrams will be analyzed from an environmental, economic and technical point of view (Garrido-Baserba *et al.* 2012).

The selection process of an alternative WWTP configuration followed by NOVEDAR_EDSS is illustrated in Figure A3.1.

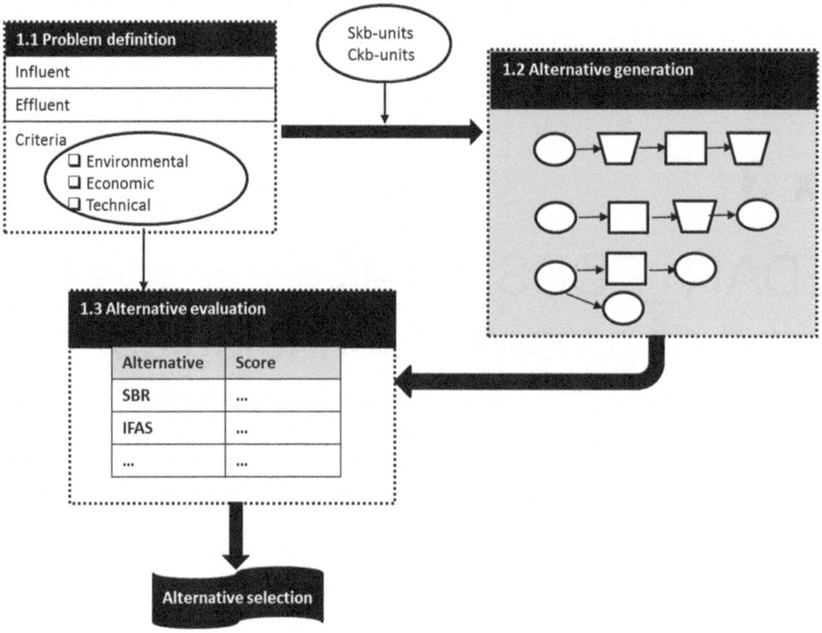

Figure A3.1 Scheme for the NOVEDAR_EDSS alternative process selection (Castillo *et al.* 2016).

A3.2 PROBLEM DEFINITION

First of all, the problem is defined. For that purpose, users introduce influent data for the specific scenario. In this screen the user must specify the influent water quality to be treated as well as the stage of development of the technologies to be considered (Figure A3.2).

After that, the effluent requirements are included. In these tab (Figure A3.3) the user must indicate the final destination of the generated effluent, it is possible to choose between: river, sensitive area and sea. Based on the option indicated by the user, the system supports the decision meeting the appropriate legislation. Thus, once the user selects the final destination of the water, the EDSS chooses the treatment options that achieve the corresponding quality requirements.

Based on user preferences, the criteria to be prioritized are selected to allow the consideration of different economic, environmental and technical aspects when comparing alternatives. In this screen (Figure A3.4) the user can decide at what degree prioritizes different objectives that can be pursued during the design of a WWTP.

Next screen is related to the Cost-Benefit Analysis (Figure A3.5). The display is characterized by two parts. One of them offers the possibility to modify some predefined data (IRR, NPV, etc.) that the EDSS will take to carry out a conventional cost-benefit analysis. The other part supports the methodology related to the environmental benefits quantification from the group of Francesc Hernández (University of Valencia).

It is possible for the user to select different sludge management strategies (Figure A3.6). Based on that, the program will recommend different sludge treatment.

In order to consider pathogenic load and target compounds two tabs (Figure A3.7 and Figure A3.8, correspondingly) are included to allow the user to indicate the concentration of pathogens and emerging contaminants that can be found in WWTP influent and the EDSS will estimate their removal. Currently, the target compounds are not yet activated in the program.

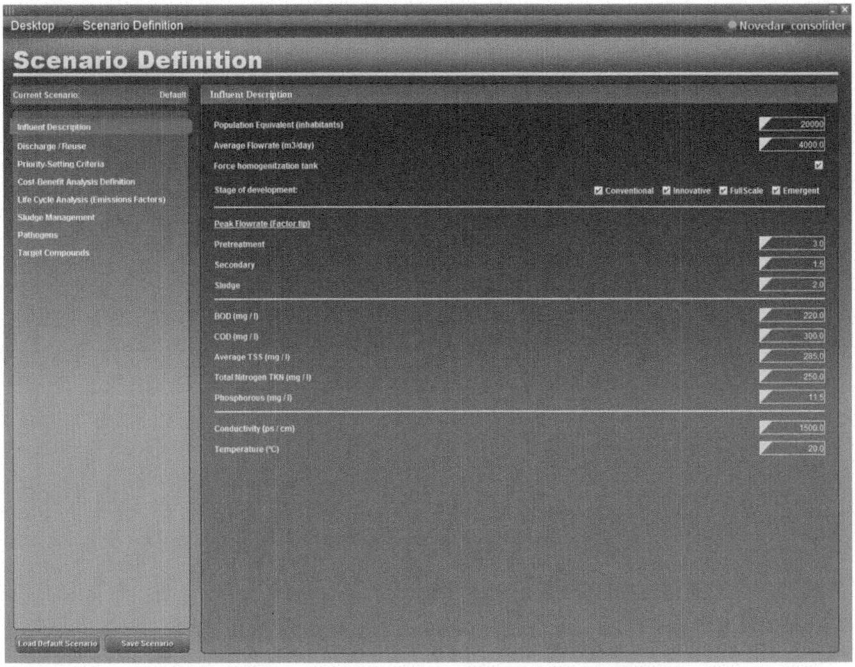

Figure A3.2 Snapshots of the scenario definition tab "Influent Information".

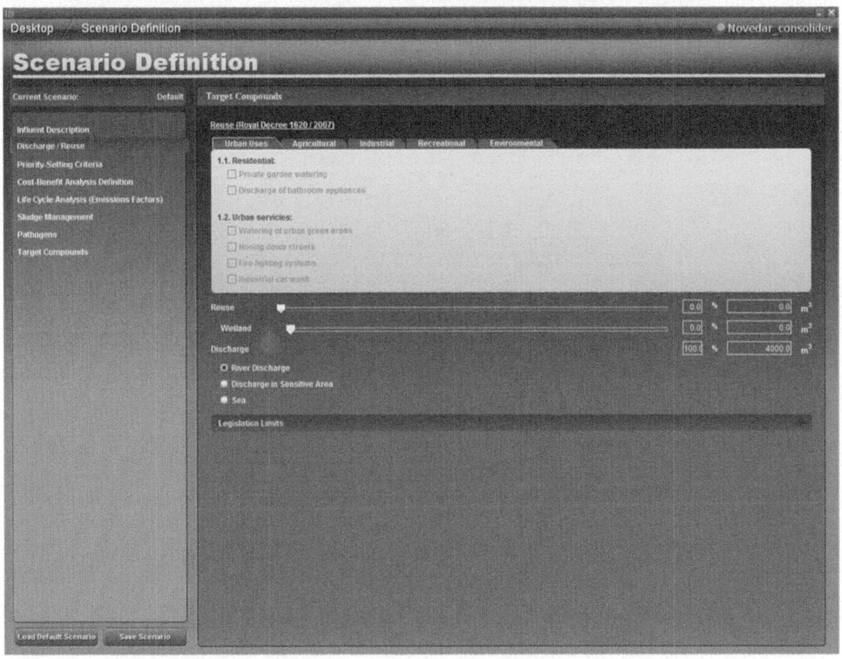

Figure A3.3 Snapshots of the scenario definition tab: effluent discharge/reuse.

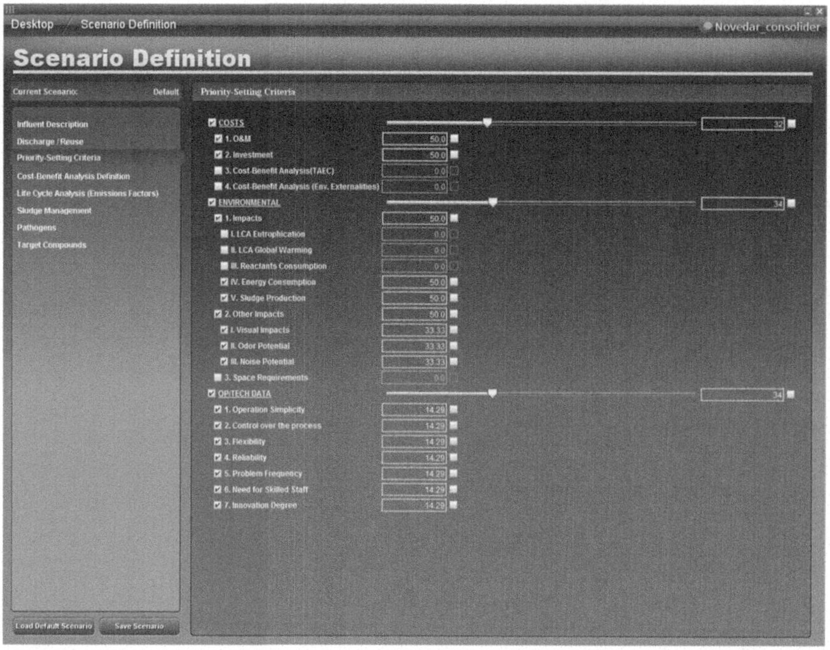

Figure A3.4 Snapshots of the scenario definition tab related to criteria and objectives prioritization.

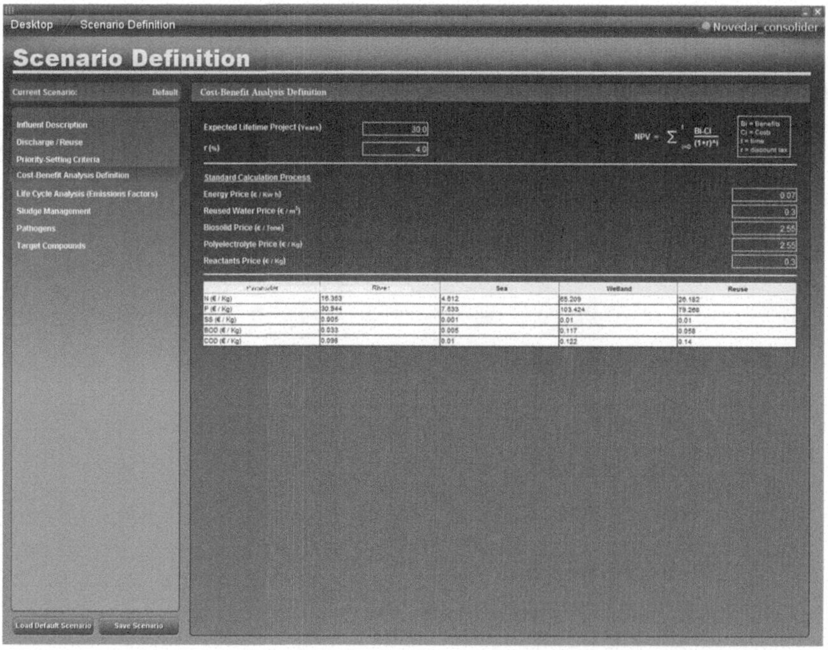

Figure A3.5 Snapshots of the scenario definition tab related to cost-benefit analysis.

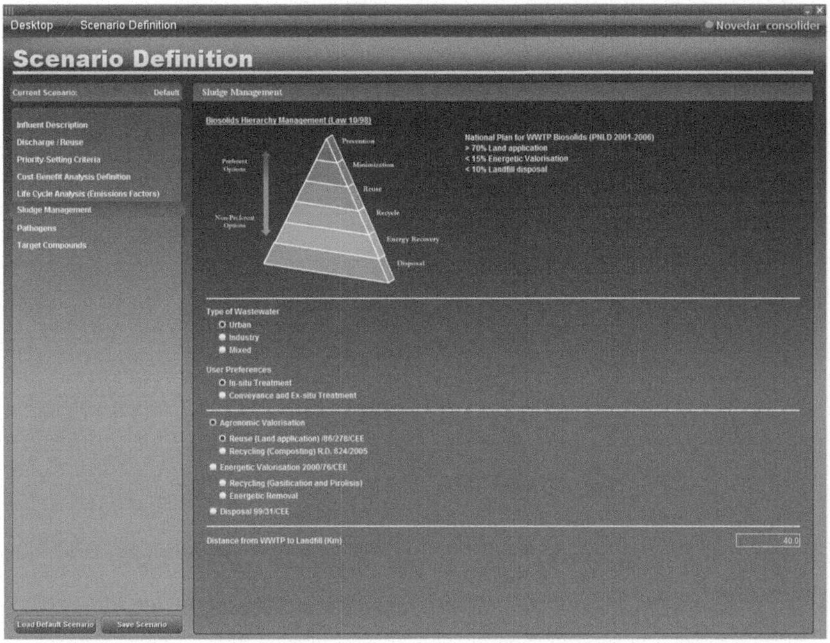

Figure A3.6 Snapshots of the scenario definition tab related to sludge management.

Figure A3.7 Snapshots of the scenario definition tab related to pathogens in the influent.

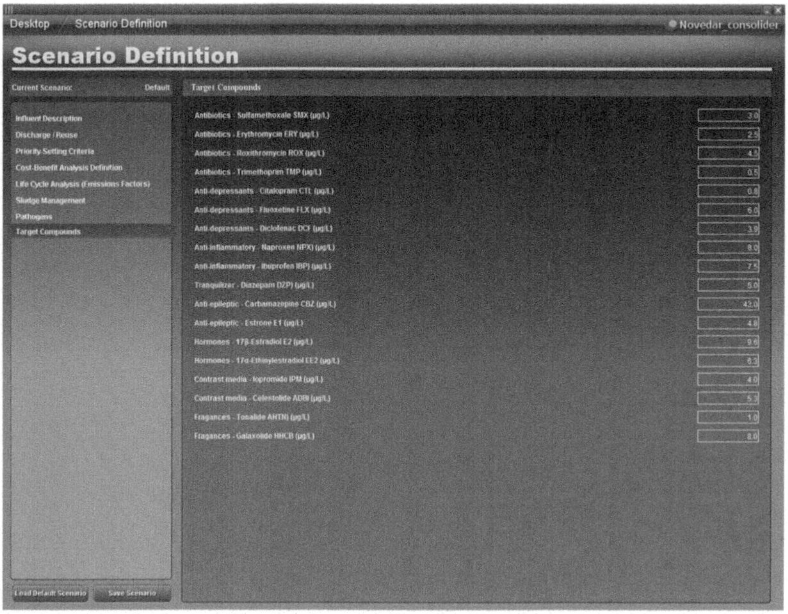

Figure A3.8 Snapshots of the scenario definition tab related to target compounds in the influent. This tab is not yet active in the program.

A3.3 ALTERNATIVE GENERATION

Once the scenario has been defined, NOVEDAR_EDSS is executed to obtain the feasible alternatives for each specific case study. In the first phase, a complete response surface of suitable WWTP network configurations is generated (Figure A3.9). Based on their capabilities, an influent range is defined for every technology. Hence, technologies are selected only when the influent data for the scenario defined is within their influent range. After that, only those process flow diagrams (PFDs) that satisfy the user's requirements are selected (Garrido *et al.* 2012). Table A3.1 presents the information included in the Skb-units for the different technologies.

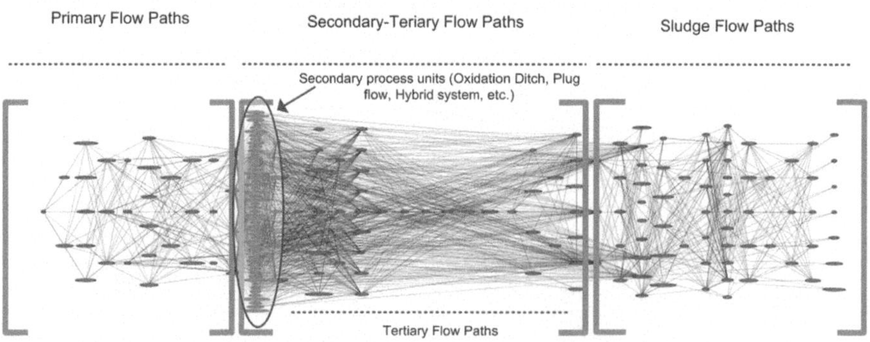

Figure A3.9 The Directed Network Structure built from the compatibility knowledge base (C-KBu) representing in a single cluster diagram all the compatibles WWTP alternatives diagrams. Unit processes corresponding to the main four parts of the WWTP flowsheet are pointed out.

Table A3.1 Formula, data and qualitative information applied to calculate values for every criterion in the knowledge base Skb-units.

Parameter	Data/Formula/Qualitative Information
Influent:	
Flow rate[1] range (m^3/d)	500–80000
Population equivalent[1] range (p.e.)	2500–500000
Effluent:	
COD removal[2] (%)	86
Phosphorus removal[2] (%)	25–70
Costs	
Investment costs[1] (for SBR)	$y = 8258.9 \cdot x^{-0.407}$ (A3.1)
O&M costs[2] (for SBR) **(x is p.e.; *y* is the total cost expressed as eur/p.e.)**	$y = 372.44 \cdot x^{-0.457}$ (A3.2)
LCA Emission factors from Ecoinvent database.	Categories considered: – Eutrophication – Global warming
Cost-Benefit Analysis (CBA) based on the Net Profit Value (NPV) Bi are benefits; Ci are costs; t is time **(30 years)** and r is discount tax **(4%)**	$NPV = \sum_{i=0}^{t} \frac{B_i - C_i}{(1+r)^i}$ (A3.3)
Impacts	
Visual Impact[1] (based on **the integration in the landscape**)	Low
Odor potential[1] (based on the treatment **process**)	High
Biogas production[1] (m^3/day) Volatile Solids Removed; based on Sludge (SLD) Production[3]; Price Biogas = 0.69 eur/m^3	$\text{SLD Production} = 0.65 \cdot (1.06 \cdot \text{SRT}) \cdot \text{Influent} \cdot \left(0.7 \cdot \frac{\text{COD Influent}}{1000}\right)$ (A3.4) $\text{Biogas Production} = \text{VSR} \cdot 1.03$ (A3.5)
Technical	
Flexibility[1]	Very high
Simplicity[3]	Low

[1]*CEDEX* (2013). [2]*Yu et al.* (1997). [3]*Metcalf & Eddy*(2003). *Parameters in the CBA can be easily modified by the user.*

Values for influent (Table A3.1) refer to the range of values for which every treatment alternative included in the database is appropriate. Taking into account these values, only those technologies able to treat the influent (defined in every scenario) are selected. As for effluent data, the range of values for every parameter corresponds to the treatment alternative removal efficiency.

In the LCA, the emission factors correspond to the Ecoinvent database, and the method applied to calculate them was developed by the Center of Environmental Science (CML) (detailed information can be found in the supplementary material).

Visual impact refers to the integration of a treatment alternative in the landscape, while odor potential considers its potential to release odors. Both, visual impact and odor potential corresponds to qualitative impacts and the evaluation range is from very low to very high, compared to the other treatment alternatives considered, based on expert criteria.

Finally, evaluation for qualitative criteria (i.e. impacts and technical) ranges from very low to very high, compared to the other treatment alternatives included in the database, based on expert criteria.

After the evaluation of the whole set of embedded treatment trains the desired output is attained. All treatment trains have been scored. Therefore, a set of outputs in the form of suitable flow diagrams is finally generated in the form of a network structure, as showed in Figure A3.10.

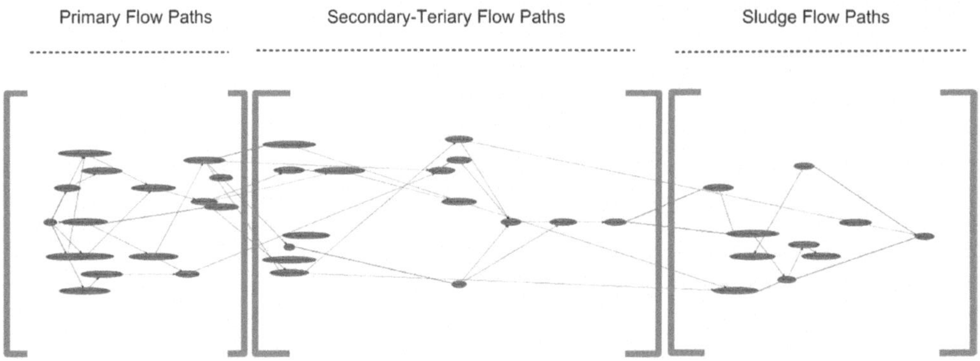

Figure A3.10 Example of treatment trains selected from the Directed Network Structure after the screening of options.

In the program a screen as showed in Figure A3.11 is presented and it is possible to go through the different operational unit and their corresponding technologies by selecting them in the screen.

A3.4 ALTERNATIVE EVALUATION

Subsequently, a multi-criteria decision analysis is applied to rank the treatment process alternatives to maximize the degree of satisfaction of the different objectives (i.e., lower environmental impact and lower investment and operational costs while obtaining proper effluent concentrations) through valuing multiple selected environmental, economic and technical criteria. Both quantitative (e.g., costs, space requirements) and qualitative (e.g., flexibility, impacts) criteria are considered. Moreover, some criteria are positive (e.g., reliability), whereas others are negative (e.g., impacts); this makes a difference in the way they are considered because higher reliability implies a higher score when prioritizing this criteria, while an alternative with higher impacts obtains a lower score for this criteria. To calculate the score for every quantitative criterion, the best score (i.e., 10) is given to the most appropriate alternative (e.g., the cheapest one), and the worst score (i.e., 0) is given to the least appropriate (e.g., the most expensive one). After that, a normalization process is applied to rank the alternatives according to their values (e.g., cost in euros) for this specific criteria between the scores 0 and 10. As for qualitative criteria, they are qualified from "very low" to "very high," and to calculate the score, they take values of 0 (very low) to 4 (very high) (therefore they take quantitative values) and the way to calculate the scores is the same as that applied for quantitative criteria.

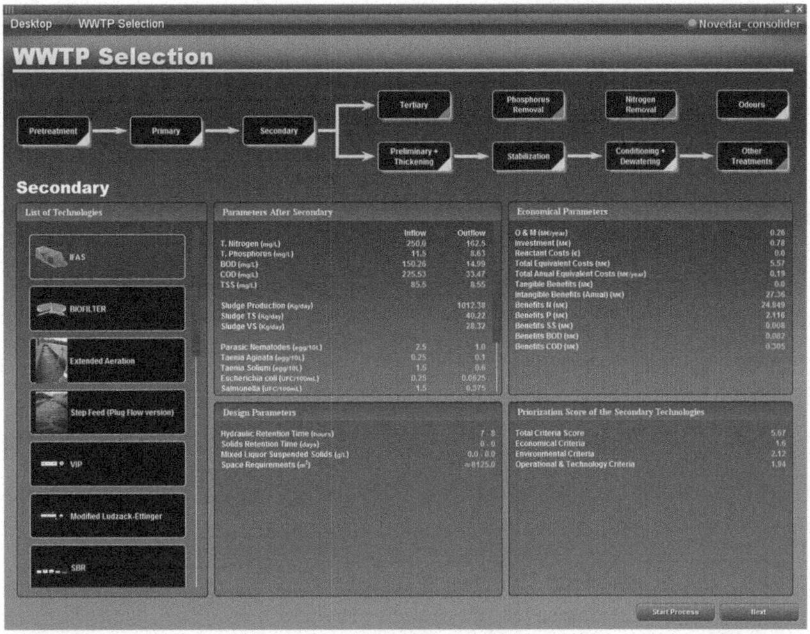

Figure A3.11 Snapshot of the treatment alternative generated.

Finally, users can select the weight for every considered criterion depending on their preferences. To obtain the total score (Equation A3.6) for each alternative in a single combined criterion, each criteria is multiplied by its appropriate criteria weight, followed by summing the weighted scores for all criteria:

$$V(X) = \sum_{i=1}^{n} W_i \cdot V_i(X_i) \tag{A3.6}$$

where "V" is the score, "W" is the weight for each criterion, "i" is the criterion, "n" is the total number of criterion and "X" corresponds to the considered treatment alternative.

Cost-benefit analysis takes into account costs (i.e. operational expenditure (OPEX) and capital expenditure (CAPEX)) as well as benefits (i.e. biogas production and reused water) for an economic assessment of different treatment process alternatives. Alternately, the cost-benefit analysis can be performed by also taking into account the environmental externalities, which are equivalent to the value of the positive externalities associated with avoiding the discharge of pollution into the environment (Molinos-Senante *et al.* 2012). LCA is an extended environmental indicator used to quantify the environmental impacts associated with a product or process throughout its entire lifecycle (ISO, 2006a; ISO, 2006b), and it considers a holistic perspective of the process system. In NOVEDAR_EDSS, both eutrophication and global warming impact categories are considered to evaluate the LCA for each alternative (Garrido-Baserba *et al.* 2014) (detailed information can be found in the supplementary material). Moreover, three different technical criteria are considered to rank the alternatives: reliability, simplicity and flexibility. The operation reliability is related to the process robustness in the face of flow variations and disturbances, while operation simplicity is related with the need of specialized staff. Finally, operation flexibility takes into account how the process can perform through different operational conditions.

Therefore, among the alternatives of the process flow diagrams obtained, a ranked list for the secondary treatment process technologies is generated based on the criteria selected.

Once a whole line of technologies have been selected, composing a complete WWTP alternative, the EDSS program displays the results that would support the decision-making. The results screen consists of four main parts (Figure A3.12):

(1) Technical part
(2) Economic
(3) Environmental
(4) Operational

Figure A3.12 Snapshot of the results interface showing the results for one specific scenario, once the water and sludge lines have been selected.

A3.5 NEW FEATURES AND CHARACTERISTICS: NOVEDARPLUS_EDSS

NOVEDARplus_EDSS is currently in stage of development, even though some of the features described in this document have been already included. An upgraded knowledge base of the EDSS has been developed to include the latest technologies and new configurations as well as new functionalities and criteria to identify more efficient WWTPs and resource recovery possibilities.

In this new EDSS the first step for scenario definition includes also the characteristics of the side-stream (Figure A3.13).

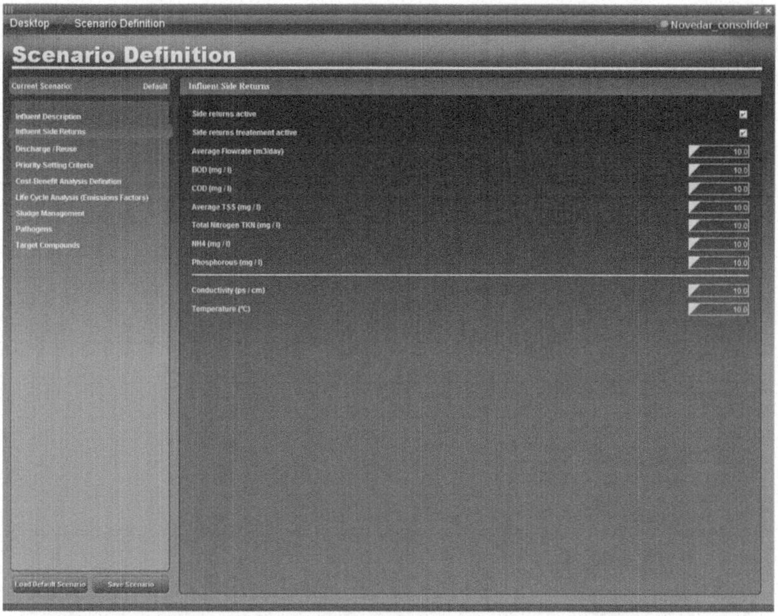

Figure A3.13 The scenario definition tab "Influent Side-stream".

When defining the user preferences, it is now possible to select criteria related with resource recovery, including: nitrogen recovery, phosphorus recovery and biogas recovery (Figure A3.14).

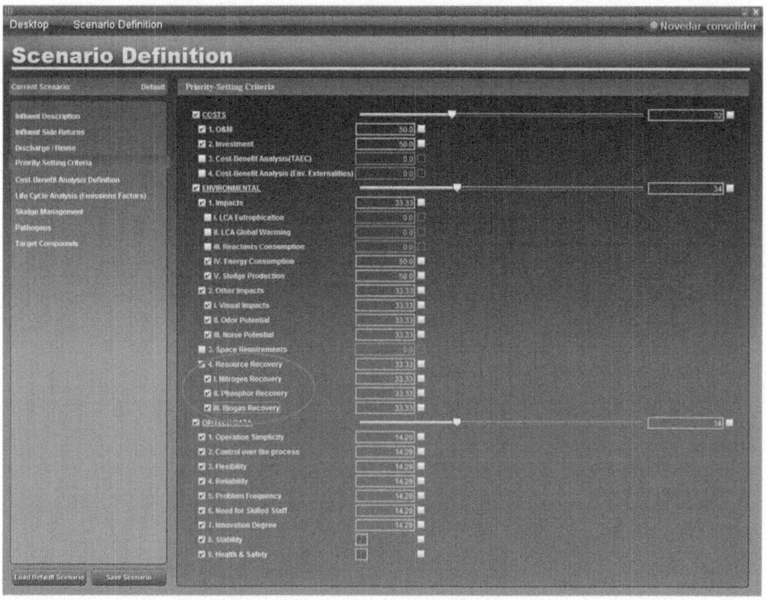

Figure A3.14 Snapshot of the scenario definition screen related to criteria and objectives prioritization, including criteria related to resource recovery.

Since the side-stream has been included, it is possible to select some new treatments and configurations developed for both biologic (Figure A3.15) and physic-chemical treatment (Figure A3.16), including resource recovery strategies.

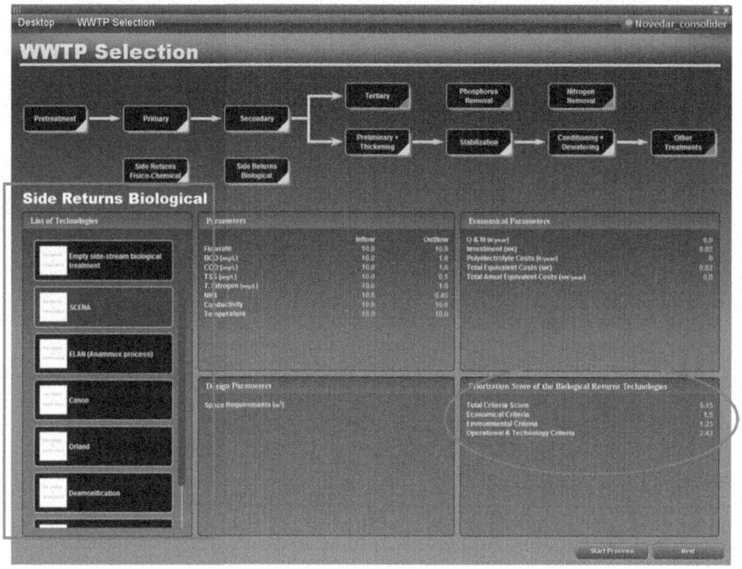

Figure A3.15 Snapshot of the generated treatments interface showing the biological treatments included in the side-stream as well as their score from the multi-criteria analysis.

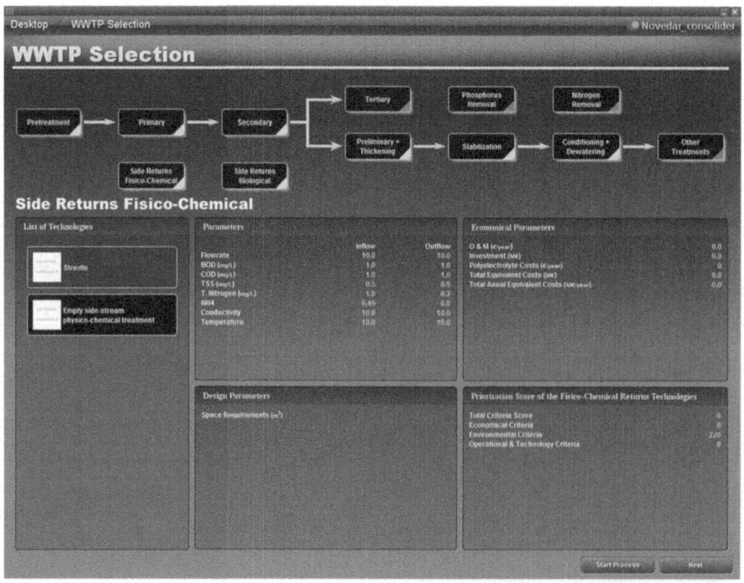

Figure A3.16 Snapshot of the generated treatments interface showing the physico-chemical treatments included in the side-stream as well as their score from the multi-criteria analysis.

In the mainstream, some resource recovery strategies can be selected (Figure A3.17).

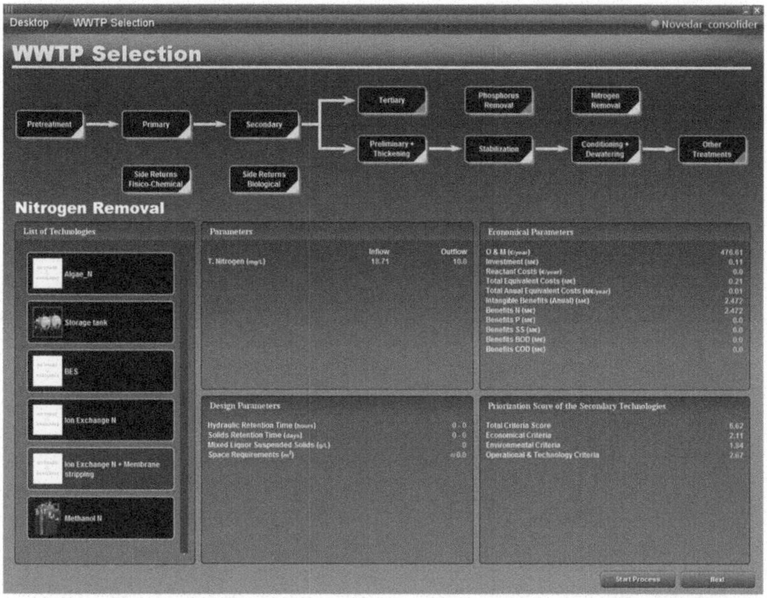

Figure A3.17 Snapshot of the generated treatments interface showing the treatments included to remove nitrogen as well as their score from the multi-criteria analysis.

Index

IWA Publishing's authorised EU representative for General Product Safety Regulations is Diane D'Arras, 15 rue Duret, 75116 Paris, France, e-mail: safety@iwap.co.uk.

Printed and bound by CPI Group (UK) Ltd, Croydon, CR0 4YY
16/06/2026
02140885-0007